DGZfP-Jahrestagung 2025

Berichtsband BB 184

Berlin, Germany
26-28 May 2025

ISBN: 979-8-3313-2083-6

Printed from e-media with permission by:

Curran Associates, Inc.
57 Morehouse Lane
Red Hook, NY 12571

Some format issues inherent in the e-media version may also appear in this print version.

Deutsche Gesellschaft Fuer Zerstoerungsfreie Pruefung (DGZfP)
Max-Planck-Str. 6
12489 Berlin
Germany

Phone: +49 30 67807-0
Fax: +49 30 67807-109

mail@dgzfp.de

Additional copies of this publication are available from:

Curran Associates, Inc.
57 Morehouse Lane
Red Hook, NY 12571 USA
Phone: 845-758-0400
Fax: 845-758-2633
Email: curran@proceedings.com
Web: www.proceedings.com

TABLE OF CONTENTS

DGZfP-Jahrestagung 2025

Neues Konzept zur Detailsensitivitätsüberwachung in der industriellen Computertomographie - das EURAMET-Projekt SensMonCT

Uwe EWERT[1], Holger ROTH[2], Simon BURKARD[3], Josephine GUTEKUNST[4], Virpi KORPELAINEN[5], Anne-Françoise OBATON[6], Ulrich NEUSCHAEFER-RUBE[7], Marko KATIC[8]

[1] Kowotest GmbH, Germany, uwe@ewert-net.de
[2] Waygate Technologies, Digital Solutions GmbH, Baker Hughes, Germany,
[3] METAS, Eidgenössisches Institut für Metrologie METAS, Switzerland,
[4] Microworks GmbH, Germany,
[5] Teknologian tutkimuskeskus VTT Oy, Finland,
[6] Laboratoire National de Métrologie et d'Essais (LNE), French National Metrology Institute (NMI), France,
[7] PTB, Physikalisch-Technische Bundesanstalt, Germany,
[8] FSB, Sveuciliste U Zagrebu, Fakultet Strojarstva I Brodogradnje, Croatia

Kurzfassung

Digitale Röntgenbildgebung und Computertomographie (CT) werden in der Industrie zur Fehlererkennung, Fehlerbewertung und zum dimensionellen Messen eingesetzt. Dies erfordert korrekte experimentelle Systemeinstellungen für eine ausreichende Sichtbarkeit und Detektierbarkeit von Unregelmäßigkeiten und Strukturelementen. Ein neuer Parameter, die Detail Detection Sensitivity (DDS – Detailnachweis-Sensitivität), wird eingeführt. Ein entsprechender Standardentwurf (WK84836-24) wurde dem ASTM E07-Ausschuss vorgelegt. Der ASTM-Leitfaden (Guide) E 1441 beschreibt drei wesentliche Funktionen für die Charakterisierung von iCT-Systemen. Dabei handelt es sich um die Contrast Discrimination Function (CDF), die Modulations-Übertragungs-Funktion (MTF) und das Contrast Detail Diagram (CDD). Die damit verbundenen Verfahren und Formeln zur Bestimmung dieser Funktionen und der DDS werden anhand von Messungen von neu entwickelten Bildgütetestkörpern mit Bohrungen von unterschiedlichem Durchmesser mit industriellen Computertomographie-Systemen (iCT) und durch Modellierung vorgestellt. Derzeit wird die DDS von iCT-Systemen von menschlichen Bedienern bewertet, was unzuverlässig und kostspielig ist. Daher werden im Rahmen des EURAMET-Projekts "SensMonCT" neue Bildgütetestkörper und rückführbare automatisierte Mess- und Überwachungsmethoden entwickelt und Verfahren zur Bewertung der DDS von iCT-Systemen und deren Standardisierung entwickelt.

Keywords:
Computertomographie, Bildgüte, Detailerkennung, Kontrastdetaildiagramm, Modulationsübertragungsfunktion, rückführbare Messung, Normen.

NDT net
Medien- und Publikationspartner

DGZfP – Jahrestagung, 26. bis 28. Mai, Berlin, 2025

Neues Konzept zur Detailsensitivitätsüberwachung in der industriellen Computertomographie - das EURAMET-Projekt SensMonCT

von
Uwe EWERT[1], Holger ROTH[2], Simon BURKHARD[3], Josephine GUTEKUNST[4], Virpi KORPELAINEN[5], Anne-Françoise OBATON[6], Ulrich NEUSCHAEFER-RUBE[7], Thomas BLUMENSATH[9], Marko KATIC[8]

[1] *KOWOTEST GmbH, Teltow, Germany* uwe@ewert-net.de
[2] *Waygate Technologies, Baker Hughes, Wunstorf, Digital Solutions GmbH, Germany*
[3]*METAS, Federal Institute of Metrology, Bern-Wabern, Switzerland ;*
[4]*Microworks GmbH, Karslruhe, Germany ;*
[5]*National Metrology Institute VTT MIKES, Espoo, Finland ;*
[6]*Laboratoire National de Métrologie et d'Essais (LNE), French National Metrology Institute (NMI), Paris, France ;*
[7]*PTB, Physikalisch-Technische Bundesanstalt, Braunschweig, Germany ;*
[8]*FSB, Sveuciliste U Zagrebu, Fakultet Strojarstva I Brodogradnje, Zagreb, Croatia ;*
[9]*University of Southampton, Institute of Sound and Vibration Research (ISVR), Highfield, UK*

Qualifizierung und Qualitätssicherung von CT-Systemen

- **Bildqualität und Bildgüteprüfkörper** (BPK) für Radiographie und CT
- **Erkennbarkeit von Löchern** und Poren durch **menschliche Auswerter** in CT-Schnittbildern
- **Contrast Detail Diagram** (CDD - ASTM E 1441) - wird erweitert
- **Modulation Transfer Function** (MTF) und relativer Kontrast in Abhängigkeit vom Lochdurchmesser (ASTM E 1695-20)
- **Kontrastunterscheidungsfunktion** (CDF - ASTM E 1695-20)
- **Neue Prüfkörper** zur Qualitätssicherung und Überwachung von CT-Scannern
  - **ASTM-Normentwurf: Direkte Bestimmung der Detail-Erkennbarkeit (DDS)** für die Computertomographie (CT), **Numerische Nachweisgrenze** (DDS_{limit}) und neues SensMonCT-Design, DGzfP-Richtlinie D7 und VDA-Richtlinie.

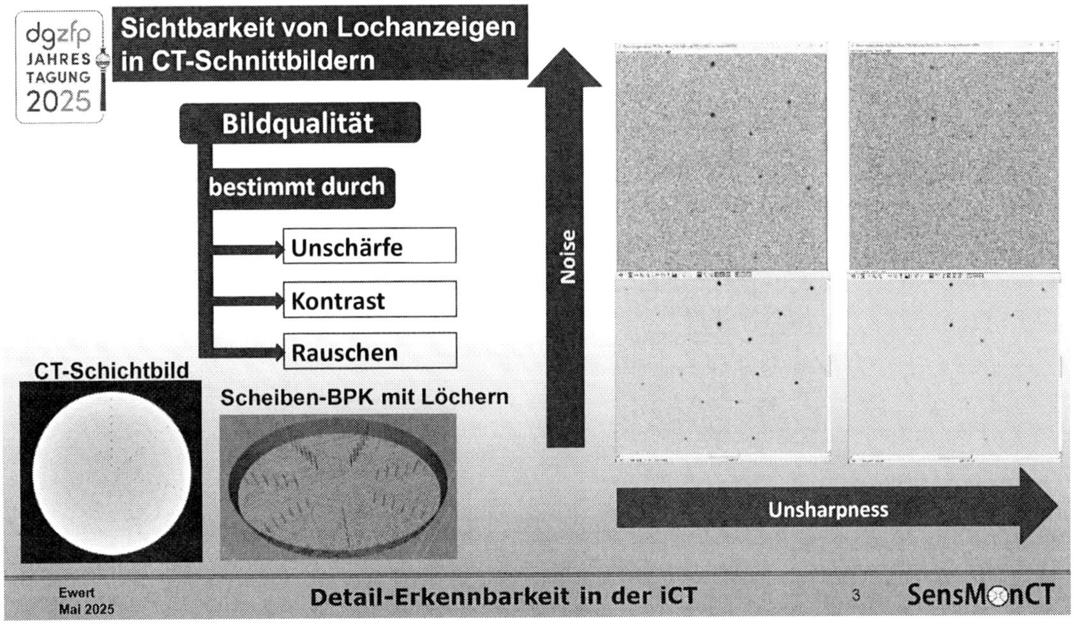

dgzfp JAHRESTAGUNG 2025
Sichtbarkeit von Lochanzeigen in CT-Schnittbildern
Bildqualität
bestimmt durch
Unschärfe
Kontrast
Rauschen
CT-Schichtbild
Scheiben-BPK mit Löchern
Noise
Unsharpness
Ewert
Mai 2025
Detail-Erkennbarkeit in der iCT
3
SensMonCT

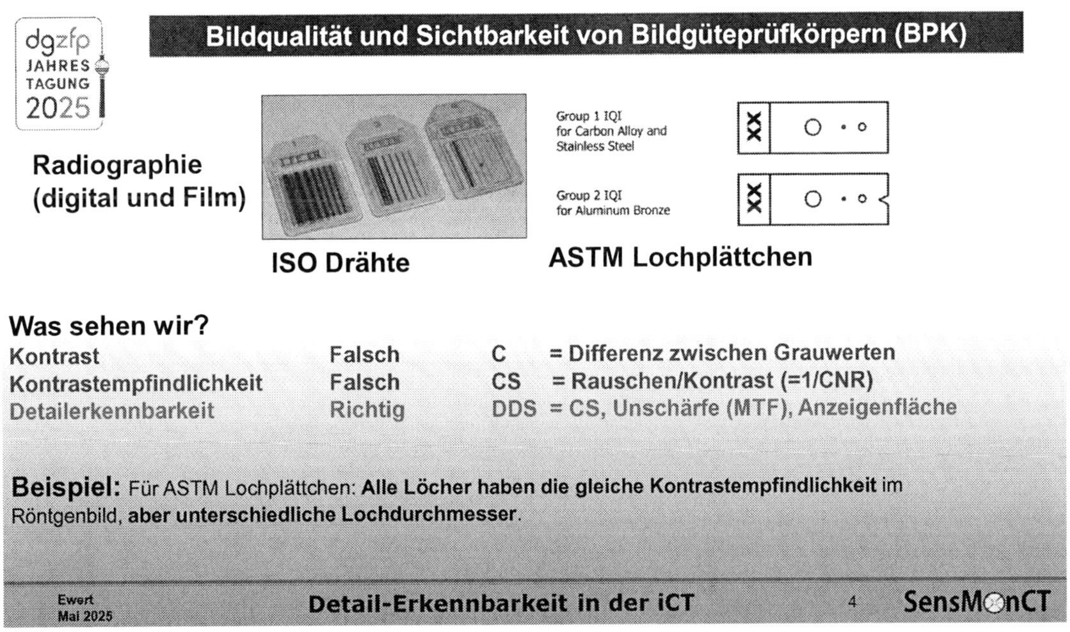

dgzfp JAHRESTAGUNG 2025
Bildqualität und Sichtbarkeit von Bildgüteprüfkörpern (BPK)
Radiographie (digital und Film)
Group 1 IQI for Carbon Alloy and Stainless Steel
Group 2 IQI for Aluminum Bronze
XX
ISO Drähte
ASTM Lochplättchen
Was sehen wir?
Kontrast Falsch C = Differenz zwischen Grauwerten
Kontrastempfindlichkeit Falsch CS = Rauschen/Kontrast (=1/CNR)
Detailerkennbarkeit Richtig DDS = CS, Unschärfe (MTF), Anzeigenfläche
Beispiel: Für ASTM Lochplättchen: Alle Löcher haben die gleiche Kontrastempfindlichkeit im Röntgenbild, aber unterschiedliche Lochdurchmesser.
Ewert
Mai 2025
Detail-Erkennbarkeit in der iCT
4
SensMonCT

Bildqualität und Sichtbarkeit von Bildgüteprüfkörpern (BPK)

Computer-Tomographie:
Diese BPKs sind nicht für CT geeignet

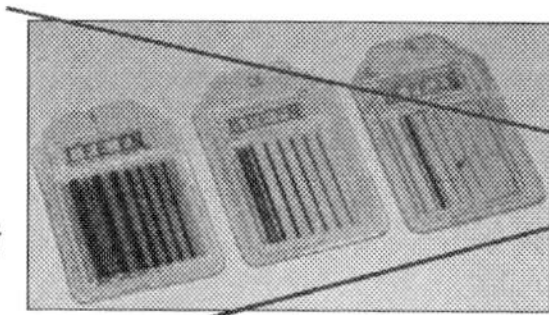

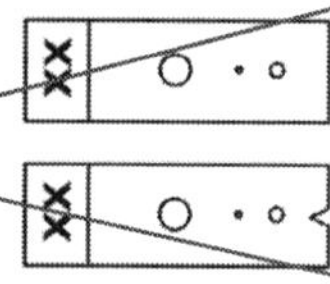

~~ISO~~ **Drähte** **ASTM Lochplättchen**

Warum?
*Bei der CT hängt der Kontrast **nicht von der Materialdicke** ab; sondern nur vom Material*
Alle IQIs zeigen den gleichen Kontrast wie das Material.

Neue BPKs für CT:

ASTM E 1695-20

Verwenden Sie Stäbe oder Scheiben

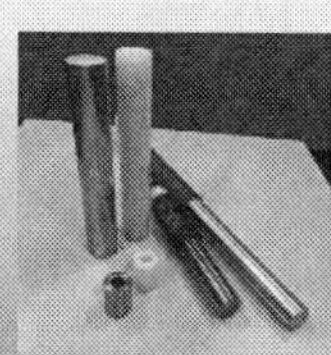

Bestimme MTF, CDF and CDD
- Modulation Transfer Function
- Contrast Discrimination Function
- Contrast Detection Diagram

Erforderliche Qualitätssicherung für die neue **ASTM E 3375-23**
Standard Practice for Cone Beam Computed Tomographic (CT) Examination

Neuer BPK für die Empfindlichkeitsmessung der CT-Detailerkennung mit Lochscheiben - Project SensMonCT -

Löcher mit unterschiedlichem Durchmesser ASTM-Ausführung

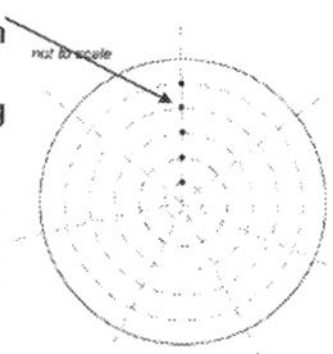

Messung mit Lochscheiben-BPK
- Überwachung der Detailerkennbarkeit (DDS).
- Was ist der kleinste Porendurchmesser (Lochdurchmesser), der erkannt werden soll/kann?

Beispiel: Al-Disk
Durchmesser: 50 mm
Löcher: 50 µm - 200 µm

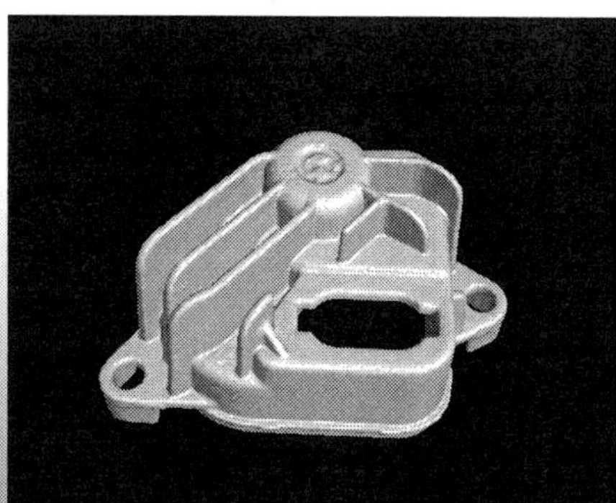

dgzfp JAHRESTAGUNG 2025

Computer-Tomographie: Modulation Transfer Function (MTF)

CT

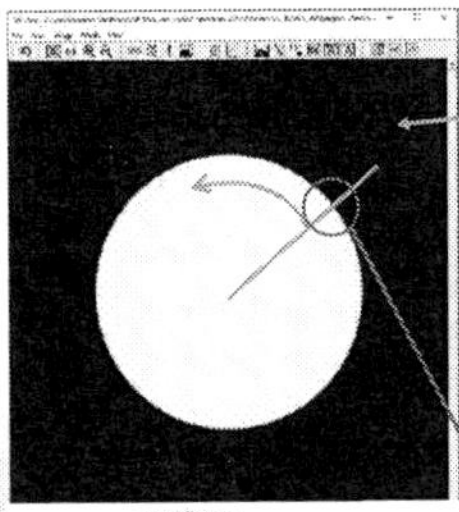

- Erklärt und definiert in ASTM E1441-19 und ASTM E 1695-20.
- Die MTF-Messung erfolgt anhand des kreisförmigen Profils der zentralen 2D-Schicht einer Scheibe oder eines Stabes.
- Die $MTF_{10\%}$ entspricht dem iSR_b^{image} der zentralen 2D-Schicht.

Kantenantwortfunktion ERF (Edge Respones Function und Linienspreizfunktion LSF

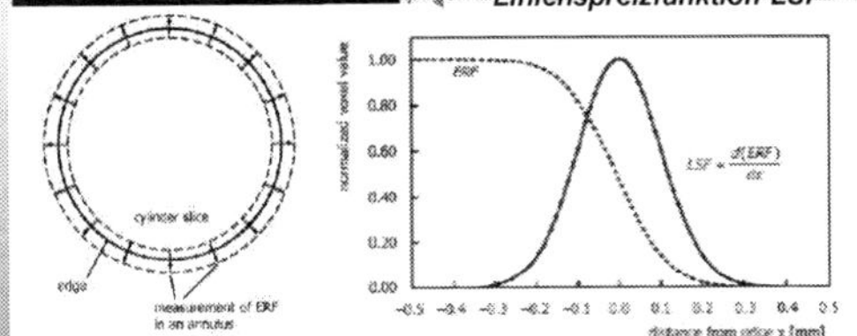

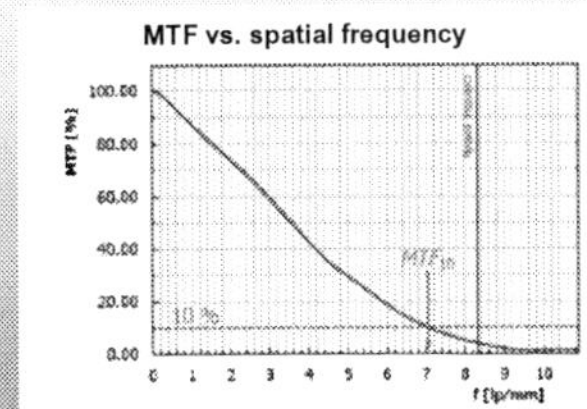

Die MTF wird als Magnitudespektrum der LSF berechnet

dgzfp JAHRESTAGUNG 2025

MTF und relativer Kontrast der Lochanzeigen

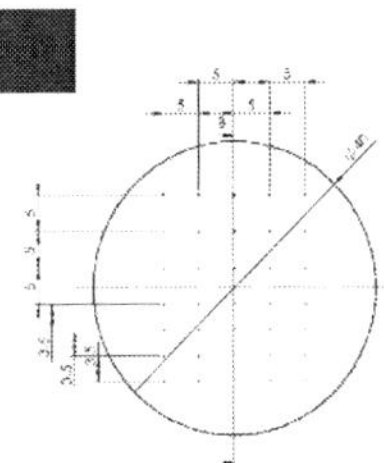

- Der gemessene relative Kontrast der Löcher liegt knapp unter der MTF(D*).
- In den Simulationen liegen die relativen Kontrastwerte genau auf der MTF-Kurve.
- D* ist der Lochdurchmesser in Voxeln.

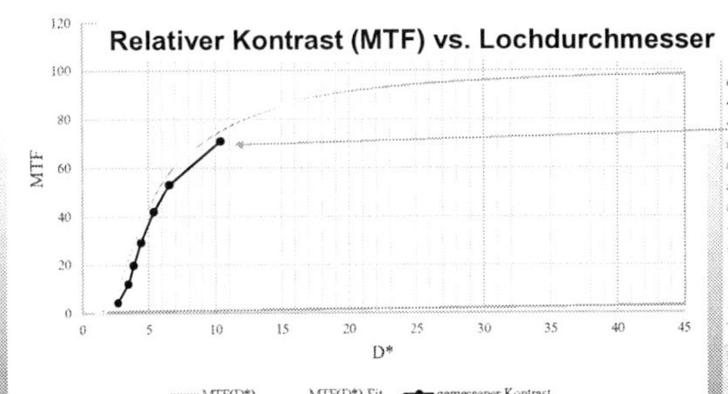

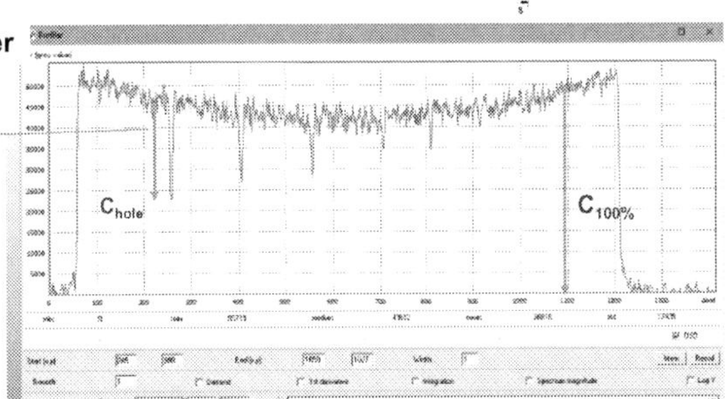

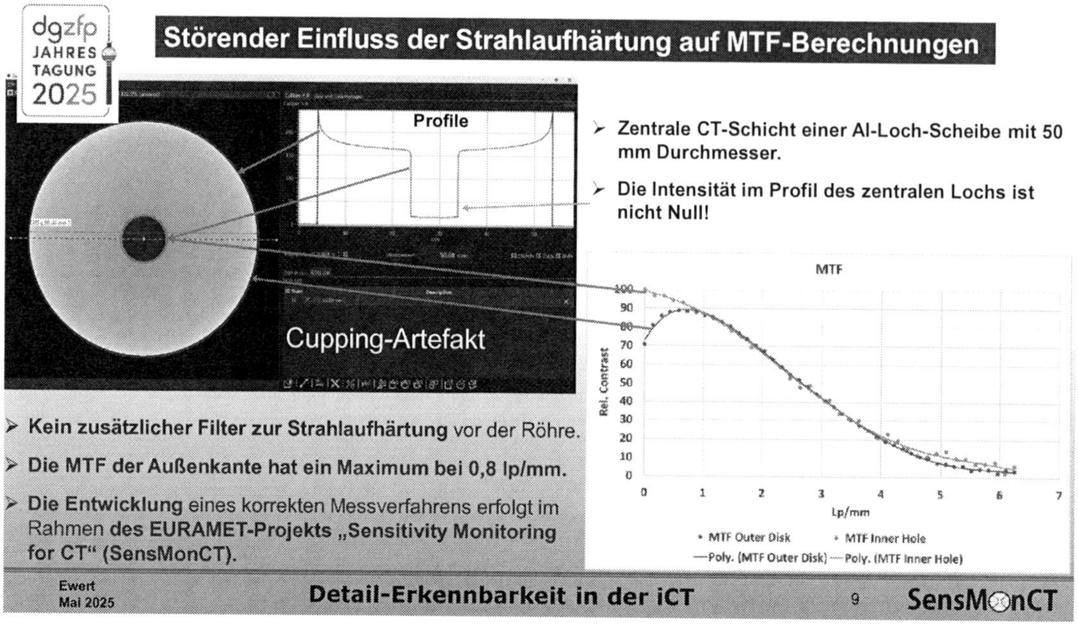
dgzfp JAHRESTAGUNG 2025
Störender Einfluss der Strahlaufhärtung auf MTF-Berechnungen
Profile
Cupping-Artefakt
➢ Zentrale CT-Schicht einer Al-Loch-Scheibe mit 50 mm Durchmesser.
➢ Die Intensität im Profil des zentralen Lochs ist nicht Null!
MTF
Rel. Contrast
Lp/mm
MTF Outer Disk
MTF Inner Hole
Poly. (MTF Outer Disk)
Poly. (MTF Inner Hole)
➢ Kein zusätzlicher Filter zur Strahlaufhärtung vor der Röhre.
➢ Die MTF der Außenkante hat ein Maximum bei 0,8 lp/mm.
➢ Die Entwicklung eines korrekten Messverfahrens erfolgt im Rahmen des EURAMET-Projekts „Sensitivity Monitoring for CT“ (SensMonCT).
Ewert
Mai 2025
Detail-Erkennbarkeit in der iCT
9
SensMonCT

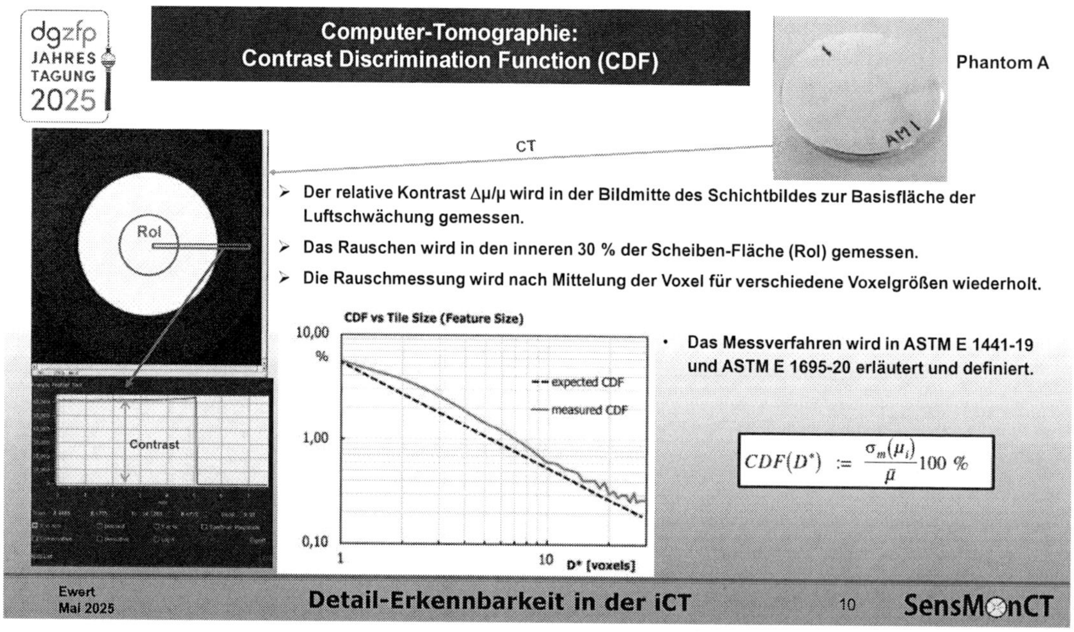
dgzfp JAHRESTAGUNG 2025
Computer-Tomographie:
Contrast Discrimination Function (CDF)
Phantom A
CT
RoI
Contrast
➢ Der relative Kontrast Δµ/µ wird in der Bildmitte des Schichtbildes zur Basisfläche der Luftschwächung gemessen.
➢ Das Rauschen wird in den inneren 30 % der Scheiben-Fläche (RoI) gemessen.
➢ Die Rauschmessung wird nach Mittelung der Voxel für verschiedene Voxelgrößen wiederholt.
CDF vs Tile Size (Feature Size)
10,00 %
1,00
0,10
1
10
D* [voxels]
expected CDF
measured CDF
• Das Messverfahren wird in ASTM E 1441-19 und ASTM E 1695-20 erläutert und definiert.
$CDF(D^*) := \frac{\sigma_m(\mu_i)}{\bar{\mu}} 100\ \%$
Ewert
Mai 2025
Detail-Erkennbarkeit in der iCT
10
SensMonCT

Computer-Tomographie: CDD-Analyse in der Rekonstruktion der zentralen Schnittebene

ASTM E1441

Achtung: ***Das Erkennbarkeitslimit (DDS_{limit}) ist nicht die Schnittstelle von MTF und CDD, wie in älteren Veröffentlichungen beschrieben****. Es handelt sich um den CDD-Wert bei 100%.*

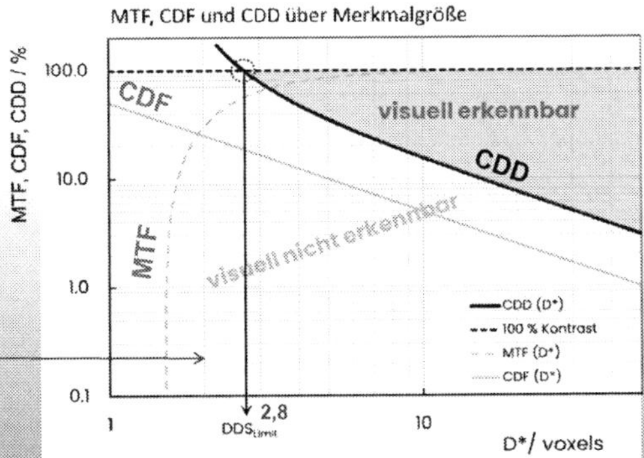

Neue Erkennbarkeitsgrenze für Lochanzeigen mit Durchmesser in Voxeln

- **Kombinierte Darstellung von CDD, CDF und MTF mit 100% CDD als DDS_{limit} bei 2,8 Voxeln.**
- **Das berechnete kleinste Loch, das gerade noch zu sehen ist, ist 70 µm groß bei 25 µm Voxelgröße.**
- **Grau markiert ist der Bereich, in dem Lochanzeigen sichtbar sind.**

$$CDD(D^*) = c \cdot \frac{CDF(D^*)}{MTF\left(\frac{1}{2D^*}\right)}$$

Der physiologische Faktor für menschliche Beobachter ist c= 3 für Zylinderlöcher

Doppelscheiben-BPK

Die Abbildung ist eine Kopie des DDS-Standardentwurfs, der bei ASTM eingereicht wurde.

Doppelscheiben-BPK zur Bestimmung der Detailerkennbarkeit bei unterschiedlichen Strahllängen.

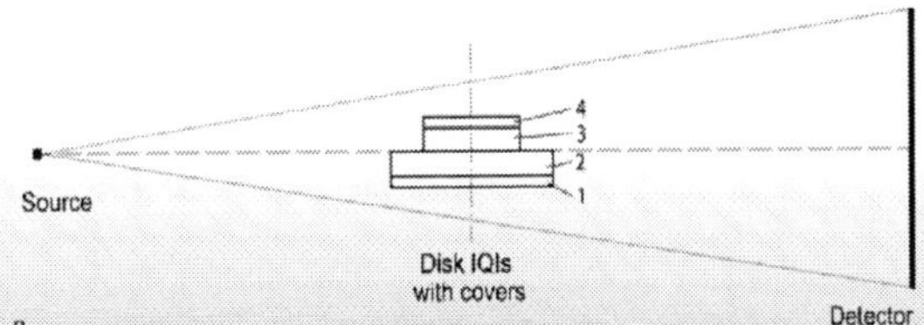

1, 4 – Scheiben ohne Löcher zum Abdecken von Lochscheiben
2, 3 – IQI-Scheiben mit unterschiedlichen Durchmessern unter Berücksichtigung der Strahllängen

Al - Doppelscheiben-BPK:
SR_b^{image} = 50 µm, Röhrenspannung 70 kV, Voxelgröße 25 µm, Vorfilter 0,5 mm Cu.

Links – Lochscheibe mit 50 mm Durchmesser – sichtbare Löcher 80 µm
Rechts – Lochscheibe mit 80 mm Durchmesser – sichtbare Löcher 130 µm

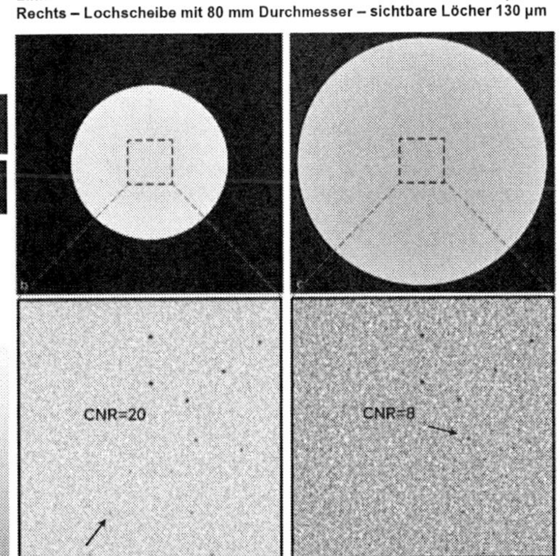

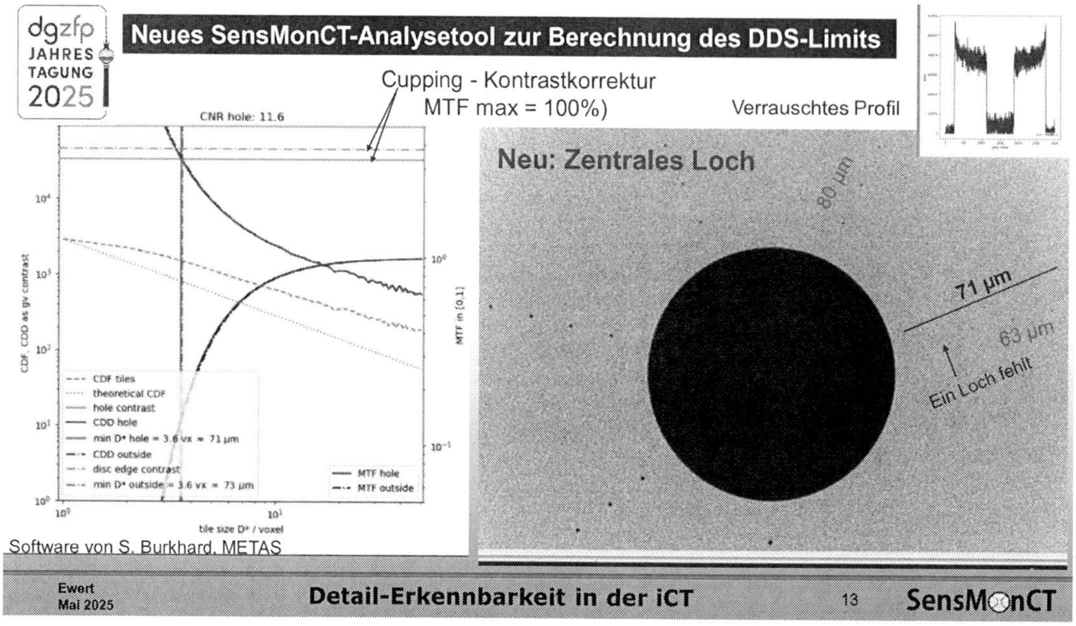
dgzfp JAHRESTAGUNG 2025
Neues SensMonCT-Analysetool zur Berechnung des DDS-Limits
Cupping - Kontrastkorrektur
MTF max = 100%)
Verrauschtes Profil
CNR hole: 11.6
CDF tiles
theoretical CDF
hole contrast
CDD hole
min D* hole = 3.6 vx = 71 µm
CDD outside
disc edge contrast
min D* outside = 3.6 vx = 73 µm
MTF hole
MTF outside
tile size D* / voxel
CDF, CDD as gv contrast
MTF in [0,1]
Neu: Zentrales Loch
80 µm
71 µm
63 µm
Ein Loch fehlt
Software von S. Burkhard, METAS
Ewert
Mai 2025
Detail-Erkennbarkeit in der iCT
13
SensMonCT

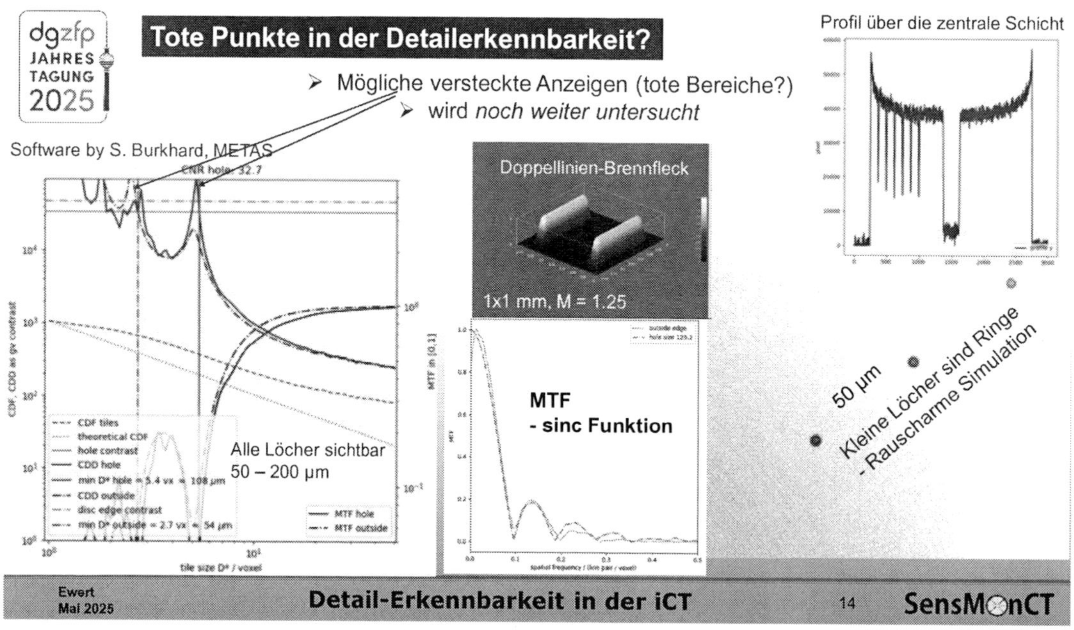
dgzfp JAHRESTAGUNG 2025
Tote Punkte in der Detailerkennbarkeit?
Mögliche versteckte Anzeigen (tote Bereiche?)
wird *noch weiter untersucht*
Profil über die zentrale Schicht
Software by S. Burkhard, METAS
CNR hole: 32.7
CDF tiles
theoretical CDF
hole contrast
CDD hole
min D* hole = 5.4 vx = 108 µm
CDD outside
disc edge contrast
min D* outside = 2.7 vx = 54 µm
MTF hole
MTF outside
Alle Löcher sichtbar
50 – 200 µm
tile size D* / voxel
CDF, CDD as gv contrast
MTF in [0,1]
Doppellinien-Brennfleck
1x1 mm, M = 1.25
MTF
- sinc Funktion
50 µm
Kleine Löcher sind Ringe
- Rauscharme Simulation
Ewert
Mai 2025
Detail-Erkennbarkeit in der iCT
14
SensMonCT

dgzfp JAHRESTAGUNG 2025

Zusammenfassung

- Bildgüteprüfkörper (BPKs) für die Radiographie (Drähte, Plattenlöcher) sind nicht für die Überwachung der Detailerkennbarkeit (DDS) von CT-Schichtbildern geeignet.
- Für die CT kann die Erkennbarkeitsgrenze (DDS_{limit}) von Löchern (und Poren) in Schnittbildern anhand von Kontrastdetaildiagrammen (CDD) bei 100% relativem Kontrast (relativ zur Luft) von Zylinderproben oder Zylinderscheiben bestimmt werden.
- Die CT-Prüfpraxis ASTM E 3375-23 verlangt die Berechnung von CDDs zur Qualitätssicherung.
- Die Erkennbarkeitsgrenze hängt von der Strahllänge im Objekt (Durchmesser der Scheibe) und der Fehlerform ab.
- Zylinderscheiben mit Löchern wurden entwickelt und werden standardisiert (ASTM, ISO), um die Erkennbarkeit (numerisch und visuell) von Löchern zu bestimmen und die Detailerkennbarkeit von CT-Systemen zu überwachen.
- Aufhärtungseffekte tragen zu einer Verschlechterung der menschlichen Erkennbarkeit bei.
- Modifizierte Zylinderscheiben und eine zugehörige Analysesoftware zur Bestimmung von DDS_{limit} werden entwickelt unter Berücksichtigung von Aufhärtungseffekten.
- Die Eignung der vorgeschlagenen Lochscheiben und der Software wird im Rahmen des SensMonCT-Projektes verifiziert.
- Das Verfahren wird in einer ASTM-Norm, einer VDA-Richtlinie und der DGZfP-Richtlinie D7 demnächst und langfristig in einer ISO-Norm der Reihe ISO 15708 beschrieben.

dgzfp JAHRESTAGUNG 2025

Danksagung
"Das Projekt (JRP 23NRM05: SensMonCTII) wurde von der European Partnership on Metrology finanziert, vom Forschungs- und Innovationsprogramm Horizon Europe der Europäischen Union und von den teilnehmenden Ländern kofinanziert."

uwe@ewert-net.de

ENDE

Fragen... ?

uwe@ewert-net.de

Non-Destructive Evaluation of Additively Manufactured Metallic Components

C. Hakan GÜR
Middle East Technical University, Ankara, TR, chgur@metu.edu.tr
[1] Department of Metallurgical and Materials Engineering
[2] Welding Technology and Non-Destructive Testing Center

Abstract. Additive manufacturing (AM), i.e., building up 3D objects by adding materials layer upon layer, produces high-value, sophisticated components that cannot be fabricated via traditional methods. Variations in raw materials and process parameters often cause problems such as dimensional inaccuracy, discontinuities, nonhomogeneous microstructure, and anisotropic behaviour. These may adversely affect the mechanical properties and functionality of the product. Thus, the qualification of AM components requires more effort to verify their structural integrity and performance. Because differences in volume, build lots, and platforms make destructive tests for qualifying AM products less successful, non-destructive evaluation (NDE) has grown in popularity. However, due to their complicated geometry and high microstructural variability, AM parts are challenging to check with traditional NDE techniques. This paper reviews the current state of the NDE approaches used for assessing and qualifying additively manufactured metallic components with a special focus on wire arc additive manufacturing.

Keywords: Metallic components, additive manufacturing, non-destructive evaluation

1. Introduction

Non-destructive Evaluation (NDE) is critical in ensuring safety, integrity, and reliability in all industries by allowing the detection of imperfections created during manufacture (in situ), between the manufacturing phases (in-line), and as the final examination of finished objects (off-line). Non-destructive methods have also been used for microstructure characterization, determination of mechanical properties, and residual stress by measuring physical parameters such as sound velocity, ultrasonic backscattering, electrical conductivity, and magnetic properties; for long-term monitoring of structures, and for estimating the remaining life.

Quality control and assurance methods have expanded as a result of industrialization. NDT was born due to requests for standardized inspections in mass production and technological advances that increased the detectability of flaws. During the 1970s and 1980s, it was realized that a greater understanding of the impacts of harsh conditions on materials and the relevance of defects in terms of failure potential in high-risk technologies such as nuclear, oil-gas, and aerospace was required. Moreover, it was requested to increase inspection reliability, link fault types and dimensions to the structural significance and potential impact on performance or structural integrity, and apply risk-based reliability evaluations. Materials, flaws, and inspection have been integrated with the help of fracture mechanics and finite element analysis. The damage tolerance criteria have been established

https://doi.org/10.58286/31268

to ensure the structural integrity. In the 1980s and 1990s, estimating the crack propagation rate and probability of detection became important tasks. Various engineering methodologies that use NDE have evolved in parallel with changes in industrial demands and the quality assurance perspective. Fitness-for-Service numerically assesses whether a component is fit for continuing use and identifies areas that require repair or replacement by gathering data from several sources, and using materials engineering, fracture mechanics, finite element analysis, statistics, and probability. Structural Health Monitoring senses variations in the signals or magnitudes of the measured parameters. Data collected from many sensors reflects the state of the structure throughout its life cycle, and gives indications for early detection of damage in passive parts in aerospace, wind turbines, and nuclear power plants. Condition-based Maintenance monitors several parameters, such as vibration or temperature in dynamic components of engines and presses, to identify any significant change that would shorten their useful life. It subsequently emerged that measurements of changes in bulk material properties, or material state, were also required due to material degradation and damage build-up. There has been interest in using data gained through NDE prognostics to predict the remaining useful life in life-cycle management approaches for aging systems, such as periodic inspections in power plants. By utilizing NDE data gathered from advanced sensors, robotic systems, wireless data transfer, and improved data processing, novel approaches to prognostics are opening up new options [1,2]. Advances in materials, microelectronics, digital technology, computers, data processing, and modeling have contributed to NDE techniques. New piezoelectric materials for high-temperature ultrasonic inspections [3]; Phased Array Ultrasonics using a probe having numerous elements pulsing individually at varied time intervals that provide 2D or 3D images for flaw detection; innovations in air-coupled ultrasonics and laser ultrasound [4,5] are some examples of the advances in ultrasonic testing. The transition from film to digital radiography systems affected data acquisition, image enhancement, data analysis, and preservation. Industrial radiography has been transformed by digital detectors and computed tomography (CT), and their combination is the most appropriate choice for 3D inspections of components [6]. Advanced mechanical scanning, coupled with data processing, has recently begun to adopt artificial intelligence to assist NDT inspectors. Simulation-assisted Automatic Defect Recognition, which uses deep learning and machine learning to reduce human interaction and error, has been incorporated into ultrasonic and radiographic tests [7].

2. Wire Arc Additive Manufacturing and Non-Destructive Evaluation

Additive Manufacturing (AM) has enabled the production of high-value, complex, and customized metallic components that would not have been possible using traditional methods. Powder bed fusion-type processes, such as Electron Beam Melting and Selective Laser Melting, offer exceptional possibilities for layer-by-layer production of metallic components with complex shapes by selectively melting the metal powder with a moving electron or laser beam. On the other hand, in Wire Arc Additive Manufacturing (WAAM), the metal wire is fed into a welding nozzle at the tip of the robotic arm, which uses an electric arc to melt the metal and deposit it in layers according to a digital 3D design. WAAM has found an area of strength in producing low-cost, large-sized parts made of high-value materials with medium complexity. It can also be used to repair and maintain damaged parts and components. The aerospace sector focuses on manufacturing complex-shaped Ti- and Ni-alloy parts by WAAM [8]. However, further research is needed for the full integration of WAAM into the aircraft design and manufacturing stages [9]. By altering the material locally throughout the WAAM process, more than one type of material can be used in a single manufactured component to create functional designs [10].

Because of the many layers and associated thermal cycles, a suitable combination of WAAM parameters (current, voltage, torch travel and wire feed speeds, interlayer temperature and dwell time, shielding gas and flow rate, torch and substrate angle, electrode diameter, contact tip-to-work distance) is required for specific design requests [11-12]. WAAM technique, in contrast to other AM technologies, does not require a vacuum atmosphere or expensive equipment and can be performed using typical welding equipment. However, machining and post-heat treatments are usually necessary to improve the surface finish and strength and to minimize anisotropy and residual stress [13,14]. Variations in raw materials and process conditions generally cause dimensional inaccuracies, discontinuities, non-uniform microstructure, and anisotropic mechanical behaviour. Mechanical tests for AM parts may be insufficiently accurate due to nonuniformity and anisotropy within the as-built volume, as well as differences in the lots. Furthermore, individual test specimens might not effectively represent the final product. A reliable and efficient methodology is required for the identification and assessment of defects, the determination of material properties, the assurance of integrity, and the qualification and certification of AM components. Closed-loop systems and in-process feedback are essential for managing process performance and variability [15]. NDE can help to tackle these problems. Several studies on non-destructive techniques in AM have been published in recent years [16-18]. As a multiscale and multi-physics coupling, WAAM technology is accompanied by higher heat input, which results in a weak ability to control a larger-sized molten pool, a more complex metallurgical process, difficulty in accurately controlling the energy, and an increased likelihood of producing defects. Simultaneously, internal stress and phase evolution are caused by the non-steady state thermal cycle during the constant scanning of the heat source. The residual stress might cause unacceptable deformation, delamination, and fatigue performance loss. Furthermore, inconsistent material composition, unstable process parameters, and fluctuations in external conditions can easily result in porosity, slag inclusions, lack of fusion, cracks, and hump [19]. Because of the layer-upon-layer characteristic, defects in a deposition layer can also affect the quality of the preceding deposition layer, and even the finished product. These defects may result in the functional failure of parts [20]. Real-time monitoring of material deposition and feedback is critical for the control and stability of WAAM processes [21,22]. Defects often generate optical, thermal, acoustic, and electromagnetic signals. Changes in such signals can be detected and analysed to provide a basis for real-time monitoring [23,24].

2.1 Optical and Thermal Methods

The optical signals can be used to analyze the quality of the deposition layer via a CCD or CMOS camera, optical emission spectrometer, etc. To manage the arc length during the GTAW-based WAAM process, Bonaccorso et al. [25] combined an image processing method and arc voltage measurement. Zhang et al. [26] developed a method for real-time monitoring of weld perturbations and defects in the pulsed GTAW process using the spectroscopy signal. They investigated the correlation between the monitoring parameters, i.e., root mean square, variance, and kurtosis of spectrum bands, and the weld defects. Zhan et al. [27] designed an online monitoring and correction system for plasma welding-based WAAM using a camera-controlled wire-feeding. Huang et al. [28] designed a portable spectrometer based on a linear CCD with real-time acquisition and processing of spectral data to investigate the GTAW of Al-alloys. They developed a classification model with algorithm optimization to provide a reliable assessment of different types of porosities. Tang et al. [29] developed an online intelligent surface defect detection system for WAAM through a deep learning algorithm. To achieve a reliable weld surface, a vision sensor was used to get the image of the shaped surface synchronously. The system was verified to be practical with high accuracy and efficiency for surface defects.

According to Sreedhar et al. [30], infrared thermography can be a practical and reliable weld-monitoring and assessment method using high frame rates and temperature resolution of infrared cameras. Temperature distributions near and around the weld pool can provide statistically distinct features in defect-free and defective regions. Offset-positioned thermal imaging of TIG welding is an effective monitoring technique for detecting porosities in the AA2219 welds. Khanzadeh et al. [31] suggested a method to detect porosities based on morphological characteristics and the temperature distribution of the melt pool's top surface. The 2D melt pool dataset collected by a pyrometer and a thermal camera was further analysed using self-organizing maps to distinguish between similar and dissimilar melt pools. This approach accurately estimated porosity location on Ti-6Al-4V thin wall specimens. Chen et al. [32] proposed a composite passive infrared thermography approach based on the pixel width curve and the stack temperature field to identify the weld bead form problems. A convolutional neural network model was used to differentiate between normal and flawed weld beads, and the system received the defect data back to perform welding and milling repairs.

2.2 Acoustic and Ultrasonic Methods

Acoustic emission (AE) signals can be used to infer the arc condition, melt pool dynamics, and material internal change in WAAM applications. Saad et al. [33] investigated the correlations between an AE signal and the welding pool modes (no-keyhole, keyhole, and cutting). The signal power spectral density was used to assess the data in the frequency domain, and a neural network was used to discriminate the keyhole mode from the cutting mode. According to Pal et al. [34], AE signals can reliably predict droplet transfer mode and defects during the arc welding process. They correlated the collected welding arc sound, current, and voltage signals in the time and frequency domains with the various process parameters and metal transfer modes. The arc sound of continuous and pulsed GMAW was also examined for different process parameters. They concluded that arc sound could be utilized to detect welding flaws. Chen et al. [35] used the arc and sound sensors to obtain information about pulsed GTAW. The voltage and sound features were extracted, and the weighted mean method was used to fuse the signal features to predict the changes in arc length. They noticed that multi-sensor information fusion technology can utilize the complementarities of different sensors and reflect variations in arc length better than single-voltage sensors. The arc length must be stable to achieve stable welding quality. The multi-sensor information fusion technique of combining voltage and sound sensors can be used to control welding torch height. To estimate the weld deposition efficiency in pulsed MIG welding, Bhattacharya et al. [36] developed artificial neural network models that used the welding sound signal in conjunction with the welding current and arc voltage signals. They used an AE sensor, a voltage sensor, and a Hall-effect sensor to monitor the efficiency of the deposition process. AE testing can also be utilized for rapid inspection of as-built parts by installing several sensors on the component to collect data on the location of discontinuities. However, further NDT techniques are necessary to obtain quantitative findings on size and depth [37].

An ultrasonic sensor has been developed to detect porosity formation during powder bed fusion manufacturing. The system was verified on the reference samples for its potential use for in situ implementation [38]. Rieder et al. [39] reported that, for non-complex AM parts, online monitoring of the surface dynamics during the build-up is possible by analysing the ultrasonic backwall signals; also, information about porosities can be obtained by evaluating the ultrasonic velocities. A coupling medium is required to permit transmission and reception of sound waves with little reflection between the probe wedge and material when using conventional UT for in situ applications in WAAM. Geng et al. [40] used an equivalent heat dissipation factor to characterize the variation of heat dissipation in WAAM. To achieve a stable process, they developed a theoretical model to optimize inter-pass temperature and heat input for each layer and verified the model on a wall part. Lopez et al. [41] presented an ultrasonic technique based on in situ detection of defects in WAAM of ER5356 Aluminium straight walls. Ultrasonic tests on walls at temperatures between room temperature and 100°C showed the satisfactory performance of the solder flux as a coupling medium, with little to no soundwave amplitude losses or noise. Finally, the effectiveness of the proposed technique was investigated using acoustic attenuation, impedance, and transmission estimations.

PAUT is suitable for the detection of WAAM defects, but with some limitations due to the level of surface roughness. Common waviness values for WAAM items can easily exceed the millimeter range, thus, machining is required before testing [42]. Javadi et al. [43] proposed a calibration method and inspection strategy for PAUT to estimate the size and shape of the lack of fusion. They manufactured a steel calibration specimen with intentionally embedded small W-carbide balls simulating the defects by depositing 20 layers via WAAM. They reported that the majority of the reflectors were detected successfully by using a 5 MHz 64-element phased array transducer. Lopez et al. [44] assessed the effectiveness of PAUT to find defects in aluminum components with several degrees of surface finish. CIVA simulations and experimental results showed that the PAUT probes can overcome the limitations caused by the surface roughness of the samples, for a maximum of 90 μm average waviness profile; also, defects ranging from 2 to 5 mm can be characterized.

Laser Ultrasonics (LU) shows potential for non-contact in-situ monitoring, even at high temperatures. Millon et al. [45] investigated the effectiveness of a LU system for detecting surface and subsurface flaws. They used Rayleigh surface waves generated with a higher amplitude than volume waves by LU. Two Nd:YAG lasers were used for ultrasound generation and detection. The models showed that a 0.2-mm spot diameter enables to generation and detection of 15 MHz ultrasound frequency and gives access to 0.1-mm size flaws. Everton et al. [46] used laser-generated surface waves to detect artificially generated defects in Ti-6Al-4V samples produced by laser Powder Bed Fusion. Another non-contacting ultrasonic transducer, EMAT (electromagnetic acoustic transducer) is less sensitive than piezoelectric transducers and can only operate at short distances from electrically conducting surfaces. Guo et al. [47] developed an in-situ monitoring system with 0.5-mm spatial resolution and 0.4-s response time using a specially designed hollow EMAT. Experiments showed that the signal power can dynamically monitor the different wire states.

2.3 Eddy Current Method

The eddy current method is suitable for detecting defects such as cracks and non-fusion pores in the final components produced by AM processes. However, since additive manufacturing causes complex temperature fields, it is difficult in situ to evaluate AM processes via eddy current [48]. He et al. [49] developed eddy current testing systems using high-sensitivity magnetic sensors with a small size. The results showed that the methods can detect the flaws in the titanium alloy sample prepared by laser AM. They also compared the results with those of a traditional inductive coil system. Gel'atko et al. [50] investigated the ability of eddy current testing to detect various artificial defects in the stainless-steel samples produced via selective laser melting. The results showed the importance of choosing the appropriate measurement regime, excitation frequency, and secondary parameters; also, defect properties such as shape, size, if they are filled with unmolten powder, or if they reach the surface.

2.4 Radiographic Methods

X-ray computed tomography (XCT) has been efficiently used in determining defect sizes, shapes, and locations. However, challenges exist, such as higher resolution for detecting finer pores to allow for higher accuracy and precision in pore measurement, and correlating porosity and pore distribution measurements to more cost-effective measurement techniques to reduce verification costs [51,52]. XCT provides information about the individual pore sizes and shapes in 3D, with the % porosity values by reconstructing a series of radiographic images [53]. It can also find main faults cost-effectively using quick scans with minimal postprocessing. The capabilities go beyond individual part inspections to process parameter optimization using small test coupons and indirect build inspections using witness specimens. Based on the extensive defect analysis of the witness specimen, this enables the evaluation of complicated parts. It has been proven that scans may be performed at various settings, for various analytical aims, and hence can be tailored to unique quality inspection workflows. In the detailed evaluation, each scan requires many hours to collect data. Synchrotron-based XCT, on the other hand, allows for rapid data collection in a matter of minutes, enhanced spatial contrast and resolution compared to micro-XCT technologies, and sub-micron voxel resolution. Advanced in situ x-ray technologies have recently been employed to generate critical insights into defect generation during the AM process. Recently, data-constrained modelling with multi-energy synchrotron X-rays was used by Xavier et al. [54] to evaluate the distribution of flaws in 316L specimens fabricated using laser metal deposition. The method enables the calculation of sub-voxel volume compositions and, as a result, the calculation of defect sizes below CT resolution. It beats previous thresholding algorithms even when used with higher resolution CT datasets (smaller voxel size), giving a dependable approach for determining the correct quantity of porosities in additively built metal parts.

3. Conclusion

Wire Arc Additive Manufacturing (WAAM) has been used across various industries for individualized medium- to large-scale structures and small-quantity productions. The generation and evolution of defects in the deposition layers can be accompanied by changes in thermal, optical, acoustic, electromagnetic, and radiographic signals which can be detected by various non-destructive evaluation (NDE) methods. The performance of NDE methods varies with the material type, surface condition, shape and position of defects, and environment. Their detection accuracy relies on the sensor's resolution and the correlation between extracted features and defects. Using a single technique or sensor cannot fully obtain defect information due to the intricate geometries and severe environment of WAAM. Thus, it is necessary to develop cutting-edge NDE techniques and multi-technique/multi-sensor systems. Besides, coupling NDE with data capturing and processing technologies, deep learning, and machine learning could open new possibilities for optimizing WAAM process.

References

[1] Vrana J (2020) NDE perception and emerging reality: NDE 4.0 value extraction. Materials Evaluation 78: 835-851.

[2] Bond LJ (2022) From nondestructive testing to prognostics: Revisited. In: Meyendorf N, Ida N, Singh R, Vrana J (eds) Handbook of nondestructive evaluation 4.0, Springer, pp 177-204.

[3] Kažys R, Voleišis A, Voleišiene B (2008) High-temperature ultrasonic transducers: Review. Ultrasound. 63: 7-17.

[4] Gebhardt W, Spies M, Arnold W (2002) Air coupled ultrasonic transducers: Design and applications. In: Maev RG (ed) Acoustical Imaging, Vol 26. Springer, Boston, pp 381-387.

[5] Zhao Y, Chen J, Sun J, Song J, Ma J, Liu S, Zhu Y (2018) Applications of laser ultrasonic technique on nondestructive testing and evaluation of materials, MATEC Web Conf., 173: 02033.
[6] Kastner J, Heinzl C (2015) X-ray computed tomography for NDT and materials characterization. In: Liu Z, Ukida H, Ramuhalli P, Niel K (eds) Integrated imaging and vision techniques for industrial inspection. Advances in computer vision and pattern recognition, Springer, London, pp 227-250.
[7] Balasubramaniam K, Gantala T, Purushothaman P, Surekha M (2022) Simulation assisted automatic defect recognition for NDE 4.0 inspection datasets. JNDE 19: 35-40.
[8] Rodrigues TA, Duarte V, Miranda RM, Santos TG, Oliveira JP (2019) Current status and perspectives on wire and arc additive manufacturing. Materials 12: 1121.
[9] Pant H, Arora A, Gopakumar GS, Chadha U, Saeidi A, Patterson AE (2023) Applications of WAAM for aerospace component manufacturing. Int J Adv Manuf Technol 127: 4995–5011.
[10] Abe T, Sasahara H (2016) Dissimilar metal deposition with a stainless steel and nickel-based alloy using wire and arc-based additive manufacturing. Precis Eng 45: 387-395.
[11] Nguyen VT, Minh PS, Uyen TMT, Do TT, Ngoc HVT, Le MT, Tien Nguyen VT (2023) WAAM technique: Process parameters affecting the mechanical properties and microstructures of low-carbon steel. *Metals, 13:* 873.
[12] Vo TH, Christelle G, Frédéric V (2023) Proposition of a model to determine the process parameters for manufacturing thick parts with WAAM technology. Production & Manufacturing Research, 11: 2240871
[13] Xia C, Pan Z, Polden J, Li H, Xu Y, Chen S, Zhang Y (2020) A review on wire arc additive manufacturing: monitoring, control and a framework of automated system, J. Manuf. Syst. 57: 31-45.
[14] Chen X, Shang X, Zhou Z, Chen S-G (2022) A review of the development status of wire arc additive manufacturing technology. Advances in Materials Science and Engineering 2022: ID 5757484.
[15] Seifi M, Salem A, Beuth J, Harrysson O, Lewandowski JJ (2016) Overview of materials qualification needs for metal additive manufacturing. JOM 68: 747-764.
[16] Lu QY, Wong CH (2018) Additive manufacturing process monitoring and control by non-destructive testing techniques: Challenges and in-process monitoring. Virtual and Physical Prototyping 13: 39-48.
[17] Koester LW, Bond LJ, Taheri H, Collins PC, 2019, Nondestructive evaluation of additively manufactured metallic parts: In situ and post-deposition. In: Froes F, Boyer R (eds), Additive manufacturing for the aerospace industry, Elsevier, pp 401-417.
[18] Gür CH (2022) Non-Destructive Methods for Assessment and Qualification of Metallic Components Additively Manufactured by Powder Bed Fusion Processes. In: e-Proceeding of Int Conf on Adv Mater Sci & Eng and High-Tech Devices Applications, ICMATSE, Ankara, 27-29 Oct 2022, no. 14, pp 22-27.
[19] Wu B, Pan Z, Ding D, Cuiuri D, Li H, Xu J, Norrish J (2018) A review of the wire arc additive manufacturing of metals: properties, defects, and quality improvement. J Manuf Process 35: 127–139.
[20] Chen X, Kong F, Fu Y, Zhao X, Li R, Wang G, Zhang H (2021) A review on wire-arc additive manufacturing: typical defects, detection approaches, and multisensor data fusion-based model. Int J Adv Manuf Technol 117: 707-727.
[21] Rodrigues TA, Duarte V, Miranda RM, Santos TG, Oliveira JP (2019) Current status and perspectives on wire and arc additive manufacturing. Mater 12: 1121.
[22] Xia C, Pan Z, Polden J, Li H, Xu Y, Chen S, Zhang Y (2020) A review on wire arc additive manufacturing: Monitoring, control and a framework of automated system. Journal of Manufacturing Systems 57: 31-45.
[23] Lopez A, Bacelar R, Pires I, Santos T, Quintino L (2017) Mapping of non-destructive techniques for wire and arc additive manufacturing inspection. In: Proc. of the 7th Int. Conf. on Mechanics and Materials in Design, Portugal, pp 1829-1844.
[24] Raut LP, Taiwade RV (2021) Wire arc additive manufacturing: A comprehensive review and research directions. J. of Materi Eng and Perform 30: 4768-4791.
[25] Bonaccorso F, Cantelli L, Muscato G (2011) An arc welding robot control for a shaped metal deposition plant: modular software interface and sensors. IEEE Trans Ind Electron 58: 3126-3132.
[26] Zhang Z, Yu H, Lv N, Chen S (2013) Real-time defect detection in pulsed GTAW of Al alloys through online spectroscopy. J Mater Process Technol 213: 1146-1156.
[27] Zhan Q, Liang, Y, Ding J, Williams S (2016) A wire deflection detection method based on image processing in wire arc additive manufacturing. Int J Adv Manuf Technol 89: 755-763.
[28] Huang Y, Wu D, Zhang Z, Chen H, Chen S (2017) EMD-based pulsed TIG welding process porosity defect detection and defect diagnosis using GA-SVM. J Mater Process Technol 239: 92-102.
[29] Tang S, Wang G, Zhang H, Wang R, 2017, An online surface defects detection system for WAAM based on deep learning. Solid Freeform Fabrication. In: Proc. of the 28th Annual International Solid Freeform Fabrication Symposium, pp 1965-1981. https://api.semanticscholar.org/CorpusID:86866657
[30] Sreedhar U, Krishnamurthy CV, Balasubramaniam K, Raghupathy VD, Ravisankar S (2012) Automatic defect identification using thermal image analysis for online weld quality monitoring. J Mater Process Technol 212: 1557-1566.
[31] Khanzadeh M, Chowdhury S, Bian L, Tschoop MA, 2017, A methodology for predicting porosity from thermal imaging of melt pools in additive manufacturing thin wall sections. In: ASME 2017 12th Int.

Manufacturing Sci. and Eng. Conf. collocated with the JSME/ASME 6th Int. Conf. on Materials and Processing *Vol. 2, Additive Manufacturing, Materials.* LA, California. June 4-8, 2017, V002T01A044.
[32] Chen X, Zhang H, Hu J, Xiao Y, 2019, A passive online defect detection method for wire and arc additive manufacturing based on infrared thermography. In: Proc. of the 30th Annual Int. Solid Freeform Fabrication Symp., pp 1497-1510.
[33] Saad E, Wang H, Kovacevic R (2006) Classification of molten pool modes in variable polarity plasma arc welding based on the acoustic signature. Journal of Materials Processing Technology 174: 127-136.
[34] Pal K, Bhattacharya S, Pal SK (2010) Investigation on arc sound and metal transfer modes for online monitoring in pulsed gas metal arc welding. Journal of Materials Processing Technology 210: 1397-1410.
[35] Chen B, Wang J, Chen S (2010) A study on application of multi-sensor information fusion in pulsed GTAW. Industrial Robot 37:168-176.
[36] Bhattacharya S, Pal K, Pal SK (2012) Multi-sensor-based prediction of metal deposition in pulsed gas metal arc welding using various soft computing models. Appl Soft Comput 12: 498-505.
[37] Li Z, Zou X, Bao F, Gu F, Ball AD (2023) A review of acoustic emission monitoring on additive manufacturing. In: Zhang H, Feng G, Wang H, Gu F, Sinha JK (eds) Proceedings of IncoME-VI and TEPEN 2021, Mechanisms and Machine Science, vol 117. Springer, Cham, pp 867-878.
[38] Slotwinski JA, Garboczi EJ, 2014, Porosity of additive manufacturing parts for process monitoring, In: AIP Conference Proceedings Vol. 1581, pp 1197-1204.
[39] Rieder H, Dillhöfer A, Spies M, Bamberg J, Hess T, 2014, Online monitoring of additive manufacturing processes using ultrasound, 11th European Conference on Non-Destructive Testing, 6-10 Oct 2014, Prag, e-Journal of Nondestructive Testing, Vol. 19, ndt.net/id=16533.
[40] Geng H, Li J, Xiong J, Lin X (2017) Optimisation of interpass temperature and heat input for wire and arc additive manufacturing 5A06 aluminum alloy. Sci. Technol. Weld. Join. 22: 472-483.
[41] Lopez AB, Sousa JP, Pragana JPM, Bragança IMF, Santos TG, Silva CMA (2022) In situ ultrasonic testing for wire arc additive manufacturing applications. *Machines 10:* 1069.
[42] Yehorov Y, da Silva LJ, Scotti A (2019) Balancing WAAM production costs and wall surface quality through parameter selection: A case study of an Al-Mg5 alloy multilayer-non-oscillated single pass wall. J. Manuf. Mater. Process 3: 32.
[43] Javadi Y, MacLeod CN, Pierce SG, Gachagan A, Lines D, Mineo C, Ding J, Williams S, Vasilev M, Mohseni E, Su R (2019) Ultrasonic phased array inspection of a wire arc additive manufactured sample with intentionally embedded defects. Addit. Manuf. 29: 100806.
[44] Lopez AB, Santos J, Sousa JP, Santos TG, Quintino L (2019) Phased array ultrasonic inspection of metal additive manufacturing parts. J. Nondestruct. Eval. 38: 62.
[45] Millon C, Vanhoye A, Obaton A-F, Penot J-D (2018) Development of laser ultrasonics inspection for online monitoring of additive manufacturing. Weld World 62: 653-661.
[46] Everton S, Dickens P, Tuck C, Dutton B, 2015, Evaluation of laser ultrasonic testing for inspection of metal additive manufacturing. In: Laser 3D manufacturing II, International Society for Optics and Photonics, Proceedings of SPIE. 9353.
[47] Guo J, Huang A, Hu R, Xu H, Yang G, Pang S (2020) An in-situ monitoring system for electron beam wire-feed additive manufacturing. Sens. Actuators A Phys 307: 111983.
[48] Chen Y, Peng X, Kong L, Dong G, Remani A, Leach R (2021) Defect inspection technologies for additive manufacturing. International Journal of Extreme Manufacturing 3:022002.
[49] He D, Wang Z, Kusano M, Kishimoto S, Watanabe M (2019) Evaluation of 3D-Printed titanium alloy using eddy current testing with high-sensitivity magnetic sensor. NDT & E International 102: 90-95.
[50] Geľatko M, Hatala M, Botko F, Vandžura R, Hajnyš J (2022) Eddy current testing of artificial defects in 316L stainless steel samples made by additive manufacturing technology. *Materials 15:* 6783.
[51] Thompson A, Maskery I, Leach RK (2016) X-ray computed tomography for additive manufacturing: A review. Meas Sci Technol 27: No. 072001.
[52] Lopez A, Bacelar R, Pires I, Santos TG, Sousa JP, Quintino L (2018) Non-destructive testing application of radiography and ultrasound for wire and arc additive manufacturing. Addit. Manuf. 21: 298-306.
[53] Du Plessis A, Yadroitsev I, Yadroitsava I, Le Roux SG (2018) X-ray microcomputed tomography in additive manufacturing: A review. 3D Printing and Additive Manufacturing 5: 227-247.
[54] Xavier MS, Yang S, Comte C, Bab-Hadiashar A, Wilson N, Cole I (2020) Nondestructive quantitative characterization of material phases in metal additive manufacturing using multi-energy synchrotron X-rays microtomography. Int J Adv Manuf Technol 106: 1601-1615.

DGZfP-Jahrestagung 2025

Röntgen als ZfP-Verfahren im Bauwesen

Sebastian SCHULZE[1]
[1] bauray GmbH, Hamburg, schulze@bauray.de

Kurzfassung

Seit Beginn des Jahrtausends haben bildgebende Verfahren eine erstaunliche Entwicklung in der zerstörungsfreien Prüfung im Bauwesen erfahren. Ultraschallecho wurde mit der Entwicklung von Punktkontaktprüfköpfen wirtschaftlich einsetzbar. Multikanal- und Breitbandantennen haben neue Einsatzmöglichkeiten von Radar im Bauwesen eröffnet. Betondeckungsmessungen sind mit nie dagewesener Geschwindigkeit und Genauigkeit möglich.

Das älteste diagnostische Verfahren aber, die Radiographie - an Materialien zuerst eingesetzt von Wilhelm Conrad Röntgen Ende des 19. Jahrhunderts - wird in der Baupraxis bislang kaum verwendet. Warum das so war, welche Entwicklungen ganz aktuell im Gange sind, und welches Potential das Verfahren besitzt, soll dieser Vortrag zeigen.

Vorgestellt wird das im Jahr 2024 überarbeitete und neu herausgebrachte Merkblatt "B 01 Mobile Durchstrahlungsprüfung im Bauwesen"- Ende der 1980er Jahre immerhin das erste Merkblatt der Bauwesen-Reihe. An vielen aktuellen Einsatzbeispielen aus der Baupraxis aus den Jahren 2023 bis 2025 wird außerdem die wiedergewonnene Praxisrelevanz belegt - von der Identifikation unverpresster Hüllrohre an Spannbetonbrücken über die Bewertung von Querschnittsverlusten an Bewehrungsstahl in situ (d.h. unterhalb der Betonoberfläche), bis hin zum Einsatz im Denkmalschutz.

Keywords: X-ray, RT, radiographic testing, non-destructive testing in civil engineering, NDT-CE, building inspection, structural inspection, tendon duct inspection, monument protection, corrosion

bauray

Röntgen als ZfP-Verfahren im Bauwesen

Dr.-Ing. Sebastian Schulze

Röntgen im Bauwesen - Motivation

bauray 2

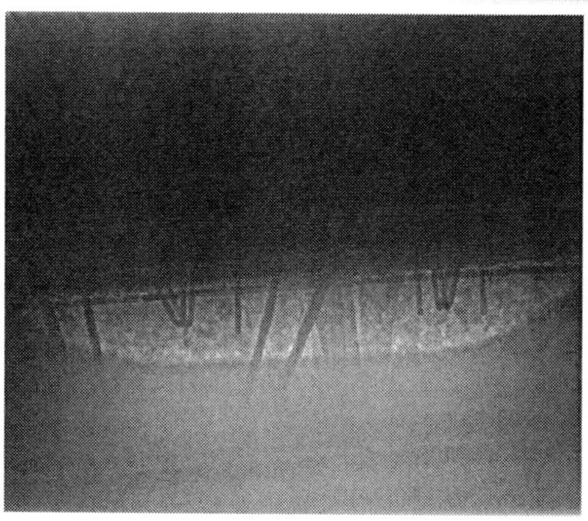

- **Bildzeichnende** Darstellung des Bauteilinnern
- **Millimetergenaue** Rekonstruktion von Einbauteilen
- **Beidseitige Zugänglichkeit** erforderlich
- **Strahlenschutz** einhalten

Röntgen als ZfP-Verfahren im Bauwesen Dr.-Ing. Sebastian Schulze

dgzfp Jahrestagung 2025 26. - 28. Mai Berlin

Neues Merkblatt B 01

bauray 3

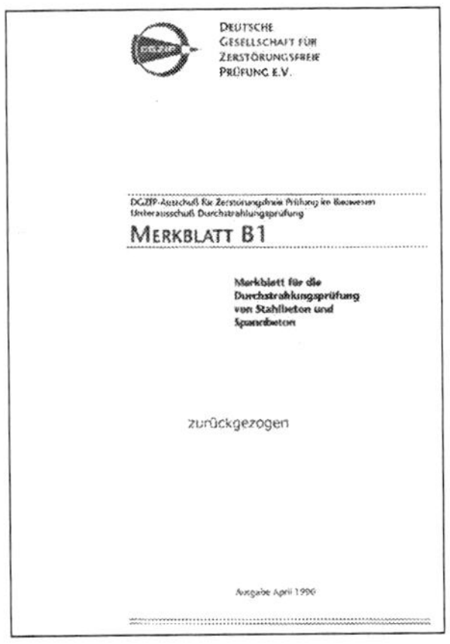

analog	digital (+analog)
Röntgen+Gamma	Röntgen (+Gamma)
Stationär (+mobil)	Mobil (+stationär)
1990	2024

dgzfp Jahrestagung 2025, 26. – 28. Mai Berlin

Röntgen als ZfP-Verfahren im Bauwesen Dr.-Ing. Sebastian Schulze

Neues Merkblatt B 01

bauray 4

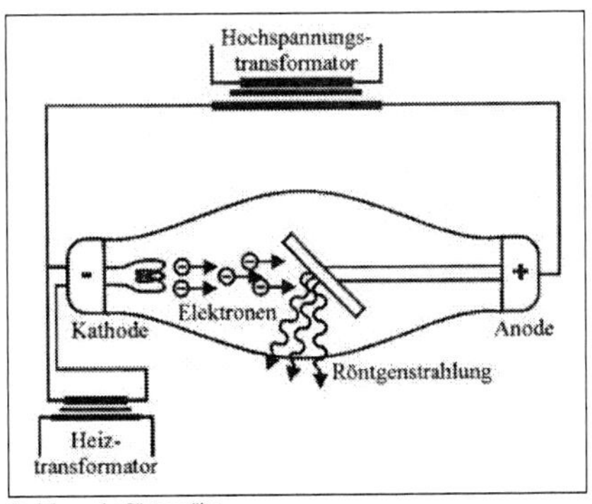

Funktionsweise Röntgenröhre

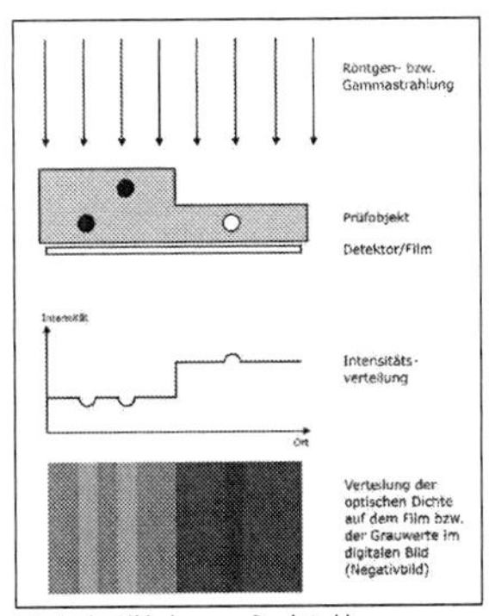

Prinzip der Bildgebung per Durchstrahlung

dgzfp Jahrestagung 2025, 26. – 28. Mai Berlin

Röntgen als ZfP-Verfahren im Bauwesen Dr.-Ing. Sebastian Schulze

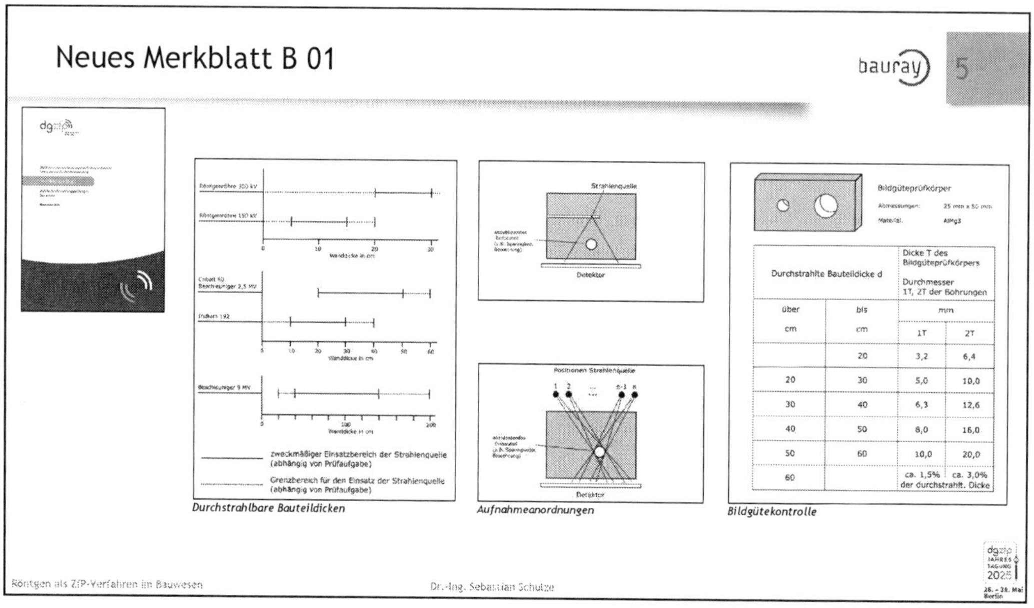
Neues Merkblatt B 01
bauray
5
Strahlenquelle
Detektor
Positionen Strahlenquelle
Detektor
zweckmäßiger Einsatzbereich der Strahlenquelle (abhängig von Prüfaufgabe)
Grenzbereich für den Einsatz der Strahlenquelle (abhängig von Prüfaufgabe)
Durchstrahlbare Bauteildicken
Aufnahmeanordnungen
Bildgüteprüfkörper
Abmessungen: 25 mm x 50 mm
Material: AlMg3
Durchstrahlte Bauteildicke d
Dicke T des Bildgüteprüfkörpers
Durchmesser 1T, 2T der Bohrungen
über cm | bis cm | mm 1T | 2T
– | 20 | 3,2 | 6,4
20 | 30 | 5,0 | 10,0
30 | 40 | 6,3 | 12,6
40 | 50 | 8,0 | 16,0
50 | 60 | 10,0 | 20,0
60 | – | ca. 1,5% | ca. 3,0% der durchstrahlt. Dicke
Bildgütekontrolle
Röntgen als ZfP-Verfahren im Bauwesen
Dr.-Ing. Sebastian Schulze
dgzfp JAHRESTAGUNG 2025
26. - 28. Mai Berlin

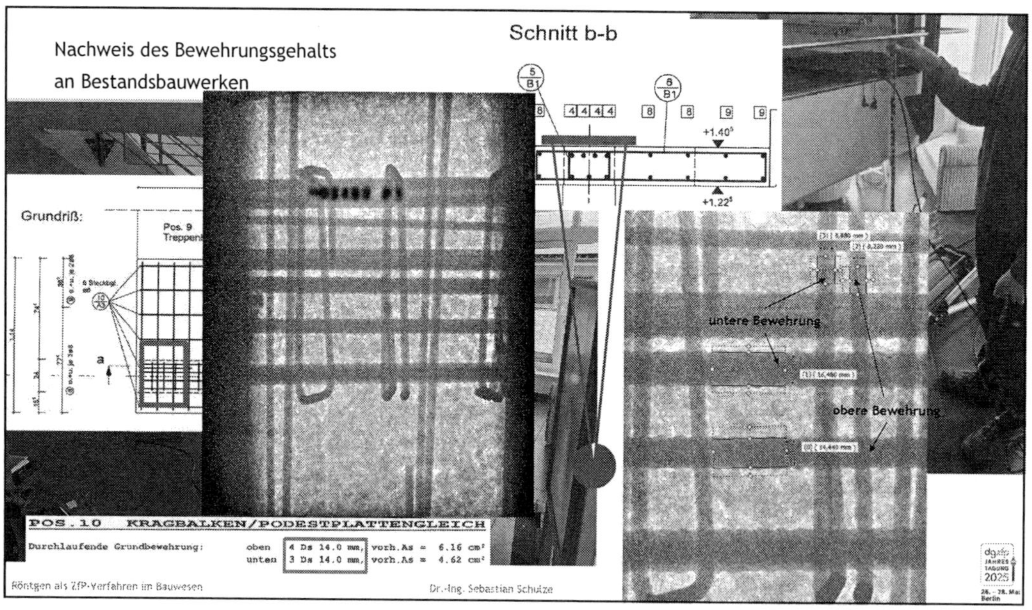
Nachweis des Bewehrungsgehalts an Bestandsbauwerken
Schnitt b-b
+1.40
+1.22
Grundriß:
untere Bewehrung
obere Bewehrung
POS. 10 KRAGBALKEN/PODESTPLATTENGLEICH
Durchlaufende Grundbewehrung: oben 4 Ds 14.0 mm, vorh.As = 6.16 cm²
unten 3 Ds 14.0 mm, vorh.As = 4.62 cm²
Röntgen als ZfP-Verfahren im Bauwesen
Dr.-Ing. Sebastian Schulze
dgzfp JAHRESTAGUNG 2025
26. - 28. Mai Berlin

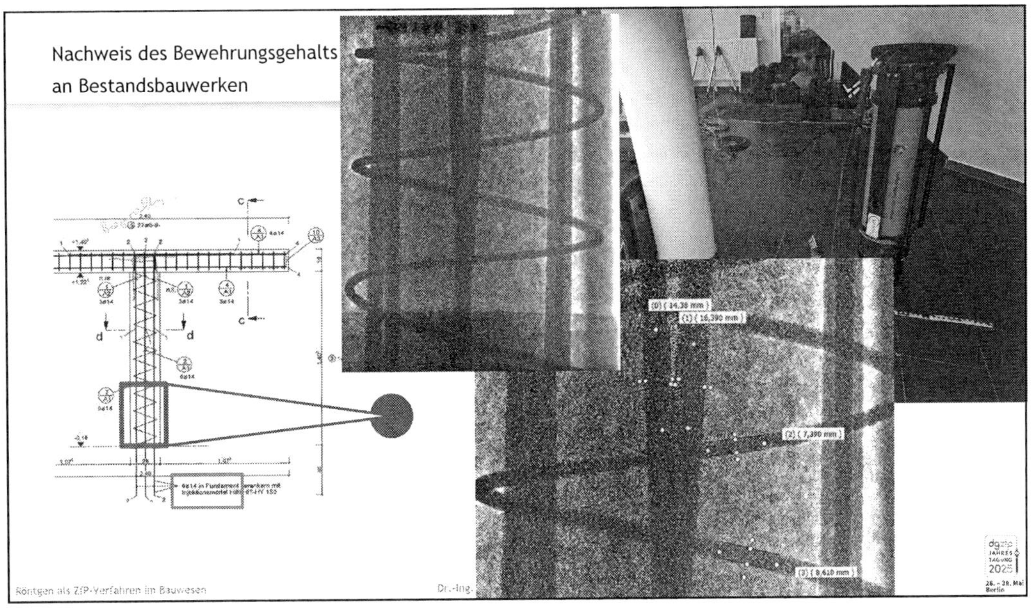
Nachweis des Bewehrungsgehalts an Bestandsbauwerken
Röntgen als ZfP-Verfahren im Bauwesen
Dr.-Ing.
dgzfp JAHRESTAGUNG 2025
26. - 28. Mai Berlin

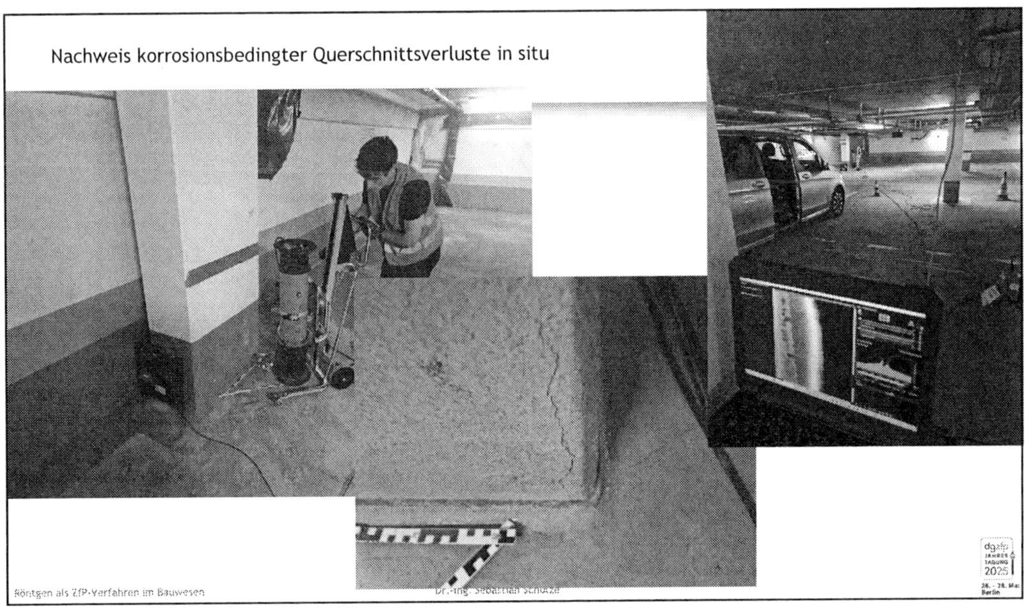
Nachweis korrosionsbedingter Querschnittsverluste in situ
Röntgen als ZfP-Verfahren im Bauwesen
Dr.-Ing. Sebastian Schulze
dgzfp JAHRESTAGUNG 2025
26. - 28. Mai Berlin

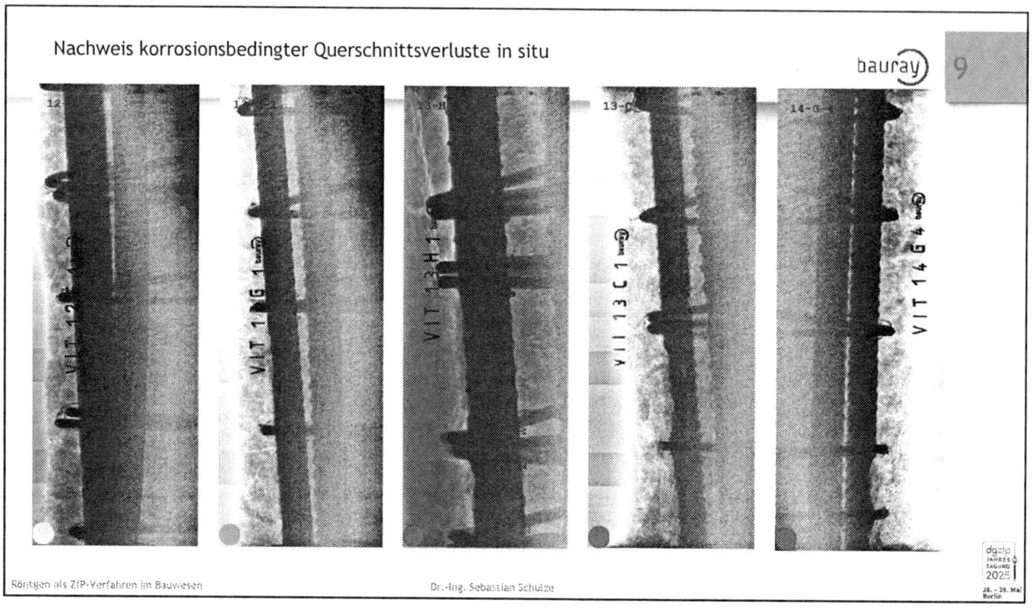
Nachweis korrosionsbedingter Querschnittsverluste in situ
bauray
9
VIT 13 H 1
VIT 13 C 1
VIT 14 G 4
Röntgen als ZfP-Verfahren im Bauwesen
Dr.-Ing. Sebastian Schulze

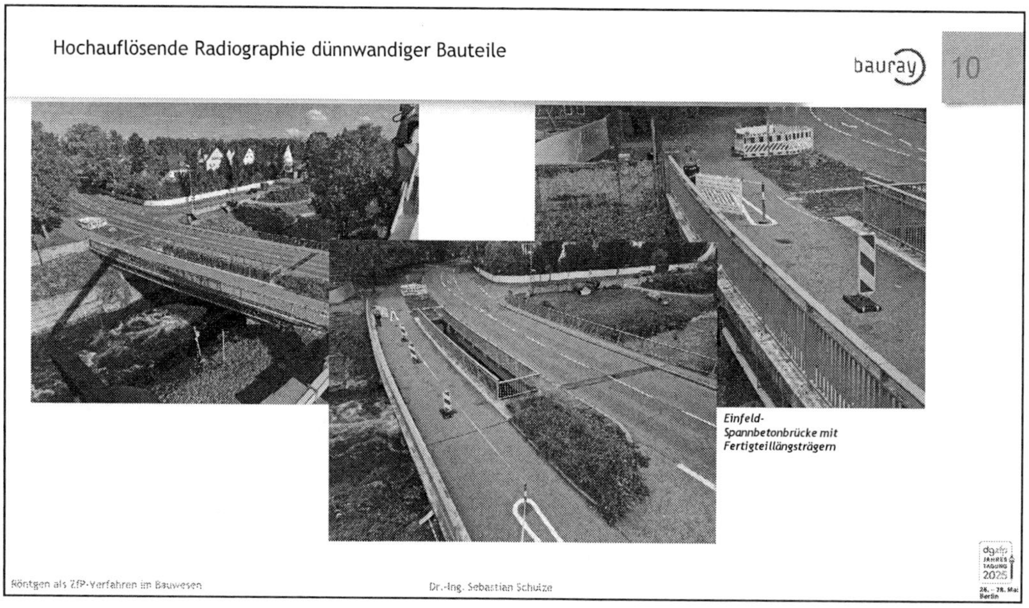
Hochauflösende Radiographie dünnwandiger Bauteile
bauray
10
Einfeld-Spannbetonbrücke mit Fertigteillängsträgern
Röntgen als ZfP-Verfahren im Bauwesen
Dr.-Ing. Sebastian Schulze

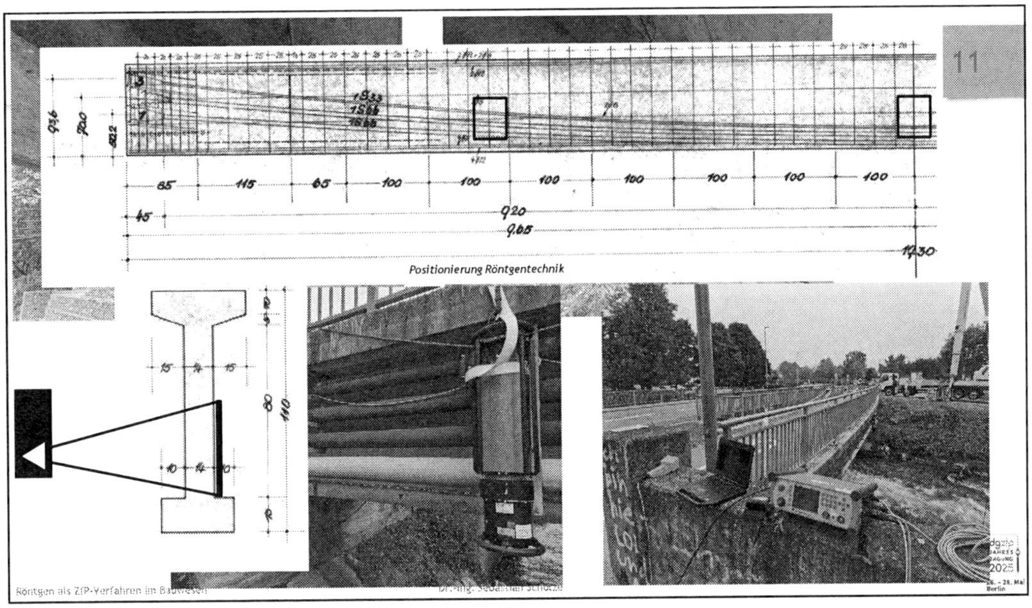
11
Positionierung Röntgentechnik
Röntgen als ZfP-Verfahren im Bauwesen
Dr.-Ing. Sebastian Schulze
dgzfp JAHRESTAGUNG 2025
26. – 28. Mai Berlin

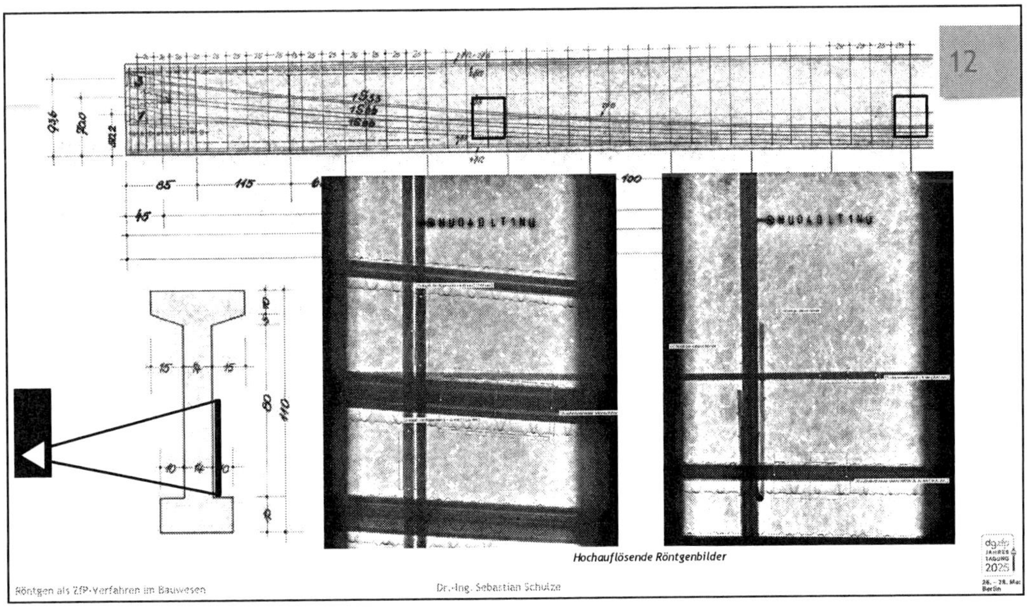
12
Hochauflösende Röntgenbilder
Röntgen als ZfP-Verfahren im Bauwesen
Dr.-Ing. Sebastian Schulze
dgzfp JAHRESTAGUNG 2025
26. – 28. Mai Berlin

Hochauflösende Radiographie dünnwandiger Bauteile

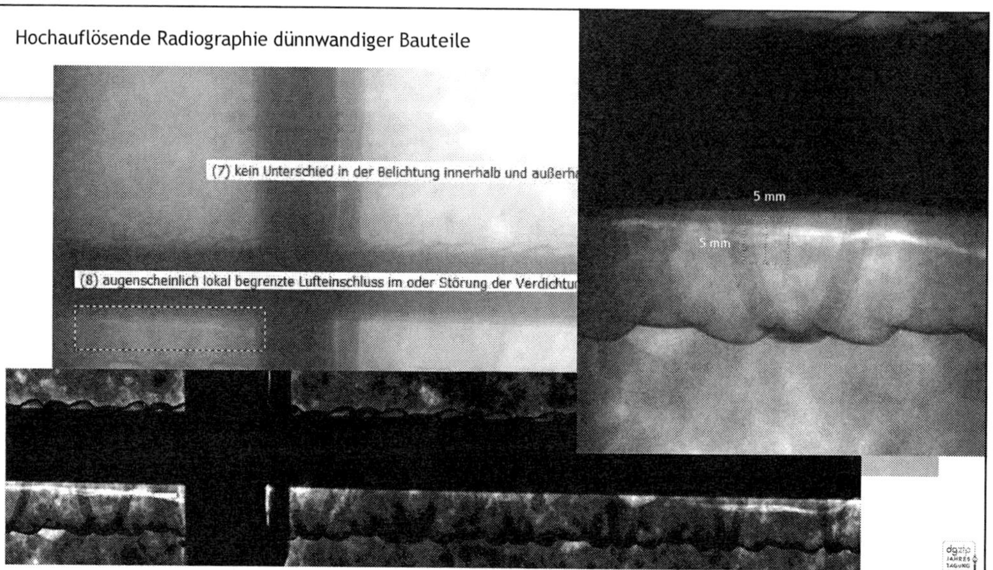

Radiographie an massiveren Bauteilen

bauray 14

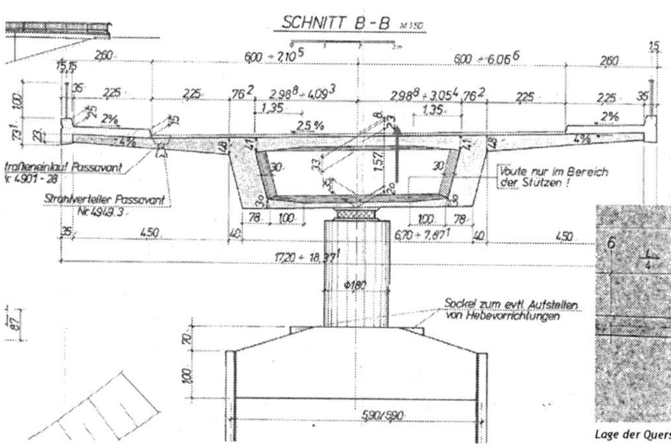

Spannbeton-Hohlkastenbrücke, Durchstrahlungsrichtung Fahrbahnplatte (d = 31 cm)

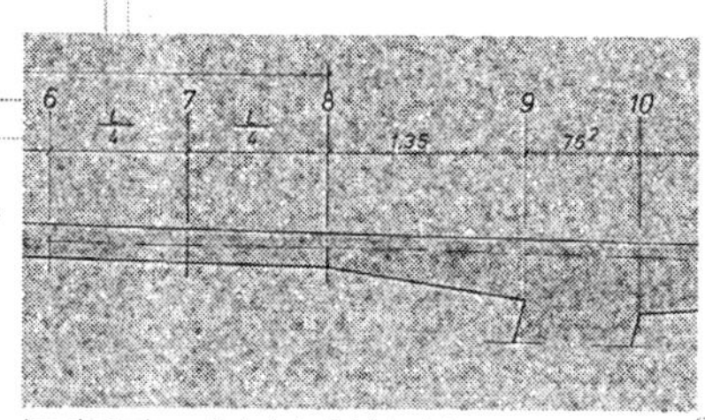

Lage der Querspannglieder in der Fahrbahnplatte

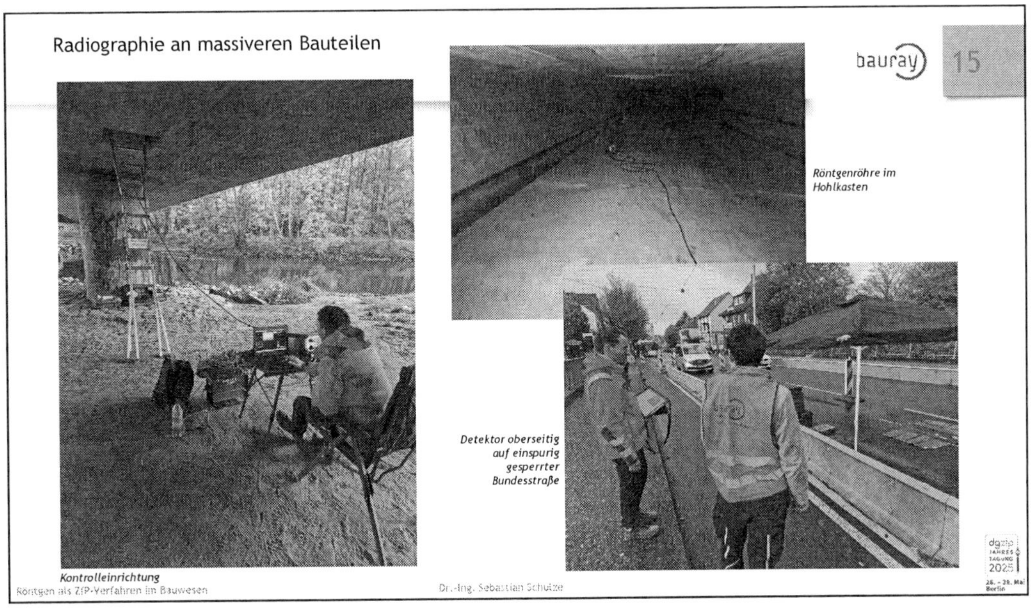
Radiographie an massiveren Bauteilen
bauray
15
Röntgenröhre im Hohlkasten
Detektor oberseitig auf einspurig gesperrter Bundesstraße
Kontrolleinrichtung
Röntgen als ZfP-Verfahren im Bauwesen
Dr.-Ing. Sebastian Schulze
2025

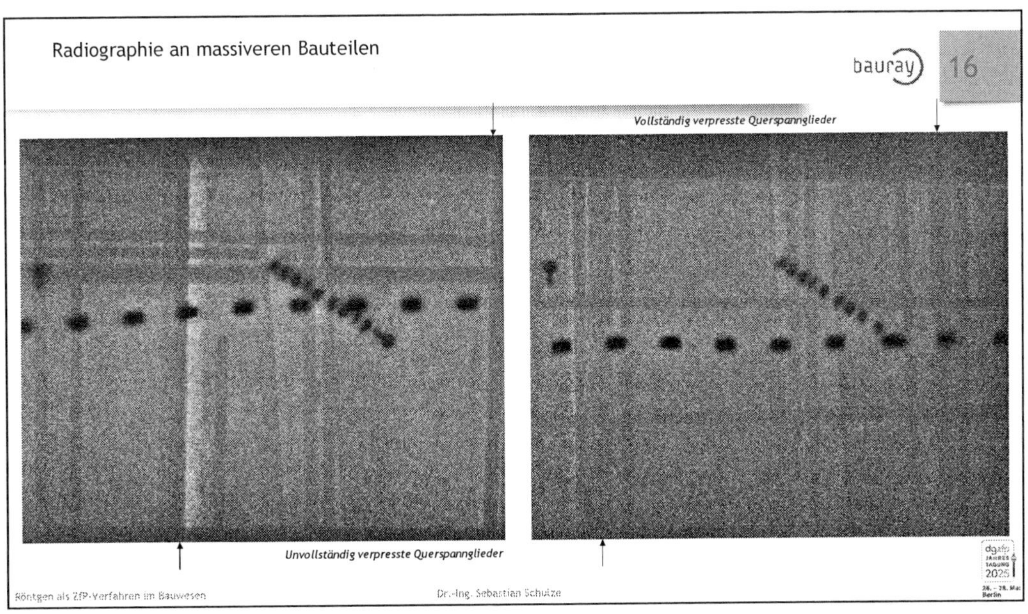
Radiographie an massiveren Bauteilen
bauray
16
Vollständig verpresste Querspannglieder
Unvollständig verpresste Querspannglieder
Röntgen als ZfP-Verfahren im Bauwesen
Dr.-Ing. Sebastian Schulze
2025

Identifikation SpRK-gefährdeter Bewehrung in situ

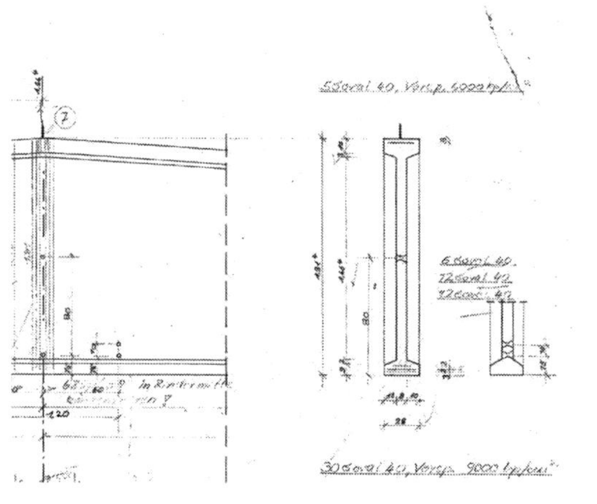

Durchstrahlung Untergurt Spannbetonträgersteg t = 9...12 cm

Identifikation SpRK-gefährdeter Bewehrung in situ

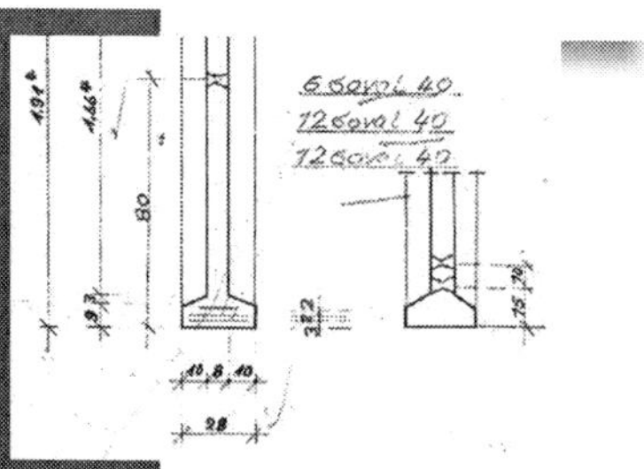

Abmessungen Sigma oval St 145/160

Bezeichnung	Oval 20	Oval 30	Oval 35	Oval 40
Nennabmessung mm	8,0 × 3,0	9,0 × 4,2	10,0 × 4,3	11,0 × 4,5
Querschnitt mm²	20	30	35	40

Aus: BAW-Brief 3/2006 „Spannungsrisskorrosion von Spannstählen"

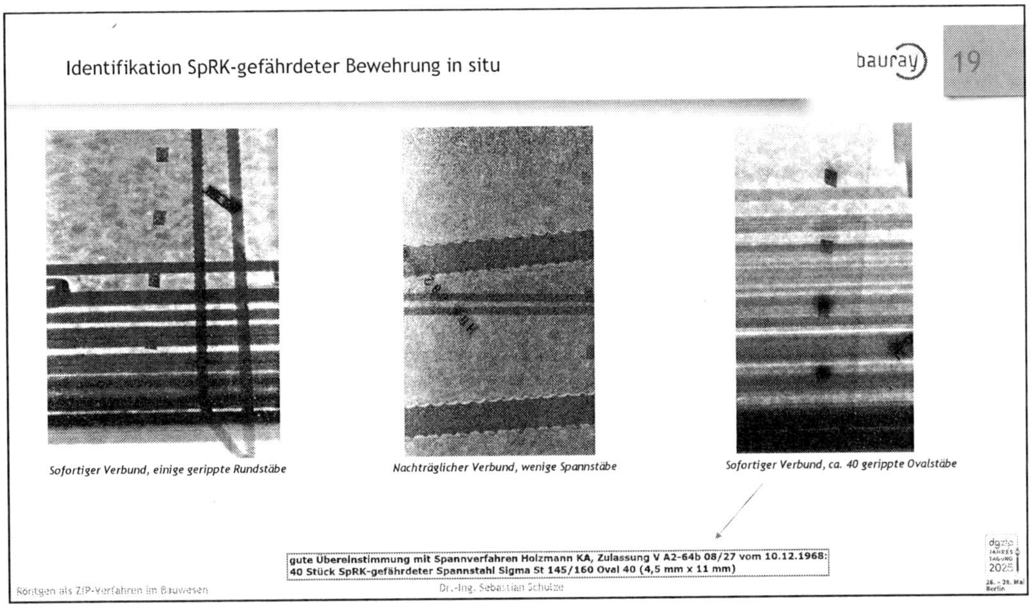

Identifikation SpRK-gefährdeter Bewehrung in situ
bauray
19
Sofortiger Verbund, einige gerippte Rundstäbe
Nachträglicher Verbund, wenige Spannstäbe
Sofortiger Verbund, ca. 40 gerippte Ovalstäbe
gute Übereinstimmung mit Spannverfahren Holzmann KA, Zulassung V A2-64b 08/27 vom 10.12.1968:
40 Stück SpRK-gefährdeter Spannstahl Sigma St 145/160 Oval 40 (4,5 mm x 11 mm)
Röntgen als ZfP-Verfahren im Bauwesen
Dr.-Ing. Sebastian Schulze
dgzfp
JAHRESTAGUNG 2025
26. – 28. Mai Berlin

Anwendung auf Bauteile > 35 cm
bauray
20
Röntgen als ZfP-Verfahren im Bauwesen
Dr.-Ing. Sebastian Schulze
dgzfp
JAHRESTAGUNG 2025
26. – 28. Mai Berlin

Anwendung auf Bauteile > 35 cm

bauray 21

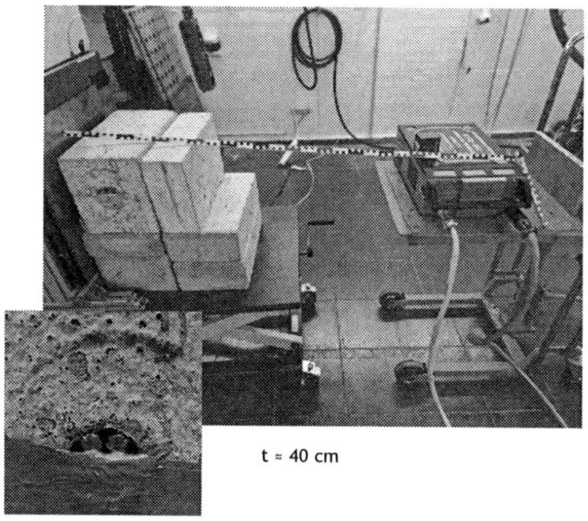

t = 40 cm

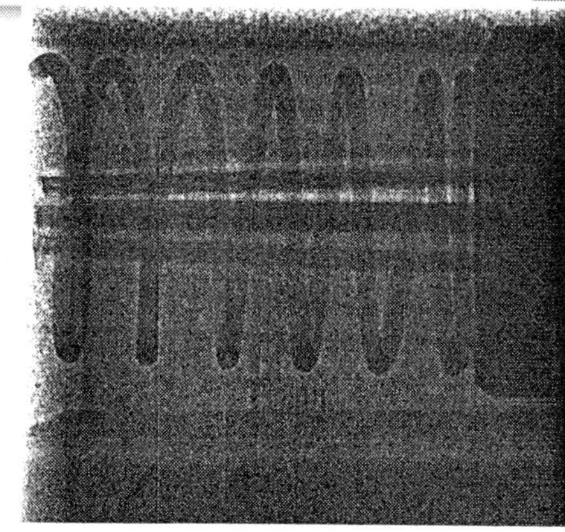

Zusammenfassung

bauray 22

- Radiographie mit wirtschaftlichem Aufwand (mobile, tragbare Eintankröhre 300 keV, realisierbarer Strahlenschutz) aktuell bis etwa 30-35 cm Beton machbar
- Künftig Einsatz bis ca. 60 cm Gesamtstärke möglich (2,5 MeV)
- Durchstrahlung von denkmalgeschützten Bauwerken und Holzbauten
- Identifikation und - bis ca. 20-25 cm - millimetergenaue Vermessung von Einbauteilen
- Keine Bauteilöffnungen zur Verifizierung erforderlich

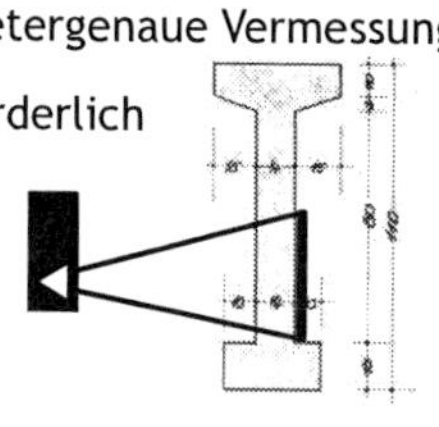

- Beidseitige Zugänglichkeit erforderlich
- Strahlenschutz i.d.R. problemlos

Vielen Dank für die Aufmerksamkeit!

Ultraschallprüfung von Betonkonstruktionen mittels Trockenkontakt-Sensorarrays: Stand der Technik und neue Ansätze

Andrey BULAVINOV, Roman PINCHUK, Andrey SAMOKRUTOV
ACS-Solutions GmbH, Saarbrücken
Kontakt E-Mail: andrey.bulavinov@acs-international.com

Kurzfassung

Die in den modernen Ultraschall-Betonprüfgeräten eingesetzten Punktkontaktsensoren (Dry-Point-Contact / DPC) weisen einzigartige Eigenschaften für die Prüfung von inhomogenen und Starkstreuenden Werkstoffen auf.

Die Ultraschallprüfung von Betonkonstruktionen hat sich als wichtige Methode zur zerstörungsfreien Materialuntersuchung etabliert. Dieser Vortrag beleuchtet den aktuellen Stand der Technik bei der Anwendung von Trockenkontakt-Sensorarrays, die eine effiziente und großflächige Prüfung ermöglichen, ohne dass ein Koppelmittel erforderlich ist. Neben der Diskussion bewährter Ansätze werden neue Entwicklungen vorgestellt, die die Signalqualität und Auflösung weiter verbessern.

Ein besonderer Schwerpunkt liegt auf der Optimierung der Sensorkonfiguration sowie der Weiterentwicklung der Algorithmen zur Datenanalyse und Bildgebung. Darüber hinaus werden innovative Ansätze zur dreidimensionalen Visualisierung von Prüfergebnissen und zur Erkennung von Materialfehlern in heterogenen Betonbauteilen präsentiert. Diese Fortschritte tragen dazu bei, die Zuverlässigkeit und Praktikabilität der Ultraschallprüfung im Bauwesen weiter zu steigern.

Keywords: Betonprüfung mit Ultraschall, Ultraschallbildgebung, niederfrequenter Ultraschall-Gruppenstrahler, Dry-Point-Contact Arrays

Acoustic Control Systems – ACS Group

Ultraschallprüfung von Betonkonstruktionen mittels Trockenkontakt-Sensorarrays: Stand der Technik und neue Ansätze

1

Inhaltsverzeichnis

2

Ultraschallprüfung von Betonkonstruktionen mittels Trockenkontakt-Sensorarrays: Stand der Technik und neue Ansätze

Niederfrequente Ultraschallwandler mit Trockenkontakt: Aufbauprinzip und mögliche Anwendungsfelder

Niederfrequente Ultraschallwandler mit Trockenkontakt: Aufbauprinzip und mögliche Anwendungsfelder

Niederfrequente Ultraschallwandler in klassischer Ausführung 25 KHz bis 1000 KHz

Niederfrequente Ultraschallwandler mit Trockenkontakt: Aufbauprinzip und mögliche Anwendungsfelder

Aufbau des Trockenkontaktwandlers

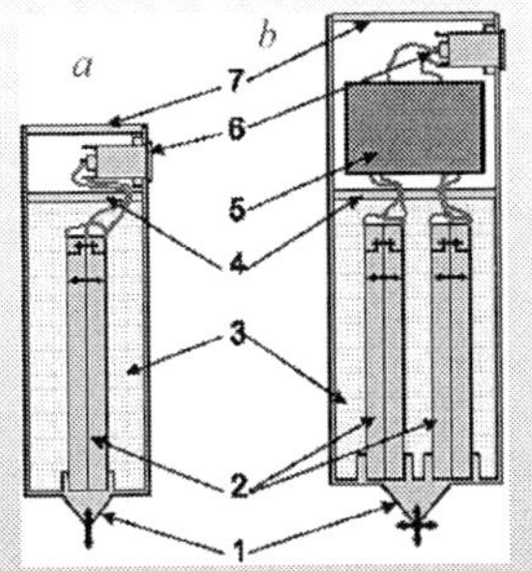

1 - Verschleißspitze
2 - Piezokeramikelemente
3 - Dämpfer
4 - Hermetische Schallwand
5 - Kommutator (orange, nur bei b)
6 – Stecker
7 - Abdeckung

Transducer	Frequency, KHz	Type	Wave mode	Damping
S1902	25	DPC	Shear	Yes
S1905	25	DPC	Longitudinal	Yes
S1844	50	DPC	Longitudinal	No
S1801	50	DPC	Shear	Yes
S1802	50	DPC	Shear	Yes
S1803	100	DPC	Longitudinal	Yes
S1806	100	DPC	Shear	Yes
S1805	150	DPC	Longitudinal	Yes
S1807	150	DPC	Shear	Yes
S1808	250	DPC	Shear	No
S1820	50 / 250	DPC	Shear	Yes
S1823	70	DPC	Long & Shear	Yes
Customized	Up to 400	DPC	Shear or long	-

 5

Niederfrequente Ultraschallwandler mit Trockenkontakt: Aufbauprinzip und mögliche Anwendungsfelder

Niederfrequente Ultraschallwandler und Wandlerarrays

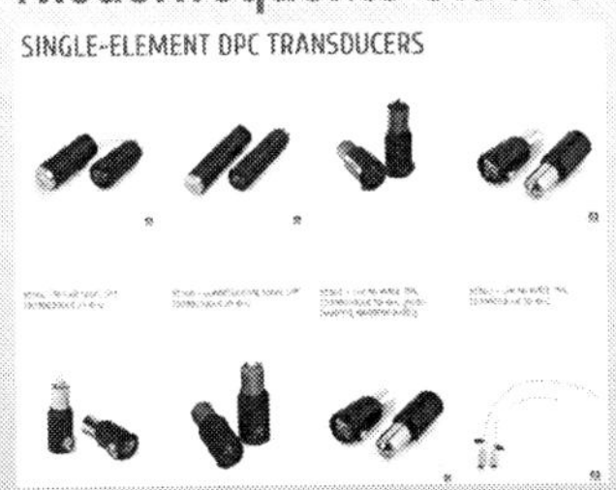

 6

Ultraschallprüfung von Betonkonstruktionen mittels Trockenkontakt-Sensorarrays: Stand der Technik und neue Ansätze

Mögliche Anwendungsgebiete der niederfrequenten Trockenkontaktwandler

 7

Mögliche Anwendungsgebiete der niederfrequenten Trockenkontaktwandler

Materialcharakterisierung durch Pulsgeschwindigkeitsmessung

Ermittlung der Druckfestigkeit

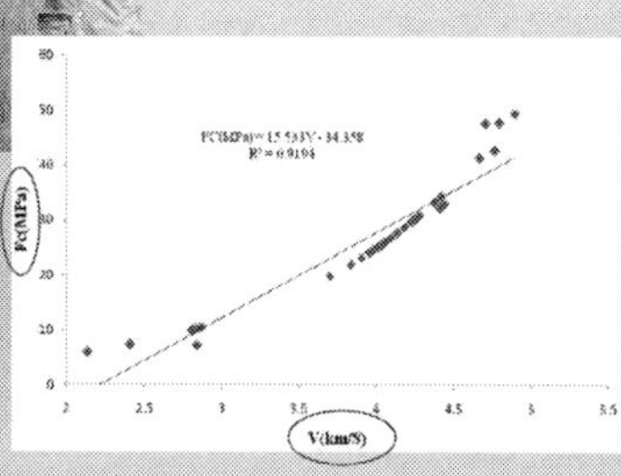

Ermittlung des Elastizitätsmoduls

Der Young-Modul E (Elastizitätsmodul) kann aus den Geschwindigkeiten der **Longitudinalwelle** (v_L) und der **Transversalwelle** (v_T) sowie der **Materialdichte** ρ berechnet werden.

Formel:

$$E = \rho \cdot v_T^2 \cdot \frac{3v_L^2 - 4v_T^2}{v_L^2 - v_T^2}$$

Variablen:

- E: Young-Modul (Elastizitätsmodul), in Pascal (Pa)
- ρ: Dichte des Materials (kg/m³)
- v_L: Geschwindigkeit der Longitudinalwelle (m/s)
- v_T: Geschwindigkeit der Transversalwelle (m/s)

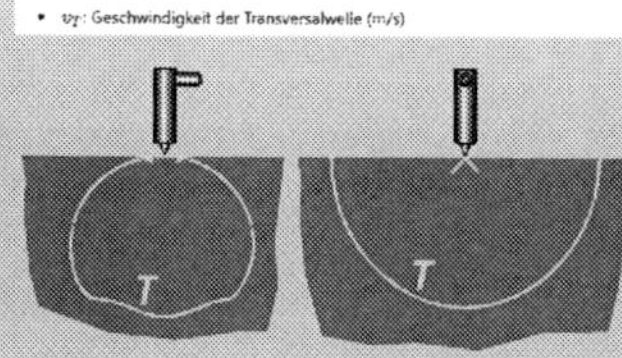

 8

Mögliche Anwendungsgebiete der niederfrequenten Trockenkontaktwandler

Bildgebende Materialprüfung

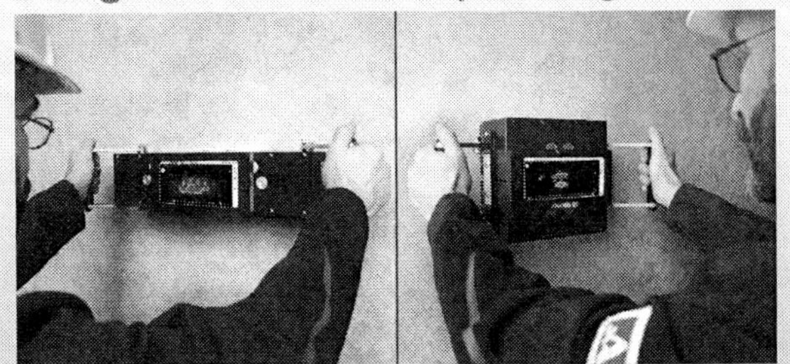

3D View in a single measurement position

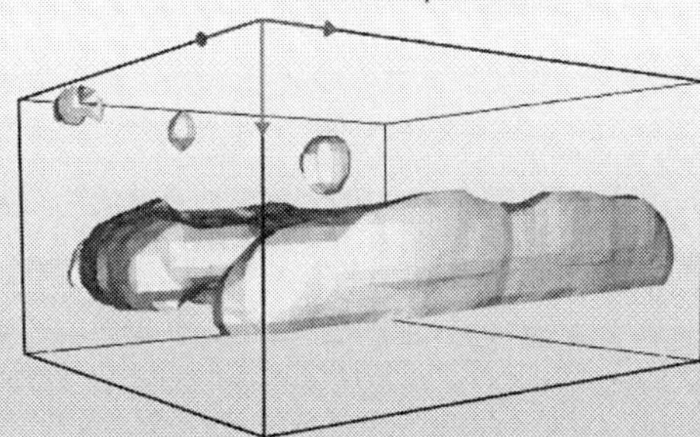

 9

Niederfrequente Ultraschallwandler mit Trockenkontakt: Aufbauprinzip und mögliche Anwendungsfelder

Structural Health Monitoring mit eingebetteten Sensoren

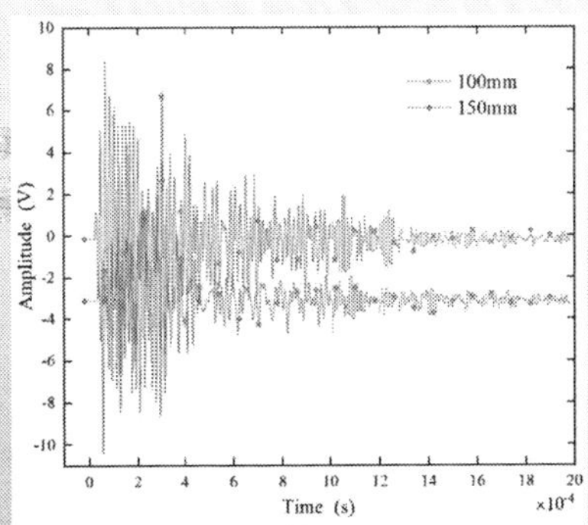

(c)

*** Coda wave based ultrasonic methods for concrete – Bundesanstalt für Materialprüfung BAM in Berlin**

 10

Ultraschallprüfung von Betonkonstruktionen mittels Trockenkontakt-Sensorarrays: Stand der Technik und neue Ansätze

Aktuelle Trends im System und Gerätebau

 11

Aktuelle Trends im System und Gerätebau

1. **Miniaturisierung**
2. **Vernetzung (Link zu Smartphone & Tablet)**
3. **Verwendung von GPU-Rechenpower zur Datenverarbeitung**
4. **Verwendung von KI & Augmented Reality**
5. **Einsatz von Drohnen / Crawlers**

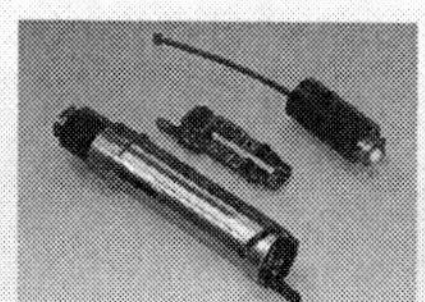

 12

Ultraschallprüfung von Betonkonstruktionen mittels Trockenkontakt-Sensorarrays: Stand der Technik und neue Ansätze

Traditionelle und „ungewöhnliche“ Prüfapplikationen von Trockenkontaktwandlern in der Bauindustrie

 13

Traditionelle und „ungewöhnliche“ Prüfapplikationen von Trockenkontaktwandlern

Wanddickenmessungen beim einseitigen Zugang

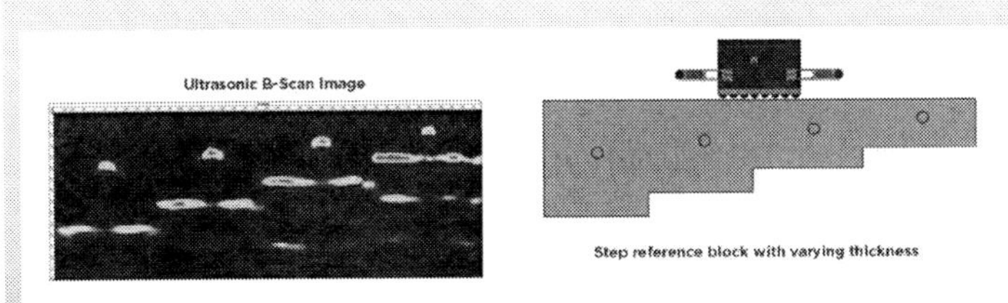

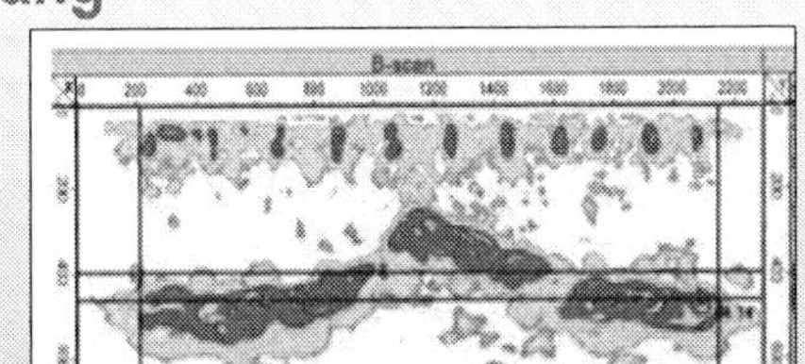

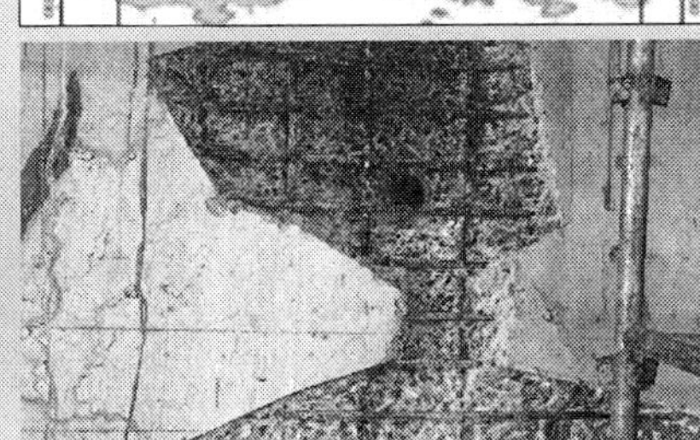

 14

Application of UT on concrete structures

Ortung von Strukturelementen (Spannglieder, Bewehrung, Anker uws)

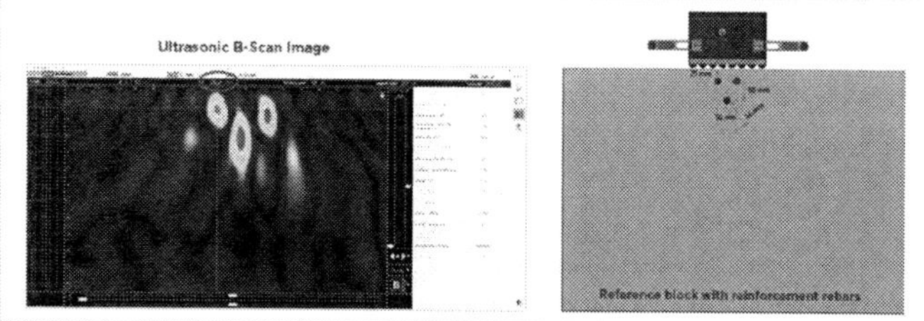

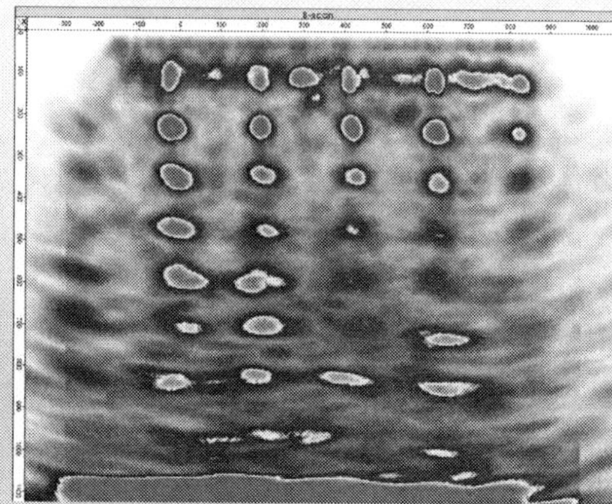

Traditionelle und „ungewöhnliche“ Prüfapplikationen von Trockenkontaktwandlern

Nachweis von Materialfehlern

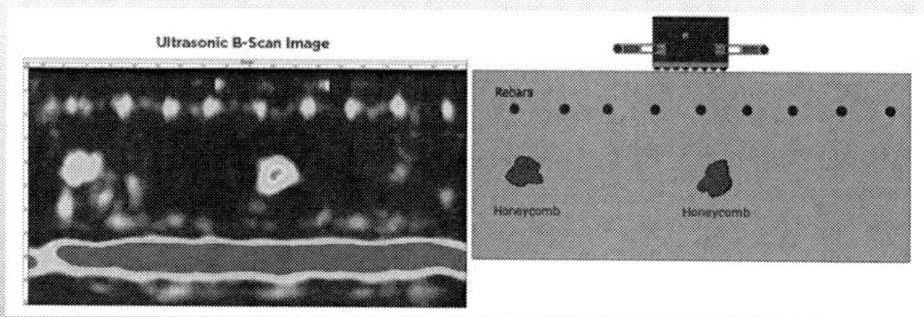

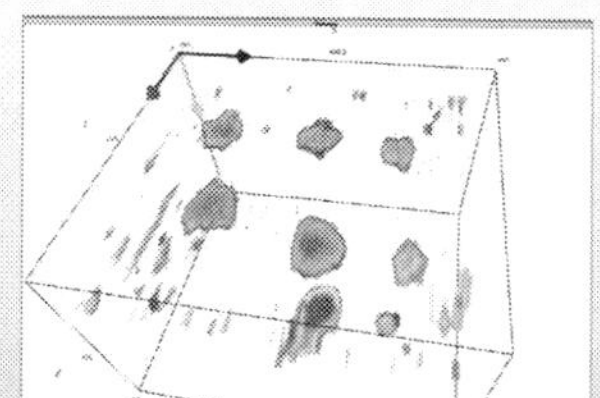

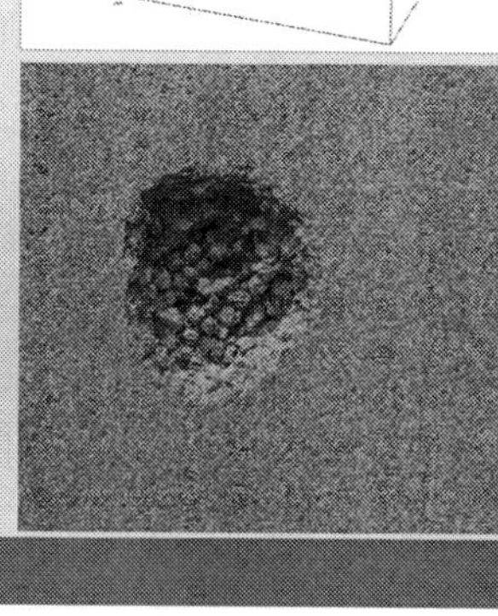

Traditionelle und „ungewöhnliche“ Prüfapplikationen von Trockenkontaktwandlern

Nachweis von Materialfehlern

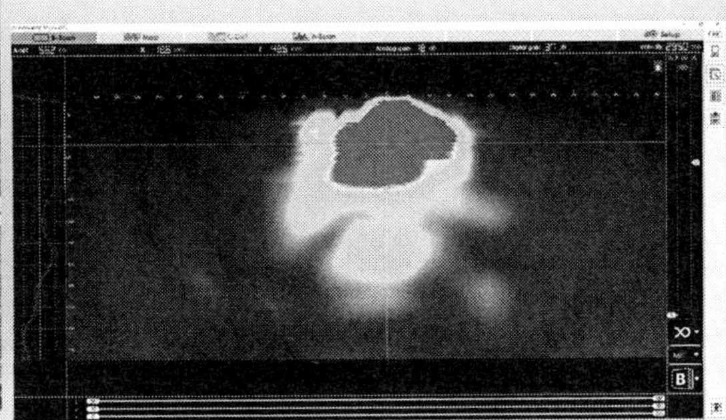

 17

Traditionelle und „ungewöhnliche“ Prüfapplikationen von Trockenkontaktwandlern

Bewertung des Verfüllungsgrads der Spannglieder

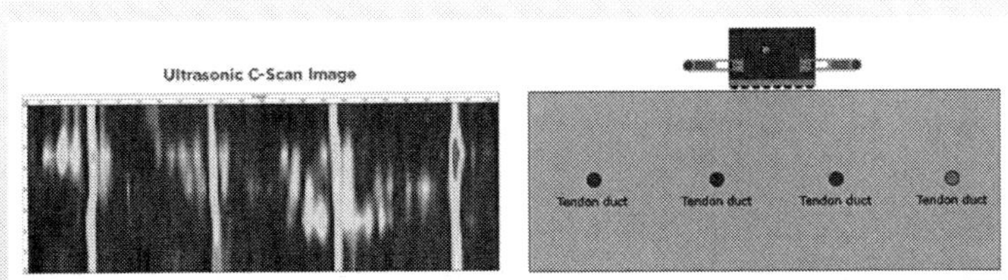

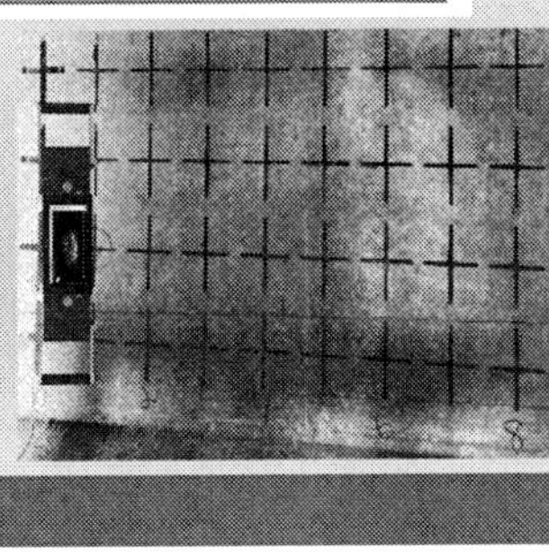

 18

Traditionelle und „ungewöhnliche“ Prüfapplikationen von Trockenkontaktwandlern

Rißtiefenvermessung

Anordnung des Prüfgerätes in Bezug auf den Rißfehler

Beugungsanzeige von der Rißspitze

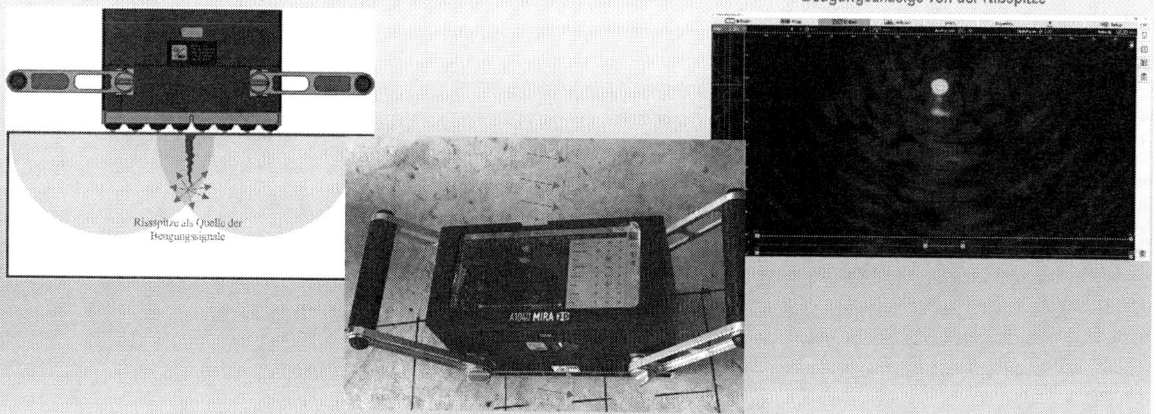

Traditionelle und „ungewöhnliche“ Prüfapplikationen von Trockenkontaktwandlern

Prüfung von feuerfesten Karbonelektroden

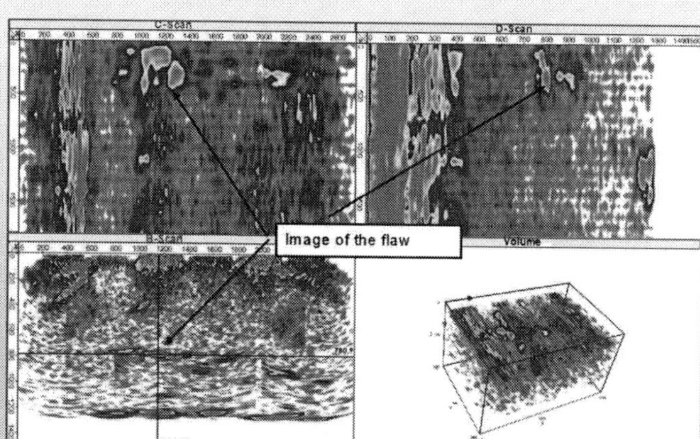

Application of UT on concrete structures

Prüfung von Parmesan-Käse

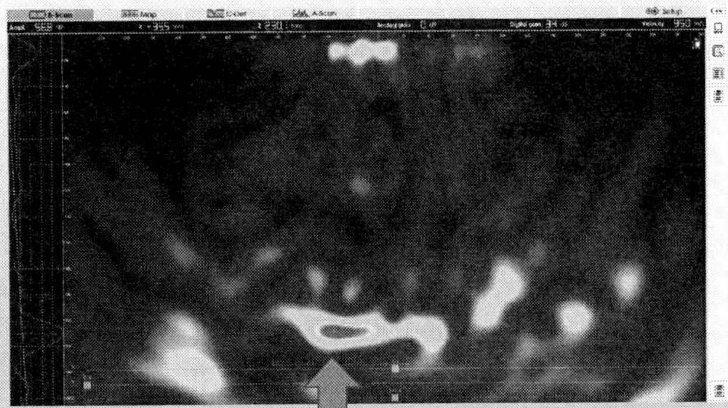

Rückwandanzeige

© Acoustic Control Systems – ACS Group 2025 21

Traditionelle und „ungewöhnliche“ Prüfapplikationen von Trockenkontaktwandlern

Holzprüfung in Durchschallung

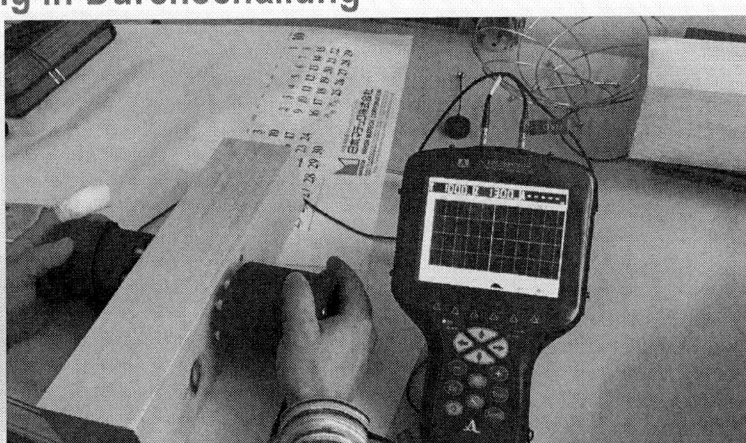

© Acoustic Control Systems – ACS Group 2025 22

Application of UT on concrete structures

Prüfung von Asphalt

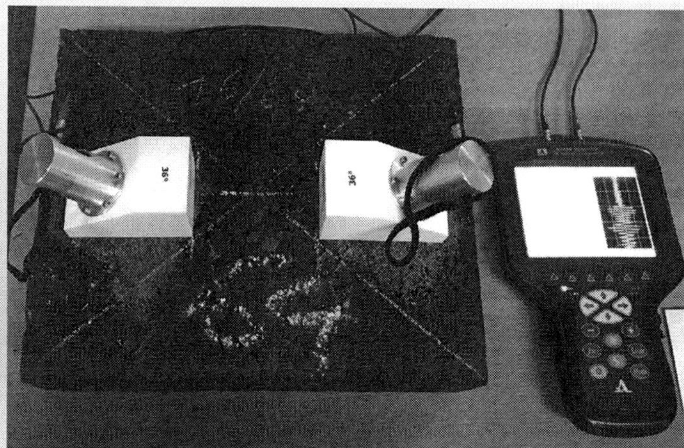

 23

Traditionelle und „ungewöhnliche“ Prüfapplikationen von Trockenkontaktwandlern

Rohrprüfung mit geführten Wellen

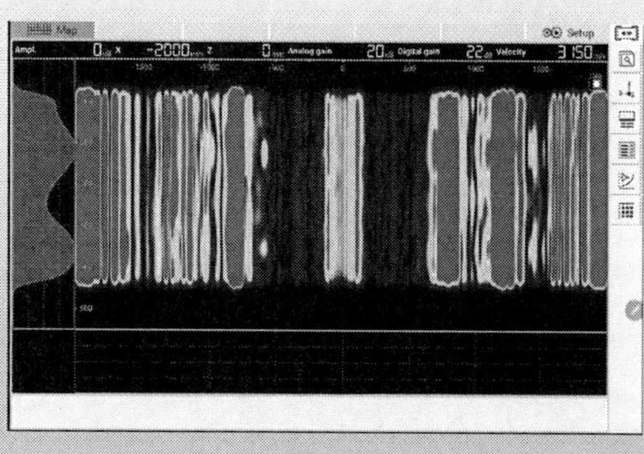

 24

Acoustic Control Systems – ACS Group

Danke für Ihre Aufmerksamkeit

www.acs-international.com

25

DGZfP-Jahrestagung 2025

Improvements in Impact Echo data quality through time signal stacking

Fabian DETHOF [1], Fabian DOLL [1], Kilian-Philipp HELDT [1], Esther KINDEL [1], Maurice KUCZ [1], Hasan STUBLLA [1], Sylvia KEẞLER [1]
[1] Helmut-Schmidt-Universität/Universität der Bundeswehr Hamburg, Hamburg, Deutschland, fabian.dethof@hsu-hh.de

Abstract. The Impact Echo method is an important non-destructive testing method for condition assessments of reinforced concrete structures such as bridges, railways, or power plants. Measurements are typically conducted using only one receiver to measure elastic waves emitted from a point-like mechanical impact. As a result, data quality shows a high degree of variability depending on the surface roughness, geometry of the concrete structure, and the degree of heterogeneity inside the concrete. To overcome this drawback, the advantages of repeated Impact Echo measurements are investigated in this study. For the measurements, a coarse grid is defined first. Within each grid cell, multiple measurements are conducted using different source and receiver locations while maintaining constant offset between them. Then these measurements are stacked in the time domain. Using a constant offset offers two key advantages: it eliminates the need to account for onset delays and ensures varying onset times for side wall reflections. This approach has proven effective for estimating the resonance frequency. However, spatial resolution is reduced due to the averaging of signals, which may limit the detectability of smaller defects.

Keywords: Impact Echo, time averaging, delay and sum, noise reduction, array

Introduction

Impact Echo (IE) is a commonly used non-destructive testing method in the field of civil engineering. Applications of this method include thickness measurements or the detection of defects such as voids [1,2], honeycombs [3], or delaminations [4]. It has also been applied for inspections of the grouting in tendon ducts [5].

During the measurements, a mechanical impact is applied and the resulting waveform is recorded a couple of centimeters away from the impact location. Measurements aim on determining the frequency of the zero-group-velocity S1 Lamb wave mode [6], which is correlated to the thickness of the investigated structure or the distance between the measurement surface and a defect, respectively. Generally, measurements are conducted using a one-source-one-receiver configuration. Therefore, data quality can very heavily depending on the surface roughness, concrete quality, or the degree of heterogeneity inside the concrete (e.g. heavy reinforcement). Another factor, which limits IE data quality is the presence of geometry effects for structures with lateral dimensions not much larger than the thickness [7,8]. As IE aims on measuring Lamb waves, this method is suited best for plate-like structures. If the lateral dimensions of the investigated structure are not significantly larger than its thickness, reflections from the side walls can introduce artifacts in the frequency spectrum, potentially leading to inaccuracies in estimating the S1 Lamb wave frequency [9]. At later stages of IE signals, these reflections excite eigenmodes of the structure [10]. For IE measurements, Eigenmodes with no amplitude modulation in thickness direction seem to be most prominent [8,10]. Different signal processing approaches, which

NDT net
Medien- und Publikationspartner

https://doi.org/10.58286/31234

utilize the short-time Fourier transform and continuous wavelet transform, have already been developed to minimize the influence of geometry effects [11,12].

Unlike for ultrasonic testing, where array devices are common to increase the signal-to-noise ratio of recorded signals, arrays have rarely been used for IE testing and their advantages have therefore not fully been investigated. Groschup and Grosse showed the advantage of repeated measurements for IE and also developed an array device (MEMS) for air-coupled IE testing [13,14]. However, as air-coupled IE signals generally contain more noise than IE signals from sensors coupled directly to the concrete, arrays coupled directly to the concrete should result in the highest data quality.

As IE sensors are generally very expensive, this study aims on presenting a simple measurement setup with one source and one receiver that can be used to simulate a small array. Measurements using this setup are then compared to measurements, which were acquired using a traditional measurement setup. Before that, the idea behind this measurement setup is presented and the coherency of IE signals is discussed.

1. Coherency of IE signals

Usually, arrays use some kind of delay-and-sum technique to maximize the signal-to-noise ratio for certain phases. Therefore, it is important to investigate how the delay needs to be applied for the S1 Lamb wave depending on the source location. Groschup and Grosse [13] have shown that the S1 Lamb wave can be measured at source-receiver offsets smaller than half the thickness of the inspected structure. Also, phase delays in this region are small. However, these observations were made based on the results of 2D simulations. As previous studies have shown that the results of 2D and 3D IE simulations can produce quite different results [15], this observation will be tested for 3D simulations.

To do so, the elastodynamic finite integration technique (EFIT) [16] is used to simulate elastic wave propagation. Here, a plate with dimensions of 1 m × 1 m × 0.2 m is used. The simulated material is assumed to be homogeneous with material parameters representing those found in concrete (P-wave velocity: 4000 m/s, S-wave velocity: 2400 m/s, density: 2300 kg/m^3). With the typical formula used for estimations of the zero-group-velocity S1 Lamb wave mode of:

$$f = \frac{\beta c_p}{2d} \tag{1}$$

a resonance frequency of f = 9530 Hz can be estimated. Here, c_p is the P-wave phase velocity, d is the thickness of the structure, and β is the Lamb wave correction factor, which depends on the Poisson's ratio of the material and has a value of about 0.953 for concrete with a Poisson's ratio of 0.2 [6]. A Gaussian pulse with a center frequency of 12 kHz is used for simulations [15]. To ensure numerical stability, the grid cell size was chosen at 1 mm. Multiple simulations were carried out with sources located along a line at x = 45 cm. Y-positions range from y = 8-92 cm in increments of 4 cm. receivers are located along the same line at y = 2-92 cm in increments of 2 cm. The recorded signals for a source located at y = 44 cm are shown in Figure 1. Here, the undisturbed S1 Lamb wave mode can be observed near the source location at times < 0.5 ms. Afterwards, this signal is masked by reflections coming from the side walls of the plate.

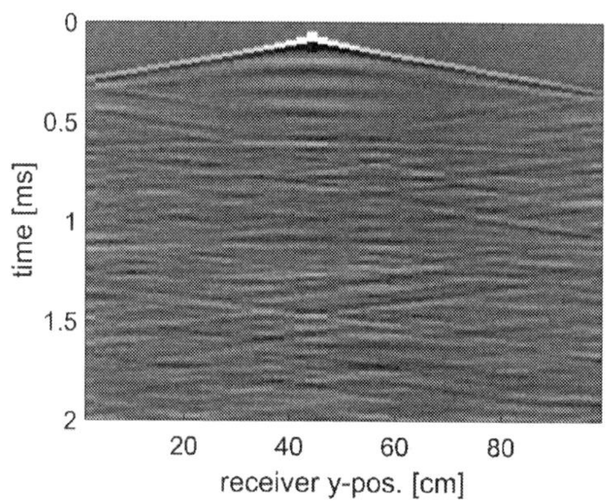

Figure 1. Recorded IE signals for a simulation with the source located at y = 44 cm.

Just from a visual evaluation of this signal, the S1 Lamb wave seems to be coherent around the source location. An IE B-scan is computed using the power spectral density to evaluate the resonance frequency and determine its spatial distribution. Figure 2 shows that the frequency peak is located at around 9.5 kHz and can be observed at distances (offsets) up to 10 cm from the source, which is half the structure's thickness. Evaluating the phase spectrum of the computed signals reveals small phase differences of less than 8° relative to the zero-offset signal, observed for all signals within a 10 cm radius from the source. Therefore, IE signals obtained in the vicinity of the source can simply be stacked in time domain without the need for a delay correction to maximize the S1 Lamb wave signal.

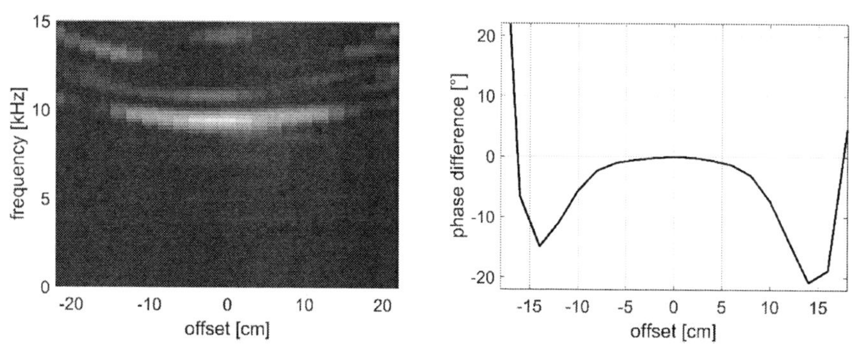

Figure 2. Computed power spectral density (left) for the simulation results shown in Figure 1. Phase difference to the zero-offset signal at the resonance frequency of 9.5 kHz (right).

From these Figures, it can be concluded that signals located closer to the source have the smallest phase difference to the zero-offset signal. Additionally, the window length, at which the undisturbed S1 Lamb wave can be observed is largest for smaller offsets (see Figure 1). However, as it is the goal to minimize the amplitude of side wall reflections, signals with larger offsets will be the focus of the measurement setup proposed in the next section.

2. Measurement setup

2.1 Reciprocity

To understand the proposed measurement setup, first the spatial assignment of IE measurements needs to be discussed. As no standards for the conduction of IE measurements exist, there are no rules whether measurements should be assigned to the source location, the receiver location, or the midpoint between the two. In geophysics, it is common that measurements are assigned to the midpoint between source and receiver. The reason for that is that space-time reciprocity applies to media with time invariant boundary conditions [17]. Spatial reciprocity implies that source and receiver locations could be swapped, and the recorded waveforms would be identical. To demonstrate spatial reciprocity for IE signals, two signals with swapped source and receiver locations at y = 36 cm and y = 44 cm (same plate as in section 1, x=45 cm), respectively, are shown in Figure 3. It can be observed that these signals are almost identical with amplitude differences being more than 10 orders of magnitudes smaller than the amplitudes themselves. The reason for these differences can likely be attributed to numerical inaccuracy during the simulations. Therefore, IE measurements will be assigned to the midpoint between source and receiver hereafter.

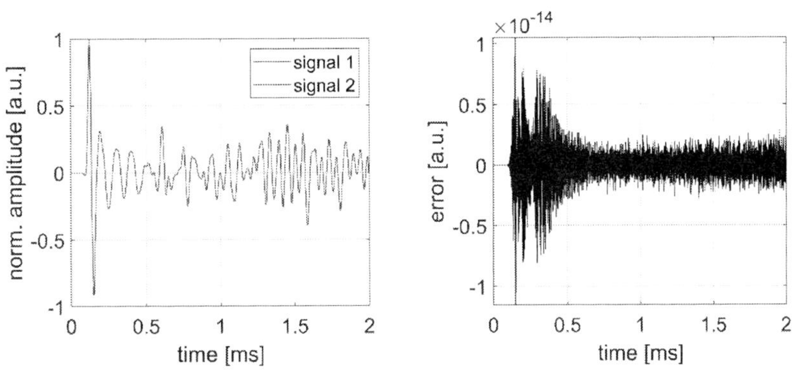

Figure 3. Two exemplary signals recorded at swapped source and receiver locations (left) and the amplitude difference (error) between these two signals (right).

2.2 Time-domain stacking

Next, the effectiveness of time-domain stacking of IE signals shall be demonstrated using simulation results. To do so, all time signals recorded within 10 cm from the source location are summed up without introducing any delay. The results are plotted against the zero-offset time signals in Figure 4. Stacking effectively reduced reflections coming from the side walls as well as the overall noise level at later stages of the signals. However, as the stacking is performed along a line of recordings, reflections coming from the third dimension like the one observable at 0.6 ms cannot be suppressed. To do so, the stacking would need to be performed across a certain area.

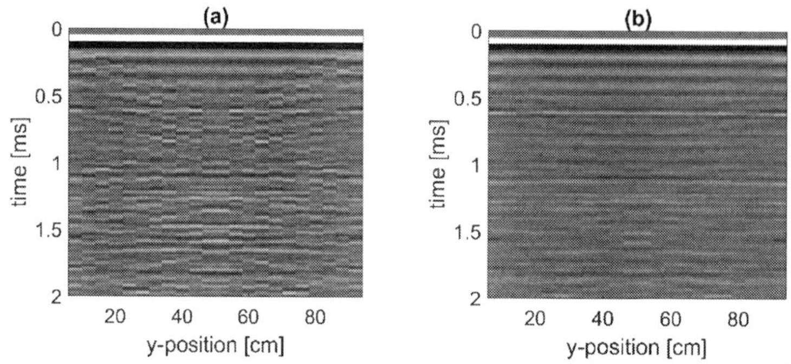

Figure 4. Comparison between simulated zero-offset time signals (a) and stacked time signals (b).

For frequency estimations, stacking of IE time signals is advantageous as well. Figure 5 shows the obtained B-scans (as power spectral density) for the signals in Figure 4. While geometry effects affect the frequency spectra near the edges of the specimen, a homogeneous and constant resonance frequency can be observed for stacked time signals. Also, the overall noise level, especially at higher frequencies, is reduced through the stacking. Please note that common midpoint gathers commonly used in seismics [18] are not effective to suppress side wall reflections, as the total distance these waves need to travel from the source to the receiver stays constant. Therefore, an averaging over different midpoints located closely together is necessary. Consequently, spatial resolution during IE measurements will be compromised in favor of achieving more accurate frequency measurements.

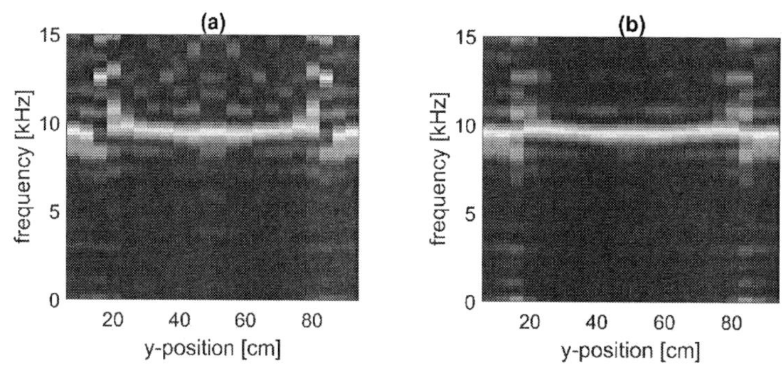

Figure 5. Comparison between power spectral densities of zero-offset time signals (a) and stacked time signals (b).

2.3 Measurement geometry

The proposed measurement setup aims to efficiently stack two-dimensional IE time signals without requiring any additional equipment beyond a standard IE device and a manual impactor. For real-world applications, the proposed measurement setup is quite simple to ensure easy applicability. First, the inspected structure is divided into smaller, square segments. Each segment has a fixed source (or receiver) location in its center. Measurements are repeated four times for each segment with the receiver (or source) located on each corner of the segment. With this setup, it can be assured that the offset between source and receiver is constant for every measurement. Therefore, it is not necessary to accurately determine the source time to have all signals aligned for the phase-coherent summation afterwards. The device-intrinsic triggers (e.g. STA/LTA trigger [19] or using the Akaike Information Criterion [20]) ensure that the signal onset is located at a fixed point in the recorded time signal. Assuming the signal onset is accurately and consistently determined for all measurements, the recorded S1 Lamb waves should be coherent, so that signals can simply be added to obtain stacked signals. A sketch of the proposed measurement setup is shown in Figure 6. Please note that due to reciprocity, the source and receiver locations can be exchanged. The size of the segments should be chosen in a way that the source receiver offset, which in this case is $a/2^{1/2}$ is less than half of the thickness, so that the S1 Lamb wave can be observed. Also, the distance between segments can be varied depending on the spatial resolution desired for IE inspections. Corner points inside the measurement grid do not have to be co-located with those from the neighboring segment as shown in the sketch.

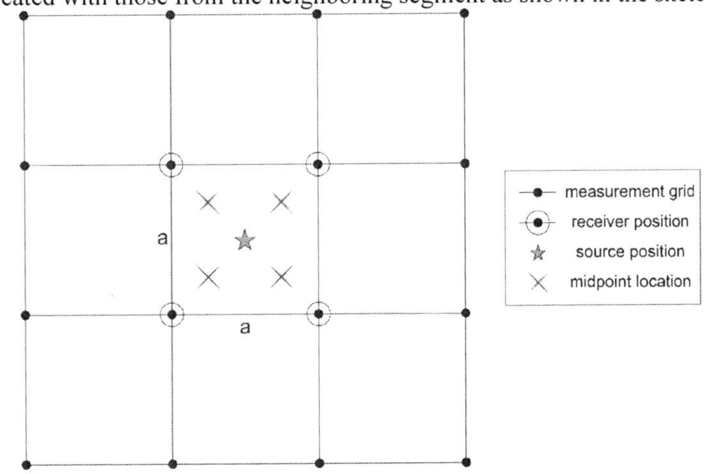

Figure 6. Proposed measurement setup for IE.

2.4 Array response function

This setup effectively allows to emulate a four-receiver array without the need of building or buying one. To measure the quality of an array to suppress other signals based on their incidence angle and phase velocity, the response function of an array can be computed [21]. The array response function (ARF) shows how effective (plane) waves coming from a certain direction can be filtered through time-domain stacking. ARFs are usually plotted depending on the direction where the signal is coming from (backazimuth) and the apparent slowness (inverse of the velocity) of a signal. It is termed "apparent" slowness, as signals might not always propagate along the measurement surface. For example, the S1 Lamb wave was found to arrive almost simultaneously at all receivers of an array during IE simulations. Therefore, the apparent velocity of this signal is infinity, which results in an apparent slowness of zero.

Figure 7a shows the ARF for the proposed measurement setup with a grid width of a=10 cm for waves with a frequency of 10 kHz. It can be observed that artifacts occur for backazimuths of 0°, 90°, 180°, and 270°, which is 45° from the receiver locations and aligns with the center of the measurement grid edges (compare to Figure 6). However, these artifacts occur for apparent slownesses > 0.0006 s/m, which corresponds to apparent velocities < 1667 m/s. As the side wall reflections occurring during IE can be assumed to propagate within the measurement surface, waves with such small apparent velocities should not really be present in IE data (for concrete structures). However, it must be mentioned that receiver locations in the proposed measurement setup are arranged along a circle with radius $a/2^{1/2}$ at even azimuthal increments. If the number of receivers is only increased by one, the side lobes can be minimized even more effectively as shown in Figure 7b.

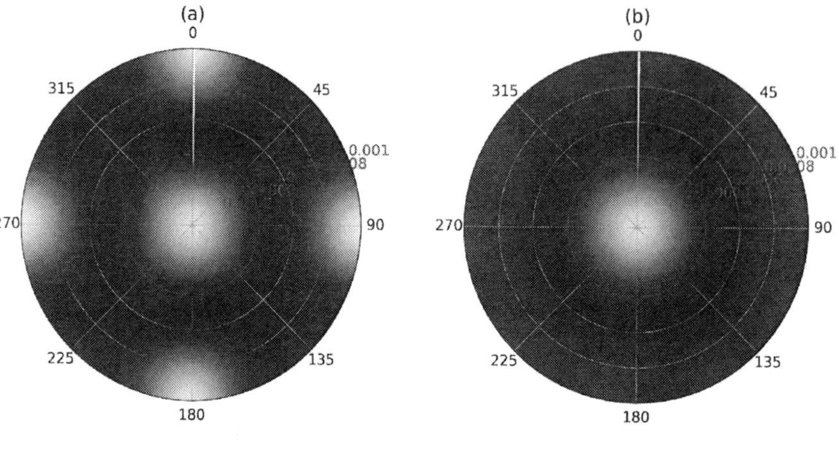

Figure 7. ARF for four receivers (a) in a circular arrangement around a common center (measurement setup from Figure 6). (b) shows the ARF if five instead of four receivers are arranged in a circle. The azimuthal axis shows the backazimuth in degrees, while the radial component gives the apparent slowness in s/m.

3. Results

The proposed measurement setup is applied for measurements on a real concrete structure. The investigated structure is a reference specimen made out of reinforced concrete. It has dimensions of 150 cm × 150 cm × 32 cm and contains 36 spherical honeycombs of different diameters in different depths [22]. It was originally designed to test the capabilities of ultrasonic testing, but IE measurements were carried out on this specimen regardless. The measurement grid contains 14×14 square grid cells with an edge length of 10 cm each, so that a minimum distance of 5 cm was kept to the nearest side wall. With four measurements carried out for each grid cell, a total of 28×28 measurements were conducted. In the post-processing, measurements were arranged according to their midpoint location. To eliminate different strengths of impacts during the measurements, all time signals were normalized to their maximum amplitude. The stacking here is performed by using a 2×2 moving average filter applied to neighboring measurements for each sample of the recorded signals.

The measured time signals can be compared to the stacked signals in Figure 8. The side wall reflections, which mask the S1 Lamb wave oscillation at early stages of the signal are effectively removed and an improved coherency between the signals can be observed. Later stages of IE signals generally show a reduced noise level and specific phases like the signal arriving at about 3 ms can be identified easier.

As the dimensions of the specimen are rather impractical for IE testing, only a short time window at the start of the time signals is used for evaluation. Here, the frequency corresponding to the largest amplitude of the short-time Fourier transform for frequencies between 2-30 kHz and times between 1.2-3.0 ms is determined for each dataset. The used time window had a length of 1.6 ms and an overlap of 95% to the next window.

The results from this frequency analysis are shown in Figure 9. Honeycomb locations are marked green. For a specimen with a thickness of 32 cm, a resonance frequency of about 6 kHz can be expected. From the results, it is evident that the reduced noise level from the time signals translates to a reduced noise level for the obtained S1 Lamb wave frequencies. Especially near the edges of the specimen, where side wall reflections superpose the S1 Lamb wave signal, some very high or very low frequencies can be observed in the raw data. Overall, the reduced noise level for the stacked time signals reduces the difficulty of data interpretation as the frequency shifts caused by honeycombs can now be identified easier. However, as spatial resolution is compromised for better data quality, indications for smaller defects could be easier overlooked due to this approach. Nevertheless, such an effect cannot be observed in the dataset evaluated here and should be investigated in future studies.

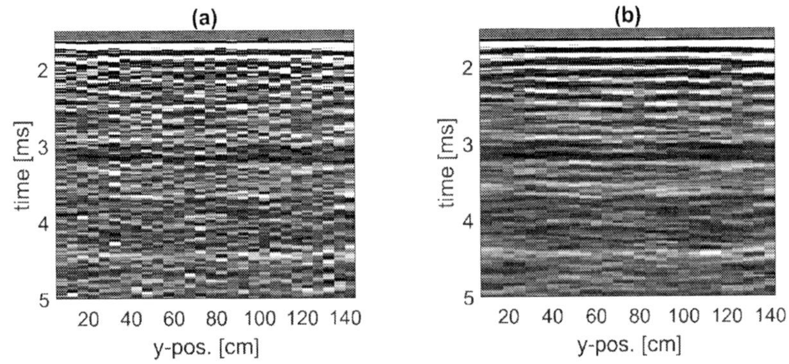

Figure 8. Raw IE time signals from measurements on a reference specimen (a) compared to the stacked time signals (b) for a line scan at x=35 cm.

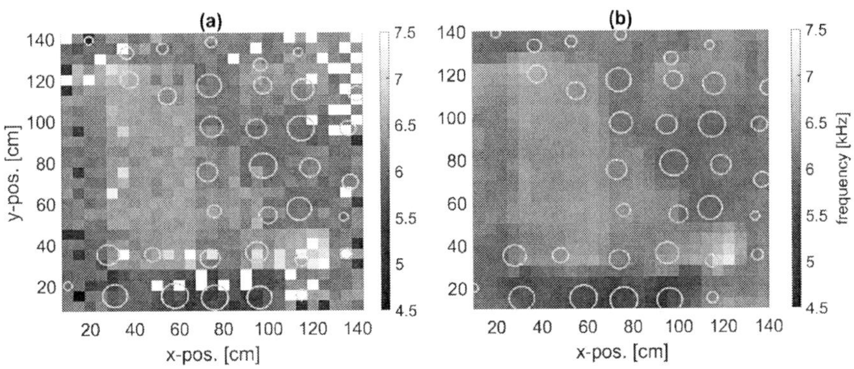

Figure 9. Resonance frequencies obtained from the raw data (a) and the stacked time signals (b).

4. Conclusions

A simple measurement setup for IE has been proposed, which utilizes time-space reciprocity to enable quasi-array measurements for IE devices, which generally consist of only one source and receiver each. Numerical simulations were used to compute the phase shift between the signals of the S1 Lamb wave mode at different distances from the source. The results revealed that the phase shift is nearly zero at distances up to half the thickness of the inspected structure. This knowledge was used to stack IE signals in time-domain using the quasi-array measurements from the proposed setup. The results show reduced noise levels in both the time and frequency domains. Especially reflections coming from the side walls of the structure can be reduced effectively leading to a higher coherency in the obtained time signals. The grid size should be chosen based on the ARF and maximum offset from the source. While the proposed setup sacrifices some spatial resolution for improved data quality, this trade-off should be carefully considered during data evaluation.

References

[1] R. Cao, M. Ma, R. Liang and C. Niu, "Detecting the void behind the tunnel lining by impact-echo methods with different signal analysis approaches," Appl. Sci., vol. 9, no. 16, p. 3280, 2019.

[2] F. Schoefs, O. Abraham and J. Popovics, "Quantitative evaluation of contactless impact echo for non-destructive assessment of void detection within tendon ducts," J. of Construct. Build. Mater., vol. 37, pp. 885-892, 2012.

[3] D. Algernon, S. Feistkorn, Y. Schiegg and B. Mühlan, "Zerstörungsfreie Detektion von Kiesnestern in Stahlbetonbauteilen," Eidgenössisches Departement für Umwelt, Verkehr, Energie und Kommunikation UVEK, 2021.

[4] S. Dorafshan and H. Azari, "Deep learning models for bridge deck evaluation using impact echo," J. of Construct. and Build. Mater., no. 263, pp. 109-120, 2020.

[5] B. Jaeger, M. Sansalone and R. Poston, "Detecting voids in grouted tendon ducts of post-tensioned concrete structures using the impact-echo method," Struct. J., vol. 93, no. 4, pp. 462-473, 1996.

[6] A. Gibson and J. Popovics, "Lamb wave basis for impact-echo method analysis," J. of Eng. Mech., no. 131(4), pp. 438-443, 2005.

[7] F. Schubert, R. Lausch and H. Wiggenhauser, "Geometrical effects on impact-echo testing of finite concrete specimens," in *Proceedings of international symposium non-destructive testing in civil engineering (NDT-CE)*, 2003.

[8] M. Sansalone and W. Streett, Impact-Echo. Nondestructive Evaluation of Concrete and Masonry, Bullbrier Press, 1997.

[9] D. Algernon, H. Wiggenhauser, and F. Schubert "A new approach for the investigation of geometry effects in impact-echo measurements-experimental results and numerical simulations," in *9th European Conference on NDT*, 2006.

[10] F. Dethof and S. Keßler, "Explaining impact echo geometry effects using modal analysis theory and numerical simulations," NDT & E Int., vol. 143, p. 103035, 2024.

[11] F. Dethof and S. Keßler, "Geometry effects in impact echo signals-Why traditional time-frequency filtering fails to remove them," in e-J. of Nondestruct. Test., *Proc. 20th World Conference on Non-Destructive Testing*, Incheon, Korea, 2024.

[12] F. Dethof and S. Keßler, "Robustes Processing für Impakt Echo Signale von kompakten Strukturen," in *DGZfP-Jahrestagung 2024*, Osnabrück, Germany, 2024.

[13] R. Groschup and C. Grosse, "Neue Ansätze zur Anwendung der Impakt-Echo-Methode," in *DGZfP-Jahrestagung 2014*, Potsdam, Germany, 2014.

[14] R. Groschup and C. Grosse, "MEMS microphone array sensor for air-coupled impact-echo," Sensors, vol. 15, no. 7, pp. 14932-14945, 2015.

[15] F. Dethof, D. Algernon, C. Thurnherr and S. Keßler, "Factors Influencing Impact Echo Simulations in Thick Concrete Structures," J. of Adv. Concr. Tech., 2024.

[16] P. Fellinger, R. Marklein, K. Langenberg and S. Klaholz, "Numerical modeling of elastic wave propagation and scattering with EFIT—elastodynamic finite integration technique," Wave motion, vol. 21, no. 1, pp. 47-66, 1995.

[17] K. Aki and P. Richards, Quantitative Seismology, Mill Valley, California: University Science Books, 2009, pp. 122-124.

[18] C. Zhang and T. Ulrych, "Estimation of quality factors from CMP records," Geophysics, vol. 67, no. 5, pp. 1542-1547, 2002.

[19] A. Tmkoczy, "Understanding and parameter setting of STA/LTA trigger algorithm," in New manual of seismological observatory practice (NMSOP), Potsdam, Germany, Deutsches GeoForschungsZentrum GFZ, 2009, pp. 1-20.

[20] C. Suhendi, M. Sudibyo, I. Erlangga and A. Arbad, "Automatic event identification from tectonic earthquakes with modified Akaike Information Criterion (mAIC)," in *IOP Conference Series: Earth and Environmental Science*, 2019.

[21] S. Rost and C. Thomas, "Array seismology: Methods and applications," Reviews of geophysics, vol. 40, no. 3, pp. 2-1 - 2-27, 2002.

[22] F. Dethof and S. Kessler, "Design of Concrete Mock-Ups with Complex Defect Scenarios Using Numerical Simulations," J. of Nondestruct. Eval., vol. 43, no. 59, 2024.

Auf dem Weg zur kontinuierlichen Korrosionsüberwachung von chemischen Anlagen im laufenden Produktionsbetrieb

Dr. Corinna Hardes [1], Dr. Dirk Treppmann [2]
[1] Evonik Operations GmbH, Marl, Deutschland, Corinna.Hardes@evonik.com
[2] Evonik Operations GmbH, Marl, Deutschland, Dirk.Treppmann@evonik.com

Kurzfassung

Chemische Produktionsanlagen sind ein Erfolgsfaktor für den Wirtschaftsstandort Deutschland und ein Versorgungsgarant für lebenswichtige Produkte z. B. der Pharma- und Nahrungsmitteltechnik. Durch die Bereitstellung von Produkten wie Membranen zur Abscheidung von grünem Wasserstoff sind chemische Anlagen längst ein wichtiges Standbein der grünen Transformation geworden. Um die Gefahren für Mensch und Umwelt zu reduzieren, ist die regelmäßige Prüfung von chemischen Produktionsanlagen und deren Druckgeräten gesetzlich vorgeschrieben. Zumeist ist dafür ein vollständiger Anlagenstillstand erforderlich. Die hiermit verbundenen Stillstandzeiten sowie die An- und Abfahrvorgänge kosten Zeit, Geld und Energieressourcen. Die diskontinuierliche Zustandsüberwachung mittels zerstörungsfreier Werkstoffprüfung (beispielsweise Phased Array und Digitale Radiografie), die im Vortrag der Autoren anhand verschiedener praxisnaher Beispiele vorgestellt wird, ist bereits ein wichtiger Meilenstein zur risikobasierten Instandhaltung. Diese bedingt jedoch den Einsatz von Fachpersonal in einer körperlich herausfordernden Arbeitsumgebung. Ein erstrebenswertes Ziel ist es, dem Grundprinzip einer Smart-Watch folgend, die Korrosionsvorgänge in Druckgeräten mit Sensortechnik im laufenden Anlagenbetrieb zuverlässig permanent zu überwachen. Dies stellt zum jetzigen Stand der Technik eine besondere Herausforderung dar, weil Produktionsprozesse nicht selten bei erhöhten Temperaturen von mehreren Hundert Grad Celsius ablaufen. Wie dieser Herausforderung begegnet werden soll, wird auch im Vortrag der Autoren aufgezeigt, die im Rahmen des BMWK-Forschungsvorhabens HotWalls auf die Entwicklung von SHM-Technologie zur kontinuierlichen Zustandsüberwachung von Druckbehältern bei erhöhter Prozesstemperatur mittels geeigneter Sensoren setzen, um Stillstandzeiten zu reduzieren und die CO2-Bilanz der Prozesse zu optimieren.

Keywords: Zustandsüberwachung, Chemische Industrie, Instandhaltung, Druckgeräte, Sensortechnik, Werkstofftechnik, Bundesministerium für Wirtschaft und Klima (BMWK)

Auf dem Weg zur kontinuierlichen Zustandsüberwachung von chemischen Anlagen im laufenden Produktionsbetrieb

Dr. Corinna Hardes, Dr. Dirk Treppmann

Dr. Corinna Hardes
Evonik Operations GmbH
Paul-Baumann-Str. 1, 45772 Marl
corinna.hardes@evonik.com

EVONIK
Leading Beyond Chemistry

Chemische Produkte von Evonik
Versorgungssicherheit für das alltägliche Leben

Hygieneprodukte
Rhamnolipid-Biotenside für umweltfreundliche Duschgels und Spülmittel

Kosmetika
Emollients für nachhaltige Kosmetika

Pharmazeutika
Lipidnanopartikel für Nukleinsäuren wie mRNA

Nahrung
Veganes Kollagen als Nahrungsergänzungsmittel

EVONIK
Leading Beyond Chemistry

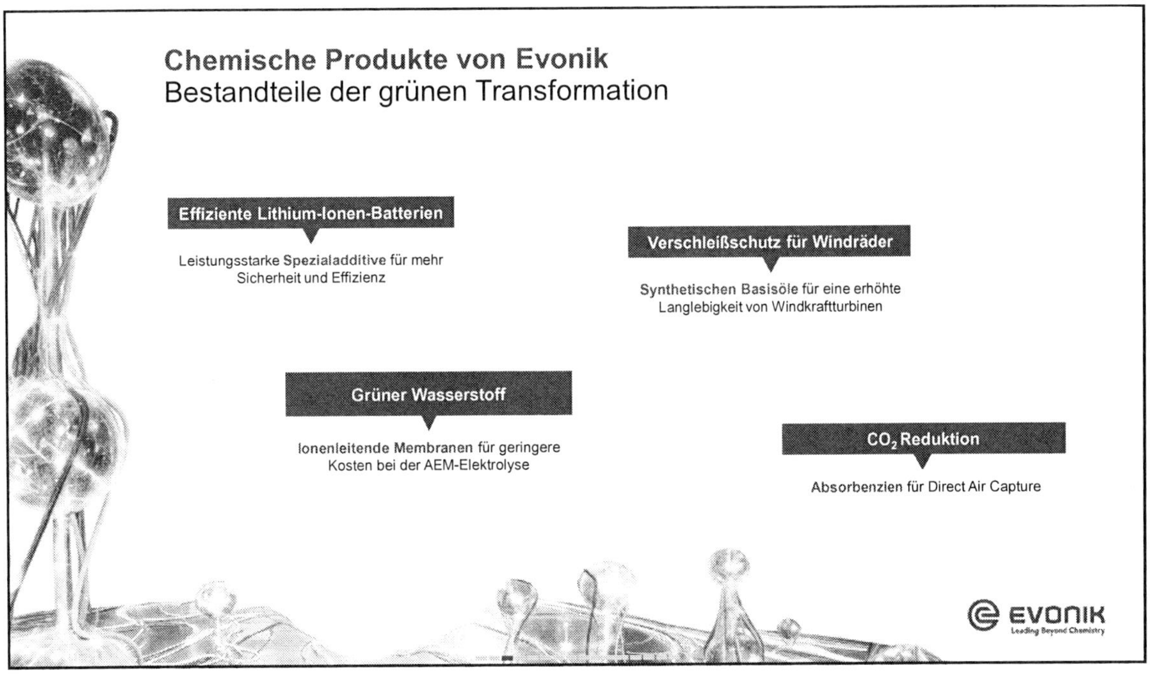
Chemische Produkte von Evonik
Bestandteile der grünen Transformation
Effiziente Lithium-Ionen-Batterien
Leistungsstarke Spezialadditive für mehr Sicherheit und Effizienz
Verschleißschutz für Windräder
Synthetischen Basisöle für eine erhöhte Langlebigkeit von Windkraftturbinen
Grüner Wasserstoff
Ionenleitende Membranen für geringere Kosten bei der AEM-Elektrolyse
CO_2 Reduktion
Absorbenzien für Direct Air Capture
EVONIK
Leading Beyond Chemistry

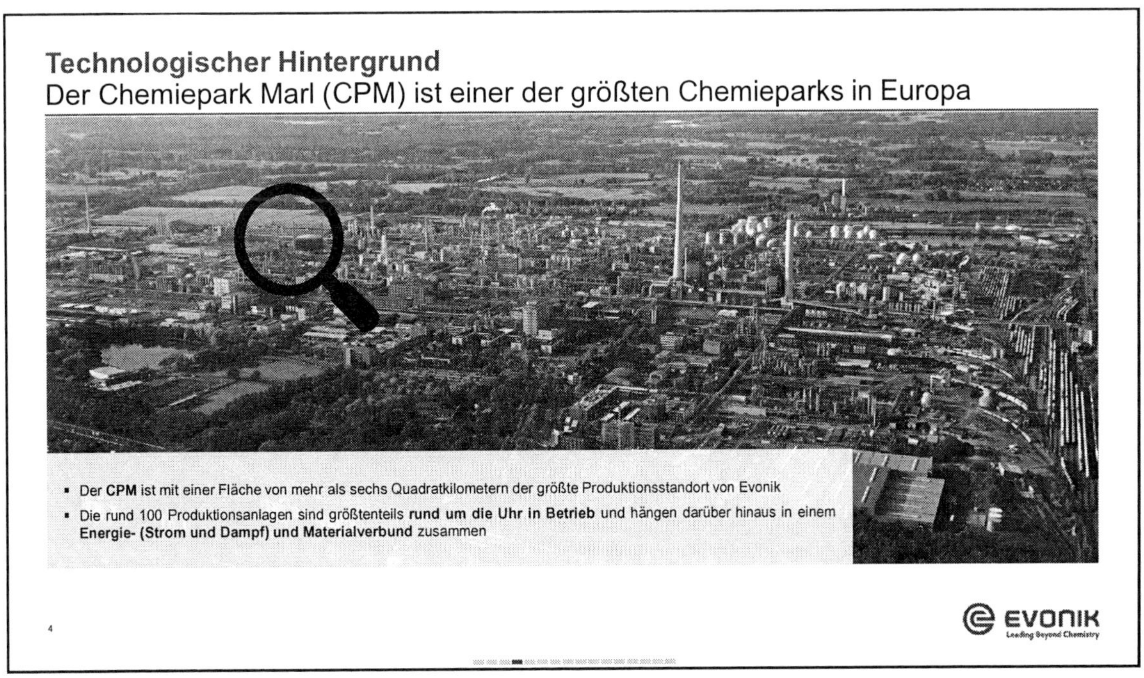
Technologischer Hintergrund
Der Chemiepark Marl (CPM) ist einer der größten Chemieparks in Europa
Der CPM ist mit einer Fläche von mehr als sechs Quadratkilometern der größte Produktionsstandort von Evonik
Die rund 100 Produktionsanlagen sind größtenteils rund um die Uhr in Betrieb und hängen darüber hinaus in einem Energie- (Strom und Dampf) und Materialverbund zusammen
4
EVONIK
Leading Beyond Chemistry

Technologischer Hintergrund
Hochkomplexe Apparate chemischer Anlagen sind oftmals überwachungsbedürftig

- In viele Produktionsprozesse sind **Druckgeräte** involviert (Wärmetauscher, Druckbehälter etc.) → hoher Druck, hohe Temperatur, **kontinuierliche** Betriebsweise
- Revisionsbedingte **Abstellungen** oder unplanmäßige **Ausfälle** Kosten **Zeit, Geld** und **energetische Ressourcen**

5

EVONIK
Leading Beyond Chemistry

Zustandsüberwachung in der chemischen Industrie
Auf dem Weg zur kontinuierlichen Überwachung im laufenden Anlagenbetrieb

Shutdown Testing
(Wasser-) Druckprüfung

Shutdown Testing
Konventionelle ZfP/ Ersatzprüfung

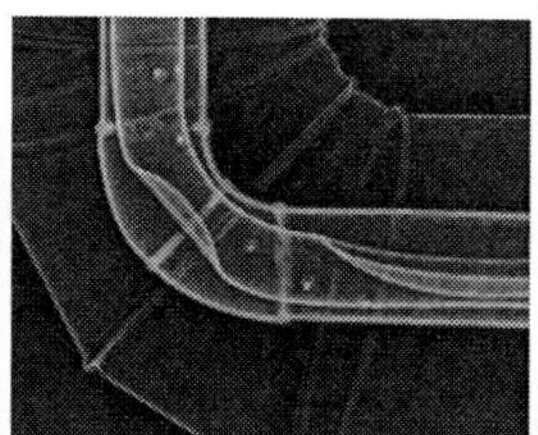

On-Stream Testing
Lokales *Corrosion-Mapping* an Druckbehältern/ Lokale digitale Radiographie an Rohrleitungen

Structural Health Monitoring (SHM)
Sensorbasierte SHM-Lösung für Korrosionsüberwachung bei erhöhter Temperatur

Diskontinuierlich mit Prüfpersonal vor Ort | Kontinuierlich

ZfP: Zerstörungsfreie Werkstoffprüfung | SHM: Structural Health Monitoring

6

EVONIK
Leading Beyond Chemistry

Zustandsüberwachung in der chemischen Industrie
Digitale Radiographie an Rohrleitungen

Radiografie am menschlichen Körper und an Rohrleitungen innerhalb von chemischen Produktionsanlagen

7

EVONIK
Leading Beyond Chemistry

Zustandsüberwachung in der chemischen Industrie
Anwendungsbeispiele: Digitale Radiographie im laufenden Betrieb

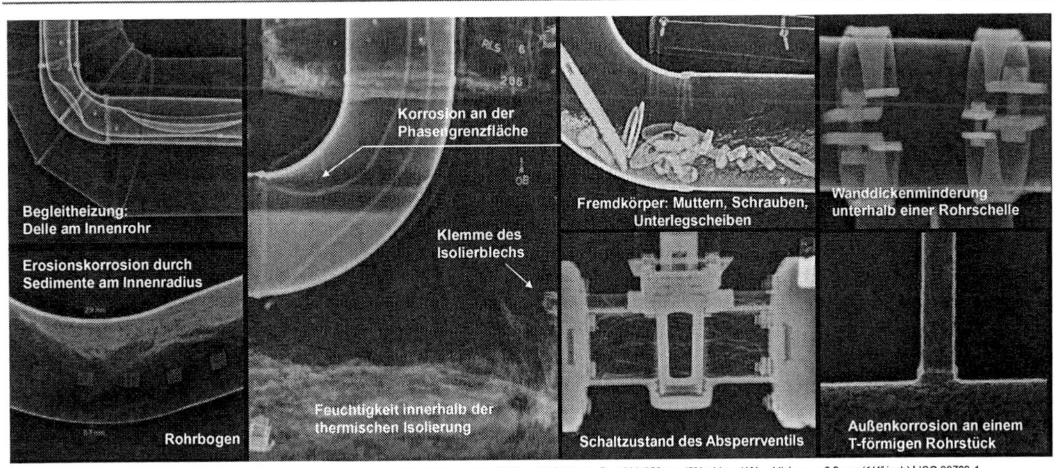

Die Wanddickenmessung mittels Durchstrahlungsprüfung im laufenden Anlagenbetrieb ist begrenzt auf Außendurchmesser D_a = 200-250 mm (9" inch) und Wanddicken s = 6.3 mm (1/4" inch) | ISO 20769-1

8

EVONIK
Leading Beyond Chemistry

Zustandsüberwachung in der chemischen Industrie
Vorbeugende Instandhaltung durch smarte Inspektionskonzepte

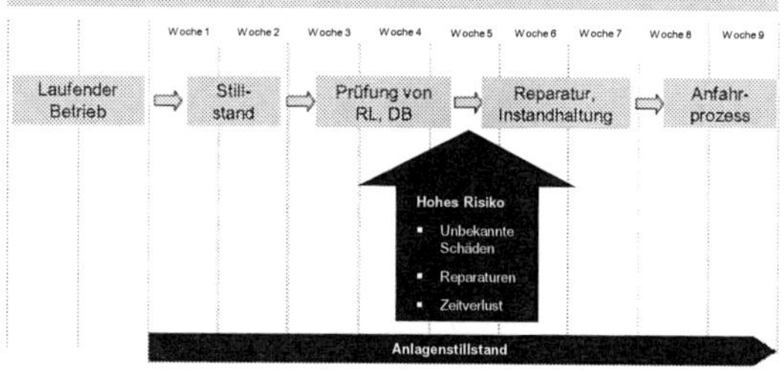

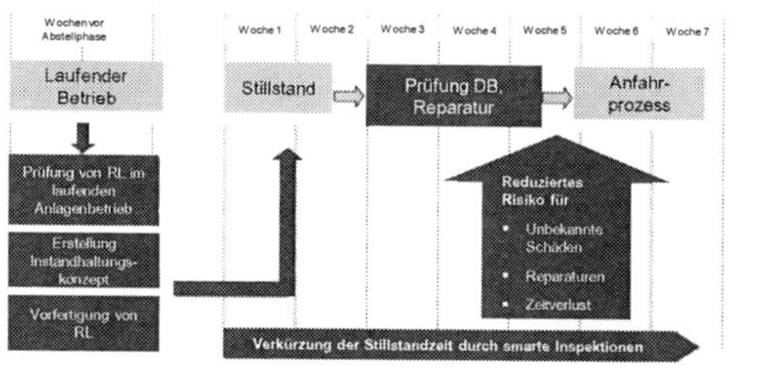

Wasserdruckprüfung mit herkömmlichen Inspektionskonzepten und ZfP
- Impliziert lange Stillstandzeiten
- Ermöglicht nur Informationen über den aktuellen Zustand
- Keine spezifische Aussage zur Lebensdauervorhersage

Risikobasierte Inspektion mithilfe von ZfP
- Schadenfrüherkennung und Lebensdauerabschätzung
- Vorfertigung defekter Rohre, Analogieschlüsse für Druckbehälter
- Ersetzt Wasserdruckprüfung

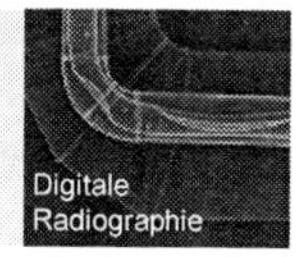

DB: Druckbehälter | RL: Rohrleitung | ZfP: Zerstörungsfreie Werkstoffprüfung

9

EVONIK Leading Beyond Chemistry

Zustandsüberwachung in der chemischen Industrie
Ultraschallprüfung mittels Phased-Array → Lokale Korrosionserscheinungen

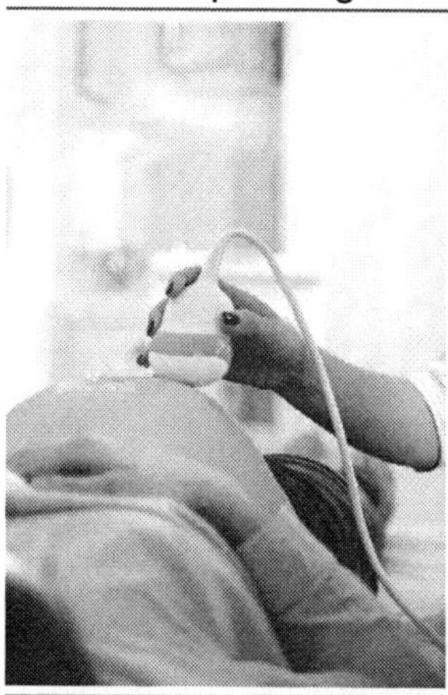

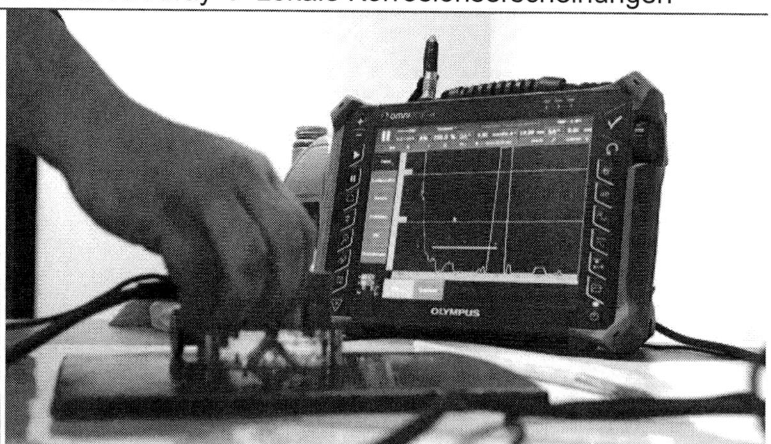

Ultraschallprüfung am menschlichen Körper und an Chemieanlagen

10

EVONIK Leading Beyond Chemistry

Zustandsüberwachung in der chemischen Industrie
Ultraschallprüfung mittels Phased-Array → Lokale Korrosionserscheinungen

- Feuchtigkeits- und Chloridioneneintrag → Lochkorrosion auf der Innenseite eines 316L-Behälters
- Phased-Array: **Außenseitiger** Scan → **Innenseitige Korrosionsdetektion** (kartographische Wanddickenverteilung)
- Deutlich verbesserte Abbildung des Behälterzustands als im Falle einer punktuellen Wanddickenmessung
- **Momentane Herausforderung:** Phased-Array Messungen im laufenden Anlagenbetrieb (Temperatur ↑)

Reale Behälterinnenwand | Radiographisches Bild | Phased-Array Scannereinheit mit *Corrosion Map* | Automatisierte Phased-Array Scannereinheit für die Schweißnahtprüfung

11

EVONIK
Leading Beyond Chemistry

Technologieentwicklung: Forschungsprojekt HotWalls (BMWK)
Kontinuierliche Zustandsüberwachung im laufenden Anlagenbetrieb

Bundesministerium für Wirtschaft und Klimaschutz

BMWK-Förderkennzeichen: 03EN4088 A-E

Bisherige Vorgehensweise bei der wiederkehrenden Prüfung von Druckbehältern

- Interne Inspektion (5-jährig), (Wasser-) Druckprüfung (10-jährig) oder Nutzung von ZfP während Stillstandphasen
- Im Gegensatz zur Rohrleitungsprüfung gibt es keine allgemein anerkannten Ersatzprüfungen für die Druckbehälterinspektion im laufenden Anlagenbetrieb

Sensorbasierte kontinuierliche Überwachung von Druckbehältern im laufenden Betrieb (HotWalls)

- **Die Konsortialpartner sollen ein HT-geeignetes sensorbasiertes** Prüfsystem für die tomographische Kartografie lokaler Korrosionsschäden entwickeln
- Langfristiges Ziel: Vollständige Substituierung der wiederkehrenden Prüfung von Druckbehältern

ZfP: Zerstörungsfreie Werkstoffprüfung | HT: Hochtemperatur | KI: Künstliche Intelligenz | FSAT: Frequency steerable acoustic transducer

12

EVONIK
Leading Beyond Chemistry

Technologieentwicklung: Forschungsprojekt HotWalls (BMWK)
Beispiel für ein industrielles Anforderungsprofil an ein Sensorsystem

Bundesministerium für Wirtschaft und Klimaschutz

BMWK-Förderkennzeichen: 03EN4088 A-E

- Anerkennung als Ersatzprüfung
- Vermeidung von Kreuzkontamination
- Eignung für gängige Apparate
- Aussage zur Wanddickenverteilung
- Erkennung verschiedener Korrosionsformen
- Einsatz im laufenden Anlagenbetrieb (Temp. ↑)
- Eignung für verschiedene Prozessbedingungen

13

EVONIK Leading Beyond Chemistry

Technologieentwicklung: Forschungsprojekt HotWalls
Rechtliche Rahmenbedingungen

Bundesministerium für Wirtschaft und Klimaschutz

BMWK-Förderkennzeichen: 03EN4088 A-E

- Jetziger Stand: Die alleinige kontinuierliche Überwachung von Druckbehältern im laufenden Anlagenbetrieb ist **rechtlich nicht als Ersatzprüfung zulässig**
- Zum jetzigen Zeitpunkt ist, nach Kenntnis der Autoren, **kein technisch präzises Sensorsystem kommerziell erhältlich**
  - → Fehlende Datenbasis für die Zuverlässigkeit des SHM-Systems
  - → **Ziel:** Langfristig soll diese Technologie weit verbreitet genutzt werden und Einzug in technische Regelwerke und Normen finden

Wirtschaftlicher Nutzen
Zeitersparnis
Ökologischer Nutzen

Beispiel für einen FSAT-Wandler [1]

[1] Marquez Reyes, O. A. ; Moll, J. ; Mohammadgholiha, M. and De Marchi, L., SN-FSAT system for single sensor ultrasonic guided wave beam steering and damage detection, 11th European Workshop on Structural Health Monitoring (EWSHM), 2024

FSAT: Frequency steerable acoustic transducer | SHM: Structural Health Monitoring

14

EVONIK Leading Beyond Chemistry

Zusammenfassung
Auf dem Weg zur kontinuierlichen Zustandsüberwachung von chemischen Anlagen

- Chemische Produktionsanlagen haben ein **hohes Gefahrenpotenzial** für Mensch und Umwelt (Korrosion, Verschleiß, Ermüdung) → gesetzlich **überwachungsbedürftig**
- Ein Anlagenstillstand kostet **Zeit, Geld und energetische Ressourcen**
- ZfP als Ersatzprüfung ist für einige Komponenten im laufenden Anlagenbetrieb möglich → Risiko: Arbeiten an heißen Oberflächen
- **Kontinuierliche** Sensorüberwachung wünschenswert
  - Herausforderungen:
    - Behältergröße (bspw. 300 m³), komplexe Struktur
    - Druck ↑ Temperatur ↑
    - Explosionsschutz, Cyber-Security

ZfP: Zerstörungsfreie Werkstoffprüfung

15

EVONIK Leading Beyond Chemistry

16

EVONIK Leading Beyond Chemistry

Drohnen als Werkzeug in der zerstörungsfreien Prüfung

Roman DENDL
Ingenieurbüro DENDL e.U., 8421 Wolfsberg im Schwarzautal, Österreich,
roman.dendl@ib-dendl.com

Kurzfassung

Drohnen können durch ihre Technologie und entsprechende Einsatzplanung wertvolle Dienste für die Qualität und Betriebssicherheit von Anlagen leisten. Speziell die zerstörungsfreie Werkstoffprüfung kann mittels diverser Drohnentechnologien unterstützt werden.

Mittlerweile sind hochwertige Drohnen standardmäßig mit 4K-Kameras ausgestattet und wenn es noch ein wenig spezieller werden soll, dann hat die Drohne auch eine Wärmebildkamera an Board. Auch der Zoomfaktor ist bei den Kameras mehr als beeindruckend. Somit können aus sicherer Entfernung detailgenaue Aufnahmen gemacht werden. Zusätzliche Soft- und Hardware erleichtern speziell die wiederkehrende Inspektion von Bauteilen. Mit der automatisierten Routineinspektion werden Objekte identifiziert und man bekommt konsistente Aufnahmen.

Spezielle Drohnen sind für Indoor-Bereiche mit einem Kollisionskäfig ausgestattet. Dieser Käfig bewirkt, dass die Drohne, obwohl sie mit Infrarotsensoren ausgestattet ist, bei einer Kollision mit einem anderen Gegenstand keinen Schaden erleidet und stabil weitergeflogen werden kann. Diese Indoor-Drohne ist ebenso mit einer 4K-Kamera und gleichzeitig mit einer Wärmebildkamera ausgestattet. Zusätzlich sind LEDs angebracht, die eine Leuchtkraft bis zu 10.000 Lumen haben. Dadurch können finstere Räume, wie z.B. Lagertanks oder Behälter, ohne zusätzliche Beleuchtung, inspiziert werden. Durch neueste Entwicklungen können solche Drohnen auch UT-Wanddickenmessungen in luftigen Höhen durchführen. Unterschiedliche Schallköpfe liefern entsprechende Ergebnisse. Ausgestattet mit einer kleinen Schleifscheibe, können diese Drohnen auch die Oberfläche bearbeiten, sodass eine normgerechte Wanddickenmessung durchgeführt werden kann. Die Einsatzmöglichkeiten speziell bei Indoor-Drohnen werden laufend verbessert und weiterentwickelt.

Der hochauflösende LiDAR-Sensor in Kombination mit dem entsprechenden Software-Algorithmus erstellt hochgenaue 3D-Karten und digitale Zwillinge der unzugänglichsten Orte mit Zentimetergenauigkeit. Somit ist nicht nur eine Inspektion von Innenräumen möglich, sondern auch deren Vermessung, wodurch neue Erkenntnisse gewonnen werden können.

Keywords:
Drohneninspektion, UT-Wanddickenmessung, visuelle Prüfung, Indoor, Wärmebild

IB-DENDL
Ingenieurbüro

Drohnen als Werkzeug in der zerstörungsfreien Prüfung

Roman DENDL

Vorstellung

IB-DENDL
Ingenieurbüro

Schweißaufsicht
Inspektionstätigkeiten
Qualitätssicherung
Zerstörungsfreie Werkstoffprüfung
u.v.m.

Drohneninspektionen
Indoor - Outdoor

Thermografie
Anlagenthermografie
Elektrothermografie
Gebäudethermografie

INGENIEURBÜRO FÜR WERKSTOFFWISSENSCHAFTEN / DI Roman Dendl 2

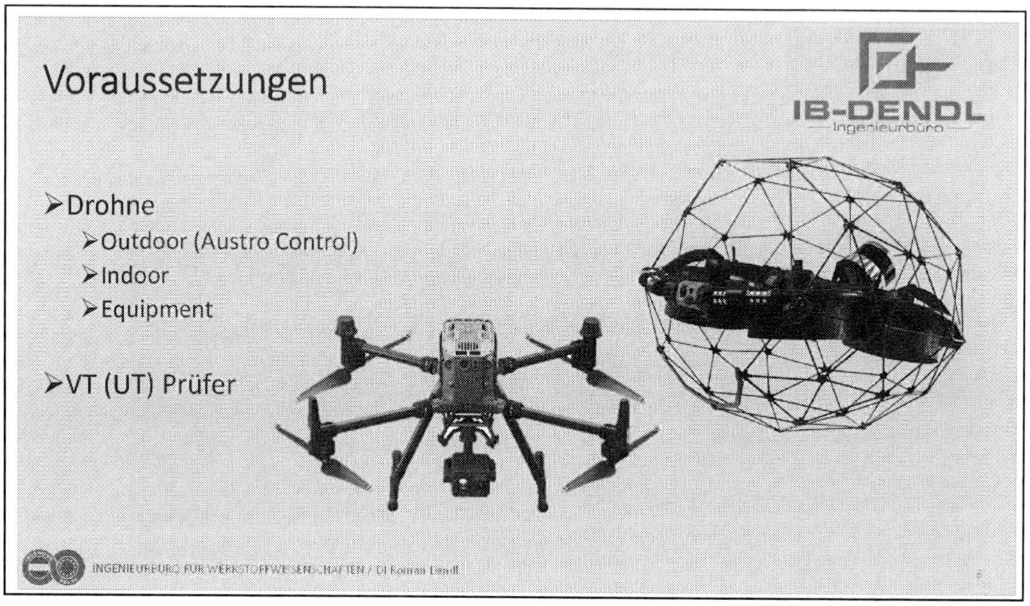
Voraussetzungen
IB-DENDL
Ingenieurbüro
➢Drohne
➢Outdoor (Austro Control)
➢Indoor
➢Equipment
➢VT (UT) Prüfer
INGENIEURBÜRO FÜR WERKSTOFFWISSENSCHAFTEN / DI Roman Dendl

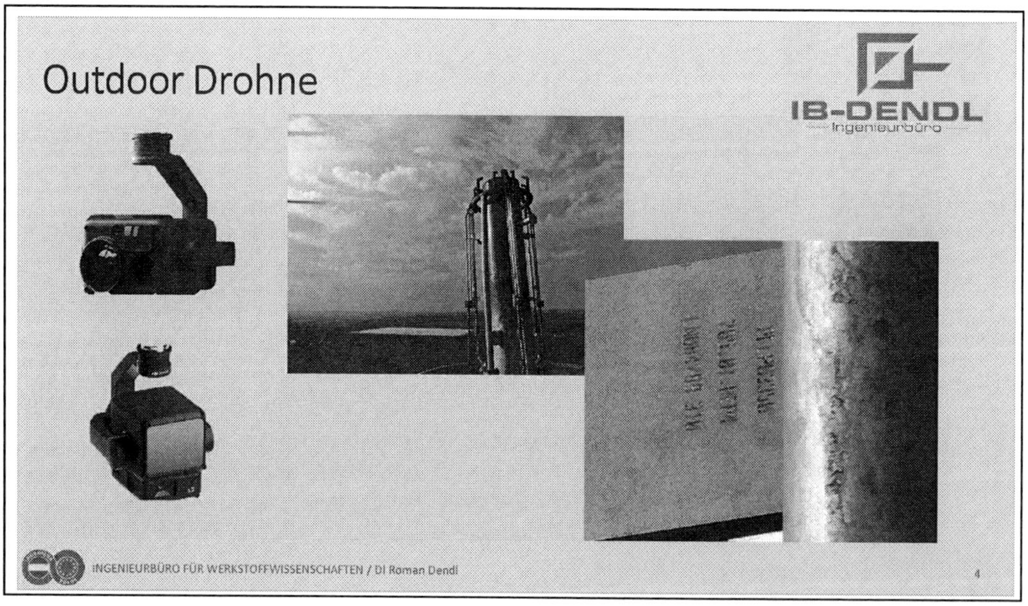
Outdoor Drohne
IB-DENDL
Ingenieurbüro
INGENIEURBÜRO FÜR WERKSTOFFWISSENSCHAFTEN / DI Roman Dendl
4

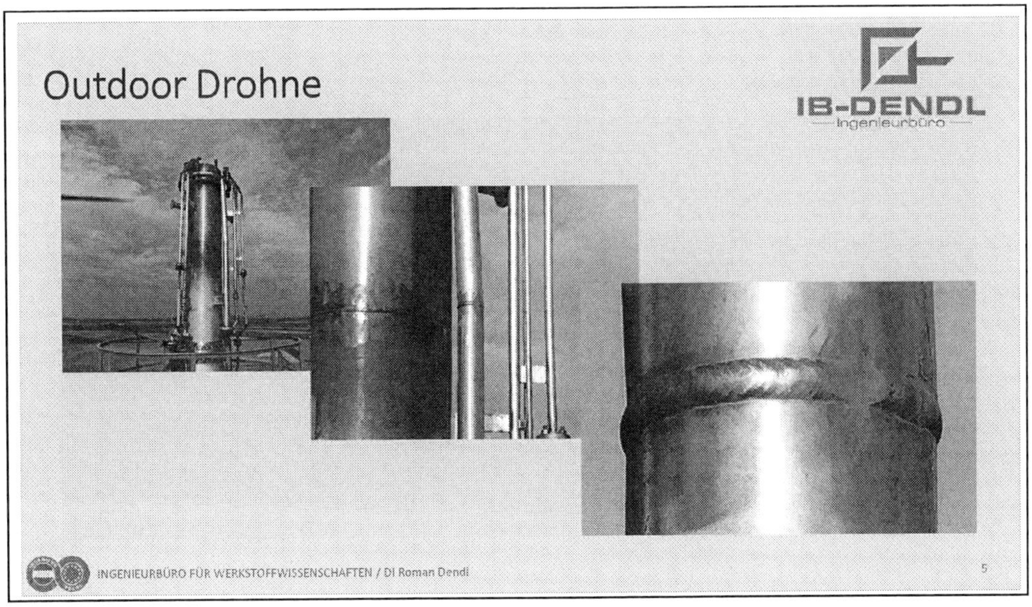
Outdoor Drohne
IB-DENDL
Ingenieurbüro
INGENIEURBÜRO FÜR WERKSTOFFWISSENSCHAFTEN / DI Roman Dendl
5

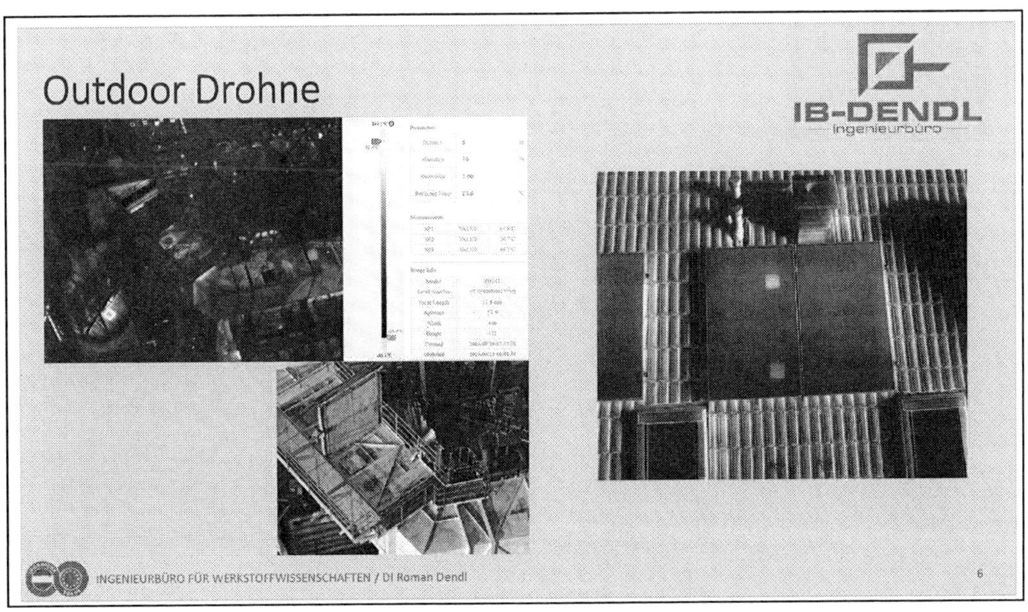
Outdoor Drohne
IB-DENDL
Ingenieurbüro
INGENIEURBÜRO FÜR WERKSTOFFWISSENSCHAFTEN / DI Roman Dendl
6

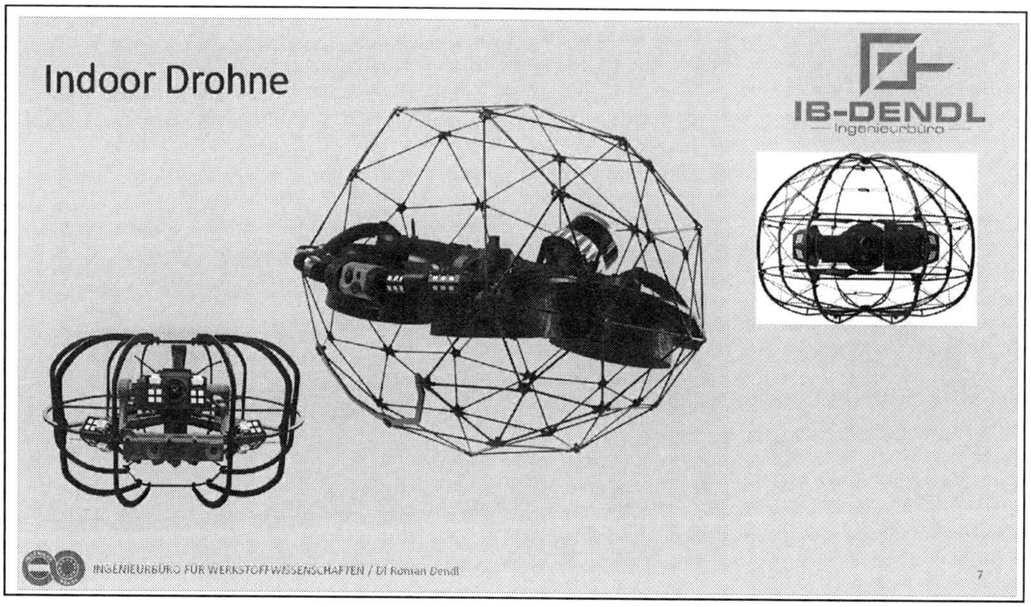
Indoor Drohne
IB-DENDL
Ingenieurbüro
INGENIEURBÜRO FÜR WERKSTOFFWISSENSCHAFTEN / DI Roman Dendl
7

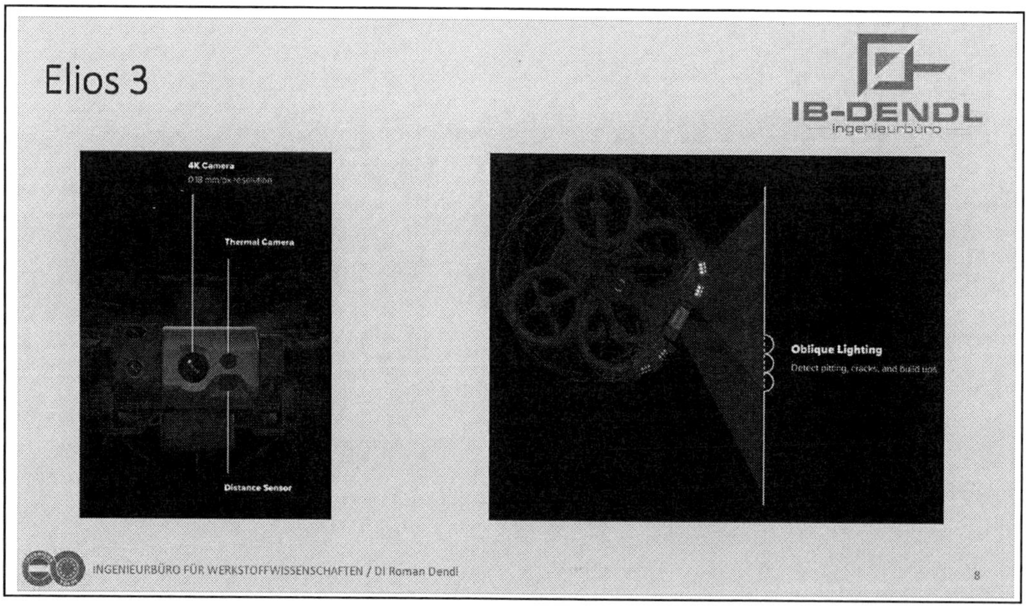
Elios 3
IB-DENDL
Ingenieurbüro
4K Camera
Thermal Camera
Distance Sensor
Oblique Lighting
Detect pitting, cracks, and build ups
INGENIEURBÜRO FÜR WERKSTOFFWISSENSCHAFTEN / DI Roman Dendl
8

Elios 3

IB-DENDL
Ingenieurbüro

INGENIEURBÜRO FÜR WERKSTOFFWISSENSCHAFTEN / DI Roman Dendl 9

Elios 3

IB-DENDL
Ingenieurbüro

INGENIEURBÜRO FÜR WERKSTOFFWISSENSCHAFTEN / DI Roman Dendl 10

Elios 3

IB-DENDL
Ingenieurbüro

INGENIEURBÜRO FÜR WERKSTOFFWISSENSCHAFTEN / DI Roman Dendl

11

Elios 3 - Beispiele

IB-DENDL
Ingenieurbüro

INGENIEURBÜRO FÜR WERKSTOFFWISSENSCHAFTEN / DI Roman Dendl

12

Elios 3 - Beispiele

IB-DENDL
Ingenieurbüro

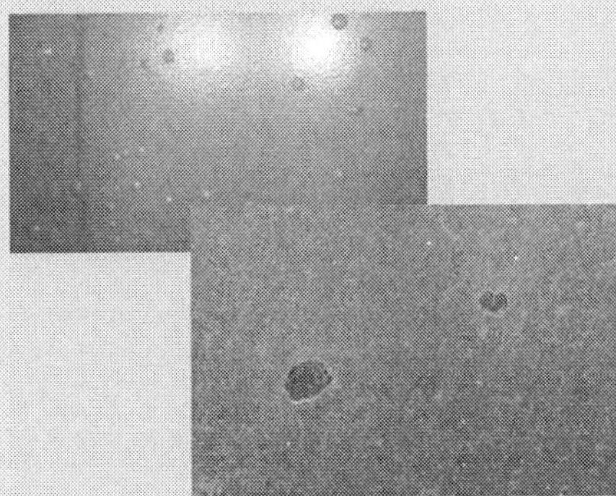

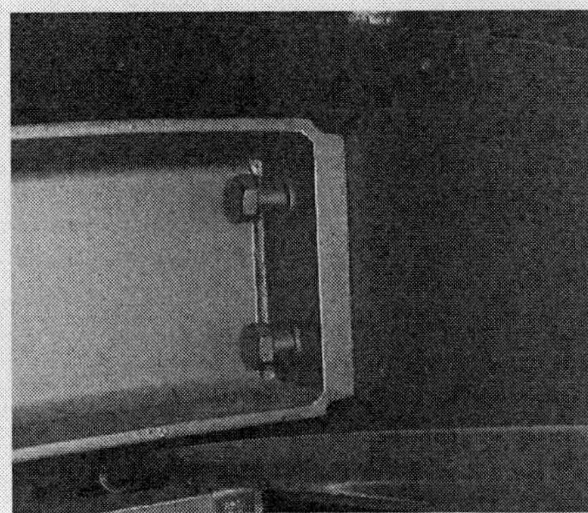

INGENIEURBÜRO FÜR WERKSTOFFWISSENSCHAFTEN / DI Roman Dendl

13

Elios 3 - Beispiele

IB-DENDL
Ingenieurbüro

INGENIEURBÜRO FÜR WERKSTOFFWISSENSCHAFTEN / DI Roman Dendl

14

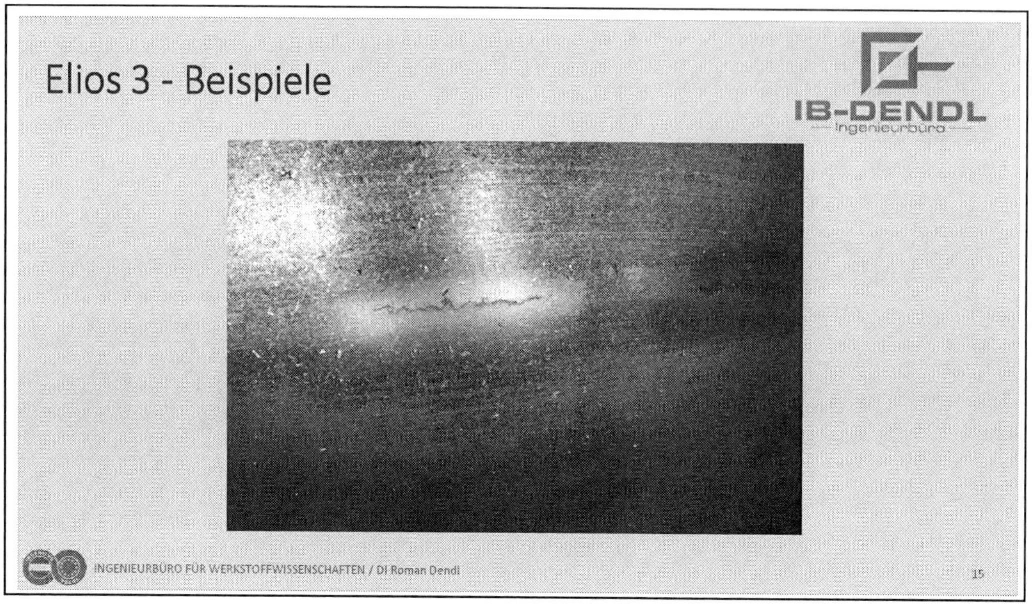
Elios 3 - Beispiele
IB-DENDL
Ingenieurbüro
INGENIEURBÜRO FÜR WERKSTOFFWISSENSCHAFTEN / DI Roman Dendl
15

Elios 3 - UT Payload
IB-DENDL
Ingenieurbüro
INGENIEURBÜRO FÜR WERKSTOFFWISSENSCHAFTEN / DI Roman Dendl
16

Elios 3 - Einsatzgrenzen

IB-DENDL
ingenieurbüro

- ➢Akku
- ➢Ex-Schutz
- ➢Starres System (mit Vorbehalt)
- ➢~~Nur~~ visuelle Prüfung
- ➢Visuelle Prüfung und UT-Wanddickenmessung

INGENIEURBÜRO FÜR WERKSTOFFWISSENSCHAFTEN / DI Roman Dendl 17

Elios 3 - UT Payload

IB-DENDL
ingenieurbüro

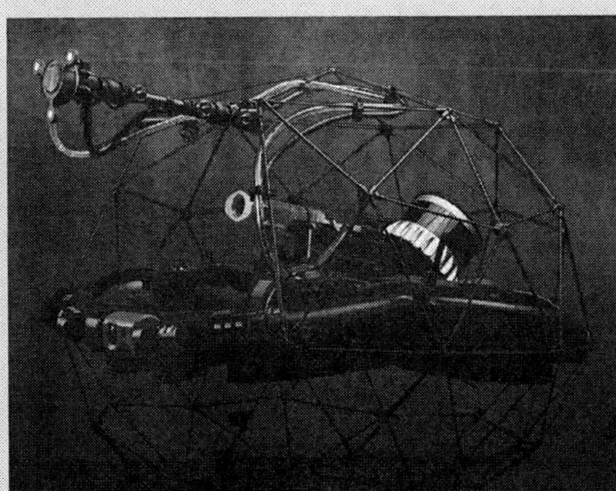

INGENIEURBÜRO FÜR WERKSTOFFWISSENSCHAFTEN / DI Roman Dendl 18

Elios 3 - UT Payload

IB-DENDL
Ingenieurbüro

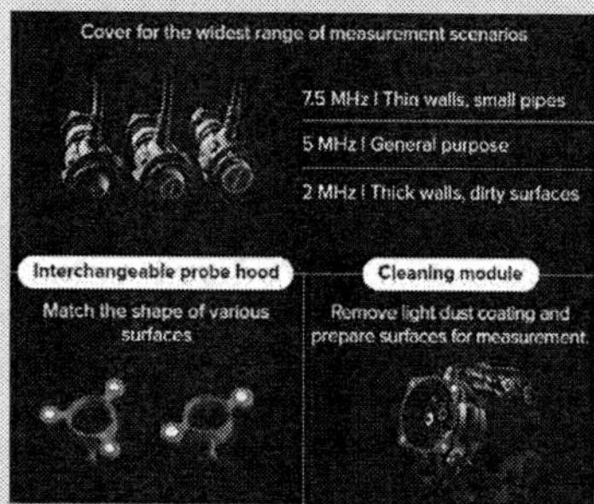

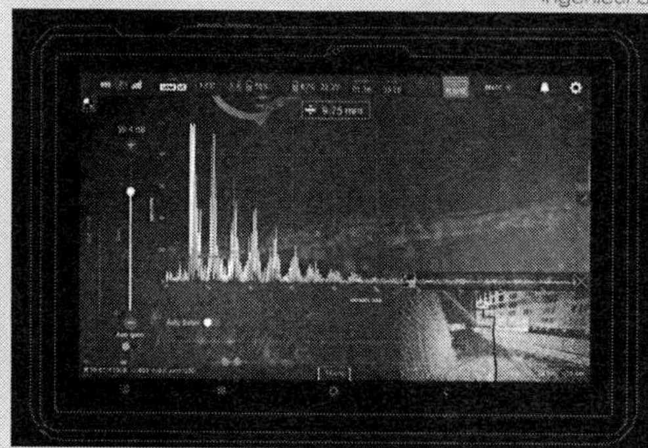

Elios 3 - UT Payload

IB-DENDL
Ingenieurbüro

Elios 3 - UT Payload

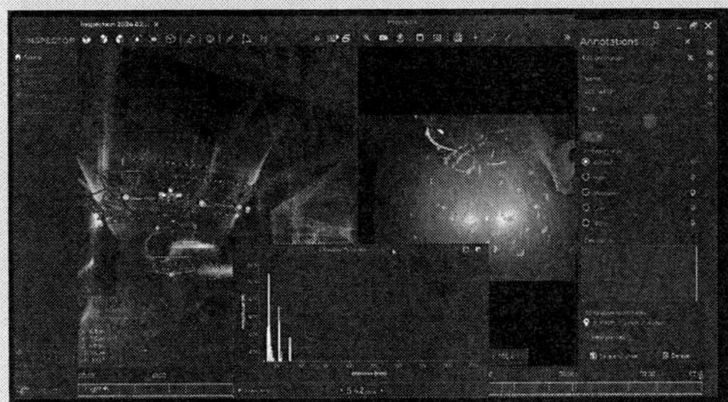

INGENIEURBÜRO FÜR WERKSTOFFWISSENSCHAFTEN / DI Roman Dendl 21

Kontaktdaten

Ingenieurbüro für Werkstoffwissenschaften und Schweißtechnik
DI(FH) Roman Dendl, M.Sc., MBA
Tel.: +43 664 2627558
Mail: roman.dendl@ib-dendl.com
www.ib-dendl.com

INGENIEURBÜRO FÜR WERKSTOFFWISSENSCHAFTEN / DI Roman Dendl 22

Digitalizing Trust: Advancing Standardization and Traceability in NDE Workflows under NDE 4.0

Srijan Tiwari[1], Krishnan Balasubramanian[2]

[1,2]Centre for Non-Destructive Evaluation (CNDE), Indian Institute of Technology Madras, Chennai, India

[1]srijantiwari10@gmail.com , [2]balas@iitm.ac.in

The Non-Destructive Testing (NDT) industry is expected to grow from $16.4 billion in 2024 to $35.4 billion by 2032, driven by increasing demands in infrastructure, energy, and manufacturing. Despite this, the industry remains constrained by fragmented data management, manual inspection reporting, and disjointed workflows across procurement, certification, and talent development.

This article highlights the need for digital transformation in NDE workflows aligned with Industry 4.0 principles. A review of industry reports and stakeholder surveys reveals critical inefficiencies: 70% of firms experience procurement delays, 60% report certification inconsistencies, and 52% of job seekers face mismatches between skills and job roles. A key issue stems from the variability in inspection reports—shaped by inspector subjectivity, device dependence, and inconsistent data capture—which hinders traceability and audit readiness.

This research advocates for the adoption of a unified, digital framework that addresses these inefficiencies through standardized inspection templates, cloud-based data traceability, and AI-powered credential validation. Compliance with global standards such as ISO 17020 and ASME Section V is emphasized. The framework promotes the use of QR codes for immutable report verification, APIs for interoperability between systems, and AI-based tools for structured CV building and role matching. Furthermore, collaboration with regulatory bodies and targeted training programs are essential to drive adoption and consistency in practice. By embedding digital infrastructure into core NDE processes, the industry can significantly improve operational efficiency, minimize errors, and enable intelligent decision-making.

Keywords: Digitalisation, NDE4.0, Report Standardization, Asset Integrity, Certification Traceability, Inspection Workflow, Industry 4.0, AI in NDT

dgzfp JAHRESTAGUNG 2025
26. – 28. Mai Berlin

Digitalizing Trust: Advancing Standardization and Traceability in NDE Workflows under NDE 4.0

Srijan Tiwari*, Krishnan Balasubramaniam, Rajan Ananthanarayanan

Center for Non-Destructive Evaluation, Indian Institute of Technology Madras, Chennai, INDIA

*srijantiwari10@gmail.com

INTRODUCTION

Why NDE 4.0?

- The NDT industry is projected to grow from **$16.4B (2024) to $35.4B (2032)**, yet struggles with manual workflows, inconsistent reports, and fragmented hiring/training ecosystems.
- **NDE 4.0** integrates digital tools like **cloud platforms, AI/ML, and standard reporting**, solving these inefficiencies and driving Industry 4.0-ready transformation.

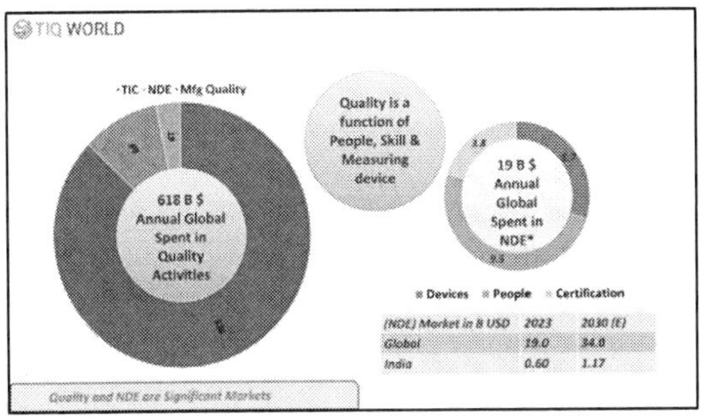

- Bridge the gap in **Procurement, Training & Talent** via a unified digital platform.
- Standardize inspection **report formats** aligned with global codes (ASME, ISO 17020).
- Enable AI-driven **validation, traceability, verification** and compliance.

MARKET RESEARCH : INDUSTRY GAPS

- **70%** firms face procurement delays
- **60%** report certification inconsistencies
- **52%** job seekers struggle with role matching
- Reports differ across inspectors, risking misinterpretation
- Lack of traceability in paper-based workflows

"Reports" Interpretation & Subjectivity?

Outcome of an Inspection is dependent on:

Person + **Data Collection** + **Part by Itself**

Data Collection → **Device Dependency**

- Inconsistent NDE reports arise due to inspector subjectivity and technique familiarity.
- This lack of standardization and traceability often leads to costly errors and Uncaptured insights

DIGITALIZED WORKFLOW – ENABLING NDE 4.0

- **Cloud-based Reporting** for traceability & access
- **Standardized Inspection Reporting** to reduce variability and improve audit trails.
- **Digital Certification Validation** to ensure verified, traceable skills.
- **Smart Talent Hiring** with AI-matched, certification-tagged profiles.
- **Workforce Credentialing**: Digital validation of certifications and AI-powered CV structuring.

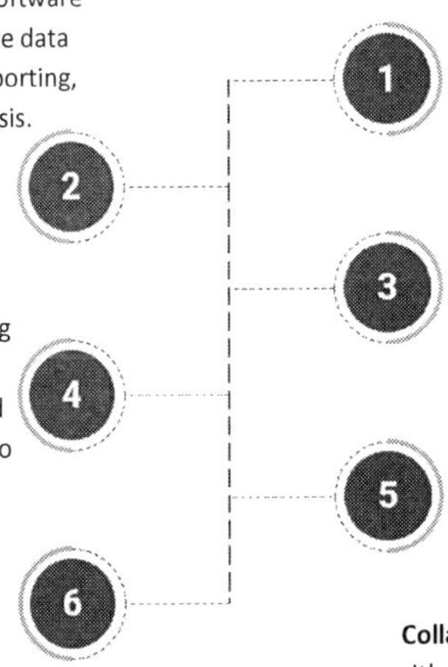

1. **Develop a Unified Framework**: Create a standardized template or framework for inspection reports that includes mandatory fields, terminologies, and formats.
2. **Adopt Digital Tools**: Implement digital platforms or software that automate data collection, reporting, and analysis.
3. **Ensure Interoperability:** Use open standards and APIs to enable seamless data sharing between different systems and stakeholders
4. **Training and Awareness:** Provide training programs for inspectors and organizations to adopt standardized practices.
5. **Regulatory Collaboration:** Work with regulatory bodies such as **NABL, ISNT,ASNT** to align standards with industry requirements.
6. **QR Codes for Data Integrity:** Usage of QR Codes to ensure the authenticity and immutability of inspection data.

REFERENCES

- ASNT Certification Report (2023)
- ISNT Hiring Trends (2023)
- Market Research Future – Global NDT Report
- ISO 17020 / ASME Sec V
- https://www.scribd.com/document/487081748/ASME-Section-V-Art-7-Magnetic-Particle-Examination accessed on March 08, 2024

Auf dem Weg zu 4D-CT in der Additiven Fertigung: Bewertung algorithmischer Grenzen und Möglichkeiten für die industrielle zerstörungsfreie Prüfung

Nadine MÖNTER[1], Katharina BLIEDTNER[1], Frank HEROLD[1]
[1] VisiConsult X-ray Systems & Solutions GmbH, 23617 Stockelsdorf, Deutschland,
f.herold@visiconsult.de

Kurzfassung

Die Computertomographie (CT) ist eine in der Industrie weit verbreitete Methode zur zerstörungsfreien Prüfung (NDT), die die Visualisierung und Analyse innerer Strukturen von Bauteilen ermöglicht. Insbesondere in der additiven Fertigung (AM), bei der Bauteile schichtweise aufgebaut werden, bietet die industrielle CT erhebliche Vorteile für die Defekterkennung und Qualitätskontrolle. In diesem Kontext können beispielsweise Einschlüsse oder Bindefehler auftreten. Traditionell beschränkt sich die Analyse solcher Defekte auf den Endzustand des Bauteils, wodurch die zeitliche Entwicklung der Defekte unbekannt bleibt.

Durch die Schätzung von Bewegungen während der Produktion oder Belastungstests wird es jedoch möglich, nicht nur Defekte zu erkennen, sondern auch deren Ursprung, zeitliche Entwicklung und die Wechselwirkungen, die zu ihrer Entstehung geführt haben, zu bestimmen. Zeitaufgelöste CT-Scans bieten somit das Potenzial, das Verhalten von Strukturen unter wechselnden Bedingungen zu visualisieren und eine geeignete Grundlage für Modelle zur Vorhersage der Defektentwicklung zu schaffen.

In unserer vorherigen Arbeit [1] wurde ein Algorithmus entwickelt, der die Bewegung zwischen einem statischen Anfangszustand und einem Endzustand modelliert, um die Rekonstruktion zeitlicher Zwischenvolumina zu ermöglichen. Dieser Artikel erweitert diesen Ansatz durch die Analyse unter verschiedenen Anforderungen, wie etwa der Größe der Bewegung, um Möglichkeiten und Grenzen zu identifizieren. Dadurch wird eine solide Basis geschaffen, um den Algorithmus auf dynamische Zustände auszudehnen und dynamische Materialeigenschaften zu bestimmen.

Keywords: 4D-Computertomographie (4D-CT), Additive Fertigung (AM), Dynamische Prozessanalyse

Förderkennzeichen:
01IY2115A-C

Auf dem Weg zu 4D-CT in der Additiven Fertigung

Bewertung algorithmischer Grenzen und Möglichkeiten für die industrielle zerstörungsfreie Prüfung

Nadine MÖNTER
Katharina BLIEDTNER, Frank HEROLD
VisiConsult X-ray Systems & Solutions GmbH
Brandenbrooker Weg 2-4, 23617 Stockelsdorf
Contact E-Mail: f.herold@visiconsult.de

DLR Projektträger
Federal Ministry of Education and Research
Funded by the European Union

Motivation & Beschreibung

In den letzten Jahren gewinnt die additive Fertigung (AM) von Metallen zunehmend an Bedeutung in der Automobil- und Luftfahrtindustrie [1]. Daher wird die damit verbundene Bauteilanalyse – insbesondere im Hinblick auf Sicherheitsaspekte – immer wichtiger und erfordert präzise und zuverlässige zerstörungsfreie Prüfmethoden (NDT). Es ist daher nicht nur entscheidend, das Vorhandensein von Defekten in einem Bauteil zu identifizieren, sondern auch deren Ursprung und zeitabhängige Entwicklung (4D) zu verstehen, um Herstellungsprozesse zu optimieren.

In unserer vorherigen Studie [2] haben wir einen dämonen-basierten Algorithmus [3,4] implementiert, der es ermöglicht, die Bewegung zwischen zwei statischen Objektzuständen zu modellieren (siehe Abb. 1). Der entwickelte Algorithmus unterliegt jedoch bestimmten Einschränkungen, wie der Wahl verschiedener Parameter, der Größe der Objektbewegung und der Größe des Objekts. Wir erweitern unsere bisherige Arbeit durch eine Parameterstudie. Diese Untersuchung ist entscheidend für ein tieferes Verständnis der Anwendbarkeit und bietet eine robuste Grundlage für die zeitaufgelöste digitale Radiographie in 2D, sowie deren Erweiterung zu einem vollständigen 4D-Algorithmus in der zerstörungsfreien Prüfung (NDT).

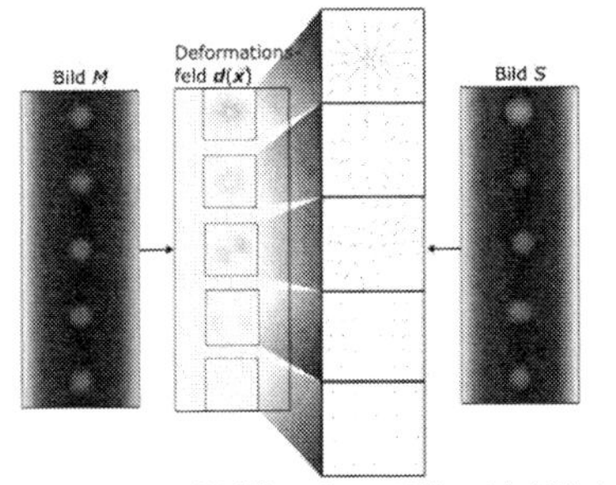

Abb. 1: Das Deformationsfeld $d(\vec{x})$ zwischen zwei Bildern wird mithilfe eines dämonen-basierten Algorithmus berechnet. **Hinweis:** Die Verschiebungsvektoren zeigen in die entgegengesetzte Richtung der Bewegung.

Methoden

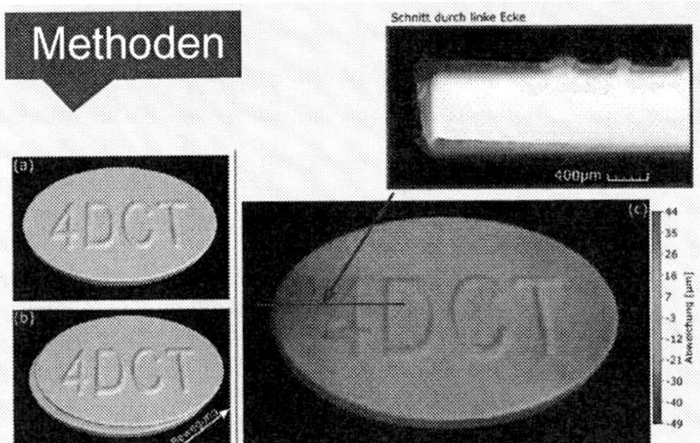

Abb. 2: Rekonstruierte Volumina. (a) Eisenplatten-Phantom unter Startbedingungen. (b) Start (unten) und Ende (oben) der Eisenplatte, die die Bewegung nach zwei Rotationen des CT-Scanners darstellt. (c) Verformte Eisenplatte nach einer einzigen Rotation. Die Farbskala veranschaulicht die Abweichung der verformten Eisenplatte von einem Referenz-Eisenplatten-Phantom.

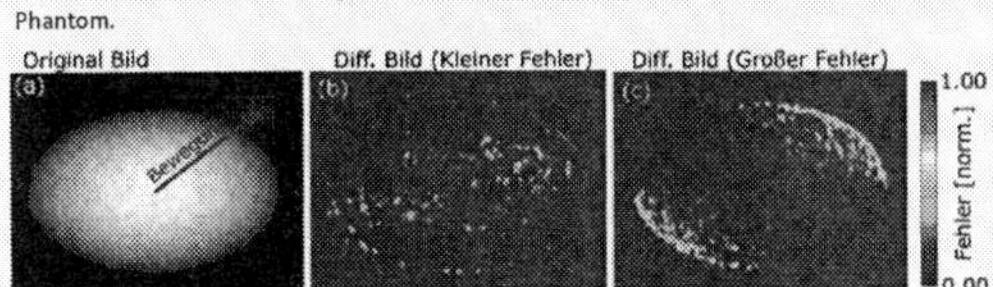

Abb. 3: (a) Simuliertes Ellipsoid, das sich in Richtung oben rechts bewegt. (b, c) Differenzbild: Endposition des Ellipsoids vs. deformiertem Ellipsoid basierend auf Parametern, die (b) einen geringen Morphing-Fehler und (c) einen hohen Morphing-Fehler aufweisen.

Dämonen-basierter Algorithmus mit 4D-CT Erweiterung

Die Objektbewegung zwischen zwei aufeinanderfolgenden Bildern wird mit einem iterativen, dämonen-basierten Algorithmus [3,4] geschätzt (siehe Abb. 1). In diesem Prozess wird ein bewegtes Bild M (Start) mithilfe eines Deformationsfeldes $\boldsymbol{d}(\vec{x})$ in ein statisches Bild S (Ende) überführt, wobei $\vec{x}$ die Pixelposition bezeichnet. Das Deformationsfeld $\boldsymbol{d}(\vec{x})$ wird iterativ durch das Vektorfeld

$$\vec{v} = -(M - S)\left(\frac{\vec{\nabla}S}{\left|\vec{\nabla}S\right|^2 + \alpha^2(M-S)^2} + \frac{\vec{\nabla}M}{\left|\vec{M}\right|^2 + \alpha^2(M-S)^2}\right), \qquad (1)$$

aktualisiert, wobei α ein in [4] eingeführter Homogenisierungsfaktor ist. Anschließend wird $\vec{v}$ mit einem Gaußfilter einer Filterkernel-Größe von $r \times r$ und einer Schrittweite s geglättet.

Um Zwischenbilder zu erzeugen, werden die Pixel von M basierend auf $\boldsymbol{d}(\vec{x})$ um eine feste Zeitschrittweite t verschoben. Die Erweiterung auf das 4D-CT-Szenario erfolgt durch $t_\nu(\tau; N) = \frac{\tau - \nu}{N}$, wobei die erforderlichen Morphing-Zeiten t_ν zweier Bilder in Abhängigkeit von ihrem Aufnahmeort (mit Blickwinkel $1 \leq \nu \leq N$), sowie der absoluten Zeit τ des Systems berechnet werden. Hierbei bezeichnet N die Anzahl der Projektionen pro Umdrehung.

Parameter Studie

Abbildung 2 veranschaulicht die Einschränkungen des dämonen-basierten Algorithmus. Das Morphing der initialen Eisenplatte in die finale Eisenplatte führt nicht zu einer vollständigen Übereinstimmung (siehe farbcodierte Abweichung in Abb. 2(c)). Die Berechnung des Deformationsfeldes wird durch den Homogenisierungsfaktor α, die Filterkernel-Größe $r \times r$ und die Schrittgröße s (vgl. Gl. (1)) beeinflusst. Diese Studie untersucht den Einfluss dieser Parameter, (s, r, α), auf den Morphing-Fehler. Darüber hinaus wird der Einfluss der Objektbewegung und -größe auf die Leistung des Algorithmus quantifiziert. Die Bedeutung der Auswahl eines optimalen Parametersatzes wird in Abb. 3 gezeigt, wo die Differenzbilder zwischen Referenz und gemorphtem Ellipsoid unterschiedliche Morphing-Übereinstimmungen aufweisen, die nur durch die Wahl der oben genannten Parameter beeinflusst werden.

Ergebnisse & Ausblick

Abbildung 4 zeigt den Morphing-Fehler eines simulierten Ellipsoids für unterschiedliche pixelweise Bewegungsschritte (Δp), abhängig von den Parametern (s, r, α), wobei α fixiert ist. Der Morphing-Fehler ist am geringsten bei der kleinsten Bewegung $(\Delta p = 1)$ unter Verwendung der Parameter $s = 1$, $r = \{1, 11, 31, 41, 51, 61, 71\}$ und $\alpha = 1$. Mit zunehmenden Bewegungsschritten steigt auch der Morphing-Fehler.

In Abb. 5 wird der Parametersatz (s, r, α) gewählt, der für unterschiedliche Bewegungsgrößen den minimalen Fehler liefert. Bei allen drei Objektgrößen beobachten wir ein sigmoidales Verhalten des Fehlers mit zunehmender Bewegung, mit einem Wendepunkt zwischen $\Delta p = 3$ und 4, was auf eine Bewegungslimitierung von drei Pixeln hinweist. Darüber hinaus zeigen größere Objekte eine größere Stabilität bei der Beibehaltung der Morphologie während des Morphings, da kleine Deformationen im Vergleich zur Gesamtobjektgröße vernachlässigbar werden.

Zukünftige Arbeiten werden die Integration von realen Messungen unter Verwendung der in dieser Parameterstudie identifizierten optimalen Parameter umfassen.

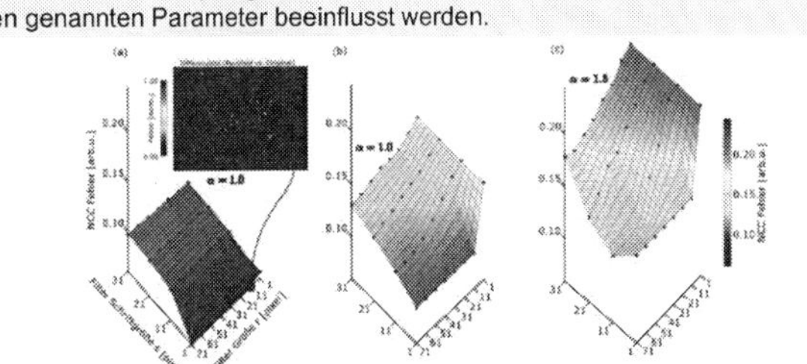

Abb. 4: *Morphing-Fehler (normalized cross correlation (NCC) Fehler) des simulierten Ellipsoids für verschiedene Pixelschrittweiten (Δp) (a)-(c), abhängig von dem Parameterset (s, r, α), wobei das best-performende $\alpha \in \{0.5, 1.0, 1.5\}$ fix gewählt wurde.*

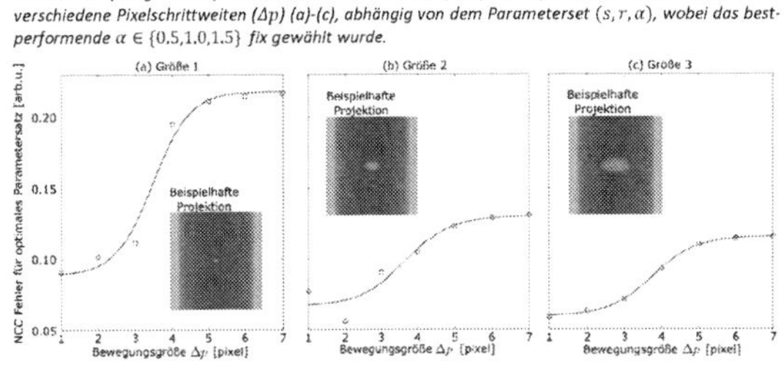

Abb. 5: Objekt- und Bewegungsgrößenanalyse. Der Morphing-Fehler (normalized cross correlation (NCC) Fehler) wird evaluiert für verschiedene Bewegungsgrößen (Δp) und größer werdende Objektgrößen (a)-(c).

[1] M. Attaran, The rise of 3-D printing: The advantages of additive manufacturing over traditional manufacturing, Business horizons **60.5** (2017): 677-688, 2017.
[2] L. Kramer, K. Bliedtner, F. Herold, 4D Computed Tomography for the Analysis of Dynamic Properties of Metallic Additive Manufactured Components, 13th Conference of Industrial Computed Tomography (iCT) 2024, 2024.
[3] H. Wang, L. Dong, J. O'Daniel, R. Mohan, A. Garden, K. Kian Ang, D. Kuban, M. Bonnen, J. Chang, R. Cheung, Validation of an accelerated 'demons' algorithm for deformable image registration in radiation therapy, Phys. Med. Biol. **50** (2005) 2887-2905, 2005.
[4] J. Ehrhardt, R. Werner, D. Säring, T. Frenzel, W. Lu, D. Low, H. Handels, An optical flow based method for improved reconstruction of 4D CT data sets acquired during free breathing, Medical Physics **34.2** (2007), 711-721, 2007.

DGZfP-Jahrestagung 2025

Röntgenprüfung und Porenklassifizierung in der additiven Fertigung: Eine systematische Studie an Al-Legierungen

Kevin EICKHOFF[1], Rafael MEINHARDT[2], Stefan GROTTKER[2], Frank HEROLD[1]
[1]VisiConsult X-ray Systems & Solutions GmbH
[2]Fraunhofer-Einrichtung für Additive Produktionstechnologien IAPT
k.eickhoff@visiconsult.de

Kurzfassung

Die additive Fertigung (engl.: *additive manufacturing* (AM)) von Metallen hat in den letzten Jahren für den Leichtbau komplexer Bauteile in der Automobil- und Luftfahrtindustrie zunehmend an Bedeutung gewonnen. Die dafür benötigten Legierungen müssen sowohl hochleistungsfähig als auch sicher sein. Um die Entwicklung neuer Werkstoffe für AM zu beschleunigen und Ressourcen zu sparen, ist eine effiziente Materialentwicklungsmethodik zur Qualifizierung neuer AM-Werkstoffe unerlässlich.
In diesem Zusammenhang eröffnet die Röntgenprüfung zahlreiche Möglichkeiten, um eine umfassende metallurgische Untersuchung und Schliffbildanalyse zu ersetzen. Insbesondere eine Porenanalyse des AM-Materials erlaubt Rückschlüsse auf die Materialeigenschaften, wie z.B. die Dichte, durch Klassifizierung und Verortung der Poren sowie deren statistische Auswertung. Mit Hilfe der Röntgenprüfung werden Probekörper schnell und reproduzierbar über ihr gesamtes Volumen analysiert, sodass Unregelmäßigkeiten innerhalb einer Legierungsprobe zuverlässig erkannt werden können. Durch den Einsatz eines Mikrofokus-Computertomographie (µCT)-Systems können auch kleine Poren mit Durchmessern von bis zu 30 µm erkannt und analysiert werden. Im Rahmen des öffentlich geförderten GeniAl-Forschungsprojekts wird in der hier vorgestellten systematischen Studie die Porosität einer $AlSi_{10}Mg$-Legierung im Laser-Pulverbett-Verfahren untersucht, um die optimalen 3D-Druckparameter zu ermitteln und eine Methodik zur schnelleren Parameterbestimmung zu identifizieren. Die erzielten Ergebnisse werden zur iterativen Optimierung des Druckprozesses und zur Verbesserung der Effizienz der Materialentwicklung verwendet.

Keywords: Additive Manufacturing (AM), Computertomographie (CT), Porenanalyse

Röntgenprüfung und Porenklassifizierung in der additiven Fertigung: Eine systematische Studie an Al-Legierungen

VCxray by VisiConsult

Motivation

In den letzten Jahren gewinnt die additive Fertigung (engl.: *additive manufacturing* (AM)) von Metallen in der Automobil-, Luft- und Raumfahrtindustrie immer mehr an Bedeutung [1]. Dadurch hat die zugehörige Bauteilanalyse - insbesondere mit Fokus auf Sicherheitsaspekte - zunehmend an Relevanz gewonnen. In diesem Zusammenhang sind die zerstörungsfreie Bewertung der Porosität im Material und deren Minimierung durch Optimierung der AM-Druckparameter von großem Interesse. Hierbei ist neben der Porosität und der entsprechend reduzierten Dichte [2] vor allem eine Klassifizierung der Poren von Bedeutung [3,4]. Diese ermöglicht es besonders durch den Einsatz von künstlicher Intelligenz (KI) einen effizienten Weg zur Optimierung der Materialentwicklung zu eröffnen [5] (siehe Abb. 1 (a)). Gegenstand des vorgestellten Förderprojekts ist die Korrelation von AM- und zugrundeliegenden Gusswerkstoffen mit Fokus auf Materialeigenschaften wie Porosität, Porenstatistik und Materialfestigkeit. Die ersten beiden Aspekte werden mit Hilfe der µCT-Technologie ermittelt, wozu neue Porenklassifizierungen benötigt bzw. bestehende Porenklassifizierungen [3] weiterentwickelt werden, welche auf die Bedürfnisse der Materialentwicklung abgestimmt sind.

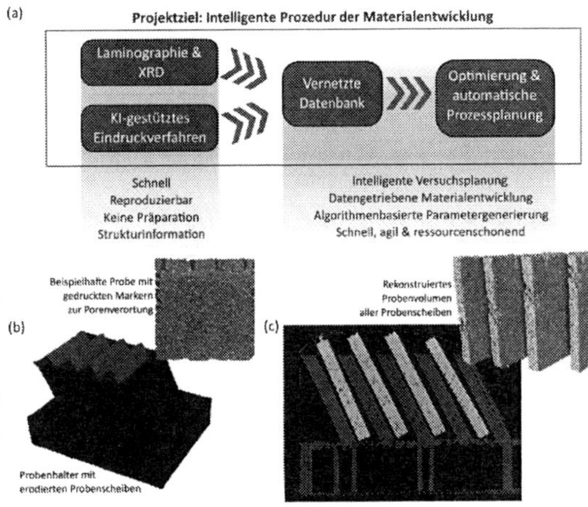

Abb. 1: (a) Illustration der Prozessabläufe für die Materialentwicklung gemäß dem Projektziel. (b) Bilder der genutzten AM-Proben aus $AlSi_{10}Mg$ mit eingedruckten Markierungen zur verlässlichen Porenverortung. (c) Zentralschnitt durch das rekonstruierte Probenvolumen sowie Illustration eines separierten Volumens der Probenscheiben.

Studie & Porenklassifizierung

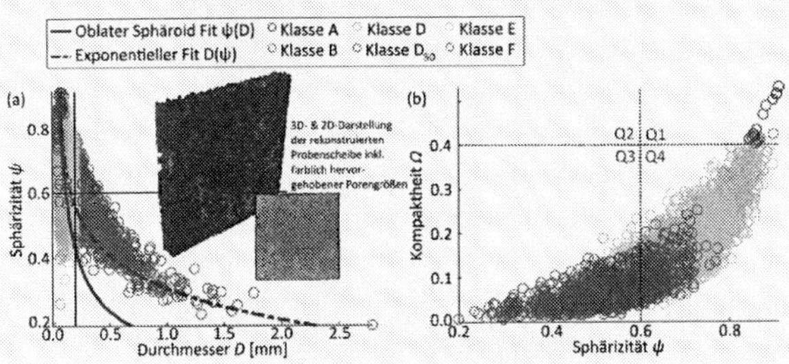

Abb. 2: Beispielhafte Porenanalyse und -klassifizierung in Form von (a) Ψ-D- und (b) Ψ-Ω-Diagrammen. Die Linien in (a) zeigen nicht-lineare Anpassungen gemäß eines Modells für einen oblaten Sphäroiden (durchgezogen) und einem exponentiellen Modell (gestrichelt). Aus letzterem ergibt sich der Grenzdurchmesser D_{lim}. Die Bilder in (a) zeigen zudem das rekonstruierte 3D-Volumen der Probe mit farblich kodiertem Porenvolumen V (blau: $V = 2.6 \cdot 10^{-5}\,mm^3$ bis rot: $V = 0.7\,mm^3$).

In der präsentierten systematischen Studie liegt der Schwerpunkt auf der µCT von AM-Proben und der damit verbundenen Porenklassifizierung. Hierzu wird eine $AlSi_{10}Mg$-Referenzlegierung verwendet, die im Laser-Pulverbett-Verfahren (PBF-LB/M) unter Verwendung verschiedener Parameter wie der Laserleistung P und der Scan-Geschwindigkeit v verdruckt wird. Vier erodierte Scheiben von der Mitte bis zum Rand des gedruckten Würfels werden mittels µCT und einer anschließenden Porenanalyse untersucht (siehe Abb. 1 (b)-(c)). Aufbauend auf grundlegenden Poreneigenschaften wie der Sphärizität Ψ und dem Durchmesser D werden unter anderem auch die Begrenzungsbox jeder Pore sowie der Randabstand und die Kompaktheit Ω berücksichtigt. Um die Porenbewertung auf die Bedürfnisse zuzuschneiden, wird die bestehende Klassifizierung [3] angepasst, indem zusätzliche Größen wie z. B. ein Grenzdurchmesser (D_{lim}) hinzugefügt werden. Letzterer stellt einen extrapolierten Porendurchmesser für eine verschwindende Porensphärizität dar, der auf einer nicht-linearen exponentiellen Anpassung des Durchmessers $D(\Psi)$ basiert und somit die beiden Eigenschaften Porendurchmesser und -sphärizität verbindet. Abbildung 2 zeigt ein Beispiel für klassifizierte (farbkodierte) Poren innerhalb eines *Ψ-D-Diagramms* (a) und *Ψ-Ω-*Diagramms (b) zusammen mit den Linien nicht-linearer Anpassungen. Durch diese Analyse gibt die durchgeführte systematische Studie sowohl Aufschluss über globale Größen wie die Porosität Φ und Porenverortung als auch über die Porenform in Abhängigkeit der Druckparameter.

Ergebnisse & Ausblick

Die Durchführung der Porenklassifizierung für jeden Druckparameter führt zu Druckparameter-Landschaften, wie in Abb. 3 beispielhaft dargestellt. Man beachte, dass die gemessenen Daten (schwarze Punkte) durch Interpolation hochskaliert werden. Die sich daraus ergebenden Parameterlandschaften ermöglichen es, den optimalen AM-Parametersatz zu finden, indem z. B. ermittelt wird, wo die Porosität Φ (vgl. Abb. 3 (a)) oder auch die Anzahl von Poren einer bestimmten Klasse (vgl. Abb. 3 (b)) zu- bzw. abnimmt. Der vorgeschlagene Grenzdurchmesser D_{lim} oben rechts in (b) zeigt ähnliche Tendenzen wie die Klassifizierung und wird als zusätzlicher Indikator verwendet. Die dargestellten Ergebnisse zeigen, dass eine auf Röntgen-µCT basierende Porositätsanalyse in Kombination mit einer angepassten und maßgeschneiderten Porenklassifizierung eine effektive Optimierung der AM-Druckparameter ermöglichen und somit sowohl Material als auch Entwicklungszeit sparen könnte.

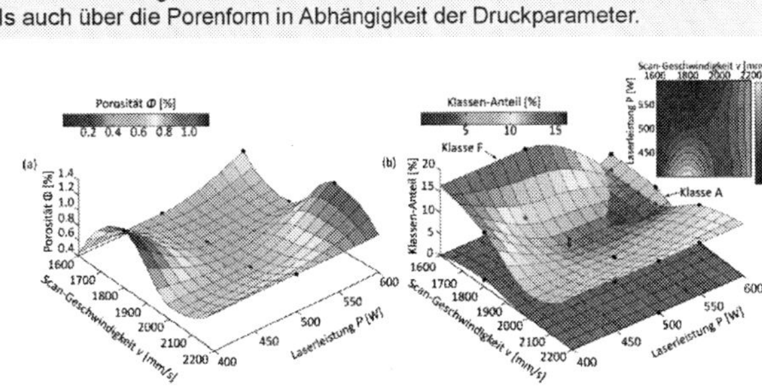

Abb. 3: Druckparameter-Landschaften der $AlSi_{10}Mg$ Proben als Funktionen der Scan-Geschwindigkeit v und der Laserleistung P bei einer Hatch-Distanz von 160 µm. Die gemessenen Daten (schwarze Punkte) wurden durch Interpolation hochskaliert. Die aufgetragenen Analyseparameter sind (a) die Porosität Φ und (b) der Anteil von Poren der Klassen A (unten) und F (oben). Das obere rechte Bild in (b) zeigt den Grenzdurchmesser D_{lim} aus der exponentiellen nicht-linearen Anpassung (vgl. Abb. 2(a)).

[1] M. Attaran, Business horizons, **60** (5), 677-688 (2017)
[2] D.-S. Shim, Journal of Materials Research and Technology, **15**, 119-134 (2021)
[3] N. Nudelis and P. Mayr, Metals, **11** (12), 1912 (2021)
[4] R. Snell et al., Jom, **72**, 101-109 (2020)
[5] N. Samadiani et al., Journal of Intelligent Manufacturing, 1-41 (2024)

Kevin EICKHOFF [1], **Rafael MEINHARDT** [2], **Stefan GROTTKER** [2], **Frank HEROLD** [1]
[1]VisiConsult X-ray Systems & Solutions GmbH, Brandenbrooker Weg 2-4, 23617 Stockelsdorf
[2]Fraunhofer IAPT, Am Schleusengraben 14, 21029 Hamburg
Kontakt E-Mail: k.eickhoff@visiconsult.de

Danksagungen: Wir bedanken uns für die finanzielle Unterstützung durch das Bundesministerium für Bildung und Forschung (BMBF) im Rahmen des Förderprojekts 01LY2301B. Wir danken der Firma Fehrmann Materials GmbH & Co. KG wird für die Zusammenarbeit und die Bereitstellung des Probenmaterials.

DLR Projektträger
GEFÖRDERT VOM Bundesministerium für Bildung und Forschung

Charakterisierung der Strömungen in der Extrusionseinheit von FFF-Druckern mittels X-ray Particle Tracking Velocimetry

Phi-Long CHUNG, Julian KATTINGER, Mike KORNELY, Julian EHRLER, Marc KREUTZBRUCK und Christian BONTEN
Universität Stuttgart, Institut für Kunststofftechnik, Stuttgart, Deutschland,
phi-long.chung@ikt.uni-stuttgart.de

Kurzfassung. Die additive Fertigung ist eine junge Fertigungstechnologie, die durch nahezu unbegrenzte Designmöglichkeiten überzeugt. Besonders in der Produktentwicklung, Prototypenfertigung und Kleinserien finden additive Fertigungsverfahren ein breites Anwendungsspektrum. Eines der verbreitesten Verfahren ist die Fused Filament Fabrication (FFF). Angesichts steigender Anforderungen an schnelle, effiziente und kostengünstige Prozesse sind Modellierungs- und Simulationsanalysen unerlässlich. Zentral ist dabei das Aufschmelz- und Strömungsverhalten des Kunststofffilaments im Druckkopf. Gelingt es, diesen Vorgang durch mathematische Modelle in Simulationsanalysen zu übertragen, könnte der Prozess bis an seine Grenzen optimiert werden. Um nun Einblicke in den laufenden Druckprozess zu gewinnen, ohne diesen zu beeinflussen, kommen Methoden der zerstörungsfreien Prüfung und der Strömungsmechanik zum Einsatz.

Die Particle Tracking Velocimetry (PTV) ist ein optisches Verfahren aus der Strömungsmechanik zur Messung dreidimensionaler Partikelbewegungen in Strömungen. Diese Messmethodik setzt jedoch transparente Medien und eine uneingeschränkte optische Zugänglichkeit für die Lichtquelle sowie die Kamerasysteme voraus. In der Röntgendurchstrahlungsprüfung stellt die optische Transparenz hingegen keine Einschränkung dar, da sie detaillierte Einblicke in das Innere von Bauteilen, wie FFF-Druckdüsen, ermöglicht. Herausforderungen ergeben sich jedoch, wenn Röntgenstrahlen den Kunststoff durchdringen, da dieser im Vergleich zu Metallen nahezu transparent erscheint. Dies erschwert die Visualisierung dynamischer Strömungsvorgänge im geschlossenen Schmelzekanal erheblich.

Diese Arbeit stellt eine neuartige Methode zur Strömungsanalyse in undurchsichtigen Medien mittels X-ray Particle Tracking Velocimetry (XPTV) vor. Am Beispiel des Strömungsverhaltens in einer FFF-Druckdüse wurden hoch abschwächende Wolframpartikel in das Filament eingebettet und in-situ untersucht. Erstmals konnte die Strömung einer Kunststoffschmelze mithilfe von XPTV sichtbar gemacht und analyisiert werden. Die vorgestellte Methodik erfasste dabei erfolgreich die ausgebildeten parabolischen Geschwindigkeitsprofile.

Keywords: Additive Fertigung, Strömungsanalyse, Durchstrahlungsprüfung, X-ray Particle Tracking Velocimetry

Medien- und Publikationspartner

https://doi.org/10.58286/31240

1. Einführung

Die präzise Erfassung und Analyse von Strömungsfeldern ist von zentraler Bedeutung in zahlreichen wissenschaftlichen und ingenieurtechnischen Disziplinen. Besonders in der Strömungsmechanik und Materialwissenschaft spielt das Verständnis der Geschwindigkeitsverteilung in komplexen Strömungen eine entscheidende Rolle. Hierzu kommen rheologische Mess- und Charakterisierungsverfahren zum Einsatz, bei denen eine kontrollierte Strömung im zu untersuchenden Werkstoff erzeugt wird. Die resultierende Materialantwort – gemessen in Form von Spannung und Schergeschwindigkeit an definierten Punkten – erlaubt Rückschlüsse auf materialabhängige Eigenschaften, ohne das vollständige Strömungsprofil erfassen zu müssen. Zur direkten Messung des Strömungsprofils werden Verfahren wie die Particle Image Velocimetry (PIV) oder die Particle Tracking Velocimetry (PTV) eingesetzt. Diese Methoden liefern hochauflösende Einblicke in die Strömungsdynamik, sind jedoch auf optisch transparente transparente Systeme beschränkt. Die X-ray Particle Tracking Velocimetry (XPTV) stellt eine neuartige Methode dar, die die Verfolgung von Partikeln in opaken Medien ermöglicht. Dadurch können erstmals Strömungsfeldern in bislang unzugänglichen, geschlossenen Systemen untersucht werden. Im Unterschied zu konventionellen optischen Verfahren kommen hierbei eine Röntgenquelle und ein Detektor anstelle von Kamerasystemen und Lichtquellen zum Einsatz. [1, 2]

Der Nutzen von Röntgenstrahlungen zur Visualisierung von Strömungen in opaken Materialien wurde in der Literatur bereits häufig aufgezeigt. Lee et al. zeigten 2003 bereits das Potenzial der Durchstrahlungsprüfung auf, indem innenliegende Strömungen visualisiert wurden [3]. Auch eine tomographische Auswertung des 3D-Raums durch die Kombination von Computertomographie und PTV wurde in ersten Arbeiten untersucht [4, 5]. Die PTV wurde seitdem genutzt, um tiefengemittelte Strömungsfelder zu visualisieren, wobei in einigen Experimenten auch Synchrotron-Strahlung genutzt wurde [6]. Aufgrund der geringen Verfügbarkeit von Synchrotron-Strahlenquellen, wurden verschiedene Ansätze zur Optimierung dieser Methode verfolgt. Zum einen wurden mehrere Röntgenquellen in Stereo-Anordnung aufgebaut, um eine räumliche Verfolgung von Partikeln zu ermöglichen [7]. Parker et al. zeigten zudem, dass Aufnahmeraten von bis zu 1 kHz möglich sind, sodass auch industriell-relevante Geschwindigkeiten mit der XPTV betrachtet werden könnten[8]. Zum aktuellen Stand der Forschung wird die XPTV jedoch primär zur Untersuchung von Medien mit bereits bekanntem Strömungsverhalten eingesetzt.

Im Rahmen dieser Arbeit werden die aktuellen Fortschritte der XPTV präsentiert und die Methodik am Beispiel eines Hot-Ends eines konventionellen 3D-Druckers vorgestellt.

2. Stand der Technik

2.1 Fused Filament Fabrication

Die Fused Filament Fabrication (FFF) ist das am weitesten verbreitete additive Fertigungsverfahren. Abb. 1a zeigt schematisch den Aufbau eines konventionellen FFF-Druckers. Dieser besteht grundlegend aus einem Portalsystem, einem Druckkopf und einer Materialzuführung. Letzteres dient dazu, das Kunststofffilament mithilfe eines mechanischen Zahnradvorschubs in die beheizte Düse des Druckkopfs zu führen. Eine Heizpatrone innerhalb des Heizblocks ist für die Bereitstellung der thermischen Energie verantwortlich, um das Filament zu aufschmelzen. Der geschmolzene Polymerstrang wird anschließend Schicht für Schicht auf die (beheizte) Bauplatte aufgetragen bis das Bauteil fertiggestellt ist. Der Druckkopf ist, wie in Abb. 1 b gezeigt, in das Hot-End und das Cold-End unterteilt. Das Hot-End beinhaltet die Düse mit einem verjüngten, konischen Abschnitt und einer Kapillare.

Gegenteilig zum Hot-End, welches die Aufgabe hat, das Filament zu plastifizieren, stellt das Cold-End sicher, dass das zugeführte Filament in einem festen Zustand bleibt bis das Filament dem Schmelzekanal zugeführt wird.

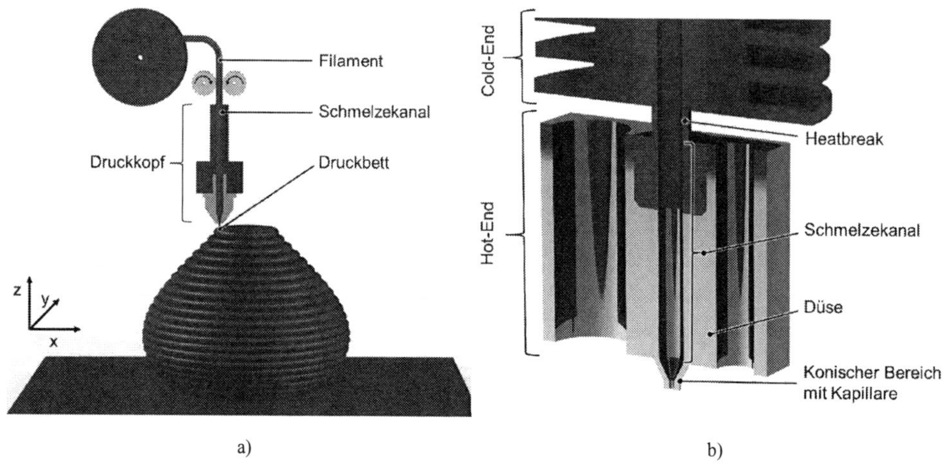

Abb. 1: a) Schematische Darstellung des Fused Filament Fabrication und b) Schnittdarstellung einer 3D-Druckdüse

2.2 Aktuelle Modelle und experimentelle Ansätze

Gegenwärtig bestehen zwei analytische Ansätze, die versuchen den Schmelzefluss im Inneren einer Düse zu beschreiben. Dabei unterscheiden sich die Ansätze in den modellierten Filamentgeschwindigkeiten. Das erste Modell nach Bellini et al. [9] geht von isothermen Bedingungen in der Düse aus und nimmt an, dass das Filament nach Eintritt in den Schmelzekanal in einem vollständig geschmolzenen Zustand vorliegt. Da dieses Modell jedoch auf der Hagen-Poiseuille-Gleichung basiert, können nur niedrige Filamentgeschwindigkeiten genau beschrieben werden. Im Gegensatz dazu bildet das zweite Modell [10] hohe Filamentgeschwindigkeiten ab, unter der Annahme, dass das zugeführte Filament spätestens im konischen Bereich der Düse aufschmilzt und dabei einen dünnen Schmelzfilm ausbildet. Während eine Vielzahl der Parameter, wie der Extrusionskraft, der Schmelzfilmdicke oder dem Druckverlust, unter Berücksichtigung dieser hohen Geschwindigkeiten mithilfe mathematischer Modelle abgebildet werden können, bestehen nach wie vor Unklarheiten bei der Modellierung des Schmelzeübergangs. Aktuelle Forschungsarbeiten [11–13] sind sich bislang uneinig über die vorherrschenden Ursachen des Abschmelzvorgangs, was zu Unterschieden in den gesetzten Randbedingungen der verschiedenen Modelle führt. Die numerisch aufgestellten Modelle sind dadurch zwar anhand der experimentellen Daten kalibriert, können jedoch auch nur für die kalibrierten Fälle präzise Vorhersagen liefern. Dadurch bleibt es weiterhin ungeklärt, wie sich der Spalt zwischen dem einströmenden Filament und der Düse füllt.

Um das Verständnis über die Aufschmelz- und Strömungsvorgänge im Schmelzekanal zu erweitern und bestehende Modelltheorien als auch Simulationen zu validieren, wurden bereits zahlreiche experimentelle Untersuchungen durchgeführt. So wurden beispielsweise Wägezellen [14] eingesetzt, um die auftretenden Extrusionskräfte aufzuzeichnen. Dadurch konnte ein plötzlicher Anstieg der Extrusionskraft bei einer kritischen Filamentgeschwindigkeit in Abhängigkeit von dem eingesetzten Werkstoff und der Düsengeometrie bestimmt werden. Eine weitere Arbeit nutzte die Thermografie [15], um die Temperaturen im Druckprozess zu überwachen. Sämtliche Methoden haben gemeinsam, dass sie nur indirekte Einblicke in das Aufschmelz- und Strömungsverhalten ermöglichen. Weiterführende Arbeiten versuchten daher den Prozess so realitätsnah wie möglich zu betrachten. Eine Methode bestand aus dem Präparieren eines Filaments mit Farbmarkern [16]. Nach der Extrusion des modifizierten Filaments wurde der Druckprozess gestoppt, das erkaltete Filament aus der Düse entnommen und die Verteilung der Farbstoffe im Filament mittels Lichtmikroskopie und Computertomographie betrachtet. Dies ermöglichte Rückschlüsse auf das zugrundeliegende Strömungsprofil [16]. Um den Aufschmelzvorgang in-situ zu betrachten, wurden Düsen aus Glas genutzt, sodass auch die Füllung des Spalts zwischen Filament und Düse betrachtet werden konnte [17]. Die dabei generierten Ergebnisse können jedoch nicht direkt auf konventionelle metallische Düsen, aufgrund unterschiedlicher Reib- und Wärmeleiteigenschaften, übertragen werden.

2.4 Optische Geschwindigkeitsmessverfahren

Die Particle Image Velocimetry (PIV) als auch die Particle Tracking Velocimetry (PTV) sind optische Geschwindigkeitsmessverfahren und stellen moderne bildgebende Messverfahren dar. Im Rahmen dieser Verfahren werden dem zu untersuchenden Fluid sog. Tracerpartikel beigefügt, welche mit einer Lichtquelle hoher Intensität beleuchtet werden. Durch die Bestrahlung einer gesamten 2D-Ebene können die enthaltenden Tracer mit einem Kamerasystem erkannt und als Grauwertbilder aufgezeichnet werden. Werden mehrere Aufnahmen der Ebene in einem zeitlichen Abstand Δt aufgezeichnet, können die Trajektorien der Partikel bei einer hohen Zeitauflösung annährend linear bestimmt werden und daraus die Geschwindigkeitsvektoren errechnet werden. Um das Strömungsfeld des Fluides nicht zu beeinflussen müssen die eingesetzten Tracerpartikel so klein wie möglich dimensioniert sein und eine ähnliche Dichte zum untersuchten Fluid besitzen. Zusätzlich muss die Lichtquelle, meistens ein Laser, die Partikel optisch erfassen können und das gestreute Licht von dem Kamerasystem aufgefangen werden [2]. Die PIV und PTV unterscheiden sich dabei grundsätzlich in ihrer Methodik als auch in der Auswertung. Während die PIV Geschwindigkeitsfelder durch die Verschiebung von Mustern in aufeinanderfolgenden Bildern mittels Kreuzkorrelation bestimmt werden, verfolgt die PTV einzelne Partikel konsekutiv über mehrere Bilder. Dies führt dazu, dass bei PIV-Analysen ein höherer Tracerpartikelanteil für die Erkennung von Verschiebungsfeldern notwendig ist. Gegenteilig dazu werden bei der PTV ein geringer Tracerpartikelanteil für die zuverlässige Erkennung von Einzelpartikeln benötigt.

3. Versuchsdurchführung

3.1 Verwendete Materialien

Im Rahmen dieser Arbeit wurde ein schlagzähmodifiziertes Polystyrol (PS, Typ PS486N, INEOS Styrolution, Frankfurt am Main) verwendet. Aufgrund der Tatsache, dass Kunststoffe Röntgenstrahlungen nur geringfügig abschwächen und dadurch ein geringer Kontrast in den Grauwertbildern besteht, können kaum bis keine Informationen über das tatsächliche Strömungsprofil extrahiert werden. Um diese physikalische Limitierung zu umgehen, wird ein Kontrastmittel benötigt, welches die Röntgenstrahlung stärker abschwächt und gleichzeitig den Kunststoff nicht hinsichtlich seiner Fließeigenschaften beeinflusst. Das eingesetzte Wolframpulver von Werth-Metall (Erfurt), mit einer Dichte von $\rho = 19.25\ \mathrm{g/cm^3}$, Ordnungszahl von 74 und einer mittleren Partikelgröße von 32 µm erfüllt diese Anforderungen vollständig. Das Wolframpulver wird über die Doppelschneckenextrusion in den Kunststoff mit einem Anteil von 0,17 M.-% eingearbeitet, um so auch eine PTV-Analyse zu ermöglichen. Das resultierende Compound wurde anschließend granuliert und mittels Einschneckenextrusion zu einem Filament mit einem Durchmesser von 1,75 mm verstreckt.

3.2 Experimenteller Aufbau

Die verwendeten Bilddaten wurden mithilfe eines in-situ Versuchsaufbaus, für den Betrieb in einem Computertomographen konzipiert, generiert (siehe Abb. 2 a).

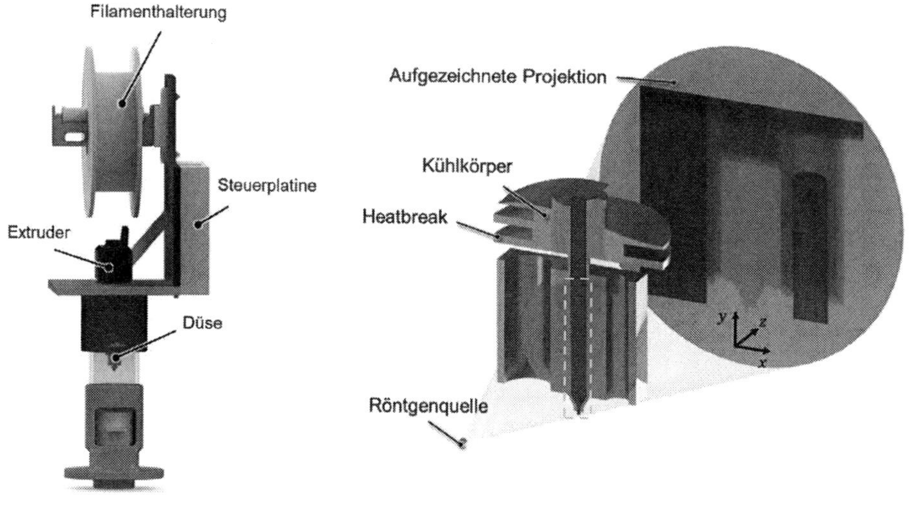

Abb. 2: a)Schematischer Aufbau des in-situ CT-Messaufbaus und b) Schnittansicht der untersuchten Düse

Der FFF-Extrusionsaufbau wurde im Probenraum des Computertomographen Yxlon FF20 CT (Comet Yxlon GmbH, Hamburg) positioniert. Die CT-Anlage ist ausgestattet mit einer Nano-fokus Transmissionsröhre (Typ: Comet FXE190.61) und einem Detektor (Typ: Varex 2530HE, Varex Imaging, Salt Lake City, Utah, USA) mit einer Speichertiefe von 16-bit. Der Aufbau ist mit kommerziell erhältlichen Bauteilen aus dem Bereich der FFF ausgestattet. Als Extruder ist das Bondtech LGX model (Värnamo, Schweden) verbaut, um kontinuierlich Filament zuzuführen. Die Anbindung zwischen Materialzuführung und der Integraldüse aus EN AW 6082 wird durch das E3D-V6 Hot-End realisiert. Aufgrund der begrenzten zeitlichen Auflösung des Detektors und den hohen Strömungsgeschwindigkeiten in der Polymerschmelze, wurden die Experimente auf Basis von Röntgendurchstrahlung durchgeführt. Die Region-of-Interest (ROI) ist in Abb. 2b durch die orange-gestrichelte Umrandung markiert. Es wurde eine geometrische Vergrößerung von ca. 7,45 erreicht. Dafür wurde der Source-to-Detector Distance auf 805 mm und der Source-to-Object Distance auf 108 mm gesetzt. In den Messungen wurde eine Strahlenleistung von 11,4 W durch eine Spannung von 190 kV und einer Stromstärke von 60 µA realisiert. Der Detektor wurde auf eine Aufnahmerate von 15 Hz gesetzt und mit einer Zeit von 67 ms belichtet. Diese Parameter zeigten ein Optimum zwischen hoher zeitlicher Auflösung und gutem Contrast-to-Noise-Verhältnis (CNR).

Die in-situ Experimente wurden bei einer konstanten Temperatur von 240 °C und variierenden Exrusionsgeschwindigkeiten zwischen 0,5 mm/s und 2 mm/s durchgeführt. Vor den Versuchen wurde der Zahnradvorschubs kalibriert.

3.3 Bilddatenaufbereitung

Um den Algorithmus der PTV anwenden zu können, müssen die aufgezeichneten Radiogramme zunächst aufbereitet werden. Dies ist notwendig, um so viele Partikel wie möglich über die Segmentierung detektieren zu können. Die Verarbeitungsschritte zur Berechnung der Trajektorien eines jeden Partikels ist in Abb. 3 dargestellt. Die PTV wurde mit der Open-Source Software Fiji durchgeführt.

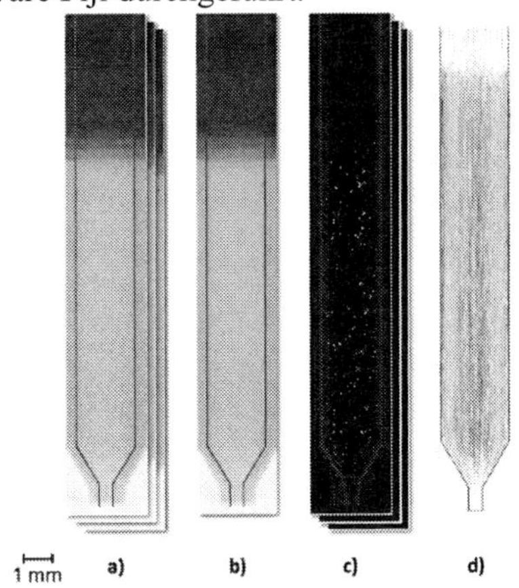

Abb. 3: Bildverarbeitungsschritte zur Analyse des Schmelz- und Fließverhaltens im FFF: a) Rohbild-Datensatz, b) Gemitteltes Hintergrundbild, c) Partikeldetektion und d) Partikel Tracking

Zunächst wurden die Rohbilder auf den ROI zugeschnitten, sodass nur der Schmelzekanal im Inneren der Düse abgebildet ist. Aufgrund des relativ geringen Kontrasts in den Rohbildern wurde ein zeitgemitteltes Referenzbild berechnet und von den einzelnen Frames subtrahiert. Anschließend wird die PTV-Analyse mit dem Fiji-Plugin TrackMate durchgeführt. Anhand der Differenzbilder müssen die einzelnen Partikel erfasst und anschließend verfolgt werden. Für die Partikeldetektion wurde der Laplacian of Gaussian (LoG)-Filter auf den Imagestack angewandt und führte im Mittel zu einer Menge von 250 detektiertern Partikeln in jedem Bildausschnitt. Nachdem die Partikel erfasst wurden, wird für die Bestimmung der Partikeltrajektorien der Linear Assignment Problem (LAP)-Tracker genutzt. Die ermittelten Informationen wurden anschließend mit Python und entsprechenden Bibliotheken ausgewertet, um die ermittelten Trajektorien zu filtern und auszuwerten. Dabei wurden Artefakte bei der Trajektorienberechnung entfernt – dies beinhaltete zum einen unrealistische Partikelpfade, die entgegen oder orthogonal zur Strömungsrichtung ermittelt wurden. Wie bei sämtlichen bildverarbeitenden Verfahren und Methoden ist auch die Bilddatensegmentierung bzw. die Algorithmen der Partikeldetektion und –verfolgung abhängig von der Bildqualität der Eingabedaten und verwendeten Parametern. Dies führt dazu, dass Partikel mit geringem CNR nur abschnittsweise erkannt werden, während der LAP-Algorithms die Bilddaten verarbeitet. In seltenen Fällen führt dies dazu, dass benachbarte Partikel fälschlicherweise anstelle des aktuell verfolgten Partikels getrackt werden und in den ausgewerteten Daten abrupte Richtungswechsel in den Trajektorien vorliegen. Um dem entgegen zu wirken, wurden die Längen aller Trajektorien bestimmt und ein Verhältnis zwischen der absoluten Trajektorienlänge und der Schmelzkanallänge gebildet. Verhältnisse unter 80 % wurden ausgeschlossen, sodass kurze Trajektorien sowie abrupte Richtungswechsel im Verlauf der Trajektorien eliminiert wurden.

4. Ergebnisse und Diskussion

Die Röntgenbilder stellen die Schnittansicht des Schmelzekanals dar, sodass Tiefeninformationen der Partikel nicht bestimmt werden können. Dennoch erlauben die Messdaten die Bestimmung zeitlicher Veränderungen in der x-y-Ebene sowie die Analyse der Häufigkeitsverteilung entlang der x-Achse, wodurch die Partikelgeschwindigkeiten in der betrachteten Ebene berechnet werden können. Während dies für den zylindrischen Schmelzkanal, in dem die erwartete Geschwindigkeitskomponente in Richtung des Düsenaustritts liegt irrelevant erscheint, ist es für die Analyse des konischen Austritts deutlich komplexer.

Eine Auswertung der Trajektorien erfolgte für alle Filamentgeschwindigkeiten zwischen 0,5 mm/s und 2,0 mm/s. Beispielhaft sind in Abb. 4 die Trajektorien der erfassten Partikel bei einer Filamentgeschwindigkeit von 1,5 mm/s dargestellt. Weiterhin wurden die Trajektorien entsprechend ihrer Geschwindigkeit in einzelne Bereich über dem normierten Durchmesser unterteilt. Die höchsten Geschwindigkeiten sind im Zentrum, während die niedrigsten Geschwindigkeiten entlang der Wand vorzufinden sind. Zudem kann festgestellt werden, dass die Partikel nicht bis in die Kapillare verfolgt werden können, wo deutlich höhere Strömungsgeschwindigkeiten zu erwarten sind. Dies liegt an den zeitlichen Auflösungsgrenzen der Systemkomponenten (Röntgenquelle und Detektor).

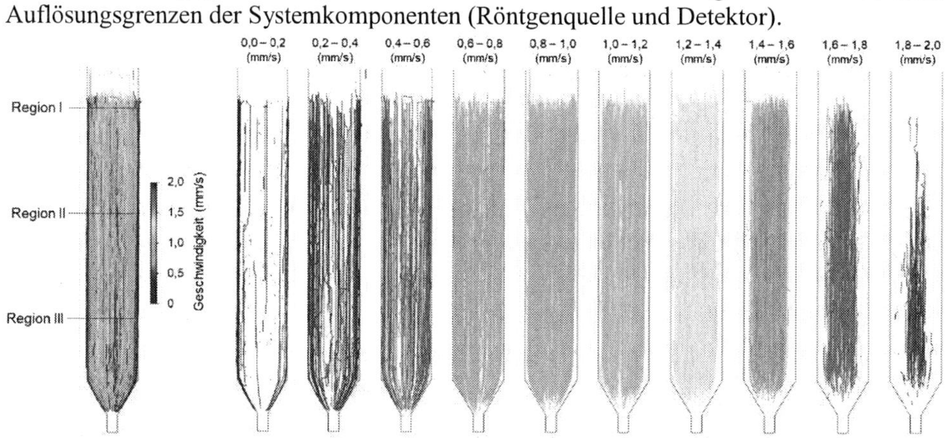

Abb. 4: Partikeltrajektorien im Schmelzekanal bei einer Geschwindigkeit von 1,5 mm/s und qualitative Aufteilung der Geschwindigkeitsbereiche über den normierten Durchmesser [18]

Abb. 5 zeigt die Histogramme der detektierten und verfolgten Partikel in den Abschnitten I bis III innerhalb des Schmelzkanals bei einer Filamentgeschwindigkeit von 1,5 mm/s. Dabei ist zu beachten, dass die Histogramme die radiale Verteilung der Partikel darstellen, die die Abschnitte I, II und III passieren. Wie von einem parabolischen Geschwindigkeitsprofil zu erwarten, ist die höchste Partikelanzahl in der Mitte vorzufinden, während eine Verlangsamung in Richtung der Wand vorliegt. Darüber hinaus deuten zwei Peaks bei x = ± 0,6 mm auf höhere Geschwindigkeiten in diesem Bereich hin, was jedoch auf das dreidimensionale Strömungsprofil und der durchstrahlten Querschnittsfläche zusammenhängt (in Abb. 6 dargestellt). Ein weiterer beobachteter Effekt ist, dass die Gesamtheit der detektierten Partikel entlang des normierten Durchmessers abnimmt, je näher der Abschnitt dem Düsenaustritt kommt. Dieser Rückgang ist auf das Abschmelzen zurückzuführen. Das Strömungsprofil wird dabei von einer Propfenströmung in Region I zu einem eher Newton'schen Profil in Region III überführt. Obwohl die mittlere Geschwindigkeit über alle Regionen hinweg konstant bleibt, weist Region III einen größeren Anteil schnell bewegender Partikel auf, die schwieriger zu tracken sind. Dies führt zu einer insgesamt abnehmenden Anzahl detektierte Partikel zwischen den Regionen I–III.

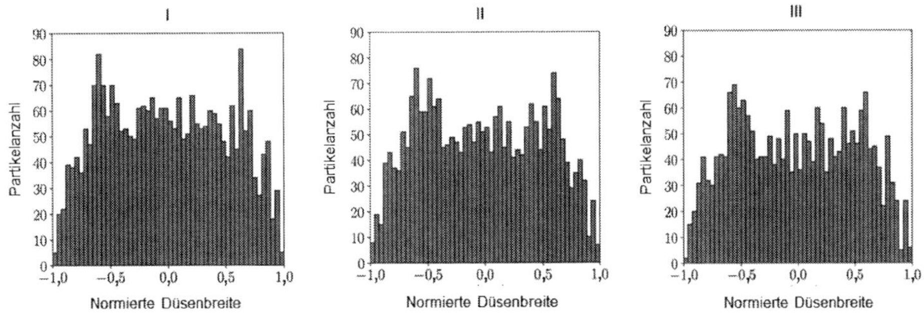

Abb. 5: Histogramm aller detekierten Partikel an den Regionen I-III bei einer Geschwindigkeit von 1,5 mm/s

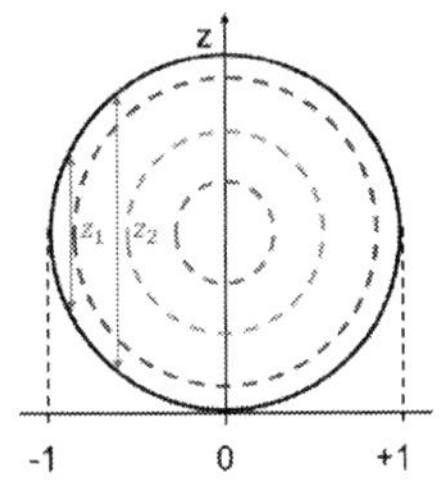

Abb. 6: Querschnitt des Schmelzekanals und schematischer Darstellung der Geschwindigkeitsbereiche.

5. Fazit und Ausblick

Im Rahmen dieser Arbeit wurde gezeigt, dass die Kombination von Durchstrahlungsprüfung und Particle Tracking Velocimetry ein noch tieferer Einblick in die komplexen Strömungsphänomene optisch unzugänglicher Systemen möglich ist, wie am Beispiel des untersuchten Hot-Ends gezeigt wurde. In den Experimenten konnte nicht nur die Polymerschmelze in-situ visualisiert werden, sondern auch einzelne Partikel verfolgt werden, um das resultierende Geschwindigkeitsprofil zu bestimmen. Trotz der Tatsache, dass der untersuchte Anwendungsfall an die technischen Grenzen des CT-Systems stößt, war es möglich die in-situ-Experimente durchzuführen, was erneut das Potenzial von XPTV als leistungsstarkes Werkzeug zur Untersuchung beliebiger Strömungen aufzeigt. In zukünftigen Arbeiten soll die Methode durch die Untersuchung verschiedener Kunststoffe weiter validiert und ein, für die Anwendung geeignetes Messsystem, verwendet werden. Da die vorgeschlagene Methode auf der additiven Fertigung basiert, werden zudem weitere mögliche Anwendungsbereiche erforscht. Möglich in der Kunststoffverarbeitung – aber nicht darauf beschränkt – könnten beispielsweise die Untersuchung von Mischprozessen in Ein- oder Doppelschneckenextrudern sein. Der Einsatz von schnelleren Detektoren und höherenergetischen Röntgenquellen öffnet potenziell die Türen zur Abbildung realitätsnaher Strömungsgeschwindigkeiten und der tomographischen Röntgen-Particle-Tracking-Velocimetry.

Danksagung

Die vorgestellten Ergebnisse wurden im Rahmen eines von der Deutschen Forschungsgemeinschaft e.V. (DFG) unter der Projektnummer 545960701 geförderten Forschungsprojekts erarbeitet.

Referenzen

[1] Heindel, T.J. X-Ray Flow Visualization: Techniques and Applications. In: *Journal of Fluids Engineering,* 2024, **146**(1). Doi:10.1115/1.4064050

[2] Herwig, H. *Strömungsmechanik.* 2. Auflage. Berlin, Heidelberg: Springer, 2006. ISBN 3-540-32443-7.

[3] Lee, S.-J. und G.-B. Kim. X-ray particle image velocimetry for measuring quantitative flow information inside opaque objects. In: *Journal of Applied Physics,* 2003, **94**(5), 3620–3623. Doi:10.1063/1.1599981

[4] Mäkiharju, S.A., J. Dewanckele, M. Boone, C. Wagner und A. Griesser. Tomographic X-ray particle tracking velocimetry. In: *Experiments in Fluids,* 2022, **63**(1). Doi:10.1007/s00348-021-03362-w

[5] Parker, J.T., J. DeBerardinis und S.A. Mäkiharju. Enhanced laboratory x-ray particle tracking velocimetry with newly developed tungsten-coated O(50 μm) tracers. In: *Experiments in Fluids,* 2022, **63**(12). Doi:10.1007/s00348-022-03530-6

[6] Fouras, A., J. Dusting, R. Lewis und K. Hourigan. Three-dimensional synchrotron x-ray particle image velocimetry. In: *Journal of Applied Physics,* 2007, **102**(6). Doi:10.1063/1.2783978

[7] Heindel, T.J., J.N. Gray und T.C. Jensen. An X-ray system for visualizing fluid flows. In: *Flow Measurement and Instrumentation,* 2008, **19**(2), 67–78. Doi:10.1016/j.flowmeasinst.2007.09.003

[8] Parker, J.T., T. Dreier, D. Nilsson und S.A. Mäkiharju. In-lab X-ray particle velocimetry for multiphase flows: Design principles and demonstration of O(1 kHz) XPV. In: *Flow Measurement and Instrumentation,* 2024, **96**, 102536. Doi:10.1016/j.flowmeasinst.2024.102536

[9] Bellini, A., S. Guceri und M. Bertoldi. Liquefier Dynamics in Fused Deposition. In: *Journal of Manufacturing Science and Engineering,* 2004, **126**(2), 237–246. Doi:10.1115/1.1688377

[10] Osswald, T.A., J. Puentes und J. Kattinger. Fused filament fabrication melting model. In: *Additive Manufacturing,* 2018, **22**, 51–59. Doi:10.1016/j.addma.2018.04.030

[11] Kattinger, J., M. Kornely, J. Ehrler, C. Bonten und M. Kreutzbruck. Analysis of melting and flow in the hot-end of a material extrusion 3D printer using X-ray computed tomography. In: *Additive Manufacturing,* 2023, **76**, 103762. Doi:10.1016/j.addma.2023.103762

[12] Serdeczny, M.P., R. Comminal, M.T. Mollah, D.B. Pedersen und J. Spangenberg. Viscoelastic simulation and optimisation of the polymer flow through the hot-end during filament-based material extrusion additive manufacturing. In: *Virtual and Physical Prototyping,* 2022, **17**(2), 205–219. Doi:10.1080/17452759.2022.2028522

[13] Serdeczny, M.P., R. Comminal, M.T. Mollah, D.B. Pedersen und J. Spangenberg. Numerical modeling of the polymer flow through the hot-end in filament-based material extrusion additive manufacturing. In: *Additive Manufacturing,* 2020, **36**, 101454. Doi:10.1016/j.addma.2020.101454

[14] Kattinger, J., T. Ebinger, R. Kurz und C. Bonten. Numerical simulation of the complex flow during material extrusion in fused filament fabrication. In: *Additive Manufacturing,* 2022, **49**, 102476. Doi:10.1016/j.addma.2021.102476

[15] Zhang, J., E. Vasiliauskaite, A. de Kuyper, C. de Schryver, F. Vogeler, F. Desplentere und E. Ferraris. Temperature Analyses in Fused Filament Fabrication: From Filament Entering the Hot-End to the Printed Parts. In: *3D Printing and Additive Manufacturing,* 2022, **9**(2), 132–142. Doi:10.1089/3dp.2020.0339

[16] Peng, F., B.D. Vogt und M. Cakmak. Complex flow and temperature history during melt extrusion in material extrusion additive manufacturing. In: *Additive Manufacturing,* 2018, **22**, 197–206. Doi:10.1016/j.addma.2018.05.015

[17] Hong, Y., M. Mrinal, H.S. Phan, V.D. Tran, X. Liu und C. Luo. In-situ observation of the extrusion processes of Acrylonitrile Butadiene Styrene and Polylactic Acid for material extrusion additive manufacturing. In: *Additive Manufacturing,* 2022, **49**, 102507. Doi:10.1016/j.addma.2021.102507

[18] Kattinger, J., M. Kornely, J. Ehrler, P.-L. Chung, C. Bonten und M. Kreutzbruck. Röntgen Particle Tracking Velocimetry als Methode der rheologischen Charakterisierung. In: *29. Stuttgarter Kunststoffkolloquium.* ISBN 978-3-9818681-4-2.

Realistische Simulation von Röntgencomputertomografie-Systemen

Anthony ORTH [1], Carsten BELLON [1], Matthias BRAUN [2], Kuan LI [3], Tamara REUTER [2], Markus BARTSCHER [3], Stefan KASPERL [4], Tino HAUSOTTE [2]

[1] Bundesanstalt für Materialforschung und -prüfung (BAM), Berlin, carsten.bellon@bam.de

[2] Lehrstuhl für Fertigungsmesstechnik, Friedrich-Alexander-Universität Erlangen-Nürnberg, Erlangen

[3] Physikalisch-Technische Bundesanstalt, Braunschweig

[4] Fraunhofer-Institut für Integrierte Schaltungen IIS, Entwicklungszentrum Röntgentechnik EZRT, Fürth

Kurzfassung

Die Entwicklung und Bewertung realitätsnaher Nachbildungen eines industriellen CT-Systems in einer Simulationssoftware waren Inhalt des WIPANO Forschungsprojektes CTSimU2 – „Realistische Simulation realer Röntgencomputertomografie-Systeme mit basisqualifizierter Simulationssoftware“. Als Voraussetzung dienen dabei Simulationssoftwares, die durch das Testframework aus dem Vorprojekt CTSimU1 basisqualifiziert wurden. Das Testframework testet die hinreichende physikalische Korrektheit und Funktionalität einer Simulationssoftware (Basisqualifizierung der Software).

Für eine realitätsnahe Nachbildung ist nicht nur die Güte der Simulationssoftware, sondern insbesondere die Güte der Parametrisierung des realen CT-Systems in der Simulationssoftware ausschlaggebend. Dabei kann das Vorgehen der Parametrisierung in vier Schritte unterteilt werden: die Datenaufnahme am realen CT-System (Schritt 1), die Auswertung der aufgenommenen Daten für die Generierung allgemeiner Parameterangaben (Schritt 2), die Übertragung der Parameter in die spezifischen Simulationssoftwares (Schritt 3) und die Validierung der resultierenden Simulationsergebnisse Simulationsparameter durch einen geeigneten Test an dimensionellen Messgrößen (Schritt 4)

Ziel des Projektes war es neben der Erarbeitung allgemeiner Methoden zur Datenaufnahme und Auswertung der Daten, die Entwicklung eines Tests, auf dessen Basis die ausreichend korrekte Simulation einer realen Anlage im Hinblick auf eine spätere Messunsicherheitsbestimmung beurteilt werden kann. Die erarbeiteten Ergebnisse sollen wie bereits im Vorprojekt CTSimU1 in einen Richtlinienentwurf für die Richtlinienreihe VDI/VDE 2630 übertragen werden. Dieser Beitrag fasst die Arbeiten und Ergebnisse des Projekts zusammen.

Keywords: Radiographic Testing (RT), Computertomografie, Simulation

Realistische Simulation von Röntgencomputertomografie-Systemen

CTSimU 2

Realistische Simulation realer Röntgencomputertomografie-Systeme mit basisqualifizierter Simulationssoftware – CTSimU2

DGZfP-Jahrestagung, 26. – 28. Mai 2025

Carsten Bellon, et al.

BAM
Bundesanstalt für Materialforschung und -prüfung

Konsortium

Forschungspartner:

- Markus Bartscher, **PTB**
- Kuan Li, **PTB**
- Carsten Bellon, **BAM**
- Anthony Orth, **BAM**
- Tino Hausotte, **FMT**
- Tamara Reuter, **FMT**
- Matthias Braun, **FMT**
- Stefan Kasperl, **EZRT**

PTB

BAM

FMT Lehrstuhl für Fertigungsmesstechnik

Fraunhofer EZRT

Industriepartner

diondo x-ray systems and services

Waygate Technologies

ZEISS

SIEMENS

SIEMENS Healthineers

VOLUME GRAPHICS

Weidmüller

werth

Assoziierter Partner:

HEITEC

Übersicht

- Die Projektreihe CTSimU
- Das Projekt CTSimU2: Ziele und Vorgehen
- Anforderungen an das digitale Modell
- Test des digitalen Modells
- Ergebnisse
- Zusammenfassung und Ausblick

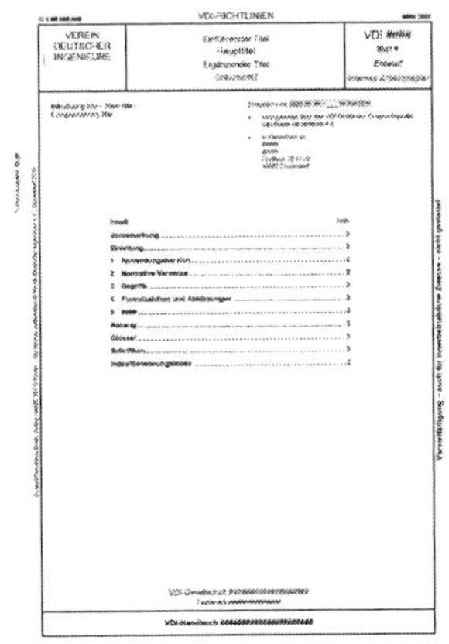

Template: Verein Deutscher Ingenieure e.V., Düsseldorf, 2021

27.05.2025 REALISTISCHE SIMULATION VON RÖNTGENCOMPUTERTOMOGRAFIE-SYSTEMEN 3

Motivation für die Projektreihe CTSimU

Es gibt verschiedene Normen und Richtlinien zur Messunsicherheitsbestimmung, aber sie

- sind durch die Komplexität des CTs nicht umzusetzen
- beachten nicht die Besonderheiten von CT-Simulationen
- sind zeit- und ressourcenintensiv

Aufgabenspezifische Messunsicherheitsbestimmung mittels Simulation
➔ ressourcen- und kostensparend

Kein standardisiertes Verfahren, um
- **Simulationssoftware zu qualifizieren,**
- **digitale Nachbildungen zu qualifizieren und**
- **am Ende eine dimensionelle Messung mit Unsicherheitsangabe durchzuführen.**

27.05.2025 REALISTISCHE SIMULATION VON RÖNTGENCOMPUTERTOMOGRAFIE-SYSTEMEN 4

Aufbau der CTSimU-Reihe

✓	CTSimU1	Schritt 1: Testframework zur Überpüfung der Simulationssoftware hinsichtlich der korrekten Nachbildung relevanter physikalischer Effekte und Funktionalitäten
aktuell	CTSimU2	Schritt 2: Werkzeugkasten zur Erstellung eines digitalen Zwillings und Test zur Überprüfung der digitalen Nachbildung
zukünftig		Schritt 3: Bestimmung der aufgabenspezifischen Messunsicherheit mittels Simulation unter Anwendung der Monte-Carlo-Methode

Funktionsweise Simulationssoftware

- Nachstellen der Strahlungs-Materie Wechselwirkung
- Nachstellen einer virtuellen Szene mit Komponenten analog zu realem CT-System
- Vorgehensweisen zur Simulation von radiographischen Projektionen:
  - Monte-Carlo-basierte Teilchenphysik-Simulationen
  - Ray-Casting Simulationen

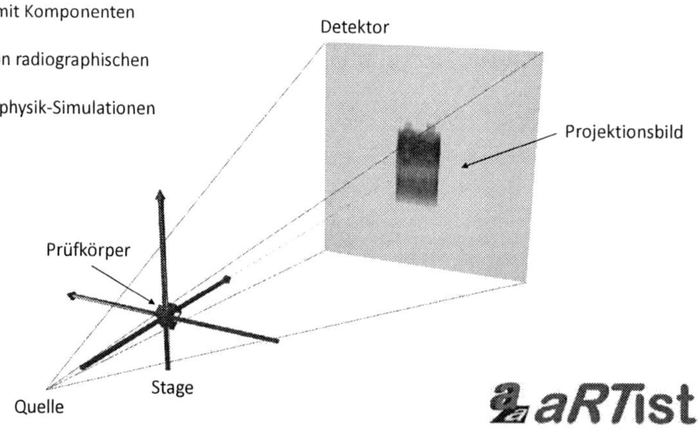

aRTist

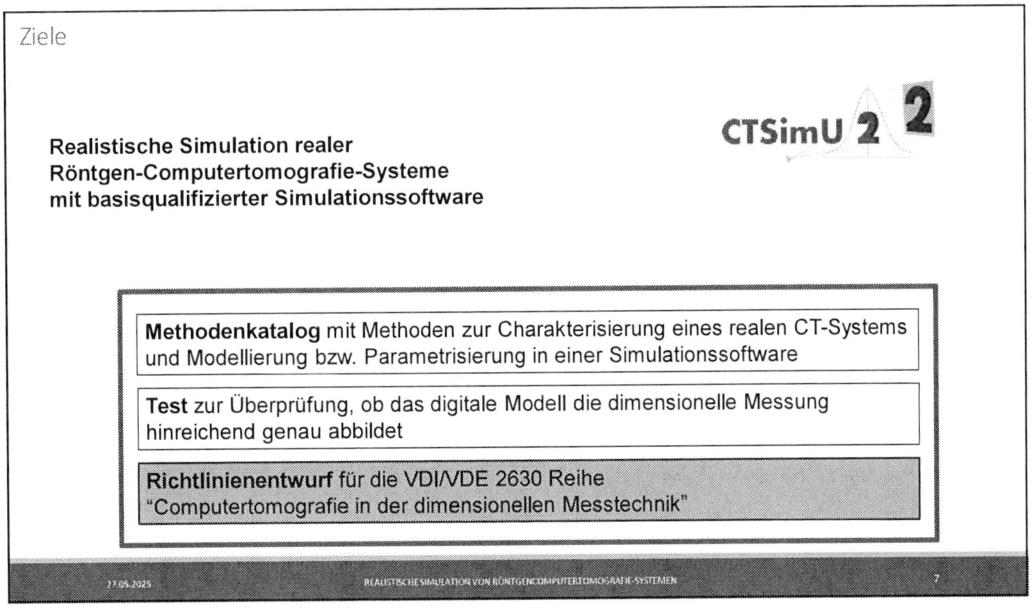
Ziele
CTSimU 2
Realistische Simulation realer
Röntgen-Computertomografie-Systeme
mit basisqualifizierter Simulationssoftware
Methodenkatalog mit Methoden zur Charakterisierung eines realen CT-Systems und Modellierung bzw. Parametrisierung in einer Simulationssoftware
Test zur Überprüfung, ob das digitale Modell die dimensionelle Messung hinreichend genau abbildet
Richtlinienentwurf für die VDI/VDE 2630 Reihe "Computertomografie in der dimensionellen Messtechnik"
27.05.2025
REALISTISCHE SIMULATION VON RÖNTGENCOMPUTERTOMOGRAFIE-SYSTEMEN
7

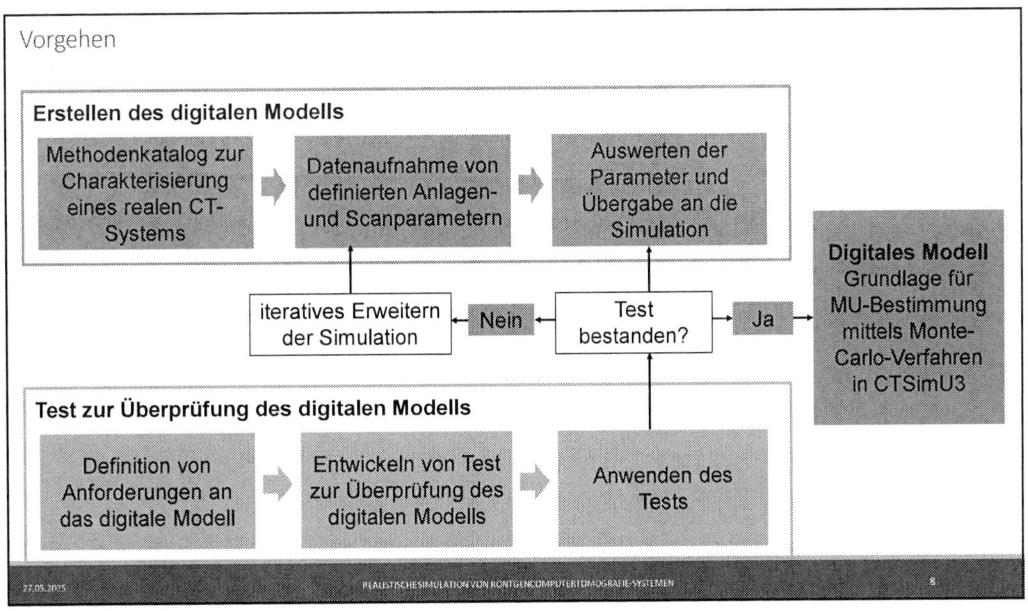
Vorgehen
Erstellen des digitalen Modells
Methodenkatalog zur Charakterisierung eines realen CT-Systems
Datenaufnahme von definierten Anlagen- und Scanparametern
Auswerten der Parameter und Übergabe an die Simulation
iteratives Erweitern der Simulation
Nein
Test bestanden?
Ja
Digitales Modell Grundlage für MU-Bestimmung mittels Monte-Carlo-Verfahren in CTSimU3
Test zur Überprüfung des digitalen Modells
Definition von Anforderungen an das digitale Modell
Entwickeln von Test zur Überprüfung des digitalen Modells
Anwenden des Tests
27.05.2025
REALISTISCHE SIMULATION VON RÖNTGENCOMPUTERTOMOGRAFIE-SYSTEMEN
8

Anforderungen an das digitale Modell

Testkonzept basierend auf modifizierter metrologischer Verträglichkeit von Wiederholmessung an kalibrierten Prüfkörpern

Anforderungen

Vergleich der Messgrößen

- Radien
- Uni-/ Bidirektionale Maße
- Abstand Mittelpunkt zu Mittelpunkt

des digitalen Modells mit den realen Messergebnissen.

Test des digitalen Modells

$$E_{n,DM} = \frac{(\bar{x}_{sim} - x_{STL}) - (\bar{x}_{mes} - x_{calib})}{\sqrt{U_{sim}^2 + U_{mes}^2}}, \text{ wenn } \frac{u_{cal}}{u_p} + \frac{u_{p,sim}}{u_p} < 2$$

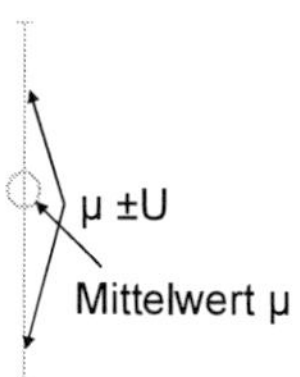

Messung
Erweiterte Messunsicherheit in Anlehnung an VDI 2630 Blatt 2.1
(Berechnet mit weniger als 20 Wiederholmessungen)

$$U_{mes} = k * \sqrt{u_{cal}^2 + u_p^2 + u_b^2}$$

Simulation
Unsicherheit durch Rauschen und Streustrahlung

$$U_{sim} = k * u_{p,sim}$$

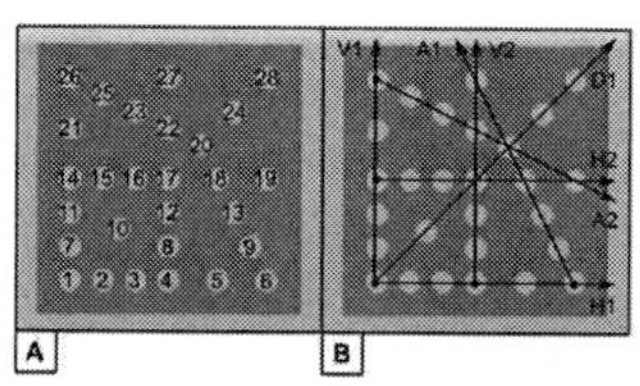

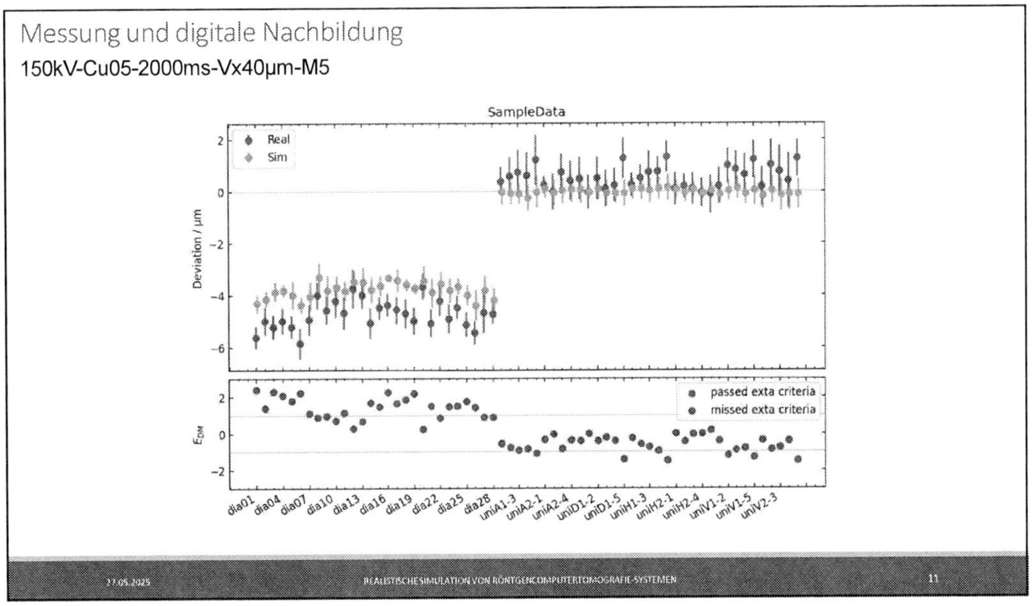
Messung und digitale Nachbildung
150kV-Cu05-2000ms-Vx40µm-M5
SampleData
Real
Sim
Deviation / µm
passed exta criteria
missed exta criteria
dia01 dia04 dia07 dia10 dia13 dia16 dia19 dia22 dia25 dia28 uniA1-3 uniA2-1 uniA2-4 uniD1-2 uniD1-5 uniH1-3 uniH2-1 uniH2-4 uniV1-2 uniV1-5 uniV2-3
27.05.2025
REALISTISCHESIMULATION VON RÖNTGENCOMPUTERTOMOGRAFIE-SYSTEMEN
11

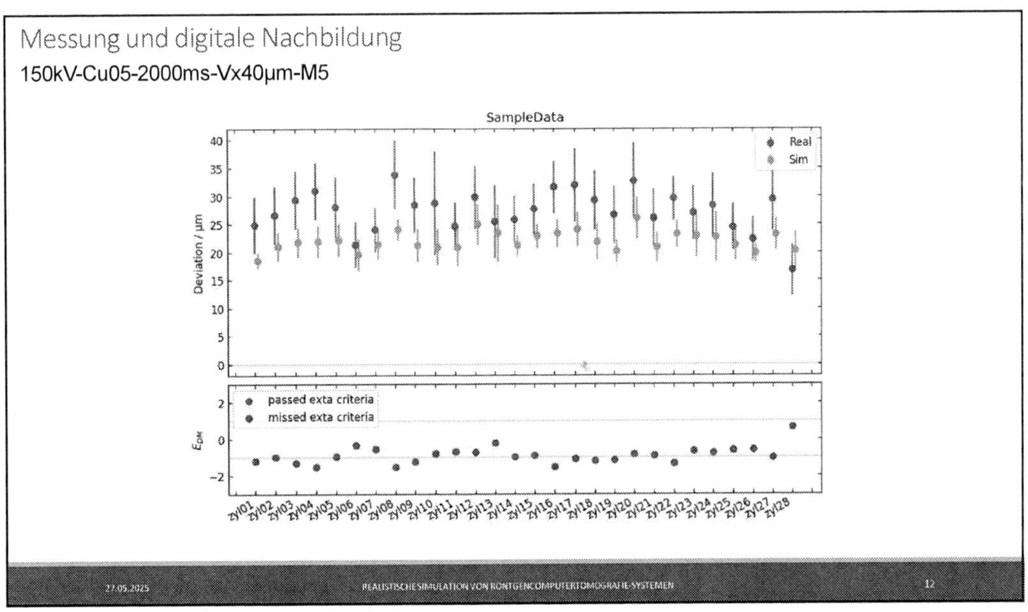
Messung und digitale Nachbildung
150kV-Cu05-2000ms-Vx40µm-M5
SampleData
Real
Sim
Deviation / µm
passed exta criteria
missed exta criteria
zyl01 zyl02 zyl03 zyl04 zyl05 zyl06 zyl07 zyl08 zyl09 zyl10 zyl11 zyl12 zyl13 zyl14 zyl15 zyl16 zyl17 zyl18 zyl19 zyl20 zyl21 zyl22 zyl23 zyl24 zyl25 zyl26 zyl27 zyl28
27.05.2025
REALISTISCHESIMULATION VON RÖNTGENCOMPUTERTOMOGRAFIE-SYSTEMEN
12

Zusammenfassung und Ausblick

Vorstellung der Projektreihe CTSimU

Ziel: Standardisiertes Verfahren zur Bestimmung der Messunsicherheit mittels Simulation

CTSimU1 (2019-2021): Erstellung eines Frameworks zur Basis-Qualifizierung einer Simulationssoftware

CTSimU2 (2022-2024): Entwicklung eines Werkzeugkastens zur Erstellung eines digitalen Zwillings und eines Tests zur Überprüfung von diesem

CTSimU3 (eingereicht): Bestimmung der aufgabenspezifischen Messunsicherheit mittels Simulation unter Anwendung der Monte-Carlo-Methode

27.05.2025 REALISTISCHE SIMULATION VON RÖNTGENCOMPUTERTOMOGRAFIE-SYSTEMEN 13

CTSimU 2

Vielen Dank für die Aufmerksamkeit!

Die dargestellten Arbeiten sind im Rahmen des WIPANO-Förderprojektes CTSimU2 (03TN0049A-L) durchgeführt worden. Die WIPANO-Förderrichtlinie wird vom Bundesministerium für Wirtschaft und Klimaschutz finanziert und vom Projektträger Jülich verwaltet.

WIPANO

Gefördert durch:

Bundesministerium für Wirtschaft und Klimaschutz

aufgrund eines Beschlusses des Deutschen Bundestages

DGZfP-Jahrestagung 2025

Einsatz der digitalen Brennfleckvermessung zur Optimierung von Röntgenröhren in Entwicklung und Produktion

Klaus Bavendiek[1], Alexander Warrikhoff, Florian Felske[2]
[1] KOWOTEST GmbH, Langenfeld, Deutschland, kb@kowotest-nord.de
[2] rtw RÖNTGEN-TECHNIK DR. WARRIKHOFF GmbH & Co. KG, Neuenhagen bei Berlin, Deutschland, rtw@rtwxray.de

Kurzfassung

Für die Abbildung und normkonforme Vermessung von Röntgenröhren Brennflecken hat die Kowotest GmbH im Jahr 2021 die digitale Brennfleck Kamera KowoSpot auf den Markt gebracht, die mit einem kleinen digitalen Detektor gestochen scharfe Abbildungen von Brennflecke im Sekundentakt ermöglicht. Neben der Lebensdauer Überwachung von Brennflecken in den bestehenden Anlagen bietet die KowoSpot auch große Vorteile bei der Entwicklung und Produktions-Überwachung von Röntgenröhren.

Das KowoSpot System wird bei der rtw RÖNTGEN-TECHNIK DR. WARRIKHOFF GmbH & Co. KG sehr erfolgreich in der Entwicklung und Produktion von MCTG Stabanoden Röntgenröhren verwendet. Diese Röntgenröhren wurden speziell für den industriellen Einsatz zur Überprüfung von Schweißnähten in Wärmetauschern entwickelt und arbeiten mit Spannungen bis zu 150kV und Brennfleckgrößen von teilweise unter 200µm.
Da die Lenkung und Fokussierung des Brennfleckes durch den Anodenstabs elektromagnetisch erfolgt, ist eine Einstellung des Brennflecks im Entwicklungs- und Produktionsprozess notwendig. Hierfür wird bei der rtw die KowoSpot X verwendet.

Die digitale Anzeige und normgerechte Auswertung des Brennflecks erleichtert und beschleunigt diesen Prozess.

Eine exakte und normgerechte Vermessung des Brennflecks und des dadurch erreichten Auflösungsvermögens ist für die Entwicklung und Produktion von Röntgenröhren bei rtw essentiell und so wird das KowoSpot System derzeit auch in den Produktionsprozess anderer Röhrentypen integriert.

Bei der Überprüfung von Rohr-Rohrboden z.B. in Wärmetauschern wird der Brennfleck in Richtung des Elektronenstrahls genutzt. Bei Radialschweißnähten wird eher der Brennfleck in einem Abnahmewinkel von 90° zum Elektronenstrahl genutzt.

Für Brennflecke <100µm kann auch die EN12543-5 mit der Messung an einer Kante verwendet werden. Dafür wird der Kowospot X so umgebaut, dass das Pinhole Element gegen ein Drahtkreuz Element ausgetauscht wird. Die Kowospot SW wird so erweitert, dass nun die beiden Kanten (senkrecht und waagerecht) automatisch erfasst und aus der resultierenden Unschärfe die Brennfleckgröße in die beiden Richtungen ermittelt und ausgegeben wird.

Keywords:
Brennfleck Vermessung, EN 12543, Wärmetauscher, Kowospot X

rtw RÖNTGEN-TECHNIK
DR. WARRIKHOFF GmbH & Co. KG

F. Felske, A. Warrikhoff

Einsatz der digitalen Brennfleckvermessung zur Optimierung von Röntgenröhren in Entwicklung und Produktion

KOWOTEST
SUPPLIERS OF EQUIPMENT FOR INSPECTION

K. Bavendiek

In der modernen ZfP werden immer häufiger neue Materialien, wie z.B. Titan oder Carbon überprüft. Dies ist z.B. bei Rohrverbindungen bei Wärmetauschern oder Turbinen nötig. Standard Röntgenröhren mit Reflexionstarget weisen eine zu große Bauform auf und können in den Rohren nicht eingeschoben werden. Röntgenröhren in Stabanodenbauform haben entweder zu große Brennflecke oder sind als offene Systeme an Vakuumpumpen gebunden. Die Herausforderung besteht darin, sehr kleinen Brennflecke der neuen Stabanoden rückführbar und standardisiert zu vermessen.

Die Firma KOWOTEST bietet seit einigen Jahren ein speziell entwickeltes, digitales Messsystem für Röntgenröhren mit Reflexionstarget an. Die entsprechenden Messadapter und Aufbauten können an vielen bestehenden Röntgenröhren direkt adaptiert werden. Die Ergebnisse werden in einem Zertifikat nach ASTM E1165 oder DIN 12543-2 ausgegeben. Mit dem System können auch sehr kleine Brennflecke unter 100µm vermessen werden. Für den Einsatz von Stabanoden fehlen bislang die Adapter sowie die Aufbauanleitung.

Kowospot X
Pin Hole Camera

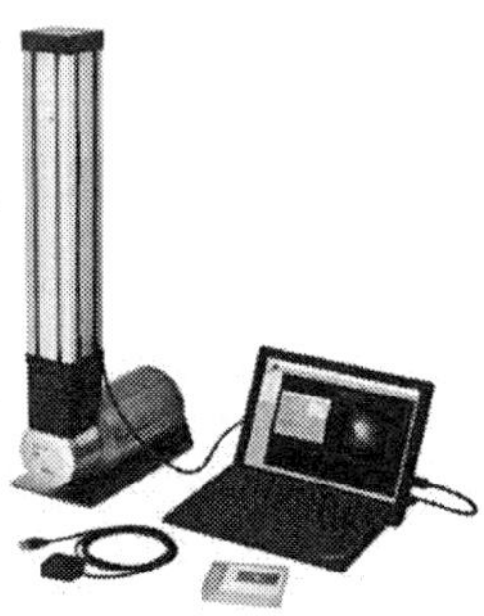

Messaufbau zum Vermessen des Brennflecks einer Stabanodenröntgenröhre

Messaufbau einer Stabanodenröntgenröhre mit dem KOWOSPOT X

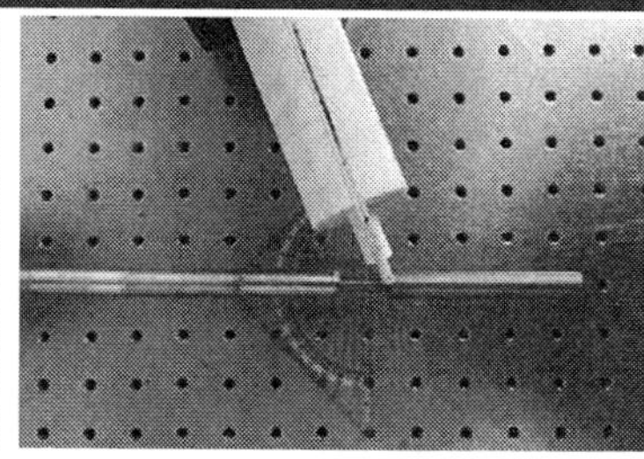

Ausrichtung des KOWOSPOT X unter 21°

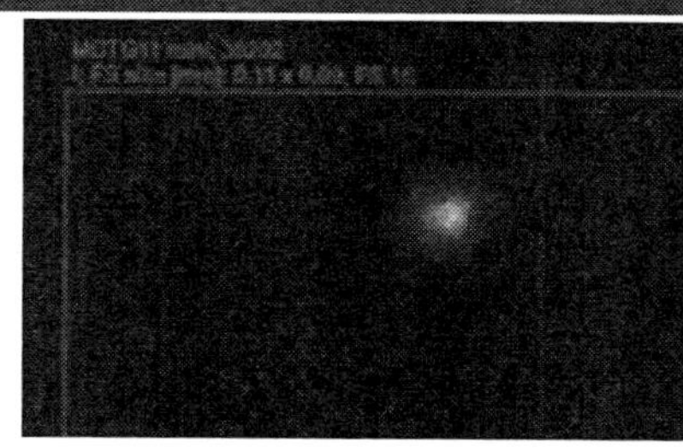

Brennfleckaufnahme einer MCTG11S bei 100kV, dargestellt und ausgewertet mit der Kowospot SW

Diskussionen

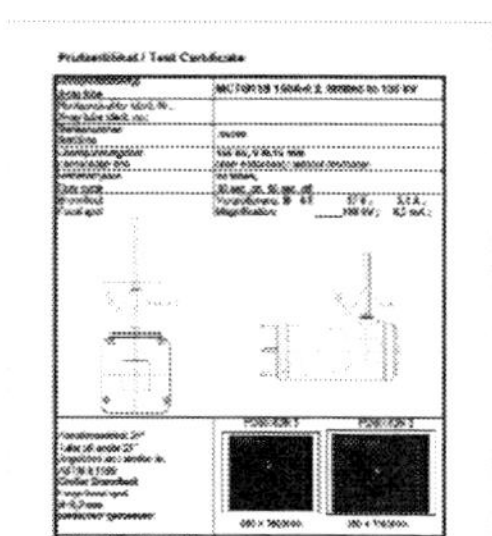

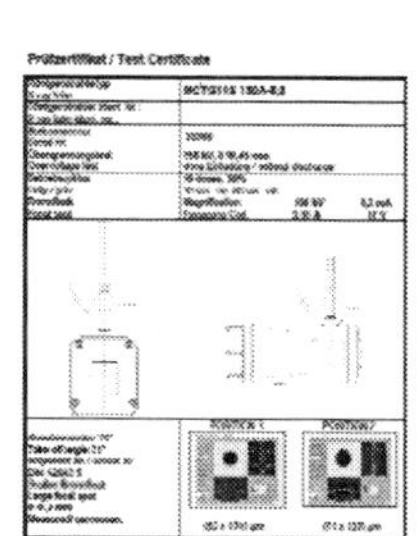

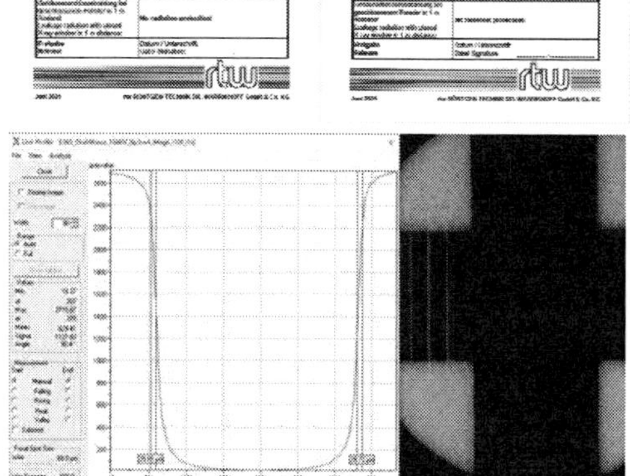

(a) Bei der Überprüfung von Rohr-Rohrboden z.B. in Wärmetauschern wird der Brennfleck in Richtung des Elektronenstrahls genutzt. Bei Radialschweißnähten wird eher der Brennfleck in einem Abnahmewinkel von 90° zum Elektronenstrahl genutzt.
In diesem Poster wird der Abnahmewinkel von 21° vorgeschlagen, um eine möglichst gute Aussage über die Brennfleckgröße bei den verschiedenen Applikationen treffen zu können. Hierzu ist die Stabanodenröntgenröhre z.B. an einem Messtisch auszurichten. Die Adapter des KOWOSPOT X können somit sehr gut auf den Brennfleck hin angepasst werden. Die Brennfleckaufnahme ist sofort auswertbar und kann in einem Zertifikat dem Kunden ausgehändigt werden. Die Messung ist damit reproduzierbar.
Das Messsystem von KOWOTEST deckt die Brennfleckvermessung über einen großen Bereich von Brennflecken von 5mm bis 50µm ab. Damit können alle sich auf dem Markt befindlichen Stabanoden vermessen werden. Beim Flachtarget wird der Abnahmewinkel von 21° vorgeschlagen.

(b) Für Brennflecke <100µm kann auch die EN12543-5 mit der Messung an einer Kante verwendet werden. Dafür wird der Kowospot X so umgebaut, dass das Pinhole Element gegen ein Drahtkreuz Element ausgetauscht wird.
Die Kowospot SW wird so erweitert, dass nun die beiden Kanten (senkrecht und waagerecht) automatisch erfasst und aus der resultierenden Unschärfe die Brennfleckgröße in die beiden Richtungen ermittelt und ausgegeben wird.
Vier verschiedene Drahtkreuz Elemente sind verfügbar für Brennfleck Größen von 5 - 10µm, von 10 - 20µm, von 20 - 40µm und von 40 - >100µm.

Zerstörungsfreie Prüfung mit Röntgen-Phasenkontrast: Materialanalyse mittels Scannertechnologie

Markus SCHNEIDER[1], Veronika LUDWIG[1], Constantin RAUCH[1], Stephan SCHREINER[1], Thilo MICHEL[1], Gisela ANTON[1], Stefan FUNK[1]
[1] ECAP, Friedrich-Alexander Universität Erlangen-Nürnberg, Erlangen
Kontakt E-Mail: markus.ms.schneider@fau.de

Kurzfassung

Die Röntgen-Phasenkontrast-Bildgebung bietet vielversprechende Möglichkeiten für die zerstörungsfreie Materialprüfung. Diese Methode ermöglicht neben der Erkennung von Defekten und Unregelmäßigkeiten auch eine detaillierte Analyse der inneren Strukturen von Bauteilen. Der vorgestellte Phasenkontrast-Scanner kombiniert zwei Interferometer, wodurch eine quantitative Untersuchung von Strukturgrößen in einem breiten Spektrum möglich wird.

Im Gegensatz zur klassischen Röntgenbildgebung, die ausschließlich auf Absorptionskontrasten basiert, erzeugt diese Bildgebungsmethode zusätzliche Informationen in Form eines differentiellen Phasenbilds und eines Dunkelfeldbilds. Durch das differentielle Phasenbild werden Objekte sichtbar, die die Wellenfront der Röntgenstrahlung ablenken, wodurch Materialübergänge erkannt werden können. Das Dunkelfeldbild ergänzt diese Analyse, indem es Kleinwinkelstreuungen detektiert, die mit herkömmlicher Bildgebung nicht sichtbar wären. Diese zusätzlichen Bildkontraste ermöglichen die Erkennung von Defekten wie Mikrorissen, Poren oder anderen strukturellen Unregelmäßigkeiten, die die Materialqualität beeinträchtigen können.

Ein entscheidender Vorteil des Scanners liegt in der Kombination der beiden Interferometer, welche auf unterschiedliche Strukturgrößen sensitiv sind. Dies eröffnet neue Möglichkeiten, um komplexe Materialien und Bauteile effizient zu untersuchen und die Qualitätssicherung zu verbessern.

Das Poster stellt den Aufbau und die Funktionsweise des Scanners vor und präsentiert erste Ergebnisse aus durchgeführten Messungen. Besonders hervorgehoben wird der zusätzliche Informationsgewinn, der durch die ergänzenden Phasenkontrastbilder im Vergleich zum herkömmlichen Absorptionsbild erzielt wird.

Keywords: Röntgen-Phasenkontrast-Bildgebung, Zerstörungsfreie Materialprüfung, Strukturgrößenbestimmung

https://doi.org/10.58286/31258

Friedrich-Alexander-Universität
Erlangen-Nürnberg

Erlangen Centre for Astroparticle Physics

Zerstörungsfreie Prüfung mit Röntgen-Phasenkontrast: Materialanalyse mittels Scannertechnologie

M. Schneider[1], V. Ludwig[1], C. Rauch[1], S. Schreiner[1], T. Michel[1], G. Anton[1], S. Funk[1]
[1] ECAP, Friedrich-Alexander-Universität Erlangen-Nürnberg, Erlangen

Röntgenstrahlung ermöglicht in der zerstörungsfreien Materialprüfung einen Blick ins Innere von Objekten. In der klassischen Radiografie entsteht der Bildkontrast zwischen unterschiedlichen Materialien durch Absorption der Röntgenstrahlung.

Die Phasenkontrastbildgebung erweitert dieses Absorptionsbild durch zwei weitere Bildmodalitäten: Dem differentiellen Phasenbild und dem Dunkelfeldbild. Der Phasenkontrast basiert auf der Brechung der Röntgenstrahlung durch das untersuchte Objekt. Dadurch werden Dichtegradienten zwischen verschiedenen Materialien sichtbar. Das Dunkelfeldbild kann poröse und fasrige Strukturen sichtbar machen, was eine Qualitätsprüfung des untersuchten Objekts erlaubt. Durch Kombination zweier Phasenkontrastscanner wird eine quantitative Strukturgrößenanalyse möglich.

Gitterbasiertes Scannersystem

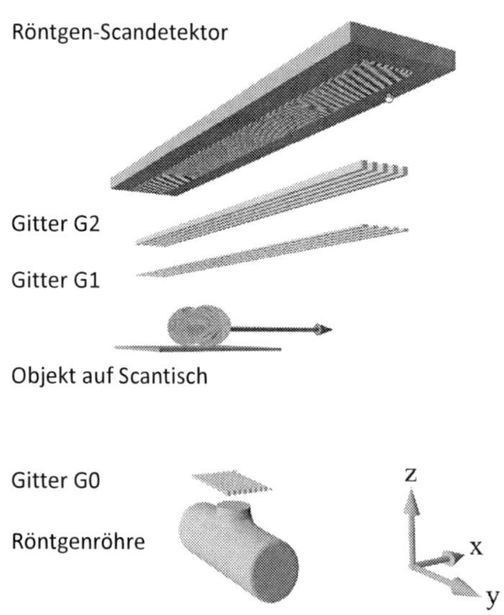

Sub-Pixel Auflösung

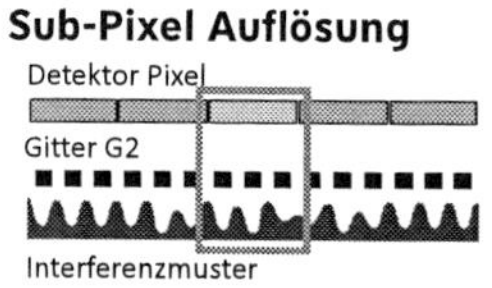

Phase-stepping Prozedur

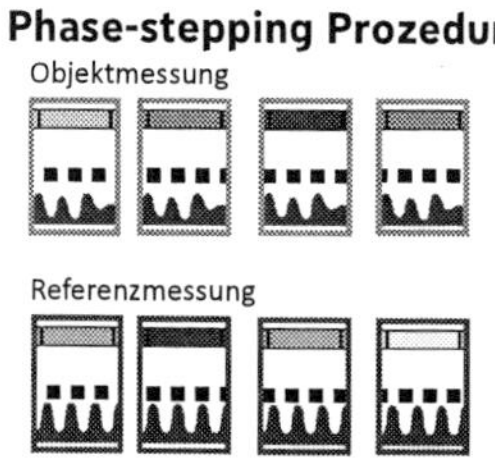

Berechnung der drei Bilder

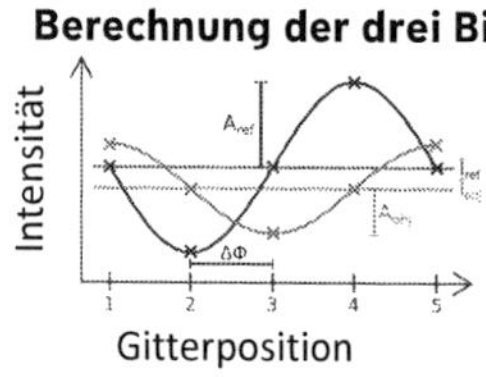

Präparierte Gummibärchen (Holz, Stecknadel, Aluminiumpulver). Das differentielle Phasenbild zeigt Übergänge zwischen verschiedenen Materialien. Das Holz und Pulver verursachen durch ihre fasrige und poröse Struktur ein starkes Dunkelfeldsignal.

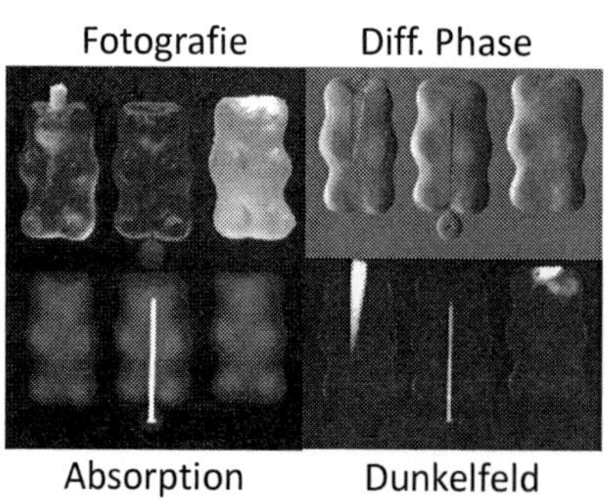

Holz in geriebenem Käse

Beschädigte Sattelstütze

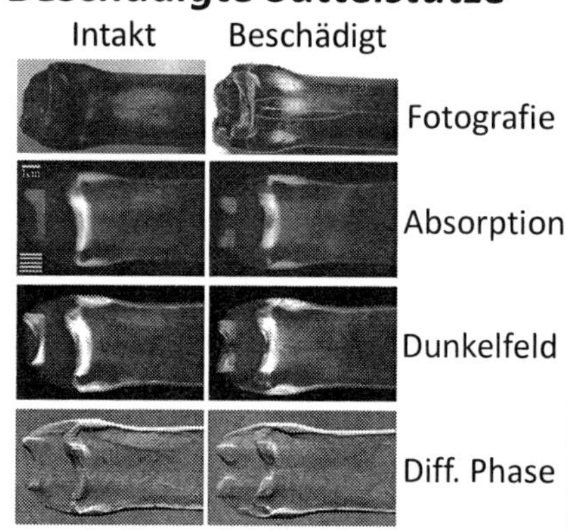

Sattelstützen hergestellt aus Carbonfasern (bereitgestellt von all ahead composites GmbH, Benzstraße 5, 97209 Veitshöchheim). Die Beschädigung ist im Dunkelfeld und der differentiellen Phase sichtbar, während sie im Absorptionsbild unsichtbar bleibt. [2]

Strukturgrößenbestimmung

Das Dunkelfeldsignal ist abhängig von der Korrelationslänge, was ein Maß für die Sensitivität darstellt. Die Korrelationslänge kann durch die Position der Probe und den zwei Phasenkontrastscannern variiert werden. Durch ein geeignetes Modell ist es möglich aus dem Verlauf des Dunkelfeldsignals in Abhängigkeit der Korrelationslänge die sub-pixel kleinen Strukturgrößen quantitativ zu bestimmen. Die Abbildung zeigt das Dunkelfeldsignal von Aluminiumpulver verschiedener Radien in Abhängigkeit der Korrelationslänge.

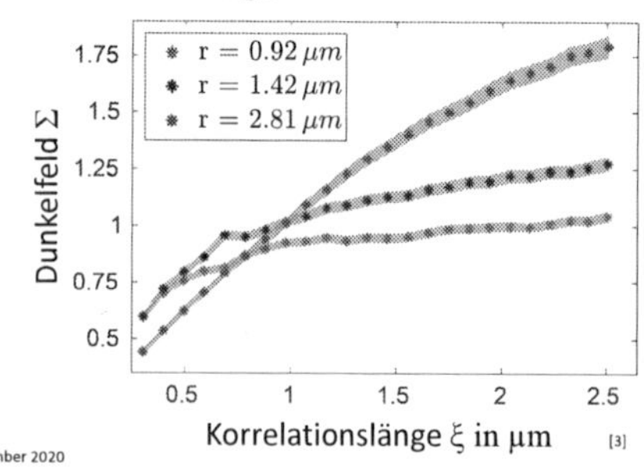

Contact: markus.ms.schneider@fau.de

[1] K. Esch: „Vergleich von Scanningmethoden des Talbot Lau Scanners anhand eines bewegten Objektes und Anwendung in der Lebensmittelkontrolle", Bachelorarbeit, FAU, September 2020
[2] V. Ludwig: „Development of image acquisition and alignment methods for X-ray dark-field and phase-contrast imaging", Dissertation, FAU, Juni 2020
[3] H. Müller: „Examination of the correlation length dependency of the dark-field in an X-ray Talbot-Lau interferometer", Praktikumsbericht, FAU, Dezember 2018

DGZfP-Jahrestagung 2025

Evaluierung und Optimierung von Rückstreuverfahren zur Analyse von Betonbahnschwellen

Dirk HOFMANN [1], Peter KRÜGER [1], Aurimas PUKÉNAS [1],
Susanne HILLMANN [2], Frank HEROLD [3]
[1] Fraunhofer Institut für Keramische Technologien und Systeme IKTS
[2] Deutsches Zentrum für Schienenverkehrsforschung beim Eisenbahn-Bundesamt, Fachbereich Sicherheit und kritische Infrastruktur DZSF
[3] VisiConsult X-ray Systems & Solutions GmbH

Kurzfassung

Schäden im Beton können nicht nur die Nutzungsdauer verringern, sondern auch zum Entgleisen von Schienenfahrzeugen führen. Die optische Beurteilung kann bei Schäden, wie Abplatzungen oder Rissen, die vordergründig auf der Oberfläche auftreten ausreichend sein. Bei tieferliegenden oder verdeckten Defekten bieten sich Verfahren, wie die Gamma- oder Röntgenstrahlung an. Für Messungen an Objekten mit einseitigem Zugang wie z.B. bei Betonbahnschwellen im eingebauten Zustand ist Compton-Rückstreuung eine Möglichkeit diese hinsichtlich ihrer inneren Rissbildung zu überwachen. Im Gegensatz zu Transmissionstechniken wird bei den Rückstreuverfahren die vom Objekt gestreute Strahlung gemessen. Dabei ist sowohl die Strahlungsquelle als auch der Detektor vor dem Messobjekt platziert. Die Anwendungsmöglichkeiten der Rückstreuverfahren sind vielfältig. Das Verfahren kann sowohl für Dichtebestimmung als auch für Radiographien herangezogen werden. Während in der Medizin die eingesetzte Strahlendosis für die Anwendbarkeit ausschlaggebend ist, sind für industrielle Anwendungen andere Parameter wichtig: Auflösung, Messtiefe, Messzeit, Messgeometrie und Kosten des Messsystems.

Der Beitrag beschreibt die Evaluation von Rückstreuverfahren zur Analyse von Betonbahnschwellen. Dabei wurden u.a. verschiedene Hardwarekonstellationen sowie variable Probekörper in einer Untersuchungsreihe verwendet und optimiert. Ergänzt werden diese Versuche durch eine Simulation zur Abschätzung von Anordnungen, Abständen, Eindringtiefen und Streukörpern.

Keywords: Computertomographie, Bahnwesen, Betonschwellen

Evaluierung und Optimierung von Rückstreuverfahren zur Analyse von Betonbahnschwellen

Fraunhofer
IKTS

D. Hofmann[1], P. Krüger[1], A. Pukėnas[1], S. Hillmann[2], F. Herold[3]
[1]Fraunhofer-Institut für Keramische Technologien und Systeme IKTS
[2]Deutsches Zentrum für Schienenverkehrsforschung beim Eisenbahn-Bundesamt, Fachbereich Sicherheit und kritische Infrastruktur DZSF
[3]VisiConsult X-ray Systems & Solutions GmbH

Einleitung

Schäden im Beton können nicht nur die Nutzungsdauer verringern, sondern auch zum Entgleisen von Schienenfahrzeugen führen. Die optische Beurteilung kann bei Schäden, wie Abplatzungen oder Rissen, die vordergründig auf der Oberfläche auftreten ausreichend sein. Bei tieferliegenden oder verdeckten Defekten bieten sich Verfahren, wie die Gamma- oder Röntgenstrahlung an.

Für Messungen an Objekten mit einseitigem Zugang wie z.B. bei Betonbahnschwellen im eingebauten Zustand, ist die Compton-Rückstreuung eine Möglichkeit diese hinsichtlich ihrer inneren Rissbildung zu überwachen. Im Gegensatz zu Transmissionstechniken wird bei den Rückstreuverfahren die vom Objekt gestreute Strahlung gemessen. Im Rahmen eines Projektes wurde die Evaluation von Rückstreuverfahren zur Analyse von Betonbahnschwellen untersucht. Dabei wurden u.a. verschiedene Hardwarekonstellationen sowie variable Probekörper in einer Untersuchungsreihe analysiert.

Stand der Technik

Im Gegensatz zu Transmissionstechniken werden bei Rückstreuverfahren die vom Objekt gestreute Strahlung gemessen. Dabei ist sowohl die Strahlungsquelle als auch der Detektor vor dem Messobjekt platziert. Die Anwendungsmöglichkeiten sind dabei sehr vielfältig. Das Verfahren kann sowohl zur Dichtebestimmung als auch für Radiographien herangezogen werden. Die erste Anwendung der Compton-Streuung wurde 1956 von Odeblad und Norhagen für die Messung der Elektronendichte vorgestellt. Während in der Medizin die freigesetzte Strahlendosis für die Anwendbarkeit ausschlaggebend ist, sind für industrielle Anwendungen andere Parameter wie Auflösung, Messtiefe, Messzeit und Messgeometrie wichtig. Das erste industrielle Rückstreu-Abbildungssystem AIDECS (engl. Automated Inspection Device for Explosive Charge in Shell, IRT Corporation) wurde 1976 für die Qualitätskontrolle von Artilleriegeschossen entwickelt. In der Folgezeit entstanden verschiedene kommerzielle Anwendungen und Versuchsstudien, wie für die Raumfahrtindustrie, die Landminensuche, Sicherheitsanwendungen (z. B. Gepäck- und Fahrzeugkontrolle) sowie für den Einsatz in der Archäologie.

[Die vollständigen Literaturangaben sind dem »Berichte des Deutschen Zentrums für Schienenverkehrsforschung Bericht 01 (2024)« zu entnehmen.].

Untersuchungen und Ergebnisse

Beim Aufbau der vorliegenden Untersuchungsreihe wurde sich am etablierten Prinzip der Lochkamera-Technik orientiert. Dabei wird das Objekt über einen Fächerstrahl bestrahlt und mittels einer Loch- oder Schlitzblende bestehend aus hochabsorptivem Material auf einen Flächendetektor als invertierte Projektion abgebildet. Die Basis des verwendeten Aufbaus bildet dabei eine 2 mm starke, mit Blei ummantelte Edelstahl-Umhausung, in der ein Flächendetektor integriert wurde. In der Umhausung befindet sich ein 3 mm großes Loch, welches während der Untersuchung verkleinert und mit unterschiedlichen Blenden beaufschlagt wurde (Abb.1).

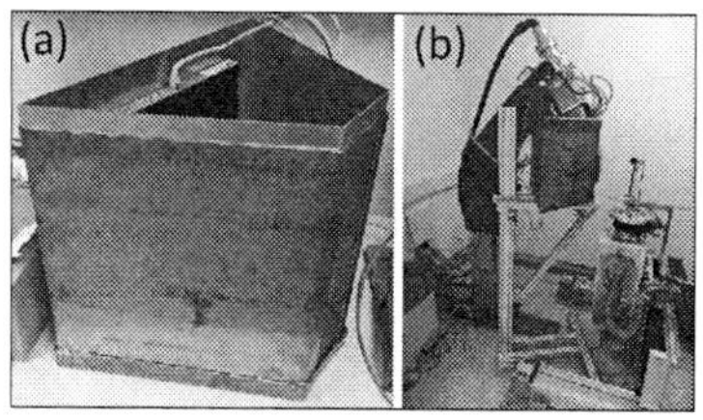

Abb 1. Aufbau der Rückstreuversuche: (a) mit 2 mm Blei verkleidete Edelstahl-Lochkamera mit Flächendetektor; (b) Gesamtaufbau mit Röhrenröhre, Lochkamera und Manipulator.

Als Untersuchungsgegenstand dienten bereits verbaute Bahnschwellen mit markanten Gebrauchsmerkmalen, wie oberflächigen Rissen, Beschädigungen und umweltbedingten Abnutzungserscheinungen. Aus diesen Schwellen wurden unterschiedliche Formate, wie Keile oder Würfel herausgeschnitten und als Prüfkörper vorbereitet.

Erste Aufnahmen zeigen sowohl den Armierungsstahl wie auch die innere Struktur, bestehend aus Poren, Rissen und Einschlüssen (Abb.2 und Abb.3). Die Aufnahmequalität bzw. die Auflösung ist sehr stark abhängig von der jeweiligen Dicke der Probengeometrie.

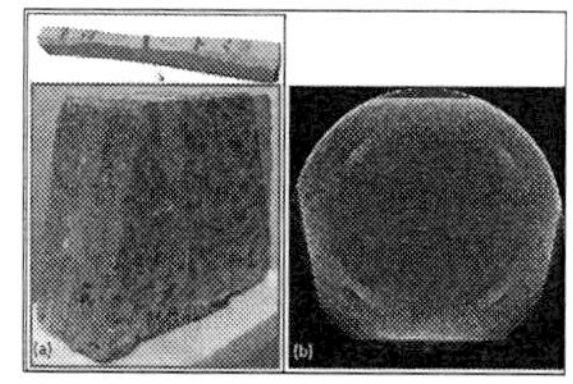

Abb.2: CT-Messung eines Prüfkörpers, ausgeschnitten aus der Mitte einer Betonbahnschwelle. (a) Prüfkörper ca. [cm] 9 x 21 x 25,
(b) Schnittbild, in dem lediglich die Armierungsstäbe zu erkennen sind. Messparameter: 210 kV, 9 W.

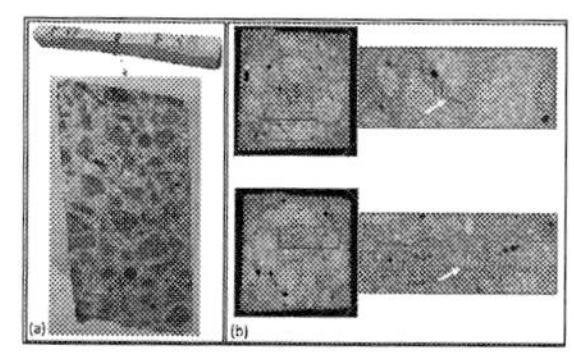

Abb.3: CT-Messung eines Prüfkörpers, ausgeschnitten aus der Mitte einer Betonschwelle: (a) ca. [cm] 9 x 9,5 x 16 Prüfkörper,
(b) Schnittbilder, auf denen die innere Struktur, Poren und Risse (gelbe Pfeile) zu erkennen sind. Röhren-/Messparameter: 195 kV, 18 W, 0,8 mm Cu-Filter.

Basierend auf den ersten Erkenntnissen wurden mithilfe des Anlagen- und Systemherstellers VisiConsult aus Stockelsdorf bzw. deren Partnerfirma DIONDO aus Hattingen weitere Aufnahmekombinationen und -konstellationen untersucht, um die Abbildungsqualität zu erhöhen. Dabei wurden zunächst höhere Röhrenspannungen sowie variierende Filterarten und -dicken gegenüber modifizierten Blendendurchmessern untersucht (Abb.4). Die deutlichsten Unterschiede konnten über das Histogramm und deren Grauwertverteilungen analysiert werden (Abb.5).

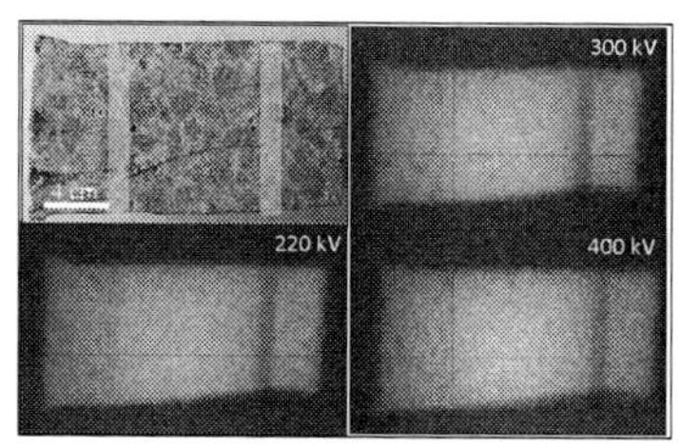

Abb.4: CT-Messung des keilförmigen Prüfkörpers mit unterschiedlichen Röhrenspannungen. Die hell-hinterlegten Bereiche auf dem Originalbild verdeutlichen die Positionen des integrierten Armierungsstahls.

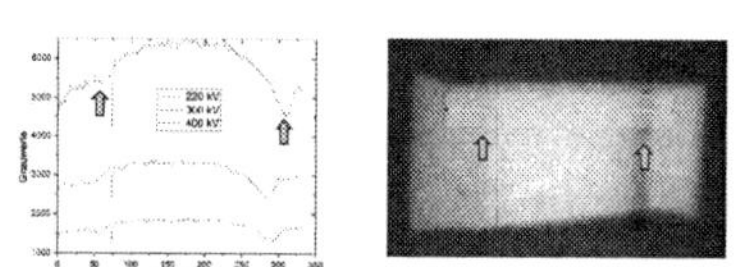

Abb 5. CT-Messung mit Histogramm-Darstellung des keilförmigen Prüfkörpers bei der jeweiligen Röhrenspannung. Der gelb-hinterlegte Bereich verdeutlicht den betrachteten Grauwertbereich

Zusammenfassung und Ausblick

Rückstreutechnologien haben sich bereits in vielen Branchen etabliert. Grundlegend hat dieses Verfahren den Vorteil das Röntgenquelle und Detektor auf derselben Seite des Messobjektes stehen. Das ermöglicht es, viele Objekte mit eingeschränkter Zugänglichkeit zu erfassen bzw. im eingebauten Zustand zu prüfen.

In der vorliegenden Untersuchungsreihe sind die Grenzen dieses Verfahrens an Betonbahnschwellen zu untersuchen. Dabei wurde sich zunächst auf die Lochkamera-Technik konzentriert, wobei der bisherige Stand der Technik bestätigt werden konnte. Es konnten innere Strukturen qualitativ unterschieden und visualisiert werden. Über eine Analyse des Grauwertspektrums konnten bei steigender Röhrenspannung auch eine quantitative Erhöhung der Grauwerte nachgewiesen werden.

Um dies auch qualitativ bewerten zu können, sollen im Weiteren Simulation genutzt werden, wobei der Gesamtaufbau und die ideale Konstellation der Einzelkomponenten numerisch bewertet werden. Zudem werden andere Detektorvarianten geprüft, die sowohl mehr Energie und somit mehr Informationen erfassen können, wie auch andere Bauformen. Ferner sollen Komponenten, wie Kollimatoren und Schlitzblenden so in den Aufbau integriert werden, um eine möglichst vollständige Abbildung der Schwelle sowie die größtmögliche Auflösung der inneren Bestandteile erreichen zu können.

Kontakt: Dr.-Ing. Dirk Hofmann
Fraunhofer-Institut für Keramische Technologien und Systeme IKTS
ZfP-Zentrum
Maria-Reiche-Str. 2, 01109 Dresden
Telefon +49 351 888 15-692
dirk.hofmann@ikts.fraunhofer.de

Deutsches Zentrum für Schienenverkehrsforschung beim Eisenbahn-Bundesamt
Projektnummer 11-17117-2680-23103

DGZfP-Jahrestagung 2025

Advanced pulse eddy current (PEC+) technology for automatic examination of steel plates for detection of hard spots

Andrej KIRIKOV [1], Alexey SHELUKHIN[2]
[1]Nordinkraft AG, Remchingen, Germany, andrej.kirikov@nordinkraft.de
[2]Nordinkraft AG, Remchingen, Germany, aleksei.shelukhin@nordinkraft.de

Abstract. Non-destructive inspection of steel plates for "hard spots" is a modern requirement imposed by most energy companies on steel plates manufacturers. The most appropriate testing method involves eddy current (EC), or pulse eddy current (PEC) techniques, which are sensitive to areas with abnormal hardness. However, a significant challenge for classical EC or PEC probes is suppressing interfering factors such as influence of residual magnetization, rolling scale, variations in lift-off, "edge effect," etc. These factors can easily involve both false indications and missing of critical defects. NORDINKRAFT AG has developed an innovative testing technique, so called „PEC+", which utilizes multifrequency analysis of specially shaped pulses. The associated algorithms, integrated into both hardware and software, effectively neutralize most negative factors, ensuring reliable inspection results.

Keywords: Hard spot, Residual magnetization, Eddy current, Multifrequency

Foreword

Hard spots are localized areas in the plate surface where the hardness exceeds the specified value. Hard spots may result in microcracks in the pipe production process. Those microcracks, developing in the pipelines, destroy them, resulting in explosions. Because of severe pipeline incidents related to hard spots which have occurred in recent years, new regulations for plates/pipes makers are being developed and appeared. Non-contact detection of hard spots, and non-contact evaluation of plates hardness are getting a technical challenge of highest importance.

1. Formulation of modern requirements to methods and equipment for non-contact examination of plates for hard spots

Below is the list of most important trending requirements related to non-contact examination of plates for hard spots. The formulations below are the result of numerous discussions with representatives of the most prominent steel plants all over the world.

(1) HS-detecting equipment shall guarantee the test of at least one surface of the plate with 100% coverage.
(2) The coverage above shall be also applied to edges and even corners of the plate. No or minimal untested zones at edges and corners is getting required.
(3) The equipment shall be able to detect shallow hard spots; their depth of penetration starts from 100.... 200 μ.
(4) The high-capacity test stations providing detailed C-scans showing exact locations and dimensions of the detected hard spots are needed.
(5) The minimal size of artificial hard spot to be detected is at least 20mm x 20 mm (trending down to 10 mm x 10 mm). The probability of detection shall be close to 100% ($P \geq 0{,}99$).

https://doi.org/10.58286/31222

(6) De-scaling of plates prior to the test is not acceptable. Rolling scale of up to 100μ thick shall not interfere the test.
(7) De-magnetization of plates before testing is unacceptable. Residual magnetization from cranes shall not influence the test results.
(8) Not only detection of hard spots, but also hardness evaluation function shall be available. It is trending now that not only a qualitative analysis, but also the quantitative one shall take place. It means, the modern equipment shall be able to measure/estimate hardness and its distribution all over the plate.
(9) Moreover, the precision of the hardness evaluation accuracy shall be generally better than one obtained with Leeb, especially for cases related to shallow defects.

The sentences (8) and (9) may need a brief explanation. The definition of hard spot above does not mean the borders of real hard spots shall always have steep gradients in their hardness. Such gradients are typical for artificial hard spots (as shown in Fig. 1, a) only. The hardness in real hard spots may grow up and go down slowly (see diagram b in Fig. 1).

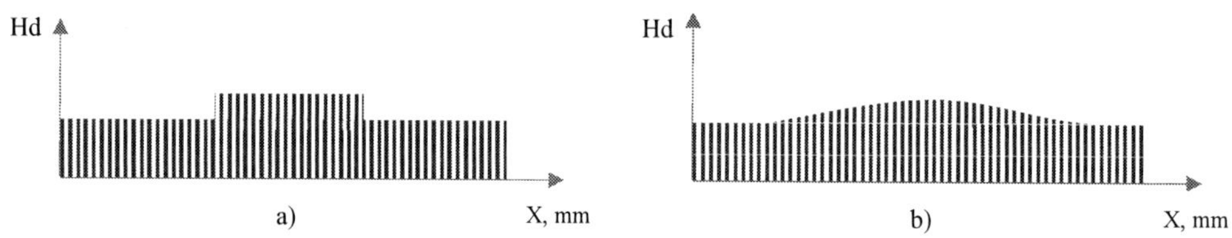

Fig. 1. Typical hardness profiles: - a – steep - in artificial HS; - b - "slow" - in real hard spots.

Therefore, all differential EC-sensors, or differential algorithms, applicable for detection of cracks, cannot detect the excessive hardness with slow gradients. Therefore, they do not meet modern requirements.

The consideration above leads to the necessity to non-contactly evaluate hardness all over the plate, regardless of whether the gradients in hardness are steep or slow.

Considering the above, the modern non-contact HS detectors shall have the function of evaluating the absolute hardness.

Moreover, according to sentence (9) the precision of non-contact evaluation of the excessive hardness shall generally be better than one available with Leeb.

Till recently Leeb has been considered as the only tool for performing practical measurements of hardness in field. However, as it will be demonstrated below, the precision and reproducibility of Leeb at examination of hard areas may be completely insufficient.

2. Why conventional approaches cannot meet the above requirements

2.1 Technique based on traditional EC-probes

It is well known the conventional EC-sensors may have ability to detect hard spots; however, they can be affected by many other factors as, for instance, a lift-off, rolling scale, residual stress, residual magnetization (actually, three independent components), edges, etc.

As we see from the diagram in Fig.1, the conventional EC operates with two available and measurable parameters only. They are:

- Amplitude A of the receiving signal.
- Phase φ of the receiving signal.

Only these two parameters may be applied both to classify hardness of the area of the plate where the EC-probe is currently on and get rid of interfering factors.

We are facing the situation when number of sources of information (just two) is substantially less than number of influencing factors (which are at least five).

Conventional EC-probe Configuration	Current	Spectrum
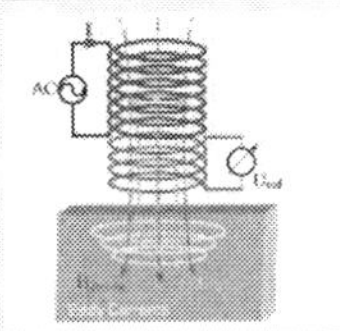	$I(t) = I_0 \times \sin 2\pi F \times t$	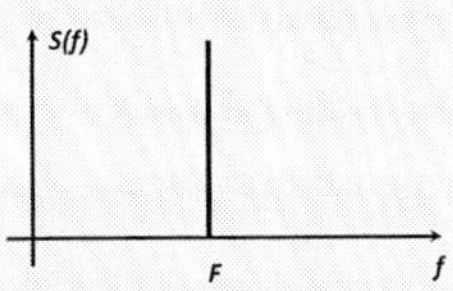

Fig.2. Conventional EC-technique, shape of the current through its transmitting coil, and its spectrum. It is clear that the only frequency F is available for analysis.

In other words, the negative factors may substantially reduce the efficiency of the conventional EC-techniques. It can hardly meet the items from at least (6) to (9) of the requirements above.

2.2 PEC-technique

Classical PEC, which physical basis is schematically represented in Fig. 3., operates with the rectangular (or almost rectangular) pulses generated by the transmitting coil.

Pulsing mode of the PEC-sensor assumes the availability of a great variety of frequencies. Potentially, they can be sources of information both for detection of hard spots and eliminating the interfering factors.

PEC-probe Configuration	Current	Spectrum
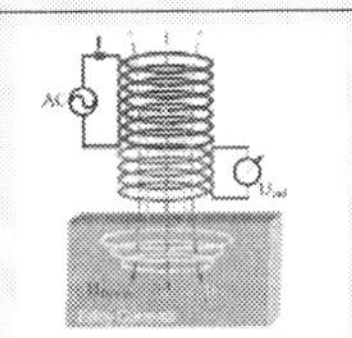	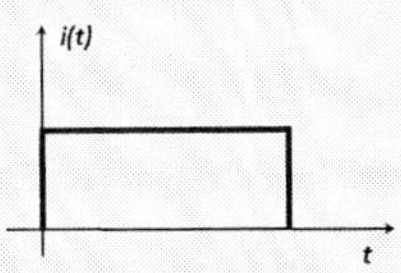	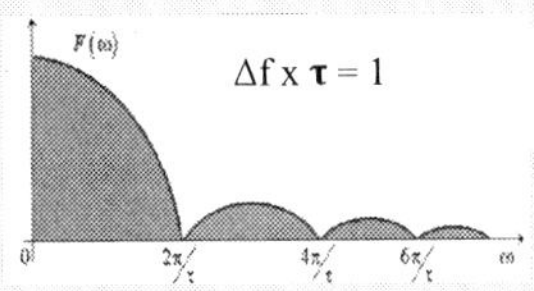

Fig.3. PEC- technique. The shape of the current through the transmitting coil is almost rectangular; the spectrum of the informative signal is concentrated in the area of low frequencies, which are dominating.

However, spectrum in Fig.2 demonstrates that most of the power of the transmitting electromagnetic pulses is concentrated in the range of low frequencies.

Low frequencies, dominating in spectrum, have their specifics. On the one hand, rolling scale is transparent for them, and cannot significantly interfere the test. On the other hand, low frequency current may penetrate too deep into the material to be sufficiently sensitive to the superficial hard spots.

Dominating of low frequencies in signal spectrum may reduce noise immunity of the equipment and decrease probability of hard spots detection. That is why the classical PEC cannot meet all the trending requirements listed in paragraph 1.

3. "PEC+" - What it is and why it is so powerful

Nordinkraft AG conceptualized advanced principles of multifrequency analysis, developed and patented the so-called "PEC+" technique. It will be shown below that this innovation enables meeting contemporary requirements providing a reliable detection of hard spots, eliminating of harmful industrial factors, and evaluation of hardness all over the plate. Of course, the precision of the evaluation may get influenced by some interfering factors; however, generally it can be acceptable, especially for situations when Leeb is not working properly.

The configuration of PEC+ technique is shown in Fig.4.

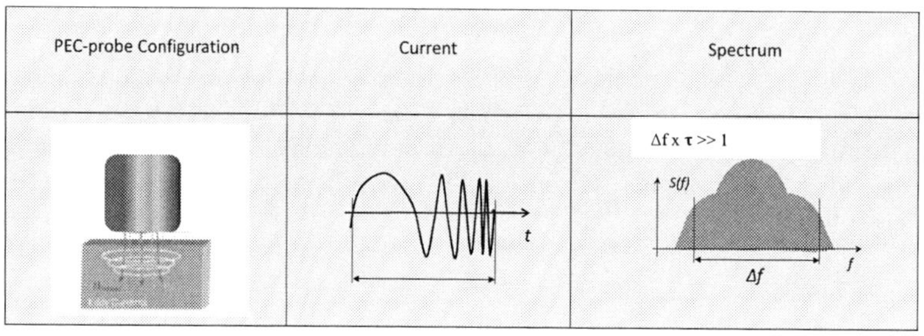

Fig. 4. PEC+. Compare to the classical PEC, the shape of the current through the transmitting coil here is more complicated; $\Delta f \times \tau >> 1$. low frequencies do not dominate.

PEC+ assumes application of signals with $\Delta f \times \tau >>1$. Reference transmitted pulse and its spectrum is shown in Fig.5. It has the following specifics:

- an extended range of frequencies available for analysis.
- low frequencies do not dominate in spectrum.
- On the contrary, the higher frequencies which are sensitive to skin-deep defects can prevail.

It means, unlike the conventional EC techniques, the PEC+ deals with signals of high complexity, containing a specific distribution of the variety of available frequencies. Analysis of spectrums of such signals allows reduction of noise, suppression of negative factors, increasing probability of detection and measuring/estimation of hardness of even skin-deep hard spots, which cannot be evaluated by Leeb.

Considering disadvantages of classical PEC technique, we came to the conclusion, that the frequencies in the range of at least 15 - 40 kHz (depth of penetration $\delta \approx 0{,}2....0{,}1$ mm) shall dominate in the spectrum.

The multifrequency analysis assumes:

- Transmitting and receiving electromagnetic pulses with TBP = $\Delta f \times \tau >> 1$.
- performing Fourier transformation of receiving signals.
- selecting a sequence of specific frequencies f1....... fN in their spectrums.
- monitoring of amplitudes A1 ......AN of the respective frequencies f1....... fN.
- calibration of working channels.
- estimation of hardness H of every portion of the plate by means of "system of basic equations".
- Corrections in basic equations on a basis of a practical test experience.

a)

b)

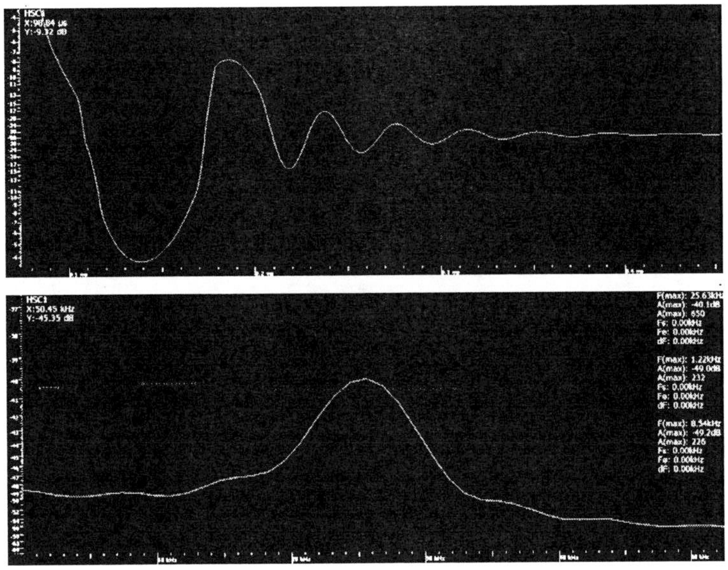

Fig.5. PEC + -sensor. (a) – A-scan - shape of the receiving signal; (b) – its Spectrum (result of the Fourier transformation).

A system of basic equations (SBE) allows estimating of hardness H in every specific area of the PEC+ sensor coupled with, and suppressing the negative factors.

The SBE has the following formal view:

$$\left.\begin{array}{l} A1 = A1(H, P1, P2, \ldots\ldots PN) \\ A2 = A2(H, P1, P2, \ldots\ldots PN) \\ A3 = A3(H, P1, P2, \ldots\ldots PN) \\ \ldots\ldots\ldots\ldots\ldots \\ A(N+1) = A(N+1)(H, P1, P2, \ldots\ldots PN) \end{array}\right\} \quad (1)$$

Here A1.... A(N +1) – amplitudes of pre-selected frequencies in spectrum; P1.....PN - parameters, related to negative factors we are going to get rid of. It is clear, that the number of equations shall be equal to the number of factors influencing spectrum, including hardness.

The idea behind (1) is based on the following pre-assumptions:

(1) The spectrum is wide enough to ensure enough conditionally independent amplitudes to separate hardness from other factors.

(2) All the amplitudes of the pre-selected sequence A1....A(N+1) are conditionally independent in terms of their correlation with hardness and other most affecting factors.

4. Practical examples and evidence of how the pec+ works

4.1 Blind test on the magnetized sample with artificial hard spots

The diagram in Fig. 6 represents the process of making artificial hard spots. The sample was manufactured by one of our customers located in Korea.

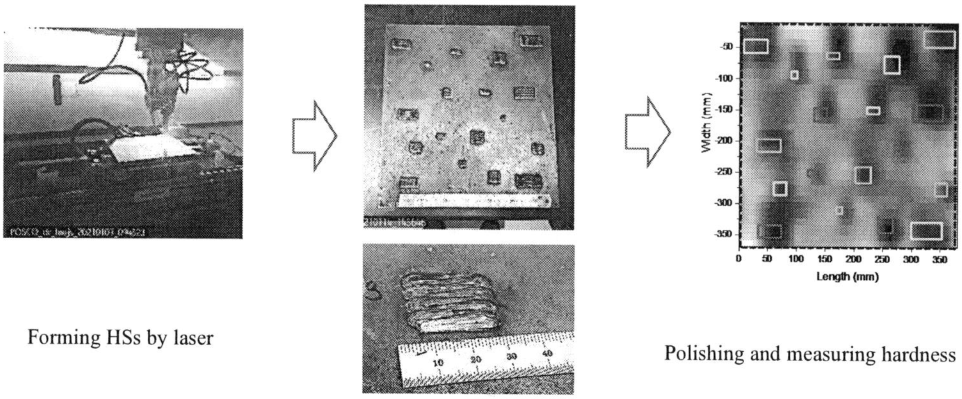

Fig. 6. Process of artificial hard spots manufacturing.

The datasheet on microhardness of the artificial hard spots and their dimensions (a) and the result of the blind test performed and reported by Nordinkraft AG (b) are represented in Fig.7.

The information on the sample was provided by the customer after the blind test of the sample was performed and reported by Nordinkraft AG.

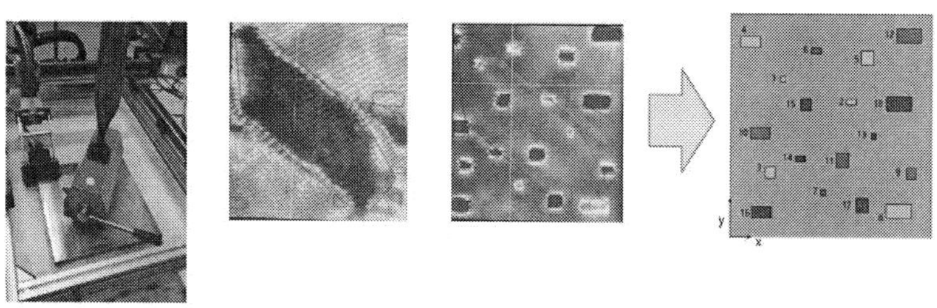

No.	Hardness Hv0.1	Defect Size (W x L) (mm)
1	230	10 x 10
2	230	10 x 20
3	230	20 x 20
4	230	20 x 40
5	230	25 x 25
6	230	25 x 50
7	260	10 x 10
8	260	10 x 20
9	260	20 x 20
10	260	20 x 40
11	260	25 x 25
12	260	25 x 50
13	[illegible]	10 x 10
14	[illegible]	10 x 20
15	[illegible]	20 x 20
16	[illegible]	20 x 40
17	[illegible]	25 x 25
18	[illegible]	25 x 50

a) b) c) d) e)

Fig. 7. Blind test and obtaining of the results; a- sample magnetization; b- result of scanning, one-frequency mode; c- result of scanning, multifrequency mode; trace of the magnet is visualized by the black line; d, e - datasheet on microhardness of the artificial hard spots, their dimensions and locations.

The results of evaluation of hardness in the same sample performed consequently by PEC+-sensor and Leeb are shown in Fig.8. The point "base Material" was selected in a defect free area of the sample.

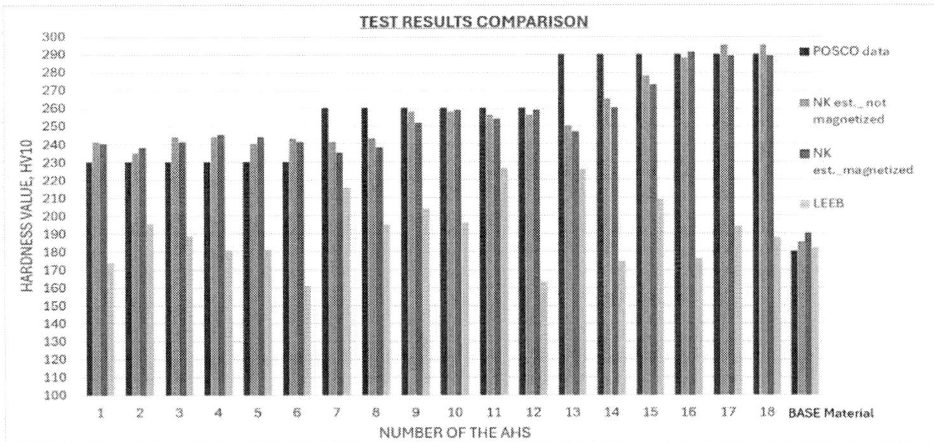

Fig.8. The results of evaluation of hardness performed consequently by Leeb and PEC-sensor. Numbers of artificial hard spots areas (from 1 to 18) are shown in a horizontal line. The last point "Base Material" belongs to a defect-free area of the sample.

Black lines demonstrate micro-hardness of the respective hard spots or base area; the data had been obtained in the customer's laboratory. Red lines represent the results of evaluation of hardness, taken by PEC+ sensor from the magnetized sample. The blue lines are the results of evaluation of hard spots by PEC+ sensor after the sample full de-magnetization. The yellow lines represent the results of measurements performed by Leeb.

Looking at the diagram in Fig. 8 the following conclusions can be made:

(1) Most of the results of evaluation of hardness, performed by PEC+ sensor, are very close to ones, obtained in the laboratory (as micro-hardness). Especially for the hard spots having dimensions of 15 mm x 15 mm and more.

(2) Magnetization of the sample does not significantly affect the results of hardness evaluation performed by PEC+ sensor.

(3) Leeb demonstrates a big deviation in all cases except for one point "Base Material". The hard spots are almost "invisible" for Leeb. The average deviation from a true hardness is about 80HV. According to the customer, their depth of the defects is about 200µ. Most probably, the hardness of such hard spots cannot be measured by Leeb.

To sum up: the diagram in Fig. 8 demonstrates a great advantage of the PEC-sensor over Leeb in terms of precision of thin hard spots hardness evaluation. It is remarkable, that the results of "Base Material" hardness measurement (an area beyond the hard spots) are almost the same. The difference between PEC-sensor and Leeb is just several HV.

4.2 Automatic testing equipment. Tests of PEC+ on a full-size plate with artificial hard spots of different sizes and locations

The following parameters were to be verified at this test:

- Probability of detection; artificial defects of different dimensions, including 10mm x 10 mm.
- Detection of artificial hard spots at edges and corners of the plate.
- False indications. Repeatability and reliability of the test.

The magnetized plate made of the material from the customer contained 31 artificial hard spots made by pinning (method and tool have been developed and patented by Nordinkraft). Hardness of those hard spots exceeded the basic hardness of the plate by about 20%.

The below results were obtained during the pre-shipment inspection of the equipment. Nordiscan-PL-HS-4500 and C-scan of the plate are represented in Fig. 9. The defects increasing a pre-set hardness level have been automatically marked in red by software.

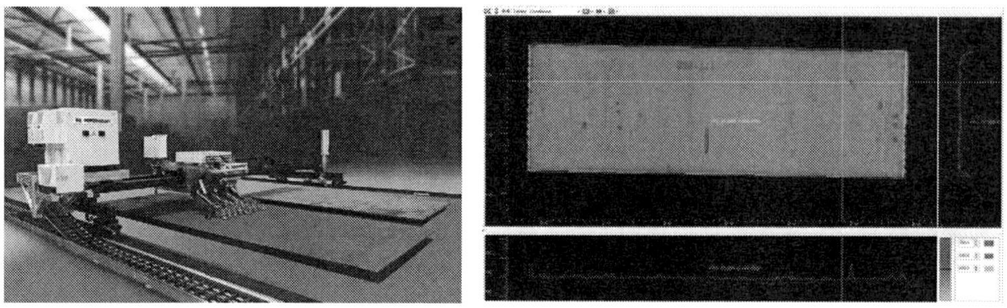

Fig.9. Nordiscan-PL-HS-4500 and C-scan of the plate. Artificial hard spots are automatically shown in red. All 31 defects are detected. No false indications.

The resolution of the scan allows reading the plate ID which had been made presumably at its production. All 31 defects we systematically detected; probability p of detection is close to 100% (p =1). False indications are very seldom – about one indication during 100 runs.

Fig.10. shows the same scan, but all the defects, having dimensions of 10 mm x 10mm are rounded by yellow. Photo of the area of the right-up corner is also represented in Fig 10, right. Two defects of 10 mm x 10mm are located at edges. The defect 15mm x 15mm is made exactly on the plate corner. All four corners of the plate have such defects.

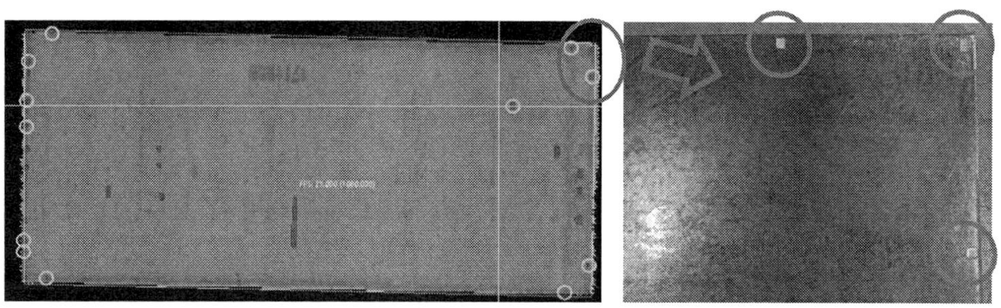

Fig.10. Left: the same scan, but all the defects, having dimensions of 10 mm x 10mm are rounded by yellow; right - photo of the right-up corner area; two defects of 10 mm x 10mm and one defect 15 mm x 15mm which is located exactly on the plate corner.

Conclusions

(1) The new method of detecting of hard spots and evaluation of plates hardness can meet most of the modern requirements formulated in paragraph 1.

(2) Automatic testing equipment based on PEC+ has already found multiple industrial applications and achieved a positive reputation. Daily field experience and feedback from the customers operating all over the world facilitate further development of the PEC+ method and its new modifications.

DGZfP-Jahrestagung 2025

Simulationsgestützte Entwicklung eines neuartigen Sensors für die Wirbelstromprüfung von Kohlenstofffasergelegen

Till SCHULZE[1,2], Henning HEUER[1,2], Dirk HOFMANN[1,2], Christian PILZ[1], Martin SCHULZE[1], Chokri CHERIF[2], Johannes MERSCH[2]
[1]Fraunhofer-Institut für Keramische Technologien und Systeme IKTS
[2]Technische Universität Dresden, Institut für Aufbau- und Verbindungstechnik der Elektronik

Kurzfassung

Kohlenstofffaserverstärkte Verbunde (CFK) werden seit Jahrzehnten erfolgreich in Industriezweigen wie Luft- und Raumfahrttechnik eingesetzt. Doch gerade in diesen Branchen sind kritische Anforderungen einzuhalten, was idealerweise eine zerstörungsfreie Prüfung voraussetzt. Das Wirbelstromprüfverfahren hat sich dafür als besonders geeignet herausgestellt.

Um einen Kohlenstofffaserverbund in verschiedenen Schichten und Tiefen hinsichtlich Fehlstellen von Luft- oder Metalleinschlüssen sowie Delaminationen exakt charakterisieren zu können, müssen die Prüfkörper mit verschiedenen Sensoren und individuellen Parametern mehrmals aufwendig gescannt werden. Je nach gewünschter Auflösung und Größe des Prüfobjektes dauert eine solche Messung mehrere Stunden. Um eine höhere Genauigkeit in Laminatdickenrichtung zu erreichen, sind oft weitere Scans mit verschiedenen Sensoren erforderlich. Um diesen Aufwand zu reduzieren, wurde ein neuer, konzentrisch aufgebauter Sensor entwickelt.

Die Auslegung solcher Sensoren erfolgte bisher experimentell oder auf Grundlage von Erfahrungen. Durch die Fortschritte in der FEM-Simulation können neuartige Sensordesigns exakt ausgelegt und numerisch entwickelt werden. Das Ziel der Simulation ist die Entwicklung einer Spule, die sensitiv auf Leitfähigkeitsänderungen in der Struktur reagiert. Bei anisotropen und heterogenen Materialien wie CFK nimmt somit die Modellierung der Struktur eine zentrale Rolle ein. Die Heterogenität der Fasern entsteht durch die höhere Leitfähigkeit im Zentrum eines Rovings im Vergleich zu seinem Randbereich. Im Modell werden zylinderförmige Fasern über eine periodische Funktion approximiert, welche die Leitfähigkeiten innerhalb der jeweiligen Schicht modelliert.

In mehreren Studien wurden Simulationen mit variierenden Spulenradien, Windungszahlen, Drahtdurchmessern und weiteren Parametern durchgeführt, um ein ideales Design zu entwickeln. Dabei wurde das Messignal der Empfängerspulen bei unterschiedlichen Leitfähigkeiten, einerseits durch die Heterogenität der Fasern und andererseits bei Auftreten von vorhandenen Fehlstellen wie Faserbrüchen und metallischen Einschlüssen ausgewertet. Die Untersuchungen ergaben Anforderungen und Randbedingungen für den Sensor, welcher erfolgreich hergestellt wurde.

Keywords: Wirbelstromprüfung, Simulation, CFK-Qualitätssicherung

Simulationsgestützte Entwicklung eines neuartigen Sensors für die Wirbelstromprüfung von Kohlenstofffasergelegen

Fraunhofer IKTS

TECHNISCHE UNIVERSITÄT DRESDEN

T. Schulze[1,2], H. Heuer[1,2], D. Hofmann[1,2], C. Pilz[1], M. Schulze[1], C. Cherif[2], J. Mersch[2]
[1]Fraunhofer-Institut für Keramische Technologien und Systeme IKTS
[2]Technische Universität Dresden, Institut für Aufbau- und Verbindungstechnik der Elektronik

Einleitung und Motivation

Kohlenstofffaserverstärkte Verbunde (CFK) werden seit Jahrzehnten erfolgreich in Industriezweigen wie Luft- und Raumfahrttechnik eingesetzt. Doch gerade in diesen Branchen sind kritische Anforderungen einzuhalten, was idealerweise eine zerstörungsfreie Prüfung voraussetzt. Das Wirbelstromprüfverfahren hat sich dafür als besonders geeignet herausgestellt.

Um einen Kohlenstofffaserverbund in verschiedenen Schichten und Tiefen hinsichtlich Fehlstellen von Luft- oder Metalleinschlüssen sowie Delaminationen exakt charakterisieren zu können, müssen die Prüfkörper mit verschiedenen Sensoren und individuellen Parametern mehrmals aufwendig gescannt werden.

Grundlegend besteht ein Wirbelstromsensor aus einer Erregerspule, welche mit einer Wechselspannung beaufschlagt wird und ein primäres Magnetfeld erzeugt. Dieses induziert Wirbelströme in einer elektrisch leitfähigen Struktur. Durch die Ströme wird ein sekundäres Gegenfeld erzeugt, welches mit von den Änderungen der Leitfähigkeit in der Struktur abhängt. Mit einer zweiten Spule (Empfänger) oder auch der Erregerspule selbst können diese Änderungen gemessen und so Fehlstellen wie z.B. Faserbrüche, Delaminationen oder Metalleinschlüsse detektiert werden.

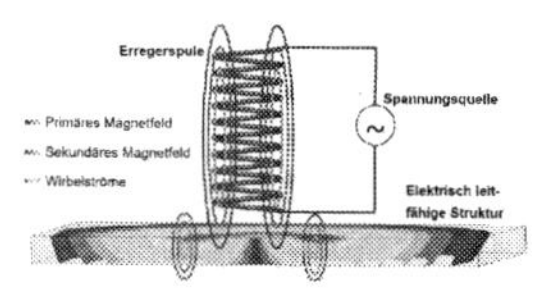

Prinzip der Wirbelstromprüfung.

Wirbelstromprüfung von Kohlenstofffasern

Kohlenstofffasern sind im Vergleich zu Metallen schwach elektrisch leitfähig. Um in diesen Materialien Wirbelströme zu induzieren, werden Prüffrequenzen größer als 1 MHz verwendet. Ein typischer Prüfvorgang ist in folgender Abbildung zu erkennen. Ein an einem multiaxialen Verfahrtisch befestigter Wirbelstromsensor wird über dem Prüfobjekt bewegt und misst in definierten Abständen die Struktur.

Prüfung eines CFK-Bauteils am Fraunhofer MPECS®.

Dabei wird jeder Position ein Intensitätswert zugewiesen, welche mit der elektrischen Leitfähigkeit korrespondiert. Somit ergibt sich folgendes Graustufenbild eines triaxialen Kohlenstofffasergeleges:

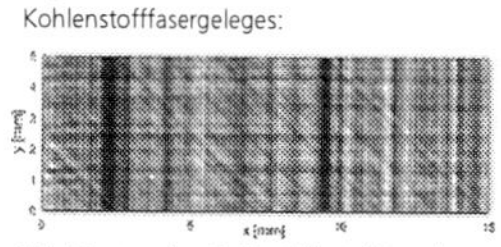

Wirbelstromscan eines triaxialen Kohlenstofffasergeleges.

Je nach gewünschter Auflösung und Größe des Prüfobjektes dauert eine solche Messung mehrere Stunden. Um eine höhere Genauigkeit in Laminatdickenrichtung zu erreichen, sind oft weitere Scans mit verschiedenen Sensoren erforderlich. Um diesen Aufwand zu reduzieren, wurde ein neuer, konzentrisch aufgebauter Sensor entwickelt.
Die Auslegung solcher Sensoren erfolgte bisher experimentell oder auf Grundlage von Erfahrungen. Durch die Fortschritte in der FEM-Simulation können neuartige Sensordesigns exakt ausgelegt und numerisch entwickelt werden.

Simulation von Wirbelströmen

Das Ziel der Simulation ist die Entwicklung einer Spule, die sensitiv auf Leitfähigkeitsänderungen in der Struktur reagiert. Bei anisotropen und heterogenen Materialien wie CFK nimmt somit die Modellierung der Struktur eine zentrale Rolle ein.

Modellierung der Faserstruktur

Die Heterogenität der Fasern entsteht durch die höhere Leitfähigkeit im Zentrum eines Rovings im Vergleich zu seinem Randbereich [1]. Im Modell werden somit eng aneinander liegende, zylinderförmige Fasern über eine analytisch beschreibbare, periodische Funktion approximiert, welche die Leitfähigkeiten innerhalb der jeweiligen Schicht entsprechend der Faserrichtung modelliert.

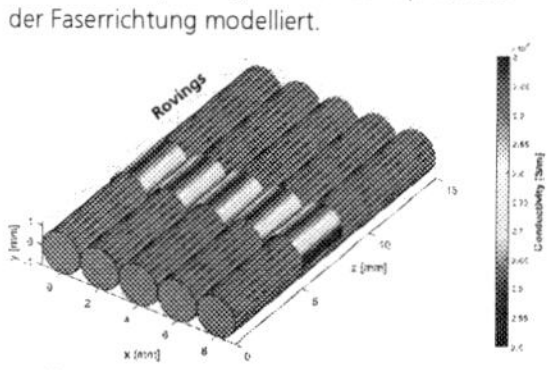

Verteilung elektrischer Leitfähigkeit im Kohlenstofffaserverbund.

Die Anisotropie einer Schicht kann mit einem Tensor für die elektrische Leitfähigkeit beschrieben werden [2]:

$$\sigma = \begin{pmatrix} \sigma_{\parallel} \cos^2\theta + \sigma_{\perp} \sin^2\theta & \frac{\sigma_{\parallel}-\sigma_{\perp}}{2}\sin(2\theta) & 0 \\ \frac{\sigma_{\parallel}-\sigma_{\perp}}{2}\sin(2\theta) & \sigma_{\parallel} \sin^2\theta + \sigma_{\perp} \cos^2\theta & 0 \\ 0 & 0 & \sigma_t \end{pmatrix}$$

Darin stehen $\sigma_{\parallel}$, $\sigma_{\perp}$ und σ_t für die elektrische Leitfähigkeit in Faserrichtung, quer Faser und in Dickenrichtung. Für den Faserwinkel der Lage steht θ.
Um die Heterogenität abzubilden, wird für die Leitfähigkeit in Faserrichtung die approximierte, periodische Funktion eingesetzt. In folgender Abbildung ist die Wirbelstromausbreitung in der obersten Schicht eines multiaxialen Kohlenstofffasergeleges qualitativ dargestellt. Dabei ist die sich über dem Zentrum der Struktur befindende Erregerspule ausgeblendet.

Wirbelstromausbreitung im mehrlagigen Kohlenstofffaserverbund.

Entwicklung des Sensors

Zur Verbesserung der Messgenauigkeit bei hohen Prüfgeschwindigkeiten wurde ein kompakter Sensoraufbau gewählt, welcher aus konzentrisch angeordneten Spulen besteht, die über jeweils individuelle Eigenschaften verfügen. Somit können bei einem einzigen Scan Wirbelstrombilder unterschiedlich aufgelöster Tiefen erzeugt werden, um daraus ein 3D-Modell des Laminats abzuleiten.

In folgender Abbildung ist die numerisch ermittelte Wirbelstromverteilung für verschiedene Konfigurationen dargestellt. Die schwarze Spule ist jeweils die Erregerspule. Dabei ist sehr gut zu erkennen, wie die Wirbelströme in unterschiedlichen Tiefen der Struktur eindringen. Somit können Sensoren simulativ optimal ausgelegt werden, um definierte Lagen und kritische Zustände zu detektieren und zu visualisieren.

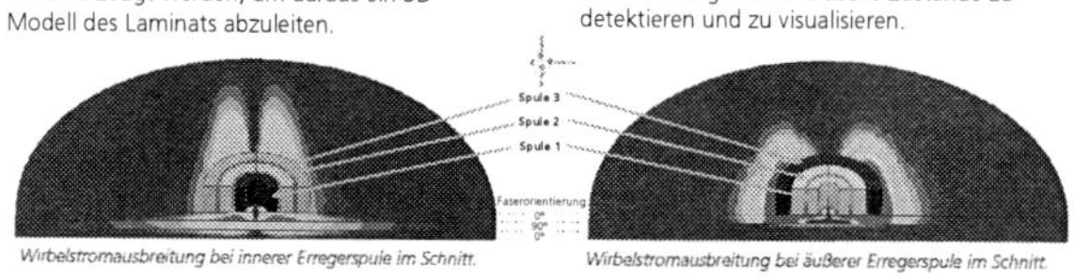

Wirbelstromausbreitung bei innerer Erregerspule im Schnitt.

Wirbelstromausbreitung bei äußerer Erregerspule im Schnitt.

Fertigung des Sensors

In mehreren Studien wurden Simulationen mit variierenden Spulenradien, Windungszahlen, Drahtdurchmessern und weiteren Parametern durchgeführt, um ein ideales Design zu entwickeln. Dabei wurde das Messignal der Empfängerspulen bei unterschiedlichen Leitfähigkeiten, einerseits durch die Heterogenität der Fasern und andererseits bei Auftreten von vorhandenen Fehlstellen wie Faserbrüchen und metallischen Einschlüssen ausgewertet.
Die Untersuchungen ergaben Anforderungen und Randbedingungen für den Sensor und dessen Fertigung. Dazu werden die Spulen nacheinander über einen Stabkern gewickelt und anschließend verdrahtet und vergossen.

Wickeln der Spulen auf Stabkern.

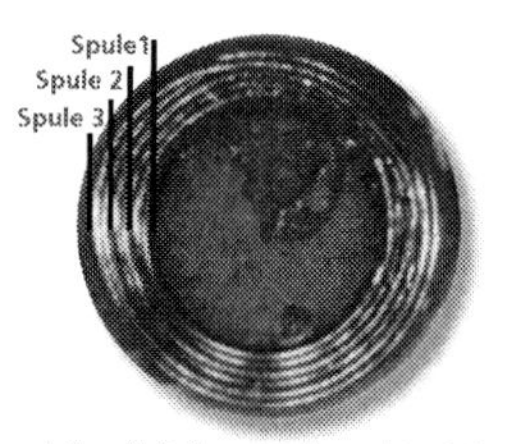

Fertig gewickelter Sensor mit drei konzentrischen Spulen.

Zusammenfassung

Zur Erhöhung der Messgenauigkeit, insbesondere der Tiefenauflösung bei hohen Prüfgeschwindigkeiten entwickelten das Fraunhofer IKTS und die TU Dresden einen konzentrisch aufgebauten, kompakten Wirbelstromsensor.
Im Rahmen dieses Projektes konnte somit die Auslegung wirtschaftlich effizient und technische nahezu vollständig durch Simulation abgesichert werden. Einige ausgewählte Vorteile des Sensor sind:

- Erhöhung der Messgenauigkeit insbesondere in der Tiefenauflösung
- Reduktion des Prüfprozesses durch einmaliges Abfahren
- Individuell einstellbare Sensitivität auf spezielle Fehlstellen (Art, Lokalisierung und Charakterisierung)
- Grundlage für die Rekonstruktion eines 3D-Modells eines CFK-Laminats durch Kombination von verschiedenen Sende- und Empfangskonfigurationen

Fazit

Die Wirbelstromprüfung etabliert sich als zerstörungsfreies Verfahren zur Qualitätssicherung von kohlenstofffaserverstärkten Kunststoffen in vielen Industriezweigen.

Durch die Fortschritte in der Simulation elektromagnetischer Felder sowie der Modellierung anisotroper und heterogener Strukturen können Sensoren besser ausgelegt und dimensioniert werden.

Die innovative Idee eines neuen Sensorkonzept vom Fraunhofer IKTS und der TU Dresden konnte somit ohne die Herstellung teurer Prototypen und aufwendiger Messreihen umgesetzt werden.

Kontakt: Till Schulze
Fraunhofer-Institut für Keramische Technologien und Systeme IKTS
Wirbelstromverfahren
Maria-Reiche-Str. 2, 01109 Dresden
Telefon +49 351 88815 556
till.schulze@ikts.fraunhofer.de

Quellen:

[1] YI, Q.; WILCOX, P.; HUGHES, R.: Modelling and evaluation of carbon fibre composite structures using high-frequency eddy current imaging. Composites 248(2023).
[2] CHENG, J.; JI, H.; QIU, J.; TAKAGI, T.; UCHIMOTO, T.; HU, N.: Novel electromagnetic modeling approach of carbon fiber-reinforced polymer laminate for calculation of eddy currents and eddy current testing signals. Journal of Composite Materials 49(2015)5, S. 617–631.

Gefördert durch
DFG Deutsche Forschungsgemeinschaft

EddyWeb – der Wirbelstromsensor für den Web-Browser

Gerhard MOOK, Yuri SIMONIN
Otto-von-Guericke-Universität Magdeburg, Deutschland, mook@ovgu.de

Kurzfassung

Smartphones, Tablets, Notebooks und Desktop-Rechner nutzen Browser, um auf externe Daten zuzugreifen. Warum nicht auch auf einen Wirbelstromsensor? Das wird mit EddyWeb möglich!

Dieser Sensor besitzt einen Miniatur-Webserver und spannt sein eigenes WLAN auf. Wenn man sich mit diesem Netz verbindet, entsteht ein Mehrfrequenz-Wirbelstromgerät, das sich für viele Prüf- und Ausbildungsaufgaben geeignet. Über eine intuitive Benutzer-Oberfläche werden alle Einstellungen vorgenommen und die Daten als Punkte in der XY-Ebene dargestellt. Neben den Standardfunktionen sind Recorder und Filter integriert. Ein Testsignalgenerator hilft, sich in die Funktionen einzuarbeiten.

Der Sensor kann elektronisch von Absolut- auf Differenzbetrieb umgeschaltet werden, so dass kein Sensorwechsel mehr erforderlich ist. Die Sensoren sind aufeinander abgeglichen, so dass die Einstellungen bei Sensorwechsel nicht verändert werden müssen.

Keywords: Wirbelstromsensor, Geräteintegration, Absolutsensor, Differenzsensor, Mehrfrequenz

OTTO VON GUERICKE UNIVERSITÄT MAGDEBURG

G. Mook, Y. Simonin
Institut für Werkstoffe | Technologien | Mechanik

EddyWeb – ein Wirbelstromsensor für den Web-Browser

Nutzbar mit vielen Endgeräten...

Android-Geräte

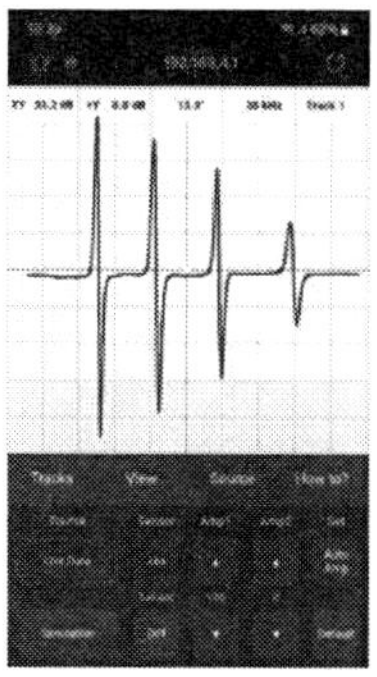

Samsung Galaxy S24 (Chrome)

iOS-Geräte

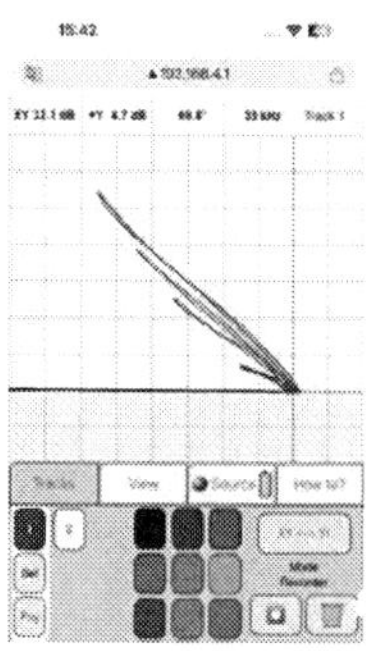

iPhone 14 Pro (Chrome)

iPad

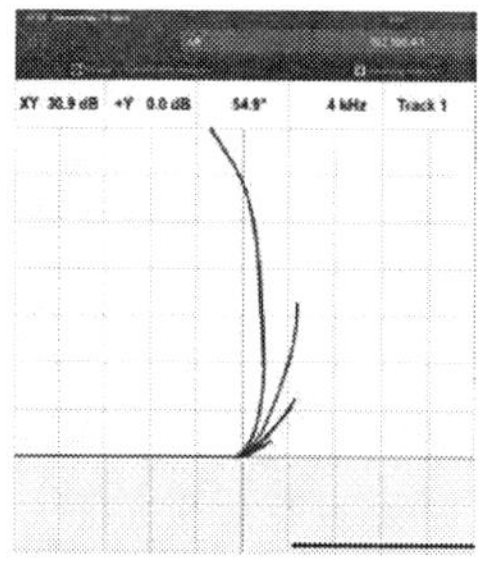

iPad (Safari)

iPhone

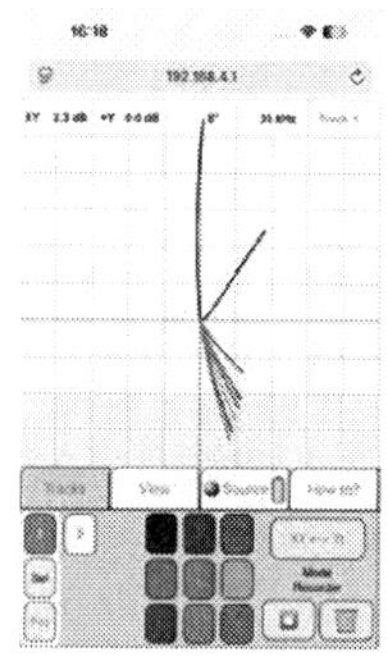

iPhone (Safari)

Windows-Geräte

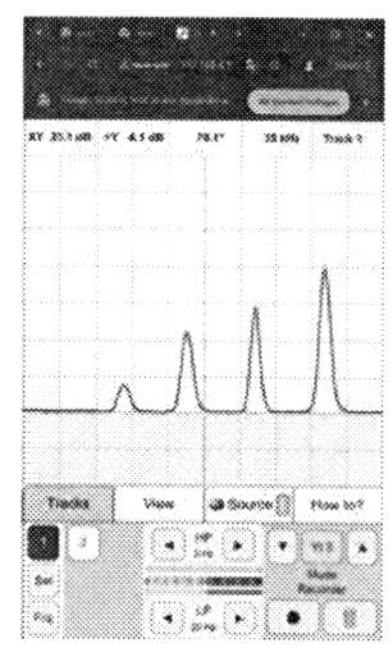

Windows-PC (Chrome)

... und Browsern

Chrome

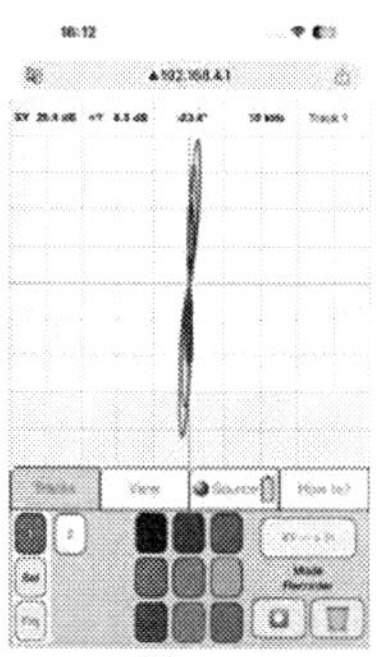

iPhone

Opera

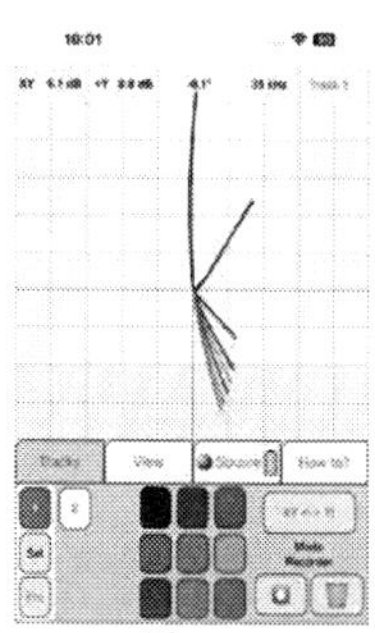

iPhone

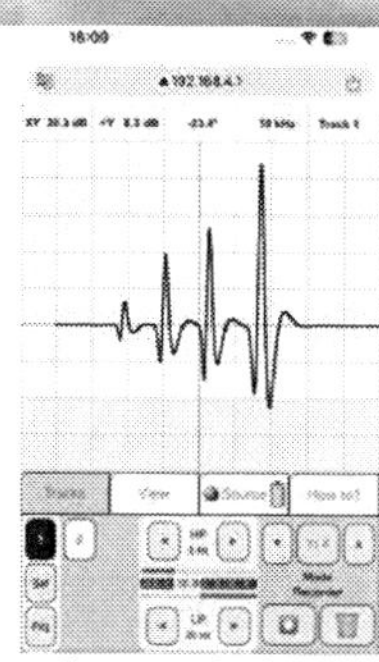

Chrome auf iPhone

Firefox

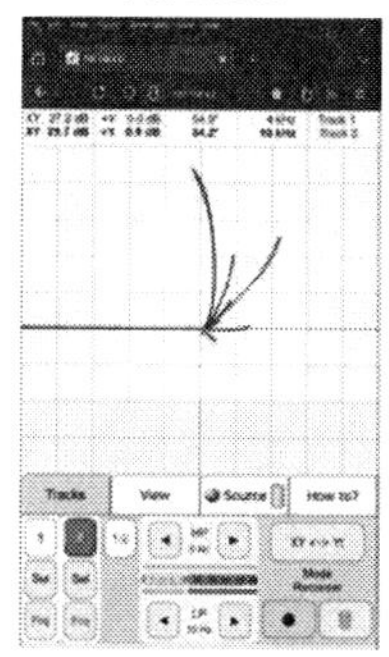

Windows-PC

Edge

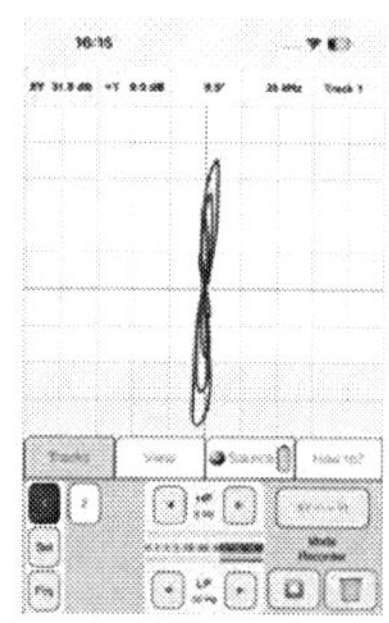

iPhone

Versionen - Bedienung - Besonderheiten

Verfügbare Sensoren

Blick auf das Sensordisplay

Kombinierter Sensor | Differenz-sensor | Absolut-sensor

Bedienung

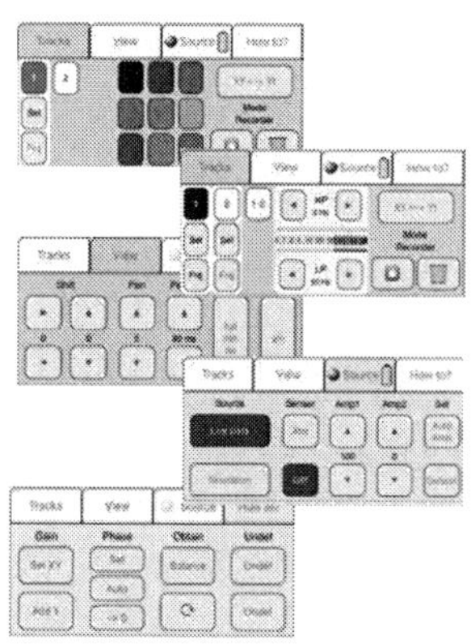

Besonderes

- Sensor mit integriertem Gerät
- Sensor spannt eigenes WLAN auf
- Absolut- und Differenzsensor elektronisch umschaltbar
- Alle Sensoren erzeugen gleiche Signale
- Zweifrequenztechnik
- Hoch- und Tiefpassfilter
- Akkubetrieb möglich
- Gespeicherte Mustersignale zum Üben ohne Testkörper
- Diverse Testkörper erhältlich
- Die Sensoren besitzen eine USB-C-Buchse zum Laden und zur kabelgebundenen Datenübertragung
- Kalibrierbar nach DIN EN ISO 155448

Kompatibilität mit EddyCation

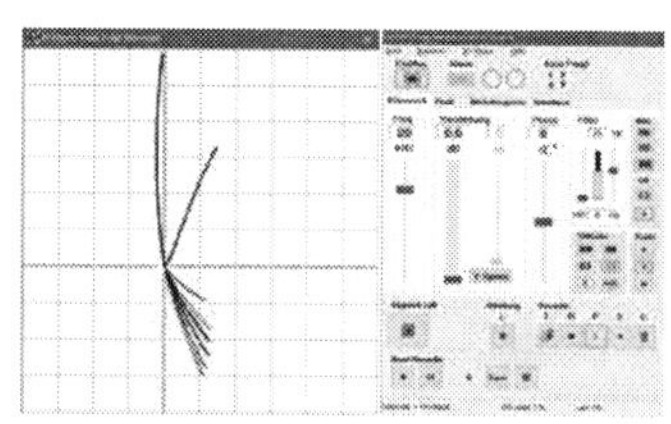

Über USB-Kabel kann der Sensor auch direkt in die bekannte EddyCation-Umgebung eingebunden werden.

Testkörper für den Grundkurs

Der magnetische Barkhausen-Effekt angewandt als standardisiertes in-line Prüfverfahren integriert in verketteten Produktionslinien

Michael MAASS, Björn REITZ
Stresstech GmbH, 56477 Rennerod, infoDE@stresstech.com

Kurzfassung. Barkhausenrauschen ist ein seit über 100 Jahren bekannter physikalischer Effekt, dessen Anwendung in der zfP-Welt noch immer ein Nischendasein fristet. Erst vor 40 Jahren haben die Gründer der Firma Stresstech damit begonnen auf Basis des magnetischen Barkhauseneffekts neue Produkte für die robuste, zerstörungsfreie Qualitätskontrolle in der industriellen Massenproduktion zu etablieren. In diesem Vortrag werden ausgehend von den physikalischen Grundlagen, zu Beginn die aktuellen Signalauswertealgorithmen und ihre Korrelationen mit verschiedenen Materialeingeschaften wie Eigenspannung, Härte und die Schleifbranddetektion dargestellt. Nachfolgend werden die Umsetzungen in moderne zfP-Prüfanlagen zur Detektion von Schleifbrand und Prüfung auf Eigenspannung in der Serienfertigung präsentiert. Ein Schwerpunkt bilden Anwendungsbeispiele aus Qualitätskontrolle in der hochvolumigen Massenproduktion mit kurzen Zykluszeiten, wie in der Komponentenproduktion der Automobilwirtschaft und der Elektromobilität gefordert.

Hierbei werden insbesondere die Themen Quantifizierung der Prüfbefunde, Inline-Prüfung, deren Verkettung und Integration in die Prozesskontrolle sowie Art und Umfang der Einsatzmöglichkeiten angesprochen.

Keywords: Barkhausenrauschen, BN, zerstörungsfreie Prüfung, zfP, elektromagnetisches Prüfverfahren, Schleifbrand-Detektion, Eigenspannungs-Prüfung, Wärmebehandlungs- und Härte-Prüfung In-line, In-Situ, automatisiert, verkettete Produktionslinie

Einführung

Das Barkhausenrauschen ist ein physikalisches Phänomen, das in ferromagnetischen Materialien auftritt. Es beschreibt sprunghafte Magnetisierungsänderungen, die durch Domänenwandbewegungen innerhalb des Materials verursacht werden. Dieses Rauschen wurde nach dem deutschen Physiker Heinrich Barkhausen benannt, der es 1919 entdeckte. Heute findet das Barkhausenrauschen Anwendung in der zerstörungsfreien Werkstoffprüfung und Charakterisierung von ferromagnetischen Materialien.

1. Grundlagen des Barkhausenrauschens [1]

Ferromagnetische Materialien verfügen im submikroskopischen Bereich über eine Vielzahl von magnetischen Domänen (Weiss´sche Bezirke), innerhalb derer die magnetischen Momente der Atome parallel ausgerichtet sind. Jede Domäne verfügt somit über eine magnetische Vorzugsrichtung die untereinander durch Domänenwände (Blochwände) getrennt sind.

https://doi.org/10.58286/31280

Solange kein magnetisches Feld anliegt, sind die magnetischen Domänen in chaotischer Unordnung, das Werkstück ist in Summe feldfrei (Abb. 1).

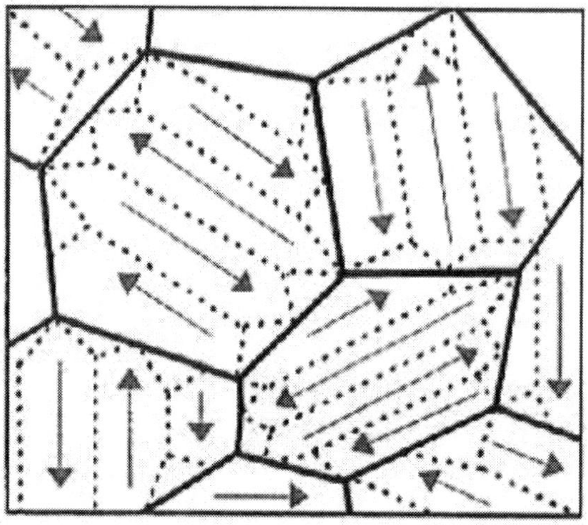

Abb. 1. Schematische Darstellung der magnetischen Domänen innerhalb eines in Summe feldfreien Werkstücks

Wirkt ein äußeres Magnetfeld auf das Werkstück, versuchen sich die magnetischen Domänen entlang der Feldlinien des äußeren Feldes auszurichten. Der Vorgang der Ausrichtung der Domänen erzwingt eine Bewegung der Domänenwände. Die Domänen, deren Vorzugsrichtung am ehesten dem äußeren Magnetfeld entsprechen, wachsen auf Kosten der entgegengesetzt orientierten Domänen. Hierbei verschiebt sich die Domänenwand. Wird das Magnetfeld kontinuierlich verstärkt, tritt schließlich Sättigung ein – im Sättigungszustand sind alle Domänen parallel zum äußeren Magnetfeld ausgerichtet. Geht das äußere Feld wieder zurück auf Null, versuchen die magnetischen Domänen wieder ihre ursprüngliche Lage einzunehmen.

Diese Bewegungen erfolgen nicht kontinuierlich, sondern in diskreten Sprüngen, da sie von Mikrostrukturen wie Ausscheidungen, Korngrenzen, Versetzungen im Kristallgitter, Eigenspannungsfeldern und kleinen Bereichen von anderen Materialphasen behindert werden. Während der Bewegung der Domänenwand wird Energie verbraucht, um diese Hindernisse zu überwinden. Die dabei auftretenden diskreten Sprünge erzeugen elektrische Impulse, die als Barkhausenrauschen gemessen werden können.

2. Physikalische Parameter und Einflussfaktoren

Die Anregung des Barkhausen-Rauschens erfolgt über elektromagnetische Wechselfelder vom niedrigen Hertz bis in den unteren Kilohertzbereich. Über den sogenannten Skin-Effekt konzentriert sich die Wirkung auf Barkhausenrausch-Prüfung auf Auswertungen im oberflächennahen Tiefenbereich bis ca. 1-2 mm Tiefe, abhängig von elektrischer Leitfähigkeit und magnetischer Permeabilität des Werkstoffs.

Die von den diskreten Sprüngen der Domänenwände hervorgerufenen elektromagnetischen Impulse, also das empfangene Barkhausenrauschen, hat ein breites Frequenzspektrum im Bereich von kHz bis zu niedrigen MHz (Abb. 2.).

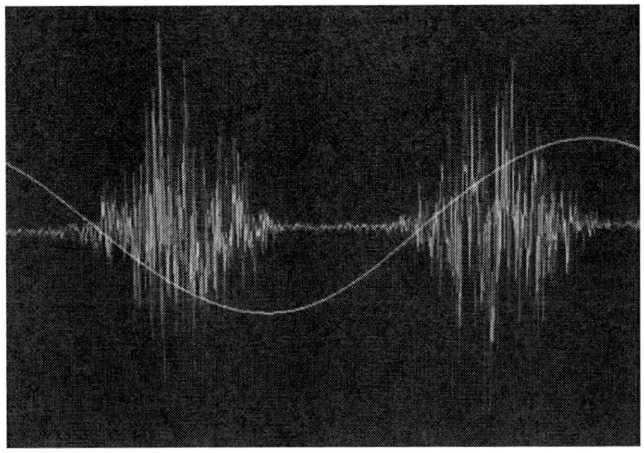

Abb. 2. Darstellung der niederfrequenten elektromagnetischen Anregung zusammen mit dem empfangenen Barkhausenrauschsignals

Zwei wichtige Materialeigenschaften beeinflussen die Intensität des Barkhausen-Rauschens.

Die eine ist das Auftreten und die Verteilung von elastischen Spannungen, die die Art und Weise beeinflussen, wie die Domänen eine bevorzugte Magnetisierungsrichtung wählen und darin „einrasten“. Dieses Phänomen der Wechselwirkung von elastischen Mateiraleigenschaften mit der Domänenstruktur und weiteren magnetischen Eigenschaften des Materials wird als „magnetoelastische Wechselwirkung“ bezeichnet. Als Ergebnis der magnetoelastischen Wechselwirkung verringern Druckspannungen die Intensität des Barkhausen-Rauschens, während Zugspannungen es in Materialien mit positiver magnetischer Anisotropie (Eisen, die meisten Stähle und Kobalt) verstärken.

Diese Tatsache kann sehr simpel und zuverlässig ausgenutzt werden, um durch die Messung der Intensität des Barkhausen-Rauschens (entsprechend dem Energieinhalt des Barkhausenrauschsignals) die Höhe von Eigenspannung zu bestimmen (Abb. 3.). Die Barkhausenrausch-Messung zeigt zudem auch die Richtung der Hauptspannungen an.

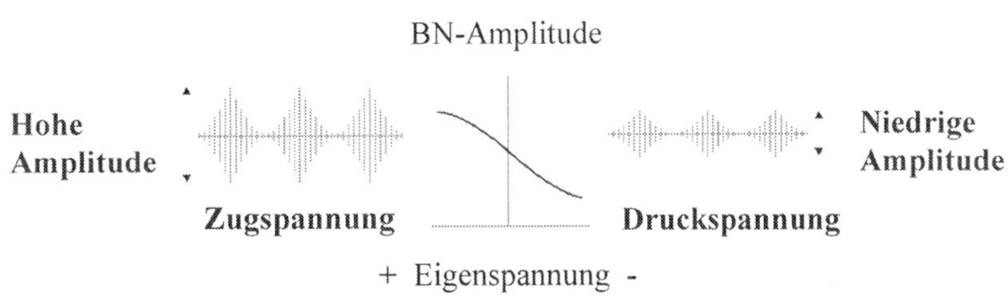

Abb. 3. Darstellung der niederfrequenten elektromagnetischen Anregung zusammen mit dem empfangenen Barkhausenrauschsignals

Eine weitere wichtige Materialeigenschaft, die mit Barkhausenrauschen bewertet werden kann, sind Mikrostruktur und Gefügeausbildung eines Prüfkörpers. Diese Materialeigenschaft lässt sich auch allgemein als Härte des Werkstoffs beschreiben: Die Intensität des Barkhausenrauschens nimmt in Gefügen mit zunehmender Härte kontinuierlich ab. Auf diese Weise liefern Barkhausenrauschmessungen einfach auswertbare Informationen über den mikrostrukturellen Zustand des Materials (Abb. 4.). Zudem wirkt sich die Mikrostruktur eines Prüfkörpers auch direkt auf die Form des Ausgangssignals aus. Beispielsweise haben hartmagnetische Materialien breitere und weichmagnetische Materialien schmalere Barkhausenrausch-Signalhüllkurven. Charakteristisch für die Auswertung eines Barkhausenrausch-Signals sind somit auch Amplitude, Spitze und Breite der Signalhüllkurven, die Informationen über die Mikrostruktur des Prüfkörpers liefern.

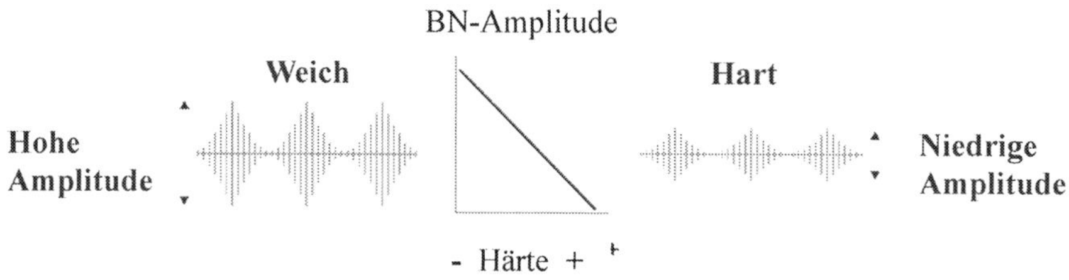

Abb. 4. Darstellung der niederfrequenten elektromagnetischen Anregung zusammen mit dem empfangenen Barkhausenrauschsignals

Zusammengefasst hängt das Barkhausenrauschen von mehreren Faktoren ab, darunter:

- ✓ **Materialeigenschaften:** Gefüge, Korngröße und chemische Zusammensetzung beeinflussen die Domänenbewegung.
- ✓ **Mechanische Spannungen:** Innere oder äußere Spannungen verändern die Magnetisierungsdynamik.
- ✓ **Externe Magnetfelder:** Die Frequenz und Stärke des angelegten Magnetfeldes bestimmen die Messbedingungen.

Somit sind Prüfkalibrierungen in Abhängigkeit vom jeweiligen Werkstoff, dem verwendeten Sensortypen und den Einstellparametern der angeregten magnetischen Wechselfelder zu erstellen.

3. Messtechnik zur zuverlässigen Prüfung mit dem Barkhausenrauschen

Die Prüfung mit dem Barkhausenrauschen erfordert

- ✓ **Spektralanalysator (Rollscan):** die mobile, einfach tragbare Prüfelektronik ermöglicht die Rauschsignal-Auswertung in verschiedenen Frequenzbereichen. Die Prüfelektronik enthält die Leistungsquelle mit verschiedenen, wählbaren Wechselfeldanregungen und einen digitalen Signalprozessor (Abb. 5.). Die Geräteausführung kann abhängig von der jeweiligen Einsatzart variieren.

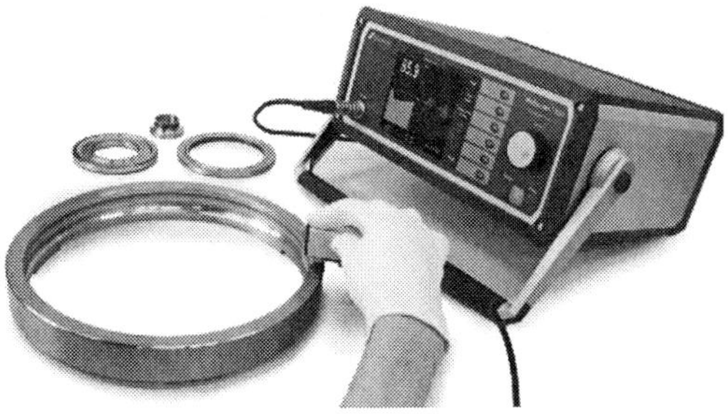

Abb. 5. Darstellung der aktuellen Standardprüfelektronik Rollscan 350

- ✓ **Barkhausenrausch-Sensor:** ein an das zu prüfende Bauteil angepasster Sensor für die zyklische Wechselfeldanregung und Aufnahme des induzierten Rauschsignals. Entsprechend der Vielzahl der Anwendungen wird eine sehr große Anzahl verschiedener Sensortypen angeboten und darüber hinaus häufig auch einzelne anwendungsspezifisch optimierte „customized“ Sensoren (Abb. 6).

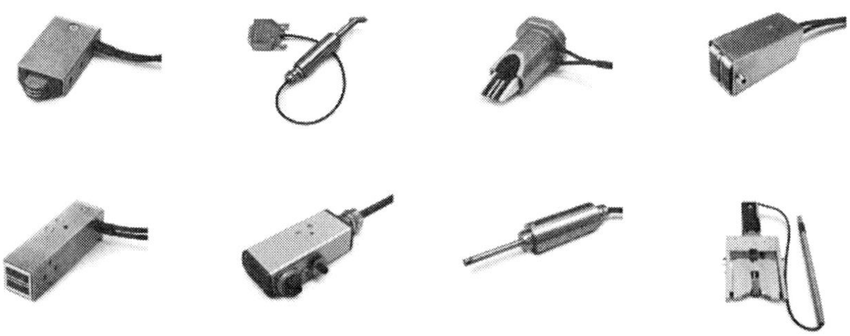

Abb. 6. Kleine Auswahl verschiedener Barkhausenrauschen-Standardsensoren

✓ **Software** für mehrdimensionale Datenerfassung **(ViewScan)** mit C-Scan Darstellung und Multiparameteranalyse für komplexe Burstanalysen **(MicroScan)** (Abb. 7.).

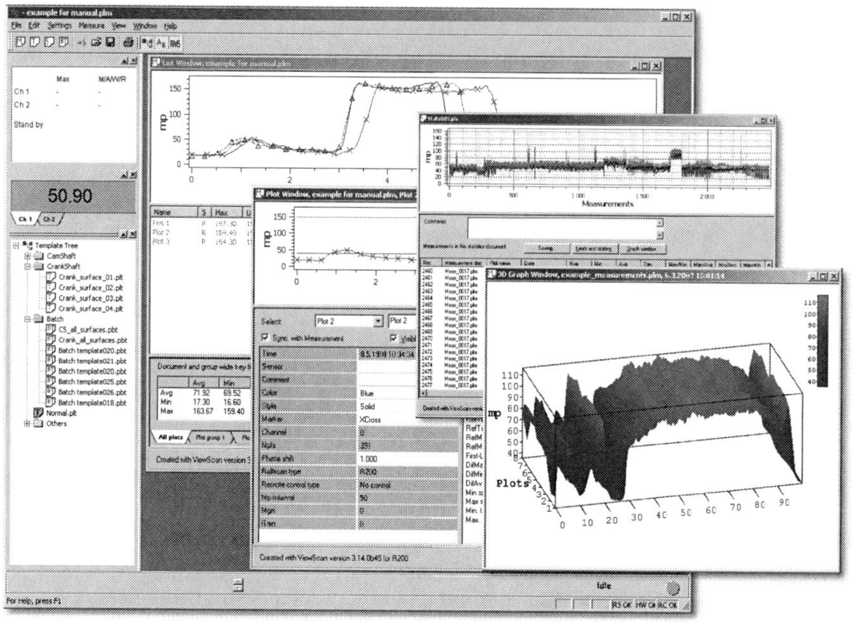

Abb. 7. PC basierte Barkhausenrausch-Auswertung ViewScan und MicorScan

✓ **Optionale Handhabungseinrichtung** für das Bauteil. Hier steht eine große Palette von Mechaniken zur Bauteilhandhabung zur Verfügung. Von einem einfachen Prüfstand, über halbautomatische Prüfmechaniken bis hin zu vollautomatischen Prüfsystemen integriert in verkettete Fertigungslinien mit Speicherung und Analyse großer Datenmengen zur langfristigen Qualitätssicherung, sowie der Anbindung an Datenbanksysteme des Kunden.
Auf die unterschiedlichen automatisierten Systemlösungen für die Linienintegration wird nachfolgend ausführlich eingegangen.

4. Industrielle Anwendungen des Barkhausenrauschen

Viele Oberflächenbearbeitungen, wie z. B. Schleifen, Hartdrehen, Einsatz-, Induktionshärten und Kugelstrahlen verursachen erhebliche Änderungen in Eigenspannung und Mikrogefügestruktur der Randzone, die mittels Barkhausenrauschen charakterisiert werden können. Die Kontrolle von Hartfeinbearbeitungen (Prüfung auf Schleifbrand) stellt eine der Hauptanwendungen des Barkhausenverfahrens dar. Prozesse wie Kriechen und Ermüdung ändern ebenfalls die Eigenspannung und Mikrostruktur und können mittels Barkhausenrauschen analysiert werden.

Praktische Anwendungen der magnetoelastischen Barkhausenrauschenmethode können allgemein in drei Kategorien eingeteilt werden:

✓ **Eigenspannungsanalyse:** Bestimmung von Eigenspannungen in gehärteten oder mechanisch bearbeiteten Bauteilen, um Rissbildungen und Verformungen vorzubeugen.

Somit können Kaltverformungsprozesse, bei denen komplexe Druckeigenspannungsverteilungen in der Randschicht entstehen, durch Barkhausenrauschen charakterisiert werden.

Mit dem Barkhausenrauschen kann beispielsweise die Deckung und Gleichmäßigkeit des Kugelstrahlprozesses überprüft werden. Damit kann das Barkhausenrauschen zur Charakterisierung und Bewertung der Effektivität des Kugelstrahlverfahrens und zur Verbesserung der Qualitätskontrolle des Kugelstrahlens verwendet werden.

- ✓ **Härteprüfung:** Das Barkhausenrauschen kann zur Bewertung des Härtegrades und Härtetiefen von Materialien verwendet werden und somit thermische Schädigungen identifizieren, die durch Bearbeitungsprozesse wie Einsatzhärten, Induktionshärten oder auch Schleifen entstehen können. Das Barkhausenrauschen kann problemlos an die Produktionsraten der meisten Fertigungslinien angepasst werden und ermöglicht eine Härtekontrolle in Echtzeit, z.B. für eine einfache Unterscheidung von gehärteten und ungehärteten Bauteilen in einer in einer Produktionslinie.

- ✓ **Sichere Detektion von Schleifbrand und Steuerung des Schleifprozesses:** Schleifbrand ist eine allgemeine Bezeichnung für thermische Schädigungen, die bei Schleifprozessen an der Oberfläche auftreten. Schleifbrand verursacht lokale Verfärbungen auf der Oberfläche und kann Oberflächenschichten erweichen oder verhärten.
  - Wenn die Temperatur über dem normalen Anlaßbereich, aber unter der Austenitisierungstemperatur (Ac3) liegt, wird ein langsames Abschrecken angewandt, um einen weichen Werkstoff zu bilden, der als neu-angelassener Martensit (over-tempered martensit OTM) bezeichnet wird. Die Bildung von OTM wird als Anlassbrand bezeichnet, das einen Zugspannungszustand erzeugt und die Härte an der Oberfläche verringert.
  - Falls die Temperatur über der Austenitisierungstemperatur (Ac3) liegt, wird eine Schnellabschreckung vorgenommen, um den nicht-angelassenen Martensit (untempered martensite UTM) auf der Oberflächenschicht des geschliffenen Werkstücks zu bilden. Die UTM-Schicht ist härter, korrosionsanfälliger, empfindlicher für Mikrorisse und zudem spröder als der Kern des Werkstücks und ändert die Oberflächenintegrität und den Oberflächenspannungszustand in den Zugzustand. Die Bildung der harten Martensitschicht (UTM) wird als Nachhärtungsbrand bezeichnet, der eine andere Form des Schleifbrandes ist.

  Da das Barkhausenrauschen sowohl bei Härte- als auch bei Spannungsänderungen die gleiche Reaktion zeigt, können verschiedene Formen von Schleifverbrennungen zuverlässig erkannt werden.
  - Somit wird das Barkhausenrauschen zum online Monitoring der Schleifprozesse eingesetzt durch die Überwachung des Zustands der geschliffenen Teile. Schleifschäden, die aufgrund von Scheibenverschleiß, falscher Vorschubgeschwindigkeit, Scheibendrehzahl und anderen Parametern auftreten, können mit der Barkhausenrauschen-Analyse sicher zu 100% erkannt werden.
  - Folglich ist die sichere 100%ige Erkennung von Schleifverbrennungen eine der häufigsten industriellen Anwendungen des Barkhausenrauschens, die von vielen Organisationen wie dem Department of the Air Force der US Navy, SAE, FAA, AST und vielen anderen anerkannt ist.

✓ **Erkennung von Oberflächendefekten durch Beschichtungen hindurch:** Die Erkennung von betriebsbedingten thermischen Beschädigungen durch Chrombeschichtungen hindurch ist eine einfache und zuverlässige Anwendung für die Prüfung mit dem Barkhausenrauschen. Eines der Haupteinsatzgebiete für die Erkennung von Oberflächendefekten durch Cr-Beschichtungen sind Untersuchungen an Fahrwerken von Flugzeugen. Bei der Überholung von Flugzeugen spart der Einsatz des Barkhausenrauschens zur Erkennung von Oberflächenfehlern durch die Verchromung mindestens eine Woche Zeit pro Teil, da das unnötige Abbeizen von Teilen wie Achsen, Zylindern und Bohrungen vermieden wird (Abb. 8.).

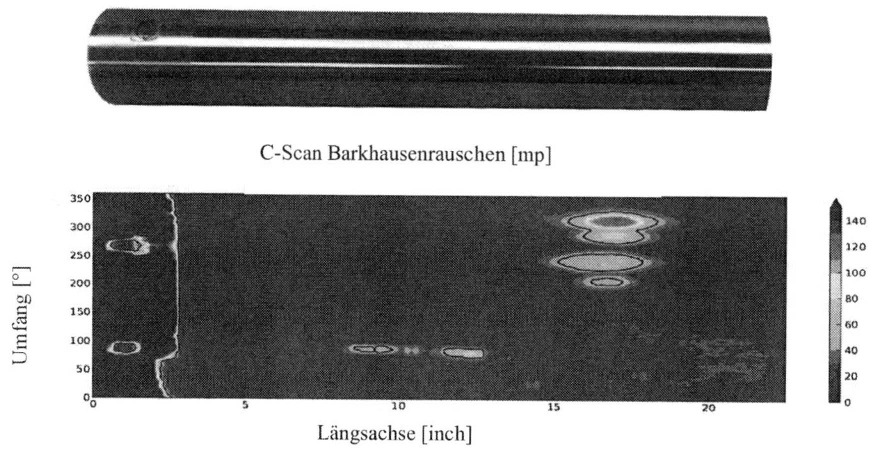

Abb. 8. Barkhausenrauschen Prüfbefund durch Cr-Beschichtung als C-Scan

5. Beispiele automatisierte Barkhausenrauschen-Prüfsysteme in Fertigungslinien

Im Folgenden werden verschiedene Beispiele automatischer, in Fertigungslinien integrierter Barkhausenrauschen-Prüfsysteme zur Erkennung von Schleifbrand- und Wärmebehandlungsfehlern dargestellt. Wie schon mit Bezug auf die Vielfalt der Sensortypen beschrieben, wurden auch für die automatisierte Serienprüfung abhängig vom Bauteil und der Aufgabenstellung unterschiedliche Systemlösungen etabliert. Roboterbasierte Lösungen werden bei einer großen Typenvielfalt bevorzugt, während bei hohen Durchsatzraten insbesondere an Wellenbauteilen wie z.B. Nockenwellen mehrere Sensoren parallel über Linearachsen zugestellt werden. Vorgestellt werden hier in-line Systeme für die Prüfung von Nockenwellen, Kurbelwellen und Zahnräder.

5.1 In-line Prüfung von Nockenwellen

Abhängig von der Anzahl der Nocken und dem geforderten Durchsatz wird eine entsprechend notwendige Anzahl von Prüfkanälen und Sensoren parallel betrieben. Abb. 9 zeigt eine 1-kanalige Lösungen für die Prüfung von Pumpennockenwellen mit zwei Nocken pro Welle. Spezifikationen hierfür waren:

✓ Pumpennockenwellen mit 2 Nocken
✓ Prüfung aller Nocken- und Lagerflächen innerhalb von ca. 60 Sek.
✓ 1-kanalige Lösung mit Nockenwellen-Sensor
✓ Große Typenvielfalt: ca. 50 versch. Nockenwellentypen
✓ Kundenspezifische Beladung manuell oder integriert in eine Roboterzelle

Abb. 8. Barkhausenrauschen Prüfbefund durch Cr-Beschichtung als C-Scan

Zahlreiche CamScan 1000 Prüfsysteme wurden weltweit an Nockenwellenfertigungen für den Truck-Bereich ausgeliefert (Abb. 9.) mit überwiegend sehr ähnlichen Prüfanforderungen:

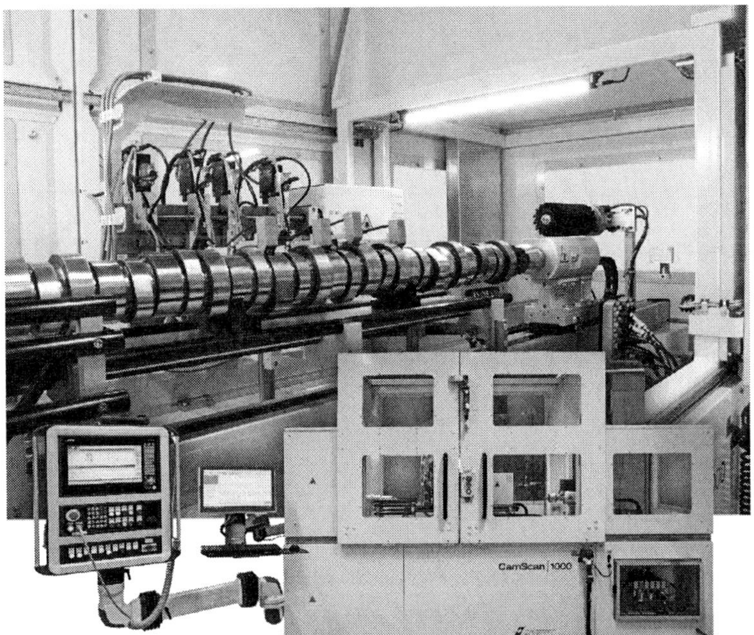

Abb. 9. Automatisiertes 4-kanaliges Prüfsystem für Nockenwellen mit bis zu 24 Nocken

- ✓ LKW-Nockenwellen mit bis zu 24 Nocken
- ✓ Identische Anlagen bei verschiedenen Kunden weltweit
- ✓ Prüfung aller Nockenflächen, Lagerflächen (z.T. auch der Stirnflächen) innerhalb von 1.5 Min.
- ✓ 4-kanalige Lösung mit 4 Nockenwellensensoren parallel im Einsatz
- ✓ Beladung automatisiert über Portal
- ✓ Die automatisierte Einstellung auf verschiedene Wellentypen erfordert typenabhängig eine automatisierte Anpassung des Abstands zwischen den 4 Nockenwellensensoren

Für ein amerikanischen Automobilhersteller im PKW-Bereich wurden Nockenwellenprüfsysteme mit voller Parallelität realisiert für eine Zykluszeit von 8s.

Bei sehr großer Typenvielfalt bieten roboterbasierte Prüfzellen eine effizientere Lösung (Abb. 10.).

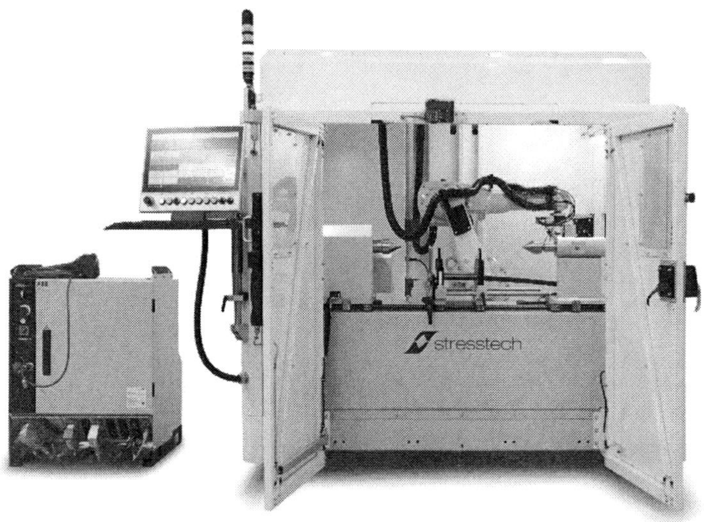

Abb. 10. Automatisiertes roboterbasiertes Prüfsystem für Nockenwellen

Im Beispiel Abb. 10 prüft die roboterbasiertes Systemlösung RoboScan S in einer Automobilproduktion in den USA mit 2 Sensoren parallel eine PKW-Nockenwelle an insgesamt 9 Messpositionen in 44s:

- ✓ PKW-Nockenwellen mit 8+1 Nocken
- ✓ Prüfung aller Nockenflächen innerhalb von 44s
- ✓ Prüfen von 1.800 Nockenwellen pro Tag in 6 Produktionslinien mit 6 Prüfzellen
- ✓ 2-kanalige Lösung mit 2 Prüfkanälen und 2 Nockenwellensensoren parallel im Einsatz, geführt vom Roboter.
- ✓ Beladung automatisiert über Roboter

5.2 In-line Prüfung von Kurbelwellen für Trucks und Marine

Abb. 11 zeigt die Standardlösung RoboScan M für zahlreiche Kurbelwellenfertigungen für die Einsatzbereiche Truck und Marine. Die Vielfalt der Kurbelwellen, welche zumeist auch unterschiedliche Sensortypen erfordert, in Verbindung mit der generellen Notwendigkeit die Prüfung der Hub- und Hauptlager mit jeweils verschiedenen Sensortypen durchführen zu müssen, führt zu dieser robotergestützten Standardlösung. Typische Prüfspezifikation für die meisten Kurbelwellenfertigungen sind:

- ✓ Kurbelwellen für LKW und Marine
- ✓ Häufig mit 7 Hauptlagern und 6 Hublagern
- ✓ Prüfung aller Hub- und Hauptlager innerhalb von 20 Min. (im Takt der Schleifmaschine)
- ✓ Automatisierter Sensorwechsel (Wechsel zwischen Sensor für Hauptlager und Hublager) während des Prüfvorgangs
- ✓ Meist ca. 5-10 verschiedene Wellentypen

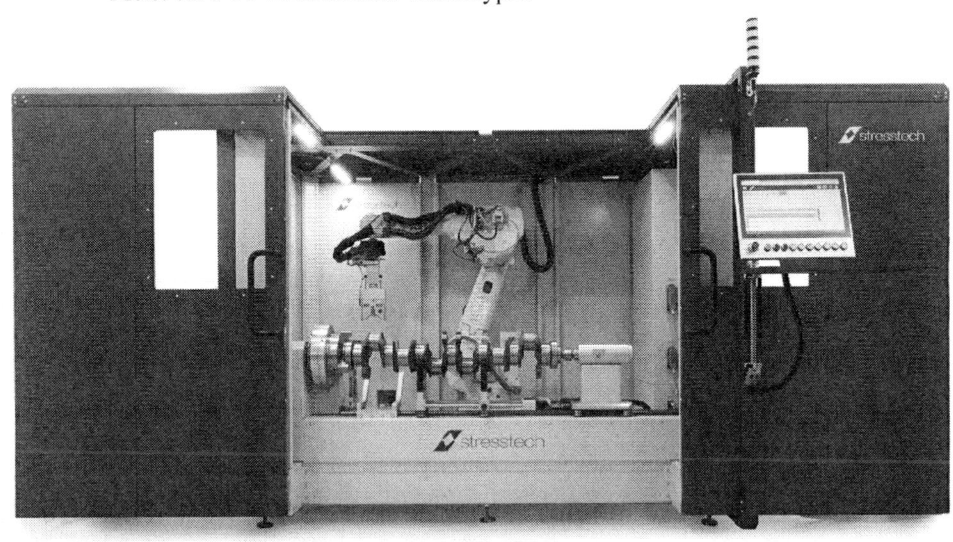

Abb. 11. Automatisiertes roboterbasiertes Prüfsystem für Nockenwellen

5.3 In-line Prüfung Zahnrädern

Die Prüfung von Zahnrädern erfordert ein sehr genaues robotergesteuertes Abscannen der Zahnradgeometrien mit speziell hierfür vorhandenen Sensoren (Abb. 12). Über die fortlaufende Miniaturisierung der Sensoren werden auch die kleinen EV Getriebekomponenten mit Modul < 1 prüfbar.

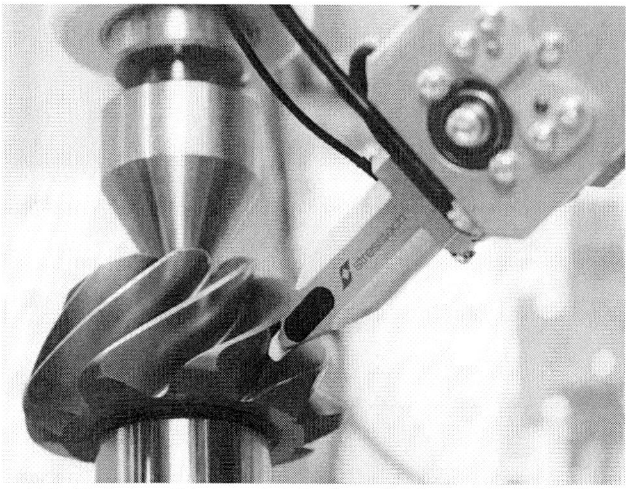

Abb. 12. Automatisiertes roboterbasiertes Prüfsystem für Zahnräder

Über eine vereinfachte Eingabesoftware werden Ihre bauteilspezifischen Parameter an die Robotersteuerung übertragen. Hierdurch ist das Einlernen von neuen Bewegungsprogrammen auch für angelernte Fachkräfte möglich. Somit wird bereits beim Einlernen sichergestellt, dass die Prüfbedingungen wie z.B. Kontaktfläche oder Anlagewinkel den Anforderungen entsprechen. Die somit erstellten Programme sind direkt mit der ViewScan-Software zur Prüfsignalauswertung und -darstellung verbunden. Für jeden zu prüfenden Zahnradtyp wird ein eigenes Messprogramm erstellt. Das entsprechende Interface der ViewScan-Software ruft dieses Programm auf, wenn die entsprechende Zahnradtyp-Vorlage ausgewählt wird.

Optional ist die EasyGear-Software für die Messung von gerad- und schrägverzahnten Stirnrädern erhältlich. Hierbei wird das Zahnrad nach Eingabe der klassischen Zahnradparameter visualisiert und das Bewegungsprogramm automatisch erstellt. Allgemeine Funktionalitäten der RoboScan und GearScan Systemlösungen

- ✓ Zahnflankensensor wird vom Roboter geführt
- ✓ Geeignet für vertikal orientierte Teile wie Hypoid-Zahnräder und Lager
- ✓ Mehrere Prüfspuren pro Zahnflanke wählbar
- ✓ Prüfung aller Zahnflanken insbesondere zur Einrichtung von Schleifmaschinen
- ✓ Prüfung von ausgewählten Zahnflanken bei Integration in die Serienfertigung (Monitoring des Schleifvorgangs)
- ✓ Zahnfußprüfung möglich

DGZfP-Jahrestagung 2025

Anwendung zerstörungsfreier Methoden zur Untersuchung von Randzoneneigenschaften

Philipp Reeh [1], Dr. Sebastian Send [1]
[1] Stresstech GmbH, Rennerod, Deutschland, philipp.reeh@stresstech.com
[1] Stresstech GmbH, Rennerod, Deutschland, Sebastian.send@stresstech.com

Kurzfassung

Oberflächenbearbeitungen wie bspw. Schleifen, Hartdrehen, Wärmebehandlung (Einsatz- und Induktionshärten) und Kugelstrahlen haben einen erheblichen Einfluss auf die Änderungen von Randzoneneigenschaften wie etwa die Eigenspannung und die Mikrogefügestruktur. Die zuverlässige und schnelle Charakterisierung solcher Randzoneneigenschaften ist notwendig, um eine Qualitätskontrolle bei der Hartfeinbearbeitung in Fertigungsprozessen sicherzustellen. In diesem Beitrag wird insbesondere die röntgenografische Eigenspannungsmessung von kristallinen Werkstoffen basierend auf der Elastizitätstheorie und dem Braggschen Gesetz beleuchtet. Ziel solcher Messungen ist u.a. eine Ermittlung von Randzoneneigenschaften, um ZfP-Verfahren wie bspw. die Barkhausenrauschen-Analyse (BNA) zu quantifizieren. Zudem können anhand der charakteristischen Beugungslinien zusätzliche Kenngrößen ermittelt werden, die mit weiteren Eigenschaften der Randzone wie etwa Mikrohärte, Härteverlauf und Korngröße, korrelieren.

Keywords: Röntgenografische Eigenspannungsmessung, Qualitätskontrolle, Mikrohärteverlauf, Korngröße

Anwendung zerstörungsfreier Methoden zur Untersuchung von Randzoneneigenschaften

Philipp Reeh[1], Sebastian Send[1]

[1]Stresstech GmbH, Konnwiese 18, 56477 Rennerod, Germany

Kontakt: philipp.reeh@stresstech.com

stresstech

Röntgenografische Eigenspannungsanalyse

Die röntgenografische Eigenspannungsanalyse von kristallinen Werkstoffen basiert auf dem Braggschen Gesetz und der Messung richtungsabhängiger Gitterdehnungen nach der $\sin^2\psi$-Methode. Die europäische Norm DIN EN 15305 [1] definiert hier die Anforderungen an die Probe und die Prüfeinrichtung. Dazu werden verschiedene Röntgendiffraktometer mit variabler Röntgenenergie eingesetzt, die Messungen von Eigenspannungen im ω-, χ- oder modifizierten χ-Modus ermöglichen. Hochsensitive Flächendetektoren auf Halbleiterbasis (2D-Detektoren) erlauben schnelle und präzise zerstörungsfreie Eigenspannungsanalysen im Sekundenbereich [2].

Braggsches Gesetz →

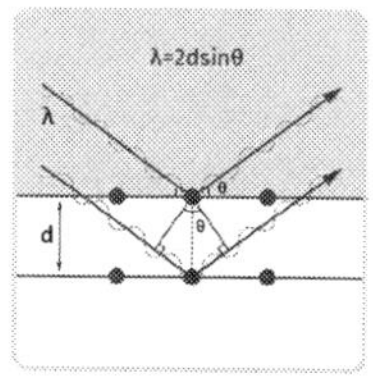

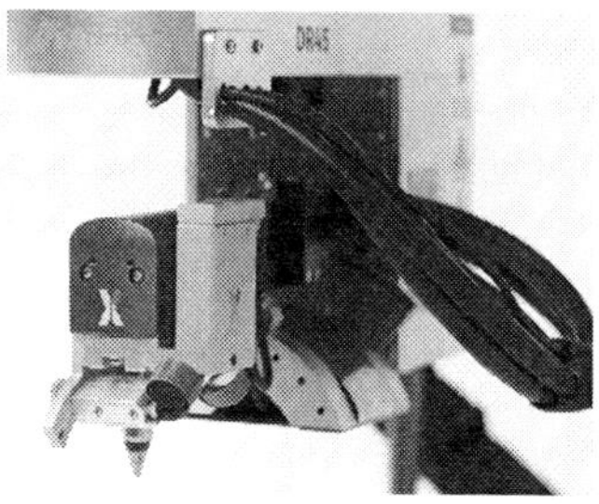

↑ Xstress DR45 Röntgendiffraktometer mit 2D-Detektoren

↓ Integration von 2D-Beugungsdaten über einen Abschnitt des Debye-Scherrer-Rings mit dem Öffnungswinkel Δγ für die röntgenografische Ermittlung der Eigenspannungen [2]

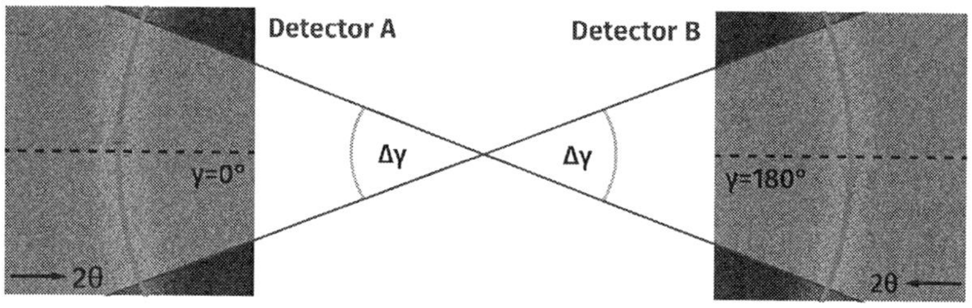

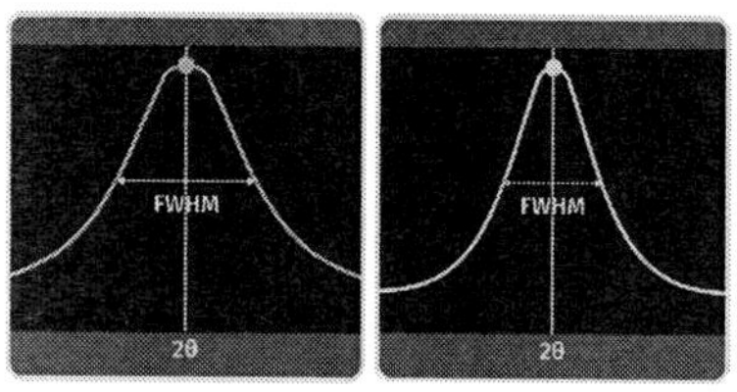

↑ Änderung der vollen Halbwertsbreite im Beugungssignal (links: hartes Material, rechts: weiches Material)

Mikrohärte und Härteverläufe

Die zur Eigenspannungsmessung ausgewerteten charakteristischen Beugungslinien können zur Korrelation mit weiteren Werkstoffeigenschaften verwendet werden. Hierzu werden die vollen Halbwertsbreiten (FWHM) der Beugungslinien bestimmt. Für stahlbasierte Materialien verhalten sich diese Halbwertsbreiten u.a. proportional zur Mikrohärte.

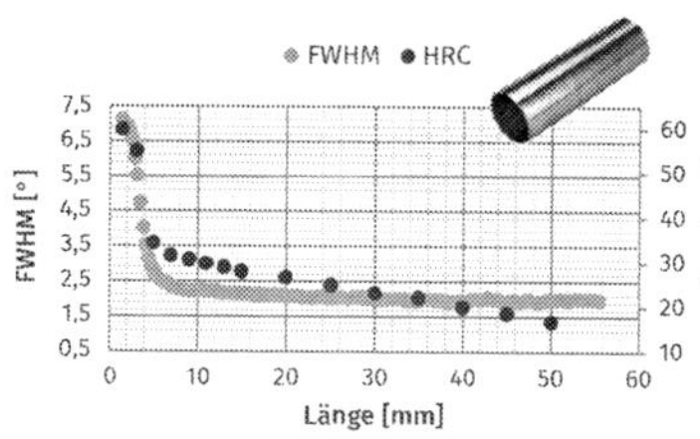

↑ Kontinuierlicher Halbwertsbreitenscan in Echtzeit über eine Zylinderrolle in axialer Richtung mittels XY-Tisch

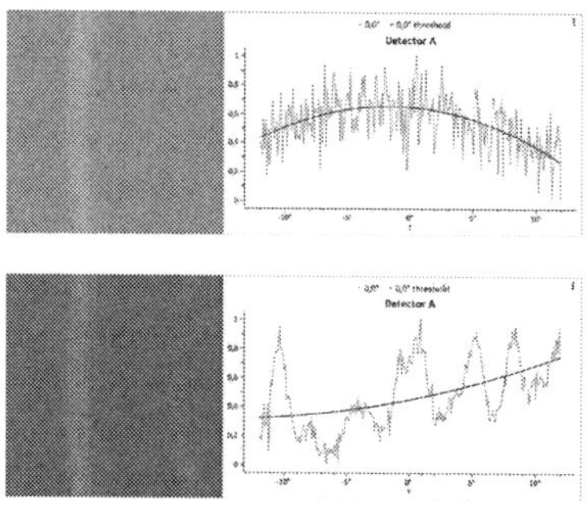

← 2D- und 1D-Intensitätsdaten (feinkörniges Material)

← 2D- und 1D-Intensitätsdaten (grobkörniges Material)

Korngröße

Im Gegensatz zur Eigenspannungsmessung beruht die röntgenografische Korngrößenbestimmung auf einer Analyse der Intensitätsverteilung entlang des Debye-Scherrer-Rings der Probe. Diese Methode wurde für WC-Pulver sowie Bauteile aus der Hartmetallindustrie erfolgreich angewendet [3].

Akkreditierung

Stresstech GmbH ist ein nach DIN EN ISO IEC 17025:2018 akkreditiertes Labor für röntgenografische Eigenspannungsmessungen und Restaustenitbestimmung.

Literatur und Quellenangaben

[1] DIN EN 15305, Zerstörungsfreie Prüfung — Röntgendiffraktometrisches Prüfverfahren zur Ermittlung der Eigenspannungen (2008)

[2] S. Send, D. Dapprich, M. Palosaari, "Fast residual stress determination by means of two-dimensional X-ray diffraction using the sin2ψ-method", Proceedings of ICRS 11 - The 11th International Conference on Residual Stresses, Nancy (2022), Article hal-04015041

[3] S. Send, R. Lehto, T. Mäkäläinen, M. Palosaari, "Non-destructive analysis of the tungsten carbide grain size by means of two-dimensional X-ray diffraction", International Journal of Refractory Metals and Hard Materials, Vol. 120 (2024), Article 106596

DGZfP-Jahrestagung 2025

Der komplexe Barkhausenrausch-Burst und die vielfältigen Korrelationen zu unterschiedlichen Materialeigenschaften – aktuelle Ansätze zu Multiparameter-Auswertungen

Rico DOBRENKO[1], Carlo Scheer[2]
[1] Stresstech GmbH, Rennerod, Deutschland, rico.dobrenko@stresstech.com
[2] Stresstech GmbH, Rennerod, Deutschland, carlo.scheer@stresstech.com

Kurzfassung

Neben der konventionellen Analyse des RMS-Wertes des Barkhausenrausch-Bursts werden Charakteristiken wie z.B. Peakposition, Peakhöhe, Halbwertsbreite oder Pulshöhenverteilung betrachtet. Anhand der verschiedenen Eigenschaften des Bursts können Korrelationen erstellt werden, um die Beeinflussung von Materialeigenschaften durch einen Prozess zu beurteilen. Somit ist es möglich, eine qualitative Aussage über den Zustand eines Werkstücks nach einem Produktionsschritt zu geben und den Prozess selbst zu überwachen.

Keywords: Barkhausenrauschen, Randzoneneigenbschaften, Prozessüberwachung, Qualitätskontrolle

Der komplexe Barkhausenrausch-Burst und die vielfältigen Korrelationen zu unterschiedlichen Materialeigenschaften – aktuelle Ansätze zu Multiparameter-Auswertungen

Rico Dobrenko[1], Carlo Scheer[2]

Konventionelle Auswertung des Barkhausenrauschens (BHR)

Das Barkhausenrauschen ist in der qualitativen Materialcharakterisierung ein fest etabliertes Verfahren zur Detektion von Gefüge- und Eigenspannungsunterschieden, wie sie zum Beispiel bei der Wärmebehandlung oder Hartfeinbearbeitung auftreten können. Dazu wird konventionell der quadratische Mittelwert (root mean square - RMS) des Barkhausenrausch-Bursts betrachtet. Unter gleichbleibenden Bedingungen wie Material und Prüfparameter beeinflussen im Wesentlichen die Eigenspannungen und die Härte die Höhe der BHR-Amplitude. Dabei gilt:

- ✓ Eine geringere Härte hat eine höhere Amplitude zur Folge
- ✓ Höhere Eigenspannungen haben eine höhere Amplitude zur Folge

Eine Trennung der beiden Charakteristiken ist nicht möglich, außerdem zeigt der RMS ein asymptotisches Verhalten bei sehr großen bzw. kleinen Werten der Eigenspannung und Härte. Um die Auswertemöglichkeiten zu erweitern, werden neben dem RMS weitere Charakteristiken des BHR-Bursts betrachtet.

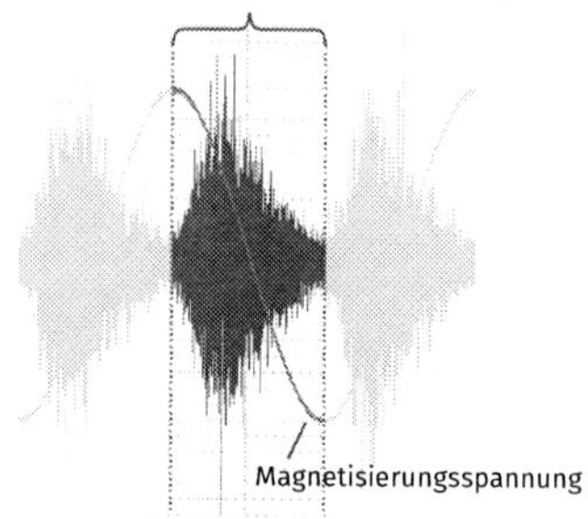

Rohsignal des Barkhausenrauschens

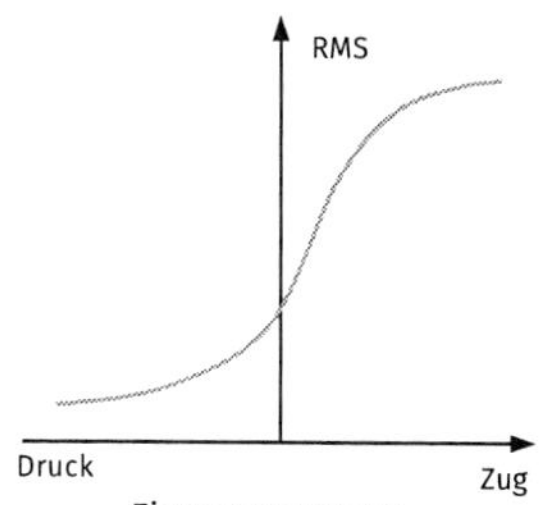

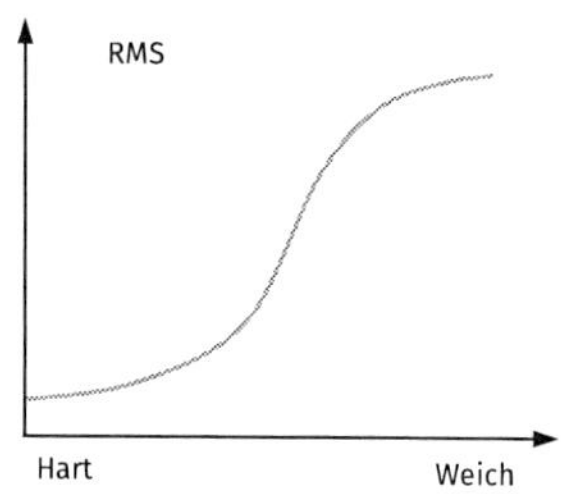

Schematische Darstellung der Zusammenhänge zwischen Eigenspannungen bzw. Härte und BHR-Amplitude:

- ✓ Höhere Eigenspannung → Höhere BHR-Amplitude
- ✓ Geringere Härte → Höhere BHR-Amplitude

Betrachtung weiterer Charakteristiken des Barkhausenrauschens (BHR)

Zur Charakterisierung des Werkstoffs können weitere Eigenschaften des Barkhausenrausch-Bursts herangezogen werden. Als besonders vielversprechend hat sich die Betrachtung der Peakposition, der Koerzitivität, des Integrals sowie des Spektrums erwiesen. In [1] wird gezeigt, dass bei zunehmender Schädigung des Werkstücks durch einen Schleifprozess der RMS nicht weiter ansteigt, andere Parameter wie Peakposition und die Koerzitivität jedoch weiter zunehmen. In [2] wird gezeigt, dass unter Verwendung verschiedener Eigenschaften des BHR-Bursts die Einhärtetiefe abgeschätzt werden kann. Die Sensitivität der Detektion von unterschiedlicher Materialeigenschaften lässt sich durch Kombination mehrerer Eigenschaften zusätzlich erhöhen.

Beispiele für Anwendungen sind:

- ✓ Wärmebehandlung
- ✓ Hartfeinbearbeitung [1]
- ✓ Einhärtetiefe [2]

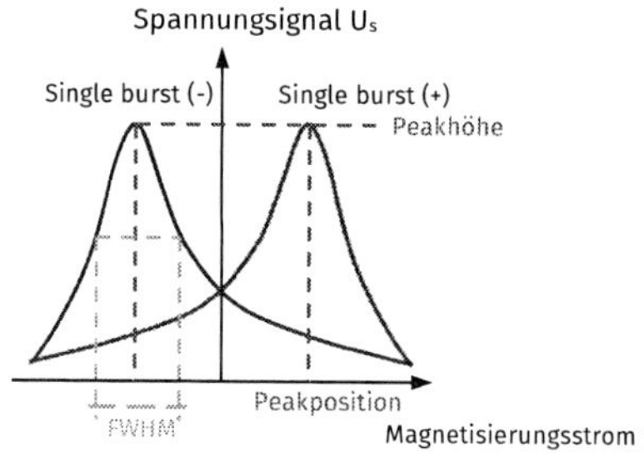

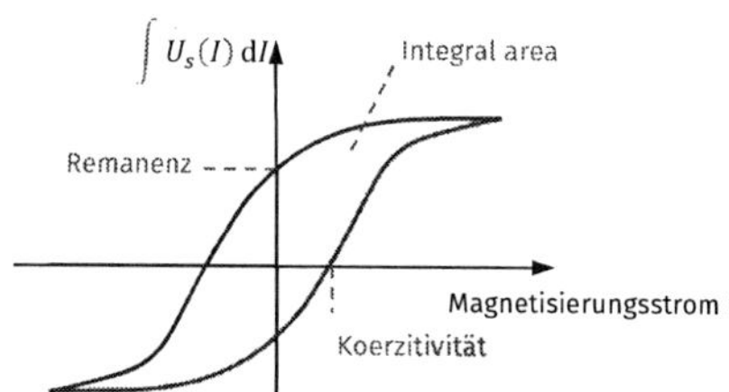

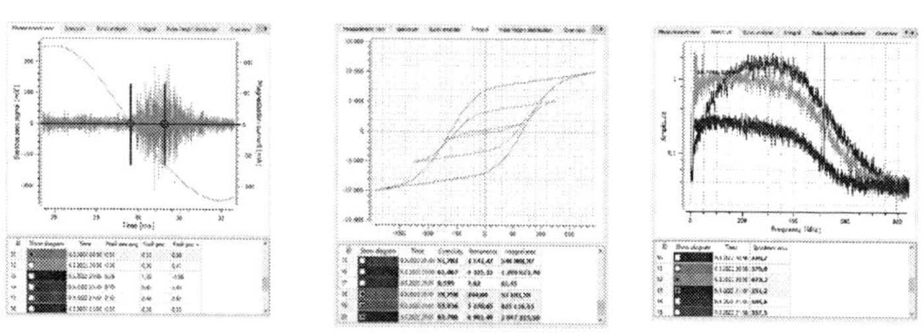

Berechnung von a) Peakposition, b) Koerzitivität, Remanenz und Integralfläche anhand der Hysteresekurve, c) Spektralanalyse

Literatur und Quellenangaben

[1] FVA-Nr. 723 II Sichere Schädigungsdetektion, Abschlussbericht erstellt von der Arbeitsgruppe FVA 723 II „Sichere Schädigungsdetektion"

[2] Send, S., Dapprich, D., Thomas, J., Suominen, L., „Non-destructive Case Depth Determination by Meansof Low-Frequency Barkhausen Noise Measurements", Journal of Nondestructive Evaluation (2018) 37:82

DGZfP-Jahrestagung 2025

Ersatzfehler: Schlüsselkomponenten der zerstörungsfreien Prüfung

Konstantin HÄRTEL[1], Kristin BEYREUTHER[1], Julia NOACK[1], Pascal KRAUS[1], Antje ZÖSCH[1],
Peter LICKSCHAT[2], Steffen WEISSMANTEL[2],
[1]imq-Ingenieurbetrieb für Materialprüfung, Qualitätssicherung und Schweißtechnik GmbH, Crimmitschau, Germany
[2]Laserinstitut Hochschule Mittweida, Mittweida, Germany

Kurzfassung

Für die Wirbelstromprüfung von Bauteilen und Komponenten werden Ersatzfehler zum Einstellen der Prüfsysteme benötigt. Bei der Rissprüfung haben sich Nuten etabliert, die mittels Erodierens hergestellt werden. Wie jedes Fertigungsverfahren hat auch das Senkerodieren Vor- und Nachteile, die zu Form- und Maßabweichungen führen können.

Um eine höhere Präzision bei der Herstellung der Ersatzfehlergeometrien zu erreichen, wurde das Laserablationsverfahren qualifiziert und weiterentwickelt, so dass nicht nur V-förmige Nuten, sondern auch U-förmige Nuten ohne thermische Gefügebeeinflussung oder Schmelzanhaftungen an den Geometriekanten erzeugbar sind. Mithilfe dieser lasergenerierter Ersatzfehler konnte der Einfluss von toleranzbedingter Maßabweichungen auf das Wirbelstromsignal systematisch untersucht werden.

Die Ergebnisse der Untersuchung zeigen, dass eine Phasendrehung des Wirbelstromsignals insbesondere im Bereich von schmalen und flachen Nuten auftritt und dass Fertigungstoleranzen einen signifikanten Effekt auf die Phasendrehung haben. Bei größeren Fehlergeometrien ist eine Phasenänderung vernachlässigbar.

Ebenso wir die Signalamplitude von den Fertigungstoleranzen beeinflusst. Im Bereich mit einer Nutbreite von kleiner als 300 µm sind Signaländerungen bei Fertigungstoleranzen vernachlässigbar. Bei Nutenbreiten größer 300 µm hingegen, haben Fertigungsungenauigkeiten starken Einfluss auf die Wirbelstromamplitude.

Bei der Betrachtung der Nuttiefe verhält es sich umgekehrt. Abweichungen vom Sollmaß im Bereich bis ca. 250 µm Tiefe haben Abweichungen vom Sollmaß einen großen Einfluss auf das Wirbelstromsignal. Ab einer Tiefe über 250 µm ist die Signalveränderung erheblich geringer und kann nahezu vernachlässigt werden.

Es ist anzumerken, dass die ermittelten Ergebnisse im hohen Maß von der eingesetzten Prüftechnik abhängig sind. Bei der Nutzung anderer Prüfsonden, -einstellungen und -geräten müssen die aufgezeigten Zusammenhänge überprüft und gegebenfalls angepasst werden.

Keywords: ZfP, Vergleichskörper, EDM, Wirbelstromprüfung, Signalentstehung

Ersatzfehler: Schlüsselkomponenten der zerstörungsfreien Prüfung

Konstantin HÄRTEL[1], Kristin BEYREUTHER[1], Julia NOACK[1], Pascal KRAUS[1], Antje ZÖSCH[1],
Peter LICKSCHAT[2], Steffen WEISSMANTEL[2],
[1]imq-Ingenieurbetrieb für Materialprüfung, Qualitätssicherung und Schweißtechnik GmbH, Crimmitschau, Germany
[2]Laserinstitut Hochschule Mittweida, Mittweida, Germany

dgzfp FÜR SICHERHEIT. JEDEN TAG.

Motivation

Bei der Wirbelstromrissprüfung von Komponenten in der Wälzlager- und Getriebetechnik werden Vergleichskörper mit Ersatzfehlern zur Einrichtung des Prüfverfahrens benötigt. Dabei werden meist Nuten in kleinsten Dimensionen eingesetzt, deren Breite und Tiefen im Bereich von ca. 50 µm bis 150 µm liegt. Auch bei der Ultraschallprüfung werden solche Ersatzfehler in Form von Flachbodenbohrungen zum Einstellen des Prüfsystems genutzt.
Gerade wenn es sich beim Prüfteil um gehärtete Bauteile handelt, ist die Herstellung solcher definierten Nuten und Bohrungen durch konventionelles Fräsen nicht möglich, so dass das Funkenerosionsverfahren angewendet wird.

Die Funkenerosion hat eine Reihe von verfahrensbedingten Nachteilen, dazu zählen die Herstellung der Elektrode, die im Größenbereich der herzustellenden Geometrie sein muss, und das für den Prozess benötigte Dielektrikum, welches einen Gefahrstoff darstellt und gesondert entsorgt werden muss. Zusätzlich erschwert der Elektrodenverschleiß während der Bearbeitung die Herstellung der kleinen Dimensionen und es können sich Schmelzrückstände an den Geometriekanten ablagern.
Da Toleranzen bei jedem Fertigungsverfahren auftreten, sollen diese Einflüsse auf das Wirbelstromsignal systematisch untersucht werden.

Qualifizierung des Laserablationsverfahren

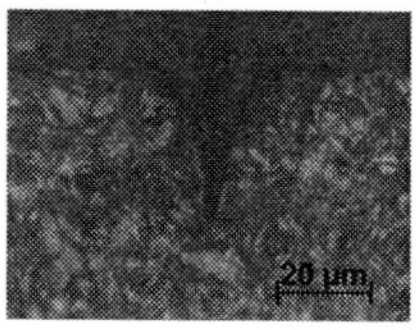

Gefügebild V-Nut geätzt, Schmelzablagerungen am Rand der Nut

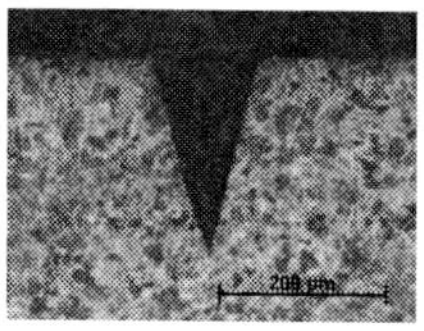

Gefügebild V-Nut geätzt

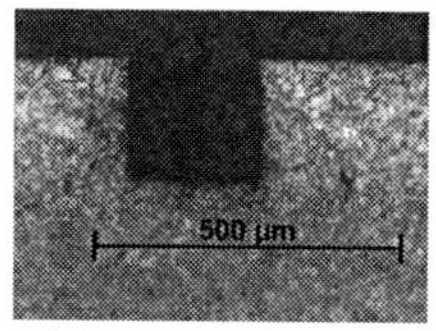

Gefügebild U-Nut geätzt

Erste Versuche:
- V-förmige Nutgeometrie mit Böschungswinkel vom ca. 70 °
- Schmelzablagerungen an den Geometriekannten
- Teilweise Gefügeveränderungen durch thermische Belastung

Weiterentwicklung:
- V-förmige Nutgeometrie mit Böschungswinkel vom ca. 70 °
- Geometrie ohne Schmelzanhaftungen
- Keine thermische Belastung der Kanten

Aktueller Stand:
- U-förmige Nutgeometrie mit Böschungswinkel vom ca. 90°
- Geometrie ohne Schmelzanhaftungen
- Keine thermische Belastung der Kanten

Einfluss der Fertigungstoleranzen auf das Prüfsignal

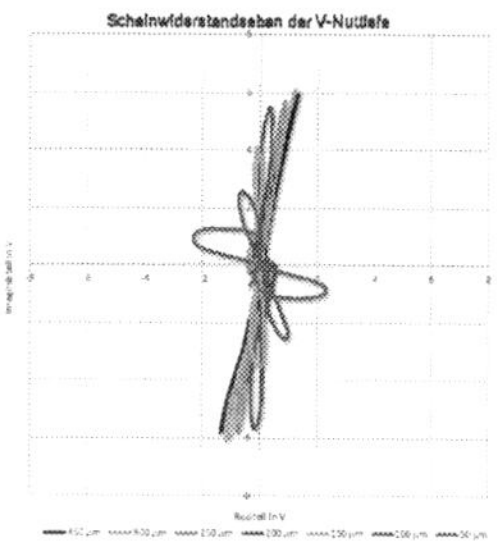

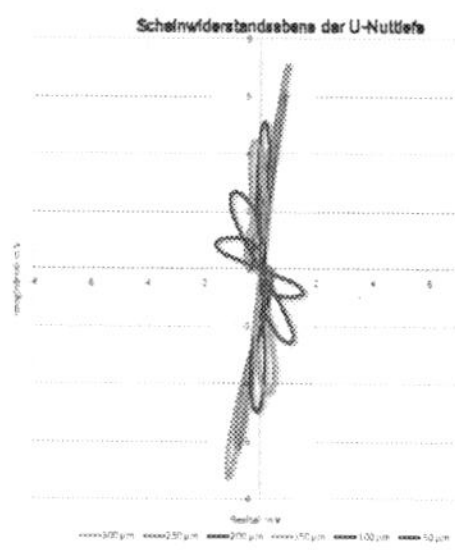

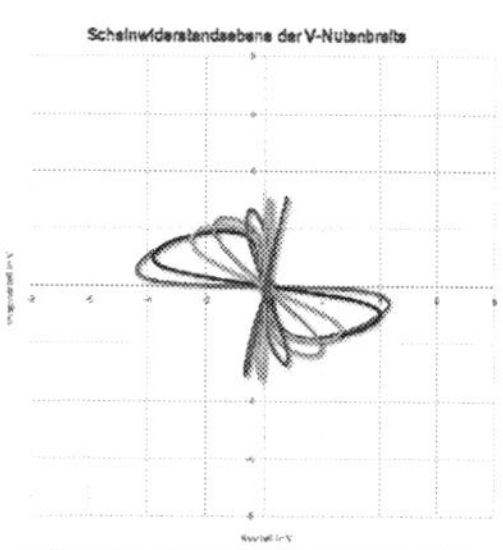

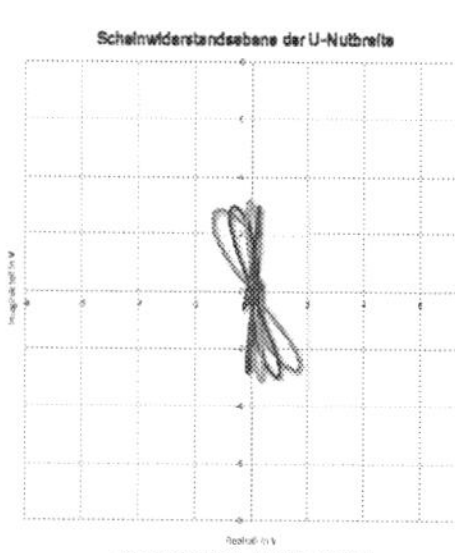

Phase:
Die Änderung der Nuttiefe und der Nutbreite wirken sich auf die Phasendrehung entgegengesetzt aus. Im Bereich von schmalen und flachen Nuten beeinflussen Geometrieabweichungen durch Fertigungstoleranzen den Phasenwinkel erheblich. Signale von sehr breiten und tiefen Nuten tolerieren Abweichungen bei der Herstellung.

Nutbreite:
Der Einfluss der Nutbreite auf die Signalamplitude ist im Bereich bis 300 µm als gering einzuschätzen. Danach steigt die maximale Elongation mit zunehmender Breite. Fertigungsabweichungen wirken sich somit erst ab einer Mindestnutbreite auf die Signalamplitude aus.

Zielgröße Breite	Toleranzgrenzen	ET Signal	Abweichung
50 ± 10 µm	40 µm	6,28 V	0,028 dB
	60 µm	6,30 V	
300 ± 10 µm	290 µm	8,79 V	0,500 dB
	310 µm	9,31 V	

Nuttiefe:
Fertigungstoleranzen bei der Herstellung der Tiefe wirken sich im Bereich bis ca. 250 µm stark auf die Signalamplitude aus. Bei sehr tiefen Nuten ist der Einfluss von Abweichungen erheblich geringer.

Zielgröße Tiefe	Toleranzgrenzen	ET Signal	Abweichung
50 ± 10 µm	40 µm	3,42 V	1,960 dB
	60 µm	4,29 V	
300 ± 10 µm	290 µm	11,85 V	0,073 dB
	310 µm	11,95 V	

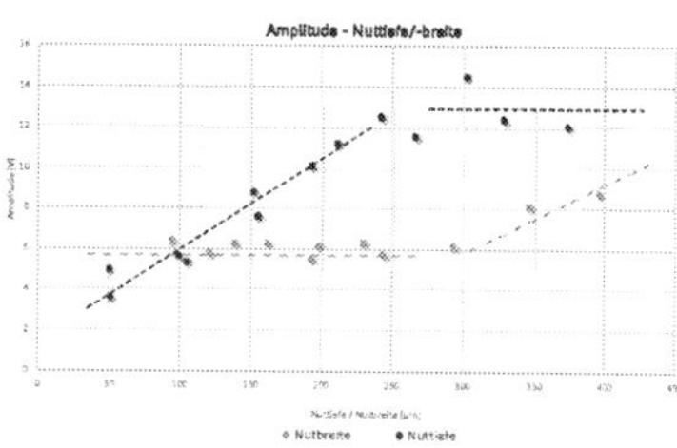

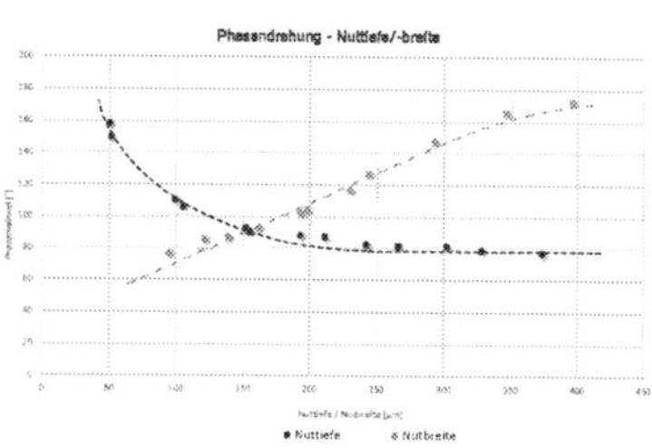

Prüfgerät: Rohmann Elotest B320
Prüffrequenz: 3 MHz
Differenzsonde: Wirkbreite 1,8 mm
Hochpass: statisch
Tiefpass: 1 kHz

In die Bestimmung der Korrelationen wurden sowohl die Signale der V-Nuten als auch die der U-Nuten einbezogen.

Danksagung

Wir bedanken uns für die finanzielle Unterstützung durch die Europäische Union und den Freistaat Sachsen.

imq Ingenieurbetrieb für Materialprüfung, Qualitätssicherung und Schweißtechnik GmbH
Laserinstitut Hochschule Mittweida
Kofinanziert von der Europäischen Union
HOCHSCHULE MITTWEIDA
Diese Maßnahme wird mitfinanziert durch Steuermittel auf der Grundlage des vom Sächsischen Landtag beschlossenen Haushaltes.

Zusammenfassung / Fazit

Die Herstellung von Ersatzfehlernuten unterliegt wie jede andere Fertigung gewissen Toleranzen. Die systematische Untersuchung hat gezeigt, dass Abweichungen in der Breite erst ab ca. 300 µm einen Einfluss auf das Fehlersignal haben. Tiefenabweichung bewirken bis 250 µm eine Veränderung der ET-Amplitude, danach ist der Einfluss vernachlässigbar. Gerade im Bereich sehr kleiner Tiefen- und Breitenausdehnungen haben Toleranzen starken Einfluss auf die Signalphase. Das sollte bei der Wahl der Sortierblende beachtet werden. Kreisblenden sind zu bevorzugen. Die ermittelten Ergebnisse hängen von der Prüftechnik ab. Bei Verwendung anderer Prüfsonden, -einstellungen oder -geräte müssen die festgestellten Zusammenhänge überprüft und gegebenenfalls angepasst werden.

www.imq-gmbh.com

DGZfP-Jahrestagung 2025

Entwicklung eines energieautarken Wirbelstromprüfsystems für die Zustandsüberwachung von Bauteilen und Bauwerken

René GANSEL[1], Stefan ZWOCH[1], Tim HARTMANN[1], Markus QUANZ[2], Armin LOHRENGEL[2], Hans Jürgen MAIER[1], Sebastian BARTON[1]

[1]Institut für Werkstoffkunde, Leibniz Universität Hannover, 30823 Garbsen, Deutschland, gansel@iw.uni-hannover.de; zwoch@iw.uni-hannover.de; t.hartmann@stud.uni-hannover.de; maier@iw.uni-hannover.de; barton@iw.uni-hannover.de

[2]Institut für Maschinenwesen, TU Clausthal, 38678 Clausthal-Zellerfeld, Deutschland, quanz@imw.tu-clausthal.de; lohrengel@imw.tu-clausthal.de

Kurzfassung. Ein zerstörungsfreies Prüfsystem, das die erfassten Daten drahtlos überträgt und mittels sog. Energy Harvesting oder Batterie betrieben werden kann, ist an Orten einsetzbar, die für eine Inspektion nicht ständig oder nur schwer zugänglich sind. Darüber hinaus kann das Prüfsystem eine kontinuierliche Überwachung gewährleisten, was die Betriebssicherheit verschiedener Bauteile und Strukturen verbessern kann.

Im Rahmen des Beitrags wird die Entwicklung eines Prototyp-Prüfsystems vorgestellt. Dieses System besteht aus einem günstigen Wirbelstromsensor, einer energieeffizienten Auswerteeinheit mit einer Ultra-Low-Power-Mikrocontrollereinheit und einer Long-Range Wide Area Network (LoRaWAN) Datenübertragungseinheit. Die drahtlose und energieeffiziente Datenübertragung über LoRaWAN ermöglicht den Zugang zum industriellen Internet of Things und damit die Zusammenführung von Daten mehrerer Sensorsysteme in einer Cloud sowie die Verarbeitung dieser Daten, um den jeweiligen Zustand ortsunabhängig zu beurteilen. Das Prüfsystem kann hinsichtlich der Funktionalität modular erweitert werden. Mittels Multiplexing können weitere Wirbelstromsensoren verwendet werden. Zudem können weitere Sensoren wie z. B. Temperatur-, Bewegungs- oder Beschleunigungssensoren ergänzt werden. Die Fähigkeit zur drahtlosen Aktualisierung erlaubt zudem eine Anpassung des Auswertealgorithmus, um z. B. zwischen konventioneller Wirbelstromprüfung und Oberwellenanalyse wechseln zu können, oder auch die Anpassung von Alarmschwellen. Somit ermöglicht das System eine Materialcharakterisierung, Fehlerprüfung und Abstandskontrolle. Anwendungsbeispiele sind das Strukturmonitoring von kritischen Stellen an Brücken oder die Überwachung von bereits vorhandenen Rissen an Schweißnähten von Windenergieanlagen sein.

Keywords: Wirbelstromprüfung, intelligente Sensoren, Internet of Things, strukturelle Zustandsüberwachung, Ermüdungszustand, Energieautarkie

1 Einführung und Motivation

Angesichts steigender Anforderungen an Sicherheit, Zuverlässigkeit und effizienten Instandhaltungsstrategien gewinnt die Bewertung der strukturellen Integrität und Betriebssicherheit von Industrieanlagen, Infrastrukturen und kritischen Komponenten zunehmend an Bedeutung. Insbesondere an schwer oder nur temporär zugänglichen Strukturen stoßen herkömmliche Inspektionsverfahren an ihre Grenzen. Das hier vorgestellte System adressiert diese

Medien- und Publikationspartner

https://doi.org/10.58286/31243

Herausforderungen durch eine innovative Verbindung von zerstörungsfreier, elektromagnetischer Prüfung und drahtloser Datenübertragung.

Ein zentrales Ziel ist die Entwicklung eines modularen Prüfsystems, das sich flexibel an verschiedene Anwendungsfälle anpassen lässt. Der Einsatz eines kostengünstigen Wirbelstromsensors in Kombination mit einer energieeffizienten Auswerteeinheit und einer Long-Range Wide Area Network (LoRaWAN) Datenübertragung schafft die Grundlage für ein robustes und integrierbares/applizierbares Monitoring-System. Zudem können durch die Nutzung von Energy Harvesting oder Batteriebetrieb auch entlegene Prüforte ohne aufwendige Energie- und Dateninfrastruktur überwacht werden. Dies ist besonders relevant in Bereichen wie dem Strukturmonitoring von Brücken [1–3] oder der Überwachung kritischer Schweißnähte an Windenergieanlagen [4, 5], wo die Anwendung herkömmlicher Prüftechniken oft aufwendig und kostenintensiv ist.

Ein weiterer wesentlicher Vorteil ist die hohe Flexibilität und Erweiterbarkeit des Systems. Neben den Signalen der Wirbelstromsensoren für die Materialcharakterisierung, Fehlerprüfung und Abstandskontrolle von kritischen Strukturen können zudem Signale von Temperatur-, Bewegungs- oder Beschleunigungssensoren erfasst und verarbeitet werden. In Abhängigkeit der Prüfparameter, der Eigenschaften und Anzahl der Sensoren und der notwendigen Leistung der Auswerteeinheit sind lediglich einfache Schaltungsanpassungen notwendig. Die Möglichkeit zur drahtlosen Aktualisierung des Auswertealgorithmus erlaubt zudem eine Anpassung an verschiedene Prüfverfahren, wie z. B. die konventionelle Wirbelstromprüfung oder die Analyse von Oberwellen. Weiterhin kann das Prüfsystem flexibel in Abhängigkeit der Verfügbarkeit kabelgebunden, mittels Batterie oder Energy Harvesting mit Energie versorgt werden. Diese Modularität eröffnet vielseitige Einsatzmöglichkeiten und unterstützt die Weiterentwicklung und Anwendung der Prüftechnologien. Einen Überblick über mögliche Wirbelstromsensoren, Auswertemöglichkeiten und die Vorgehensweise bei der drahtlosen Datenübertragung liefert Abbildung 1.

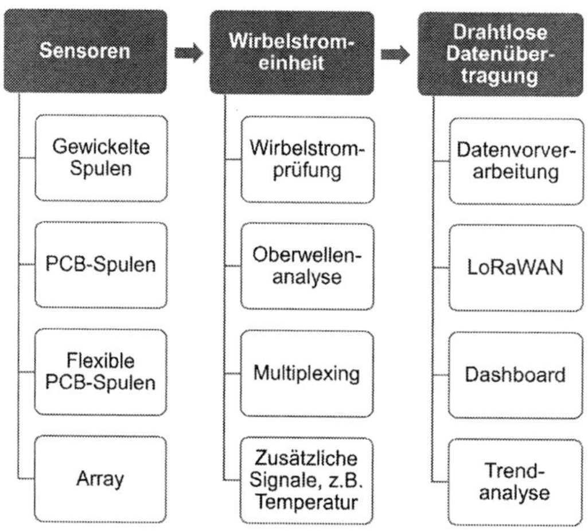

Abb. 1. Darstellung der Flexibilität und Erweiterbarkeit des Prüfsystems

Im Zentrum der Motivation steht also der Bedarf, herkömmliche Prüftechniken durch eine moderne, energieeffiziente und drahtlos vernetzte Lösung zu ergänzen. Ziel ist es, die Vorteile eines kontinuierlichen und flexiblen Monitorings zu nutzen, um durch präzisere Diagnosewerkzeuge eine prädiktive, zielgerichtete und wirtschaftliche Wartung zu ermöglichen. Der Beitrag stellt daher nicht nur eine beispielhafte technische Realisierung eines solchen Prüfsystems vor, sondern beleuchtet auch dessen Potenziale für zukünftige Anwendungen im Bereich der zerstörungsfreien Prüfung und das daraus resultierende Verbesserungspotenzial der Betriebssicherheit.

2 Monitoringkonzepte auf Basis energieautarker Wirbelstromprüfsysteme

2.1 Belastungsmonitoring am Beispiel einer Zahnwelle

Hochleistungsfähige Maschinenelemente sind zentrale Bestandteile moderner technischer Systeme und spielen in zahlreichen industriellen Anwendungen eine entscheidende Rolle. Eine wesentliche Herausforderung ist hierbei das Belastungsmonitoring, das nicht nur die mechanische Integrität der Komponenten sicherstellt, sondern auch prädiktive Wartungsstrategien ermöglicht und damit hilft, ungeplante Ausfallzeiten zu vermeiden. Die zustandsorientierte Wartung ist ein systematischer Ansatz, der die Erkenntnisse aus der Belastungsüberwachung nutzt, um geeignete Maßnahmen zu empfehlen. Dieser Ansatz gliedert sich in drei grundlegende Schritte: die periodische oder kontinuierliche Erfassung von Daten, deren Verarbeitung und die anschließenden Wartungsentscheidungen [6].

Welle-Nabe-Verbindungen gehören zu den am stärksten beanspruchten Maschinenelementen im Antriebsstrang und sind meist zentral im Kraftfluss angeordnet. In dem zugrundeliegenden Forschungsprojekt „Lastsensitive Zahnwelle mit sensorischem Werkstoff“ des DFG Schwerpunktprogramms (SPP 2305) „Sensorintegrierende Maschinenelemente als Wegbereiter flächendeckender Digitalisierung“ wird ein Belastungsmonitoring an einer Zahnwelle aus metastabilem austenitischen Edelstahl realisiert. Das Gesamtziel des Projekts besteht darin, ein autonomes Sensorsystem in das Maschinenelement Zahnwelle zu integrieren, ohne dabei die Primärfunktion des Systems zu beeinträchtigen.

Um eine zustandsorientierte Wartung zu realisieren, werden hierbei Werkstoffsensoren in Verbindung mit Wirbelstromsensoren und einer Auswerteeinheit zur Detektion mechanischer Überbelastung an einer Zahnwelle eingesetzt. Die Datenerfassung erfolgt dabei über die Wirbelstromprüfung des Werkstoffsensors und die Datenverarbeitung auf der Auswerteeinheit mittels low-power Mikrocontrollereinheit (MCU). Das Funktionsprinzip des Werkstoffsensors beruht auf einer permanenten Gefügeumwandlung von paramagnetischem Austenit zu ferromagnetischem Martensit in Abhängigkeit von der mechanischen Belastung. Die Realisierung der Belastungserfassung und Informationsspeicherung durch das Sensormaterial, das diese Funktion ohne externe Energieversorgung ausführt und das Auslesen in größeren Intervallen ermöglicht, führt zu einem besonders energieeffizienten Belastungsmonitoring. Weitere Informationen zu den Voraussetzungen eines Werkstoffsensors und dessen Umwandlungsverhalten unter mechanischer Belastung sind in [7, 8] zu finden. Die Energieversorgung der Auswerteeinheit wird durch eine integrierte DC-Maschine im Generatorbetrieb realisiert [9]. Die erfassten und vorverarbeiteten Prüfdaten werden über eine drahtlose LoRaWAN-Übertragungskette an den Endanwender übermittelt, siehe [10] für die Konzeptdarstellung. Basierend auf dem Trendverlauf der cloudbasierten Visualisierung der Daten über ein Dashboard können diese weltweit über einen Browser abgerufen und Wartungsentscheidungen getroffen werden.

Ein Demonstratoraufbau der sensorintegrierten, autonomen Zahnwelle ist in Abbildung 2 inklusive einer Detailansicht der integrierbaren Auswerteeinheit zu finden.

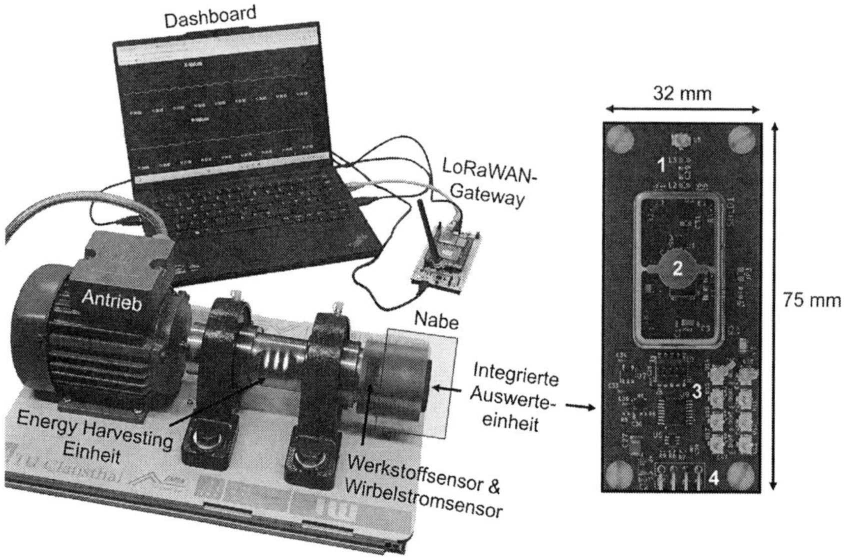

Abb. 2. Demonstratoraufbau der sensorintegrierten, autonomen Zahnwelle mit integrierter Energy Harvesting Einheit, Auswerteeinheit, inhärentem Werkstoffsensor, appliziertem Wirbelstromsensor und LoRaWAN-Übertragungskette sowie Detailansicht der Auswerteeinheit mit 1: Radiofrequenz-Signalpfad, 2: Mikrocontrollereinheit, 3: Analogschaltung und 4: Energieversorgung

Wie in Abbildung 2 dargestellt, wurde die Zahnwelle als Hohlwelle ausgeführt, um die Energy Harvesting Einheit und die Auswerteeinheit zu integrieren. Die Breite der Auswerteeinheit beträgt 32 mm und die Länge 75 mm.

Darüber hinaus ist die Position der Nabe angedeutet. Der Werkstoffsensor und der Wirbelstromsensor sind außerhalb des Verbindungsbereichs im Zahnfuß positioniert. Die Sensorkabel und bei Bedarf auch die PCB-Antenne können über das Kopfspiel der Nabe aus der Zahnwelle bzw. der Verbindung herausgeführt werden. Ein Belastungsmonitoring der Zahnwelle wird mittels Wirbelstromprüfung des Werkstoffsensors und der anschließenden Übertragung der Daten realisiert. Die drahtlose Übertragung der vorverarbeiteten Daten an das LoRaWAN Gateway, von dort an den LoRa-Server und schließlich an den Anwendungsserver und den MQTT-Broker ermöglicht eine Visualisierung der Daten beim Endbenutzer. In diesem Kontext erfolgt die Visualisierung des Real- und Imaginärteils der Wirbelstromsignale über ein Dashboard entsprechend der Darstellung auf dem Laptop in Abbildung 2. Erfolgt eine Überbelastung der Zahnwelle, so resultiert eine Martensitbildung im Werkstoffsensor und die Wirbelstromsignale weisen eine Änderung auf, die direkt im Dashboard ersichtlich ist.

Eine Gegenüberstellung der Signalqualität des entwickelten MCU-basierten Wirbelstromsystems mit einem industriellen Wirbelstromprüfsystem wurde in [10] anhand eines typischen Kalibrierkörpers für die Materialcharakterisierung durchgeführt. Trotz der deutlich kompakteren Bauweise und der geringeren Signalamplitude erzeugt das MCU-basierte Wirbelstromprüfsystem eine ähnliche Signalqualität wie das industrielle Prüfsystem.

Neben der Signalqualität ist die Energieeffizienz der Auswerteeinheit ein wesentlicher Aspekt im Kontext der Autonomie. Die Energieeffizienz ist wichtig, um im Batteriebetrieb eine möglichst lange Laufzeit des Gesamtsystems zu ermöglichen und eine vom Betriebszustand unabhängige, vollständige Abdeckung des Leistungsbedarfs mittels Energy Harvesting gewährleisten zu können. Insbesondere die Sensitivität der Wirbelstromsensoren und der hierfür notwendige Erregerstrom sowie die statische Verlustleistung der Mikroelektronik und das gewählte Prüfintervall besitzen einen wesentlichen Einfluss auf die Energieeffizienz.

Überdies bietet hinsichtlich der Modularität der Auswerteeinheit der eingesetzte Multiplexer die Möglichkeit, in kürzester Zeit Daten von bis zu vier Wirbelstromsensoren bzw. Werkstoffsensoren zu erfassen. Darüber hinaus steht an der Auswerteeinheit eine zweite analoge Schnittstelle zur Verfügung, um z. B. Temperaturdaten oder Beschleunigungsdaten zu erfassen. Multiplexer mit mehr Kanälen oder zusätzliche analoge Schnittstellen sind möglich.

Um die entsprechenden Daten versenden zu können, muss eine drahtlose Verbindung hergestellt werden. In öffentlichen Netzen ist die Übertragungszeit jedoch begrenzt. Der sogenannte Duty Cycle beschreibt eine rechtlich geregelte Begrenzung der Sendezeit von Geräten in einem bestimmten Funkband. Im verwendeten 868 MHz-Band liegt sie je nach Frequenz zwischen 0,1 und 10 %. Im Folgenden wird von der häufig geltenden Übertragungszeit von 1 % ausgegangen [11]. Ein Übertragungsvorgang dauert ca. 1,5 s. Dies bedeutet, dass innerhalb einer Stunde ca. 24 Übertragungen möglich sind. Entsprechend können in öffentlichen Netzen ca. alle 150 s Messdaten versendet werden.

Im kontinuierlichen Messbetrieb beträgt die gemessene Leistungsaufnahme der Auswerteeinheit bei Raumtemperatur 48 mW. Für die Übertragung im Low-Power-Modus im Bereich 868 bis 915 MHz bei 14 dBm und 3,3 V Spannungsversorgung ist eine Leistungsaufnahme bei Raumtemperatur von 86 mW zu erwarten [12]. Daraus ergibt sich eine Leistungsaufnahme von 134 mW bei kontinuierlicher Messung und drahtloser Datenübertragung. Durch die permanente Speicherung der Belastungsdaten in den Werkstoffsensoren ist eine Betriebsweise mit periodischer Messung möglich, die zu einer deutlich geringeren Leistungsaufnahme führt. Nach mindestens 148 s des Prüfsystems im Energiesparmodus werden entsprechend eines seriellen Ablaufs maximal 0,5 s für die Wirbelstromprüfung und 1,5 s für die Datenübertragung benötigt. Um die Leistungsaufnahme weiter zu reduzieren, wird im Energiesparmodus die Analogschaltung in den Shutdown-Modus und die MCU in den STOP 2-Modus [12] versetzt. Infolgedessen beträgt die durchschnittliche Leistungsaufnahme des gesamten Systems über einen vollständigen Messzyklus von 150 s etwa 1,34 mW, was einer Reduzierung der durchschnittlichen Leistungsaufnahme um 99 % entspricht.

Die Dauer des Systems im Energiesparmodus kann in Abhängigkeit des Belastungsfalls erhöht werden. Ist das System z. B. für 300 s im Energiesparmodus, so liegt eine durchschnittliche Leistungsaufnahme von etwa 0,75 mW vor. In der nachfolgenden Tabelle 1 wird unter Einbezug der zwei unterschiedlichen Messzyklen eine grobe Schätzung der Betriebslaufzeit für eine Lithium Thionylchlorid Batterie und für einen Superkondensator bei Raumtemperatur gegeben. Zudem sind weitere Charakteristika der zwei Energieversorgungsarten beispielhaft gegenübergestellt.

Tabelle 1. Beispielhafte Gegenüberstellung der Betriebslaufzeit für unterschiedliche Energieversorgungsarten

Merkmale	Batterie	Superkondensator
Beispiel	Saft LS 14500 AA [13]	SDBH3R0L1061320 [14]
Energieinhalt	33 696 Ws	43,75 Ws (bei 3 V → 0,5 V)
Schätzung Laufzeit bei 1,34 mW	Ca. 291 Tage	Ca. 9,1 Stunden
Schätzung Laufzeit bei 0,75 mW	Ca. 520 Tage	Ca. 16,2 Stunden
Betriebstemperaturbereich	-60 °C bis +85 °C	-40 °C bis +105 °C
Wiederaufladbar	Nein	Ja in Sekunden bis Minuten
Größe	Durchmesser = 14,4 mm, Länge = 50 mm	Durchmesser = 12,5 mm, Länge = 20 mm
Selbstentladung	Sehr gering	Hoch

Die Tabelle 1 zeigt, dass das prototypische Prüfsystem mit lediglich einer Batterie für den langfristigen Einsatz geeignet ist. Bei der Verfügbarkeit eines Energy Harvesting Systems kann z. B. ein Superkondensator eingesetzt werden. In Bezug auf die Zahnwelle und den verfügbaren Bauraum unter- oder oberhalb der Platine sind entweder eine Batterie oder zwei parallel geschaltete Kondensatoren möglich, wodurch die Laufzeit ohne Zwischenladung der Kondensatoren verdoppelt werden kann. Der Kondensator fungiert dabei als Pufferspeicher für Betriebszustände, in denen die Drehzahl zu gering ist und der Generator nicht die notwendige Energie liefert. Zu beachten ist, dass beim Entladen der Energiespeicher, insbesondere bei Kondensatoren, ein Spannungsabfall auftritt, der einen zusätzlichen DC-DC-Wandler erfordert, um z.B. das 3,3 V Spannungsnetz für die MCU stabil zu halten. Im Falle einer vollständigen Entladung des Kondensators ist das Prüfsystem in der Lage, sich bei erneuter Energieversorgung durch die Energy Harvesting Einheit automatisch neu zu starten und die drahtlose Datenübertragung mittels LoRaWAN wieder herzustellen. Darüber hinaus bietet das vorgestellte, prototypische MCU-basierte Prüfsystem auf Basis der bereits erarbeiteten Erfahrungswerte Optimierungspotenzial hinsichtlich der Leistungsaufnahme, Kompaktheit und Robustheit gegenüber mechanischen und thermischen Einwirkungen.

2.2 Rissüberwachung an strukturkritischen Schweißnähten

Eine weitere Anwendungsmöglichkeit des MCU-basierten Prüfsystems kann die Rissüberwachung an Großstrukturen sein. Um das Risswachstum in Schweißverbindungen kontinuierlich zu überwachen und frühzeitig Maßnahmen ergreifen zu können, werden verschiedene Monitoring-Systeme eingesetzt. Ein ausführlicher Überblick über den Einsatz von zerstörungsfreien Prüfverfahren zur Zustandsüberwachung von Brücken ist in [1–3] und von Windenergieanlagen in [4, 5] gegeben.

Zu den vorhandenen Technologien in Bezug auf eine kontinuierliche Zustandsüberwachung gehören akustische Schallemissions-Überwachungssysteme [2, 15], die während des Risswachstums freigesetzte Schallwellen erfassen, sowie Risswachstumssensoren, die auf Dehnungsmessstreifen [2, 16] oder anderen elektronischen Sensoren basieren. Ergänzende Maßnahmen umfassen regelmäßige manuelle Überprüfungen und visuelle Inspektionen, die zusammen mit den technologischen Ansätzen eine umfassende Überwachung gewährleisten sollen. Ein auf der Wirbelstromprüfung basierendes Monitoring-Konzept ist in der Anwendung bisher nicht bekannt [2, 5].

Im Labormaßstab wurde in [17] das Potenzial einer MCU-basierten Auswerteeinheit in Kombination mit flexiblen PCB-Spulen für eine Rissdetektion untersucht. Die Aspekte der Energieeffizienz und drahtlosen Datenübertragung wurden hierbei nicht betrachtet. Während der zyklischen Ermüdung einer Kreuzprobe aus Baustahl wurde evaluiert, ob mit dem Prüfsystem eine Überwachung der Risslänge möglich ist. Über die Änderung der Spulenimpedanz konnte eine Risslänge berechnet werden [17]. Flexible PCB-Sensoren bieten zudem das Potenzial der dauerhaften Applikation, auch auf geometrisch anspruchsvollen Bauteilen [17, 18].

Das entwickelte MCU-basierte Wirbelstromprüfsystem zeichnet sich durch seine Modularität aus und ermöglicht die Anpassung an die jeweiligen Anforderungen der Prüfaufgabe mit geringem Aufwand. So kann es beispielsweise notwendig sein, den Analog-Digital-Converter, den Digital-Analog-Converter und den Tiefpassfilter anzupassen, wenn eine höhere Prüffrequenz erforderlich ist. Bei signifikant unterschiedlichen Sensoreigenschaften kann es erforderlich sein, die Operationsverstärker anzupassen, um entsprechende Ausgangsströme zu realisieren. Darüber hinaus besteht die Möglichkeit, die Anzahl an Wirbelstromsensoren über den eingesetzten Multiplexer und die Anzahl an zusätzlichen Sensoren, wie beispielsweise Temperatursensoren, über die Anzahl an analogen Kanälen anzupassen.

Soll nun mit diesem System ein sog. Hot-Spot an einer Schweißverbindung oder ein bereits vorliegender Riss überwacht werden, kann das System positionsnah an die Struktur appliziert werden. Wächst ein Riss in den Bereich des Sensorwirkbereichs, so ändern sich die Wirbelstromsignale und eine Änderung wird im Dashboard ersichtlich. Eine Energieversorgung ist über Batterien oder über das Energy Harvesting z. B. von Solarenergie denkbar.

Um das Potenzial von PCB-Sensoren für die Rissüberwachung zu evaluieren, wurde ein Ø 8 mm PCB-Sensor mit jeweils 5 Wicklungen auf der Erreger- und Messspule (siehe Abbildung 3 und 4) für die Wirbelstromprüfung von zwei unterschiedlichen nutbehafteten Kalibrierkörpern verwendet. Der Sensor wurde an eine Linearachse befestigt und jeweils mittig auf die Nuten zu bewegt. Die Verfahrrichtung liegt parallel zur Nut, um das Risswachstum mittig in Richtung der Spule zu simulieren. Hinsichtlich der Prüffrequenz wurde eine typische Frequenz für die Schweißnahtprüfung von 200 kHz (siehe z. B. [19]) und eine Frequenz von 1 MHz aus Gründen der Energieeffizienz verwendet. Für die Machbarkeitsstudie wurde ein industrielles Wirbelstromprüfsystem verwendet. Der Lift-Off-Effekt wurde mittels Phasenrotation auf die positive Imaginärteil-Achse gelegt. Der Einfluss der Nut äußert sich somit primär im Realteil, weshalb dieser im Folgenden betrachtet wird.

In Abbildung 3 ist die Änderung des Realteils für die Prüffrequenzen 200 kHz und 1 MHz in Abhängigkeit des Verfahrweges dargestellt. Der Verfahrweg wird der Mitte des Ø 8 mm PCB-Sensors zugeordnet. Die zwei schwarzen Kreuze markieren die Positionen bei dem sich die Spulen des PCB-Sensors geometrisch etwa an der Kante der Nut befinden. Die maximale Nuttiefe von ca. 2 mm befindet sich in der Mitte. Die Länge der Nut wird durch den in der Abbildung 3 rot gekennzeichneten Bereich dargestellt. Anhand der Signaländerung kann festgestellt werden, dass die Nuten detektiert werden konnten und die höchste Änderung auftritt, wenn die Nutlänge vollständig in der Sensorwirkbreite liegt. Erste Signaländerungen in Richtung des positiven Realteils treten auf, wenn der PCB-Sensor in die Richtung der Nut bewegt wird. Signaländerungen in Richtung des negativen Realteils treten überdies auf, sobald ein Teil der Nut unterhalb des PCB-Sensors liegt. Welche Signaländerungen resultieren, sofern der Riss nicht mittig durch den Sensor wächst, muss in weiterführenden Untersuchungen evaluiert werden. Darüber hinaus ist in Abbildung 3 zu erkennen, dass das Signal-Rausch-Verhältnis für beide Prüffrequenzen ähnlich ist.

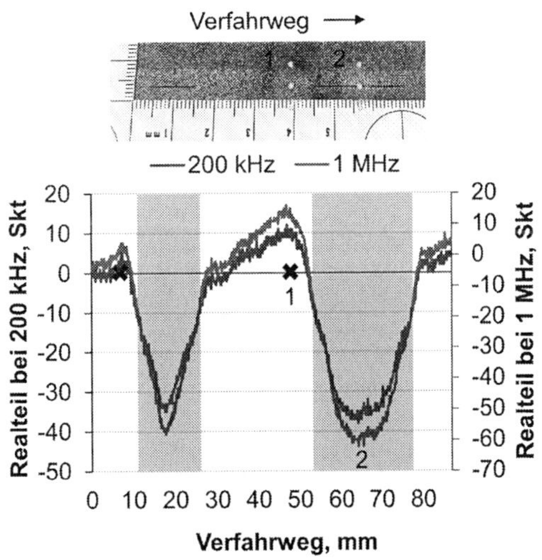

Abb. 3. Vergleich der Wirbelstromsignaländerung bei 200 kHz und 1 MHz bei der parallelen Überfahrt der zwei Nuten zur Simulation eines Risswachstums mittig durch den applizierten Wirbelstromsensor

Neben der Überwachung der Risslänge sollte anhand einer Unterpulverschweißnaht mit unterschiedlichen Nuttiefen, die entlang der Wärmeeinflusszone positioniert sind, evaluiert werden, ob zudem die Risstiefe überwacht werden kann. Hierfür wurde das identische Prüfsystem mit einer Frequenz von 1 MHz verwendet. Der PCB-Sensor wurde ebenfalls mittels Linearachse entlang der Wärmeeinflusszone bzw. parallel zu den Nuten verfahren, wie im oberen Bild der Abbildung 4 verdeutlicht. Die Nuten wurden erosiv mittels Wasserstrahlschneiden erzeugt und haben eine Länge von 15 mm sowie einen Abstand von 10 mm zueinander. Weitere Informationen zur Nuterzeugung mittels Wasserstrahlschneiden sowie zu den Wirbelstromprüfergebnissen dieser sind in [20] zu finden.

Anhand der Signaländerung während der Überfahrt des PCB-Sensors über die Nuten in Abbildung 4 ist zu erkennen, dass eine erste Änderung bei einer Nuttiefe von 0,2 mm erfolgt. Mit steigender Nuttiefe ist eine steigende Änderung des Realteils zu beobachten.

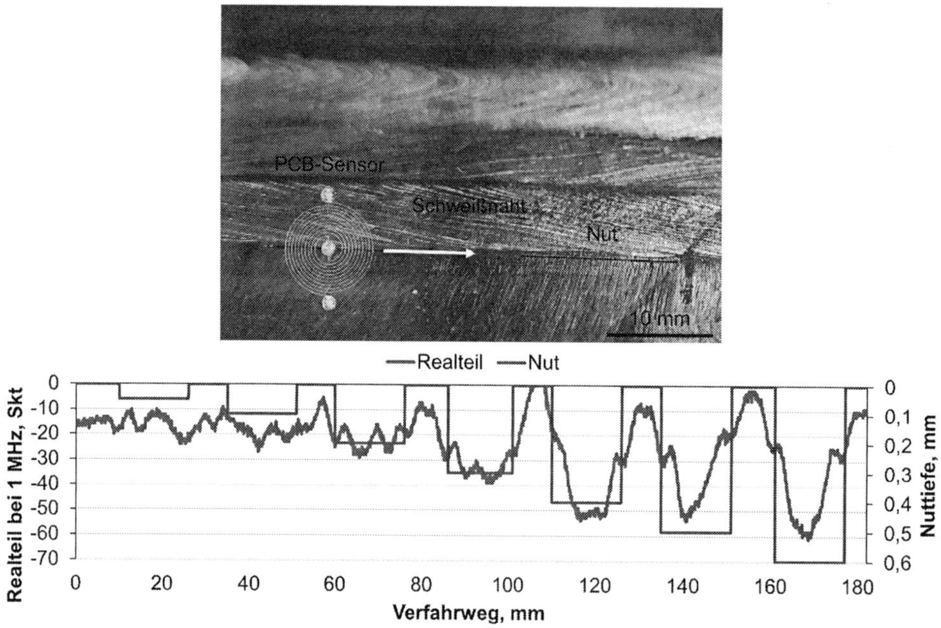

Abb. 4. Darstellung des PCB-Sensors und der Verfahrrichtung über die Nuten sowie Änderung des Realteils in Abhängigkeit der Nuttiefe

Zusammenfassend kann festgestellt werden, dass bei der Befundung eines Risses die Rissüberwachung hinsichtlich der Länge mithilfe eines PCB-Sensors als realistisch einzustufen ist. Ist die Risslänge größer als die Sensorwirkbreite so erfolgt keine Änderung in den Wirbelstromsignalen. Dies impliziert, dass ein Sensorarray von mehreren nebeneinander angeordneten PCB-Sensoren notwendig ist. Nachfolgend kann hingegen die Risstiefe weiterhin überwacht werden. Inwiefern diese zwei Parameter anhand der Wirbelstromsignale differenziert werden können, muss anhand von weiterführenden Untersuchungen erarbeitet werden.

3 Zusammenfassung

Im Rahmen des Beitrags wurde die Entwicklung eines integrierbaren Prototyp-Wirbelstromprüfsystems anhand des Anwendungsbeispiels der Zahnwelle vorgestellt. Dieses System besteht aus einem kostengünstigen Wirbelstromsensor, einer energieeffizienten Auswerteeinheit mit einer Ultra-Low-Power-Mikrocontrollereinheit und einer Long-Range Wide Area Network (LoRaWAN) Datenübertragungseinheit. Zudem kann die Energieversorgung über eine Batterie oder eine Energy Harvesting Einheit realisiert werden.

Darüber hinaus wurde die Flexibilität und die Modularität des Prototyp-Prüfsystems aufgezeigt. Die mehrfache Verwendung von leicht zu applizierenden PCB-Spulen und energieautarken Wirbelstromprüfsystemen, welche die Daten drahtlos in einem LoRaWAN-Netzwerk übertragen, kann somit ein ortsunabhängiges Monitoring anhand eines übersichtlichen Dashboards ermöglichen. Hieraus ergeben sich neben dem Bauteilmonitoring Möglichkeiten zur Überwachung von Hot-Spots oder bereits vorliegenden Rissen an Großstrukturen wie Brücken oder Windenergieanlagen.

Danksagung

Diese Studie wurde gefördert durch die Deutsche Forschungsgemeinschaft (DFG) - Projektnummer 466760574 - Projekttitel „Lastsensitive Zahnwelle mit sensorischem Werkstoff". Das Projekt ist Teil des Schwerpunktprogramms 2305 mit der Projektnummer 441853410. Die Autoren danken der DFG für die finanzielle Unterstützung.

Referenzen

[1] A. M. Abdallah; R. A. Atadero; M. E. Ozbek: A State-of-the-Art Review of Bridge Inspection Planning: Current Situation and Future Needs, Journal of Bridge Engineering 27, 2022, 2, S. 1–20, DOI:10.1061/(ASCE)BE.1943-5592.0001812.

[2] S. Taheri: A review on five key sensors for monitoring of concrete structures, Construction and Building Materials 204, 2019, S. 492–509, DOI:10.1016/j.conbuildmat.2019.01.172.

[3] S. Kashif Ur Rehman; Z. Ibrahim; S. A. Memon; M. Jameel: Nondestructive test methods for concrete bridges: A review, Construction and Building Materials 107, 2016, S. 58–86, DOI:10.1016/j.conbuildmat.2015.12.011.

[4] Y. Yang; F. Liang; Q. Zhu; H. Zhang: An Overview on Structural Health Monitoring and Fault Diagnosis of Offshore Wind Turbine Support Structures, Journal of Marine Science and Engineering 12, 2024, 3, S. 1–17, DOI:10.3390/jmse12030377.

[5] M. Civera; C. Surace: Non-Destructive Techniques for the Condition and Structural Health Monitoring of Wind Turbines: A Literature Review of the Last 20 Years, Sensors, 2022, 22 (4), S. 1–52, DOI:10.3390/s22041627.

[6] A. K. Jardine; D. Lin; D. Banjevic: A review on machinery diagnostics and prognostics implementing condition-based maintenance, Mechanical Systems and Signal Processing 20, 2006, 7, S. 1483–1510, DOI:10.1016/j.ymssp.2005.09.012.

[7] R. Gansel; M. Quanz; A. Lohrengel; H. J. Maier; S. Barton: Qualification of Austenitic Stainless Steels for the Development of Load-Sensitive Material Sensors, Journal of Materials Engineering and Performance 33, 2024, 17, S. 9004–9016, DOI:10.1007/s11665-024-09287-9.

[8] R. Gansel; C. Heinrich; A. Lohrengel; H. J. Maier; S. Barton: Development of Material Sensors Made of Metastable Austenitic Stainless Steel for Load Monitoring, Journal of Materials Engineering and Performance 33, 2024, 23, S. 13570–13582, DOI:10.1007/s11665-024-09910-9.

[9] M. Quanz; R. Gansel; S. Barton; A. Lohrengel: Energy Harvesting in rotierenden Maschinenelementen, Mitteilungen aus dem Institut für Maschinenwesen der Technischen Universität Clausthal 48, 2024, S. 89–93, DOI:10.21268/20241023-10.

[10] R. Gansel; S. Zwoch; M. Quanz; G. Schäfer; A. Lohrengel; H. J. Maier; S. Barton: Development of an energy-autonomous eddy current sensor system for in-situ component monitoring, e-Journal of Nondestructive Testing 29, 2024, 6, S. 1–8, DOI:10.58286/29942.

[11] J. Haxhibeqiri; E. de Poorter; I. Moerman; J. Hoebeke: A Survey of LoRaWAN for IoT: From Technology to Application, Sensors 18, 2018, 11, S. 1–38, DOI:10.3390/s18113995.

[12] STMicroelectronics: Datenblatt STM32WL55xx, STM32WL54xx, DS13293 Rev 5, 2022.

[13] Saft: Datenblatt Lithium AA Batterie LS14500, Document No. 31064-2-0923, 2023.

[14] YMIN: Datenblatt Superkondensator SDBH3R0L1061320, V1.0.0.202205, 2024.

[15] B. Weihnacht; B. Frankenstein; T. Gaul; R. Neubeck; L. Schubert: Monitoring of welded seams on the foundations of offshore wind turbines, Insight - Non-Destructive Testing and Condition Monitoring 59, 2017, 2, S. 72–76, DOI:10.1784/insi.2017.59.2.72.

[16] M. He; X. Bai; R. Ma; D. Huang: Structural monitoring of an onshore wind turbine foundation using strain sensors, Structure and Infrastructure Engineering 15, 2019, 3, S. 314–333, DOI:10.1080/15732479.2018.1546325.

[17] E. Gasparin; G. Santi; A. Nussbaumer: Eddy current crack monitoring system for structural health monitoring (SHM) applications, In: Proceedings of the 68th International Institute for Welding (IIW) Annual Assembly and International Conference, Helsinki, Finland, 2018, S. 1084–1090.

[18] E. Brun; P.-J. Cottinet; A. Pelletier; B. Ducharne: Printed Eddy Current Testing Sensors: Toward Structural Health Monitoring Applications, Sensors 23, 2023, 19, S. 1–15, DOI:10.3390/s23198345.

[19] R. Gansel; H. J. Maier; S. Barton: Detection and Characterization of Fatigue Cracks in Butt Welds of Offshore Structures Using the Eddy Current Method, Journal of Nondestructive Evaluation, Diagnostics and Prognostics of Engineering Systems 6, 2023, 2, S. 1–8, DOI:10.1115/1.4056313.

[20] R. Gansel; M. Mlinaric; T. Hassel; H. J. Maier; S. Barton: Erzeugung von künstlichen Rissen mittels der Wasserstrahltechnologie, e-Journal of Nondestructive Testing 28, 2023, 10, S. 1–11.

DGZfP-Jahrestagung 2025

Untersuchungen an prähistorischen Randleistenbeilen

Oliver MICHAEL[1], Sebastian DIECK[1], Markus, WILKE[2], Jan-Heinrich BUNNEFELD[3], Christian-Heinrich WUNDERLICH[3], Thorsten HALLE[2], Harald MELLER[3]
[1] DeltaSigma Analytics GmbH, Magdeburg
[2] Otto-von-Guericke-Universität, Magdeburg
[3] Landesamt für Denkmalpflege und Archäologie Sachsen-Anhalt, Halle (Saale)

Kontakt-E-Mail: oliver.michael@deltasigma.de

Kurzfassung

Die Herstellung von Werkzeugen aus Metallen ist ein bedeutender Schritt in der Menschheitsgeschichte. Die Untersuchung derartiger Objekte liefert tiefgreifende Einblicke in das Wissen um die Materialverarbeitung und die Fertigkeiten in der jeweiligen Epoche und Region. Bronzezeitliche Randleistenbeile sind in dieser Hinsicht aufschlussreiche Fundstücke, da ihre Gebrauchsfähigkeit bestimmter Eigenschaften – hier die Kombination aus Härte und Duktilität – bedarf. Diese Gebrauchseigenschaften resultieren aus einer bestimmten Elementzusammensetzung und einer darauf angepassten Verarbeitungsmethodik. Die chemische Zusammensetzung wie auch die Herkunft der verwendeten Legierungselemente ist bei bronzezeitlichen Beilen zumeist gut erforscht. Insbesondere bei älteren Fundstücken zeigt sich, dass die Zusammensetzung der Metalle zumeist nachteilig ist, um die erwünschten Gebrauchseigenschaften zu erzielen. Hier muss also eine adäquate Nachbearbeitung der Gussstücke erfolgt sein.

Neue Erkenntnisse zur Zusammensetzung, der Mikrostruktur und den mechanischen Eigenschaften sollen helfen die Art und Weise der Herstellung zu klären. Im Rahmen der vorliegenden Untersuchungen wurden Beile unterschiedlichen Alters zunächst mithilfe modernster Methoden werkstofftechnisch analysiert. Im zweiten Schritt wurde eine zerstörungsfreie Prüfroutine entwickelt und die Ergebnisse mit der werkstofftechnischen Analyse korreliert. Dadurch können die wesentlichen Charakteristika der Objekte ermittelt werden, ohne dass eine Beprobung notwendig ist. Damit soll ein größerer Probenumfang analysiert werden, um die Erkenntnisse statistisch abzusichern und um weitere Fundkomplexe zu erweitern.

Im Rahmen der Studie konnte ermittelt werden, wie die Beile vom Gießen bis zum fertigen Produkt verarbeitet wurden. Aus den Erkenntnissen der Studie leitet sich ab, dass die Hersteller dieser Beile über ein fundiertes und tiefgreifendes, über lange Zeit erworbenes Verständnis der Materialverarbeitung verfügt haben müssen.

Keywords: Bronzezeit, Randleistenbeile, Schmieden, Metallographie, Wirbelstrommessung

Medien- und Publikationspartner

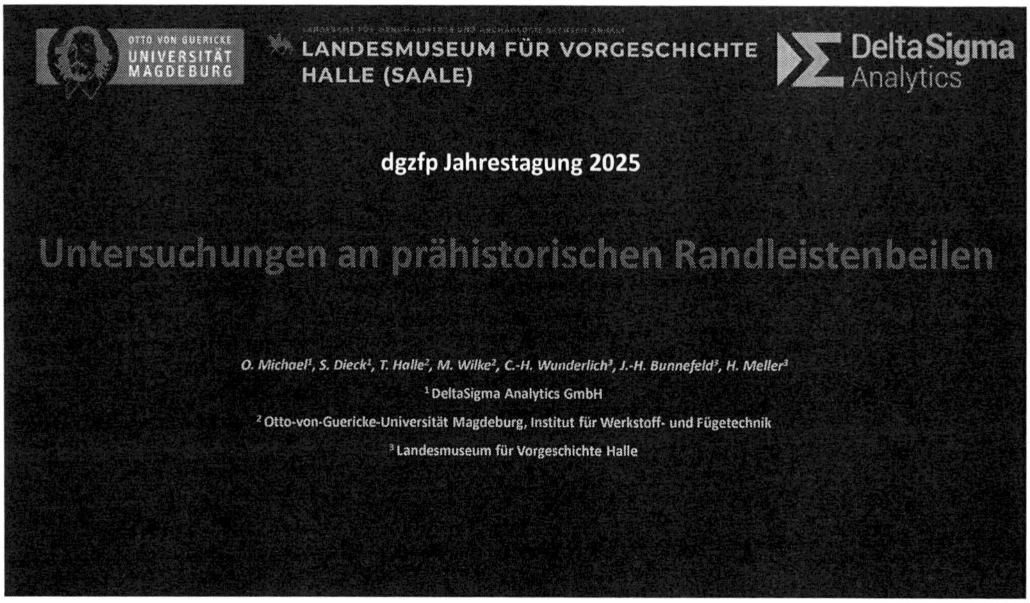
OTTO VON GUERICKE
UNIVERSITÄT
MAGDEBURG
LANDESMUSEUM FÜR VORGESCHICHTE
HALLE (SAALE)
DeltaSigma
Analytics
dgzfp Jahrestagung 2025
Untersuchungen an prähistorischen Randleistenbeilen
O. Michael[1], S. Dieck[1], T. Halle[2], M. Wilke[2], C.-H. Wunderlich[3], J.-H. Bunnefeld[3], H. Meller[3]
[1] DeltaSigma Analytics GmbH
[2] Otto-von-Guericke-Universität Magdeburg, Institut für Werkstoff- und Fügetechnik
[3] Landesmuseum für Vorgeschichte Halle

1

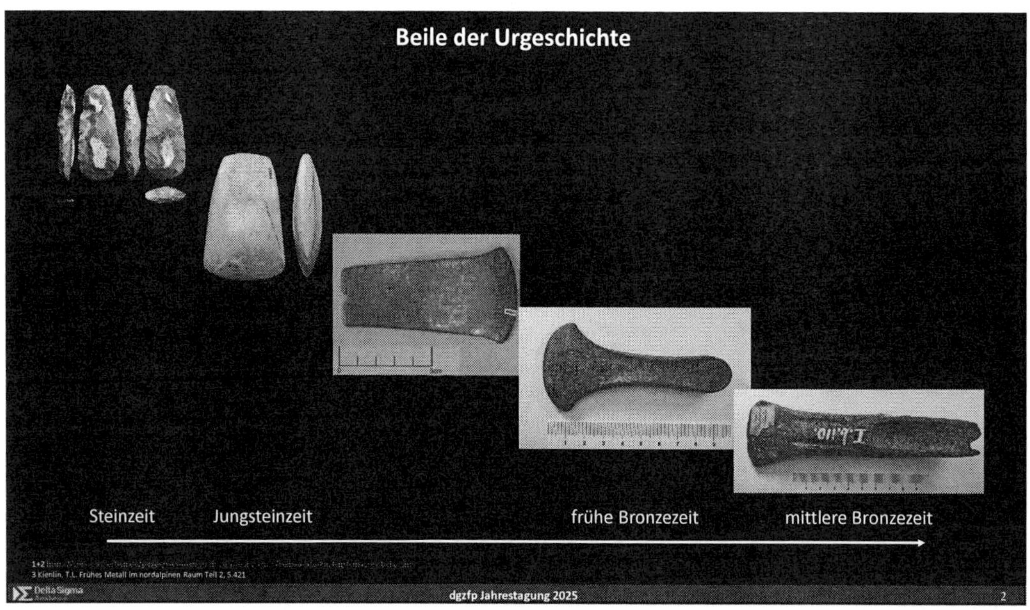
Beile der Urgeschichte
Steinzeit
Jungsteinzeit
frühe Bronzezeit
mittlere Bronzezeit
3 Kienlin, T.L. Frühes Metall im nordalpinen Raum Teil 2, S.421
dgzfp Jahrestagung 2025
2

2

Untersuchungsobjekte

- 10 Beile

Probenbezeichnung	Fundort	Datierung
P1	Beichlingen	Bz B1/2 1550-1450 v. Chr.
P2	Kanena III (Halle)	A1b/A2a 1975-1775 v. Chr.
P3		
P4	Dieskau III	A1b/A2a 1975-1775 v. Chr.
P5		
P6		
P7	Benndorf	Bz A2c-B1 1625-1500 v. Chr.
P8	Bennewitz	A2b 1775-1625 v. Chr.
P9		
P10	Halberstadt	etwa Bz B2-D 1450-1250 v. Chr.

Delta Sigma dgzfp Jahrestagung 2025 3

3

Untersuchungsobjekte

- 10 Beile

Probenbezeichnung	Fundort	Datierung
P1	Beichlingen	Bz B1/2 1550-1450 v. Chr.
P2	Kanena III (Halle)	A1b/A2a 1975-1775 v. Chr.
P3		
P4	Dieskau III	A1b/A2a 1975-1775 v. Chr.
P5		
P6		
P7	Benndorf	Bz A2c-B1 1625-1500 v. Chr.
P8	Bennewitz	A2b 1775-1625 v. Chr.
P9		
P10	Halberstadt	etwa Bz B2-D 1450-1250 v. Chr.

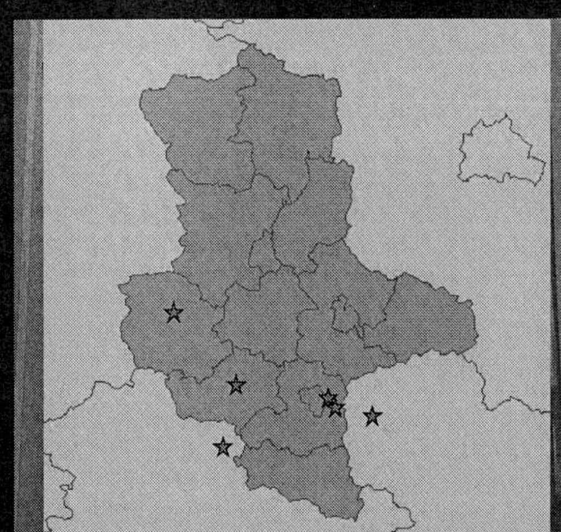

Delta Sigma dgzfp Jahrestagung 2025 4

4

Untersuchungsziele

- Fragestellung: Herstellung und Verarbeitung (Schneide, Grundkörper, Randleisten)
    - Endkonturnaher Guss mit minimaler Bearbeitung oder ausgiebiges Ausschmieden aus einem Rohling
- Teil I: Mikrostruktur-Eigenschaftsbeziehungen
    - Beprobung
    - metallographische Präparation
    - werkstofftechnische Analyse der Mikrostruktur-Eigenschaftsbeziehungen
    - Restauration
- Teil II: zerstörungsfreie Prüfung
    - Entwicklung einer Prüfroutine zur zerstörungsfreien Datenerhebung
    - Basis: Kennwerte aus Teil I
    - Studie im größeren Umfang (100%?)

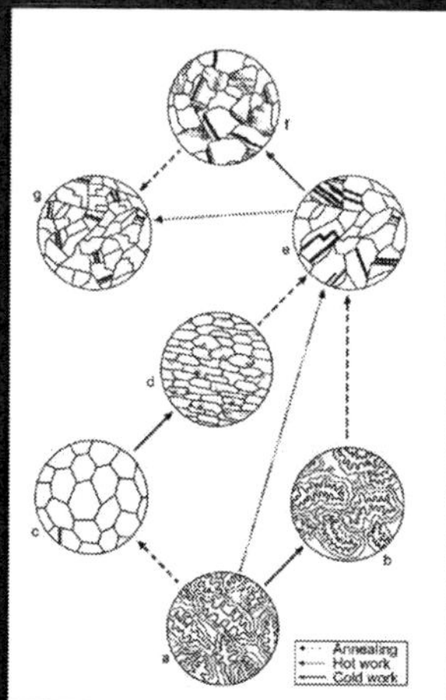

Kienlin, T.L. Frühes Metall im nordalpinen Raum Teil 1, S.46

DeltaSigma dgzfp Jahrestagung 2025 5

5

Metallographie

- Aufschlüsselung der Mikrostruktur-Eigenschafts-Beziehungen
- Herstellung von Mikroschliffen

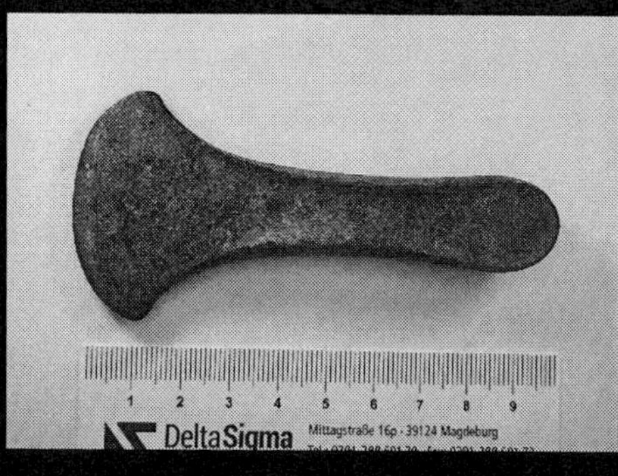

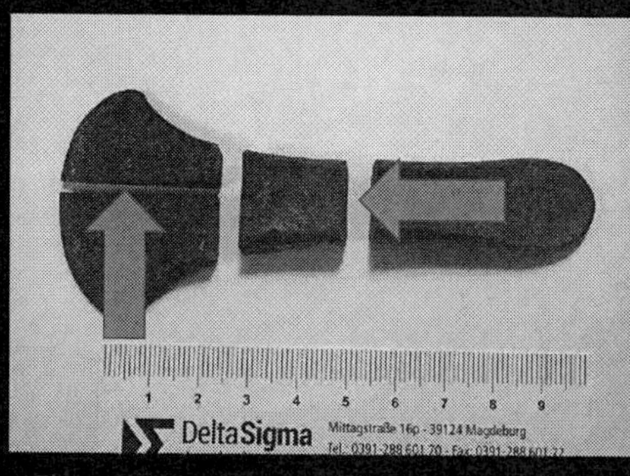

DeltaSigma dgzfp Jahrestagung 2025 6

6

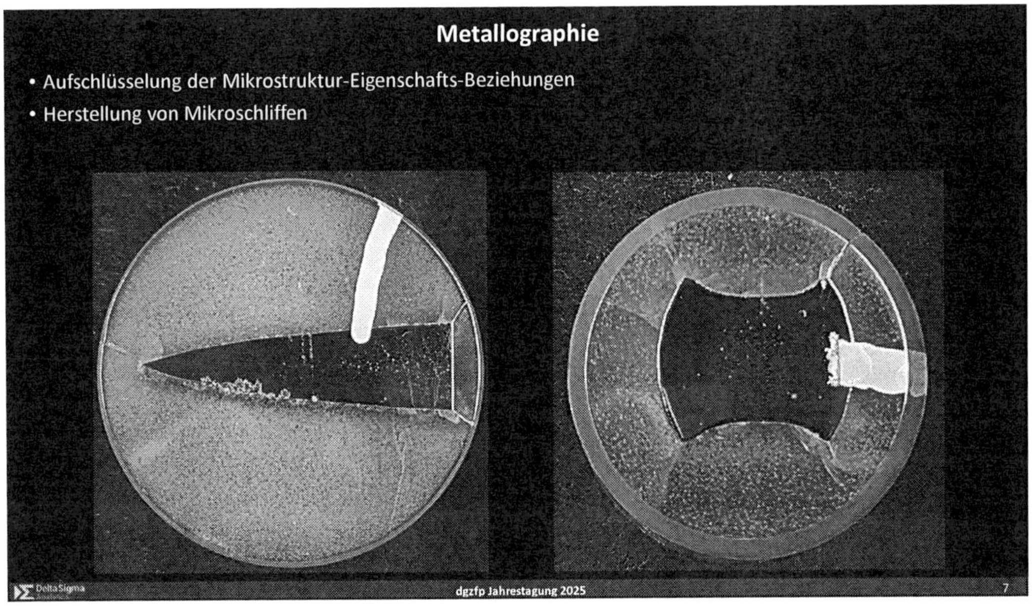
Metallographie
• Aufschlüsselung der Mikrostruktur-Eigenschafts-Beziehungen
• Herstellung von Mikroschliffen
Delta Sigma
dgzfp Jahrestagung 2025
7

7

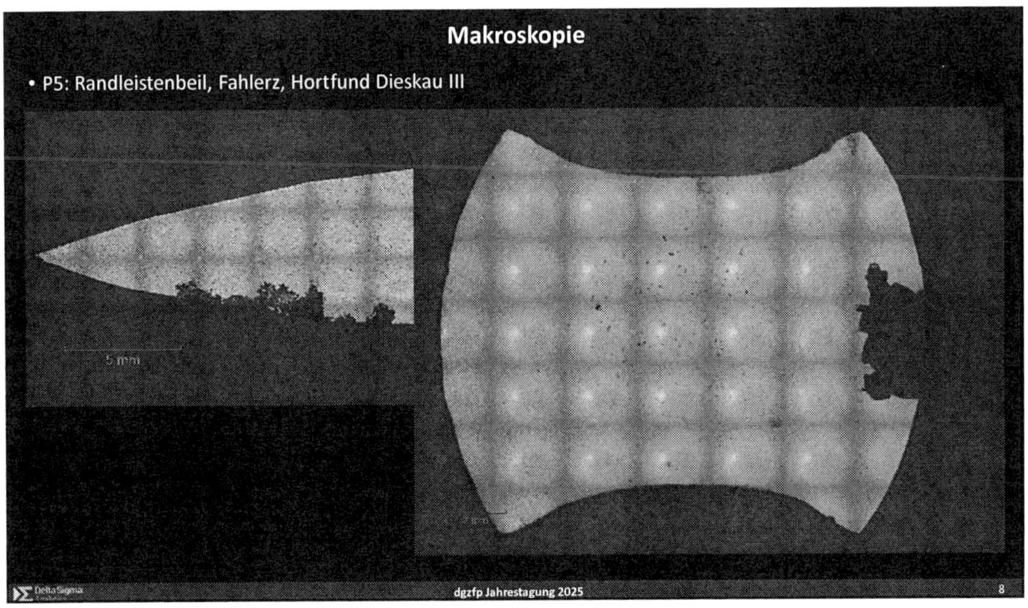
Makroskopie
• P5: Randleistenbeil, Fahlerz, Hortfund Dieskau III
5 mm
Delta Sigma
dgzfp Jahrestagung 2025
8

8

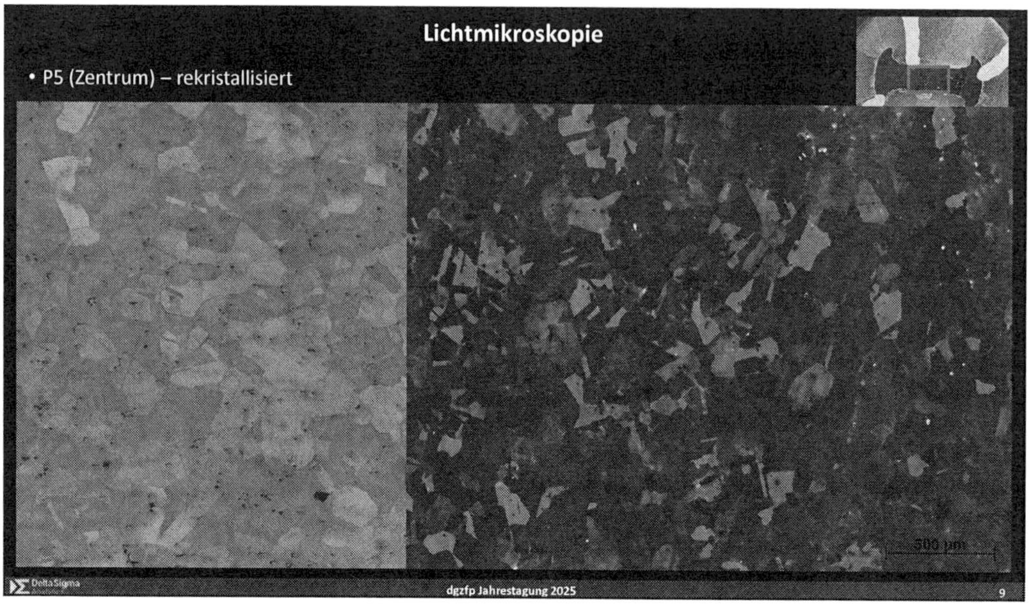
Lichtmikroskopie
• P5 (Zentrum) – rekristallisiert
500 µm
dgzfp Jahrestagung 2025
9

9

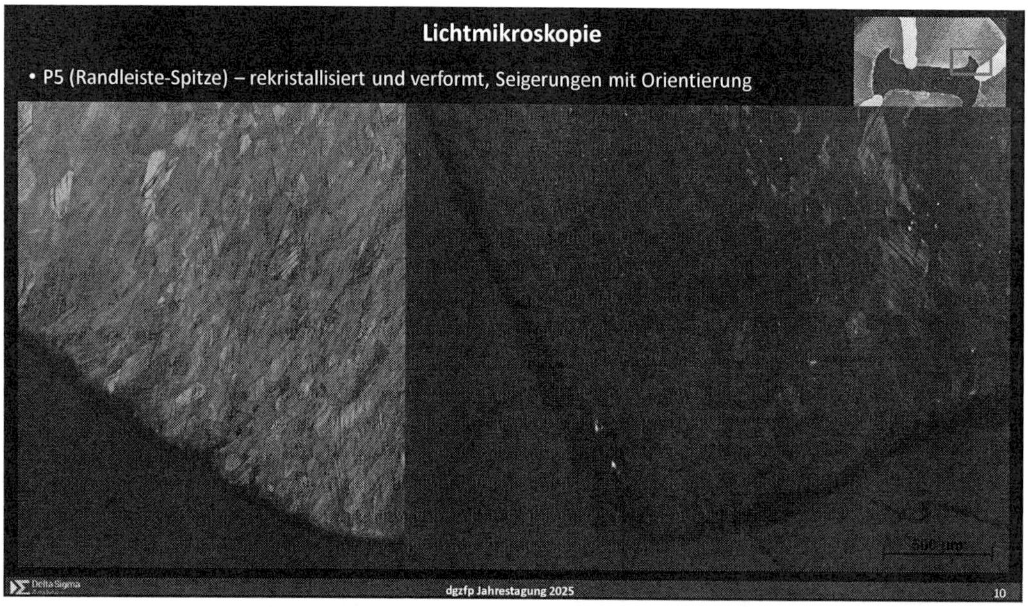
Lichtmikroskopie
• P5 (Randleiste-Spitze) – rekristallisiert und verformt, Seigerungen mit Orientierung
500 µm
dgzfp Jahrestagung 2025
10

10

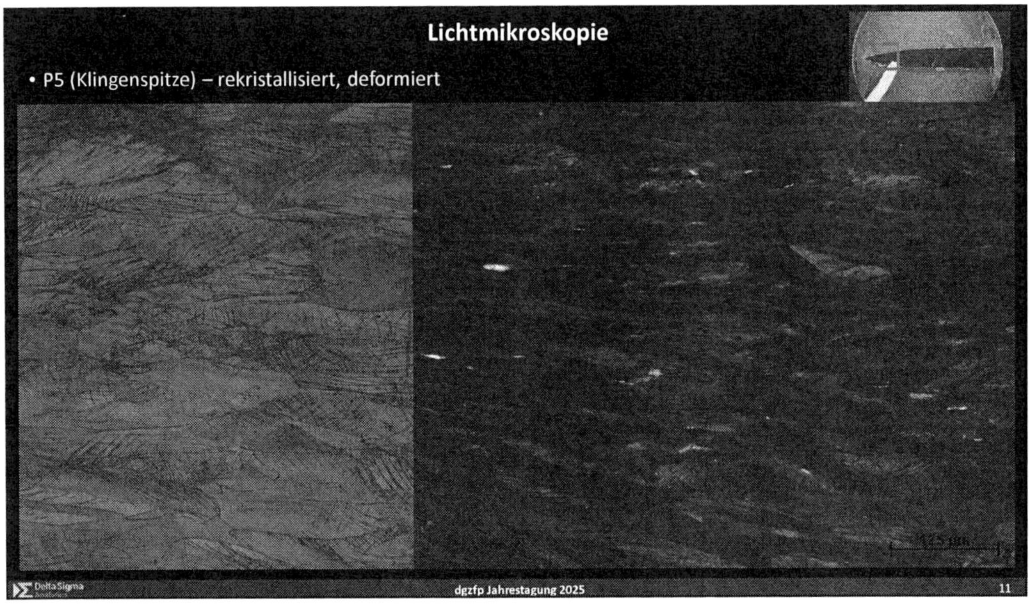
Lichtmikroskopie
• P5 (Klingenspitze) – rekristallisiert, deformiert
Delta Sigma
dgzfp Jahrestagung 2025
11

11

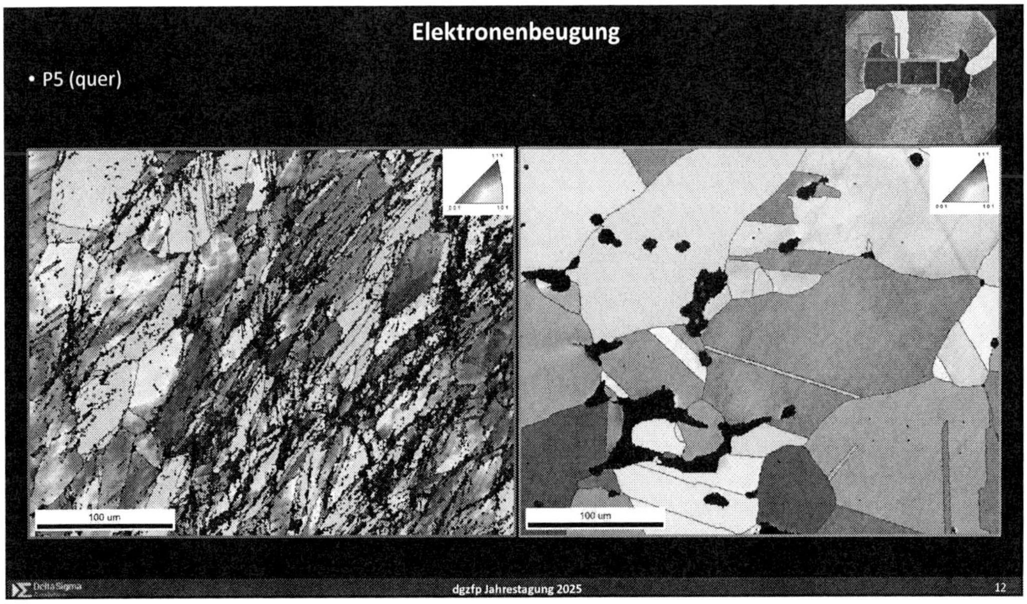
Elektronenbeugung
• P5 (quer)
100 um
100 um
Delta Sigma
dgzfp Jahrestagung 2025
12

12

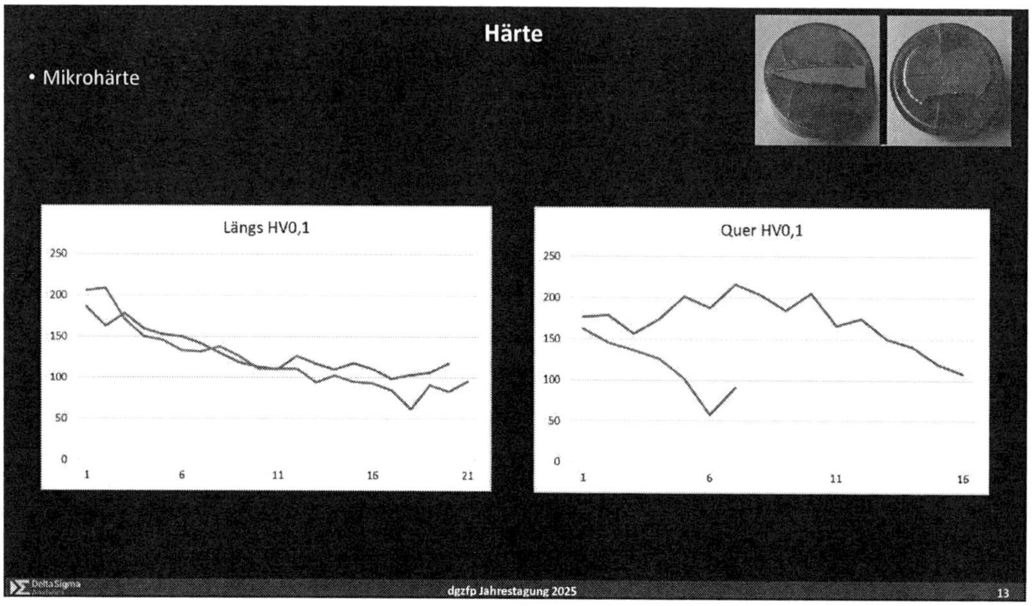
Härte
• Mikrohärte
Längs HV0,1
250
200
150
100
50
0
1
6
11
16
21
Quer HV0,1
250
200
150
100
50
0
1
6
11
16
dgzfp Jahrestagung 2025
13

13

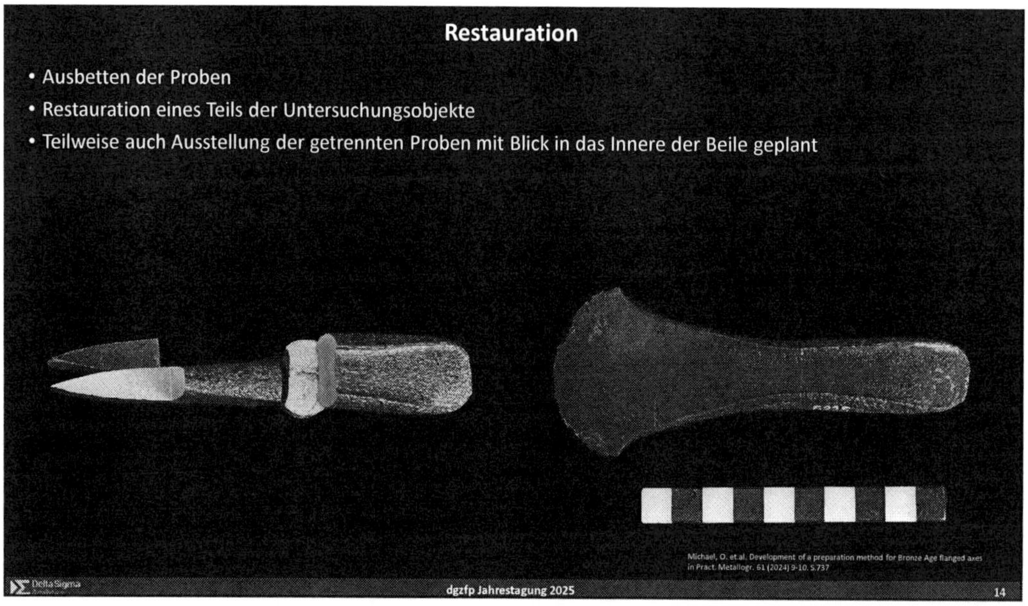
Restauration
• Ausbetten der Proben
• Restauration eines Teils der Untersuchungsobjekte
• Teilweise auch Ausstellung der getrennten Proben mit Blick in das Innere der Beile geplant
Michael, O. et.al. Development of a preparation method for Bronze Age flanged axes in Pract. Metallogr. 61 (2024) 9-10. S.737
dgzfp Jahrestagung 2025
14

14

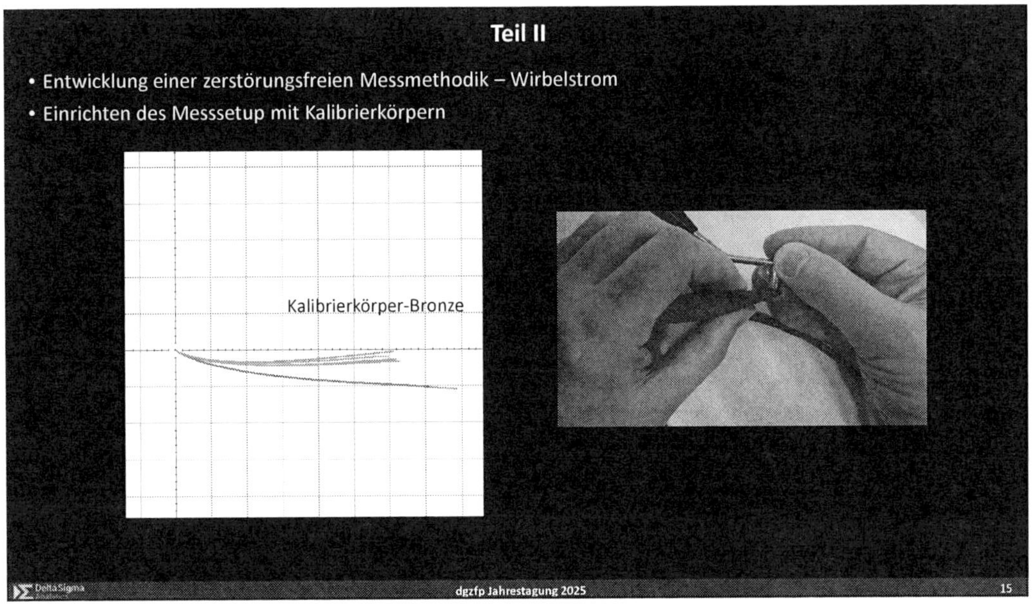
Teil II
• Entwicklung einer zerstörungsfreien Messmethodik – Wirbelstrom
• Einrichten des Messsetup mit Kalibrierkörpern
Kalibrierkörper-Bronze
dgzfp Jahrestagung 2025
15

15

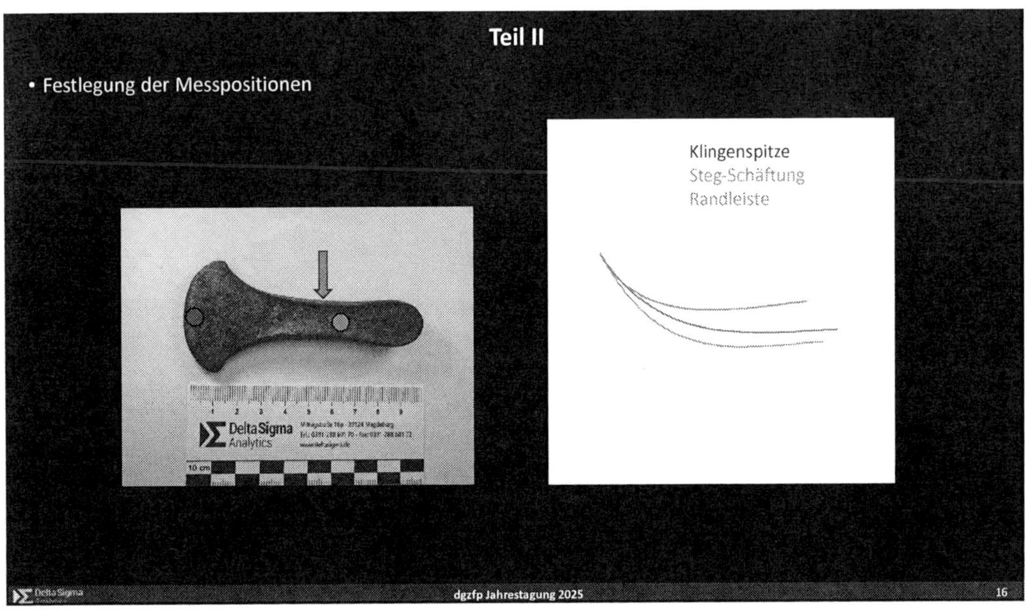
Teil II
• Festlegung der Messpositionen
Klingenspitze
Steg-Schäftung
Randleiste
DeltaSigma Analytics
10 cm
dgzfp Jahrestagung 2025
16

16

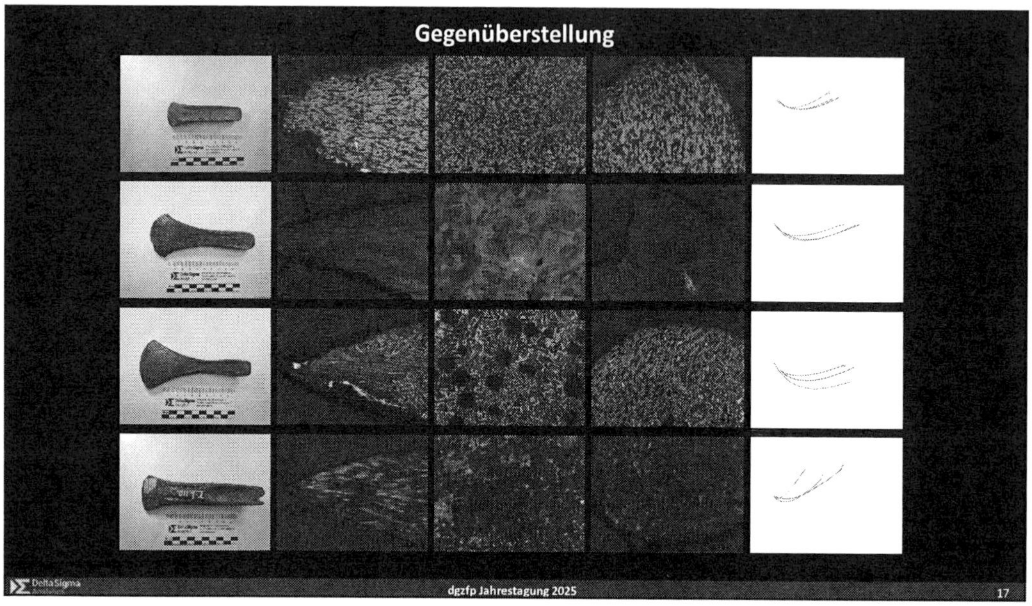

17

Zusammenfassung

- Untersuchungsobjekte: Randleistenbeile der frühen und mittleren Bronzezeit
- Untersuchungen: Metallographie
  - minimalinvasiv zerstörend aber rekonstruierbar
  - Tieferer Einblick in das Gesamtobjekt und dessen Herstellung
- Ergebnisse:
  - Herstellung: Guss → Schneide / Randleisten geschmiedet (entgegen der aktuellen Lehrmeinung für Objekte dieser Zeit in Mitteldeutschland)
  - mehrstufige thermomechanische Behandlung
  - signifikant unterschiedliche Qualität / Eigenschaften
- Korrelation der Mikrostrukturen mit den Wirbelstrommessungen ermöglicht zerstörungsfreie Einordnung weiterer Beile

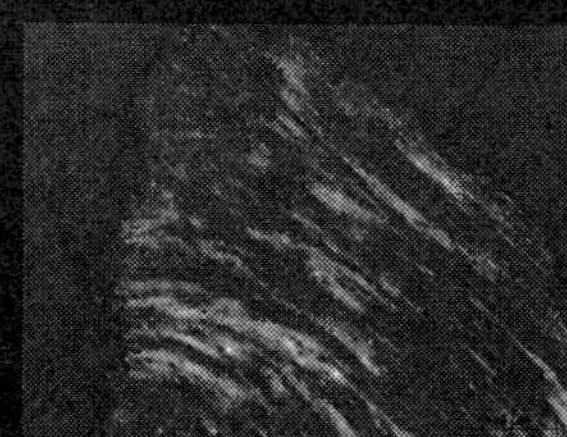

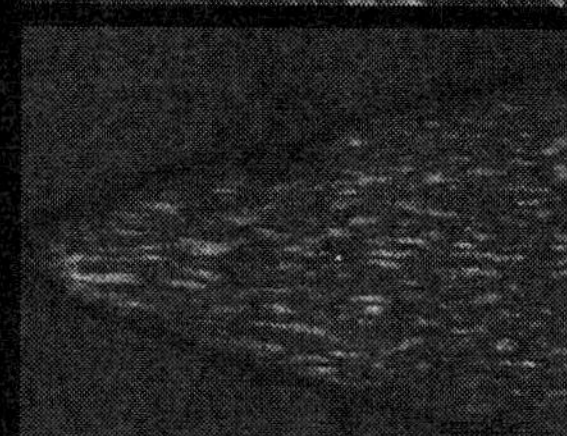

dgzfp Jahrestagung 2025 18

18

Danksagung
LANDESMUSEUM FÜR VORGESCHICHTE
HALLE (SAALE)
C.-H. Wunderlich, J.-H. Bunnefeld, H. Meller
OTTO VON GUERICKE
UNIVERSITÄT
MAGDEBURG
T. Halle, M. Wilke

19

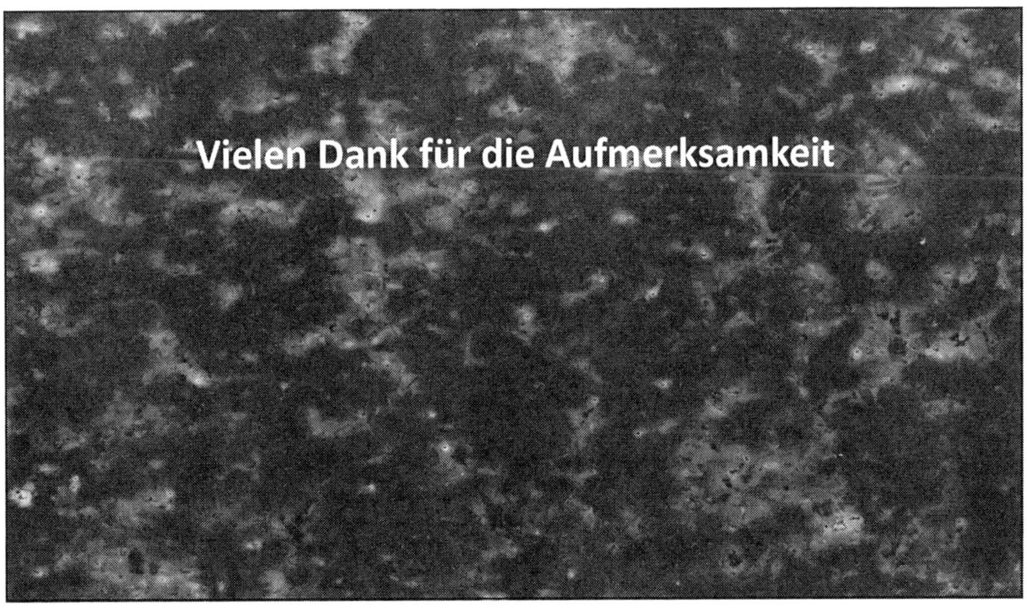
Vielen Dank für die Aufmerksamkeit

20

Ihr Partner für
Werkstofftechnik
DeltaSigma
Analytics
Ansprechpartner:
Oliver Michael
Tel.: 0391-288 601 71
Mail: oliver.michael@deltasigma.de
DeltaSigma Analytics GmbH
Mittagstraße 16 p
39124 Magdeburg
Tel.: 0391-288 601 71
Fax: 0391-288 601 72
Internet: www.deltasigma.de

21

DGZfP-Jahrestagung 2025

Ermittlung und Interpretation der Eigenschaften historischer Werkstoffe aus Sicht der Beanspruchbarkeit – eine Auswahl an Beispielen aus der Praxis

Christoph GAJDA[1], Steffen WAGNER[1]
[1]SLV Halle GmbH

Kurzfassung. Die natürliche Alterung metallischer Werkstoffe setzt der Beanspruchbarkeit von Bauteilen aus aktueller Betrachtung Grenzen, die zum Zeitpunkt der Herstellung nicht im Fokus liegen konnten. Die sich daraus ableitenden Schädigungsmechanismen wirken oft zusätzlich zu den ohnehin durch die Betriebslast hervorgerufenen Beanspruchungen. Insbesondere die bestehenden Stahlbrücken im deutschen Schienen- und Straßennetz unterliegen gegenwärtig einer kritischen Altersstruktur, wobei zu beachten ist, dass sich etwa zwei Drittel der Eisenbahnbrücken im „befriedigenden Bauwerkzustand" befinden (Zustandsnote 2,0 bis 2,4) [1]. Das sich daraus ergebende Spannungsfeld aus hoher Verkehrslast, schwierigen metallurgischen und mechanisch-technologischen Eigenschaften und teurem Neubau stellt die meisten Anwender vor extreme Herausforderungen und deckt einen hohen Untersuchungs- und Inspektionsbedarf an Bestandsbauwerken auf [2]. Dabei ist oft nicht sicher, um welchen Werkstoff es sich konkret handelt oder gar welche Eigenschaften er für eine Nachrechnung des Bauwerkes mit sich bringt. Die sich daraus ergebenden Reparatur- und Ertüchtigungsmöglichkeiten zur Erhöhung der Restnutzungsdauer sind ebenfalls schwer abschätzbar. Der Grund hierfür liegt in der historischen Herstellungsart, der Erschmelzung, der Vergießungsart sowie der stattgefundenen Alterung, zu denen oft noch ein nicht unerhebliches Maß an Ermüdungsbeanspruchung mit hoher Werkstoffausnutzung kommt.

Keywords: Altstahl, Schweißeignung, Schweißmetallurgie, Tragwerksbewertung, Historie, Denkmalschutz

1 Einführung

Stähle, deren Herstellung vor der Mitte des 20. Jahrhunderts lag, nehmen aus heutiger Sicht eine exponierte Stellung im Stahlbau ein. Das betrifft vor allem Bestandsbauwerke, deren Nutzungseigenschaften infolge einer Sanierung für die Zukunft sichergestellt bleiben sollten oder deren Werkstoffe im Zusammenhang mit wiederverwendeten Bauteilen ertüchtigt werden müssen. Vor allem Stahlwerkstoffe, deren Herstellung vor Einführung des Sauerstoffaufblasverfahrens, d. h. vor 1952, stattfand, gelten als alterungsempfindlich und zeichnen sich durch eingeschränkte Eignungen hinsichtlich ihrer schweißtechnischen Verarbeitung aus. Derartige Stähle werden im aktuellen Kontext als Altstähle bezeichnet. Mit fortschreitendem Alter kommen für die Denkmalsbewertung der Gegenwart immer mehr Stahlbauwerke, darunter auch solche, die sich nicht offensichtlich als derartige zu erkennen zeigen (z. B. Stahlbetonbauten in Skelettbauweise), in Betracht, als erhaltenswürdig eingestuft zu werden. Ihre Instandhaltung erfordert Fachwissen und Sorgfalt, soll der historische Charakter unverfälscht erhalten bleiben. Bereits im Jahr 1964 wurde dem durch die Verabschiedung der Charta von Venedig Rechnung getragen [3]. Doch nicht allein aus kultur- oder industriehistorischer Sicht sind Altstähle präsent, auch im Bereich der Nutzbauten stellen sie gegenwärtig einen Großteil des Bestandes dar. So sind ca. 50 % der Brücken im Schienennetz der Deutschen Bahn derzeit älter als 80 Jahre, davon wiederum 80 % älter als 100 Jahre [4]. Auch im Straßennetz sind ca. 6.000 Brückenbauwerke im Alter ab 80 Jahren zu finden. Für ihren Erhalt und Instandhaltung sind ein Großteil der

https://doi.org/10.58286/31278

öffentlichen Aufwendungen des Bundes bereitgestellt [2, 5]. Typische Anwendungen, in denen heute noch Altstähle zu finden sind u. a. Brücken, Türme, Masten und Hallen.

Die kategorische Ablehnung einer schweißtechnischen Verarbeitung dieser Altstähle beruht teilweise auf der nicht geklärten und durch entsprechende Untersuchungen allenfalls als „bedingt“ einzustufende Schweißeignung der Werkstoffe. Kann durch metallographisch-analytische Untersuchungen nachgewiesen werden, dass der Werkstoff prinzipiell keine untypischen schweißtechnischen Unregelmäßigkeiten erwarten lässt, so ist die Schweißeignung dennoch als „bedingt“ einzuschätzen, da keine verbindlichen Herstellerangaben bezüglich Stahlsorte und – Gütegruppe mehr herangezogen werden können.

Darüber hinaus konnte im Rahmen des in diesem Aufsatz zusammengefassten FuE-Vorhabens nachgewiesen werden, dass durch geeignete Nachbehandlungsverfahren die Eigenschaften von Altstählen verbessert und damit die Einsatzmöglichkeiten erweitert werden können. Die bisherige kategorische Ablehnung schweißtechnischer Instandsetzungsmaßnahmen an Altstählen wird durch diese Möglichkeiten des nachträglichen Ertüchtigens in Frage gestellt. Die dabei entstehenden Kosteneinsparungen gegenüber der aktuellen Vorgehensweise sind beträchtlich.

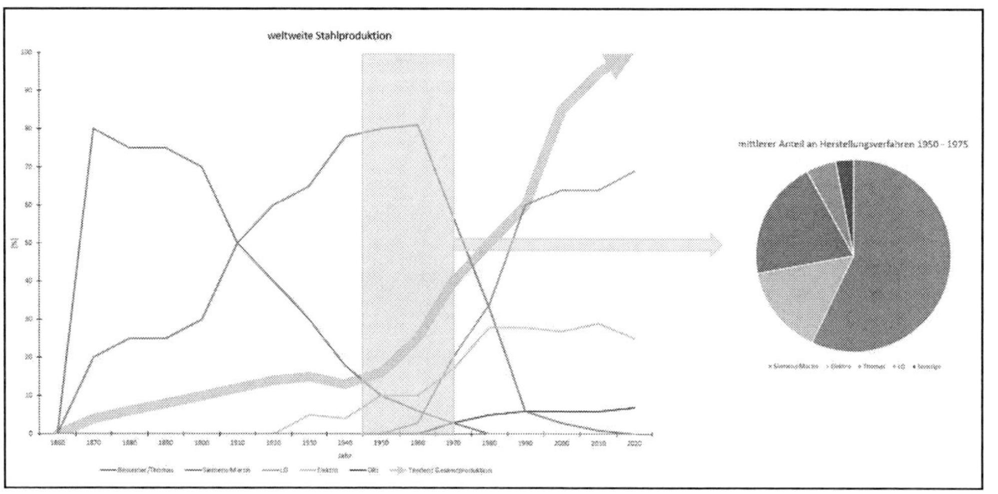

Abb. 1: Anteil der Herstellungsverfahren an der Weltproduktion an Stahl, Quelle: world steel association [6]

Abb. 1 gibt die Anteile der Herstellungsverfahren von Stahl im Zeitraum zwischen 1860 und der Gegenwart qualitativ wieder. Dies sind im Wesentlichen auch die Werkstoffe, die aktuell als Altstähle gelten und prinzipiell Fragen hinsichtlich ihrer Eigenschaften sowie ihrer weiteren Verwendung aufwerfen. Aus dieser Grafik wird deutlich sichtbar, welchen Anteil heute Werkstoffe historischer Fertigung haben. Dabei spielt das Herstellungsjahr des betreffenden Bauwerkes eine bedeutende Rolle. Während bis in die 1940er Jahre die nach dem Thomasverfahren erschmolzenen Flussstähle den Großteil der Werkstoffe ausmachen, können ab den 1950er Jahren verstärkt Siemens-Martin- oder Elektrostähle als heutige Altstähle vorgefunden werden. Bei letzteren hängt der metallurgische Zustand und damit die allgemeine Beanspruchbarkeit stark von der Art der Vergießung (dem Beruhigungszustand bzw. Desoxidationsgrad) ab. Unabhängig davon bleibt die Tatsache, dass die Werkstoffe natürlichen Alterungsvorgängen ausgesetzt sind, die ihre Eigenschaften stark beeinflussen und damit ihre heutige Nutzung im bauaufsichtlichen Bereich in Frage stellen. Altstahluntersuchungen sind aktuell keine Seltenheit mehr und erfordern ein besonderes Interesse bei Anwendern und Bauwerksprüfern, da sie mit modernen Stählen oft nicht vergleichbar sind bzw. ihre Vergleichbarkeit durch Untersuchungen eruiert werden muss. Der aktuelle Bundesverkehrswegeplan sieht bis zum Jahr 2030 vor, etwa

69 % der öffentlichen Aufwendungen für Instandhaltung und Reparatur bestehender Altsubstanz auszugeben [2]. Gleichzeitig ist festzustellen, dass ca. 50 % aller bestehenden stählernen Verkehrstragwerke vor 1950 erbaut wurden. Daraus ergibt sich eine problematische Situation für die Infrastruktur, da bis zum Abschluss der vorgesehenen Maßnahmen nicht alle Verkehrswege durch Neubauten ersetzt werden können. Zusätzlich kommen hohe finanzielle Aufwendungen eines umfangreichen Neubauprogramms dazu und bekräftigen dadurch die Forderung, Maßnahmen in den Erhalt bestehender Substanz zu stecken. Zurückliegende Forschungsvorhaben sowie regelmäßig wiederkehrende Altstahluntersuchungen in der Werkstoff- und Bauteilprüfung sind eine Referenz an dieses aktuelle Thema. So soll im Folgenden auf charakteristische Eigenschaften von Altstählen sowie ihre Abgrenzung zu modernen Werkstoffen eingegangen werden.

Abb. 2: Hbf Halle (Saale) (links) [7], genietete Stützen aus historischem Flussstahl (rechts)

Nachfolgend soll auf Grundwerkstoffuntersuchungen zur Charakterisierung des verwendeten Stahls eingegangen werden. Die vorgestellten Ergebnisse beziehen sich dabei auf im Rahmen eines Forschungs- und Entwicklungsvorhabens (49MF200119 [8]) erzeugten Werte, die aus dem historischen Bauwerk des Hallenser Hauptbahnhofes entnommen wurden (siehe Abb. 2).

2 Werkstofftechnische Untersuchungen

2.1 Schweißbarkeit und Schweißeignung von Altstählen

Soll eine Konstruktion aus unlegierten Baustählen mittels Schmelzschweißen gefügt werden, ist dazu die Betrachtung der „Schweißbarkeit" dieser Baugruppe zu beachten. Dieser Begriff bezeichnet einen zusammengesetzten Begriff, der die fertigungstechnologischen, werkstofftechnischen und konstruktiven Bedingungen der auszuführenden Schweißverfahren berücksichtigt. In der (zurückgezogenen) DIN 8528-1 [9] wird diese Verknüpfung deutlich. Abb. 4 zeigt den zusammengesetzten Begriff der Schweißbarkeit als Schnittmenge der Teilbedingungen zum Schmelzschweißen nach DIN 8528-1 [9]. Demzufolge kann in Zusammenhang mit Altstahluntersuchungen nie die Schweißbarkeit des untersuchten Werkstoffes festgestellt werden, sondern lediglich dessen Schweißeignung, da die Aspekte der Schweißmöglichkeit sowie der Schweißsicherheit nur durch die Gesamtheit der an der Konstruktion vorliegenden Bedingungen bestimmt werden.

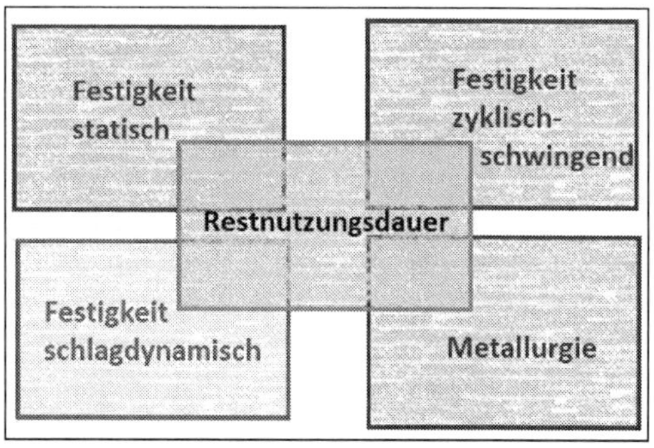

Abb. 3: Zusammenhänge zur Abschätzung der Restnutzungsdauer [10]

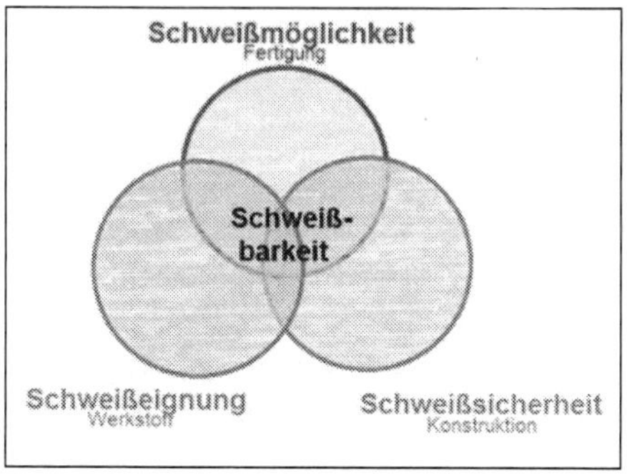

Abb. 4: Begriff der Schweißbarkeit in Anlehnung an DIN 8528-1 [9]

Im Sinne des zusammengesetzten Begriffes der Schweißbarkeit kann die Schweißeignung des Werkstoffes als vom Werkstoff vorgegebener Teilaspekt angesehen werden. Nicht zuletzt wird die Schweißeignung von Altstählen häufig durch werkstofftechnische Untersuchungen charakterisiert, wenn im Zuge von Sanierungsmaßnahmen an historischen Bau- oder Massenstählen gearbeitet werden und in bestehende Schweiß- oder Nietkonstruktionen eingegriffen werden muss. In der ebenfalls zurückgezogenen DIN 8528-2 [11] findet sich diesbezüglich eine Entscheidungsmatrix zur Beurteilung der Schweißeignung unlegierter Baustähle im Sinne der damalig geltenden Werkstoffnorm DIN 17 100 [12] anhand der Neigung des Werkstoffes zu unterschiedlichen Schädigungsereignissen beim Schweißen. Diese sind:

- Sprödbruchneigung
- Alterungsneigung
- Härtungsneigung
- Seigerungsverhalten
- Anisotropie

In der Entscheidungshilfe nach DIN 8528-2 [11] wird davon ausgegangen, dass bei Beachtung der genannten Einflussgrößen eine Schweißeignung gegeben ist, wenn keine oder möglichst wenig der genannten Faktoren auftreten. Tabelle 1 zeigt die Einflussgrößen für verschiedene Werkstoffe nach dem Verständnis der in DIN 8528-2 [11] dargestellten Schweißeignungsbewertung. Hinsichtlich der Sprödbruchneigung wird davon ausgegangen, dass eine Mindestkerbschlagarbeit von 28 J bei einer Prüftemperatur von -20°C gegeben sein muss, um eine Versprödung in der Wärmeeinflusszone (WEZ) auszuschließen und damit eine vollumfängliche Schweißeignung vor dem Hintergrund der Sprödbruchneigung zu gewährleisten. Alle davon abweichenden Ergebnisse von Kerbschlagbiegeversuchen würden die Schweißeignung zumindest negativ beeinflussen, so die Aussage der Norm. Eine ähnliche Bewertungsmethode wird hinsichtlich der Alterungsneigung angewandt. Es wird davon ausgegangen, dass die alterungsfördernden Bedingungen gleichzeitig die Schweißeignung negativ beeinflussen. Dazu zählen insbesondere unberuhigt vergossene Stähle der (damaligen) Gütegruppe 2. Bezüglich der Härtungsneigung wird davon ausgegangen, dass diese anhand des Kohlenstoffgehaltes aus der Stückanalyse abgeschätzt werden kann. Mit zunehmender Auf- bzw. Einhärtbarkeit des Werkstoffes wird die Schweißeignung schlechter. Das Seigerungsverhalten soll laut DIN 8528-2 [11] anhand des Desoxidationsgrades beurteilt werden. Finden sich die Desoxidationselemente Silizium und/oder Aluminium (beruhigt / sonderberuhigt) in der Stückanalyse wieder, kann davon ausgegangen werden, dass es keine Entmischungserscheinungen im Werkstoff gibt und daher eine Schweißeignung grundsätzlich gegeben ist. Unter der Berücksichtigung der damals geltenden Stahlgütegruppen gelten Stähle der Gütegruppen 1 und 2 als bedingt bis nicht schweißgeeignet. Dem Anisotropieverhalten (Richtungsabhängigkeit der mechanisch-technologischen Eigenschaften) wird unterstellt, dass es ebenfalls negative Einflüsse auf die Schweißeignung hat, da insbesondere das Verformungs- und Plastifizierungsverhalten in Dickenrichtung (und damit das Abbauen von Schrumpfspannungen beim Erstarren einer Schweißnaht) eine wichtige Rolle zur Bewertung der Schweißeignung spielt. Es ist dabei nicht abhängig von der Stahlsorte, sondern von der Erzeugnisdicke und dem Umformgrad.

Ist die Zuordnung zu einer vor 1980 erzeugten Stahlsorte möglich, eignen sich die in Tabelle 1 zusammengestellten Informationen für eine bessere Bewertung deren Schweißeignung [11]. Eine zunehmende Anzahl an Kreuzen bedeutet wachsende Schwierigkeiten und damit einen höheren Aufwand bei der schweißtechnischen Fertigung.

Die ebenfalls in Tabelle 1 aufgeführten Stähle St 50, St 60 und St 70 unterscheiden sich von den übrigen Werkstoffen darin, dass ihre Eigenschaften wesentlich vom Kohlenstoffgehalt abhängen und sie weitgehend im Maschinenbau eingesetzt werden. Ihre schweißtechnische Verarbeitung ist kohlenstoffabhängig mindestens stark eingeschränkt bzw. nicht mehr gegeben.

Tabelle 1: Schweißeignung unlegierter Baustähle, die vor 1980 erzeugt wurden auf Grundlage historischer Regelwerke [11]

Stahlsorte (Gütegruppe)	**Desoxidation-art** [1)]	**Behandlungs-zustand** [2)]	**Sprödbruch-neigung**	**Alterungs-neigung**	**Härtungs-neigung**	**Seigerungs-verhalten**
St 37-3	RR	N	—	—	—	—
St 46-3	RR	N	—	—	—	—
St 52-3	RR	N	—	—	x	—
St 37-3	RR	U	x	x	—	—
St 46-3	RR	U	x	x	—	—
St 52-3	RR	U	x	x	x	—
RSt 34-2	R	U, N	xx	x	—	—
RSt 37-2	R	U, N	xx	x	—	
RSt 46-2	R	U, N	xx	x	—	—
USt 34-2	U	U, N	xx	x	—	x
USt 37-2	U	U, N	xx	x	—	x
St 42-3	RR	N	—	—	xx	—
St 42-3	RR	U	x	x	xx	—
RSt 37-1	R	U, N	xxx	xx	—	—
RSt 34-1	R	U, N	xxx	xx	—	—
RSt 42-2	R	U, N	xx	x	xx	—
USt 37-1	U	U, N	xxx	xx	—	xx
USt 34-1	U	U, N	xxx	xx	—	xx
USt 42-2	U	U, N	xx	x	xxx	xx
St 33-2	—	—	xxx	x	xxx	xx
RSt 42-1	R	U, N	xxx	xx	xxx	—
USt 42-1	U	U, N	xxx	xx	xxx	xx
St 33-1	—	—	xxx	xx	xxx	xx
St 50-2	R	U, N	xxx	x	xxx	—
St 50-1	R	U, N	xxx	xx	xxx	—
St 60-2	R	U, N	xxx	x	xxx	—
St 70-2	R	U, N	xxx	x	xxx	—
St 60-1	R	U, N	xxx	xx	xxx	—

1) Kennbuchstaben für die Desoxidationsart: RR – besonders beruhigt (FF), R – beruhigt (FN), U – unberuhigt (FU)

2) Kennbuchstaben für den Behandlungszustand: U – warmgeformt unbehandelt (+AR), N – normalgeglüht (+N)

2.2 *Untersuchungen an nicht mehr benötigtem Material des Hallenser Hauptbahnhofes*

2.2.1 *Vorbetrachtungen*

Im Rahmen von Umbaumaßnahmen am Hauptbahnhof Halle (Abb. 2) konnte Restmaterial einer Dachkonstruktion aus dem Jahre 1901/1902 gewonnen werden. Die daran angestellten Untersuchungen sollen nachfolgend beispielhaft für grundhafte Altstahl-Grundwerkstoffuntersuchungen zusammengefasst werden.

Üblicherweise sollten die analytischen und technologischen Prüfungen zur Schweißeignungsbewertung unter Laborbedingungen an aus dem betreffenden Bauwerk mechanisch entnommenen Proben mit kalibrierten und metrologisch rückgeführten Geräten sowie Standards erfolgen (Metrologie – Wissenschaft des Messens). Dafür spezialisierte Prüflabore verfügen aufgrund ihrer Akkreditierung über sehr umfangreiche Erfahrungen für die Gewährleistung dieser Forderungen. Sie können anhand ihrer, durch die „Deutsche Akkreditierungsstelle" DAkkS vergebene Registriernummer beginnend mit „D" für Deutschland gefolgt von „PL" für Prüflabor und einer mehrstelligen Zahl (z. B. D-PL-17369-02-00 für das Prüflabor der SLV Halle GmbH) erkannt werden. Der genaue Umfang der jeweiligen Akkreditierung und damit die Kompetenz des Prüflabors für die notwendigen Prüfungen sind auf der Internetseite der DAkkS einsehbar.

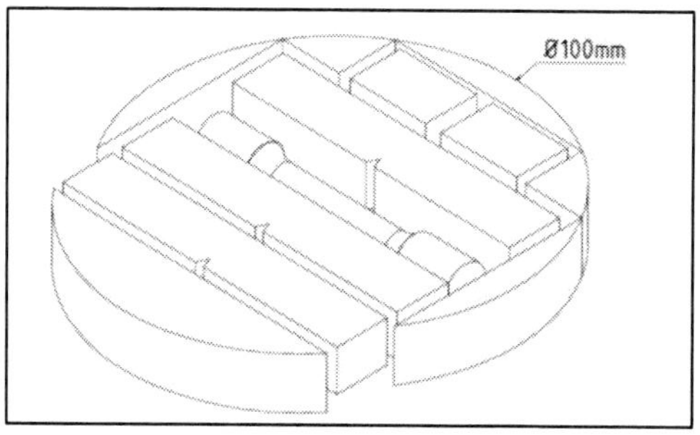

Abb. 5: Beispiel für einen Bohrkern und den daraus entnehmbaren Proben

Die erforderliche Probenentnahme sollte mittels Kernbohrungen (Abb. 5), am besten unter Flüssigkeitskühlung erfolgen. Dafür stehen spezielle, magnetisch am Bauwerk selbsthaftende Fräsgeräte zur Verfügung. Zu beachten ist, dass eine thermische Probenentnahme grundsätzlich nicht zulässig ist, da bereits während des damit verbundenen Aufheizens und Abkühlens kritische Werkstoffschädigungen, wie Risse und Versprödungen, auftreten können. Das Herausarbeiten von Proben mittels Brennschnitten würde somit ebenfalls eine zuvor auszuführende Schweißeignungsanalyse (besser „Brennschnitteignungsanalyse") erforderlich machen.

Das in Abb. 5 dargestellte Beispiel eines Bohrkerns mit einem Durchmesser von 100 mm zeigt die mögliche Positionierung von daraus entnehmbaren Proben für

- den Kerbschlagbiegeversuch (drei Proben),
- den Zugversuch (eine Probe),
- für die chemische Analyse (eine Probe) und
- die metallographische Untersuchung (mindestens eine Probe).

Die verbleibenden Reststücke in den Randbereichen können darüber hinaus für weitere metallographische Schliffe genutzt werden, z. B. zum Nachweis möglicher Anisotropien in Abhängigkeit von der Walzrichtung. Durch die Entnahme kleinerer Bohrkerne ist es möglich, die damit verbundene partielle Schädigung von Stahlkonstruktionen zu verringern. Dabei ist jedoch zu bedenken, dass daraus weniger Proben gefertigt werden können.

2.2.2 Erhöhung der Reproduzierbarkeit der Schweißeignungsbewertung

Bei komplexen Ingenieurbauwerken, wie z. B. Stahlbrücken oder -hallen ist es in der Regel nicht möglich, aus allen Einzelkomponenten Proben zu entnehmen. Zur Erhöhung der Reproduzierbarkeit der Ergebnisse bereits erfolgter Laboruntersuchungen bieten sich ergänzende Vorortuntersuchungen im Bereich von Bauwerksbereichen an, die zur Probenentnahme durch Kernbohrungen nicht geschädigt werden dürfen, können oder sollen. Dabei empfiehlt sich die in Abb. 6 beschriebene, nicht invasive, Vorgehensweise. Im Rahmen der Bewertung der dabei erhaltenen Ergebnisse muss unbedingt beachtet werden, dass diese auf der (sachgerecht beschliffenen) Oberfläche gewonnen wurden und nicht über den Querschnitt der zu untersuchenden Bauwerkskomponente. Somit sind die besonderen Eigenschaften von beruhigt und unberuhigt vergossenen Stählen (möglich vorhandene reine oberflächige „Speckschichten" und geseigerte Kernbereiche) zu berücksichtigen. Dieses kann bei Kenntnis des Baujahres relativ sicher aus den Ergebnissen der chemischen (Oberflächen-) Analyse abgeleitet werden. So weisen bei vor 1970 errichteten Bauwerken Kohlenstoffgehalte unter 0,1 %, (relativ) geringe Schwefel- und Phosphorgehalte sowie fehlende Silizium- und Aluminiumgehalte in der Regel auf eine nicht vorhandene Desoxidation, also einen unberuhigt vergossenen Zustand hin.

(ergänzende) nicht invasive Vorortuntersuchungen zur Schweißeignungsbewertung	
örtliches Beschleifen der zu prüfenden Oberflächenbereiche mit **SiC-freien** Schleifmitteln	
Bestimmung der chemischen Analyse	**Bestimmung der Härte**
transportables Emissionsspektrometer	mobiles Härteprüfgerät
Beachtung der Legierungs- und Begleitelemente: C, Si, Mn, P, S, Al, N	Ermittlung der (Oberflächen-) Härte nach dem UCI-Verfahren nach DIN 50 159-1 [13]
Bewertung der Wirkungen der Elemente auf die Schweißeignung	Umwertung der Härtewerte (HV oder HB) in Zugfestigkeiten.

Abb. 6: Nicht invasive Vorortuntersuchungen zur Schweißeignungsbewertung nach SCHUSTER [14]

Sollen anstelle mobiler Emissionsspektrometer, die nach wie vor immer noch relativ groß und schwer sind, handliche ED-RFA-Spektrometer (ED-RFA = energiedispersive Röntgenfluoreszenzanalyse) für die (Oberflächen-)Analyse verwendet werden, ist zu bedenken, dass mit diesen Geräten (bedingt durch ihr physikalisches Messprinzip mittels Röntgenstrahlen) leichte Elemente mit Ordnungszählen kleiner 11, wie z. B. Kohlenstoff und Bor, NICHT nachweisbar sind. Bei Einsatz eines solchen Geräts kann die Bewertung der Schweißeignung jedoch über die Elemente Silizium und Aluminium erfolgen, was insbesondere Aussagen zur Art der Desoxidation zulässt. Darüber hinaus erlauben sie (geräteabhängig) auch die Bestimmung von Phosphor und Schwefel mit ausreichender Genauigkeit. Damit eignen sich solche Geräte im Rahmen von Schweißeignungsuntersuchungen nur eingeschränkt. In Zukunft könnten hier weiterentwickelte Handlaserspektrometer Abhilfe bringen, welche in der Lage sind, alle für eine Schweißeignungsbewertung relevanten Elemente zu bestimmen. Natürlich sind die in diesem Abschnitt genannten Prüfungen auch an kleinen Bohrkernen (Durchmesser ≈ 20 mm) im Labor möglich. Das hat den Vorteil, dass die dem Bauwerk zerstörend entnommenen Proben nicht so groß sind und dieses damit zu stark geschädigt wird. Werden diese über ihren Radius geteilt, erlauben sie an ihren beiden Hälften jeweils eine spektrometrische Analyse und einen metallographischen Querschliff. Auf dessen Fläche kann darüber hinaus die Werkstoffhärte bestimmt werden, welche ihrerseits Aussagen zur Zugfestigkeit zulassen. Die Interpretation der dabei erhaltenen Prüfergebnisse kann dann wiederum mit den in eingangs dargestellten Abhängigkeiten erfolgen.

Abb. 7: Ermittlung der chemischen Zusammensetzung an den Bogenscheitelgelenken des Chemnitzer Viaduktes (SLV Halle)

Abb. 8: Mobiles Funkenspektrometer mit Ar-Spülung und digitaler Auswertung

Allen Methoden zur in-situ-Bestimmung der chemischen Zusammensetzung ist die lediglich oberflächige Prüflokalisierung gemein, was die Interpretation der Ergebnisse teilweise komplex werden lässt und vergleichende stationäre Untersuchungen im Labor erfordert. Weiterhin ist die Bestimmung leichter Elemente nur bedingt möglich. So können z B. Kohlenstoff mittels Röntgenfluoreszenzanalyse (RFA) und Laserspektrometrie sowie Stickstoff mittels mobiler F-OES nicht bestimmt werden.

2.2.3 Untersuchungsergebnisse

Die hier vorgestellten Altstahluntersuchungen im Rahmen von Sanierungsarbeiten oder Bauwerksbewertungen ermöglichen einen Überblick über die nutzbaren analytisch-technologischen Prüfverfahren, um eine vollständige Werkstoffcharakterisierung unter minimalinvasivem Eingriff in das Bauwerk zu ermöglichen. Begleitende Beanspruchungsmessungen im Zuge eines Bauwerksmonitoring lassen in spezifischen Fällen auch einen Vergleich der durch die Betriebslast eingebrachten Schädigung und dem Ermüdungswiderstand des Werkstoffes im geschweißten Zustand zu, wobei die vorgestellten Untersuchungen in diesem Fall um entsprechende Wöhlerlinien ergänzt wurden, die nicht Bestandteil dieses Aufsatzes sind. Die werkstofftechnischen Untersuchungen erfolgten an Stützen aus einem historischem Flussstahl der Sorte St 44 (Abb. 2, Abb. 9, Abb. 10), wobei die Proben direkt aus dem Bauteil entnommen wurden.

Die Ergebnisse der metallographischen Makro- und Mikroschliffuntersuchungen des Werkstoffs sind anhand ausgewählter Beispiele in Abb. 11 bis Abb. 14 dargestellt.

Die mikroskopische Untersuchung der polierten Schliffebene wies einen hohen Verunreinigungsgrad des Gefüges durch zahlreiche nichtmetallische Seggregate nach. Aus den Detailaufnahmen geht hervor, dass es sich hierbei um Mangansulfide und Mikroschlackeeinschlüsse handeln könnte.

Abb. 9: Stützenabschnitt mit Kennzeichnung der Probenentnahme des „Bohrkern 1“[1]

Abb. 10: Probenentnahme mithilfe einer Magnetbohrmaschine

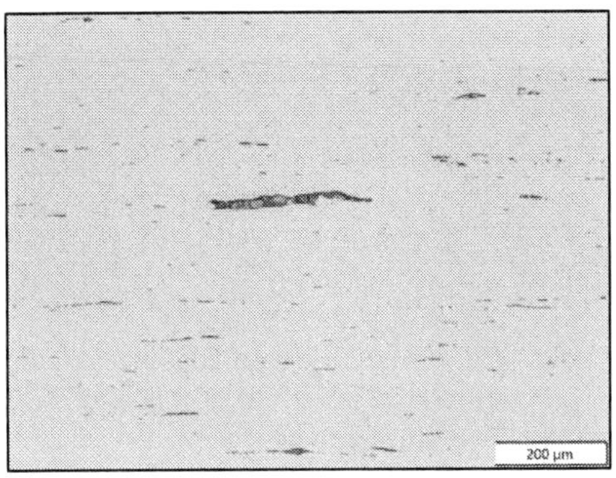

Abb. 11: Detailaufnahme des Mikroschliffs in poliertem Zustand mit nichtmetallischen Bestandteilen. Schlackeeinschlüssen sowie Mangansulfidausscheidungen. [15]

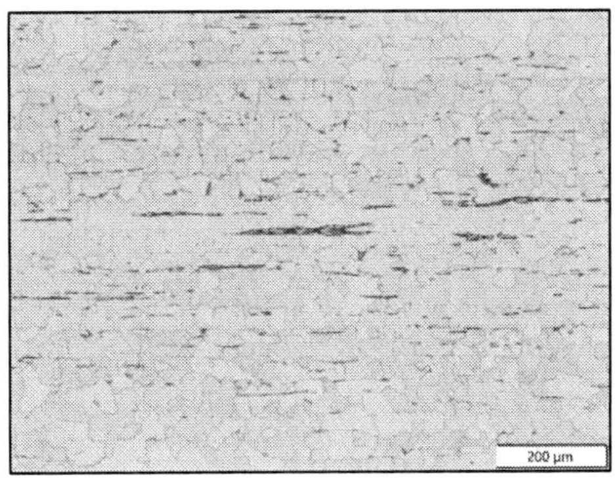

Abb. 12: Detailaufnahme des Mikroschliffs in geätztem Zustand. Ferritisches Gefüge mit geringen Anteilen an Perlit sowie zeiligen nicht-metallischen Einschlüssen und Ausscheidungen von MnS [15]

Das Mikrogefüge (geätzt mit HNO_3) zeigte sich nahezu fast vollferritisch mit vereinzelten Perlitinseln, wie es für unlegierte Massenstähle des frühen 20. Jahrhunderts mit mittleren Festigkeitseigenschaften typisch ist. Der vorliegende historische Werkstoff weist im Makroschliff eine charakteristische „Speckschicht“ an den Probenrändern, sowie eine ausgeprägte Seigerungszone im Werkstoffinneren auf, wonach der Desoxidationsgrad als unberuhigt eingeschätzt werden kann. Im Weiteren lassen sich anhand der vorliegenden Makroschliffe die im Werkstoff durch Alterungsvorgänge stattgefundenen Gefügeveränderungen beurteilen. Abb. 13 lässt die durch diesen Alterungsvorgang hervorgerufenen FRY‘schen Kraftwirkungsfiguren deutlich erkennen.

[1] Die vollständige Prüfmatrix mit allen Probenbezeichnungen kann dem Abschlussbericht des Forschungsvorhabens 49MF200119 entnommen werden. Sämtliche in diesem Aufsatz dargestellten Ergebnisse von Bohrkernuntersuchungen sind exponiert.

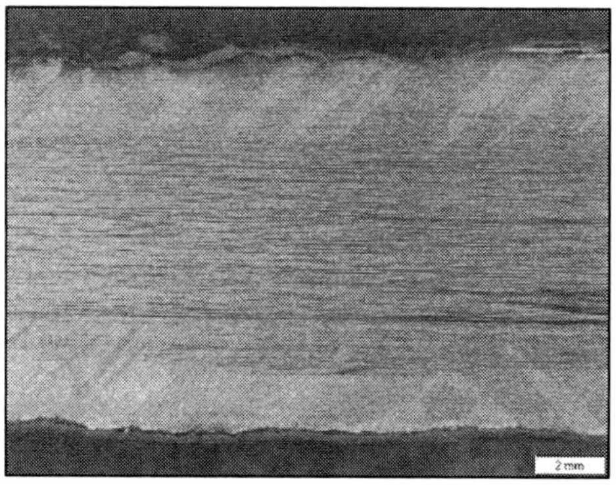

Abb. 13: Übersichtsaufnahme des Makroschliffs mit erkennbarer „Speckschicht“ an den Proben-rändern und Kraftwirkungsfiguren. [15]

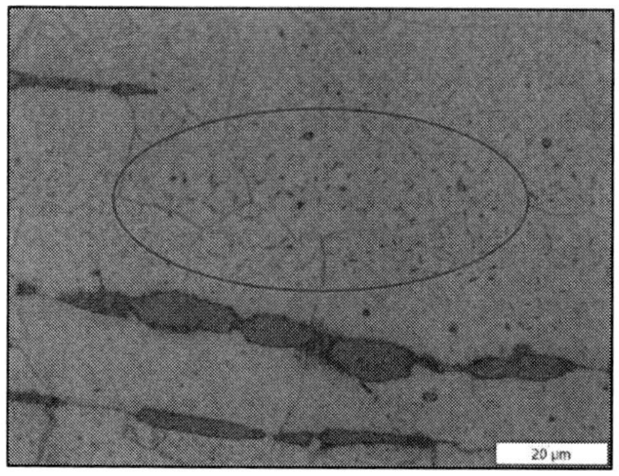

Abb. 14: Detailaufnahme des geätzten Mikroschliffs mit Eisennitridnadeln (rot). [15]

Anhand des geätzten Mikroschliffes konnten diese durch stickstoffhaltige Ausscheidungen entstandenen nadeligen Eisennitride zudem eindeutig nachgewiesen werden. Dies lässt auf eine durch Stickstoff hervorgerufene Alterungsversprödung schließen. Die Mittelwerte der Ergebnisse der Analyse können Tabelle 2 entnommen werden. Als sehr auffällig und auf einen herstellungsbedingt kritischen Verunreinigungsgrad zurückzuführen gilt der Gehalt an Schwefel (S).

Tabelle 2: Ergebnisse der funkenspektrometrischen Analyse

Proben-bez.	**Chemische Zusammensetzung (Stückanalyse) in Masse-%**													
	%C	**%P**	**%S**	**%N**	**%Mn**	**%Si**	**%Al**	**%Ti**	**%Cr**	**%Ni**	**%Cu**	**%Nb**	**%V**	**%Mo**
G21-601-BK1-CA	0,154	0,040	0,117	0,0011	0,42	0,01	0,001	0,001	0,06	0,13	0,28	0,001	0,002	0,01
G21-601-BK6-CA	0,132	0,039	0,120	0,0006	0,42	0,01	0,001	0,001	0,06	0,12	0,27	0,001	0,002	0,01
G21-601-BK8-CA	0,154	0,040	0,117	0,0011	0,42	0,00	0,001	0,001	0,01	0,13	0,28	0,001	0,002	0,01
G21-601-BK7-CA	0,051	0,138	0,112	0,0113	0,52	0,01	0,001	0,001	0,06	0,07	0,08	0,001	0,002	0,01

Für die Beurteilung einer möglichen Alterungsversprödung erfolgten Kerbschlagbiegeversuche nach Charpy gemäß der DIN EN ISO 148-1 [16]. Dafür wurden je drei ISO-V-Proben in Längsrichtung aus den Bohrkernen des Gurts der ersten Dachstütze sowie zwei Proben aus den Bohrkernen des Knotenblechs herausgearbeitet. Die Durchführung der Kerbschlagbiegeversuche erfolgte bei den Temperaturen +20 °C, 0 °C und -20 °C. Die Ergebnisse der Versuchsreihe sind in Tabelle 3 zusammengefasst.

Abb. 15 zeigt die Untersuchungsergebnisse in graphischer Auswertung mit Angabe der Übergangstemperatur $T_{ü27}$, welche im vorliegenden Fall bei ca. 12°C liegt. Hierbei ist zu berücksichtigen, dass diese nicht direkt mit der Einsatztemperatur gleichgesetzt werden kann, sofern das zu bewertende Prüfobjekt nach aktuell gültigem Bemessungsregelwerk (nach)bewertet werden soll.

Tabelle 3: Übersicht der ermittelten Kerbschlagwerte [15]

Pos.	Proben-bez.	Prüftemperatur [°C]	Probenbreite [mm]	Probenhöhe [mm]	Prüfquerschnitt [cm²]	KV in J	Mittelwert KV in J
DS-I-OG	G21-601-BK2-KV1	RT	10	10	1	24,9	35
	G21-601-BK2-KV2	RT	10	10	1	48,3	
	G21-601-BK2-KV3	RT	10	10	1	31,3	
	G21-601-BK3-KV4	0	10	10	1	12,9	14
	G21-601-BK3-KV5	0	10	10	1	15,4	
	G21-601-BK3-KV6	0	10	10	1	13,6	
	G21-601-BK4-KV7	-20	10	10	1	7,8	7
	G21-601-BK4-KV8	-20	10	10	1	6,5	
	G21-601-BK4-KV9	-20	10	10	1	6,7	
DS-IV-St	G21-601-BK8-KV1	RT	10	10	1	33,7	37
	G21-601-BK8-KV2	RT	10	10	1	39,4	
	G21-601-BK8-KV3	0	10	10	1	7,1	8
	G21-601-BK9-KV4	0	10	10	1	8,1	
	G21-601-BK9-KV5	-20	10	10	1	4	5
	G21-601-BK9-KV6	-20	10	10	1	4,5	

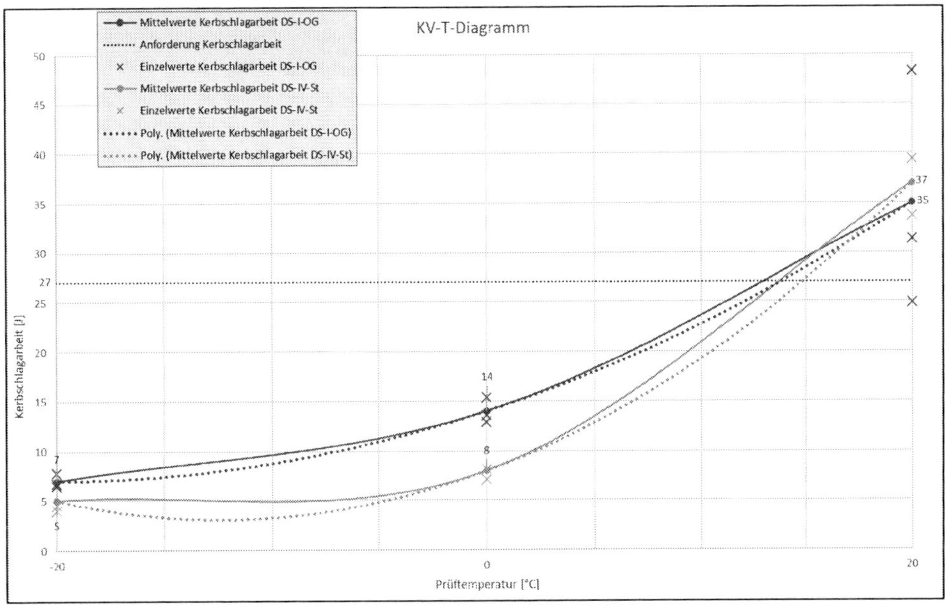

Abb. 15: Graphische Darstellung der Kerbschlagarbeitswerte [15]

Anhaltspunkte für die Nachbewertung von Zähigkeitskennwerten an aus dem Bauwerk stammenden Proben können dem aktuellen Bemessungsregelwerk nur näherungsweise entnommen werden. So wird bei der Werkstoffauswahl von Tragwerken, die nach europäischem Regelwerk (EC3) zu bemessen sind (Stahlgüteauswahl nach DIN EN 1993-1-10 [17]), dieser Umstand in Abhängigkeit von Ausnutzungsgrad und Bezugstemperatur berücksichtigt. Diese Werte sind für in-situ-Bewertungen als gegeben anzusehen. Es sind für verschiedene im bauaufsichtlichen Bereich eingesetzte unlegierte Stähle die maximal zulässigen Erzeugnisdicken angegeben, bis zu denen von einer minimalen Sprödbruchgefahr auszugehen ist. Tabelle 4 zeigt die Übersicht aus dem Eurocode 3, Teil 1-10 [17]. Für die Nachbewertung von Brückentragwerken kann gemäß den allgemeinen Bemessungsgrundlagen im Ingenieurbau eine Bezugstemperatur von -20°C angenommen werden, bei der auch die Kerbschlagbiegeversuche durchzuführen sind.

Hierbei gelten folgende Definitionen:

Bezugstemperatur: niedrigste zu bemessende Einsatztemperatur inklusive Sicherheitskonzept(e)

Übergangstemperatur: Prüftemperatur, bei der ein Übergang von Spröd- zu Verformungsbruch mit definiertem Anteil an Verformungsbruchfläche zu erwarten ist und/oder ein bestimmter Wert der Kerbschlagarbeit in J erreicht wird (meist bezogen auf 27 J Kerbschlagarbeit).

Ausnutzungsgrad: Spannungsniveau als Anteil der Nennstreckgrenze $f_y(t)$
$\sigma_{Ed1} = 0{,}75\ f_y(t)$ [MPa]
$\sigma_{Ed2} = 0{,}50\ f_y(t)$ [MPa]
$\sigma_{Ed3} = 0{,}25\ f_y(t)$ [MPa]

Tabelle 4: Auszug aus DIN EN 1993-1-10, Tabelle 2.1 [17]: Größte Erzeugnisdicken in mm (der Ausnutzungsgrad σ_{Ed} darf linear interpoliert werden).

Stahlsorte		KV		Bezugstemperatur T_{Ed} °C																				
Stahlsorte	Stahlgütegruppe	bei T °C	J_{min}	$\sigma_{Ed} = 0{,}75\,f_y(t)$							$\sigma_{Ed} = 0{,}50\,f_y(t)$							$\sigma_{Ed} = 0{,}25\,f_y(t)$						
				10	0	−10	−20	−30	−40	−50	10	0	−10	−20	−30	−40	−50	10	0	−10	−20	−30	−40	−50
S235	JR	20	27	60	50	40	35	30	25	20	90	75	65	55	45	40	35	135	115	100	85	75	65	60
	J0	0	27	90	75	60	50	40	35	30	125	105	90	75	65	55	45	175	155	135	115	100	85	75
	J2	−20	27	125	105	90	75	60	50	40	170	145	125	105	90	75	65	200	200	175	155	135	115	100
S275	JR	20	27	55	45	35	30	25	20	15	80	70	55	50	40	35	30	125	110	95	80	70	60	55
	J0	0	27	75	65	55	45	35	30	25	115	95	80	70	55	50	40	165	145	125	110	95	80	70
	J2	−20	27	110	95	75	65	55	45	35	155	130	115	95	80	70	55	200	190	165	145	125	110	95
	M,N	−20	40	135	110	95	75	65	55	45	180	155	130	115	95	80	70	200	200	190	165	145	125	110
	ML,NL	−50	27	185	160	135	110	95	75	65	200	200	180	155	130	115	95	230	200	200	200	190	165	145
S355	JR	20	27	40	35	25	20	15	15	10	65	55	45	40	30	25	25	110	95	80	70	60	55	45
	J0	0	27	60	50	40	35	25	20	15	95	80	65	55	45	40	30	150	130	110	95	80	70	60
	J2	−20	27	90	75	60	50	40	35	25	135	110	95	80	65	55	45	200	175	150	130	110	95	80
	K2,M,N	−20	40	110	90	75	60	50	40	35	155	135	110	95	80	65	55	200	200	175	150	130	110	95
	ML,NL	−50	27	155	130	110	90	75	60	50	200	180	155	135	110	95	80	210	200	200	200	175	150	130
S420	M,N	−20	40	95	80	65	55	45	35	30	140	120	100	85	70	60	50	200	185	160	140	120	100	85
	ML,NL	−50	27	135	115	95	80	65	55	45	190	165	140	120	100	85	70	200	200	200	185	160	140	120
S460	Q	−20	30	70	60	50	40	30	25	20	110	95	75	65	55	45	35	175	155	130	115	95	80	70
	M,N	−20	40	90	70	60	50	40	30	25	130	110	95	75	65	55	45	200	175	155	130	115	95	80
	QL	−40	30	105	90	70	60	50	40	30	155	130	110	95	75	65	55	200	200	175	155	130	115	95
	ML,NL	−50	27	125	105	90	70	60	50	40	180	155	130	110	95	75	65	200	200	200	175	155	130	115
	QL1	−60	30	150	125	105	90	70	60	50	200	180	155	130	110	95	75	215	200	200	200	175	155	130
S690	Q	0	40	40	30	25	20	15	10	10	65	55	45	35	30	20	20	120	100	85	75	60	50	45
	Q	−20	30	50	40	30	25	20	15	10	80	65	55	45	35	30	20	140	120	100	85	75	60	50
	QL	−20	40	60	50	40	30	25	20	15	95	80	65	55	45	35	30	165	140	120	100	85	75	60
	QL	−40	30	75	60	50	40	30	25	20	115	95	80	65	55	45	35	190	165	140	120	100	85	75
	QL1	−40	40	90	75	60	50	40	30	25	135	115	95	80	65	55	45	200	190	165	140	120	100	85
	QL1	−60	30	110	90	75	60	50	40	30	160	135	115	95	80	65	55	200	200	190	165	140	120	100

Es ist zu erkennen, dass der geprüfte Werkstoff des Dachträgers insbesondere bei tiefen Temperaturen zu sprödem Bruchversagen neigt. Mit Ausnahme des Bohrkerns 2 ist eine geringe Streuung der Versuchsergebnisse festzustellen. Abb. 16 bis Abb. 18 zeigen charakteristische Bruchflächen der Proben bei den unterschiedlichen Prüftemperaturen. So weisen die Proben bei Raumtemperatur noch deutliche Anteile an Verformungsbruch auf. Die hell-glänzenden Bereiche charakterisieren dabei die kristallinen Anteile der Bruchflächen und sind die Folge eines spröden Materialverhaltens. Mit Abnahme der Prüftemperatur zeigen die Bruchflächen zunehmend geringere Verformungen, dahingegen sind die kristallinen Anteile infolge eines Sprödbruches hier deutlich ausgeprägter. Aufgrund der geringen Kerbschlagarbeitswerte bei -20 °C wurden die Bruchflächen der entsprechenden Proben zusätzlich im Rasterelektronenmikroskop (REM) untersucht. Wie anhand Abb. 19 zu erkennen ist, weist die Bruchmorphologie große Spaltbruchflächen mit ausgeprägten transkristallinen Spaltfacetten auf. Diese sind charakteristisch für ein sprödes Materialversagen.

Abb. 16: Bruchfläche der Probe bei RT

Abb. 17: Bruchfläche der Probe bei 0°C

Abb. 18: Bruchfläche der Probe bei -20°C

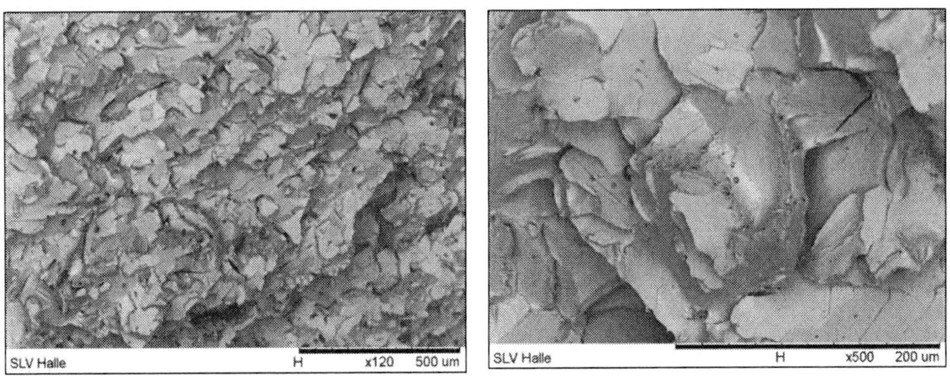

Abb. 19: Übersichtsaufnahme einer Kerbschlagprobenbruchfläche bei - 20°C im Rasterelektronenmikroskop. Spaltbruchfläche als typisches Anzeichen für sprödes Werkstoffverhalten infolge Alterung (links Übersicht, rechts Detail)

Die Ermittlung der Festigkeits- und Verformungseigenschaften sind nicht Bestandteil dieses Aufsatzes, da sie in Bezug auf die Schweißeignung eine untergeordnete Rolle spielen [8, 11, 14, 23].

Anhand der Ergebnisse der durchgeführten analytischen und mechanisch-technologischen Untersuchungen kann der Werkstoff dem unlegierten Baustahl DIN EN 10 025-2-S235JR nach aktuellem Verständnis, bzw. im historischen Kontext der Stahlsorte St 37 oder St 44 zugeordnet werden. Gleichzeitig lassen sich anhand der ermittelten Bruchdehnung Aussagen über das Plastizierungsvermögen und damit über das Ermüdungsverhalten des Werkstoffs treffen. So weist der vorliegende Altstahl eine Bruchdehnung A über 20 % auf, weswegen davon ausgegangen werden kann, dass ein ausreichendes Formänderungsvermögen vorliegt, um strukturbedingte Spannungskonzentrationen an Kerben durch lokale Plastizierung und ausreichendes Rissauffangvermögen abzufangen.

Gemäß des WTA-Merkblattes E-7-3-05/D [18] empfiehlt sich das Vorwärmen, um eine mögliche Wasserstoffversprödung durch Restfeuchtigkeit zu vermeiden, obwohl der niedrige Kohlenstoffgehalt des Grundwerkstoffes dies nicht explizit verlangen würde. Zudem konnte in Untersuchungen zu der Thematik der Ertüchtigung von Altstählen [19, 23] herausgearbeitet werden, dass die Zähigkeitseigenschaften von Werkstoffen mit geringen Kerbschlagarbeitswerten unterhalb von 10 J durch spezielle Wärmebehandlungsverfahren wieder revitalisiert werden können, wie beispielsweise durch das Normalisierungsglühen oder dem Zähigkeitsglühen. Sollte eine schweißtechnische Verarbeitung notwendig werden, wird die Prüfung möglicher thermischer Revitalisierungen für den untersuchten Werkstoff dringend empfohlen.

Ferner empfiehlt sich die Wahl eines Schweißverfahrens mit möglichst geringem Einbrand, wie beispielsweise das Lichtbogenhandschweißen (111). Um ein ausreichend zähes Schweißgut zu sowie eine bessere Verschlackung der Verunreinigungen des Stahls zu erreichen, sollten zudem grundsätzlich basisch umhüllte Stabelektroden verwendet werden. Diese eignen sich aufgrund des geringen Einbrandes vor allem hinsichtlich der festgestellten Seigerungszonen im Werkstoffinneren, die angesichts des hohen Verunreinigungsgrades und der daraus resultierenden unerwünschten Effekte nicht angeschmolzen werden sollten. Gleichzeitig sollte ein besonderes Augenmerk auf die Wahl der Schweißnahtart gelegt werden. So sind Beanspruchungen in Dickenrichtung, z.B. durch Verwendung von Kehlnähten, grundsätzlich zu vermeiden, da die Gefügezeiligkeiten mit einer hohen Neigung zur Lamellarrissigkeit einhergehen. Neben der Ableitung des Herstellungsverfahrens konnten anhand der durchgeführten chemischen Analysen zudem Aussagen über den Desoxidationsgrad des vorliegenden Thomasstahls getroffen werden. Dabei lässt die geringe Konzentration der Desoxidationselemente Silizium und Aluminium auf eine unberuhigte Vergießungsart des Werkstoffes schließen. Wie die Kerbschlagbiegeversuche zeigten, brachte der vorliegende Flussstahl vor allem bei tiefen Temperaturen Kerbschlagarbeitswerte, die deutlich unter den in modernen Regelwerken geforderten Werten von 27 J lagen. Zudem konnten anhand der metallographischen Gefügeuntersuchungen sowohl Eisennitridnadeln im Mikroschliff als auch die charakteristischen FRY'schen Kraftwirkungslinien im Makroschliff festgestellt werden.

Bestandteil der Untersuchungen waren weiterhin in-situ-Wärmebehandlungen mittels Induktion, um das Bauteil vor Ort per mobilem Energieeintrag induktiv zu erwärmen und dabei eine Normalisierungsglühung vorzunehmen (Abb. 20). Die für diese Art der Wärmebehandlung notwendige Haltezeit kann jedoch mittels handelsüblicher Induktionstechnologie, wie sie beispielsweise zum induktiven (Oberflächen-)Härten angewendet wird, nicht aufgebracht werden, sondern musste mittels einer statischen Regelung nach dem Prinzip eines Zweipunkt-Reglers erzeugt und oszillierend gehalten werden [8, 19, 23]. Aus diesem Grund wurde ein Regelkreis aufgebaut, der aus dem Probeblech als Regelstrecke, dem Induktor als Regler, einem Strahlungspyrometer als Messstelle sowie einem statischen Vergleichsglied in Form der jeweiligen Austenitisierungstemperatur T_A bestand. Letztere wurde dabei so gewählt, dass eine ausreichende Diffusion von C-Atomen stattfinden kann, ohne dabei ein zu starkes Austenitkornwachstum zu provozieren. Im Ergebnis konnten auf diese Weise die Wärmeeinflusszonen geschweißter Prüfstücke lokal normalisiert werden. Aus vorangegangenen FuE-Vorhaben ist bekannt, dass der Einfluss einer nachträglichen Kornfeinung (durch Normalisieren) sehr positiven Einfluss auf die Anhebung der Kerbschlagarbeitswerte hat [19]. Daraus resultierend wurde zunächst von einer gleichzeitigen Anhebung des lokalen Plastizierungsvermögens ausgegangen.

Abb. 20: induktive lokale Normalisierung der WEZ (links), normalisiertes Mikrogefüge der WEZ (rechts)

3 Zusammenfassung

In Zusammenhang mit den durch das in diesem Aufsatz auszugsweise vorgestellte FuE-Vorhaben [8, 19, 23] gewonnenen Untersuchungsergebnissen kann geschlussfolgert werden, dass der Gehalt an Begleitelementen ebenso wie der Grad an stattgefundener Alterungsversprödung (hervorgerufen durch Langzeitbeanspruchungen unterhalb der Streckgrenze) die Sprödbruchempflindlichkeit bei Raumtemperatur und tiefen Temperaturen erheblich steigern und somit für Konstruktionen aus solchen Werkstoffen zum erhöhten Risiko der Systemunzuverlässigkeit oder gar zur Gefährdung der Tragwerksicherheit führen können. Es konnte nachgewiesen werden, dass dieser Effekt hingegen reduziert werden kann, wenn durch entsprechend sachkundige Ertüchtigungsmaßnahmen Revitalisierungen der werkstofflichen Eigenschaften vorgenommen werden. So konnten die Kerbschlagarbeitswerte, welche die Sprödbruchneigung aus prüftechnischer Sicht verkörpern, signifikant durch nachträgliche gezielte Wärmebehandlungen erhöht werden. Perspektivisch erscheint an dieser Stelle ein Abgleich mit Methoden der Bruchmechanik zielführend, da ebenfalls nachgewiesen werden konnte, welchen Einfluss der Wärmebehandlungszustand (und damit der revitalisierte Alterungsgrad) auf das globale Rissauffangvermögen bei schweißtechnischer Verarbeitung hat. Der entsprechende Nachweis aus Sicht der Ermüdungsfestigkeit erfolgte in den Dauerschwingversuchen unter den Gesichtspunkten der Betriebsfestigkeit.

Abb. 21 zeigt zeilig ausgeschiedene Mangansulfide auf der Bruchfläche, wie sie im Zuge einer energiedispersiven Röntgenspektroskopie (EDS/EDX) am Rasterelektronenmikroskop detektiert werden konnten. Diese galten innerhalb der Bruchfläche als spröd versagende Bereiche, während der überwiegend ferritische Matrix-Werkstoff ein duktiles Schwingbruchbild zeigte. Abschließend zum Thema Ermüdungsfestigkeit gealterter Baustähle lässt sich zusammenfassen, dass eine nachweisbare Versprödung durch zeilenartige Mangansulfide nicht zwangsläufig mit einer Herabsetzung der Schwingfestigkeit einhergehen muss. Die Fähigkeit, örtliche Spannungskonzentrationen, wie sie geometriebedingt an den Nahtübergängen bei Schweißverbindungen entstehen, können durch lokal begrenztes plastisches Verformungsverhalten des Werkstoffes abgefangen werden, solange die kritische Beanspruchungsgeschwindigkeit zum Erreichen der Spaltbruchspannung nicht überschritten wird. Dieser Sachverhalt stellt eine neuartige Erkenntnis über das Ermüdungsverhalten von alterungsversprödeten Stählen unter zyklischer Beanspruchung dar, da das Spaltbruchkriterium nur von der Beanspruchungsgeschwindigkeit, nicht aber von der Höhe und Dauer der zyklischen Beanspruchung abhängig ist. So konnte trotz zahlreicher zeiliger nichtmetallischer Seggregate im Mikrogefüge und damit einhergehender Anisotropie ein duktiles Ermüdungsverhalten mit örtlicher Verfestigung, wie sie durch das Vorhandensein von Mikrorastlinien und Gleitlinien bei fraktographischer

Untersuchung der Schwingbruchflächen nachgewiesen werden konnte, beobachtet werden. Eine nachträgliche Kornfeinung der WEZ wirkt sich dabei positiv auf das lokale Plastizierungsvermögen aus.

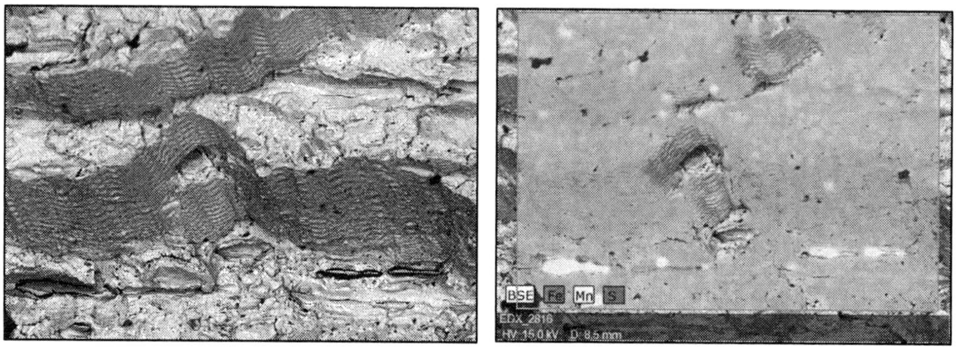

Abb. 21: Bruchoberfläche an Schwingprobe (links), Ergebnisse der EDX-Analyse im REM (rechts)

Ausgehend von diesem Sachverhalt lag der Fokus auf der Evaluierung des Einflusses dieser Werkstoffschädigung auf die zyklische und schlagdynamische Beanspruchbarkeit, da vermutet wurde, dass durch die Herabsetzung des Formänderungsvermögens auch die Fähigkeit des Werkstoffes unter hohen Beanspruchungen zu plastizieren reduziert worden war. Dabei sollte sich herausstellen, dass diese Vermutung für hohe Beanspruchungsgeschwindigkeiten auch zutreffend ist. Wie Versuche zum Einfluss der Beanspruchungsgeschwindigkeit auf die Festigkeitseigenschaften zeigten, kann für unlegierte Baustähle ein Plastizierungsverlust durch lokales Verfestigen, d. h. unter minimaler Zunahme der statischen (Dehn-)Festigkeitswerte, beobachtet werden, welches mit einer Reduzierung des Rissauffangvermögens einhergeht [20]. Insbesondere bei tiefen Temperaturen wird dieser Einfluss deutlich verstärkt, wie es schematisch Abb. 22 zeigt. Die örtliche Dehngrenze des Werkstoffes steigt bei sinkender Temperatur so lange an, bis sie die kritische Beanspruchung in Form der Spaltbruchspannung schneidet [21].

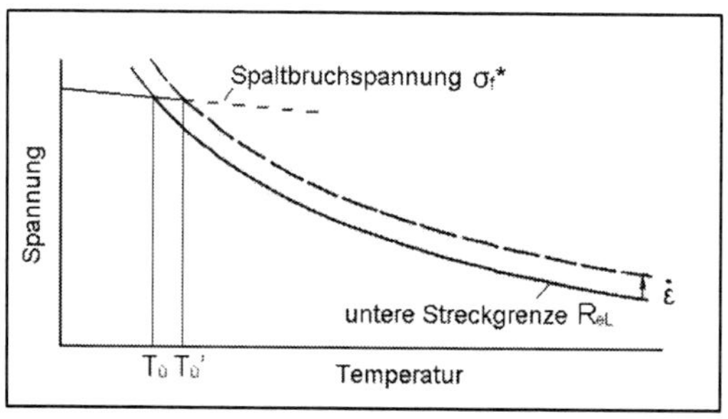

Abb. 22: Tieftemperaturverhalten unlegierter Baustähle mit ausgeprägter Streckgrenze (schematisch) nach [21]

Der durch vorangegangene Forschungsvorhaben zur Verfügung stehenden Literatur kann entnommen werden, dass die Mehrzahl der bislang untersuchten untersuchten Altstähle in allen betrachteten Kerbfällen gleichwertige Eigenschaften gegenüber einem Neumaterial aufweisen [15, 19, 22]. Dies bekräftigt die Hypothese, dass die Ermüdungsfestigkeit von Bauteilen Altstählen generell als metallurgisch-unabhängig gelten kann, da ausreichend hohes lokales Plastizierungsvermögen vorhanden ist und damit die Sprödbruchneigung aufgrund der Alterungsmechanismen größeren Einfluss auf die Bewertung der Restnutzungsdauer hat. Entscheidend für die Ergebnisse der Dauerschwingversuche war die geometrische Kerbe in Abhängigkeit vom jeweils vorliegenden Kerbfall.

Tabelle 5: Ausgewählte Kennwerte der WEZ für den untersuchten Werkstoff St 44*, ohne Ergebnisse von Ermüdungsfestigkeitsuntersuchungen [15, 23]

Entnahme	Wärmebehandlung	Streckgrenze R_{eH} [Mpa]**	Bruchdehnung A [%]**	Schwingfestigkeit $\Delta\sigma_C$ [Mpa]**	Kerbschlagarbeit K [J]**
Hbf Halle Dachstütze ca. 1900	unbehandelt	248	26	184	9
	Ofenglühung T_A = 900°C v_A = 1,4 K/s	264	37	178	70
	Induktion T_A = 950°C v_A = 1,9 K/s	-	-	254	52

*teilweise unter Berücksichtigung vorangegangener Themenstellungen und Voruntersuchungen am gleichen Werkstoff [23]
**Prüfung bei Raumtemperatur

Wie sich in den Untersuchungen zur induktiven Vor-Ort-Wärmebehandlung zeigte, existiert eine gute Korrelation zwischen der eingebrachten Energie zum Austenitisieren des Werkstoffes und dem Normalisierungseffekt bzw. den damit einhergehenden verbesserten Verformungs- und Zähigkeitseigenschaften. So konnte bewiesen werden, dass das Normalisierungsglühen bei Verwendung einer geeigneten thermischen Regelungstechnologie vollständig normalisierte Gefügebereiche erzeugt, welche zur Anhebung des lokalen Plastizierungsvermögens und damit zur Verschiebung von Anrissorten hin zu unkritischeren Bereichen führen. Tabelle 5 zeigt in diesem Zusammenhang die ermittelte Korrelation zwischen Wärmebehandlungszustand und Kerbschlagarbeit in der Wärmeeinflusszone für den untersuchten Kerbfall. Durch Verwendung von in den Feldverstärker des Flächeninduktors integrierte Glasfaser-Pyrometern kann gleichzeitig der technologisch bedingte Einfluss der durch die oberflächige Verzunderung verfälschten Strahlungsintensität eliminiert werden. Es besteht die Möglichkeit, vor Ort unter Baustellenbedingungen thermische Revitalisierungen der durch Sprödbruch gefährdeten Bereiche der potenziellen WEZ durchzuführen. Durch Schwingfestigkeitsuntersuchungen konnte gezeigt werden, dass derartig ertüchtigte Schweißnähte im vollen Einklang mit den aktuellen Bemessungsanforderungen für Schweißverbindungen unter zyklischer Belastung stehen.

4 Schrifttum

[1] Bundesanstalt für Straßenwesen (BASt):
Zustandsnoten der Brücken.
Stand: September 2023

[2] Bundesministerium für Verkehr und digitale Infrastruktur:
Bundesverkehrswegeplan 2030.
Entwurf März 2016.
Berlin: 2016-03

[3] International Council on Monuments and Sites (ICOMOS):
International Charter for the Conservation and Restoration of Monuments and Sites (The Venice Charter 1964).
Venedig: 2nd International Congress of Architects and Technicians of Historic Monu ments, 1964

[4] Lorenz, W.:
Bausystem und Tragwerk Stahl.
http://denkmalpraxismoderne.de/bausystem-und-tragwerk-stahl/
BTU Cottbus: Onlineveröffentlichung, 2018

[5] Geißler, K.:
Bewertung bestehender Stahlbrücken.
Berlin: Ernst u. Sohn, Stahlbaukalender, 2006

[6] Worldsteel Association:
World Steel Figures 2018.
Brüssel: World Steel association, 2018

[7] Skrzypczak, D.:
Teure Rekonstruktion: Kehren die Türme auf den Hauptbahnhof zurück?
Mitteldeutsche Zeitung (MZ), Lokalausgabe Halle (Saale) 28.09.2018

[8] SLV Halle GmbH.:
Bewertung der Betriebsfestigkeit von reparaturgeschweißten Altstahlkonstruktionen unter besonderer Beachtung ihrer Schweißeignung.
Abschlussbericht zum FuE-Vorhaben 2023
Halle (Saale): Förderkennzeichen: 49MF200119

[9] DIN 8528-1
Schweißbarkeit – metallische Werkstoffe, Begriffe
Ausgabe: 1973-06

[10] Gajda, Ch.:
Evaluierung von Ertüchtigungsmethoden reparaturgeschweißter und alterungsversprödeter Bauteile aus Altstählen durch gezielte Wärmebehandlungsmaßnahmen.
Vortrag in Berlin zum Verbandstag des Verbandes Innovativer Unternehmen e. V. 01.12.2016.
Berlin: Bundesministerium für Wirtschaft und Energie, 2016

[11] DIN 8528-2
Schweißbarkeit – Schweißeignung der allgemeinen Baustähle zum Schmelzschweißen
Ausgabe: 1975-03

[12] DIN 17 100
Allgemeine Baustähle – Gütevorschriften
Ausgabe: 1966-09

[13] DIN 50 159-1:
Metallische Werkstoffe – Härteprüfung nach dem UCI-Verfahren – Teil 1: Prüfverfahren.
Ausgabe: 2015-01

[14] Schuster, J., Gajda, Ch.:
Ein Blick in die Geschichte: Schweißeignung von Stählen für den Stahlbau, Teil 4: Sichere und reproduzierbare Bewertung.
Der Praktiker 10/2020, Düsseldorf, 75 (2020)

[15] Bäumel, L.:
Bewertung der zyklischen Beanspruchbarkeit alterungsversprödeter Flussstähle unter Berücksichtigung ihrer metallurgischen Eigenschaften.
Masterarbeit: Universität Kassel, 2023

[16] DIN EN ISO 148-1:
Metallische Werkstoffe – Kerbschlagbiegeversuch nach Charpy.
Ausgabe: 2017-05

[17] DIN EN 1993-1-10:
Eurocode 3: Bemessung und Konstruktion von Stahlbauten – Teil 1-10: Stahlsortenauswahl im Hinblick auf Bruchzähigkeit und Eigenschaften in Dickenrichtung.
Ausgabe: 2010-12

[18] WTA Merkblatt E-7-3-05/D:
Sanierung historischer Stahl- und Gusskonstruktionen nach WTA I – Schweißen von Altstahl.
Ausgabe: 2005

[19] SLV Halle GmbH:
Erhöhung der Restnutzungsdauer ruhend oder zyklisch beanspruchter Bauteile aus Altstählen durch gezielte metallurgisch-thermische Behandlungsmaßnahmen.
Abschlussbericht zum FuE-Vorhaben 2019
Halle (Saale): Förderkennzeichen: VF 160024

[20] Böttcher, Ch., et. al.:
Einfluss der Beanspruchungsgeschwindigkeit auf das mechanische Verhalten von unlegiertem Baustahl – Teil 1.
Stahlbau, Verlag Ernst und Sohn, Berlin, 72 (2003) 11, S. 778-787

[21] Dahl, W., u.a.:
Werkstoffkunde Stahl. Band 1: Grundlagen
Düsseldorf: Verein Deutscher Eisenhüttenleute. Springer Verlag, 1984

[22] Doliva, S:
Einfluss des Reparaturschweißens auf vorwiegend ruhende und zyklische Beanspruchung von Bauteilen aus Altstählen und deren Restnutzungsdauer in Anlehnung an die DB RIL 805.
Dissertation: Brandenburgische Technische Universität Cottbus-Senftenberg, 2016

[23] SLV Halle GmbH:
Einfluss des Reparaturschweißens auf vorwiegend ruhende und zyklische Beanspruchbarkeit von Konstruktionen und Bauteilkomponenten aus Altstählen.
Abschlussbericht zum FuE-Vorhaben 2014
Halle (Saale): Förderkennzeichen: MF120072

DGZfP-Jahrestagung 2025

Künstliche Intelligenz in der digitalen Radiographie

Lennart SCHULENBURG [1]
[1] VisiConsult X-ray Systems & Solutions GmbH, Stockelsdorf, Deutschland,
l.schulenburg@visicosnult.de

Kurzfassung. Die Integration von KI in die digitale Radiographie und andere Verfahren verändert die zerstörungsfreie Prüfung (ZfP), indem sie die Genauigkeit der Fehlererkennung und die Effizienz der Prüfung verbessert. Algorithmen für maschinelles Lernen und Deep Learning automatisieren die Bildanalyse und detektieren dabei Anomalien, die menschliche Prüfer möglicherweise übersehen.

Beispiele aus der Praxis in der Luft- und Raumfahrt, der Automobilbranche und dem Energiesektor zeigen Verbesserungen in Bezug auf Geschwindigkeit, Zuverlässigkeit und Kosteneinsparungen, da weniger Pseudoausschuss die Nacharbeit reduzieren. Dieses Paper skizziert wichtige Strategien für die Datenerfassung, das Modelltraining und die nahtlose Integration von KI in bestehende Arbeitsabläufe. Es zeigt, wie wichtig die interdisziplinäre Zusammenarbeit zwischen KI-Entwicklern, ZfP-Fachleuten und Branchenvertretern ist, um zuverlässige, qualitativ hochwertige Ergebnisse zu gewährleisten.

Durch den Einsatz von KI können Unternehmen potentiell bessere Prüfqualität erreichen und sich einen Wettbewerbsvorteil verschaffen, indem sie mittels vorausschauender Wartung und Echtzeitüberwachung die Effizienz erhöhen. Darüber hinaus ermöglicht die KI-gesteuerte Inspektion schnelle und genaue Ergebnisse, die die Produktqualität verbessern, Ressourcen optimieren und neue Möglichkeiten in der Fertigung eröffnen, was letztlich eine Innovation im Sinne Industrie 4.0-Umgebungen ist.

Keywords: Digitales Röntgen, Künstliche Intelligenz, Automatisierung, KI, Digitalisierung

Einführung

In der modernen industriellen Fertigung ist die Gewährleistung der strukturellen Integrität und Zuverlässigkeit von Komponenten entscheidend für die Prozesssicherheit und Produktionseffizienz. Zerstörungsfreie Prüfverfahren (ZfP) spielen seit langem eine zentrale Rolle bei der Qualitätskontrolle, da sie es den Herstellern ermöglichen, interne Defekte in Materialien und Baugruppen zu erkennen. Els ein Schlüsselverfahren hat sich die digitale Radiographie als grundlegendes Werkzeug für die Bewertung der inneren Merkmale industrieller Komponenten herauskristallisiert, insbesondere in Sektoren wie der Luft- und Raumfahrt, der Automobilindustrie und dem Energiesektor. Trotz ihrer Wirksamkeit ist die konventionelle Röntgenprüfung jedoch nach wie vor weitgehend von der menschlichen Interpretation abhängig, was sie anfällig für Subjektivität, Inkonsistenz und ermüdungsbedingte Fehler macht.

Um diese Einschränkungen zu überwinden, wurde die automatische Fehlererkennung (ADR) entwickelt, bei der Computer-Vision-Techniken zur Verbesserung der Fehlererkennungsgenauigkeit eingesetzt werden. Frühe ADR-Systeme stützten sich auf herkömmliche Bildverarbeitungsalgorithmen, die aufgrund ihrer Abhängigkeit von

https://doi.org/10.58286/31220

vordefinierten Regeln und ihrer begrenzten Anpassungsfähigkeit an unterschiedliche Fehlertypen und Bildgebungsbedingungen eingeschränkt waren. Mit dem Aufkommen der künstlichen Intelligenz (KI), insbesondere des Deep Learning, hat sich im Bereich der ADR ein Paradigmenwechsel vollzogen, der ausgefeiltere, datengesteuerte Fehlererkennungsmechanismen ermöglicht. Convolutional Neural Networks (CNNs), eine Untergruppe von Deep-Learning-Modellen, haben bemerkenswerte Erfolge bei der Erkennung komplexer Defektmuster in Röntgenbildern gezeigt und übertreffen herkömmliche Methoden in Bezug auf Geschwindigkeit, Präzision und Zuverlässigkeit.

Trotz des transformativen Potenzials der KI-gestützten ADR ist ihre Einführung in der industriellen Röntgeninspektion nicht ohne Herausforderungen. Die Implementierung von Deep-Learning-basierten Systemen erfordert umfangreiche kommentierte Datensätze für das Training, die aufgrund der proprietären Natur industrieller Fehlerdaten und der Seltenheit bestimmter Fehlertypen schwer zu beschaffen sein können. Darüber hinaus müssen Probleme im Zusammenhang mit der Modellverallgemeinerung, der Interpretierbarkeit und der Integration in bestehende Fertigungsabläufe angegangen werden, um die Effizienz von KI-basierten Prüfsystemen zu maximieren.

Dieser Beitrag untersucht die Entwicklung der digitalen Röntgeninspektion in industriellen Anwendungen, die Rolle der künstlichen Intelligenz bei der automatischen Fehlererkennung und die Hauptursachen für Interpretationsfehler bei von Menschen durchgeführten Inspektionen. Indem wir Parallelen zur etablierten Forschung im Bereich der medizinischen Bildinterpretation ziehen, untersuchen wir, wie menschliche kognitive Einschränkungen, Ermüdung und Verzerrungen zu Fehlern bei der Fehlererkennung beitragen. Darüber hinaus erörtern wir die aktuellen Herausforderungen bei der Implementierung von KI-gesteuerten ADR-Systemen und skizzieren zukünftige Forschungsrichtungen, die darauf abzielen, die Zuverlässigkeit, Skalierbarkeit und Erklärbarkeit von KI-gestützter Röntgeninspektion zu verbessern.

Das Ziel dieses Papers ist es, das transformative Potenzial von KI in der industriellen Qualitätskontrolle aufzuzeigen und gleichzeitig die technischen und praktischen Überlegungen anzusprechen, die für einen erfolgreichen Einsatz notwendig sind. Da die Industrie grundsätzlich auf Automatisierung und KI-gesteuerte Lösungen setzt, wird die Integration intelligenter Fehlererkennung in die Röntgeninspektion eine immer wichtigere Rolle bei der Steigerung der Fertigungseffizienz, der Reduzierung von Fehlern und der Gewährleistung höchster Standards für Produktqualität und -sicherheit spielen.

1. Warum AI für die ZfP relevant ist

1.1 Geschichte und Entwicklung der digitalen Radiographie

Während des größten Teils des 20. Jahrhunderts wurde die industrielle Radiographie ausschließlich manuell durchgeführt. Inspektoren prüften Film- oder Echtzeitbilder auf Anzeichen von Defekten. Die analoge Filmradiographie war zwar zuverlässig, schloss aber eine sinnvolle Automatisierung aus; die gesamte Fehlererkennung und -beseitigung hing von der menschlichen Interpretation ab. Ein wichtiger Wendepunkt war das Aufkommen der Computerradiographie (CR) und digitaler Detektoren in den 1980er Jahren. Die Digitalisierung bildete die Grundlage für die computergestützte Bildverarbeitung und frühe ADR-Algorithmen und ermöglichte die ersten Versuche zur automatischen Erkennung von Fehlern in Röntgenbildern. Durch die Umwandlung der Röntgenergebnisse in digitale Daten konnten die Hersteller Filter und andere Bildverarbeitungstechniken anwenden, um potenzielle Fehlerbereiche hervorzuheben, was mit Film allein nicht möglich war.

Die ersten ADR-Implementierungen in den 1990er und 2000er Jahren waren oft regelbasiert oder basierten auf klassischer Bildverarbeitung. Automatisierte Röntgen-Inspektionssysteme in Großserienindustrien (z. B. Automobilgießereien) verwendeten beispielsweise Schwellenwerte und Mustererkennung, um häufige Fehler (Lunker, Porosität usw.) in Aluminiumgussteilen und Schweißnähten zu erkennen. Diese Systeme verbesserten zwar den Durchsatz, erforderten jedoch eine umfangreiche Abstimmung durch ZfP-Experten und waren im Allgemeinen auf bestimmte Anwendungen mit gut charakterisierten Fehlertypen beschränkt. In Großserienmärkten wie der Automobilherstellung wurden diese klassischen ADR-Lösungen jedoch unentbehrlich, um die Produktionsgeschwindigkeit zu erreichen. Damit wurde ein Präzedenzfall geschaffen: Durch ADR und Automatisierung war es möglich, sich wiederholende Prüfaufgaben in großem Umfang zuverlässig zu erledigen, solange die Teile und Fehlerarten bekannt und konsistent waren.

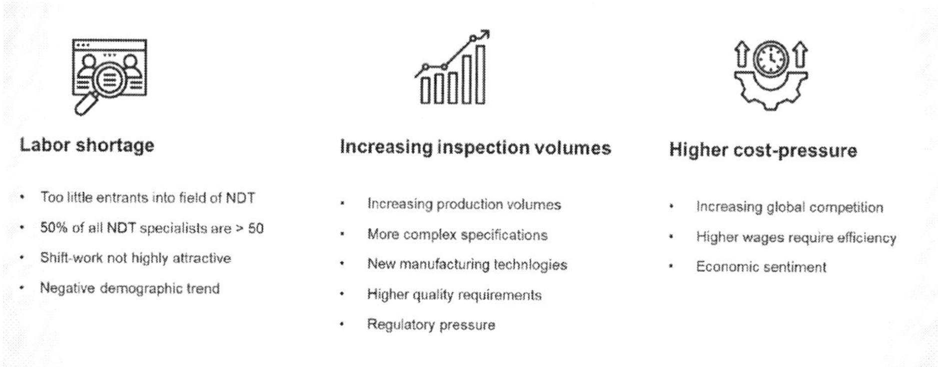

Abb. 1. Wirtschaftliche Treiber, die den Bedarf von Automatisierung fördern (Copyright VisiConsult)

In den letzten zehn Jahren hat sich die Entwicklung von ADR mit dem Aufkommen des maschinellen Lernens und insbesondere des Deep Learning beschleunigt. Traditionelle ADR-Algorithmen hatten mit komplexen oder variierenden Indikationen zu kämpfen und erforderten eine manuelle Programmierung für jedes neue Teil oder jeden neuen Fehlertyp. KI versprach einen flexibleren, datengesteuerten Ansatz. Als große Datensätze digitaler Röntgenbilder zur Verfügung standen und die Rechenleistung zunahm, begannen Forscher damit, neuronale Netze und andere Modelle zu trainieren, um Defekte direkt aus Bilddaten zu erkennen. Heutzutage zeigen auf Deep Learning basierende ADR-Prototypen eine überragende Fähigkeit zur Erkennung und Klassifizierung von Fehlern bei unterschiedlichen Bauteilgeometrien und Bildgebungsbedingungen und übertreffen damit die Leistung von manuell erstellten Algorithmen. Die automatisierte Fehlererkennung mit Deep Learning kann eine hohe Segmentierungsgenauigkeit auf echten Röntgenbildern erreichen, was ein Niveau an Präzision und Generalisierung widerspiegelt, das mit früheren Methoden nicht erreicht werden konnte. Diese KI-getriebene Entwicklung greift nun auch auf sicherheitskritische Bereiche über. Branchen mit traditionell geringerem Bilddurchsatz (z. B. Luft- und Raumfahrt, Öl und Gas) beginnen, ADR zu übernehmen, da ihre Volumina steigen und KI ihren Wert beweist; es besteht eine große Nachfrage nach intelligenten Inspektionslösungen, die menschliche Inspektoren unterstützen oder bestimmte Bewertungsschritte in diesen Sektoren automatisieren können.

1.2 Ursachen für Interpretationsfehler in der digitalen Radiographie

Der Prozess der Interpretation von Röntgenbildern ist von Natur aus komplex und erfordert ein hohes Maß an Fachwissen. Menschliche Bediener, die mit der Fehlererkennung beauftragt sind, stehen vor verschiedenen Herausforderungen, die die Genauigkeit beeinträchtigen können. Das Verständnis der Hauptursachen für Interpretationsfehler ist sowohl für die Verbesserung manueller Prüfverfahren als auch für die Verbesserung von KI-gestützten Erkennungssystemen von entscheidender Bedeutung.

Eine der Hauptfehlerquellen sind Auslassungsfehler, die auftreten, wenn ein Bediener einen Defekt entweder ganz übersieht oder nicht lange genug auf den interessierenden Bereich fixiert, um ihn richtig zu erkennen. Untersuchungen in der medizinischen Bildgebung haben gezeigt, dass Such-, Erkennungs- und Entscheidungsfehler zu diesen Fehlern beitragen. Ein Erkennungsfehler liegt vor, wenn der Bediener den Fehler zwar sieht, ihn aber nicht als signifikant erkennt. Fehler bei der Entscheidungsfindung treten auf, wenn ein Prüfer sich auf einen potenziellen Fehler konzentriert, ihn aber letztlich als Artefakt oder irrelevantes Merkmal falsch identifiziert. Aufmerksamkeit und Wahrnehmung spielen ebenfalls eine entscheidende Rolle bei der Fehlererkennung. Das menschliche Sehsystem ist so konzipiert, dass es sich immer nur auf einen begrenzten Teil eines Bildes konzentriert, und Inspektoren verlassen sich bei der Analyse von Röntgenbildern oft auf erlernte Muster und Erwartungen. Diese selektive Fokussierung kann zu Unaufmerksamkeitsblindheit führen, bei der offensichtliche Fehler übersehen werden, nur weil der Prüfer nicht aktiv nach ihnen gesucht hat.

Kognitive Verzerrungen erhöhen das Fehlerrisiko zusätzlich. „Anker-fehler“ treten zum Beispiel auf, wenn ein Prüfer sich einen ersten Eindruck von einem Bauteil verschafft und diesen Eindruck nicht korrigiert, wenn ihm neue Beweise vorgelegt werden. Framing-Effekte können die Fehlererkennung beeinflussen, indem sie die Erwartungen eines Inspektors aufgrund der Formulierung einer Inspektionsanfrage beeinflussen. Die Verfügbarkeitsverzerrung führt dazu, dass sich Inspektoren an kürzliche Fehlertrends erinnern, so dass sie erwartete Probleme eher erkennen, während sie weniger häufige Anomalien übersehen. Ermüdung ist ein weiterer wichtiger Faktor für Interpretationsfehler, insbesondere in Branchen, die eine kontinuierliche Prüfung von Komponenten erfordern. Studien haben gezeigt, dass längere Zeiträume der Fehleranalyse die Genauigkeit eines Inspektors verringern können, was sowohl zu falsch positiven als auch zu falsch negativen Ergebnissen führt. Auch eine hohe Arbeitsbelastung und ein hoher Geschwindigkeitsdruck bei der Auswertung können die Qualität beeinträchtigen, da die Bediener gezwungen sein können, Inspektionen im Eiltempo durchzuführen, um die Produktivitätsziele zu erreichen. Ablenkungen und externe Unterbrechungen spielen ebenfalls eine Rolle. Untersuchungen in der Radiologie haben gezeigt, dass selbst kurze Ablenkungen, wie z. B. die Beantwortung eines Telefonanrufs vor der Wiederaufnahme einer Untersuchung, die Fehlerquote erheblich erhöhen können. Industrielle Röntgeninspektoren arbeiten oft in Umgebungen, in denen Multitasking erforderlich ist, was die Wahrscheinlichkeit von Interpretationsfehlern aufgrund von geteilter Aufmerksamkeit erhöht.

1.3 Optimierungspotential durch KI

1.3.1 Verbesserung der Erkennungsgenauigkeit

Ein Hauptvorteil der Integration von KI in die digitale Radiographie ist die deutlich verbesserte Leistung bei der Fehlererkennung. KI-Modelle (insbesondere neuronale Netze, die auf großen Datensätzen trainiert wurden) haben das Potenzial, Defekte mit höherer

Genauigkeit und Konsistenz zu erkennen als herkömmliche manuelle Methoden. Sie sind in der Lage, schwache oder kleine Anzeichen - wie winzige Risse, Porosität oder Einschlüsse - zu erkennen, die selbst von erfahrenen Augen unter Zeitdruck übersehen werden könnten. Durch das Erlernen der subtilen Merkmale von Fehlern direkt aus den Bilddaten verringert die auf Deep Learning basierende ADR das Risiko unentdeckter Fehler und verbessert so die allgemeine Zuverlässigkeit der Inspektionen. In der Praxis haben Röntgeninspektoren aus der Luft- und Raumfahrt berichtet, dass KI-gestützte Systeme selbst komplexe Unregelmäßigkeiten in kritischen Komponenten (z. B. Turbinenschaufel-Gussteile oder Flugzeugteile aus Verbundwerkstoffen) aufzeigen können, für deren Erkennung früher mehrere Prüfer erforderlich waren. Die Konsistenz der KI-Algorithmen bedeutet auch weniger Fehlalarme: Wenn die Software richtig trainiert ist, ist die Wahrscheinlichkeit geringer, dass sie im Vergleich zu einem übervorsichtigen Menschen auf nicht defekte Anzeichen hinweist, so dass die Rate der falsch-positiven Ergebnisse sinkt und die Aufmerksamkeit nur auf das Notwendige gelenkt wird. Dieses verbesserte Signal-Rausch-Verhältnis bei der Erkennung führt zu geringeren Nacharbeits- und Ausschusskosten, da echte Fehler frühzeitig erkannt und Fehlalarme minimiert werden. Die Systemfähigkeiten von KI-Systemen müssen sorgfältig überwacht und vor und nach dem Einsatz erprobt werden.

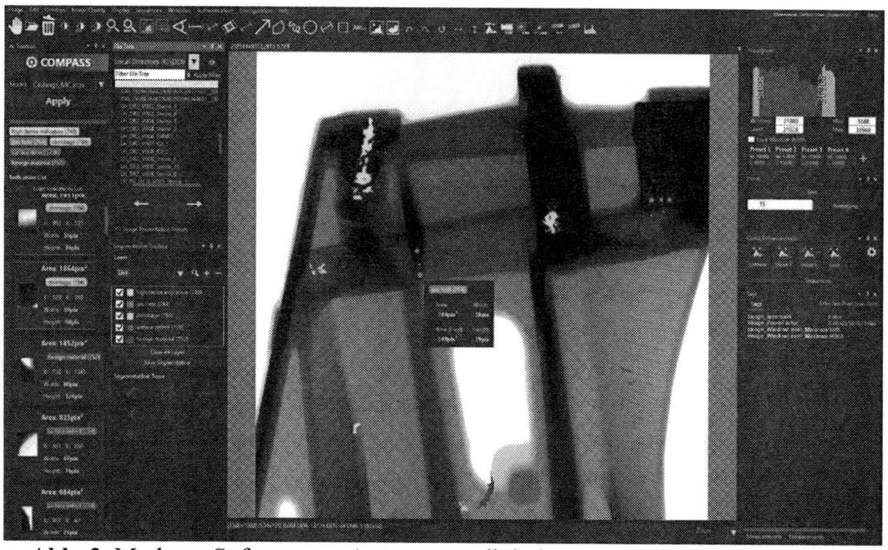

Abb. 2. Moderne Software zur Auswertung digitaler Röntgenbilder mit KI-Funktionen (Copyright VisiConsult)

1.3.2 Geschwindigkeits- und Produktivitätsgewinne

Die Automatisierung der Bildauswertung führt zu erheblichen Effizienzvorteilen. KI-gesteuertes ADR kann digitale Röntgenbilder in Sekundenschnelle analysieren und ermöglicht die Identifizierung von Fehlern während der Inspektion in Echtzeit oder nahezu in Echtzeit. Dies ist für Szenarien mit hohen Stückzahlen oder In-Line-Prüfungen von großer Bedeutung. In der Automobilproduktion zum Beispiel, wo pro Schicht Hunderte von Gussteilen oder Schweißnähten geröntgt werden, kann ein KI-System sofort Bilder mit Hinweisen von denen trennen, die unbedenklich sind, was die Arbeitsbelastung der menschlichen Prüfer drastisch reduziert und den Durchsatz beschleunigt. Eine sofortige Analyse unterstützt auch die Rückkopplungsschleifen in der Fertigung - wenn sich ein bestimmter Fehler abzeichnet, kann die Produktionslinie früher angepasst werden. Fallstudien haben gezeigt, dass KI-basierte ADR-Module zuverlässig Live-Röntgenbilder verarbeiten, um Defekte wie z. B. Gasporosität in Aluminiumgussteilen zu erkennen, so dass fehlerhafte Teile sofort vom Band genommen werden können, was die Ausbeute in der Produktion verbessert. Durch die Automatisierung der Interpretation ermöglicht KI den ZfP-Fachleuten, von einem seriellen, arbeitsintensiven Arbeitsablauf zu einem parallelen, schnellen Arbeitsablauf überzugehen, bei dem das System jedes Bild im Hintergrund triagiert oder sogar vollständig analysiert. Diese Produktivitätssteigerung hält nicht nur mit den modernen Produktionsraten Schritt, sondern trägt auch zur Senkung der Prüfkosten bei.

1.3.3 Verringerung von menschlichem Versagen und Fehlerquellen

Die Integration von KI wirkt wie ein zweites Paar Augen, das die bekannten menschlichen Faktoren bei der Röntgeninspektion abschwächt. Die wiederholte Überprüfung von Bildern kann den Prüfer ermüden, und Faktoren wie Unaufmerksamkeit oder subjektive Voreingenommenheit können dazu führen, dass Hinweise übersehen werden. Ein KI-System ermüdet nicht und verliert nicht den Fokus - es wendet dieselben Erkennungskriterien gleichmäßig auf Hunderte von Bildern an. Auf diese Weise fängt es potenzielle falsch-negative Ergebnisse ab (Fehler, die ein Mensch übersehen könnte) und liefert eine neutrale Bewertung, die nicht von individuellen Vorurteilen beeinflusst wird. Die Forschung in verwandten Bereichen hat Parallelen zur medizinischen Radiographie gezogen, wo die computergestützte Erkennung Radiologen dabei hilft, Übersehensfehler zu vermeiden; in ähnlicher Weise kann die KI-Unterstützung bei der zerstörungsfreien Prüfung Auslassungsfehler, Wahrnehmungsfehler und Unstimmigkeiten zwischen verschiedenen Prüfern erheblich verringern. Eine Industriestudie ergab, dass die Einführung eines KI-Tools zur Unterstützung von ZfP-Analysten bei der Überprüfung von CT-Röntgenaufnahmen von komplexen 3D-gedruckten Teilen für die Luft- und Raumfahrt als wirksames Sicherheitsnetz diente: Die KI wies auf mehrere subtile Anzeichen hin, die von Menschen zunächst übersehen wurden, und verhinderte so potenzielle Ausfälle im Betrieb. Durch die Entlastung der Inspektoren (die sich nun auf die kritischsten oder unklarsten Fälle konzentrieren können) verbessert die KI nicht nur die Erkennung, sondern verlängert auch die effektive Arbeitszeit der Mitarbeiter, da sie weniger ermüden. Dieser kollaborative Ansatz von Mensch und KI erhöht die Gesamtzuverlässigkeit, ohne den Menschen aus dem Spiel zu lassen.

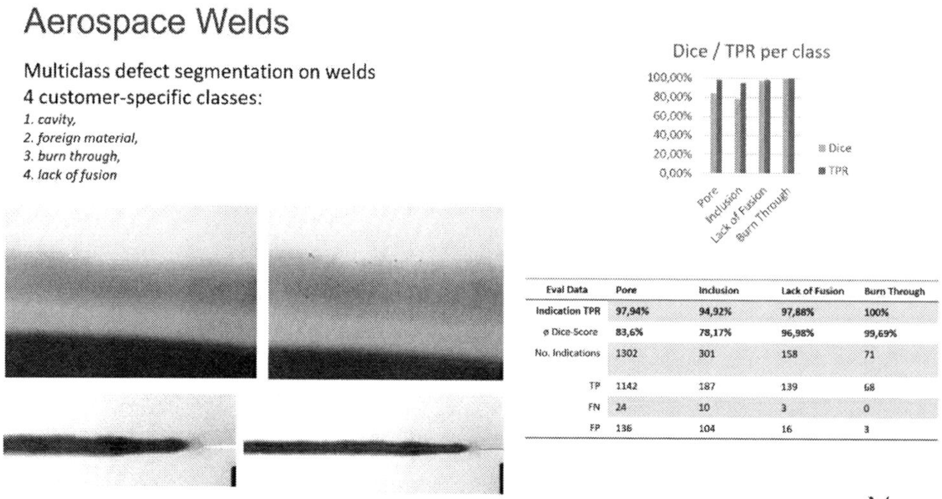

Eval Data	Pore	Inclusion	Lack of Fusion	Burn Through
Indication TPR	97,94%	94,92%	97,88%	100%
ø Dice-Score	83,6%	78,17%	96,98%	99,69%
No. Indications	1302	301	158	71
TP	1142	187	139	68
FN	24	10	3	0
FP	136	104	16	3

Abb. 3. Erkennungsraten eines modernen AI-Systems für dünne Schweißnähte in der Luft- und Raumfahrt (Copyright VisiConsult)

1.3.4 Standardisierung und Wissensbewahrung

KI-gesteuerte ADR-Systeme wenden, sobald sie trainiert sind, einheitliche Kriterien auf alle Bilder an. Dies trägt zur Standardisierung der Qualitätsannahme über Schichten und Standorte hinweg bei. In globalen Industrien ist es eine Herausforderung sicherzustellen, dass ein "OK" oder "Nicht OK" in einem Bauteil auch anderswo auf die gleiche Weise beurteilt wird - KI kann das kollektive Fachwissen der besten Inspektoren in ein Modell kodieren, das überall eine einheitliche Analyse liefert. Darüber hinaus unterstützt die KI-Integration den Wissenserhalt: Das System "merkt" sich, wie ein Fehler aussieht, selbst wenn erfahrene Mitarbeiter in den Ruhestand gehen oder die Belegschaft wechselt. Im Grunde genommen wird ein gewisses Stammeswissen über Fehlermuster in den trainierten Modellen institutionalisiert. Dieser Vorteil ist schwerer zu quantifizieren, hat aber langfristige Auswirkungen auf die Aufrechterhaltung der Qualität.

1.3.4 Strategischer und wirtschaftlicher Nutzen

Unternehmen, die KI in der radiografischen ZfP einsetzen, können Wettbewerbsvorteile erzielen. Eine verbesserte Fehlererkennung im ersten Durchgang führt zu einer höheren Produktqualität und weniger Qualitätsproblemen in der Folgezeit. Schnellere Inspektion und automatische Berichterstellung verkürzen die Zeit von der Produktion bis zur Auslieferung. Die Kosteneinsparungen durch das frühzeitige Erkennen von Defekten (Vermeidung von kostspieligen Nacharbeiten oder Ausfällen vor Ort) und die Optimierung des Arbeitsaufwands der Prüfer sind erheblich. Durch den Einsatz von KI können sich Unternehmen auch als Technologieführer präsentieren; die Einhaltung strenger Qualitätsstandards mit automatischer Rückverfolgbarkeit kann ein Verkaufsargument für Kunden sein. Darüber hinaus legt die Integration von ADR den Grundstein für künftige Innovationen: Die Dateninfrastruktur und die Algorithmen, die heute für ADR eingesetzt werden, können sich in Zukunft zu vorausschauenden Wartungssystemen und vollständig intelligenten Inspektionslinien entwickeln.

2. Herausforderungen bei der KI-gestützten ADR-Implementierung

Trotz der vielversprechenden Möglichkeiten ist die Implementierung von KI-gesteuerter ADR in der industriellen Radiographie nicht ohne Herausforderungen. ZfP-Experten müssen technische, betriebliche und sogar kulturelle Hürden überwinden, um diese Systeme erfolgreich einzusetzen.

2.1 Datenverfügbarkeit und -qualität

Leistungsstarke KI-Modelle erfordern große Mengen repräsentativer Trainingsdaten, was in der zerstörungsfreien Prüfung einen großen Engpass darstellt. Im Gegensatz zu anderen Technologiebranchen sind ZfP-Daten (Röntgenbilder von Teilen) oft geschützt und spärlich. Unternehmen zögern, Bilder von Defekten aus Gründen des geistigen Eigentums und der Sicherheit weiterzugeben, was zu begrenzten offenen Datensätzen führt. Der Aufbau eines firmeneigenen Datensatzes erfordert die Erfassung und sorgfältige Kennzeichnung von Tausenden von Bildern mit und ohne Defekte - ein arbeitsintensiver Aufwand, der sich über Monate oder Jahre erstrecken kann. Darüber hinaus müssen die Daten die Variabilität der Materialien, Dicken und Fehlertypen abdecken, auf die die KI stoßen wird. Viele ADR-Projekte scheitern, weil die anfänglichen Trainingsdaten einfach nicht umfassend genug waren, was zu Modellen führt, die im Labor gut funktionieren, aber bei neuen Fällen versagen. Die Datenqualität ist ebenfalls von entscheidender Bedeutung: Inkonsistente Bildgebungsparameter oder Fehler bei der Beschriftung können den Lernprozess in die Irre führen. Neuere Ansätze versuchen, dieses Problem durch die Erzeugung synthetischer Röntgenbilder zu lösen, um die Trainingsdaten zu ergänzen. So haben Forscher beispielsweise Simulationspipelines entwickelt, um realistische Röntgenbilder mit implantierten Defekten zu erzeugen, die von den statistischen Merkmalen realer Bilder abgeleitet sind. Es hat sich gezeigt, dass diese synthetischen Datensätze, wenn sie mit einem kleineren Satz echter Röntgenbilder kombiniert werden, die Modellgenauigkeit erhöhen und gleichzeitig die Kosten für die Datenerfassung überschaubar halten. Synthetische Daten führen jedoch das Problem der Domänenlücke ein - Modelle, die nur auf simulierten Bildern trainiert wurden, lassen sich aufgrund subtiler Unterschiede möglicherweise nicht sofort auf reale Bilder übertragen, was eine sorgfältige Validierung und häufig eine Feinabstimmung mit realen Daten erfordert. Zusammenfassend lässt sich sagen, dass die Beschaffung von ausreichend hochwertigen Daten (und die Möglichkeit, diese sicher zu teilen oder zu synthetisieren) eine der größten Herausforderungen im Bereich der KI-basierten ADR darstellt.

2.2 Integration in bestehende Arbeitsabläufe

Die Einführung von KI in einen gut etablierten ZfP-Arbeitsablauf kann störend sein. Prüfverfahren in der Luft- und Raumfahrt oder der petrochemischen Industrie unterliegen beispielsweise strengen Normen und sind durch jahrelange Praxis zertifiziert. Ein KI-System muss nahtlos integriert werden, damit es die Produktion nicht unterbricht oder eine umfassende Umschulung des Personals erfordert. Das bedeutet, dass die ADR-Software idealerweise in bestehende digitale Röntgensysteme und -software integriert werden sollte (unter Berücksichtigung von Standards wie DICONDE für Bilddaten) und die Ergebnisse in einem für die Inspektoren vertrauten Format darstellen sollte. Die Gewährleistung der Kompatibilität mit verschiedenen Detektortypen, Bildformaten und Archivierungssystemen kann nicht trivial sein. Darüber hinaus müssen Latenzzeiten und Rechenanforderungen berücksichtigt werden: Wenn die Analyse in Echtzeit erfolgen soll, benötigt das System möglicherweise leistungsstarke GPUs vor Ort oder Edge-Computing-Geräte, um Verzögerungen bei der Entscheidungsfindung zu vermeiden. Einige Implementierungen entscheiden sich für eine Cloud-basierte Verarbeitung, was jedoch Bedenken hinsichtlich der Datenübertragungsgeschwindigkeit und der Sicherheit sensibler Prüfdaten aufwirft. Außerdem stellt sich die Frage nach der Skalierbarkeit - ein Modell mag für einen Teiletyp funktionieren, aber der Einsatz in mehreren Produktlinien bedeutet eine Umschulung oder Feinabstimmung auf neue Fehlertypen, was einen anhaltenden Aufwand bei der Datenverwaltung und Modellpflege erfordert. Ohne sorgfältige Planung kann eine ADR-Initiative schwer aufrechtzuerhalten sein, insbesondere wenn sich der Fertigungsprozess oder die geprüften Designs ändern (was ein kontinuierliches Lernen oder eine regelmäßige Neuvalidierung der KI erfordert). Eine wirksame Integration erfordert daher eine enge Zusammenarbeit zwischen KI-Entwicklern, ZfP-Experten und Produktionsingenieuren, um maßgeschneiderte Lösungen zu finden, die den betrieblichen Anforderungen gerecht werden.

2.3 Menschliche Faktoren und Akzeptanz

Selbst die präziseste KI wird keinen Nutzen bringen, wenn die menschlichen Inspektoren ihr nicht vertrauen oder sie nicht richtig einsetzen. Das Vertrauen der Prüfer in KI-Empfehlungen zu gewinnen, ist eine große Herausforderung. Erfahrene ZfP-Fachleute stehen einem "Black Box"-Algorithmus möglicherweise skeptisch gegenüber, insbesondere wenn dieser gelegentlich harmlose Merkmale markiert oder einen atypischen Fehler übersieht. Es kann zu Widerständen kommen, weil man befürchtet, dass die Automatisierung die Rolle des Prüfers schmälern könnte. Daher ist es wichtig, KI als Hilfsmittel und nicht als Ersatz für das menschliche Urteilsvermögen einzusetzen. Der Trend in der Branche geht in Richtung "Assisted Defect Recognition", bei der die KI Hinweise hervorhebt und der Mensch die endgültige Entscheidung trifft. Mit diesem erweiterten Ansatz behält der Prüfer die Kontrolle und steht im Einklang mit den sicherheitskritischen Zertifizierungsanforderungen, die häufig eine menschliche Aufsicht über die endgültigen ZfP-Ergebnisse vorschreiben. Dennoch ist die Aufrechterhaltung der Aufsicht von entscheidender Bedeutung; ein übermäßiges Vertrauen in die KI (Selbstgefälligkeit) könnte gefährlich sein, wenn die Inspektoren beginnen, KI-Entscheidungen abzustempeln. Schulungsprogramme und Änderungsmanagement sind erforderlich, um die Partnerschaft zwischen Mensch und KI zu kalibrieren: Die Prüfer müssen die Fähigkeiten der KI und die Fehlerarten verstehen. Die Visualisierung der Ergebnisse kann dabei helfen - zum Beispiel, indem farbige Markierungen auf dem Röntgenbild eingeblendet werden, wo die KI einen Fehler vermutet, zusammen mit einer Vertrauenskennzahl. Dadurch erhält der Bediener einen Einblick, warum die KI eine Indikation vorschlägt. Die Bemühungen um erklärbare KI für die zerstörungsfreie Prüfung zielen darauf ab, die Transparenz weiter zu verbessern, indem Gründe für eine Fehlerklassifizierung angegeben werden (z. B. Hervorhebung bestimmter Texturmerkmale), die das Vertrauen der Benutzer stärken können. Letztendlich ist es das Ziel, die KI wie einen Junior-Inspektor zu behandeln, der nie müde wird - ihr Beitrag ist wertvoll, aber der zertifizierte Inspektor bleibt die Autorität, wodurch die Akzeptanz im täglichen Gebrauch sichergestellt wird.

2.3 Regulierung und Qualifikation

Die Einführung der künstlichen Intelligenz in die ZfP wirft Fragen in Bezug auf Vorschriften und Normen auf. Traditionell müssen die Ergebnisse der Durchstrahlungsprüfung von zertifiziertem Personal der Stufe II/III gemäß Normen wie ASME, ASTM, EN usw. bewertet werden. Wie passt eine KI in dieses System? Derzeit gibt es keine expliziten Normen für die Qualifizierung von ADR-Systemen. Frühe Anwender müssen ihr KI-System oft intern validieren (z. B. durch den Vergleich seiner Entscheidungen mit anerkannten Referenzergebnissen an einer großen Stichprobe), um Auditoren oder Kunden zufrieden zu stellen. Regulierungsbehörden und Branchengruppen beginnen, Rahmenbedingungen für die Qualifizierung von KI-Tools zu erörtern - zum Beispiel die Festlegung von Leistungsmaßstäben (Erkennungswahrscheinlichkeit, Falschanrufquote), die ein ADR-System bei repräsentativen Fehlern erfüllen muss, oder die Forderung, dass bei bestimmten kritischen Inspektionen ein Mensch anwesend sein muss. Bis solche Richtlinien formalisiert sind, gehen viele Unternehmen vorsichtig vor: Sie setzen KI in einer beratenden Funktion ein und lassen die endgültigen Bewertungen weiterhin nach dem üblichen Code durchführen. In sicherheitskritischen Branchen (Luft- und Raumfahrt, Nuklearindustrie) wird jede Änderung der Inspektionsmethodik genau geprüft; der Nachweis der Zuverlässigkeit einer KI für alle Fehlerarten und Bildbedingungen im Anwendungsbereich ist eine hohe Hürde. Eine weitere Überlegung ist die Haftung - wer ist verantwortlich, wenn eine KI einen Fehler übersieht, der später einen Ausfall verursacht? Diese rechtliche Unklarheit führt zu einem konservativen Ansatz beim Einsatz von KI in der zerstörungsfreien Prüfung. Die Bewältigung dieser Herausforderung wird wahrscheinlich eine branchenweite Zusammenarbeit zur Entwicklung von Standards und Best Practices erfordern. In der Tat ist ein Wissensaustausch über Gremien wie ASNT, ASTM-Ausschüsse und ZfP-Konferenzen im Gange, um zu kodifizieren, wie und wann KI-basierte ADR für den Einsatz qualifiziert werden kann. Sobald diese Normen entstehen, werden sie den Weg für eine breitere Akzeptanz ebnen, indem klare Anforderungen an die Dokumentation von Trainingsdaten, die Leistungsvalidierung und die regelmäßige Requalifizierung von KI-Systemen in der ZfP stellt.

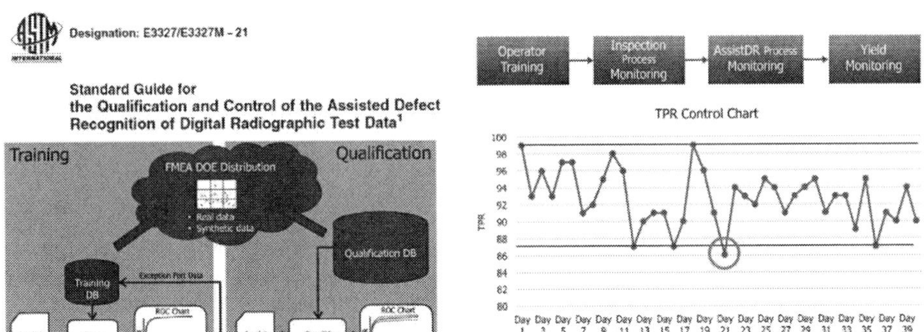

Abb. 4. Qualifizierung und Leistungsverfolgung eines KI-Systems im Zeitverlauf (Copyright VisiConsult)

2.3 Technische Limitationen

Trotz großer Fortschritte haben KI-Modelle technische Grenzen, die Praktiker erkennen müssen. Die meisten trainierten Modelle sind bereichsspezifisch - ein Netzwerk, das für die Erkennung von Porosität in Aluminiumgussteilen trainiert wurde, kann nicht ohne Weiteres fehlende Verschmelzung in Röntgenbildern von Stahlschweißnähten erkennen. Eine allgemeine KI, die für alle Materialien und Defekttypen funktioniert, ist immer noch ein ungelöstes Problem, das entweder große einheitliche Datensätze oder adaptive Algorithmen erfordert. Außerdem können Bildartefakte (Rauschen, Streuung, Strahlungswinkeleffekte) die KI verwirren, wenn sie in den Trainingsbeispielen nicht vorhanden waren. Studien haben gezeigt, dass die Qualitätsparameter von Röntgenbildern, wie Signal-Rausch-Verhältnis (SNR) und Kontrast, einen messbaren Einfluss auf die Leistung von Deep Learning bei der Fehlererkennung haben. Bei einem Modell, das auf hochauflösenden Bildern trainiert wurde, kann die Genauigkeit sinken, wenn es mit Aufnahmen von schlechterer Qualität konfrontiert wird (z. B. unterbelichtete Bilder oder starke Streuung in dickeren Schnitten). Das bedeutet, dass die Aufrechterhaltung konsistenter Bildgebungsbedingungen wichtig ist, wenn ADR oder eine andere Bildverbesserungsvorverarbeitung zur Normalisierung der Eingaben verwendet wird. Darüber hinaus sind einige Defekttypen von Natur aus schwierig - z. B. können Gussteile mit komplexer Geometrie aufgrund von überlappenden Strukturen falsche Anzeigen erzeugen, die eine KI als Defekte interpretieren könnte, wenn sie den 3D-Kontext nicht versteht. Es wird aktiv daran geforscht, solche Probleme zu entschärfen, z. B. durch die Verwendung von 3D-Kontext aus der Computertomographie als Grundlage für die 2D-ADR oder durch die Verwendung von Röntgenbildern aus mehreren Winkeln, um Mehrdeutigkeiten zu vermeiden. Schließlich ist das Paradoxon "99 % Genauigkeit ist nicht genug" in der zerstörungsfreien Prüfung deutlich spürbar: Selbst wenn eine KI eine 99 %ige Erkennungsrate erreicht, kann das 1 %ige Versäumnis katastrophal sein, wenn es sich um einen kritischen Fehler handelt. Daher wird häufig Redundanz (mehrere Modelle oder eine menschliche Kontrolle) eingebaut. Ein bemerkenswerter praktischer Ansatz besteht darin, dass anstelle einer vollautomatischen Entscheidung für jedes Bild KI eingesetzt wird, um große Stapel von Röntgenbildern zu sichten und die verdächtigsten Bilder schnell für eine menschliche Überprüfung zu markieren. Auf diese Weise wird sichergestellt, dass wahrscheinlich fehlerfreie Bilder nicht die Zeit der Inspektoren in Anspruch nehmen, während jedes Bild, das die KI als auch nur geringfügig anomal einstuft, die Aufmerksamkeit des Menschen erhält und so die seltenen Fehler auffängt. Techniken wie diese werden eingesetzt, um die Stärken und Schwächen der KI in industriellen Umgebungen praktisch auszugleichen.

3. Künftige Trends und Entwicklungen

Die kommenden Jahre versprechen bedeutende Fortschritte bei der KI-gesteuerten ADR für die digitale Radiographie, da sich sowohl die Technologie als auch die Branchenpraktiken weiterentwickeln. ZfP-Fachleute können mehrere wichtige Trends erwarten.

3.1 Prädiktive Analytik und ZfP 4.0

Über die Identifizierung von Defekten in einzelnen Bildern hinaus wird KI zunehmend eingesetzt werden, um aus Röntgendaten übergeordnete Erkenntnisse abzuleiten. Ein zu erwartender Trend ist die Integration von ADR-Ergebnissen in vorausschauende Wartungssysteme. Ein KI-System könnte das Auftreten und die Eigenschaften von Defekten im Laufe der Zeit verfolgen und vorhersagen, wann Prozessabweichungen oder

Anlagenverschleiß zu Defektspitzen führen könnten. Damit wird die ZfP von einer reinen Qualitätskontrollfunktion zu einem Werkzeug für die vorbeugende Instandhaltung, das den Grundsätzen von Industrie 4.0 entspricht. Große Mengen von Inspektionsdaten werden, sobald sie aggregiert sind, zu einer wertvollen Ressource für die Trendanalyse. KI-gesteuerte Analysen könnten beispielsweise aufzeigen, dass eine bestimmte Schweißmaschine nach einer bestimmten Anzahl von Stunden Mikrorisse zu produzieren beginnt, so dass eine Wartung erforderlich wird, bevor ein größeres Qualitätsproblem auftritt. Ein weiterer Aspekt ist die Echtzeit-Überwachung: In intelligenten Fabriken sind Inline-Radiographie-Einheiten geplant, bei denen KI nicht nur Defekte anzeigt, sondern auch in Echtzeit mit den Fertigungssystemen kommuniziert, um Probleme zu beheben (z. B. Anpassung der Gießparameter, wenn bei den letzten Teilen Porosität festgestellt wird). Im Grunde genommen entwickelt sich die zerstörungsfreie Prüfung von einem eigenständigen Qualitätssicherungssystem zu einem vernetzten Element des digitalen Fertigungssystems.

3.2 Verbesserte Algorithmen und Modelle

An der technischen Front treibt die Forschung die Möglichkeiten der KI in der Radiographie weiter voran. Wir werden wahrscheinlich ausgefeiltere neuronale Netzwerkarchitekturen sehen (und möglicherweise den Einsatz von Transformatoren oder anderen neuen KI-Modellen), die die Nuancen von Röntgenbildern besser erfassen. Eine Möglichkeit sind multimodale und 3D-Daten: die Kombination von 2D-Röntgenbildern mit 3D-Computertomographiedaten oder Simulationsmodellen, um der KI einen umfassenderen Kontext zu bieten. Eine andere Möglichkeit ist die unbeaufsichtigte Erkennung von Anomalien - statt Beispiele für jeden Defekt zu benötigen, werden Modelle auf vielen "guten" Bildern trainiert und lernen, alles zu erkennen, was anomal aussieht. Dies könnte das Long-Tail-Problem neuartiger oder seltener Fehlertypen lösen. Es gibt bereits Prototypen von Systemen, die Autoencoder oder generative adverserial Netze verwenden, um ungewöhnliche Anzeichen zu erkennen, ohne dass sie vorher explizit darauf trainiert wurden. Erklärbare KI ist ebenfalls ein Schwerpunkt: Künftige ADR-Software könnte eingebaute Erklärungsmodule enthalten, die zum Beispiel aufzeigen, welche Merkmale eines Bildes zu einer Fehlerklassifizierung geführt haben, um den Benutzern zu helfen, dem System zu vertrauen und es zu optimieren. Darüber hinaus werden kontinuierliche Verbesserungen bei der Bilderfassung (höher auflösende Detektoren, schnellere Bildgebung und fortschrittliche Algorithmen zur Bildverbesserung) der KI bessere Daten liefern und ihre Leistung steigern.

3.3 Datenzentrierte und simulationsgestützte Entwicklung

Das Paradigma der KI-Entwicklung verschiebt sich von der modellzentrierten zur datenzentrierten KI, insbesondere bei industriellen Anwendungen, wo Daten sehr wertvoll sind. Künftige ADR-Anstrengungen werden sich stark auf die Zusammenstellung optimaler Trainingsdatensätze konzentrieren - die Auswahl der richtigen Mischung von Fehlertypen, die Verwendung von Erweiterungen, um die Klassen auszugleichen, und die kontinuierliche Aktualisierung von Datensätzen, wenn neue Rückmeldungen aus der Praxis eintreffen. Die Erzeugung synthetischer Daten wird alltäglicher werden, mit digitalen Zwillingen und Monte-Carlo-Röntgensimulationsmaschinen, die Tausende von unterschiedlichen Bildern erzeugen, um Szenarien abzudecken, die physisch schwer zu erhalten sind. Wir können mit ausgefeilteren Simulationen von Röntgenbildern rechnen, die realistisches Rauschen, Artefakte und sogar zufällige Fehlerformen enthalten, um die Bereichslücke zu minimieren. Darüber hinaus werden Techniken wie die Domänenanpassung (das Erlernen der

Verallgemeinerung von Modellen von synthetischen auf reale Bilder) Fortschritte machen.

3.4 Breite Anwendung und allgemeine Standards

In dem Maße, wie sich die Erfolgsgeschichten häufen, werden immer mehr Branchen von Pilotprojekten zum vollständigen Einsatz von KI in der Radiographie übergehen. Wir beobachten bereits die Einführung in Sektoren wie dem Triebwerksbau in der Luft- und Raumfahrt (für die automatisierte Schweißnahtprüfung), der Öl- und Gaspipelineprüfung (für die Auswertung tausender Röntgenbilder von Rundnähten) und der Energieerzeugung (Prüfung von Turbinengussteilen und Kesselrohrschweißnähten). In einer Zusammenarbeit zwischen VisiConsult und GKN Aerospace wurde beispielsweise ein ADR-System für Laserschweißnähte in Bauteilen der Luft- und Raumfahrtindustrie vorgestellt, bei dem ein Deep-Learning-Ansatz mit mehreren Modellen (einschließlich eines Modells zur Segmentierung von Schweißbereichen) eingesetzt wurde, um schnelle Schlussfolgerungen mit minimalen Fehlalarmen zu erzielen, so dass es für den Produktionseinsatz geeignet ist. Solche Fallstudien zeigen, dass KI die strengen Qualitätsanforderungen kritischer Branchen erfüllen kann. Infolgedessen werden wir wahrscheinlich erleben, dass Normungsorganisationen die Verwendung von ADR formalisieren. Gremien wie ASTM, ASME, EASA und ISO arbeiten an der Einführung von Richtlinien für die Schulung und Qualifizierung von ADR-Systemen (einige Vorarbeiten wurden bereits veröffentlicht). Nationale Luftfahrtbehörden sowie Öl- und Gasregulierungsbehörden könnten empfohlene Praktiken für den Einsatz von KI-gestützter Radiographie herausgeben und diese schließlich in die Vorschriften aufnehmen, wenn das Vertrauen ausreichend ist. Diese Standardisierung wird die Akzeptanz fördern, da sie Unsicherheiten beseitigt und den Endnutzern klare Kriterien für die Akzeptanz bietet. Außerdem wird sich die Entwicklung der Arbeitskräfte verändern: ZfP-Zertifizierungen könnten Module zur Überwachung und Interpretation von KI enthalten, um sicherzustellen, dass die nächste Generation von Inspektoren mit einem KI-Assistenten genauso gut umgehen kann wie mit dem Lesen eines Films.

4. Zusammenfassung und Schlussfolgerungen

Die KI-gesteuerte automatische Fehlererkennung steht an der Spitze der technologischen Trends der digitalen Radiographie in der zerstörungsfreien Prüfung. Wie bereits dargelegt, hat die Entwicklung von der frühen Bildverarbeitung bis hin zu hochentwickelten Deep-Learning-Algorithmen ein neues Niveau an Fehlererkennungsleistung und betrieblicher Effizienz ermöglicht. Die ZfP-Profis von heute beginnen, KI nicht als Ersatz für ihr Fachwissen zu nutzen, sondern als leistungsfähiges Werkzeug, das ihre Fähigkeiten erweitert - indem es kleinere Defekte erkennt, überwältigende Bildmengen verarbeitet und datengestützte Erkenntnisse liefert, die zuvor unerreichbar waren. Fallstudien aus der Automobilindustrie, der Luft- und Raumfahrt und dem Energiesektor zeigen, dass KI-basiertes ADR bei durchdachter Implementierung die Inspektionsergebnisse erheblich verbessern kann, indem es die Zahl der übersehenen Fehler und Fehlalarme reduziert und gleichzeitig die Arbeitsabläufe beschleunigt. Die Vorteile - erhöhte Zuverlässigkeit, Konsistenz und Kosteneinsparungen - sind greifbar und haben frühe Anwender als Qualitäts- und Technologieführer auf dem Markt positioniert. Gleichzeitig erfordert der erfolgreiche Einsatz von KI in der ZfP die Bewältigung praktischer Herausforderungen: die Sicherstellung einer ausreichenden Anzahl qualitativ hochwertiger Daten, die Integration mit bestehenden Prozessen, die Aufrechterhaltung der menschlichen Aufsicht und die Einhaltung der sich entwickelnden Zertifizierungsstandards. Diese Herausforderungen werden von der ZfP-Gemeinschaft aktiv durch Innovation (z. B. synthetische Datengenerierung), Zusammenarbeit (Erfahrungsaustausch über Konferenzen und Arbeitsgruppen) und einen klaren Fokus auf die Ergänzung und nicht den Ersatz des menschlichen Prüfers angegangen. mit.

Referenzen

1. Hena, B. et al. (2024). "Towards Enhancing Automated Defect Recognition (ADR) in Digital X-ray Radiography Applications: Synthesizing Training Data through X-ray Intensity Distribution Modeling for Deep Learning Algorithms."

2. Inspenet (2023). The Integration of AI in Industrial Digital Radiography – Industry article on AI applications, benefits and challenges

3. VisiConsult (2023). Key Trends in Industrial Imaging– NDT 4.0 blog discussing Assisted Defect Recognition and automation in inspections

4. Benuzzi, S. et al. (2025). Mitigation of human factor in CT post-processing of AM parts – 14th iCT Conf. Paper, describing an AI tool to assist CT image analysis

5. ASNT Webinar (Schulenburg, 2023). Applying AI in Digital Radiography – Webinar highlighting ADR in automotive and aerospace, and discussing implementation challenges

6. Tyystjärvi, T. et al. (2023). Deploying Machine Learning for Radiography of Aerospace Welds – Case study (Trueflaw/GKN Aerospace) using multi-model ADR for weld inspection

7. Fraunhofer SCS (2022). AI for defect detection in castings (Project AI-SAR) – Project note on using simulated and real data for ADR on aluminum rims

8. Quality Digest (2003). Digital Radiography's Progress – Article on radiography evolution, noting CR enabled automation in the 1980s

DGZfP-Jahrestagung 2025

Digitale Repräsentation und Vernetzung von NDE-Informationen und -Systemen

Frank LEINENBACH[1], Dharma PANCHAL [2]
[1] Fraunhofer-Institut für Zerstörungsfreie Prüfverfahren IZFP, Campus E3.1, 66113 Saarbrücken, Deutschland, frank.leinenbach@izfp.fraunhofer.de
[2] Fraunhofer-Institut für Zerstörungsfreie Prüfverfahren IZFP, Campus E3.1, 66113 Saarbrücken, Deutschland, dharma.panchal@izfp.fraunhofer.de

Kurzfassung. Die digitale Transformation erfordert eine signifikante Anpassung der NDE-Systeme, um den steigenden Anforderungen an Integrationsfähigkeit und Interoperabilität gerecht zu werden. In der Vergangenheit beschränkten sich die Systeme oft auf einfache Steuerungsbefehle und Zustandsanzeigen. Heutzutage ist jedoch eine vollständige digitale Repräsentation der NDE-Systeme notwendig, die detaillierte Beschreibungen der Teilelemente, Parametersets und Ergebnisbeschreibungen umfasst. Dies verringert potenzielle Fehlinterpretationen und verbessert die Datenpflege und -aufbereitung, erfordert aber auch eine standardisierte Strukturierung und die Verfügbarkeit von Informationen in digitaler Form. Es ist wichtig, ein Datenmodell zu entwickeln, das sich dynamisch an die individuellen Anforderungen von NDE-Systemen und -Datensätzen anpassen und gegebenenfalls zukünftig erweitern kann. Technologischer Fortschritt, insbesondere durch Standards wie OPC UA und die Asset Administration Shell, ermöglicht eine nahtlose Vernetzung und den sicheren Austausch von Informationen in einem IIoT-Umfeld.

In diesem Manuskript werden technische Herausforderungen, Lösungsansätze und erste Implementierungen zur digitalen Darstellung von NDE-Informationen sowie deren Vernetzung in einem Datenökosystem vorgestellt.

Keywords: Digitale Repräsentation, NDE-Systeme, Interoperabilität OPC UA, Asset Administration Shell

1. Bedarf an digitalen Repräsentationen in der NDE

Ein wesentlicher Faktor zur Umsetzung aktueller Digitalisierungsstrategien ist die Bereitstellung relevanter Informationen durch entsprechende Systeme und Geräte. Dies bezieht sich sowohl auf die intelligenten Teilnehmer in einem Netzwerk als auch auf die Darstellung und Struktur der geteilten Informationen. Um diese Aussage einzuordnen, ist es notwendig, einen Blick auf die Netzwerkstrukturen der dritten industriellen Revolution zu werfen, die in vielen Fabriken heute noch den Standard bilden. Diese digitale Revolution war geprägt von der Implementierung intelligenter Systeme in der Produktion, was den Übergang von der Fließbandfertigung zu starren Automatisierungsanlagen und Robotik zur Folge hatte. Gleichzeitig entstand ein Bedarf an Kommunikationstechnologien, um Prozesse in der Fertigung zu koordinieren. Dabei wurden Kommunikationshierarchien etabliert, die durch die Automatisierungspyramide veranschaulicht werden. Die Ebenen dieser Pyramide repräsentieren unterschiedliche Aufgaben sowie die Echtzeitanforderungen und das erwartete Datenvolumen. So wird beispielsweise auf der ERP-Ebene eine „weiche" Echtzeitanforderung von Minuten bis Stunden erwartet, während auf der Aktuator-Sensor-Ebene eine „harte" Echtzeitanforderung von wenigen Millisekunden besteht. Diese vereinfachte Darstellung der Ausgangssituation zu Beginn der vierten industriellen

https://doi.org/10.58286/31226

Revolution konfrontiert aktuelle Aktivitäten mit mehreren Herausforderungen: Wie können Subsysteme Informationen austauschen, wenn sie auf derselben Hierarchieebene sind, unterschiedliche Kommunikationsprotokolle verwenden oder nicht einheitliche Datenstrukturen aufweisen? Bisher wurden diese Probleme größtenteils umgangen. Der Datenaustausch auf derselben Hierarchieebene wurde in der Regel vermieden und direkt in der Planung der Netzwerke über den Controller als zentrale Schnittstelle implementiert. Wenn ein direkter Austausch dennoch erforderlich war, wurden separate Verbindungen hergestellt.

Heute stehen diese Lösungsansätze im Widerspruch zu den Zielen von Industrie 4.0 und den Anforderungen an IIoT-Netzwerke (Industrial Internet of Things), die eine Aufweichung der Grenzen der genannten Hierarchieebenen und die direkte Verfügbarkeit relevanter Informationen zwischen den Teilnehmern eines Netzwerks vorsehen. Die Vielfalt und Inflexibilität der Kommunikationsstruktur einer Automatisierungspyramide wird dabei durch die Umsetzung einer sogenannten serviceorientierten Architektur (SOA) ersetzt. Dieser Begriff ist noch nicht eindeutig standardisiert, wird jedoch in der Regel durch die Definition der Organization for the Advancement of Structured Information Standards (OASIS) beschrieben, die ein SOA als ein Paradigma zur Organisation und Nutzung verteilter Fähigkeiten definiert, die ggf. verschiedenen Eigentumsbereichen unterliegen. Um diesen Schritt zu erreichen, müssen ideale Dienste bestimmte Anforderungen erfüllen:

- Ein Dienst ist in sich geschlossen und kann unabhängig genutzt werden.
- Ein Dienst ist über ein Netzwerk verfügbar.
- Ein Dienst hat eine klar definierte, veröffentlichte Schnittstelle.
- Ein Dienst ist plattformunabhängig.
- Ein Dienst ist in einem Verzeichnis registriert.
- Ein Dienst ist dynamisch gebunden.

Diese Anforderungen sind derzeit in der Praxis oft nicht vollständig erfüllt, zeigen jedoch deutlich, mit welchen neuen Herausforderungen Entwickler konfrontiert werden, wenn ein System in einem IIoT-Netzwerk eingebunden werden muss. Insbesondere die Verfügbarkeit im Netzwerk, die beschriebene Schnittstelle und die Plattformunabhängigkeit verdeutlichen den Rückgang des Individualismus der Hersteller hin zu einer Standardisierung und Vereinheitlichung der Kommunikationstechnologien sowie den Bedarf an einer semantischen Darstellung der verfügbaren Informationen. Diese technischen Anforderungen müssen gleichzeitig auch den rechtlichen Anforderungen des EU-Data Act [1] genügen. Die Regelungen umfassen die Verpflichtung, Produktdaten und damit verbundene Servicedaten für den Nutzer zugänglich zu machen, sowie die Gewährleistung der Interoperabilität zwischen verschiedenen Systemen und Diensten. Darüber hinaus muss sichergestellt werden, dass standardisierte Protokolle und Datenformate implementiert werden, um den Nutzern einen einfachen Datenaustausch zwischen verschiedenen Diensten innerhalb einer SOA-Umgebung zu ermöglichen. Die Einhaltung dieser Anforderungen gilt auch für die Integration von Verfahren zur zerstörungsfreien Prüfung und deren Systeme. Die Bereitstellung eines einzelnen Prozesses als Dienst stellt dabei kein Hindernis dar. Aktuelle Systeme wie Ultraschall-, Wirbelstrom- oder Thermografieverfahren haben bereits Netzwerkschnittstellen und könnten semantisch beschrieben und als geschlossenes System autonom genutzt werden. Damit erfüllen viele NDE-Systeme bereits die technischen Voraussetzungen zur Integration in IIoT-Netzwerke, allerdings steht derzeit eine standardisierte Beschreibung nicht zur Verfügung. Andere Verfahren, wie beispielsweise die Eindringprüfung, werden in der Regel manuell durchgeführt und erfordern zusätzlich ein Digitalisierungskonzept. Dies bedeutet jedoch nicht, manuelle Prozesse zu automatisieren, sondern vielmehr, den Prozess digital im Sinne einer serviceorientierten Lösung abzubilden.

Um dies zu erreichen, ist es notwendig, nicht nur die Vielfalt der NDE-Systeme, sondern auch deren Automatisierungsgrad zu berücksichtigen. Infolgedessen sind nachfolgende Anforderungen zu erfüllen, um ZFP-Systeme im Sinne einer SOA in ein IIoT-Netzwerk einzubinden:

- Einheitliche semantische Beschreibung verschiedener NDE-Systeme
- Darstellung als unabhängigen, autonom nutzbaren Dienst
- Digitale Abbildung manueller Prozesse
- Berücksichtigung des Automatisierungsgrads
- Nahtlose Kommunikation zwischen verschiedenen Systemen
- Eindeutige Identifizierung

Im Kontext der bereits erwähnten Anforderungen ist es wichtig, bestehende Standards und die Standardisierung bezüglich der Umsetzung dieser Verfahren zu berücksichtigen. Dieses Detailniveau bietet jedoch auch Chancen, da es die Grundlage für eine präzise Beschreibung bildet. Dies macht NDE-Verfahren zu idealen Datenquellen für IIoT-Netzwerke. Nachfolgend werden Technologien und Beispiele vorgestellt, die die genannten Anforderungen erfüllen.

2. Technologien zur digitalen Repräsentation

Die Beschreibung von Systemen innerhalb eines IIoT-Netzwerks erfordert eine digitale Repräsentation [2]. Diese bezieht sich auf die Art und Weise, wie Informationen, Daten und Konzepte in einem digitalen Format dargestellt und verarbeitet werden. Die Repräsentation ermöglicht es digitalen Geräten, Daten effizient zu speichern, zu analysieren und zu übertragen. Mit der fortschreitenden Entwicklung von Technologien und Medien wächst auch die Relevanz und Komplexität der digitalen Repräsentation, was neue Herausforderungen und Möglichkeiten für Forschung und Anwendung mit sich bringt. In der Industrie ist die digitale Repräsentation ein zentrales Konzept, das es Unternehmen ermöglicht, physische Objekte, Prozesse und Systeme in einer virtuellen Umgebung abzubilden. Diese Abbildung umfasst nicht nur die physikalischen Eigenschaften von Maschinen und Anlagen, sondern auch deren dynamische Verhaltensweisen, Betriebszustände und Interaktionen mit anderen Komponenten. Durch die digitale Repräsentation können Unternehmen wertvolle Einblicke in ihre Produktionsabläufe gewinnen, was zu einer verbesserten Effizienz, Qualität und Flexibilität führt. Die digitale Repräsentation spielt auch eine entscheidende Rolle bei der Entwicklung von digitalen Zwillingen, die virtuelle Abbilder von physischen Assets darstellen. Diese digitalen Zwillinge ermöglichen es, Simulationen durchzuführen, verschiedene Szenarien zu testen und fundierte Entscheidungen auf der Grundlage von Datenanalysen zu treffen. Zudem fördern sie die Zusammenarbeit zwischen verschiedenen Abteilungen und Stakeholdern, indem sie eine gemeinsame Datenbasis bereitstellen. Technologien wie OPC UA (Open Platform Communications Unified Architecture), MQTT (Message Queuing Telemetry Transport) und die Asset Administration Shell (AAS) unterstützen die Umsetzung dieser digitalen Repräsentationen, indem sie den sicheren und effizienten Austausch von Informationen zwischen Maschinen, Systemen und Anwendungen ermöglichen.

MQTT ist ein offenes Netzwerkprotokoll für die Machine-to-Machine-Kommunikation und spielt eine wichtige Rolle als leichtgewichtiges Nachrichtenprotokoll im Bereich der digitalen Darstellung und Vernetzung von Informationen und Systemen [3]. Dies erfolgt durch den Informationsaustausch mithilfe einer Publish/Subscribe-Architektur, die durch einen zentralen MQTT-Broker koordiniert wird, was eine effiziente

Datenverteilung unter mehreren Clients ermöglicht. Durch den zentralen Broker können auch Systeme mit geringer Hardwareleistung an die Architektur angebunden werden. Gleichzeitig verfügt MQTT als leichtgewichtiges Protokoll nicht über semantische Modelle oder Standards zur Interpretation seiner Daten. Die Beschreibung der Information wird damit zwar eindeutig, bleibt aber stets, gemessen an anderen Standards, individuell. Als Fazit eignet sich MQTT daher oft für Systeme mit geringer Hardwareleistung und die Integration von Altsystemen.

Ein weiterer relevanter Standard für die industrielle Kommunikation stellt OPC UA dar, der eine flexible und skalierbare Plattform für den Austausch von Daten und Informationen bietet. Ein wichtiges Merkmal ist dabei die Plattformunabhängigkeit. Das bedeutet, dass OPC UA-Anwendungen auf einer Vielzahl von Geräten und in unterschiedlichen Umgebungen ausgeführt werden können, einschließlich PCs, Embedded-Systemen und Cloud-Plattformen. OPC ist darüber hinaus protokollunabhängig, sodass der Standard als Framework auf diverse Kommunikationsschnittstellen aufgesetzt werden kann [4]. Ein weiteres Merkmal ist die Skalierbarkeit, wodurch OPC UA sowohl für kleine, isolierte Anwendungen als auch für große, komplexe Automatisierungsarchitekturen geeignet ist. Konkret bedeutet dies, dass die serviceorientierte Architektur es ermöglicht, verschiedene Dienste für den Datenaustausch, die Überwachung und die Steuerung bereitzustellen, unabhängig von der Anzahl der angeschlossenen Geräte oder der Komplexität der Kommunikation. Ein weiterer Aspekt der Skalierbarkeit von OPC UA ist das hierarchische Datenmodell, das es ermöglicht, komplexe Datenstrukturen einfach darzustellen. Mithilfe von standardisierten Informationsmodellen können Unternehmen spezifische Daten und deren Beziehungen zueinander definieren.

Neben MQTT und OPC UA soll hier die AAS als Option für die digitale Repräsentation genannt werden. Die AAS ist eine standardisierte digitale Darstellung von physischen oder digitalen Assets im Rahmen von Industrie 4.0. Sie stellt eine semantische Beschreibung von Objekten dar, die von der International Digital Twin Association (IDTA) als standardisierter digitaler Zwilling beschrieben wird [5]. Als digitaler Zwilling strukturiert die AAS-Daten mithilfe von Teilmodellen, die die Eigenschaften und Funktionen eines Assets definieren. Diese Struktur ermöglicht einen nahtlosen Datenaustausch und Interoperabilität zwischen Systemen und Herstellern und erleichtert die effiziente Integration über verschiedene Plattformen hinweg. Die verwalteten Werte eines in der AAS dargestellten Objekts können entweder offline als separate Instanz oder direkt am Objekt abgerufen werden. Diese Optionen zielen darauf ab, die Integration und den Austausch von auf diese Weise beschriebenen Systemen innerhalb bestehender Infrastrukturen zu standardisieren und den Integrationsaufwand zu minimieren. Im Vergleich scheinen insbesondere OPC UA und AAS ähnliche Ziele zu adressieren, wobei beide Technologien eigene Akzente setzen. So setzt die AAS den Fokus auf die semantische Beschreibung, die in verschiedenen Formaten implementiert werden kann, einschließlich der Speicherung in Datenbanken, als Teil eines Datenmodells und in Dateiformaten. Die AAS zwingt keine spezifische Kommunikationsstruktur auf, sondern verlässt sich auf die Interaktion mit Middleware wie BaSyx, die ihrerseits wieder auf OPC UA basieren kann [6]. OPC UA hingegen hat einen stärkeren Fokus auf die Interoperabilität und Funktionen des Kommunikationsframeworks, bietet aber auch eigene semantische Beschreibungen. Obwohl diese Spezifikationen nicht direkt von der AAS unterstützt werden, können bestehende semantische Modelle und Funktionen beider Technologien in die jeweils andere Technologie integriert werden, um den Austausch von Informationen zu ermöglichen [7].

MQTT, OPC UA und die AAS sind drei bedeutende Technologien im Kontext der Industrie 4.0, die unterschiedliche, aber komplementäre Funktionen erfüllen. Nachfolgende Sektionen werden den Fokus auf OPC UA und AAS legen, da diese für komplexere semantische Strukturierungen geeignet sind.

3. OPC UA und Companion Specification in der NDE

Ein wesentlicher Ausgangspunkt zum Einsatz von OPC UA in der NDE stellen die sogenannten Companion Specifications dar. Diese Spezifikationen bieten eine standardisierte semantische Beschreibung von Prozessen und fördern die Interoperabilität zwischen unterschiedlichen Systemen und Akteuren. Diese werden in Arbeitsgruppen innerhalb der OPC Foundation erstellt, in denen Experten aus Industrie und Wissenschaft gemeinsam solche Informationsmodelle entwickeln. Für NDE-Systeme befindet sich derzeit eine Companion Specification im Rahmen der Joint Working Group Non-Destructive Evaluation (JWG NDE) in Entwicklung. Dieses Modell soll die spezifischen Anforderungen und Herausforderungen der NDE-Technologien abdecken und die Integration in digitale Produktionsumgebungen erleichtern. Dies erfordert jedoch einen gewissen Kompromiss: Zum einen muss das Informationsmodell die allgemeinen Herausforderungen eines NDE-Verfahrens adressieren, zum anderen ist es aber unmöglich, alle spezifischen Informationen aller diversen NDE-Verfahren in einem Modell zu vereinen. Um diesem Problem zu begegnen, gibt es zwei Strategien. Einerseits sind Spezifikationen im Allgemeinen nicht starr, sondern unterliegen regelmäßigen Anpassungen, um die Flexibilität und Aktualität der Informationen zu gewährleisten. Andererseits ist es beim Einsatz von OPC UA üblich, die Informationsmodelle der Companion Specification um eigene individuelle Erweiterungen zu verfeinern. Die JWG NDE verfolgt daher die Entwicklung eines allgemeinen Informationsmodells, das die grundlegenden Aspekte der NDE beschreibt und durch eigene spezifische Submodelle individuell ergänzt werden kann. Die nachfolgende Abbildung stellt grafisch den angedachten Umfang des Informationsmodells dar. Das geplante Modell adressiert somit die digitale Abbildung des Prüfprozesses vom Eingang des Prüfauftrags bis zur Ausgabe der Prüfergebnisse und wird dabei u.a. die Verwaltung der Prüfanweisungen, die Beschreibung der Prüfsysteme sowie die Darstellung der Ergebnisse umfassen.

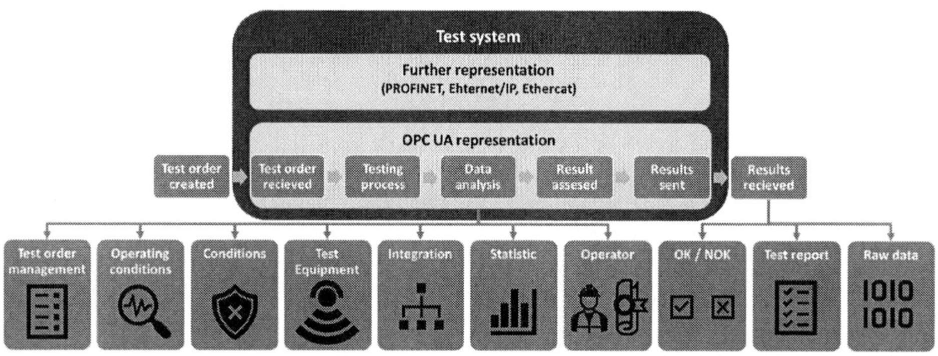

Abb. 1. Aktuelle grafische Darstellung des adressieren Umfangs des Informationsmodells der Companion Specification of OPC UA for Non-Destructive Testing - Part 1: Testing System

Auch wenn derzeit noch keine Companion Specification für NDE-Systeme existiert, können diese bereits jetzt von OPC UA profitieren, da die Implementierung des OPC UA Frameworks selbst von den Hardwareschnittstellen abhängig ist. So können bereits etablierte Schnittstellen in ein individuelles Modell überführt und in OPC UA genutzt werden. Hierfür bieten sich bspw. PROFINET, Ethernet/IP, EtherCAT oder allgemein TCP/IP an. Zudem können auch andere OPC UA Companion Specifications als temporäre Basis verwendet werden. Hier sei auf die VDMA 40100-1 „OPC UA for Machine Vision - Teil 1: Steuerung, Konfigurationsmanagement, Rezeptmanagement, Ergebnismanagement" verwiesen, die sich besonders für den Betrieb bildgebender NDE-Prozesse eignet. Konkrete Beispiele für die

Anwendung von OPC UA in NDE-Systemen zeigen sich sowohl in Forschungsvorhaben als auch bereits in verfügbaren Sensorsystemen. So ist die Verbindung von NDE-Prozessen mit OPC UA ein Forschungsthema im Projekt nexAMo [8], das vom Bundesministerium für Wirtschaft und Klimaschutz (BMWK) im Rahmen der Fördermaßnahme „Digitalisierung von Fahrzeugherstellern und der Zulieferindustrie" unterstützt wird. Ziel des Projekts ist die Entwicklung einer hochflexiblen und frei konfigurierbaren Matrixproduktion, die eine beispiellose Flexibilität in der Fertigung ermöglicht. Ein Teilziel des Projekts besteht darin, ein kompaktes CT-System für die Integration in eine digitale Industrie 4.0-Umgebung zu entwickeln. Das CT-System soll in der Lage sein, gemäß dem digitalen Auftrag nur bestimmte Detailregionen zu scannen und einen Prüfbericht gemäß digitalen Prüfanforderungen zu erstellen, der dann vollautomatisch archiviert wird. Dieser Prozess soll mittels OPC UA realisiert werden. Ein Beispiel für erste verfügbare Prüfverfahren mit OPC UA findet sich am Fraunhofer IZFP. Hier wird der OPC UA-Standard verwendet, um die Integration von Systemen in Produktionsanlagen zu erleichtern und die Anforderungen der Industrie 4.0 zu erfüllen. Die inspECT-PRO-Plattform für Wirbelstromprüfungen nutzt OPC UA zur Sortierinspektion, um Materialfehler und mechanische Eigenschaften zu erkennen. In diesem Fall wurde ein selbst definiertes Modell verwendet [9].

4. Einsatz der Verwaltungsschale für NDE-Applikationen

Auch wenn zur semantischen Beschreibung insbesondere OPC UA und AAS als komplementäre Technologien mit ähnlichem Fokus behandelt werden, so ist die Philosophie für den Einsatz einer Verwaltungsschale doch eine andere. Semantische Modelle werden dabei nicht durch Arbeitsgruppen erstellt, sondern individuell durch Endanwender oder unabhängige Gruppen für deren jeweilige Anwendung gestaltet. Diese Submodelle können dann über die IDTA verbreitet werden, wobei jedes Modell einen Bewertungsprozess durchläuft, in dem das Modell auch bei Bedarf angepasst wird. Konkret bedeutet dies, dass die Interoperabilität durch die technologischen Grundlagen der Verwaltungsschalen gewährleistet wird. Für NDE-Systeme bedeutet dies, dass für den Einsatz der AAS stets der Bedarf der Applikation im Vordergrund steht, wobei ein Blick auf veröffentlichte Submodelle die Implementierungen ggf. erleichtern kann. Diese liefern eine detaillierte Beschreibung der verschiedenen Aspekte eines Assets. Ein Beispiel für die Anwendung einer Verwaltungsschale findet sich im Forschungsprojekt EMOTION [10]. Ein Teilziel des Vorhabens war die Entwicklung einer detaillierten digitalen Darstellung des 3D-Druckers, die eine Prozessüberwachung ermöglicht. Dies bedurfte der Entwicklung und Implementierung verschiedener Submodelle, die den 3D-Drucker digital abbilden. Dabei konnten der überwiegende Teil direkt von der IDTA [11], sowie aus dem Projekt Interopera [12] bezogen werden. Lediglich das „Operation Submodel" musste für die Umsetzung selbst gestaltet werden. Nachfolgend sind alle enthaltenen Submodelle inklusive einer kurzen Beschreibung aufgelistet:

- **Digital Nameplate Submodel**: Informationen des Herstellers und der Lieferanten des 3D-Druckers.
- **Technical Data Submodel**: Technischen Spezifikationen des 3D-Druckers wie bspw. mechanische Eigenschaften und Software-Version
- **Operation Submodel**: Metadaten über alle Vorgänge (Funktionen) des 3D-Druckers, wie bspw. das Starten/Stoppen der Maschine
- **Time Series Submodel**: Metadaten für die verknüpfte Zeitreihendatenbank (InfluxDB)
- **Predictive Maintenance Submodel**: Metadaten im Zusammenhang mit der vorausschauenden Wartung

Die finale digitale Repräsentation des 3D-Druckers wurde in eine SOA implementiert, die aus fünf verschiedenen Schichten besteht. Diese wird in Abbildung 2 dargestellt. Die Datenpipeline beginnt auf dem Field Layer, der den 3D-Drucker sowie eine Reihe von Sensoren enthält. Die Sensoren senden in Echtzeit qualitätsrelevante Datensätze der Maschine an die Kommunikationsschicht mittels MQTT. Die dritte Ebene wird Service Layer genannt und beinhaltet kritische Dienste wie Influx DB, in der die Echtzeitdaten in einem Zeitreihenformat archiviert werden. Der Predictive Maintenance Service hingegen nutzt die Echtzeitdaten von Influx DB, um Anomalien im Betrieb des Druckers zu erkennen. Darüber kommt der Digital Twin Layer, in dem sich die AAS mit den Untermodellen befindet, sowie der User Layer, der u.a. Informationen mittels Dashboards visualisiert und eine Prozesssteuerung erlaubt.

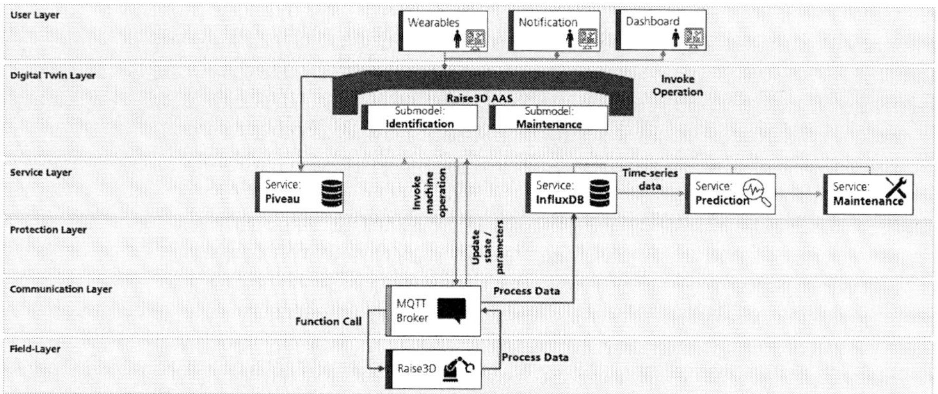

Abb. 2. Darstellung der Datenpipeline des 3D-Druckers inklusive der Einordnung der Verwaltungsschale

Sobald der Benutzer die Operation „Start Machine“ im Operation Submodel über die AAS-Benutzeroberfläche auslöst, startet der 3D-Drucker. Mit dem Start werden Sensordaten kontinuierlich an den Influx DB Service gesendet und die Metadaten im Time Series Submodel des AAS gespeichert. Gleichzeitig nutzt der Predictive Maintenance Service diese Zeitseriendaten. Bei der Erkennung von Anomalien stoppt der Service die Maschine, um mögliche Schäden am Gerät zu vermeiden. Informationen über die Anomalie werden anschließend im Predictive Maintenance Submodel des AAS veröffentlicht. Neben der beschriebenen Anwendung wird in dem gezeigten Beispiel durch den Piveau-Service eine Anbindung der AAS an weitere Datenbanken, in diesem Fall eine Piveau-Datenbank zur Metadatenerfassung, ermöglicht. Im konkreten Fall erforderte dies eine zusätzliche Konvertierung der Informationen vom AAS-Format nach DCAT-AP (Data Catalogue Vocabulary Application Profile).

5. Zusammenfassung

Innerhalb dieses Berichtes wurde die Bedeutung digitaler Repräsentationen und die Vernetzung von NDE-Informationen und -Systemen im Kontext der Industrie 4.0 beschrieben und relevante Technologien vorgestellt. Die fortschreitende digitale Transformation erfordert eine tiefgreifende Anpassung der bestehenden NDE-Systeme, um den steigenden Anforderungen an Integrationsfähigkeit und Interoperabilität gerecht zu werden. Die Untersuchung zeigt, dass die Übergabe einfacher Steuerungsbefehle und Zustände in der heutigen Zeit nicht mehr ausreicht; stattdessen müssen detailliertere digitale

Repräsentationen implementiert werden, die alle relevanten Aspekte eines NDE-Systems abdecken. Dabei kommt Technologien wie OPC UA und AAS eine Schlüsselrolle bei der Schaffung von interoperablen und adaptiven NDE-Systemen zu. Diese Technologien ermöglichen es, komplexe Datenstrukturen zu verwalten und eine nahtlose Kommunikation zwischen verschiedenen Systemen zu gewährleisten. Die Companion Specifications, die im Rahmen von OPC UA entwickelt werden, bieten dabei eine standardisierte semantische Grundlage, die es verschiedenen Akteuren in der NDE-Branche ermöglicht, effizient zusammenzuarbeiten und ihre Systeme zu integrieren. Die vorgestellten Fallstudien im Projekt EMOTION zur Implementierung der AAS für 3D-Drucker verdeutlichen die praktischen Anwendungsmöglichkeiten dieser Technologien. Sie zeigen, wie eine digitale Repräsentation von NDE-Systemen die Prozessüberwachung verbessern kann und langfristig die Effizienz von Wartungsprozessen durch die Verknüpfung mit weiteren Services innerhalb einer SOA optimiert wird.

Für die Zukunft ist es entscheidend, die Flexibilität und Anpassungsfähigkeit dieser Technologien weiter zu fördern. Eine kontinuierliche Entwicklung und Anpassung der digitalen Modelle wird notwendig sein, um den dynamischen Anforderungen der Industrie gerecht zu werden. Darüber hinaus sollten neue Standards und Protokolle entwickelt werden, die die Interoperabilität zwischen unterschiedlichen NDE-Systemen und -Datenbanken fördern. Der Austausch von Informationen und die Vernetzung von Systemen müssen dabei immer im Einklang mit den rechtlichen Rahmenbedingungen, wie dem EU-Data Act [1], erfolgen.

Referenzen

[1] Europäische Union. (2023). Verordnung (EU) 2023/2854 des Europäischen Parlaments und des Rates vom 13. Dezember 2023 über harmonisierte Vorschriften für einen fairen Datenzugang und eine faire Datennutzung sowie zur Änderung der Verordnung (EU) 2017/2394 und der Richtlinie (EU) 2020/1828 (Datenverordnung).

[2] Paolone, Gaetanino, Danilo Iachetti, Romolo Paesani, Francesco Pilotti, Martina Marinelli, and Paolino Di Felice. "A Holistic Overview of the Internet of Things Ecosystem." IoT 3, no. 4 (2022): 398–434.

[3] Silva, Daniel, Liliana I. Carvalho, José Soares, and Rute C. Sofia. "A Performance Analysis of Internet of Things Networking Protocols: Evaluating MQTT, CoAP, OPC UA." Applied sciences 11, no. 11 (2021): 4879.

[4] Nast, Michael, Hannes Raddatz, Frank Golatowski, and Christian Haubelt. "A Novel OPC UA PubSub Protocol Binding Using MQTT for Sensor Networks (MQTT-SN)." 2024 IEEE 29th International Conference on Emerging Technologies and Factory Automation (ETFA), 2024, 1–4.

[5] Barnard, Alfred. "Digital Twin Frameworks: A Comparative Study of AAS, OPC UA, and AML." 2024 International Conference on Electrical, Computer and Energy Technologies (ICECET, 2024, 1–6.

[6] Schnicke, Frank, Thomas Kuhn, and Pablo Oliveira Antonino. "Eclipse BaSyx DataBridge – Datenintegration Einfach Machen: Skalierbare Datenintegration Mit Verwaltungsschalen Durch Eclipse BaSyx." Atp-Edition : automatisierungstechnische Praxis 65, no. 5 (2023): 50–58.

[7] Neubauer, Michael, Lukas Steinle, Colin Reiff, Samed Ajdinović, Lars Klingel, Armin Lechler, and Alexander Verl. "Architecture for Manufacturing-X: Bringing Asset Administration Shell, Eclipse Dataspace Connector and OPC UA Together." Manufacturing letters 37 (2023): 1–6.

[8] BMWK. "Zukunftsinvestitionen Fahrzeughersteller und Zulieferindustrie". Aufgerufen am 25.03.2025. https://www.bmwk.de/Redaktion/DE/Publikationen/Industrie/zukunftsinvestitionen-fahrzeughersteller-und-zulieferindustrie.pdf?__blob=publicationFile&v=13.

[9] Fraunhofer. "OPC UA-fähige Sortierprüfung". Aufgerufen am 25.03.2025. https://www.vision.fraunhofer.de/de/technologien-anwendungen/technologien/zerstoerungsfreie-pruefung/wirbelstrom-sortierpruefsystem.html

[10] Webauftritt der Leitprojekt EMOTION. Aufgerufen am 25.03.2025. https://www.fraunhofer.de/de/forschung/fraunhofer-initiativen/fraunhofer-leitprojekte/emotion.html

[11] Bereitstellung von Submodellen auf der Webseite des IDTA. Aufgerufen am 25.03.2025. https://industrialdigitaltwin.org/en/content-hub/submodels

[12] Bereitstellung der Teilmodelle im Projekt Interopera. Aufgerufen am 25.03.2025. https://interopera.de/landkarte/

ZFP: Der Weg vom "Workmanship Standard" zu NDE 4.0 und darüber hinaus

Norbert MEYENDORF [1], Leonard J. BOND [2]
[1] Chemical and Materials Engineering, University of Dayton, Dayton, OH, USA,
Email: nmeyendorf1@udayton.edu
[2] Aerospace Engineering, Iowa State University, Ames, IA, USA

Kurzfassung.
Der Begriff „NDE 4.0" wurde im Herbst 2016 geprägt und ist inspiriert durch „Industrie 4.0". „Industry 4.0" ist ein Begriff, der technische Innovationen beschreibt, die wir derzeit sowohl in der Industrie als auch in der Gesellschaft als Ganzes sehen. Jede Verbesserung der Produktionstechnologien erforderte immer eine Verbesserung der Methoden zur Überprüfung der Produktqualität und -integrität, die eine Bewertung ermöglichen. Im Englischen werden die Fortschritte historisch durch verschiedene Abkürzungen beschrieben: NDT, NDE, QNDE, NDE 4.0 oder auch SHM, ISHM, MA-NDE. In Zukunft können wir eine ganzheitliche Überwachung komplexer technischer Produkte wie Autos, oder Flugzeuge erwarten, analog zum menschlichen Gesundheitssystem. Dies bedeutet, die enormen Datenmengen zu nutzen, die durch 3D-Volumen-NDE und SHM generiert werden, um mithilfe von KI und intelligenten Diagnosemethoden (QNDE) unter Berücksichtigung von Komponentenlastdaten Prognosen über die zukünftige Integrität der Produkte zu erstellen (Prognostik).

Keywords: NDT, NDE, QNDE, NDE 4.0, Smart Manufacturing, Smart NDE, SHM, Artificial Intelligence, Neural Networks, cyber-physical transformation

Einführung

Der Begriff NDE 4.0 wurde erstmals vor etwa 8 Jahren, im Jahr 2016, verwendet [1,2,3,4]. Kritiker behaupten jedoch, er beschreibe nichts Neues, sondern sei lediglich ein neues Label, das aus Marketinggründen eingeführt wurde. Wir werden jedoch zeigen, dass NDE 4.0 tatsächlich etwas anderes beschreibt. Der Begriff „NDE 4.0" beschreibt den nächsten Schritt oder die nächste Revolution in der zerstörungsfreien Prüfung als Reaktion auf die Revolution in Industrie und Gesellschaft. Kennzeichnend hierfür ist Industrie 4.0, ein Begriff, der technische Innovationen beschreibt, die aktuell sowohl in der Industrie als auch in der Gesellschaft allgemein stattfinden. Dies wird durch intelligente Fertigung, die Vernetzung von Systemen, die zunehmende Komplexität von Produkten und die Nachfrage nach Echtzeit-Überwachungen vorangetrieben. Die zerstörungsfreie Prüfung wird durch Digitalisierung, Konnektivität und Datenanalyse sowie durch den Einsatz künstlicher Intelligenz (KI), digitaler Zwillinge und Simulation, sowie innovative (advanced) zerstörungsfreie Methoden und Prüfwerkzeuge revolutioniert. Dies fördert die Vereinfachung von Prozessen, die bessere Nutzung von NDE-Daten und die Entwicklung der zerstörungsfreien Prüfung von einem Kosten- zu einem Wertschöpfungsfaktor und die

https://doi.org/10.58286/31241

Integration der zerstörungsfreien Prüfung in die digital transformierende (cyber-physical transformation) Umgebung.

1. Entwicklung von NDT/NDE

Klassisch wurden die Prüfmethoden zur Erkennung und Lokalisierung von Defekten meist am Ende des Produktionsprozesses eingesetzt. Sie wurden als zerstörungsfreie Prüfung (ZfP englisch NDT) bezeichnet und ihre Anwendung expandierte in den 1940er Jahren rasant. Auch die Bruchmechanik [5] entwickelte sich in den 1940er Jahren rasant. Diese Fortschritte führten zu neuen Ansätzen in der Strukturkonstruktion und der sicheren Lebensdauerabschätzung, die dann in neue Geräte, Vorschriften und Normen einflossen.

In den 1960er und 1970er Jahren gab es bedeutende technologische Fortschritte, unter anderem in der Verteidigungs-, Luft- und Raumfahrt-, Nuklear-, Öl- und Gasförderung sowie der Petrochemischen Industrie, die die Entwicklung der zerstörungsfreien Prüfung vorantrieben. Damals erkannte man auch, dass es keine ausreichende wissenschaftliche Grundlage für eine quantitative ZfP gab! Sowohl in Europa als auch in Nordamerika wurden umfangreiche Forschungs- und Entwicklungsarbeiten initiiert, und mehrere große Forschungsprogramme und -zentren wurden eingerichtet [6]. Das Ergebnis waren fortschrittlichere und verbesserte Prüfmethoden, die es ermöglichten, mögliche Defekte zu erkennen, Größe, Form, Ausrichtung und physikalische Eigenschaften von Anomalien zu messen und verschiedene Arten der Materialcharakterisierung zu ermöglichen. Diese Messungen, zusammen mit dem Übergang von analogen zu digitalen Messgeräten seit etwa dem Jahr 2000, liefern heute wichtige Informationen zur Bewertung von Komponenten und Herstellungsprozessen und geben Feedback, das zur Anpassung der Prozessparameter zur Verbesserung der Produktionsqualität genutzt werden kann. Die neuen und verbesserten Messmethoden zur Sicherstellung der anfänglichen Produktqualität wurden als NDE (Nondestructive Evaluation) [7] bezeichnet.

Ein quantitativerer Ansatz, die sogenannte Quantitative ZfP (QNDE), wurde entwickelt, um Daten für die Lebensdauerbewertung von Strukturen zu liefern und Bruchmechanik und ZfP zu kombinieren. Solche Bewertungen dienen dazu, zu entscheiden, ob ein Defekt von kritischer Art und Größe für ein Bauteil oder eine Struktur ist, die bestimmten Belastungen ausgesetzt ist. Die Defekteigenschaften und -belastungen sind Daten, die als Input für die Prognostik verwendet werden können, eine Bewertung zur Vorhersage der zukünftigen Lebensdauer eines Bauteils im Einsatz [6].

Neben NDT und NDE während der Fertigung wuchs das Interesse an der Überwachung von Defekten in alternden Strukturen, die Belastungen wie Ermüdung, Kriechen oder Korrosion ausgesetzt sind. Im letzteren Fall entwickelte sich auch eine verwandte Aktivität namens Structural Health Monitoring (SHM). Obwohl Daten aus NDT /NDE zu einem bestimmten Zeitpunkt erhoben werden und die Daten aus SHM kontinuierlich über einen bestimmten Zeitraum überwacht werden, ergänzen sich die bereitgestellten Informationen.

Mit der zunehmenden Nutzung von NDT, NDE und SHM wurde es notwendig, Messgrößen zur Beschreibung der Messleistung bereitzustellen. Angesichts der inhärenten Variabilität der Mikrostruktur von Werkstoffen und der darin enthaltenen Anomalien lassen sich Leistung und Zuverlässigkeit am besten statistisch bewerten, üblicherweise mithilfe der sogenannten Nachweiswahrscheinlichkeit (Probability of Detection, POD) [8].

Neben Fortschritten in der wissenschaftlichen Basis der NDE haben moderne Digitalisierungs- und Computerkapazitäten eine Umstellung der NDE-Instrumentierung von analogen auf digitale Einheiten ermöglicht. Die Fortschritte bei computergestützten Digitaltechnologien sowie neue Messumformer und Detektoren ermöglichten neue

Prüfverfahren wie die digitale Radiographie und Ultraschall-Phased-Array-Prüfung, die empfindlichere, zuverlässigere und wiederholbarere Prüfergebnisse liefern. Diese Technologien haben zudem den Aufbau von NDE-Datenbanken und in der Folge verschiedene hochentwickelte Datenintegrations- und -analysetechniken ermöglicht.

Bereits 1992 zeigten Versuche, dass die digitale Röntgenbildverarbeitung das Potenzial hat, mehr Schweißfehler zu erkennen und korrekt zu klassifizieren als typische menschliche Prüfer [9]. In den 1990er Jahren wurden neuronale Netzwerke in der zerstörungsfreien Prüfung getestet. Diese können als eine frühe Form der künstlichen Intelligenz (KI) betrachtet werden, obwohl sie hauptsächlich in Laborumgebungen eingesetzt wurden [10] und ihre Fähigkeiten durch die damals verfügbaren Computer und Software begrenzt waren. Damals suchte die NDE-Branche eher nach automatisierten Prüftechniken, die in den Produktionsprozess integriert wurden, um Fehler unter Einsatz fortschrittlicherer Datenverarbeitung frühzeitiger zu erkennen. In den 1990er und frühen 2000er Jahren basierten die in der Industrie eingesetzten Prüfungen auf der Implementierung starrer Standardverfahren, obwohl Roboter oder CNC-Maschinen zur Sensor- (Prüfkopf) Manipulation verstärkt eingesetzt wurden. Die Daten waren in der Regel statistischer Natur und wurden bei der Herstellung und Prüfung einer großen Anzahl ähnlicher Teile gewonnen. Außerdem wurden zu dieser Zeit alle Ergebnisse halbautomatischer NDE-Prüfungen von einem Prüfer bewertet.

2. Heutige Anforderungen, NDE 4.0 und Prognostik

Die aktuellen Prüfvorschriften und -standards für NDT/NDE erfordern im Allgemeinen immer noch die Einbindung des Menschen den Prüfprozess. Dennoch besteht in den heutigen cyber-physischen Produktionssystemen, in denen Maschinen potenziell miteinander kommunizieren, Daten austauschen und die Produktionsparameter modifizieren können, das Potenzial, individuell gestaltete Produkte bereitzustellen, die den festgelegten Qualitätsstandards entsprechen. Diese Art von Prozess stellt die zerstörungsfreie Prüfung vor zwei besondere Herausforderungen:

1. Qualitätssicherung und -prüfung müssen in diese intelligente und selbstlernende flexible Produktionsumgebung integriert werden.
2. Entscheidungen zur Bewertung der Signifikanz von Anomalien müssen für Einzelteile getroffen werden können, ohne dass eine konventionelle Prüfung auf Grundlage eines Codes oder Standards für eine Vielzahl identischer Teile erforderlich ist. Möglicherweise muss ein selbstkorrigierender Fertigungsprozess eingesetzt werden, der erkannte Mängel beseitigt.

Solche neuen Methoden müssen sowohl das Problem der inversen Merkmalscharakterisierung als auch die Bewertung dieser Merkmale im Sinne der Vorhersage der Auswirkungen einer Anomalie auf die Lebensdauer eines Bauteils unter erwarteten Betriebsbelastungen lösen und anschließend die Automatisierung von Annahme-/Ablehnungs- oder Reparaturentscheidungen ermöglichen.

Diese Art von Bewertungsprozess ist seit einigen Jahren geplant, die Umsetzung bleibt jedoch eine Herausforderung [11]. Die Erkennung, Charakterisierung und Bewertung einer Anomalie erfordert die in Abbildung 1 dargestellten Schritte. Informationen über den Ausgangszustand die Teilegeometrie, die lokalen Materialeigenschaften und die Fähigkeit zur Erkennung und Charakterisierung von Anomalien sind bei der Fertigung erforderlich. Für die Lebensdauerbewertung müssen diese Daten in einen Bewertungsprozess integriert werden. Dieser umfasst die Verwendung einer Modellfamilie zur Vorhersage des Schadensverlaufs in einer Betriebsumgebung (Stressoren) sowie das Verständnis der Materialversagensmodelle und Versagenskriterien, um eine erwartete Komponentenlebensdauer zu ermitteln. Im Komponentenlebensdauermanagement werden neben der anfänglichen zerstörungsfreien Prüfung (NDE) auch laufende regelmäßige Prüfungen mittels NDE und kontinuierliche Messungen mittels Strukturüberwachung (SHM) durchgeführt. Zu den größten Herausforderungen gehören die Entwicklung der notwendigen deterministischen Modelle für die Komponentenlebensdauerbewertung und die Frage, wie sowohl die anfänglichen NDE-Daten als auch die Daten aus regelmäßigen Prüfungen und SHM mit gängigen Metriken kombiniert werden können, die die „Defektmerkmale" (Art, Form, Größe und Ausrichtung) definieren und mit geeigneten Belastungsdaten (Stressorendaten) integriert werden können, um eine zuverlässige Bewertung der Komponentenlebensdauer zu ermöglichen. Solche Bewertungen müssen auch die statistischen Aspekte der POD, angewendet auf Daten aus NDE und SHM, sowie die Prognose mit verschiedenen Stressoren- und Versagensmodellen berücksichtigen. Die vollständige Implementierung des in Abbildung 1 dargestellten Prozesses ist aufgrund unvollständiger Modelle und/oder unzureichender Daten bestenfalls schwierig und kann in allen außer den einfachsten Fällen unmöglich sein.

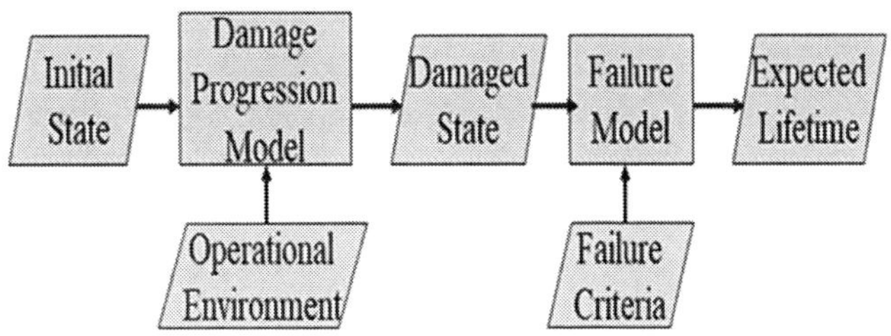

Abbildung 1: Konzeptioneller Prozess zur Bewertung der Komponentenlebensdauer [11]

Im Kontext von NDE 4.0 werden verschiedene Datenverarbeitungsansätze vorgeschlagen. Dazu gehört der Einsatz künstlicher Intelligenz (KI) sowie Modelle für Echtzeitdiagnose und -prognose. Beispiele werden entwickelt und beispielsweise bereits auf Verbundwerkstoffe angewendet. Die verschiedenen Modelle und die damit verbundenen Unsicherheiten wurden von Elenchezhian et al. [12] untersucht. Kognitive Sensorsysteme für NDE 4.0 werden derzeit mithilfe von KI-Einbettung evaluiert, zusammen mit der erforderlichen Validierung und Qualifizierung, beispielsweise von Valeske et al. [13]. Ein kritischer Punkt, der für die Implementierung von zentraler Bedeutung ist, ist die Bereitstellung geeigneter Trainingsdaten; die Erfüllung dieses Bedarfs ist noch in Arbeit.

Zusammenfassung

Ausgewählte zusammenfasende Bilder des Vortrages sind als Anhang beigefügt

References

[1] Meyendorf N, "NDE for the 21st century: industry 4.0 requires NDE 4.0", Proceedings Volume 10171, Smart Materials and Nondestructive Evaluation for Energy Systems 2017; 101710H (2017), https://doi.org/10.1117/12.2263326 , Event: SPIE Smart Structures and Materials + Nondestructive Evaluation and Health Monitoring, 2017, Portland, Oregon, United States

[2] Meyendorf NGH, Bond LJ, Curtis-Beard J, Heilmann S, Pal S, Schallert R, Scholz H, Wunderlich C. : "NDE 4.0 – NDE for the 21st century – the internet of things and cyber physical systems will revolutionize NDE", 15th Asia Pacific conference for non-destructive testing (APCNDT 2017), Singapore, 2017.

[3] Singh R, Vrana J, "NDE 4.0 – why should 'I' get on this bus now?" CINDE J. 41(4):6–13, 2020

[4] Vrana J, Singh R, "NDE 4.0 – a design thinking perspective" J NDE. 2020; 40:4. https://doi.org/10.1007/s10921-020-00735-9

[5] Cotterell, B. "The past, present, and future of fracture mechanics" Engineering Fracture Mechanics 69 (2002) 533–553

[6] Donald O. Thompson, EVOLUTION OF QNDE'S CORE INTERDISCIPLINARY SCIENCE AND ENGINEERING BASE, AIP conference proceedings, AIP Conf. Proc. 1211, 3 (2010); doi: 10.1063/1.3362421
[7] NDT and NDE, https://www.nde-ed.org/About/NDTandNDE.xhtml, CNDE Webseite, Iowa State University
[8] E. Lindgren, D. Forsyth, J. Aldrin and F. Spencer, "Probability of detection, " in ASM Handbook, Vol 17, NDE of Materials, Ed. A. Ahmad and L.J. Bond, ASM Int. p 8-22.
[9] Regina Pohle: "Manual and Automated Detection of Weld Defects in Radioscopic Images" PhD Thesis, Magdeburg, 1994
[10] G Dobmann, M Kröning, W Theiner, H Willems, "Nondestructive characterization of materials (ultrasonic and micromagnetic techniques) for strength and toughness prediction and the detection of early creep damage" Nuclear Engineering and Design, Volume 157, Issues 1–2, 1 July 1995, Pages 137-158
[11] Bond, L.J. and Meyendorf, N.G. (2019) NDE and SHM in the age of Industry 4.0., in Structural Health Monitoring 2019: enabling intelligent life-cycle health management for industry internet of things (IIOT), Edited F-Y, Chang and F. Kopsaftopoulos, (Proc. 12th International Workshop on Structural Health Monitoring, Sept. Stanford Univ., CA) DEStech Publications, Inc. [published online, memory stick, and book 2019] http://www.dpi-proceedings.com/index.php/shm2019 (Book - Vol 1) p 3-15.
[12] Muthu Ram Prabhu Elenchezhian, Vamsee Vadlamudi, Rassel Raihan, Kenneth Reifsnider and Eric Reifsnider, "Artificial intelligence in real-time diagnostics and prognostics of composite materials and its uncertainties—a review " Smart Mater. Struct. 30 (2021) 083001 (25pp).
[13] Bernd Valeske, Ralf Tschuncky, Frank Leinenbach, Ahmad Osman, Ziang Wei, Florian Römer, Dirk Koster, Kevin Becker, and Thomas Schwender, "Cognitive sensor systems for NDE 4.0: Technology, AI embedding, validation and qualification tm – Technisches Messen 2022; 89(4): 253–277

Anhang: Ausgewählte zusammenfassende Bilder aus der Präsentation

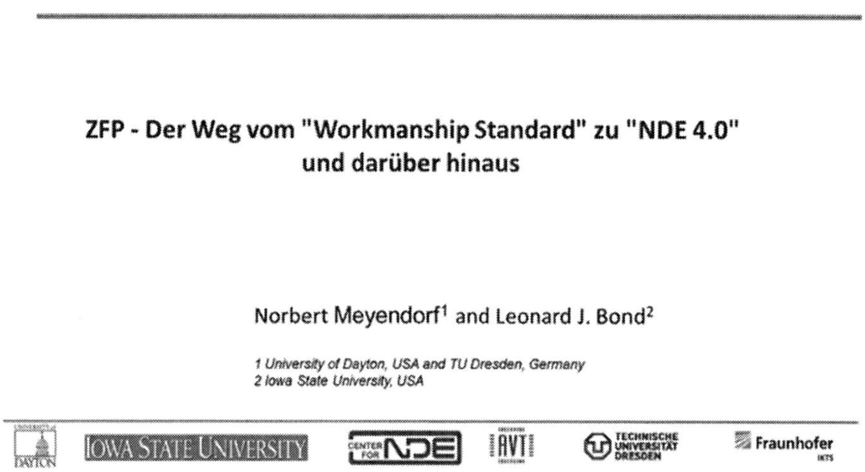

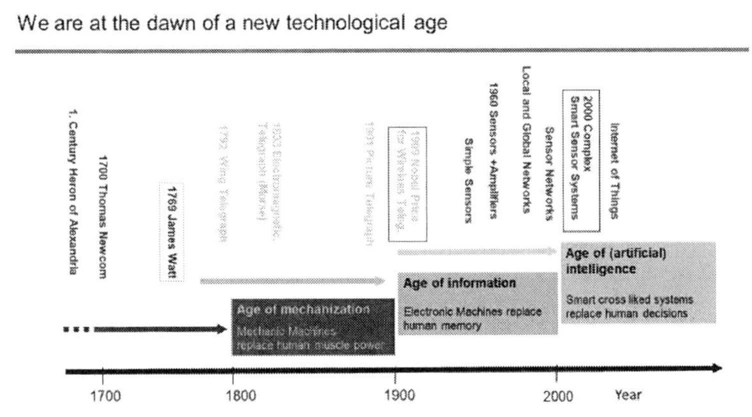

Norbert Meyendorf: Dresden Airport Seminar, 2006

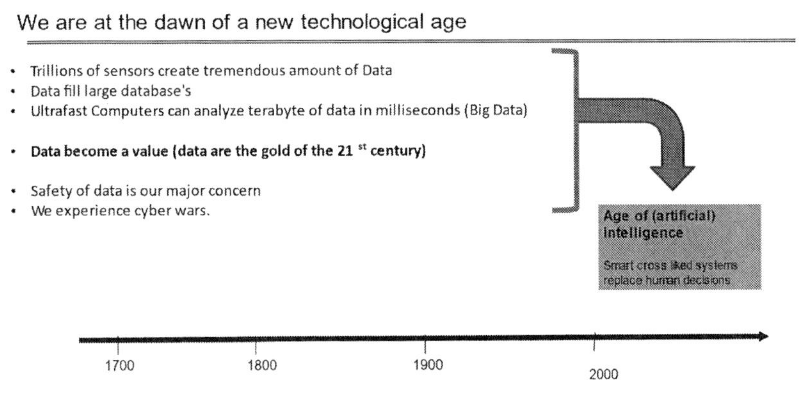

Leonard Bond: WCNDT, 2024

Die Entstehung von "NDE 4.0"

Im Sommer 2016 diskutierten wir intensiv in Berlin in dem IKTS Netzwerk PROGNET (Volmerstrasse 9) und dem etwa 100 Meter entfernten DGZfP Headquarter (ganz besonders mit Herrn Holstein) über die Anforderungen an die ZFP im Umfeld von Industry 4.0.

Im Ergebnis stand der Slogan: „ **Industry 4.0 requires NDE 4.0**". Soll heißen, wir benötigen eine Weiterentwicklung der ZfP/ NDE.

Als Ergebnis erschien im Herbst 2016 die Vorankündigung des SPIE Beitrages:

„NDE FOR THE 21ST CENTURY: INDUSTRY 4.0 REQUIRES NDE 4.0."

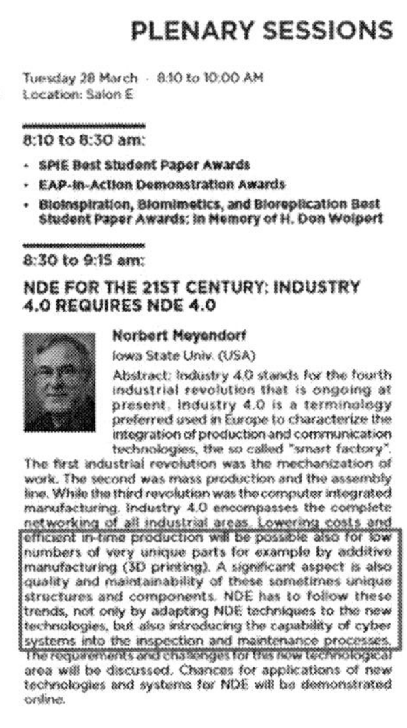

PLENARY SESSIONS

Tuesday 28 March · 8:10 to 10:00 AM
Location: Salon E

8:10 to 8:30 am:

- **SPIE Best Student Paper Awards**
- **EAP-in-Action Demonstration Awards**
- **Bioinspiration, Biomimetics, and Bioreplication Best Student Paper Awards: In Memory of H. Don Wolpert**

8:30 to 9:15 am:

NDE FOR THE 21ST CENTURY: INDUSTRY 4.0 REQUIRES NDE 4.0

Norbert Meyendorf
Iowa State Univ. (USA)

Abstract: Industry 4.0 stands for the fourth industrial revolution that is ongoing at present. Industry 4.0 is a terminology preferred used in Europe to characterize the integration of production and communication technologies, the so called "smart factory". The first industrial revolution was the mechanization of work. The second was mass production and the assembly line. While the third revolution was the computer integrated manufacturing. Industry 4.0 encompasses the complete networking of all industrial areas. Lowering costs and efficient in-time production will be possible also for low numbers of very unique parts for example by additive manufacturing (3D printing). A significant aspect is also quality and maintainability of these sometimes unique structures and components. NDE has to follow these trends, not only by adapting NDE techniques to the new technologies, but also introducing the capability of cyber systems into the inspection and maintenance processes. The requirements and challenges for this new technological area will be discussed. Chances for applications of new technologies and systems for NDE will be demonstrated online.

NDT/E History –

****Based on the ideas of Jan Achenbach, Don Thomson and Bruce Thomson and Leonard Bond* ***

- NDT: started as a "workmanship standard" looking for anomalies, but with limited detection capability
- NDT+: -- started in the 1940's better defect and materials characterization, 0 defect strategy
- 1960 to 70 advanced facture mechanics
- NDE: 1970-1990 – inspection in service "living with cracks", POD
- QNDE: -- started 1974, incudes measurement about type, shape orientation, size of defect and the properties of the material – driver being need for enhanced reliability (e.g. aerospace, nuclear, defense, oil/gas)
- 1990 - 2000 scanning large areas and SHM / condition based maintenance
- 2000- 2010 Full digitalized inspection systems, 3D imaging, NDE databases
- Smart NDE - NDE 4.0 - Today: Merging of NDE, SHM and modelling, networking of NDE and production devices, AI to support data analysis and decisions
- Future: Self organizing NDE. Holistic "healthcare" system for complex engineering products.

NDT/E Historical view of developments

NDE 4.0 -- Today Cyber physical transformation

- Merging of NDE and SHM , NDE modelling and Digital Twin
- Networking of NDE and production devices, Remote NDE
- Artificial Intelligence (AI) to support decisions

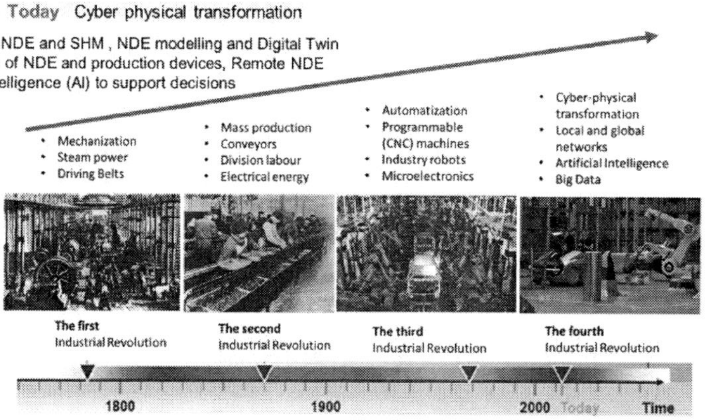

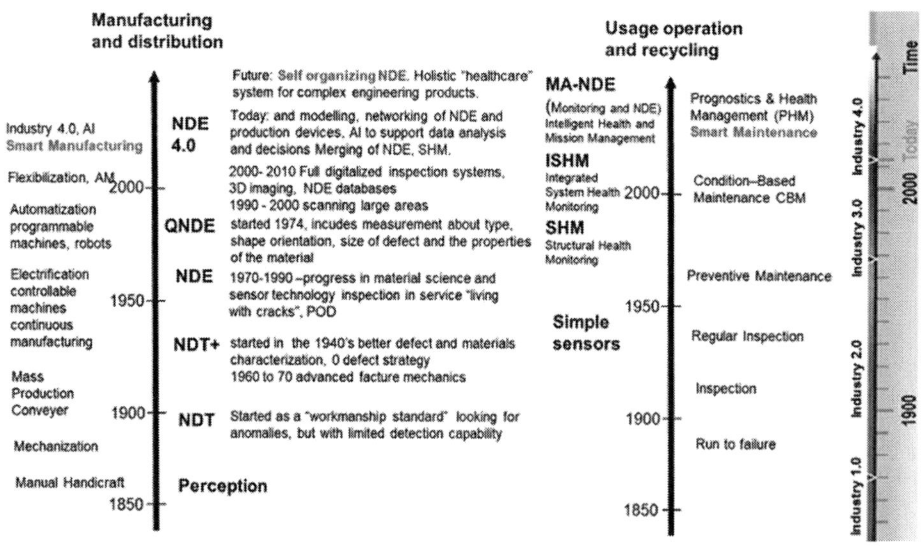

Herausforderungen für ZFP in dem Industry 4.0 Ecosystem

Neue Materialien (Miro- and Nano strukturierte Materialien)

Neue Komponenten (Mikroelektronik und mikromechanische Bauteile)

Neue Fertigungstechnologien AM, Laser Plasma ....

NDE Integration in Smart Factory

Additive Fertigung individueller Komponenten - keine Serien

Individuell Produktion nach Wünschen der Kunden

Neu Sensoren and Elektronik

Veränderung in der Entwicklung von NDE-Systemen

IOWA STATE UNIVERSITY · CENTER FOR NDE · AVT · TECHNISCHE UNIVERSITÄT DRESDEN · Fraunhofer IKTS

NDE 4.0 - NDE new trends and tools

1. **Cyber- physical transformation of NDE**
   - 3D volume data creation and managing of large files
   - Component live data files
   - Management of Big Data
   - AI supported inspection and decision making
   - Digital Twins and NDE planning and Interpretation based on modelling
   - Communicating between part and machine and NDE system
2. **Advanced NDE Methods and**
3. **Advanced Inspection tools (e.g. drones)**
4. **Remote NDE to Include competence not available onside**
5. **Make NDE available for anybody by low cost sensing and hand held devices**
6. **Solving new inspection problems for all day service**
7. **Real time monitoring of structure integrity**
8. **Prognostics**

NDT/E Historical view of developments

Smart NDE (NDE 4.0+) - Future Self organizing NDE (+SHM).

- Holistic "healthcare" system for complex engineering products.

Monitoring, diagnostics and prognostics (SHM)

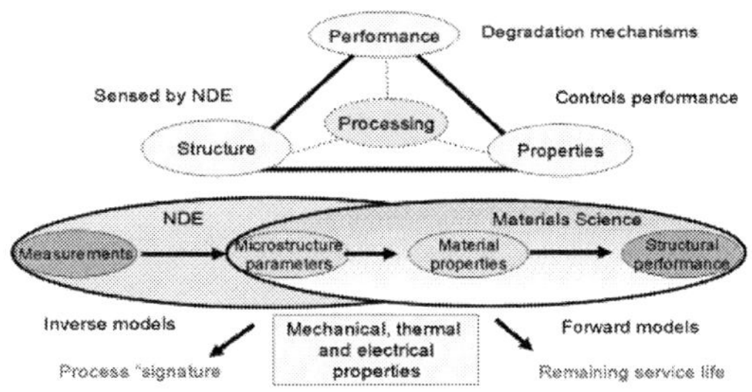

*after Ensminger & Bond (2011 & 2024)

Learning from Medicine

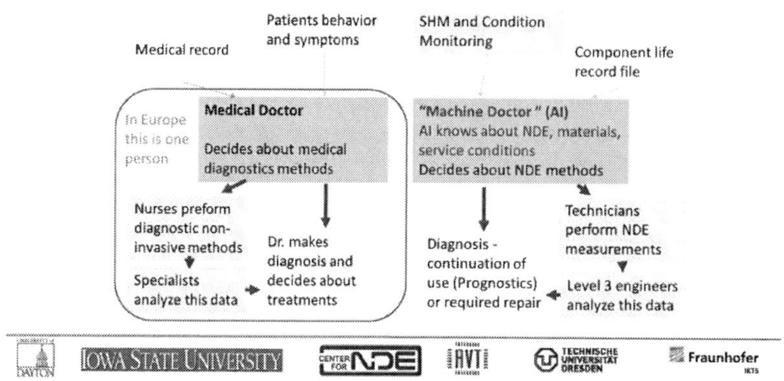

Ingredients of Smart NDE (NDE 4.0)

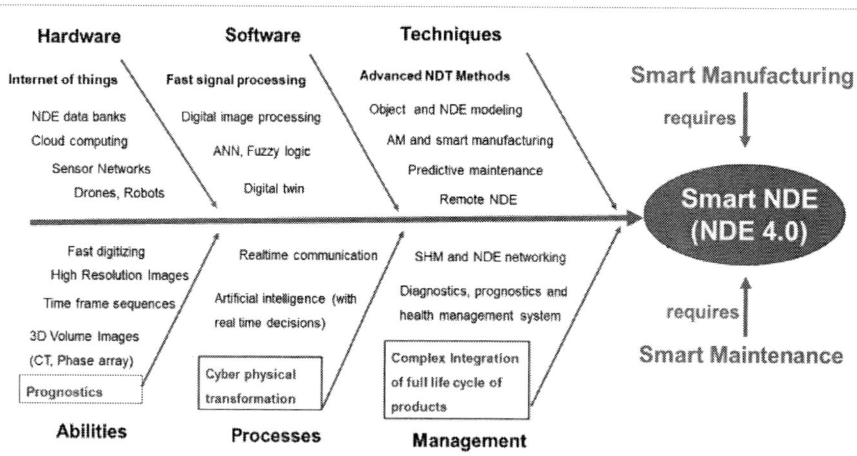

Complex Integration of full life cycle of products

(design, production, inspection (NDE), life monitoring (SHM), recycling.....)

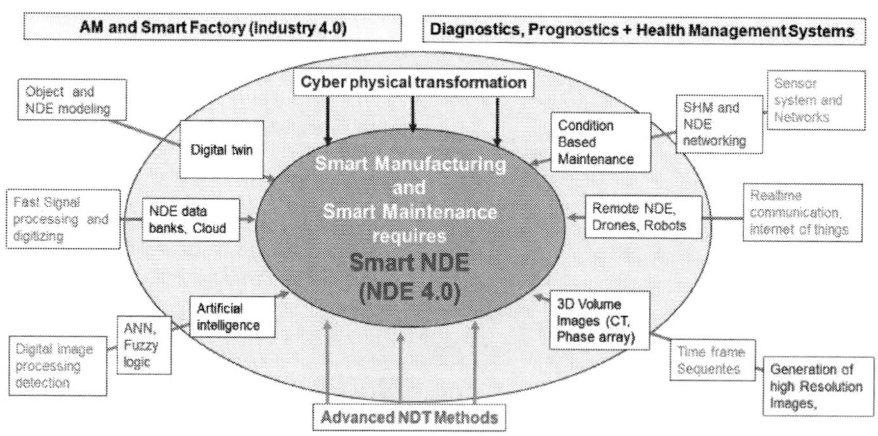

Ultraschall-Bildgebung zur Fehlerrekonstruktion in unidirektional Faser-verstärkten Kompositen unter der Annahme elliptischer Anisotropie

Martin SPIES [1]
[1] Baker Hughes Digital Solutions GmbH, Frankfurt am Main
E-Mail: martin.spies@bakerhughes.com

Kurzfassung. Bildgebende Verfahren wie die Synthetische Apertur Fokus Technik (SAFT) werden seit Jahren zur Fehlerabbildung in isotropen Materialien eingesetzt. Ihre Anwendung bei der Prüfung anisotroper Materialien erfordert die Berücksichtigung spezifischer Phänomene, insbesondere der Richtungsabhängigkeit der Ultraschallgeschwindigkeiten. Um die Anwendung eines komplexen Berechnungsalgorithmus zu umgehen, wurde basierend auf der Annahme elliptischer Anisotropie eine geschlossene Formel zur Berechnung der Gruppengeschwindigkeiten für Longitudinalwellen abgeleitet. Auf dieser Basis wurde ein modifizierter SAFT-Algorithmus implementiert und anhand der Rekonstruktion von Modellfehlern in einem Kohlefaser-verstärkten Testblock validiert.

Keywords: Ultraschall, Prüfkopf, Modellierung, Bildgebung, Fehlerrekonstruktion

Einführung

Eine der wichtigsten Aufgaben der Ultraschallprüfung im Hinblick auf die quantitative zerstörungsfreie Prüfung besteht in der Bestimmung der Position, der Form, der Größe und der Orientierung von Materialfehlern. Diesbezüglich werden bildgebende Verfahren wie die Synthetische Apertur Fokus Technik (SAFT) zur Fehlerabbildung in isotropen Materialien seit Jahren, auch im industriellen Umfeld, eingesetzt. Ihre Anwendung bei der Prüfung anisotroper Materialien erfordert jedoch die Berücksichtigung spezifischer Phänomene wie etwa der Richtungsabhängigkeit der Ultraschallgeschwindigkeiten und der dadurch im Vergleich zu isotropen Materialien veränderten Schallfeldcharakteristiken. Darüber hinaus kann die Richtungsabhängigkeit der Gruppengeschwindigkeiten in der Regel nicht mit einfachen Zusammenhängen beschrieben werden.

Um die Anwendung eines komplexen Algorithmus zu umgehen, wurde eine geschlossene Formel für die räumliche Abhängigkeit der Gruppengeschwindigkeiten der Quasi-Longitudinalwellen abgeleitet, basierend auf der Annahme sogenannter elliptischer Anisotropie. Dieser Beitrag stellt die theoretischen Grundlagen dazu in Kürze vor, die zur Implementierung eines modifizierten SAFT-Algorithmus genutzt wurden, der die Wellenausbreitungsphänomene in unidirektional faserverstärkten Kompositwerkstoffen (FVK) berücksichtigt. Es werden Ergebnisse der Rekonstruktion von Modellfehlern in einem

https://doi.org/10.58286/31231

Kohlefaser-verstärkten Testblock gezeigt, auch im Vergleich zu einem bisher genutzten, nicht-approximierenden SAFT-Algorithmus. Die Relevanz für weitere bildgebende Verfahren wie beispielsweise der Total Focusing Method (TFM) wird diskutiert.

1. Theoretische Grundlagen

1.1 Der SAFT-Algorithmus

Die Grundlagen der Ultraschall-Bildgebung mittels Synthetischer Apertur Fokus Technik sind wohlbekannt [1-4], so dass hier nur ein kurzer Überblick gegeben wird. Im Wesentlichen können die unkorrigierten, an verschiedenen Prüfkopfpositionen aufgezeichneten HF-Daten kombiniert werden, um einen einzigen Prüfkopf mit einer großen ‚synthetischen' Apertur zu simulieren, der in allen Tiefen fokussiert werden kann.

Im Folgenden wird eine geeignete Formulierung für die Verarbeitung von Daten, die mittels linearer Scans eines einzelnen Impuls-Echo-Prüfkopfes aufgezeichnet wurden, unter Berücksichtigung anisotroper Materialeigenschaften gegeben. Angenommen, der Scan zeichnet digitalisierte HF-Amplitudenwerte entlang der Materialoberfläche an Punkten mit den Raumkoordinaten $\underline{\mathbf{R}}'_i$ auf. Die aufgezeichneten zeitabhängigen Signale (A-Bilder) werden dann mit $S(\underline{\mathbf{R}}'_i, t)$ bezeichnet, wobei t das Zeitintervall zwischen dem aufgezeichneten Datenpunkt im A-Scan und dem Anregungsimpuls ist. Das für das Innere des Materials unter Anwendung des SAFT-Algorithmus zu ermittelnde Bild kann dann berechnet werden als

$$I(\underline{\mathrm{R}}_j) = n^{-1} \sum_{i=1}^{n} S(\underline{\mathrm{R}}_i', t_{ij}) \ , \tag{1}$$

wobei n die Anzahl der akquirierten Signale ist, $\underline{\mathbf{R}}_j$ bezeichnet die Raumpositionen innerhalb des Inspektionsvolumens und

$$t_{ij} = \frac{2 \mid \underline{\mathrm{R}}_i' - \underline{\mathrm{R}}_j \mid}{c_\alpha(\widehat{\underline{\mathrm{R}}_i' - \underline{\mathrm{R}}_j})} \tag{2}$$

ist die Laufzeit des Ultraschallsignals von der Prüfkopfposition $\underline{\mathbf{R}}'_i$ zur Position $\underline{\mathbf{R}}_j$ und zurück. Der Betrag der richtungsabhängigen Gruppengeschwindigkeit c_α ist hier eine Funktion des Einheitsvektors der Richtungsdifferenz ($\underline{\mathbf{R}}'_i$-$\underline{\mathbf{R}}_j$). α bezeichnet den betrachteten Wellenmodus, d.h. die Quasi-Longitudinal- (qL) oder die Quasi-Transversalwellen ($qT_{1,2}$), die sich in anisotropen Materialien ausbreiten. Diese werden als ‚quasi'-Wellen bezeichnet, da ihre Polarisation bei Ausbreitung in Nicht-Symmetrierichtungen des Materials nicht rein longitudinal bzw. nicht rein transversal orientiert ist.

1.2 Ermittlung der Gruppengeschwindigkeit in transversal-isotropen Medien

In diesem Beitrag betrachten wir unidirektional-faserverstärkte Kompositwerkstoffe, die sich durch das Modell transversaler Isotropie beschreiben lassen. Diese Medien besitzen eine Rotationssymmetrieachse entlang der Fasern, deren Ausrichtung wir mit dem Vektor $\underline{\mathbf{a}}$ kennzeichnen. Zur Materialcharakterisierung benötigen wir die fünf elastischen Konstanten C_{11}, C_{33}, C_{44}, C_{66} und C_{12} sowie die Dichte ρ. Der elastische Tensor ist im Folgenden in Matrix-Schreibweise (Voigt-Notation) dargestellt für den Fall, dass die Rotationssymmetrieachse in x-Richtung verläuft. Zum Vergleich ist die elastische Matrix für ein isotropes Medium gezeigt. Bei Isotropie gilt die lineare Abhängigkeit C_{11}=C_{12}+2C_{44}, bei transversaler Isotropie gilt die Abhängigkeit C_{33}= C_{23}+2C_{44}.

$$\begin{pmatrix} C_{11} & C_{12} & C_{12} & 0 & 0 & 0 \\ C_{12} & C_{11} & C_{12} & 0 & 0 & 0 \\ C_{12} & C_{12} & C_{11} & 0 & 0 & 0 \\ 0 & 0 & 0 & C_{44} & 0 & 0 \\ 0 & 0 & 0 & 0 & C_{44} & 0 \\ 0 & 0 & 0 & 0 & 0 & C_{44} \end{pmatrix} \quad \begin{pmatrix} C_{11} & C_{12} & C_{12} & 0 & 0 & 0 \\ C_{12} & C_{33} & C_{23} & 0 & 0 & 0 \\ C_{12} & C_{23} & C_{33} & 0 & 0 & 0 \\ 0 & 0 & 0 & C_{44} & 0 & 0 \\ 0 & 0 & 0 & 0 & C_{66} & 0 \\ 0 & 0 & 0 & 0 & 0 & C_{66} \end{pmatrix}$$

Abb. 1 Elastische Matrix in Voigt-Notation - links: isotrop, rechts: transversal-isotrop.

Wir betrachten in diesem Beitrag die Prüfung eines transversal-isotropen Kompositwerkstoffs in Kontakttechnik mittels quasi-Longitudinalwellen (qL), deren Phasengeschwindigkeiten berechnet werden können gemäß

$$V_{qL}^2 = \frac{2A}{\varrho\left[B - (B^2 - 4A)^{1/2}\right]}, \tag{3}$$

wobei die Größen A und B gegeben sind durch [5]

$$\begin{aligned} A &= C_{33}C_{66} + (\underline{a} \cdot \hat{\underline{K}})^2 \left[C_{33}(C_{11} - 2C_{66}) - C_{12}(C_{12} + 2C_{66})\right] \\ &\quad + (\underline{a} \cdot \hat{\underline{K}})^4 \left[(C_{11} - C_{66})(C_{66} - C_{33}) + (C_{12} + C_{66})^2\right], \end{aligned} \tag{4}$$

$$B = C_{33} + C_{66} + (\underline{a} \cdot \hat{\underline{K}})^2 (C_{11} - C_{33}). \tag{5}$$

Der Gruppengeschwindigkeitsvektor berechnet sich gemäß

$$\underline{c}_{qL} = V_{qL}\left[\hat{\underline{K}} + F_{qL}\,(\underline{a} \cdot \hat{\underline{K}})\left(\underline{a} - (\underline{a} \cdot \hat{\underline{K}})\hat{\underline{K}}\right)\right], \tag{6}$$

Die Größe F_{qL} enthält die elastischen Konstanten, die Faserrichtung $\underline{\mathbf{a}}$ sowie den Einheitsvektor in Einschallrichtung $\hat{\underline{K}}$ [5].

Die Auswertung von Gleichung (1) unter Nutzung von Gleichung (2) erfordert die Bestimmung der Wellenvektorrichtung $\hat{\underline{K}}$, die mit einer Gruppengeschwindigkeitsrichtung entlang der Raumrichtung

$$\hat{\underline{R}}_{ij} \equiv \widehat{(\underline{R}_i' - \underline{R}_j)}$$

verknüpft ist. Gleichung (6) beschreibt den Gruppengeschwindigkeitsvektor als Funktion der $\hat{\underline{K}}$-Richtung, so dass die Umkehrung dieser Beziehung erforderlich ist. Aufgrund der – bedingt durch den Term F_{qL} - komplexen mathematischen Struktur von Gleichung (6) gelingt die analytische Inversion jedoch nicht. In der Regel muss diese Inversion für anisotrope Materialien numerisch durchgeführt werden. Analytisch gelingt sie nur für transversal-isotrope Medien im Fall der rein transversal polarisierten Transversalwellen (SH-Wellen) [6] sowie für die quasi-Longitudinalwellen im Spezialfall elliptischer Anisotropie, wie im nächsten Abschnitt ausgeführt wird.

Ein effizientes $\hat{\underline{K}}(\hat{\underline{R}}_{ij})$-Auswerteschema kann auf folgende Weise implementiert werden [7]. Das numerische Programm durchläuft die durch einen azimutalen Winkel θ und einen polaren Winkel ϕ in Schritten von z.B. 0.25° parametrisierten Wellenvektorpositionen und bestimmt die zugehörigen Gruppengeschwindigkeitsvektoren, deren Richtungen wiederum entsprechend parametrisiert sind. Die den Vektor $\hat{\underline{K}}$ charakterisierenden Winkel θ und ϕ werden als Array gespeichert, wobei die für die Gruppengeschwindigkeitsvektoren $\underline{c}$ ermittelten azimutalen und polaren Winkel als Array-Indizes verwendet werden. Bei der Auswertung von Gleichung (2) werden dann die polaren und azimutalen Winkel für eine bestimmte Richtung $\hat{\underline{R}}_{ij}$ ($\equiv \hat{\underline{c}}$) verwendet, um die zugehörige $\hat{\underline{K}}$-Richtung aus dem gespeicherten Array herauszusuchen. Besondere Vorsicht ist in Richtungen geboten, in denen die Gruppengeschwindigkeit mehrwertig ist, wie unter Umständen im Fall der quasi-Transversalwellen - hier werden mehrere Look-Up-Arrays erzeugt und angewendet [8].

1.3 Ermittlung der Gruppengeschwindigkeiten unter der Annahme elliptischer Anisotropie

Um eine direkte räumliche Abhängigkeit der qL-Gruppengeschwindigkeiten herzuleiten, betrachten wir folgenden Spezialfall, der zum Modell der elliptischen Anisotropie führt

$$(C_{12} + C_{66})^2 = (C_{11} - C_{66})(C_{33} - C_{66}) \; . \tag{7}$$

Daraus ergibt sich die Phasengeschwindigkeit zu

$$V^2_{qL,ell} \;=\; \frac{1}{\varrho}\left[C_{33} + (C_{11} - C_{33})(\underline{a} \cdot \underline{\hat{K}})^2\right] \; . \tag{8}$$

Der Gruppengeschwindigkeitsvektor resultiert dann in einer wesentlich einfacheren Form gemäß

$$\underline{c}_{qL,ell} \;=\; (\varrho\, V_{qL,ell})^{-1}\left[C_{33}\,\underline{\hat{K}} + (C_{11} - C_{33})(\underline{a} \cdot \underline{\hat{K}})\,\underline{a}\right] . \tag{9}$$

Die Gleichungen (8) und (9) enthalten nur noch die beiden elastischen Konstanten C_{11} und C_{33}, welche die Schallgeschwindigkeiten der Longitudinalwellen bei Einschallung in Faserrichtung bzw. senkrecht dazu charakterisieren.

Da der (Energie-) Beitrag zum Wellenfeld an einem Beobachtungspunkt $\underline{R}$ durch die Gruppengeschwindigkeit charakterisiert wird, gilt für die entsprechenden Einheitsvektoren

$$\underline{\hat{R}} = \underline{\hat{c}}_{qL,ell}, \quad |\underline{\hat{R}}| = |\underline{\hat{c}}_{qL,ell}| = 1 \; . \tag{10}$$

Aus Gleichung (9) folgt

$$\underline{\hat{R}} \;=\; \frac{C_{33}\,\underline{\hat{K}} + (C_{11} - C_{33})(\underline{a} \cdot \underline{\hat{K}})\underline{a}}{\left[C_{33}^2 + (C_{11}^2 - C_{33}^2)(\underline{a} \cdot \underline{\hat{K}})^2\right]^{1/2}} \; , \tag{11}$$

und

$$(\underline{a} \cdot \underline{\hat{R}}) \;=\; \frac{C_{11}(\underline{a} \cdot \underline{\hat{K}})}{\left[C_{33}^2 + (C_{11}^2 - C_{33}^2)(\underline{a} \cdot \underline{\hat{K}})^2\right]^{1/2}} \; . \tag{12}$$

Die Kombination der Gleichungen (11) und (12) liefert eine direkte Beziehung zwischen $\underline{\hat{K}}$ und $\underline{\hat{R}}$ gemäß

$$\underline{\hat{K}} \;=\; \frac{C_{11}\,\underline{\hat{R}} + (C_{33} - C_{11})(\underline{a} \cdot \underline{\hat{R}})\,\underline{a}}{\left[C_{11}^2 + (C_{33}^2 - C_{11}^2)(\underline{a} \cdot \underline{\hat{R}})^2\right]^{1/2}} \; , \tag{13}$$

damit folgt dann aus Gleichung (9) der Betrag der Gruppengeschwindigkeit als Funktion von $\underline{\hat{R}}$ gemäß

$$c_{qL,ell}(\underline{\hat{R}}) \;=\; \left[\frac{C_{11}C_{33}}{\varrho\,(C_{11} + (C_{33} - C_{11})(\underline{a} \cdot \underline{\hat{R}})^2)}\right]^{1/2} \; . \tag{14}$$

Gleichung (14) wird nachfolgend bei der SAFT-Rekonstruktion zur Berechnung der Laufzeiten der Ultraschallsignale gemäß Gleichung (2) genutzt.

1.4 Auswertung des Rekonstruktionsvolumens

Im Prinzip stellt die Summation nach Gleichung (1) eine einfache Überlagerung der empfangenen Ultraschall-HF-Amplituden an denjenigen Materialpositionen dar, an denen die detektierten Signale gemäß den Ultraschall-Laufzeiten erzeugt worden sein könnten. Bei der Bildberechnung muss also das Materialvolumen berücksichtigt werden, das von dem jeweiligen Prüfkopf beleuchtet wird. Bei isotropem Material kann dies mit der bekannten

Formel für den Divergenzwinkel des Schallfeldes [9] geschehen. In anisotropen Medien hängt das Prüfkopf-Schallfeld jedoch stark von den elastischen Materialeigenschaften und vom Einschallwinkel relativ zu den Materialsymmetrieachsen ab. Das Schallfeld wird daher mittels Generalisierter Punktquellensynthese [10,11] berechnet und bei der Auswertung von Gleichung (1) berücksichtigt.

2. SAFT-Rekonstruktion in unidirektional Faser-verstärkten Kompositwerkstoffen

2.1 Untersuchter FVK-Testblock

Referenz [12] berichtet über Untersuchungen an zwei FVK-Testblöcken, bei denen die Fasern unter 15° und 75° zur Oberflächennormalen ausgerichtet sind. Diese Testkörper wurden mit einer Reihe von Zylinderbohrungen von der Seite und Flachbodenbohrungen (FBB) unterschiedlicher Tiefe und Durchmesser versehen. Die elastischen Konstanten dieses transversal-isotropen Verbundwerkstoffs sind gegeben durch C_{11}=125.5 GPa, C_{33}=15.1 GPa, C_{12}=8.0 GPa, C_{44}=3.4 GPa und C_{66}=6.5 GPa, die Dichte beträgt ρ=1.6 g/cm^3.

Aufgrund der geringen Abmessungen des Testblocks mit der Ausrichtung der Fasern unter 75° waren die Ergebnisse der SAFT-Rekonstruktion wenig aussagekräftig, da lediglich die mittig im Testkörper platzierten FBB detektiert und rekonstruiert werden konnten. Daher konzentrieren wir uns in diesem Beitrag auf den in Abbildung 2 gezeigten Testkörper mit der Ausrichtung der Fasern unter 15°.

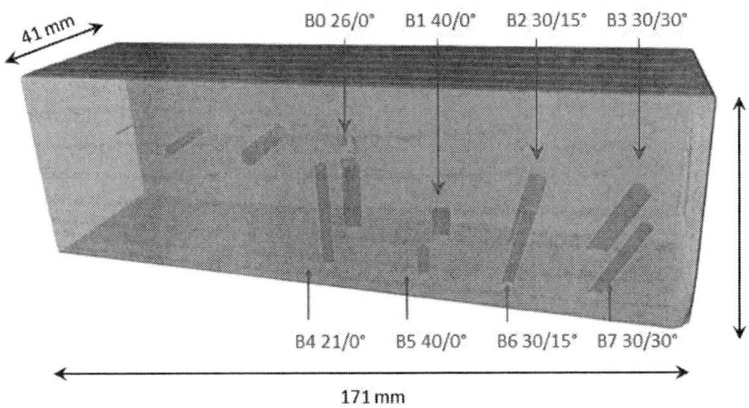

Abb. 2 FVK-Testblock mit Tiefenlage und Ausrichtung der FBB (B0-B3: Ø 5 mm, B4-B7: Ø 3 mm). Die drei Zylinderbohrungen haben eine Länge von 10 mm (Ø 1 mm) und 20 mm (Ø 3 und 5 mm).

2.2 Vergleich der exakt und der näherungsweise berechneten Gruppengeschwindigkeiten

Abbildung 3 zeigt den Vergleich der mittels Gleichungen (6) und (14) berechneten Gruppengeschwindigkeiten für die Raumrichtungen von 0° bis 90° bei einer Ausrichtung der Fasern entlang der x-Achse. Die maximale Abweichung von etwa 10 % ergibt sich für eine Richtung von 20° zur x-Achse und entspricht der Ungenauigkeit bei der Ermittlung der elastischen Konstanten.

2.3 Vergleich der SAFT-Rekonstruktionsergebnisse

Die Impuls-Echo-Messungen wurden mit einem Senkrechtprüfkopf mit einer Mittenfrequenz von 2.25 MHz und einem Durchmesser von 6.3 mm (Krautkrämer MSW-QC 2,25) durch-

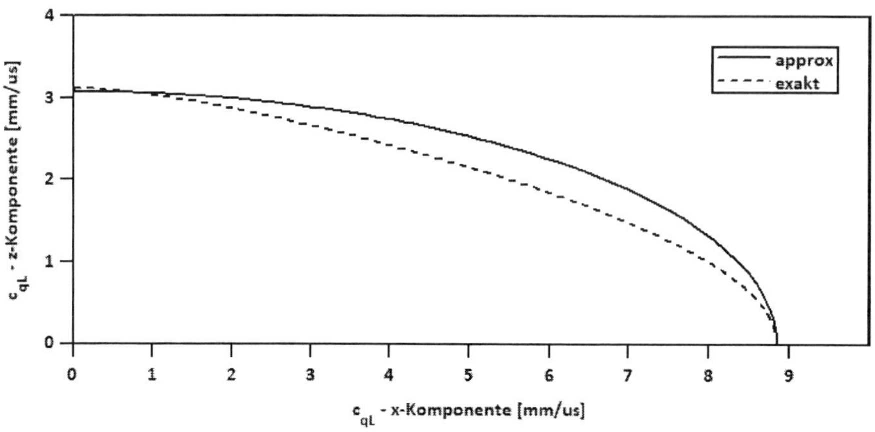

Abb. 3 Richtungsabhängige *qL*-Gruppengeschwindigkeit berechnet mittels Gleichungen (6) und (14).

geführt, der (Quasi-) Longitudinalwellen erzeugt. In Kontakttechnik wurden Impuls-Echo-HF-Daten entlang aufeinanderfolgender linearer Scans erfasst und als A-Scans aufgezeichnet. Der Prüfkopf wurde in Schritten von 0.3 mm entlang der x-Richtung positioniert (554 Schusspositionen) mit einem Spurabstand von 1.0 mm entlang der y-Richtung (38 Linienscans)[1]. Als Ergebnis dieser Scans ist in Abb. 4 beispielhaft das B-Scan-Bild für die Scanlinie Nr. 29 dargestellt, die über die Mitte der 5 mm FBB verläuft. Man erkennt das starke Eintrittsecho sowie das variierende Rückwandecho, das zum Teil durch die Bohrungen abgeschwächt wird. Die unterschiedlichen Laufzeiten des Rückwandechos sind auf die variierende Dichte der Fasern in diesem dicken Komposit zurückzuführen.

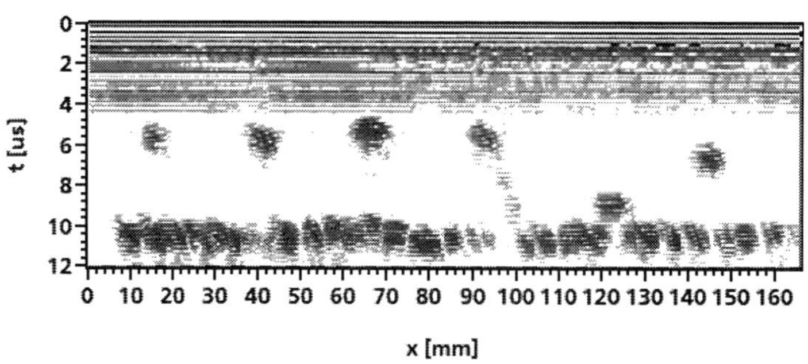

Abb. 4 B-Bild der Scan-Linie Nr. 29 mit den Anzeigen der Zylinderbohrungen links und der Flachbodenbohrungen in unterschiedlicher Tiefe rechts.

Bei einem isotropen Material kann anhand der B-Bilder unter Kenntnis der Schallgeschwindigkeit direkt die Fehlertiefe und -position ermittelt werden. Bei anisotropen Materialien ist dies jedoch aufgrund der richtungsabhängigen Gruppengeschwindigkeiten und des Effekts der Schallfeldablenkung nicht möglich. In diesem Fall ergeben sich die Tiefen und Positionen der Defekte aus den entsprechenden SAFT-Rekonstruktionen, die in Abbildung 5 für die Scan-Linie 29 dargestellt sind.

Es ist zu beachten, dass die geneigte FBB am rechten Rand der Probe aufgrund der Schallfeldablenkung nicht getroffen wird. Vergleicht man die SAFT-Rekonstruktion, die mit dem approximierenden Algorithmus zur Berechnung der Gruppengeschwindigkeiten durchgeführt wurde, mit dem bisher genutzten Algorithmus [12], so sind allenfalls marginale

---

[1] Der für diese Untersuchung verwendete HF-Datensatz wurde im Rahmen meiner Tätigkeit am Fraunhofer-Institut für Zerstörungsfreie Prüfverfahren IZFP durch meinen damaligen Kollegen Herrn Winfried Jager, inzwischen im Ruhestand, aufgenommen

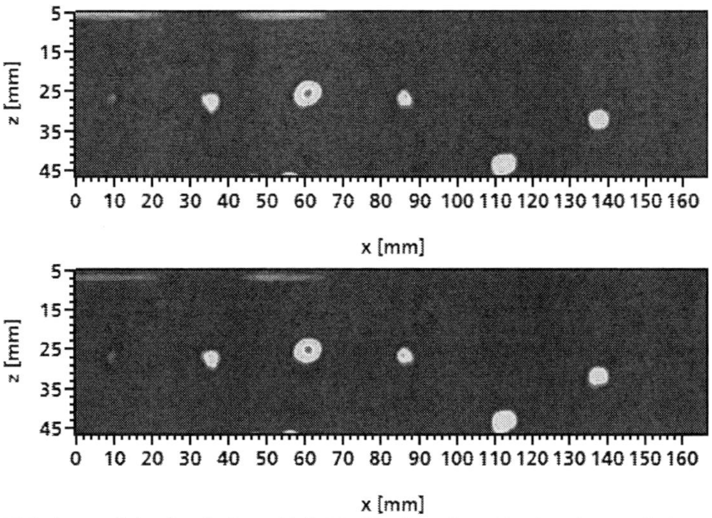

Abb. 5 SAFT-Rekonstruktion für die Scan-Linie Nr. 29 mit exakter (oben) und approximierender (unten) Berechnung der räumlich abhängigen Gruppengeschwindigkeiten.

Unterschiede zu erkennen. Ein quantitativer Vergleich ist in Abbildung 6 anhand des Lateralprofils in 26 mm Tiefe dargestellt mit den Anzeigen der drei Zylinderbohrungen und der FBB in dieser Tiefe. Abbildung 7 schließlich zeigt die Rekonstruktion der 3 mm Flachbodenbohrungen, deren Position und Tiefenlage ebenfalls korrekt wiedergegeben werden. Zur besseren Darstellung der Fehleranzeigen wurden der Bereich des Eintrittsechos und der Rückwand in den Abbildungen 5 und 7 ausgeblendet.

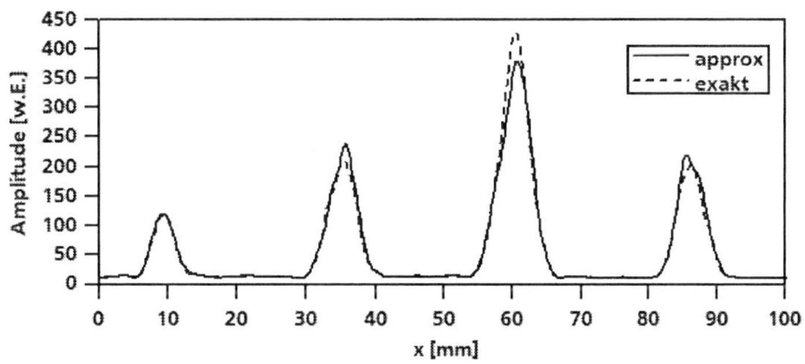

Abb. 6 Lateralprofil in 26 mm Tiefe ermittelt aus den beiden SAFT-Rekonstruktionen (Abb. 5). Der maximale Amplitudenunterschied beträgt 1 dB (5 mm Zylinderbohrung bei x = 60 mm).

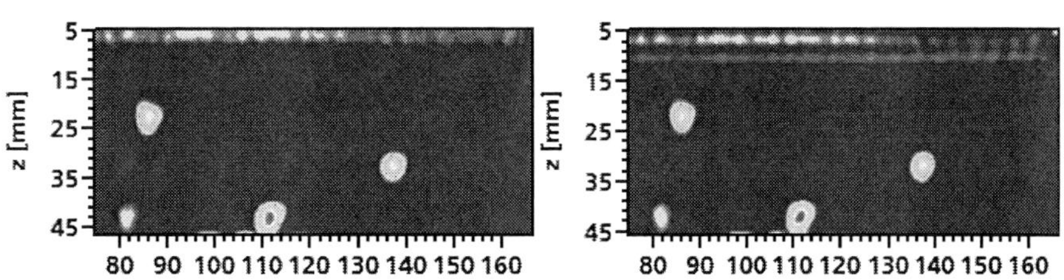

Abb. 7 SAFT-Rekonstruktion für die Scan-Linie Nr. 7 mit exakter (links) und approximierender (rechts) Berechnung der räumlich abhängigen Gruppengeschwindigkeiten.

4. Zusammenfassung

Im Rahmen dieser Untersuchung wurde unter der Annahme elliptischer Materialanisotropie eine Näherungsformel zur direkten analytischen Berechnung der räumlichen Abhängigkeit der Gruppengeschwindigkeiten für Longitudinalwellen abgeleitet. Damit wird der Einsatz eines numerischen Berechnungsalgorithmus vermieden, der die Energieausbreitung den zugehörigen Phasenrichtungen zuordnet. Dies bedeutet insbesondere einen Rechenzeitvorteil, da der bisher bei der SAFT-Rekonstruktion eingesetzte numerische Algorithmus die Raumrichtungen hochdiskretisiert durchschreiten muss, um diese Zuordnung korrekt auszuführen.

Die unter der Annahme elliptischer Anisotropie durchgeführten SAFT-Rekonstruktionen von Modellfehlern in einem FVK-Testblock zeigen lediglich marginale Unterschiede zu dem in [12] genutzten SAFT-Algorithmus. Das hier beschriebene Berechnungsschema auf Basis von Gleichung (14) besitzt damit großes Potenzial für weitere bildgebende Verfahren wie beispielsweise der Total Focusing Method (TFM). Hier wirkt sich insbesondere der genannte Rechenzeitvorteil positiv aus, da im Gegensatz zum SAFT-Algorithmus die Array-Elemente bei TFM nicht im Impuls-Echo-Modus betrieben werden, sondern im Sende- und Empfangsmodus. Dies erfordert die Ermittlung der je nach Elementanzahl des verwendeten Array-Prüfkopfes mehr oder weniger umfangreichen Energieausbreitungspfade zwischen dem jeweiligen Sende-Element, den Raumpunkten im Rekonstruktionsvolumen und den Empfangselementen.

Danksagung

Besonderer Dank gilt meinem Kollegen, Herrn Dr. Stephan Falter, Baker Hughes Digital Solutions GmbH, Hürth, für den sorgfältigen Review des Manuskripts.

Referenzen

[1] D.W. Prine; Synthetic Aperture Ultrasonic Imaging; in Proc. of the Eng. Appl. of Holography Symposium 287, Society of Photo-optical Instrumentation Engineers (1972)

[2] S.R. Doctor, T.E. Hall, L.D. Reid; SAFT - The Evolution of a Signal Processing Technology for Ultrasonic Testing, NDT International, Vol. 19 (1986), pp. 163-167

[3] M.S. Hughes, D.K. Hsu, F. Margetan et al.; Application of SAFT on Composites, in *Review of Progress in QNDE*, eds. D.O. Thompson and D.E. Chimenti, Plenum Press, New York, Vol. 11 (1992), pp. 1413-1419

[4] M. Spies, H. Rieder, A. Dillhöfer et al.; Synthetic Aperture Focusing and Time-of-Flight Diffraction Ultrasonic Imaging – Past and Present; J. Nondestruct. Eval., Vol. 31 (2012), pp. 310-323

[5] M. Spies; Elastic Waves in Homogeneous and Layered Transversely Isotropic Media: Plane Waves and Gaussian Wave Packets. A General Approach; J. Acoust. Soc. Am., Vol. 95 (1994), pp. 1748-1761

[6] M. Spies; Space-Time Far-Field Representation of Green's Function for Cross-Plane Shear Waves in General Transversely Isotropic Media; J. Acoust. Soc. Am., Vol. 102 (1997), pp. 733-740

[7] M. Spies; Kirchhoff Evaluation of Scattered Elastic Wavefields in Anisotropic Media; J. Acoust. Soc. Am., Vol. 107 (2000), pp. 2755-2759

[8] M. Spies; Modeling Transient Radiation of Ultrasonic Transducers in Anisotropic Materials Including Wave Attenuation, in *Review of Progress in QNDE*, Vol. 21, eds. D.O. Thompson and D.E. Chimenti, Plenum Press, New York (2002), pp. 807-814

[9] J. Krautkrämer, H. Krautkrämer; *Ultrasonic Testing of Materials*, Springer, Berlin (1990)

[10] M. Spies; Transducer-Modeling in General Transversely Isotropic Media via Point-Source-Synthesis. Theory; J. Nondestr. Eval., Vol. 13 (1994), pp. 85-99

[11] M. Spies; Drei Jahrzehnte Generalisierte Punktquellensynthese – Rück- und Ausblick; in diesem Berichtsband

[12] M. Spies, W. Jager; Synthetic Aperture Focusing for Defect Reconstruction in Anisotropic Media; Ultrasonics, Vol. 41 (2003), pp. 125-131

Untersuchung von mikromechanischen Schädigungsmechanismen in amorphen Thermoplasten mittels nichtlinearem Ultraschall

Marius REITINGER [1], Julian EHRLER[1], Marc KREUTZBRUCK [1]

[1] Institut für Kunststofftechnik, Universität Stuttgart, Stuttgart, Deutschland, marius.reitinger@ikt.uni-stuttgart.de

Kurzfassung. Kunststoffe sind aufgrund ihrer Vielseitigkeit, Verarbeitbarkeit und Leichtbaupotenziale in vielen Industriezweigen unverzichtbar. Doch die zunehmende Optimierung von Bauteilen, die möglichst nah an ihrer mechanischen Belastbarkeitsgrenze ausgelegt werden, erhöht sich das Risiko eines, durch kleinste Schädigungen und Ermüdungserscheinungen initiierten Bauteilversagens.

Irreversiblen Schädigungen im Nanometerbereich wie Crazing oder Scherfließen treten bei Kunststoffen typischerweise zuerst im nichtlinear-viskoelastischen Bereich während der Deformation auf. Unter anhaltender oder steigender Belastung entwickeln sich aus diesen mikromechanischen Deformationen Risse, die sich im Bauteil weiter ausbreiten. Dadurch können selbst kleinste Schädigungen und Ermüdungserscheinungen eine große Auswirkung auf die Zuverlässigkeit und Lebensdauer der Bauteile haben. Deshalb ist es entscheidend, Methoden zur frühzeitigen Detektion nichtlinearer Schädigungen zu entwickeln.

Die Ultraschallprüfung, speziell nichtlineare Ultraschallverfahren stellen hier eine vielversprechende, zerstörungsfreie Methode dar. Damit ist es möglich, die Veränderung des nichtlinearen akustischen Verhaltens des Materials und somit von nichtlinearen Schädigungen zu analysieren.

Dabei ist das auftretende Ermüdungsverhalten im nichtlinearen-viskoelastischen Bereich wie die Bildung von Crazing zu analysieren. Hierfür werden Prüfkörper durch eine definierte stufenweise Zugbelastung von mikromechanischen Deformationen bis hin zu plastischen Verformungen geschädigt. Anschließend ermöglichen Messungen mittels nichtlinearem Ultraschall die Bestimmung des Nichtlinearitätsfaktors. Der Vergleich der hierbei ermittelten Werte zeigt eine klare Tendenz zu steigenden Nichtlinearitätsfaktoren bei zunehmender Belastung. Es konnten eindeutige Schädigungen bei Dehnungen nahe der Streckdehnung festgestellt werden, die deutlich im nichtlinear-viskoelastischen Bereich liegen. Außerdem ist auch eine eindeutige Ermittlung von plastischen Verformungen mit dieser Methode möglich.

Keywords: Bauteilprüfung, Bruchmechanik, Ermüdung, Mikromechanik, zerstörungsfreie Prüfung, nichtlinearer Ultraschall, Kunststoffe

https://doi.org/10.58286/31236

1. Einleitung

Kunststoffe sind in verschiedensten Industriezweigen unverzichtbar und finden aufgrund ihrer Vielseitigkeit und guten Verarbeitbarkeit in unzähligen Anwendungsbereichen Verwendung. [1] Die zunehmenden Anforderungen an Bauteile, die mit möglichst geringem Materialeinsatz möglichst nahe an ihrer maximalen mechanischen Belastbarkeit ausgelegt sind, machen sie potenziell anfälliger für kleinste Ermüdungserscheinungen und somit für ein frühzeitiges Versagen. Die Untersuchung und das Verständnis der hierbei auftretenden mikromechanischen Schädigungsmechanismen bietet die Möglichkeit, dieses frühzeitige Versagen vorherzusagen und entsprechend im Konstruktionsprozess von Kunststoffbauteilen zu berücksichtigen.

Mikromechanische Schädigungsmechanismen von Kunststoffen

Anders als bei metallischen Werkstoffen, die oft ein deutlich einfacheres Ermüdungsverhalten aufweisen, ist die Betrachtung von Kunststoffen wesentlich komplexer. Je nachdem, durch welche Beanspruchungsart es zu einem Bruch kommt, wird zwischen Gewalt- oder Schwingbruch unterschieden, wobei Zweiteres in die Kategorie der Ermüdungsbrüche fällt. Bei der Versagensart eines Bruches gehen Rissbildungen voraus, welche wiederum aus mikromechanischen Fehlstellen hervorgehen, weshalb in makroskopische und mikroskopische Bruchmerkmale unterschieden wird. Auf mikroskopischer Ebene sind besonders das Crazing und das Scherfließen, welche beide ein Teil der mikromechanischen Deformationsmechanismen sind, von Bedeutung. Vor allem amorphe Polymere weisen oftmals ein solches Deformationsverhalten auf. [2]

Scherfließen, auch Shear-Yielding genannt, beginnt auf mikroskopischer Ebene und kann zum Versagen des Bauteils führen. Eine Besonderheit ist, dass sich Scherbänder nur bei bestimmten Dehngeschwindigkeiten und Temperaturen bilden können und das Volumen dabei konstant bleibt. Bei teilkristallinen Kunststoffen geschieht das Scherfließen durch das Abgleiten der Ketten. Zu Beginn sind es die amorphen Bestandteile, die sich verstrecken, darauffolgend die kristallinen Bereiche, die sich in Richtung der Beanspruchung orientieren. [2 – 5]

In Abbildung 1 sind die Vorgänge beim Scherfließen bei amorphen (rechts) und teilkristallinen (links) Thermoplasten dargestellt.

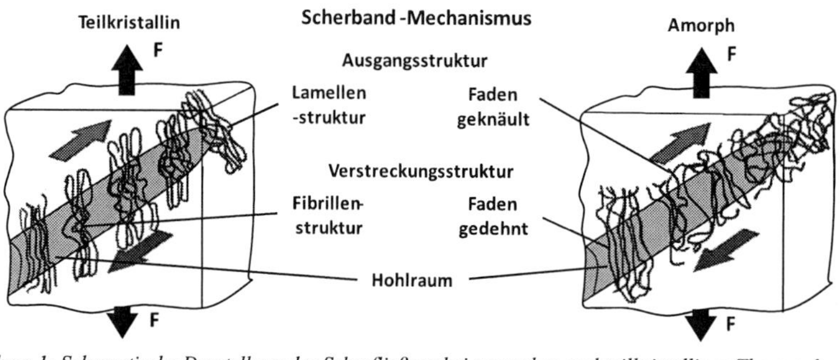

Abbildung 1: Schematische Darstellung des Scherfließens bei amorphen und teilkristallinen Thermoplasten [3]

Eine wie in Abbildung 1 rein axiale aufgebrachte Zugspannung führt mit dem Maximum der Schubspannung unterhalb von einem 45° Winkel dazu, dass die Hohlräume der Scherbänder in einem Bereich zwischen 30° und 45° auftreten. Bei ausreichend vorhandenen Scherbändern sind diese ebenfalls an der Oberfläche des Materials sichtbar. [4]

Das Wachstum solcher Scherbänder schreitet voran, solange eine ausreichend große Spannung vorhanden ist und die Verformung ihr Maximum nicht überschritten hat. Ist dieses Maximum überschritten, entsteht an dem Scherband ein Riss, der letzten Endes zum Bruch führt. [6]

Ermüdungserscheinungen können nicht nur in Form von Scherbändern auftreten, sondern auch in Form von Craze-Bildung. Die Bildung von Crazes ist, ähnlich wie die Scherband-Bildung, die Vorstufe eines Spannungsrisses und führt ebenfalls dazu, dass immer größer werdende Risse entstehen, bis es zum kompletten Bauteilversagen kommt. Mithilfe des lokalen Abgleitens bei Beanspruchung von Makromolekülen lässt sich die Entstehung von Crazes, auch Mikrorisse genannt, erklären. Sowohl amorphe als auch teilkristalline Thermoplaste weisen diese typischen Mikrorisse auf.

Bei amorphen Kunststoffen sind Crazes oftmals bereits mit bloßem Auge erkennbar (Abbildung 2, links). Sie können dabei innerhalb eines Bauteils oder ausschließlich an der Oberfläche auftreten.

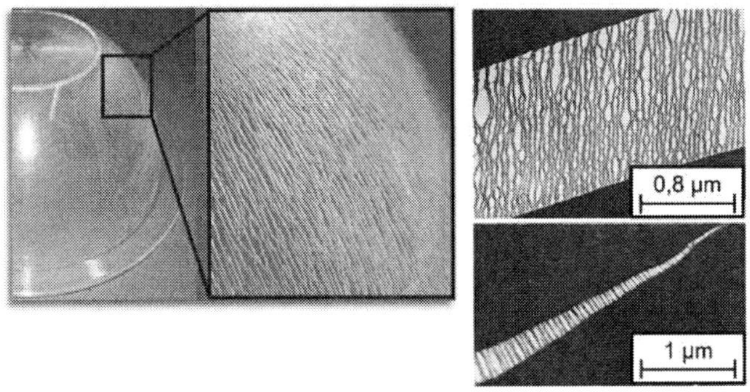

Abbildung 2: Craze-Bildung innerhalb eines Bauteils (links) und Elektronenmikroskop-Aufnahme von Crazes (rechts) [8]

Innerhalb der Crazes entstehen aufgrund plastischer Verformung sogenannte Fibrillen (Abbildung 2, rechts), welche aus Molekülbündeln bestehen. Diese Fibrillen erstrecken sich über den Mikroriss und nehmen bis zu 60 % des Volumens innerhalb des Crazes ein. Weiter sind die Fibrillen in Richtung der aufgetretenen Belastung orientiert und können dabei immer noch eine gewisse Last ertragen. Jedoch ist der Werkstoff in diesem Zustand irreversibel geschädigt, was zu einer Veränderung des viskoelastischen Verhaltens des Materials führt. Die Bildung von Crazes wird dabei von verschiedenen Faktoren wie den Materialeigenschaften, den Umgebungsbedingungen und der Art der Belastung beeinflusst. [2, 7]

Bei einer erhöhten Beanspruchung kommt es zu einem Aufbrechen der Fibrillen, was zum Versagen der Crazes und letztlich zur Rissbildung führt. [2, 8, 9] Im Gegensatz zu der Scherband-Bildung treten Crazes orthogonal zur Beanspruchungsrichtung auf und führen zu einer positiven Volumenänderung durch die Bildung des Craze-Hohlraumes.

Nichtlinearer Ultraschall

Zur Erzeugung von linearem Ultraschall werden lineare Schwingungen und Wellen verwendet. Durch beispielsweise Risse innerhalb des Materials ist keine kontinuierliche Impedanz mehr vorhanden. Die Streuung am Riss, der wie eine Grenzfläche wirkt, verringert daher die Amplitude der reflektierten Welle. Weiter ist es durch den Riss möglich, dass sich die Phase und die Laufzeit der reflektierten Welle ändern. Aufgrund der Gültigkeit des Hooke'schen Gesetzes lässt sich der Riss mithilfe des Signals und der linearen Elastizität detektieren. [10]

In nichtlinearen Werkstoffen, wie Kunststoffen, besteht kein linearer Zusammenhang zwischen Eingangsgröße und Ausgangsgröße. Damit erzeugt eine lineare Sinus-Anregung keine reine sinusartige Schwingung mehr, sondern es sind Anteile von nichtlinearen Schwingungen vorhanden. Diese Abweichung vom eingebrachten Sinus ist ein Maß für die Nichtlinearität und damit ein Indikator für Risse bzw. Mikroschädigungen. Die nichtlinearen Effekte entstehen an diesen Fehlstellen beispielsweise durch die Änderung der Plastizität des Materials oder durch nichtlineare Reibungsverluste. Durch diese wird die nichtlineare Ultraschallwelle verzerrt und es entstehen die sogenannten höheren harmonischen Schwingungen, die ein Teil der Energie der Grundschwingung übertragen bekommen. [10 – 12] Ein Maß für die auftretende Nichtlinearität ist der Nichtlinearitätsfaktor. Dieser wird aus der Beziehung zwischen den Nichtlinearitäten und den entstehenden höher harmonischen Schwingungen hergeleitet und ergibt sich zu

$$\beta = \frac{A_2}{{A_1}^2}$$

für die zweite harmonische Schwingung bzw.

$$\gamma = \frac{A_3}{{A_1}^3}$$

für die dritte harmonische Schwingung, wobei A_n die Amplitude der n-ten harmonischen Schwingung ist.

Durch Auswertung des Nichtlinearitätsfaktor lassen sich nichtlineare Fehlstellen erkennen. Grundsätzlich kann diese Methode ebenfalls zur Bestimmung des linearen- und nichtlinearen-viskoelastischen Bereichs verwendet werden.

Ziel

Das Ermüdungsverhalten von Kunststoffen stellt eine Herausforderung bei der langfristigen Verwendung von Kunststoffbauteilen dar. Die Erkennung dessen ist für die Zuverlässigkeit und Sicherheit solcher Bauteile von besonderer Bedeutung. Die Untersuchung mittels nichtlinearem Ultraschall stellt dabei eine vielversprechende Möglichkeit zur Erkennung von vorzeitigen Ermüdungsverhalten wie der Bildung von Crazes dar. Insbesondere die Craze-Bildung stellt eine Herausforderung dar, da sie in der Regel meist schwer zu erkennen ist und eine genaue Charakterisierung erfordert.
Deshalb ist das Ziel die mikromechanischen Schädigungen in amorphen Thermoplasten, welche in einem direkten Zusammenhang mit Mikrorissen und letztendlich einem Bruchversagen stehen, mittels nichtlinearem Ultraschall zu detektieren. Die Ergebnisse sollen dazu beitragen, ein grundlegend besseres Verständnis der mikromechanischen Schädigungsmechanismen von Kunststoffen zu erlangen.

2. Methodik

Werkstoffauswahl und Prüfkörperpräparation

Als Werkstoff wurde ein Polycarbonat PC1220 der Firma Lotte Chemical, Seoul, Südkorea verwendet. Hierbei handelt es sich um einen amorphen Thermoplasten, der eine vergleichsweise hohe Streckdehnung im Bereich von 6 - 7 % besitzt und damit einen größeren nichtlinearen Bereich aufweist. Daraus wurden Standard-Zugprüfkörper nach DIN EN ISO 20753 Typ A1 mit einer Dicke von 6 mm mittels einer Spritzgießmaschine vom Typ Allrounder 370s der Fa. Arburg, Loßburg, spritzgegossen.

Mechanische Charakterisierung

Die Versuche zur definierten dehnungsabhängigen Belastung bzw. Schädigung wurden auf einer Universalprüfmaschine des Typs UPM 1455 der Fa. Zwick-Roell, Ulm, durchgeführt. Dabei wurde eine Prüfgeschwindigkeit von 5 mm/min gewählt. Die Messung der Dehnung erfolgte im nichtlinearen Bereich mithilfe von Feindehnungsaufnehmern und danach über den Traversenweg der Prüfmaschine. Die Belastung erfolgte dabei stufenweise bis zu einer maximalen Dehnung von 10 %. Nach jeder Dehnungsstufe wurde eine Ultraschallmessung durchgeführt. Zusätzlich wurden Versuche bis zum Bruch durchgeführt, um die mechanischen Eigenschaften zu charakterisieren und den relevanten Messbereich zu identifizieren.

Ultraschallmessungen

Die Ultraschallmessungen wurden in einem Wasserbad mit dem in Abbildung 3 abgebildeten Prüfaufbau durchgeführt. Dabei wurde mit einer vorher festgelegten Schrittweite von 0,5 mm die Messpositionen über die gesamte Prüfkörperlänge angefahren. Bei der Durchführung der Messungen wird ein Mittelwert aus 32 Messungen pro Messstelle gebildet, um unter anderem das elektrische Rauschen des Signals zu unterdrücken.

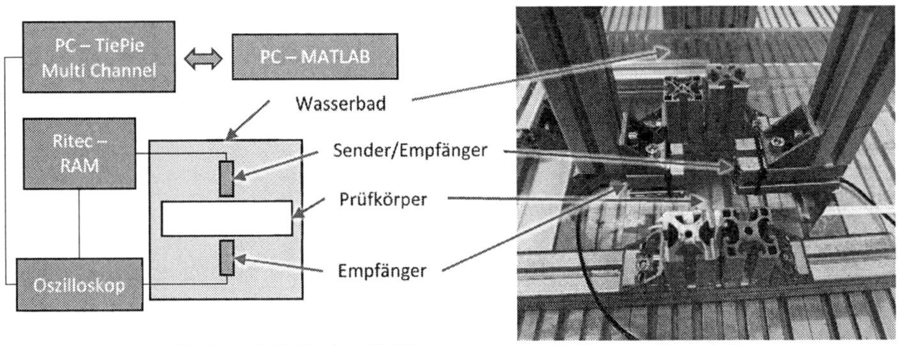

Abbildung 3: Prüfaufbau für Messungen mittels nichtlinearem Ultraschall

Für die Ultraschallmessung wird zur Erzeugung und zum Erfassen des nichtlinearen Ultraschalls ein Hochleistungsmesssystem RAM-5000-SNAP der Firma Ritec, Warwick, USA genutzt. Dieses ist speziell für ein nichtlineares akustisches Materialverhalten ausgelegt und ermöglicht unter anderem die Entstehung von harmonischen Schwingungen im Prüfkörper, die über die Schallamplituden, die ein Spektrum von geringen und hohen Werten abdecken, beobachtet werden können. Die Signale werden durch einen Sender/Empfänger-Prüfkopf des Typs Krautkrämer L5ME der Firma Waygate Technologies, Hürth, Deutschland mit einer Frequenz von 2 MHz erzeugt und einem reinen Empfänger-Ultraschallsensor des Typs TS 6 PB 4-20 P25 (Frequenz - 20 MHz, Wandler - 6 mm im Durchmesser, Fokus - 50 mm im Wasser) der Firma Karl-Deutsch, Wuppertal, Deutschland erfasst.

Zur Erfassung der Signale wurde ein Oszilloskop vom Typ Handyscope HS6 DIFF der Firma TiePie engineering, Sneek, Niederlande verwendet. Es wurde mit einer Abtastfrequenz von 500 MHz bei einer Aufnahmelänge von 50.000 Sample-Punkten gemessen, was den Vorteil der gleichzeitigen Aufzeichnung von Reflexions- und Transmissionssignalen bietet. Mit einem Trigger wurde die Aufnahme der Signaldaten zwischen dem Oszilloskop und dem Ritec RAM-5000-SNAP-System synchronisiert. Die Signale werden in der Software Multi Channel® von TiePie engineering dargestellt und im Anschluss in MATLAB, Mathworks, Natick, USA ausgewertet.

3. Ergebnisse

Für die Untersuchung der mikromechanischen Schädigungen ist besonders der Übergang von linearem zu nichtlinearem Werkstoffverhalten bzw. unterhalb der Streckgrenze relevant. In Abbildung 4 sind die Ergebnisse der Zugversuche bis zum Bruch dargestellt.

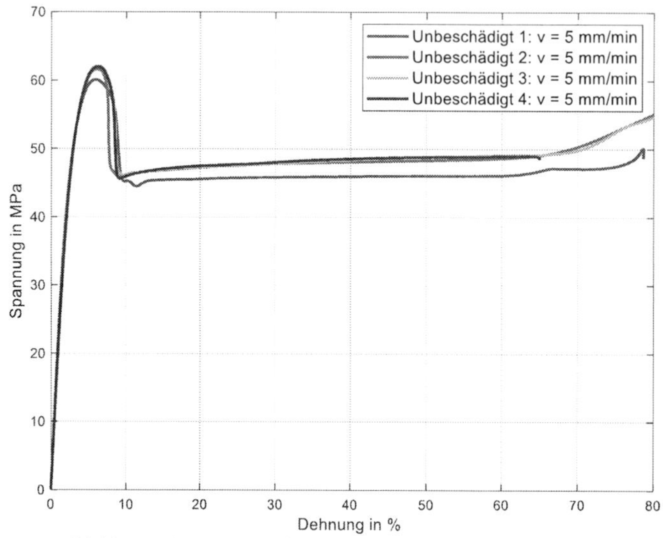

Abbildung 4: Spannungs- Dehnungs-Diagramm von PC bis zum Bruch

Die Ergebnisse zeigen, dass die Streckgrenze bei etwa 6 % Dehnung liegt. Oberhalb der Streckgrenze kommt es dann zu plastischen Verformungen, welche in Form von Einschnürungen sichtbar sind. Daraus ergibt sich die Betrachtung des Bereichs bis 6 % Dehnung. In Abbildung 5 sind Ergebnisse der Auswertungen der Nichtlinearitätsfaktoren für diesen Bereich, bei einer harmonischen Frequenz von 6 MHz, aufgezeigt.

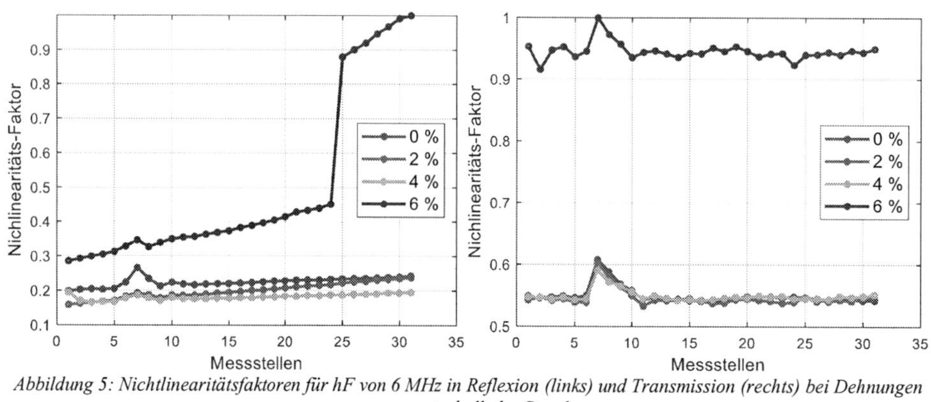

Abbildung 5: Nichtlinearitätsfaktoren für hF von 6 MHz in Reflexion (links) und Transmission (rechts) bei Dehnungen unterhalb der Streckgrenze

Sowohl bei der Auswertung von Transmission als auch Reflexion sind für den Dehnungsbereich von 0 % bis 4 % keine bzw. nur geringfügige Änderungen des Nichtlinearitätsfaktors festzustellen. Erst bei einer Erhöhung der Dehnung auf 6 % kommt es zu einem sprunghaften Anstieg des Nichtlinearitätsfaktors. Dieser Sprung konnte in weiteren Versuchen reproduziert werden. Dies lässt darauf schließen, dass es im Dehnungsbereich zwischen 4 % bis 6 % zum Auftreten von irreversiblen mikromechanischen Schädigungen kommt. Der Hochpunkt im Bereich der siebten Messstelle lässt sich auf einen kleinen Kratzer in der Oberfläche des Zugstabes zurückführen, was besonders in der Transmission erkennbar ist. Deshalb ist dieser Peak auch bei allen Dehnungsstufen erkennbar.

Für einen Vergleich der Ergebnisse bei verschiedenen harmonischen Frequenzen sind diese in Abbildung 6 mit gemittelten Nichtlinearitätsfaktoren in Reflexion (links) und Transmission (rechts) gegeneinander aufgetragen.

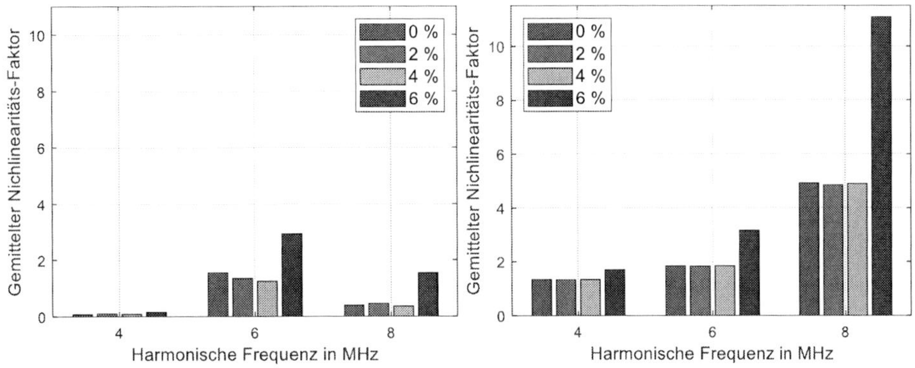

Abbildung 6: Gemittelte Nichtlinearitätsfaktoren für Reflexion (links) und Transmission (rechts)

Da der Sprung innerhalb der Kurve bei 6 % Dehnung aus Abbildung 5 links nicht bei allen Zugstäben auftritt, wird in diesem Fall der Mittelwert der Reflexion bis zur Messstelle 24, welche sich vor dem Sprung befindet, gebildet. Der zuvor beschriebene Verlauf mit ähnlichen Werten für den Nichtlinearitätsfaktor von 0 % bis 4 % und einer großen Erhöhung bei 6 % Dehnung spiegelt sich auch hier wieder. Mit dem Zusatz der dritten hF zeigt sich für 6 % Dehnung vor allem in der Transmission ein noch wesentlich größerer Unterschied von einer Verdopplung gegenüber den drei anderen Dehnungen.

Für eine bessere Interpretation der Ergebnisse wurden zusätzlich die Ergebnisse der Dehnungsstufen oberhalb der Streckgrenze betrachtet (Abbildung 7).

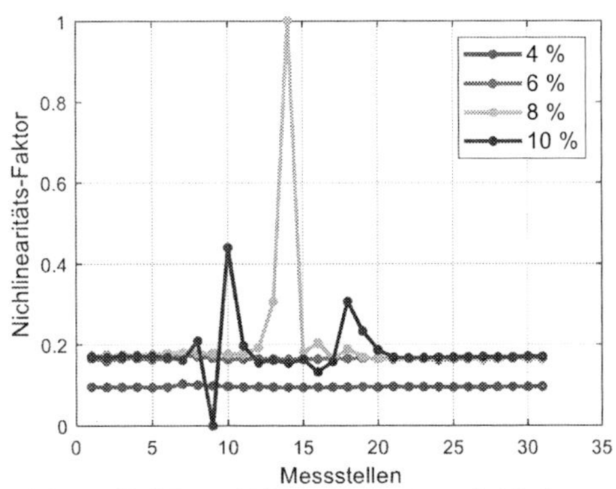

Abbildung 7: Nichtlinearitätsfaktoren für hF von 6 MHz in Transmission bei Dehnungen oberhalb der Streckgrenze

Hier werden nur die Ergebnisse in Transmission betrachtet, da sich durch das Einschnüren des Zugstabs die Oberfläche so verändert, dass die Auswertung des Reflexionssignals nicht mehr durchführbar ist, da sich das Fehler- und Eintrittsecho im Signal für diesen Bereich überlagern. Dadurch ist die Trennung des Fehlerechos für jeden Messpunkt nicht mehr eindeutig möglich. Es ist eindeutig zu erkennen, dass bei Dehnungen, die größer als Streckdehnung sind und in den Einschnürbereich oder darüber hinaus gehen, sehr markante Ausschläge des Nichtlinearitätsfaktors entstehen. Ein derartiges Ergebnis ist für eine Dehnung oberhalb der Streckdehnung zu erwarten, da es durch die Einschnürung zu einer sichtbaren Längenänderung des Zugstabs und auch zu Schäden auf molekularer Ebene kommt, die zu einer Veränderung des akustischen Verhaltens und damit zu einer Erhöhung des Nichtlinearitätsfaktors führen.

4. Fazit und Ausblick

Mithilfe der nichtlinearen Ultraschallmessungen und dem daraus bestimmten Nichtlinearitätsfaktor konnte ein deutlicher Sprung im Dehnungsbereich zwischen 4 % und 6 % festgestellt werden. Dieser ist unterhalb der Streckgrenze und bestätigt die Veränderung des akustischen Verhaltens innerhalb des Materials durch die mechanische Schädigung. Er ist außerdem ein deutlicher Hinweis auf auftretende mikromechanische Schädigungen in diesem Dehnungsbereich. Außerdem konnten plastische Verformungen mit der Methode eindeutig bestimmt werden.

In Zukunft soll der Bereich zwischen 4 % und 6 % Dehnung genauer betrachtet werden, um die auftretenden mikromechanischen Schädigungsmechanismen genauer zu untersuchen. Außerdem sollen weitere Werkstoffe untersucht werden, um durch den Vergleich des nichtlinearen akustischen Verhaltens verschiedener Werkstoffe weitere Erkenntnisse zu gewinnen.

5. Referenzen

[1] BONTEN, C. Kunststofftechnik. Einführung und Grundlagen. 3. Auflage. München: Hanser, 2020. ISBN 978-3-446-46471-1

[2] HUBER, D. Charakterisierung von Spröd/Zäh-Übergängen in PP-Compounds mittels instrumentierter Schlagbiege- und Durchstoßversuche. Masterarbeit, 2009.

[3] Polymer Service GmbH Merseburg. Scherbandbildung – Lexikon der Kunststoffprüfung, 2023 [Zugriff am: 30. Dezember 2023]. https://wiki.polymerservice-merseburg.de/index.php/Scherbandbildung

[4] Polymer Service GmbH Merseburg. Deformationsmechanismen – Lexikon der Kunststoffprüfung, 2023 [Zugriff am: 30. Dezember 2023]. https://wiki.polymerservice-merseburg.de/index.php/Deformationsmechanismen

[5] BEERBAUM, H. Ermittlung strukturbezogener bruchmechanischer Werkstoffkenngrößen an Polyethylen-Werkstoffen. Dissertation, 1999.

[6] FRIEDRICH, K. Scherbänder und Bruch in amorphen und teilkristallinen Polymeren. In: Colloid and Polymer Science, 1981, 259(2), 190–201. Doi:10.1007/BF01381760

[7] DAHLMANN, R., E. HABERSTROH und G. MENGES. Menges Werkstoffkunde Kunststoffe. 7. Auflage. München: Hanser, 2019. Hanser eLibrary. ISBN 9783446460867.

[8] EHRENSTEIN, G.W. Polymer-Werkstoffe. Struktur - Eigenschaften - Anwendung. 3. Auflage. München: Hanser, 2011. Hanser eLibrary. ISBN 9783446429673.

[9] GREIN, C. Relations entre la structure et les propriétés mécaniques de polypropylènes modifés choc: Lausanne, EPFL, 2005. https://www.researchgate.net/publication/37450338_Structure_et_mecanismes_de_microdeformation_de_polymethylmethacrylates_renforces_au_choc

[10] WANG, R., Q. Wu, K. Xiong, H. Zhang and Y. Okabe. Evaluation of the matrix crack number in carbon fiber reinforced plastics using linear and nonlinear acousto-ultrasonic detections. In: Composite Structures, 2021, 255, 112962. Doi:10.1016/j.compstruct.2020.112962

[11] GRELLMANN, W., S. SEIDLER und V. ALTSTÄDT, Hg. Kunststoffprüfung. 3. Auflage. München: Hanser, 2015. Hanser eLibrary. ISBN 9783446443907.

[12] POTTER, J.N., A.J. CROXFORD and P.D. WILCOX. Nonlinear ultrasonic phased array imaging. In: Physical review letters, 2014, 113(14), 144301. Doi:10.1103/PhysRevLett.113.144301

DGZfP-Jahrestagung 2025

Luftgekoppelte Schallgeschwindigkeitsmessung in Transmission

Yury GOLITSYN[1], Christoph PIENTSCHKE[1], Ralf STEINHAUSEN[1]
[1] Forschungszentrum Ultraschall, Halle (Saale), Deutschland, yury.golitsyn@fz-u.de

Kurzfassung. Die luftgekoppelte Ultraschallprüfung ist eine zerstörungsfreie und berührungslose Methode, um unter anderem Verbundmaterialien und Fügeverbindungen zu prüfen. Bisher wird meist in Transmissionsanordnung gearbeitet, wobei in der Regel die Amplitude des transmittierten Signals ausgewertet wird. Direkte Laufzeitmessungen sind hingegen nicht praktikabel, weil die Pulse zeitlich ausgedehnt sind und durch die Schallschwächung der Signalbeginn nicht verlässlich feststellbar ist.

Im Beitrag wird ein Verfahren vorgestellt, welches mit Hilfe eines heuristischen Ansatzes durch Referenzmessungen an Luft die Laufzeiten phasenrichtig bestimmen kann. Durch vergleichende angekoppelte Schallgeschwindigkeitsmessungen wird das Verfahren verifiziert. Prüfbeispiele an technischen Keramiken illustrieren die Besonderheiten des Verfahrens und zeigen Anwendungspotentiale. Die für jeden Scanpunkt durchgeführte Laufzeitmessung kann genutzt werden, um für die Darstellung ein Mapping der ermittelten Schallgeschwindigkeiten heranzuziehen. Im Vergleich mit den herkömmlichen C-Bildern zeigt sich, dass die Methodik in der Lage ist, Interferenz-Artefakte der Anzeige zu vermeiden. Da die Schallgeschwindigkeit von der Elastizität und der Dichte der Materialien abhängig ist, erschließt das Verfahren quantitative lokale Qualitätsaussagen zum Beispiel über die Porosität von Keramiken.

Keywords: Luftgekoppelter Ultraschall, Schallgeschwindigkeit, Transmissionsanordnung, technische Keramiken, Qualitätskontrolle

1 Einführung

Technische Keramiken finden breite Anwendung. Ein wichtiges Beispiel sind Hochtemperaturmaterialien, die u. a. zur Auskleidung von Öfen genutzt werden. Bei der Herstellung von technischen Keramiken nimmt die Bedeutung der Qualitätssicherung wie in den meisten Bereichen der Produktion kontinuierlich zu. Zerstörungsfreie Prüfverfahren werden gewünscht, die bestenfalls im Produktionstakt inline testen können. Die Luftultraschallprüfung bietet hierfür ein Verfahren, das diesen Ansprüchen gerecht wird und sich schon in verschiedenen Anwendungsbereichen bewährt hat.[1] Für die Beurteilung der Qualität sind quantitative und physikalisch basierte Aussagen besonders wertvoll. In üblicher Transmissionsmessung mit Ultraschall werden jedoch meist vergleichende Aussagen zur Amplitude getroffen. Hiermit können sehr gut Ungänzen wie Lufteinschlüsse detektiert werden. Für den Hersteller technischer Keramik sind darüber hinaus quantitative Aussagen zur Porosität bzw. Dichte eine wertvolle Prüfaussage. Die Schallgeschwindigkeit stellt einen physikalischen Zugang zur Dichte bereit, weil sie mit dieser und dem Elastizitätsmodul verknüpft ist.[2] Die angekoppelte Messung der Schallgeschwindigkeit ist ein Standardmessverfahren der Ultraschalltechnik. Jedoch ist ihre luftgekoppelte Bestimmung bisher durch die Besonderheiten der Luftschallprüfung nicht ohne weiteres möglich. Grund dafür sind die sehr langen Pulse, welche nach dem Durchgang durch die schallschwächende Probe keinen klaren Beginn haben. Wir

Medien- und Publikationspartner

 https://doi.org/10.58286/31244

haben einen heuristischen Ansatz entwickelt, um diese Schwierigkeiten für praktisch relevante Probentypen zu überwinden. Damit verfügen wir über ein Verfahren zur berührungslosen, ortsaufgelösten Erfassung der Schallgeschwindigkeiten.

Wir stellen das Verfahren in Abschnitt 2 vor, illustrieren das Vorgehen in Abschnitt 3 und zeigen erste Beispiel aus der Prüfpraxis in Abschnitt 4. In Abschnitt 5 fassen wir unsere Ergebnisse zusammen und geben einen Ausblick auf weitere Entwicklungsaufgaben.

2 Verfahren

2.1 Transmissionsprüfung

Die Luftultraschallprüfung findet üblicherweise in Transmissionsanordnung statt, wie sie schematisch in Abb. 1a dargestellt wird. Dabei befindet sich die Probe zwischen Sender und Empfänger. Die empfangene Zeitreihe der Ultraschallsignale – in der Luftultraschallprüfung als A-Bilder bezeichnet – werden ausgewertet, siehe Abb. 1b. Durch simultane Scanbewegung von Sender und Empfänger wird die Fläche der Probe lateral gescannt. Die entstehenden Ultraschallpulse werden hinsichtlich ihres Maximums ausgewertet. Diese werden ortsaufgelöst farbcodiert dargestellt, was auf die sogenannten C-Bilder führt. Es kann auch die zeitliche Lage der Maxima ermittelt werden und dann als D-Bilder dargestellt werden. Allerdings sind direkte Laufzeitmessungen in Transmissionsanordnung nicht möglich. Grund dafür ist, dass die genutzten Ultraschallpulse zeitlich sehr ausgedehnt sind. Zudem sorgt die stets deutliche Schallschwächung im Schallweg dafür, dass die phasenrichtige Auswertung der Ultraschallpulse nicht ohne weiteres möglich ist.

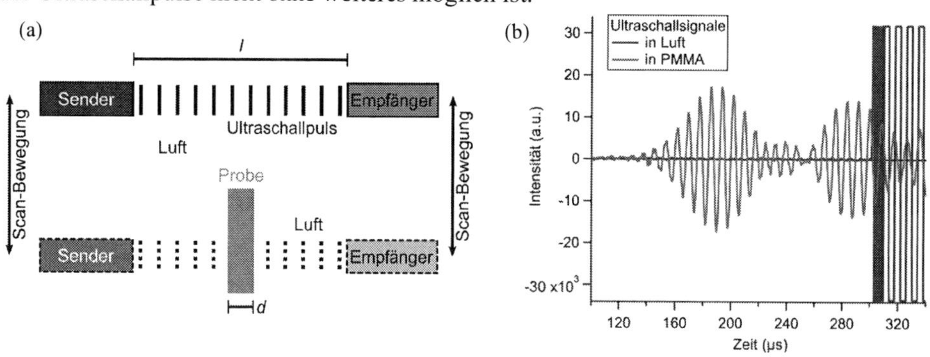

Abb. 1. (a) Anordnung zur Messung der Schallgeschwindigkeit. (b) Ultraschallsignale (A-Bild) bei 125 kHz in Luft und in PMMA.

2.2 Verfahren zur phasenrichtigen Bestimmung der Laufzeit

Die Schwierigkeiten der Laufzeitmessung werden durch das entwickelte Verfahren überwunden. Anhand von Abb. 1a wird das Vorgehen erläutert. Üblicherweise haben die Proben eine plattenförmige Geometrie mit parallelen Ein- und Austrittsflächen für den Schall, so dass die Dicke konstant ist. Der Schallpfad steht senkrecht auf den Flächen. In dieser Messanordnung wird eine Referenzmessung an Luft vorgenommen, indem für einen Punkt ohne Probe das A-Bild aufgenommen wird, siehe Abb. 1b. Die Probe wird gescannt und danach die für jeden Scanpunkt aufgenommenen A-Bilder analysiert.

Zwischen der Referenzmessung in Luft und der Messung durch die Probe ist wegen der unterschiedlichen Schallgeschwindigkeiten ein Laufzeitunterschied zu erwarten. Da Luft eine um etwa eine Größenordnung kleinere Schallgeschwindigkeit c_L im Vergleich zur

Schallgeschwindigkeit der Probe c_P aufweist, ist die Laufzeit in der Luft t_L meist deutlich länger als jene durch die Probe t_p. Analog zu Verfahren in Immersionstechnik [3] wird der Laufzeitunterschied $\Delta t = t_L - t_P$ ausgewertet. Für die Schallgeschwindigkeit in einer Probe mit der Dicke d gilt dann

$$c_p = \frac{d \cdot c_L}{d - \Delta t \cdot c_L}. \tag{1}$$

Bei der praktischen Anwendung ist zu berücksichtigen, dass der Laufzeitunterschied nicht direkt zu messen ist, weil die Pulsdauer groß und der Pulsbeginn wegen der Schallschwächung der Probe unklar ist. Aus diesem Grund wird für Δt ein Schätzverfahren genutzt. Hierbei werden die jeweils ansteigend durchlaufenen Nullstellen der A-Bilder ausgewertet. Ausgegangen wird von einer klaren Nullstelle des Referenzpulses in Luft. Ein Schätzwert für den Startpunkt des Schallpulses durch die Probe wird anhand des A-Bildes ermittelt. Hier wird eine ansteigend durchlaufene Nullstelle in der Nähe eines Zeitpunktes gesucht, bei dem die Amplitude deutlich höher als das Rauschlevel des Signals ist. Die Zeitdifferenz $\Delta t'$ zwischen den beiden Zeitpunkten ist um ein ganzzahliges Vielfaches der Periodendauer kleiner als die gesuchte Zeitdifferenz Δt. Deswegen führt der Ansatz

$$\Delta t = \Delta t' + \frac{m}{f} \text{ mit } m = 0,1,2,\ldots \tag{2}$$

auf die gesuchte Zeitdifferenz. Dabei wird m schrittweise geändert und die resultierende Schallgeschwindigkeit der Probe mit Hilfe der Mittenfrequenz f der Ultraschallwandler berechnet. In der Praxis zeigt sich, dass es nur *eine* plausible Lösung in einem relativ großen Toleranzbereich um die Schallgeschwindigkeit gibt. Die durch die Methode verursachte Mehrdeutigkeit in der Bestimmung der Schallgeschwindigkeit beeinträchtigt die praktischen Anwendungen mit technischer Keramik nicht. Es zeigt sich (vgl. Abschnitte 3 und 4), dass die Unterschiede zwischen den resultierenden Geschwindigkeiten deutlich größer sind als der für die Beurteilung der Qualität notwendige Messbereich.

Die Betrachtung der Messunsicherheiten zeigt [4], dass die relative Messunsicherheit der ermittelten Schallgeschwindigkeit größenordnungsmäßig um einen Faktor c_P/c_L höher als die Messunsicherheit von Probendicke bzw. Schallgeschwindigkeit der Luft ist. Aus diesem Grund muss die Probendicke genau und gegebenenfalls auch ortsaufgelöst bekannt sein. Die Schallgeschwindigkeit der Luft wird über eine Temperaturmessung mit aus der Literatur bekannten Formeln [5] verlässlich bestimmt. Insgesamt kann damit eine absolute systematische Messunsicherheit im Bereich von 5 Prozent erreicht werden. Abweichungen innerhalb der Probe sind in der Regel genauer bestimmbar.

3 Demonstration des Vorgehens an einem Prüfbeispiel

3.1 Messplatz und Messbedingungen

Die Messungen wurden mit dem Luftultraschallprüfsystem SONOAIR (SONOTEC GmbH Halle, Deutschland) vorgenommen. Die Messungen erfolgten bei 125 kHz bzw. 75 kHz (Prüfköpfe CF075 bzw. CF125, SONOTEC). Der Sender-Empfänger-Abstand wurde jeweils entsprechend der Nahfeldlänge der Wandler gewählt. Zur Schallerzeugung wurden die Wandler mit fünf aufeinanderfolgenden Rechteckpulsen mit der Amplitude von 400 V angeregt. Das empfangene Signal wurde rauscharm verstärkt. Aus den transmittierten Ultraschallsignalen wurde mit einer Zeitblende ein früher Ausschnitt des Ultraschallsignals gewählt. Dieser wurde so gewählt, dass nur ein direkt vom Sender zum Empfänger laufender Schallpuls ausgewertet wird. Reflektierte Signale, die im A-Bild bei späteren Zeiten auftreten, werden somit ausgeblendet. Die Scanauflösung wurde je nach Probe gewählt. Die nachfolgend präsentierten Prüfbilder (C-Bild, D-Bild, Geschwindigkeits-Mapping) sind jeweils in mm skaliert.

3.2 Herkömmliche Prüfbilder

Das Verfahren zur Bestimmung der Schallgeschwindigkeit wird anhand einer quaderförmigen PMMA-Probe mit den Abmessungen 100×100×72 mm^3 demonstriert. Die Messung erfolgte bei einer Mittenfrequenz von 125 kHz. Der Abstand zwischen Wandler und Probe betrug 15 mm. Die Scanauflösung betrug 2×2 mm^2. Das PMMA (Polymethylmethacrylat, Plexiglas) weist eine geringe Schallschwächung auf, sodass trotz einer Probendicke von 72 mm Ultraschallsignale mit einem guten Signal-zu-Rausch-Verhältnis bei einer Signalverstärkung von 70 dB gemessen werden konnten.

Beispielhaft sind detektierten Ultraschallsignale in Luft und in Probe in der Abb. 1b dargestellt. Aus den Signalen wird das Maximum bestimmt und für jeden Scanpunkt farbcodiert als sogenanntes C-Bild dargestellt (Abb. 2a). Trotz der gewählten frühen Zeitblenden weisen die resultierenden C-Bilder starke Intensitätsschwankungen auf. Diese deuten wir als Interferenzmuster der vom durchgehenden Schall angeregten stehenden Wellen in der Probe. Die Interferenzmuster im C-Bild hängen von der Wellenlänge des Schalls im Material und der Geometrie des Prüfkörpers ab und erschweren die Auswertung des Prüfbildes bezüglich der möglichen Defekte und Ungänzen im Material.

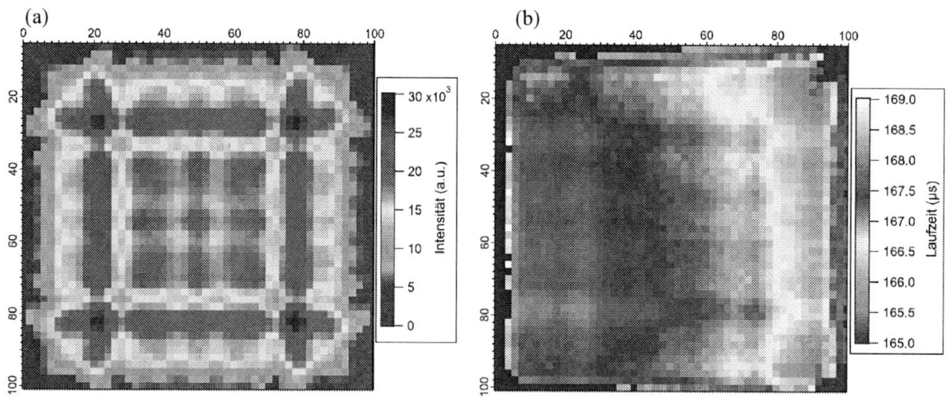

Abb. 2. Messung der Schallgeschwindigkeit in PMMA. (a) Farbcodierte Signalintensitäten (C-Bild), (b) Signallaufzeiten (D-Bild).

Eine weitere Möglichkeit, die transmittierten Signale graphisch darzustellen, bietet das sogenannte D-Bild an, bei dem die Phase des Ultraschallsignals ausgewertet wird. In der Abb. 2b ist die zeitliche Position einer ansteigend durchlaufenen Nullstelle im Bereich zwischen 165 µs und 169 µs abgebildet. Die phasenrichtige Auswertung der Nullstellen ist im Allgemeinen genauer als die Auswertung der zeitlichen Lage eines Signalmaximums. Im Vergleich zum C-Bild weist diese Darstellung geringere Interferenzen, dafür aber einen kontinuierlichen Gradienten auf. Ein solcher Gradient kann sowohl durch eine Dickenänderung der Probe als auch durch eine Änderung der Schallgeschwindigkeit im Material zustande kommen. Da die Schallgeschwindigkeit in der Luft etwa eine Größenordnung kleiner als im PMMA ist, werden Laufzeitunterschiede im Wesentlichen von der geänderten Luftstrecke bestimmt.[4] Aus diesem Grund werden Änderungen der Materialdicke mit dem luftgekoppelten Verfahren mittels Luftultraschall in der Regel sensitiver erfasst als bei einer Laufzeitmessung in Ankopplung.

3.3 Ergebnisse der luftgekoppelten Schallgeschwindigkeitsmessung

Die mittels des entwickelten Algorithmus ermittelten Schallgeschwindigkeiten sind in Abb. 3a farblich codiert dargestellt. Dabei wurden die Randbereiche des C-Bildes nicht ausgewertet, was anhand der Achsenskalierung zu erkennen ist. Das Schallgeschwindigkeits-Mapping weist keine Interferenzmuster, aber wie das D-Bild einen Gradienten auf. Dieser ist durch die mit einem Messschieber bestimmte Dickenänderung der Probe im Bereich von 0,5 mm zu erklären. Die Verteilung von Schallgeschwindigkeiten im ausgewerteten Bereich der Probe ist als Histogramm in der Abb. 3b abgebildet. Der Mittelwert und die Standardabweichung betragen (2799 ± 65) m/s. Eine gleichartige Messung in Querrichtung zur ersten Messung (Dicke 100 mm) mit höherer Verstärkung führte auf eine Schallgeschwindigkeit von (2795 ± 152) m/s. Erwartungsgemäß stimmen diese Geschwindigkeiten überein. Um diese Ergebnisse mit einer unabhängigen Messung zu verifizieren, wurde die Schallgeschwindigkeit in der Mitte der Probe mit angekoppeltem Ultraschall im Impuls-Echo-Verfahren bei 1 MHz bestimmt. Die Messung ergab einen Wert von 2750 m/s, was innerhalb der Standardabweichungen der beiden anderen Messungen liegt und mit Tabellenwerten üblichen PMMAs (je nach Sorte 2,61 km/s bzw. 2,75 m/s) verträglich ist.[2] Das Messverfahren mit luftgekoppeltem Ultraschall liefert somit verlässliche absolute Werte der Schallgeschwindigkeit.

Als weitere Testprobe für den entwickelten Schallgeschwindigkeits-Algorithmus

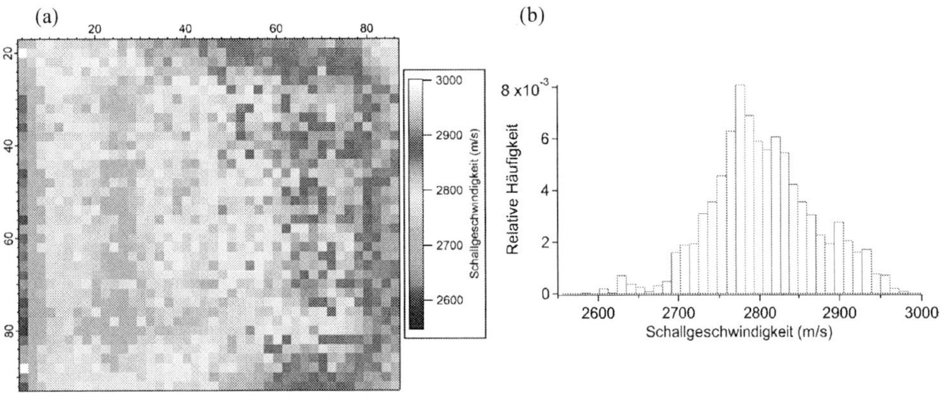

Abb. 3. Analyse der PMMA-Probe aus Abb. 2 mit der neu entwickelten Methode zur Messung der Schallgeschwindigkeit. Schallgeschwindigkeits-Mapping (a) und (b) Verteilung von Schallgeschwindigkeiten in der Probe.

wurde eine Aluminium-Platte der Dicke 20,1 mm verwendet. Die Messung der Schallgeschwindigkeit mittels Impuls-Echo-Verfahren bei 2 MHz Ultraschall in Ankopplung ergab einen Wert von 6345 m/s in der Mitte der Platte. Die Luftultraschallmessung erfolgte wieder bei 125 kHz und lieferte einen über die gesamte Platte gemittelten Wert von (6422 ± 191) m/s, welcher in einer guten Übereinstimmung mit der angekoppelten Messung liegt und kompatibel mit einem Literaturwert von 6,42 km/s.[2]

Anhand der beiden Testproben wird die Funktionsweise des in Abschnitt 2.2 skizzierten Algorithmus verdeutlicht. Entsprechend Formel (2) gibt es für einen zeitlich ausgedehnten Luftschallpuls mehrere Laufzeiten und somit verschiedene Schallgeschwindigkeiten in der Probe. Wenn diese scannend erfasst werden, ergeben sich daraus Geschwindigkeitsverteilungen. Diese sind für die untersuchte PMMA- und Aluminium-Probe in Abb. 4 zusammengestellt.

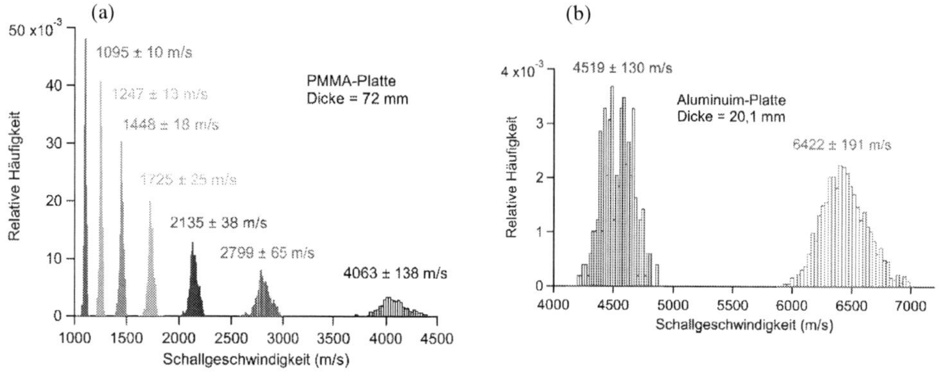

Abb. 4. Histogramme für die Geschwindigkeitsbestimmung in der (a) PMMA-Platte und (b) Aluminium-Platte bei verschiedenen Laufindizes *m* in Formel (2). Die Histogramme umfassen deutlich voneinander getrennte Geschwindigkeitsbereiche.

Gemäß Formel (1) hängen Laufzeit Δt und die Schallgeschwindigkeit der Probe c_P nichtlinear zusammen. Je höher die aus dem Algorithmus ermittelte Schallgeschwindigkeit ist, desto höher ist der Einfluss von kleinen Laufzeitänderungen. Die Streuung der gemessenen Laufzeiten wirkt sich damit stärker aus. Die Peaks bei niedrigen Schallgeschwindigkeiten weisen entsprechend eine deutlich schmalere Verteilungsbreite als die höheren auf. Außerdem nehmen aufgrund der Nichtlinearität die Abstände zwischen den Histogrammen zu. Deutlich ist zu erkennen, dass bei grober Kenntnis der korrekten Schallgeschwindigkeit eine eindeutige Identifizierung der richtigen Verteilung gelingt. Für praktisch relevante Anwendungsfälle führt der Algorithmus somit zu einer eindeutigen Lösung.

4 Praktische Prüfbeispiele

Für technische Keramiken verspricht das vorgestellte Verfahren besonders aussagekräftige Ergebnisse, weil hier Aussagen zur Porosität besonders wichtig für die Qualitätsaussage sind. Aus der Signalamplitude (C-Bildern) ist diese – auch aufgrund der diskutierten Interferenzeffekte – nicht sehr gut für eine weitgehend automatisierte Bewertung geeignet. Die ortsaufgelöste Geschwindigkeitsmessung verspricht hier einen deutlichen Mehrwert gegenüber den bisherigen luftgekoppelten Prüfverfahren.

4.1 Keramische Probe mit keilförmiger Geometrie

Zunächst wird eine keilförmige keramische Probe mit einem bekannten linearen Dickengradienten betrachtet. Die Abmessungen betrugen $400\times225\times(87\text{–}101)$ mm^3, wo der letztere Wert die Änderung der Dicke entlang der langen Probenseite darstellt. Das Dickenprofil wurde in den Algorithmus implementiert, indem die Probendicke in Abhängigkeit von der Scanposition zwischen der minimalen und maximalen Dicke linear interpoliert wurde. Die Messung erfolgte mit einem Scanraster von 5×5 mm^2 bei einer Ultraschallfrequenz von 75 kHz. Der Abstand zwischen Sender und Empfänger betrug 154 mm.

Die Abb. 5a zeigt das C-Bild der keilförmigen Probe. Die Probendicke nimmt von links nach rechts zu und führt zur kontinuierlichen Abnahme der detektierten Schallintensität. Das zugehörige D-Bild (Abb. 5b) weist anstelle eines einzigen abnehmenden Gradienten einen komplexen Verlauf auf, bedingt durch den Sprung von einer Laufzeitperiode zur nächsten. Somit sind die Intensitäts- und Laufzeitbilder bei solchen Probengeometrien schwer zu interpretieren in Bezug auf die Homogenität und mögliche Ungänzen.

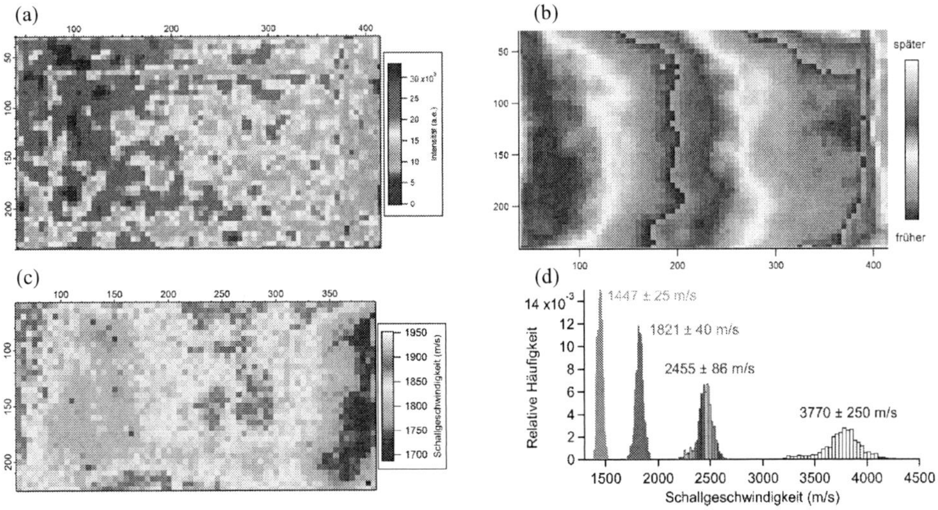

Abb. 5. Messung der Schallgeschwindigkeit in einer keilförmigen keramischen Probe (a) C-Bild, (b) D-Bild, (c) Schallgeschwindigkeits-Mapping und (d) Histogramme der ermittelten Schallgeschwindigkeit.

Dieser Nachteil ist bei dem Schallgeschwindigkeits-Mapping (Abb. 5c) nicht vorhanden. Das Diagramm zeigt eine relativ homogene Verteilung von Schallgeschwindigkeiten in der Probe, frei von Gradienten. Die mittels Algorithmus berechnete mittlere Schallgeschwindigkeit im Material beträgt (1821 ± 40) m/s. Dieser Wert kommt dem vom Keramikhersteller mit angekoppeltem Ultraschall bei ähnlicher Frequenz gemessenen Bereich zwischen 1600 m/s und 1700 m/s am nächsten. Die Histogramme zu den verschiedenen Schallgeschwindigkeiten in Abb. 5d zeigen, dass in diesem Bereich nur ein einziges Histogramm zu finden ist, welches einen großen Abstand zu den benachbarten Lösungsmöglichkeiten hat. Für praktische Belange ist die Messung somit eindeutig.

4.2 Keramische Probe mit planarer Geometrie

Im ersten Beispiel für eine technische Keramik betrachten wir eine Platte konstanter Dicke mit den Abmessungen 410×380×45 mm^3. Die 75 kHz-Wandler wurden mit einem Abstand von jeweils 30 mm zur Plattenoberfläche eingesetzt. Die Scanauflösung beträgt 5×5mm^2 und die Signalverstärkung 72 dB. In die Platte sind oberflächlich ein Pfeil und eine Probennummer eingeprägt.

Das C-Bild der keramischen Platte ist in der Abb. 6a dargestellt. Im Vergleich zum PMMA weist die keramische Platte nur einen schmalen Interferenzstreifen am Rand auf. Das liegt daran, dass die porösen Keramiken eine deutlich höhere Schallschwächung aufweisen. Die Interferenzmuster entstehen nur entlang von Materialkanten. Eine andere Ausprägung deutet erfahrungsgemäß auf Risse in axialer Richtung (parallel zur Schallausbreitungsrichtung) hin (keine Abbildung). Weiterhin kann man anhand des C-Bildes schließen, dass die Platte homogen und frei von Defekten oder Ungänzen ist. Der Bereich mit einer niedrigeren Intensität mittig an der oberen Kante ist auf den eingeprägten Pfeil zurückzuführen. Eine weitere Signatur der Oberflächentopologie stellt die eingedrückte Probennummer mittig an der linken Kante dar, welche aber aufgrund der gewählten Scanauflösung im C-Bild kaum

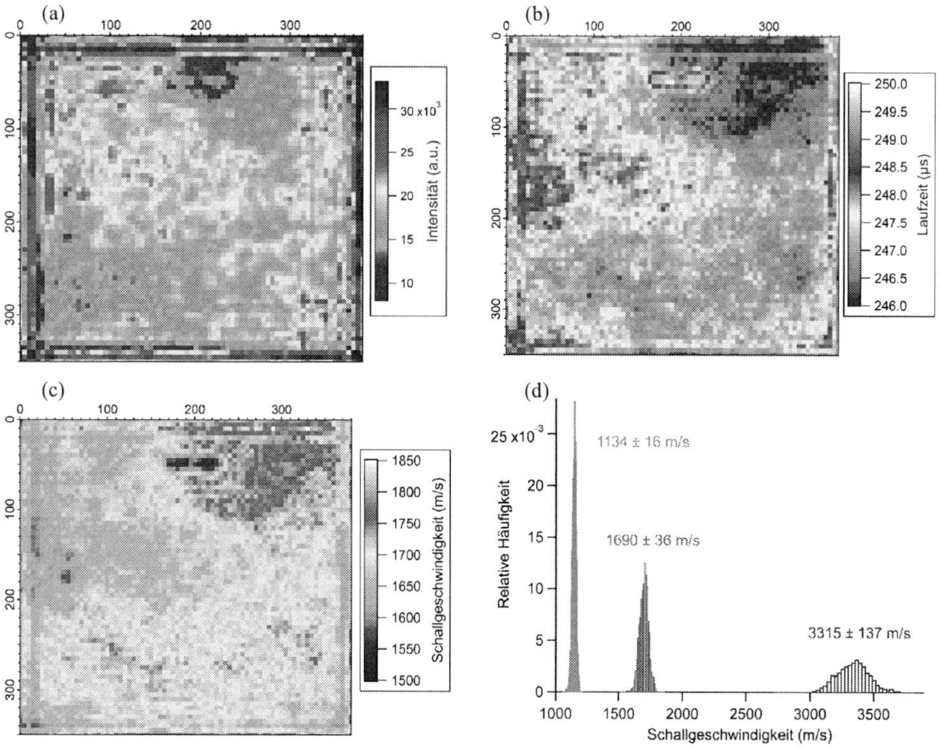

Abb. 6. Messung der Schallgeschwindigkeit in einer keramischen Platte planarer Geometrie. (a) C-Bild, (b) D-Bild, (c) Schallgeschwindigkeits-Mapping und (d) Histogramme der Schallgeschwindigkeit.

zu erkennen ist. Der von oben nach unten kontinuierlich verlaufende Farbgradient ist auf das leicht ungleichmäßige Pressen beim Herstellungsprozess der Platte zurückzuführen.

Das zugehörige D-Bild, welches die zeitliche Position einer ansteigend durchlaufenen Nullstelle abbildet, ist in der Abb. 6b dargestellt. Die Oberflächentopologie (Plattennummer links und Pfeil oben) ist in dieser Darstellungsart deutlich besser zu erkennen, weil es sehr sensitiv auf Dickenänderungen der Probe reagiert. Weiterhin ist zu erkennen, dass die Bereiche höherer Signalamplitude im C-Bild (links mittig) eine längere Laufzeit im D-Bild aufweisen und somit den Stellen mit einer geringeren Materialdicke entsprechen.

Die Abb. 6c stellt die mittels Algorithmus ermittelten Schallgeschwindigkeiten in der gesamten keramischen Platte dar. Erneut können die topologischen Effekte eindeutig erkannt werden. Das Geschwindigkeits-Mapping entspricht in guter Näherung dem Laufzeitbild, wobei die Bereiche mit der hohen Schallgeschwindigkeit den mit einer geringen Laufzeit zugeordnet werden. Insgesamt konnten für diese Probe drei Nullstellen im A-Bild ausgewertet werden. Die zugehörigen Histogramme (Abb. 6d) umfassen deutlich voneinander getrennte Geschwindigkeitsbereiche.

Da sich die Schallgeschwindigkeiten der dicken Proben in Ankopplung nicht messen ließen, kann hier keine eindeutige Zuordnung getroffen werden. Die dadurch entstehende Mehrdeutigkeit bei der Bestimmung von Schallgeschwindigkeiten ist für praktische Anwendungen der Qualitätskontrolle in der Regel jedoch nicht relevant, solange die Abweichungen innerhalb eines Histogramm-Peaks liegen, was im Beispiel einfach zu gewährleisten ist.

4.3 Keramische Probe mit Ungänzen

Anhand einer keramischen Probe der Abmessungen 410×380×45 mm^3 wurde die Bildgebung von Ungänzen näher untersucht. Die Messungen fanden bei 75 kHz statt. Das Scanraster wurde zu 5×5 mm^2 gewählt.

Die Abb. 7 zeigt die verschiedenen Prüfbilder. C-Bild (Abb. 7a), D-Bild (Abb. 7b) und Geschwindgkeitsmapping (Abb. 7c) zeigen jeweils größere Inhomogenitäten, wobei diese im D-Bild und Geschwindigkeitsmapping am klarsten heraustreten. Auch der topologische Kontrast (Prägefeil) kommt in diesen Abbildungen am stärksten zum Ausdruck. Das Histogramm der ermittelten Schallgeschwindigkeiten (Abb. 7d) wird wegen der großen Inhomogenitäten unsymmetrisch.

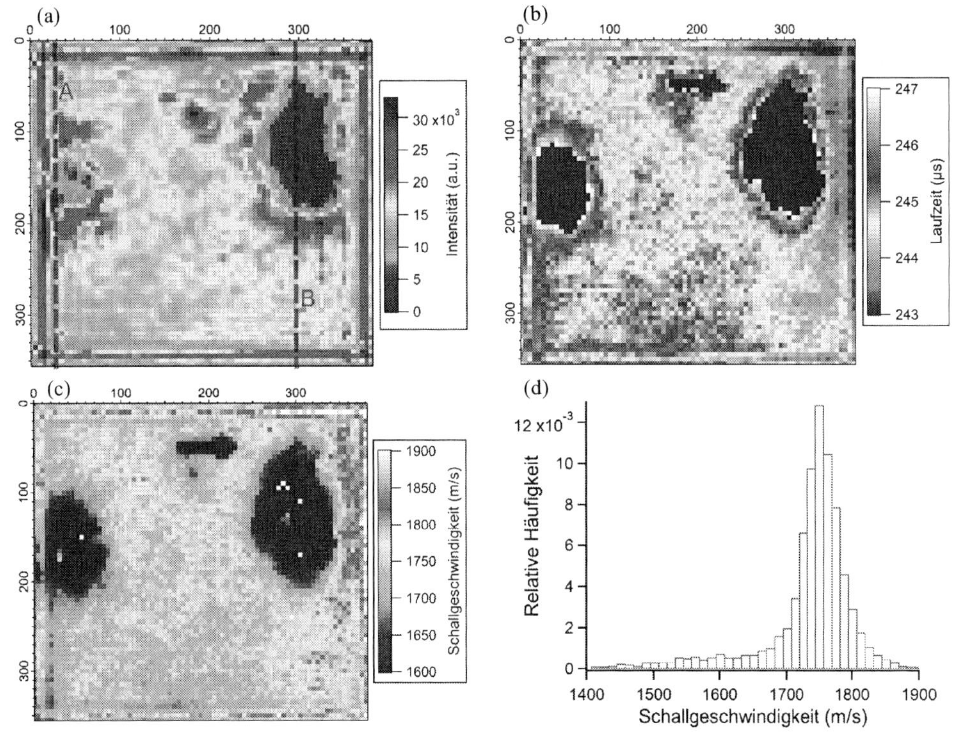

Abb. 7. Keramische Probe mit Ungänzen. (a) C-Bild mit eingezeichneten Schnitten A und B, (b) D-Bild, (c) Schallgeschwindigkeits-Mapping und (d) Histogramm für die Bestimmung der Schallgeschwindigkeit.

Die Probe wurde nach der zerstörungsfreien Prüfung entlang zweier Schnitte aufgesägt (A und B in Abb. 7a). Dabei zeigten sich Lagen mit unverdichteter Keramik, sogenannte Presslagenfehler (Abb. 8). Die in der Prüfung aufgefundenen Inhomogenitäten lassen sich klar diesen Ungänzen zuordnen.

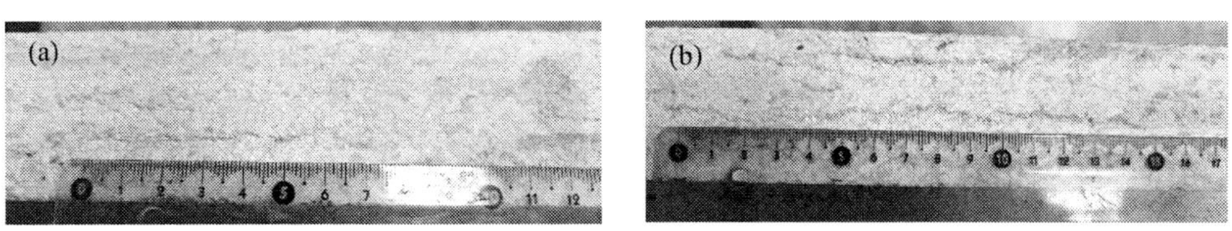

Abb. 8. Aufgeschnittene Probe. Die Durchschallungsrichtung ist die Dickenrichtung senkrecht zum Lineal. Die in den Prüfbildern (Abb. 7) auffindbaren Auffälligkeiten entlang der Schnitte (A) und (B) korrespondieren mit auffindbaren Presslagenfehlern in den Teilabbildungen (a) bzw. (b).

5 Zusammenfassung und Ausblick

Ein Verfahren zur phasenrichtigen Laufzeitbestimmung mit luftgekoppeltem Ultraschall wurde entwickelt. Dabei werden Nulldurchgänge eines Signals durch die Probe mit einem Referenzsignal in Luft verglichen. Weil die Luftultraschallpulse zeitlich ausgedehnt sind und die Startzeiten der Probenpulsen wegen der Schallschwächung nicht eindeutig bestimmbar sind, wird mittels eines heuristischen Ansatzes die korrekte Laufzeit ermittelt. Die dadurch entstehende Mehrdeutigkeit ist für praktische Anwendungen der Qualitätskontrolle in der Regel nicht relevant. Anhand von Verifikationsmessungen wurde die absolute Genauigkeit der Methode nachgewiesen. Die Schallgeschwindigkeit der Luft (vorwiegend beeinflusst durch die Lufttemperatur) und die Kenntnis der Probendicke sind relevante Einflussfaktoren bei der Bestimmung der Schallgeschwindigkeit der Proben. An Prüfbeispielen technischer Keramik wurde die Eignung der Methode zum Auffinden von Ungänzen und zur Materialcharakterisierung gezeigt. Das Verfahren ermöglicht einerseits Aussagen zur Signallaufzeit und daraus resultierend der Schallgeschwindigkeit bei inhomogenen Proben. Andererseits sind sehr genaue Angaben zur Probendicke möglich, wenn die Schallgeschwindigkeit im Material konstant ist. Ein weiterer Vorteil der Methode gegenüber der Kontakttechniken besteht darin, dass es sich um ein scannendes Verfahren handelt. Störende Interferenzartefakte der C-Bilder treten in den ermittelten Geschwindigkeits-Mappings nicht auf.

Zukünftige Arbeiten zielen darauf, aus den Unterschieden der gemessenen Schallgeschwindigkeiten auf die Porosität der Keramik zurückzuschließen, was spezielle Probenkörper erfordert. Weitere Fragen betreffen die Verbesserung des Algorithmus, welche durch ein vertieftes Verständnis des Frequenzgehaltes der ausgewerteten Ultraschallpulse möglich erscheinen. Die gezeigten Ergebnisse demonstrieren die Brauchbarkeit des Ansatzes und belegen die Anwendbarkeit für praktisch relevante Prüfaufgaben. Somit verspricht das inlinefähige und berührungslose Verfahren valide quantitative Aussagen zur Materialqualität von technischen Keramiken.

6 Danksagung

Die vorgestellten Ergebnisse entstanden im Rahmen des Forschungsprojektes „Ertüchtigung der Luftultraschallprüfung zur Material-Charakterisierung“ (ELMaC). Dieses wurde aufgrund eines Beschlusses des Deutschen Bundestages vom Bundesministerium für Wirtschaft und Klimaschutz im Förderprogramm INNO-KOM unter dem Förderkennzeichen 49MF220202 gefördert. Wir danken herzlich für die Unterstützung.

Unserem ehemaligen Kollegen Dr. Mario Kiel verdanken wir die Anregung und erste Ansätze zur vorgestellten Charakterisierungsmethode.

7 Referenzen

[1] W. Essig et al., “Air-coupled Ultrasound – Emerging NDT Method,” ZfP-Zeitung, vol. 173, pp. 32-43, no. Feb, 2021 Verfügbar unter http:/www.ndt.net/?id=25761, Letzter Zugriff 28. März 2025.

[2] J. D. N. Cheeke, Fundamentals and Applications of Ultrasonic Waves, 2nd ed., Boca Raton, FL, USA: CRC Press, 2021.

[3] K. A. Wear et al., “Interlaboratory Comparison of Ultrasonic Backscatter Coefficient Measurements From 2 to 9 MHz”, J. Ultrasound Med., vol. 24, no. 9, pp. 1235-1250, 2005. DOI. 10.7863/jum.2005.24.9.1235

[4] R. Steinhausen et al., “Scanning the Speed of Sound with Air-Coupled Ultrasound in Through-Transmission Technique”, submitted to IEEE Open Journal of Ultrasonics, Ferroelectrics, and Frequency Control, in Revision.

[5] Calculator for the speed of sound in humid air (2018) [Online]. Available: http://resource.npl.co.uk/acoustics/techguides/speedair/. Letzter Zugriff 28. März 2025.

Prüfung der Verklebung von Stegen mit Gurten in Rotorblättern, Modelluntersuchungen

Johann HINKEN[1]
[1] fitm Hinken Consult, Magdeburg, Deutschland, johann.hinken@fitm.de

Kurzfassung

Bei der Herstellung von Rotorblättern werden in der Regel zwei separat gefertigte Halbschalen zusammengeklebt. Die Klebestellen zwischen den Gurten und Stegen sind nach dem Zusammenfügen nicht mehr einsehbar. Die zerstörungsfreie Überprüfung der Klebstellen auf Lufteinschlüsse von außen ist mit herkömmlichen Verfahren der ZfP, wie z.B. Ultraschallprüfung, schwierig bzw. nicht möglich. Hier wird an einem Modell gezeigt, dass diese Prüfung mit Mikrowellen gut möglich ist. - Der Gurt wird durch ein Sandwich von 39 mm dickem Balsaholz mit zwei 2 mm dicken Schichten aus glasfaserverstärkten Kunststoff (GFK) nachgebildet sowie der Fuß des zu verklebenden Steges durch eine 10 mm dicke GFK-Platte und der dazwischenliegende Kleber durch austauschbare 3D-gedruckte 10 mm dicke PLA-Platten mit verschiedenen eingebrachten Flachbodenvertiefungen. Die handelsübliche Balsa-Schicht besteht dabei wie üblich aus einzelnen miteinander verklebten Blöcken, die holztypisch auch Jahresringe aufweisen. - Es wird im scannenden Reflexionsverfahren von oben durch das GFK/Balsa/GFK-Sandwich hindurch getestet, um Lufteinschlüsse im darunter liegenden PLA-Layer zu erkennen. Die Herausforderung ist dabei, die Anzeigen zu unterdrücken, die aus den natürlichen Unregelmäßigkeiten im Balsa-Bereich stammen. Dafür wurde eine linsenähnliche Mikrowellenantenne verwendet, die auf die Tiefenlage der PLA-Schicht fokussiert. Die Antenne ist an ein Reflektometer angeschlossen, das bei der ISM-Frequenz 24,125 GHz arbeitet. Die Signalauswertung erfolgt auf einem PC. – Die Größen der eingebrachten Flachbodenvertiefungen reichten von 30 mm x 30 mm x 3 mm bis 10mm$^{\Phi}$/1,0 mm. Im vorliegenden Fall konnten simulierte Lufteinschlüsse von 20 mm x 20 mm x 2 mm und 10 mm$^{\Phi}$/2,5 mm sicher erkannt werden. Die Gleichmäßigkeit des Balsaholzes bestimmt die Erkennbarkeit..

Keywords: ZfP, Mikrowellenprüfung, Rotorblätter, Balsaholz, Sandwich

DGZfP Jahrestagung Berlin 2025

Prüfung der Verklebung von Stegen mit Gurten in Rotorblättern, Modelluntersuchungen

Johann H. Hinken [1)]

1) fitm Hinken Consult, Breitscheidstrasse 17, D-39114 Magdeburg

@microwavetesting

fitm Hinken Consult

Rotorblätter von Windkraftanlagen

- ...werden in zwei Halbschalen hergestellt, die dann miteinander verklebt werden.

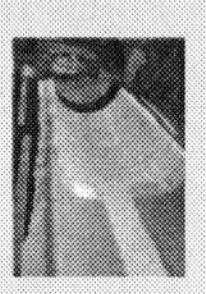

- Kleine Modellausführung:

Prüfaufgabe: Verklebung

- Die Prüfung hat von außen zu erfolgen

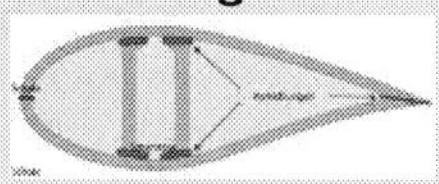

Herausforderung

- Wanddicke bis zu 50 mm
  Wand besteht häufig aus Sandwich GFK/Balsaholz/GFK,
  GFK: glasfaserverstärkter Kunststoff
- Ultraschallprüfung entfällt wegen zu hoher Dämpfung im Balsaholz

- Mikrowellenprüfung durch Balsaholz hindurch ist möglich, jedoch entstehen parasitäre Anzeigen durch Jahresringe und weitere Inhomogenitäten im Balsaholz. Diese sind zu unterdrücken

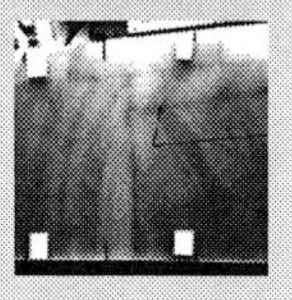

Lösungsansatz

- Scannende Mikrowellen-Reflexionsprüfung mit einer Antenne, die auf die zu untersuchende Kleberschicht in der Tiefe fokussiert >>>

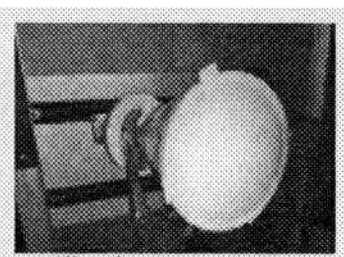

Rillenhorn mit Linse

Modell des Prüfobjektes

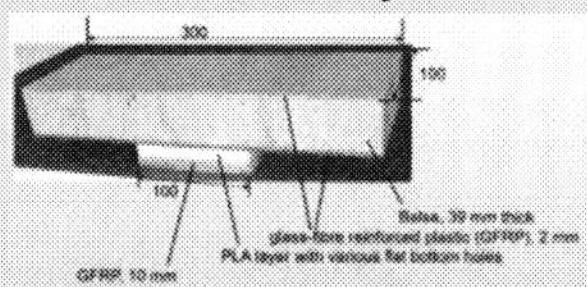

- Austauschbare PLA-Platten mit unterschiedlichen Flachbodenvertiefungen als künstlichen Fehlern:

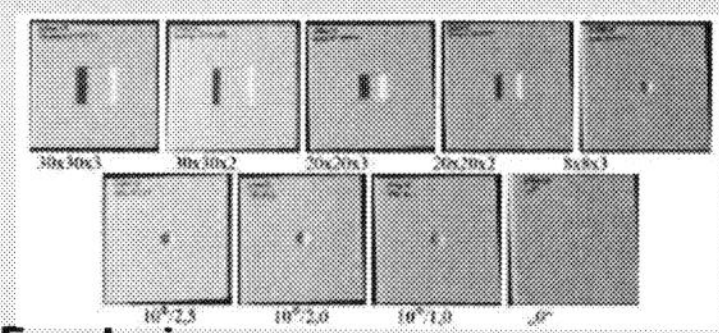

Ergebnisse

- ...als Roh- und als Schwellenwertanzeigen:

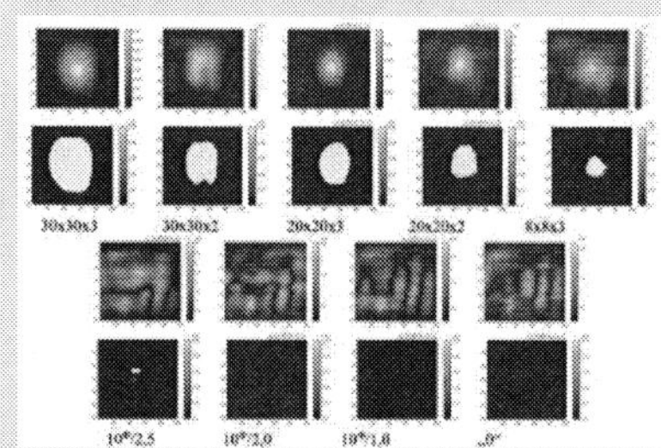

- Lufteinschlüsse bis herunter zu 10 mmΦ/2,5 mm können auf diese Weise erkannt werden. Kleinere Einschlüsse werden durch die parasitären Anzeigen aus dem Balsaholz verdeckt.

Kontakt:
Johann H. Hinken
fitm Hinken Consult
Breitscheidstr. 17
D-39114 Magdeburg

Tel.: +49 (0) 391-8868 129
Mobil: +49 (0) 171-2053208
E-Mail: johann.hinken@fitm.de
www.fitm.de

Mikrowellen-Handprüfung an Komponenten von Rohrleitungen

Johann HINKEN[1]
[1]fitm Hinken Consult, Magdeburg, Deutschland, johann.hinken@fitm.de

Kurzfassung

Rohre für den Transport von Gasen und Flüssigkeiten bestehen mitunter aus elektrisch-nichtleitendem Material wie Polyethylen (PE) oder glasfaserverstärktem Kunststoff (GFK). Diese sind mit Mikrowellen prüfbar. Prüfbedarf besteht vor allem beim Herstellen von Rohrverbindungen und im Betrieb, wobei Schäden durch unachtsamen Umgang sowie durch chemischen und abrasiven Abtrag entstehen können. Der Prüfbedarf besteht zumeist nicht an den geraden, langgestreckten Abschnitten der Rohre sondern an geometrischen Besonderheiten wie z.B. Krümmern oder Verbindungsstellen. Dabei können Rohrscanner zumeist nicht eingesetzt werden. Eine Handprüfung ist nötig. – Hier wird ein Mikrowellenhandprüfgerät vorgestellt und insbesondere sein Einsatz an Krümmern und Verbindungsstellen. Es wird bei einer Frequenz von 24 GHz im Reflexionsverfahren geprüft. Die Signaldarstellung erfolgt in der komplexen Ebene des Reflexionsfaktors, ganz ähnlich wie in der Wirbelstromprüfung. Die primäre Mikrowellenantenne ist ein handelsüblicher offener Hohlleiter mit WR42-Querschnitt (10,67 mm x 4,32 mm) und ebenem Flansch. Um den Einfluss von Liftoff-Änderungen und Verwackeln zu vermeiden, werden diesem Flansch Adapter aufgeschoben, die der jeweils vorliegenden Geometrie angepasst sind. Sie sind im 3D-Druck in PLA mit anschließender Leitlack-Metallisierung hergestellt. Sie können damit für ein auftretendes Prüfproblem schnell verfügbar sein. – Einzelheiten zur Handhabung und zu den Möglichkeiten dieser Handprüfung werden an Bespielen erläutert. Dazu gehören die Prüfung von geraden Rohren und Krümmern, Berücksichtigung der Polarisation sowie die Darstellungsunterschiede von oberflächennahen und tiefliegenden Fehlern.

Keywords: ZfP, Mikrowellenprüfung, Rohrkomponenten, Handprüfung

DGZfP Jahrestagung Berlin 2025

Mikrowellen-Handprüfung an Komponenten von Rohrleitungen

Johann H. Hinken [1)]

1) fitm Hinken Consult, Breitscheidstrasse 17, D-39114 Magdeburg

@microwavetesting

fitm Hinken Consult

Kunststoffrohre

- ...werden für den Transport von Gasen und Flüssigkeiten eingesetzt. Beispiele:

Chemieanlage mit Rohren aus glasfaserverstärktem Kunststoff (GFK)

Stumpfverschweißte PE-Rohre

- Prüfbedarf insbesondere an geometrischen Unregelmäßigkeiten wie Rohrkrümmern und Verschweißungen.
- Dafür sind automatische Scaneinrichtungen nur bedingt geeignet.
- \>>> Handprüfung ist erforderlich.
  Mikrowellenprüfung ist zweckmäßig.

Mikrowellenhandprüfgerät

- Darstellung des komplexen Reflexionsfaktors auf dem Oszilloskop
- Prüffrequenz 24,125 GHz (im ISM-Band)

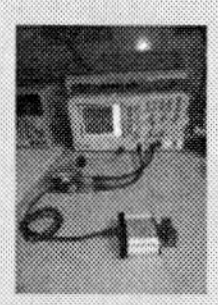

- Antenne: Hohlleiter mit WR42-Querschnitt (10,67 mm x 4,32 mm), offener ebener Standardflansch

Kontakt: Johann H. Hinken
fitm Hinken Consult
Breitscheidstr. 17
D-39114 Magdeburg
Tel.: +49 (0) 391-8868 129
Mobil: +49 (0) 171-2053208
E-Mail: johann.hinken@fitm.de
www.fitm.de

Adapter

- Aufsetzen und sicheres Verschieben auf ebener Oberfläche ist problemlos. >> klare Defektanzeige

- Aufsetzen und Verschieben auf gekrümmten Oberflächen führt zu Liftoff-Änderungen und Verwackeln mit verschmierten und nicht-reproduzierbaren Anzeigen.
- \>>> Oberflächenangepasste Adapter erforderlich zum Aufschieben auf den Standardflansch

Beispiele

- ...mit künstlichem Defekt: Polyamidplättchen 4 mm x 4 mm x 0,8 mm an Innenseite der Rohre.
- GFK-Rohr, Außendurchmesser 50 mm, Wandstärke 8 mm mit konkavem zylindrischen Adapter:

- GFK-Krümmer, Außendurchmesser an den Enden 65 mm, Wandstärke 7,5 mm mit konkavem sphärischen Adapter:

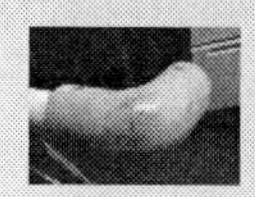

- Stumpfverschweißte PE-Rohre, Außendurchmesser 90 mm, Wanddicke 9 mm, Defekt unter Schweißraupe. Konkaver zylindrischer Adapter mit Aussparung für die Schweißraupe

Fazit

- Bei der Mikrowellenhandprüfung verhindern oberflächenangepasste Adapter das Verwackeln.
- Sie sind durch 3D-Druck schnell für einzelne Prüfaufgaben verfügbar.

Expertise oder KI – Was hilft uns in der ZfP?

Daniel KANZLER [1], Marija BERTOVIC [2]

[1] AV-NDT, Berlin, KanzlerD@av-ndt.com
[2] Bundesanstalt für Materialforschung und -Prüfung, Berlin, Marija.Bertovic@bam.de

Kurzfassung.
Die zunehmende Integration von Automatisierung und KI in die zerstörungsfreie Prüfung (ZfP) verändert nicht nur die Prüfprozesse selbst, sondern auch die Art und Weise, wie Entscheidungen getroffen werden. Während technische Systeme die Fehleranfälligkeit verringern und die Datenverarbeitung unterstützen können, bleibt die letztendliche Verantwortung beim Menschen. Dieser Beitrag untersucht die Rolle der Intuition in der Entscheidungsfindung und analysiert typische Fehleinschätzungen mithilfe der Prospekttheorie und Erkenntnissen aus der kognitiven Psychologie. Zudem wird aufgezeigt, wie fundierte Entscheidungen in KI-gestützten ZfP-Prozessen unterstützt werden können – durch Schulung, erklärbare Systeme, nutzerzentrierte Gestaltung, geeignete Metriken und eine gezielte Aufgabenverteilung zwischen Mensch und Technik.

Keywords: Zerstörungsfreie Prüfung (ZfP), Entscheidungsfindung, Künstliche Intelligenz (KI), Kognitive Verzerrungen, Zuverlässigkeit, POD, Mensch-Maschine-Interaktion, Standardisierung

Einführung

Jeder kennt die Geschichten von Feuerwehrleuten, die ohne ersichtlichen Grund plötzlich ein brennendes Haus verlassen – nur Sekunden, bevor es einstürzt. Oder den Elternteil, der instinktiv die richtige Bewegung macht, um sein Kind vor dem Sturz vom Klettergerüst zu bewahren. Diese Intuition hat uns über Jahrmillionen hinweg vor Gefahren geschützt und unser Überleben gesichert – ein Grund, stolz darauf zu sein. Doch was genau ist Intuition, und warum sollten wir darüber nachdenken – gerade in einer Zeit, in der künstliche Intelligenz, Automatisierung und maschinelle Intelligenz immer mehr an Bedeutung gewinnen? Oft wird dabei unbewusst die Gefahr des Jobverlusts und die potenzielle Sinnlosigkeit des Menschen in einer zunehmend technisierten Welt heraufbeschworen.

Es ist nicht davon auszugehen, dass der Mensch vollständig aus den Prozessen der Produktion, Wartung und des Infrastrukturbauwerks verdrängt wird – zumindest nicht in absehbarer Zeit. Sollte es dennoch irgendwann so weit kommen, dann wohl erst lange nach unserer Zeit. Doch schon heute steht die moderne zerstörungsfreie Prüfung (ZfP) vor drängenden Herausforderungen, die sofortige Lösungen erfordern. Grundsätzlich gibt es verschiedene Anforderungen der modernen ZfP, die uns tagtäglich begegnen (**Abb 1**). Um diese Anforderungen zu lösen, gibt es jedoch eine Reihe von Werkzeugen, mit denen an einer Erfüllung der Anforderungen gearbeitet werden kann. Zu den Anforderungen zählen die

https://doi.org/10.58286/31237

zunehmende Komplexität von Bauteilen sowie neue Fertigungsverfahren wie die additive Fertigung. Besonders besorgniserregend ist jedoch der schrumpfende Erfahrungspool – ein Problem, das sich angesichts des rückläufigen Interesses an naturwissenschaftlichen Fächern bei Schülern [1] weiter verschärfen wird.

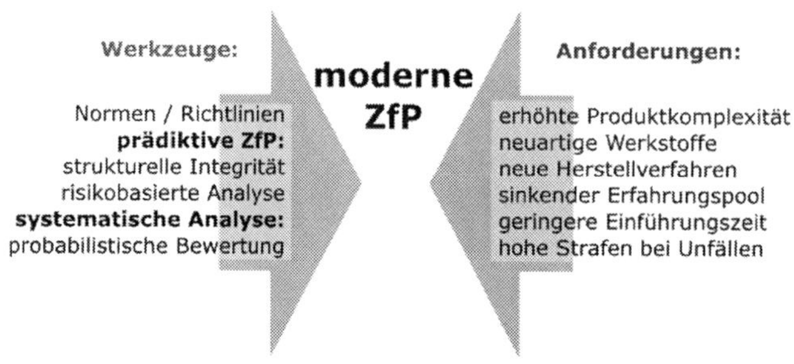

Abb 1. Aktuelle Herausforderungen in der ZfP

Auf den ersten Blick scheint die Entwicklung rund um künstliche Intelligenz (KI) und Automatisierung die Lösung für diese Herausforderungen zu sein. Der „einfache" Prüfende wird überflüssig (wenngleich nicht in jedem Bereich der ZfP), und das Problem des fehlenden Erfahrungswissens wäre damit gelöst. Oder etwa nicht?

Die Realität sieht anders aus: In jeder Branche gibt es Fälle, in denen KI versagt [2][3]. Selbst Validierungssysteme, wie sie etwa durch den *AI Act* (Verordnung über künstliche Intelligenz) [4] gefordert werden, stoßen an ihre Grenzen – Benchmarksysteme etwa scheitern [5]. Am Ende bleiben die Menschen als entscheidende Instanz über das KI-System. Doch sind sie wirklich in der Lage, diese Verantwortung zu übernehmen? Und sollten sie ihre Entscheidungen – wie so viele im Alltag – auf ihre Intuition stützen? [6]

Eines ist sicher: Die Rolle der Prüfenden verändert sich grundlegend. Sie sind nicht mehr nur dafür verantwortlich, klassische Prüfungen durchzuführen und zu dokumentieren. Mit der Einführung von KI und Automatisierung in der ZfP übernehmen sie neue Aufgaben – wie das Verstehen, Hinterfragen und Validieren automatisierter Entscheidungen. Auch wenn viele Systeme immer intelligenter werden, bleibt der Mensch die letzte Instanz im Prozess. Damit steigt nicht nur die Verantwortung, sondern auch die Komplexität der zu treffenden Entscheidungen.

Diese Aufgaben erfordern mehr als nur technisches Wissen oder Erfahrung. Prüfende tragen zunehmend die Verantwortung dafür, automatisierte Prozesse zu bewerten, Systemgrenzen zu erkennen und geeignete Maßnahmen abzuleiten. Wie in [7] beschrieben, wird es immer wichtiger, Menschen gezielt auf diese neue Rolle vorzubereiten – nicht nur durch Schulungen, sondern auch durch eine ganzheitliche nutzerzentrierte Gestaltung der Mensch-Maschine-Interaktion. Das bedeutet u.a.: verständliche, erklärbare und intuitive Systeme, angemessene Aufgabenverteilung zwischen Menschen und KI und klare Verantwortlichkeiten.

Genau hier setzt dieser Artikel an: Anstatt alle Aspekte dieses Wandels zu behandeln, konzentrieren wir uns gezielt auf einen Kernbereich – die menschliche Entscheidungsfindung. Wir fragen: Wie treffen Prüfende Entscheidungen in einer zunehmend automatisierten Umgebung? Welche Rolle spielen Intuition, Erfahrung und kognitive Verzerrungen? Und was brauchen sie, um in ihrer neuen Rolle gute Entscheidungen zu treffen?

1. Schachgroßmeister vs. Trader – wo Intuition funktioniert und wo nicht

Jeder, der gerne Schach spielt, kennt die besondere Dynamik des Blitzschachs. Diese Variante erfordert schnelle Entscheidungen, denn ein ganzes Spiel muss innerhalb weniger Minuten abgeschlossen sein. Nachdenken ist nur in besonders kritischen Stellungen möglich – der Rest basiert auf Intuition. Dennoch dominieren meist erfahrene Spieler. In diesem einfachen Fall lässt sich Intuition als die Summe der gesammelten Erfahrung im Spiel verstehen. Großmeister verfügen über das meiste Wissen und die höchste Spielpraxis – kein Wunder also, dass Magnus Carlsen auch „Mit"-Blitzweltmeister wurde [8]. Die Quintessenz: Intuition ist wichtig und hilfreich, selbst bei komplexen Aufgaben.

Ein Gegenbeispiel liefert der Aktienhandel. Der stereotype Trader à la *Wolf of Wall Street* mag sich zwar hervorragend für die Kinoleinwand eignen, doch in der Realität führt das kurzfristige Kaufen und Verkaufen von Aktien auf reiner Intuition selten zu langfristigem Reichtum [9][10]. Hier ist Intuition eher hinderlich, da sie zu impulsiven Entscheidungen verleitet, die nicht unbedingt rational oder strategisch vorteilhaft sind.

Bleiben wir noch kurz in der Welt von Gewinn und Verlust. Als ZfP-Experten betrachten wir uns meist als risikoavers – als Menschen, die versuchen, Risiken zu minimieren. Tatsächlich sind wir jedoch sowohl risikoavers (*risk-averse)* als auch risikofreudig (*risk-seeking*), je nach Situation. Diese Erkenntnis wurde durch Daniel Kahneman [9, 10, 11] wissenschaftlich untermauert und führte zur Entwicklung der *Prospekttheorie*, wie in **Abb 2**. Der Grad der Risikofreudigkeit wird häufig als v gekennzeichnet. Die Theorie besagt, dass Menschen in Gewinnsituationen eher vorsichtig agieren und einen sicheren Gewinn bevorzugen, anstatt ein höheres Risiko für einen potenziell größeren Gewinn einzugehen. Umgekehrt neigen wir dazu, risikofreudiger zu handeln, wenn Verluste drohen – wir sind eher bereit, ein gewisses Risiko in Kauf zu nehmen, um eine Niederlage abzuwenden. Die unterschiedlichen Gewinne und Verluste werden dabei individuell gewichtet (mit dem Faktor π), abhängig von unserem *Referenzpunkt* – also davon, wie viel Geld oder Ressourcen uns aktuell zur Verfügung stehen. Die Höhe des potenziellen Gewinns oder Verlusts beeinflusst ebenfalls unsere Entscheidungsfindung.

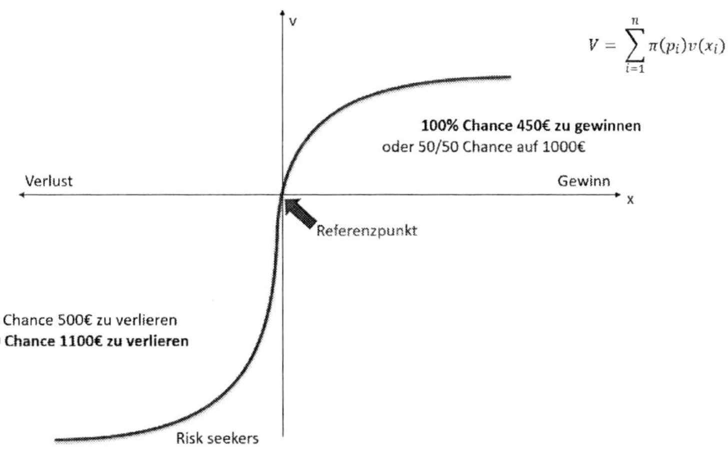

Abb 2. Prospekttheorie nach Kahneman

Nun stellt sich die Frage: In welcher Welt bewegt sich die ZfP? Daniel Kahneman führte das Konzept der „*probabilistischen Versicherung*" als Gedankenexperiment ein [9]. Dabei handelt es sich um eine Versicherung, die im Schadensfall nicht garantiert zahlt, sondern nur mit einer gewissen Wahrscheinlichkeit. Auch wenn diese Situation zunächst theoretisch erscheint, zeigt sie klare Parallelen zur ZfP. Denn auch die ZfP kann Defekte nur unter bestimmten Bedingungen zuverlässig aufspüren – und damit potenzielle Ausfälle nur mit einer gewissen Wahrscheinlichkeit verhindern.

2. Die kontra-intuitive Welt des Qualitätsprüfers

Das Konzept der "Risikosuche im Fall von Verlusten" ist etwas, das wir eigentlich nicht auf die ZfP übertragen möchten. Dennoch zeigen Beispiele aus dem Qualitätsmanagement, dass ähnliche Überlegungen in der Praxis eine Rolle spielen – wie etwa die Fälle von OceanGate, Boeing und Tesla [12, 13]. Auch die oft gehörte Aussage „ZfP kostet nur Geld" folgt diesem Gedankenansatz.

Dennoch ist klar: Die ZfP ist ein essenzielles Werkzeug im Qualitäts- und Risikomanagement. Ihre zentrale Funktion besteht darin, die Ausfallwahrscheinlichkeit von Komponenten zu reduzieren, indem sie den aktuellen Zustand eines Bauteils beschreibt. Wird beispielsweise ein Riss entdeckt, kann rechtzeitig reagiert werden, wodurch das Versagensrisiko signifikant sinkt.

Ein Blick in die DIN EN 16991 [14] bestätigt diese Rolle der ZfP bei der Reduktion von Ausfallwahrscheinlichkeiten. In einer typischen Risikomatrix, siehe **Abb 3**, verschiebt sich der Arbeitspunkt auf der y-Achse entsprechend nach unten. Allerdings wird die oft als „teuer" wahrgenommene ZfP nur dann eingesetzt, wenn sie ökonomisch sinnvoll ist. In der weiteren Analyse werden daher geringe Auswirkungen von Ausfällen zunächst außer Acht gelassen. Ein interessanter Aspekt ist der „schwarze Bereich", der in klassischen Risikomatrizen nicht explizit dargestellt wird, aber durch den juristischen Tatbestand der Fahrlässigkeit abgedeckt ist. Der Fall von OceanGate [12] liefert ein tragisches Beispiel für die Folgen von Vernachlässigungen in sicherheitskritischen Bereichen. Gegen klare Regeln und Normen und die berechtigte Kritik von Sicherheits- und Qualitätsmanagementleuten wurde widersprochen, was zu dem weltbekannten Unfall des Unterseeboots geführt hat.

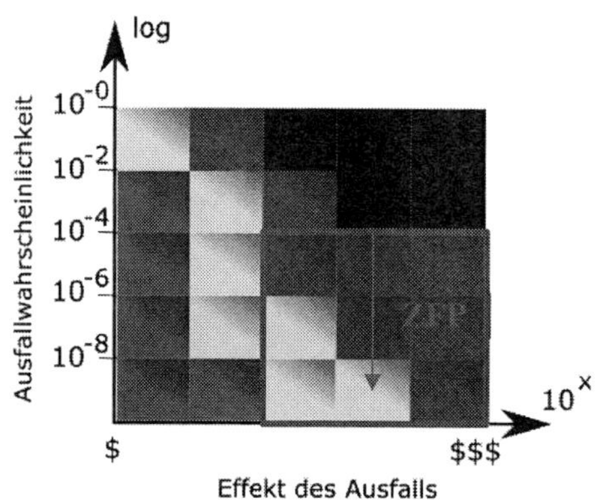

Abb 3. Risikomatrix für die ZfP

Darüber hinaus gibt es wichtige Unterschiede in der Darstellung von Risiko:

- Die Ausfallwahrscheinlichkeit wird üblicherweise auf einer logarithmischen Skala abgebildet. Technische Ausfälle treten in absoluten Zahlen sehr selten auf, sodass eine lineare Darstellung wenig aussagekräftig wäre.
- Die Auswirkungen von Ausfällen hingegen folgen oft einer exponentiellen Skala. Große technische Katastrophen – wie Flugzeugabstürze – können extreme Folgen

nach sich ziehen, sei es durch dramatische Kursverluste an der Börse oder massive Einbrüche in den Verkaufszahlen.

Diese Zusammenhänge verdeutlichen, dass die ZfP weit mehr ist als eine bloße „Kostenstelle“ – sie ist ein strategisches Instrument zur Risikominimierung und kann im schlimmsten Fall existenzbedrohende Konsequenzen verhindern.

2.1 Entscheidungsfindung als Schlüsselaufgabe in der ZfP

Die zunehmende Automatisierung in der ZfP verändert nicht nur die Prüfprozesse, sondern auch die Art und Weise, wie Entscheidungen getroffen werden. Wie in einer Studie über den Einsatz der automatisierten Fehlererkennung in der ZfP betont wird, stellen neue Technologien auch neue Anforderungen an das Prüfpersonal [21, 22] – insbesondere im Hinblick auf ihre Rolle als Entscheidungsträger in einer zunehmend digitalisierten Umgebung [7].

Entscheidungen in der ZfP betreffen nicht nur die Auswahl geeigneter Verfahren, sondern auch die Bewertung von Prüfergebnissen, die Einschätzung von Risiken und die Ableitung sicherheitsrelevanter Maßnahmen. Gleichzeitig bleibt die Entscheidung unter Unsicherheit eine zentrale Herausforderung – technisch, organisatorisch und menschlich.

2.2 Kognitive Verzerrungen in der ZfP

Obwohl wir Menschen mit einer unglaublichen Intuition ausgestattet sind, ist unser Entscheidungsprozess ein ausgeklügelter, aber oft fehlerhafter Mechanismus. Zahlreiche Studien zeigen, dass Menschen unter Unsicherheit und Zeitdruck zu kognitiven Verzerrungen und systematischen Fehleinschätzungen neigen [9][10][15][16][17]. Im Folgenden sind einige Beispiele für diese kognitiven Verzerrungen und wie sie zu einer verzerrten Entscheidungsfindung in der ZfP führen können aufgelistet.

- Risikofreudiges Verhalten bei drohenden Ausfällen: Die Prospekttheorie [9, 11] zeigt, dass Menschen in Verlustsituationen tendenziell risikofreudiger handeln. In der ZfP geht es primär darum, negative Risiken – also mögliche Ausfälle – zu vermeiden. Dies kann dazu führen, dass Entscheidungsträger trotz potenzieller Gefahren eher bereit sind, Risiken einzugehen, anstatt präventiv zu handeln.
- Der Möglichkeitseffekt (engl. *possibility effect*) [9]: Aufgrund der äußerst niedrigen Ausfallwahrscheinlichkeiten tritt der sogenannte Möglichkeitseffekt auf. Dabei werden Wahrscheinlichkeiten systematisch unterschätzt, insbesondere wenn eine ähnliche Situation noch nie zuvor aufgetreten ist. Paradoxerweise kann dieser Effekt auch dazu führen, dass kleine Wahrscheinlichkeiten überschätzt werden, sobald ein solcher Fall doch einmal eintritt. Für beide Fälle kann die Carolabrücke in Dresden [18] herangezogen werden. Erste die Unterschätzung der Ausfallwahrscheinlichkeit vor dem Unfall und dann nach Ausfall eine allgemeine Überschätzung der Gefährdungslage in der Bevölkerung.
- Die Verfügbarkeitsheuristik (engl. *availability heuristic*) [9]: Diese kognitive Verzerrung beschreibt die Tendenz, die Wahrscheinlichkeit eines Ereignisses anhand der Leichtigkeit abzuschätzen, mit der uns ähnliche Fälle in den Sinn kommen. Im Kontext der ZfP bedeutet das: Die Folgen eines Ausfalls sind schwer greifbar, solange dieser nicht bereits eingetreten ist. Eine vertiefte Diskussion zu diesem Phänomen findet sich unter anderem in [19][20].
- Automatisierungsverzerrung (engl. *automation bias*) [9]: kognitive Tendenz, automatisierte Systeme oder KI-Ergebnisse ohne Überprüfung zu akzeptieren, anstatt

sie kritisch zu hinterfragen. Bei der ZfP kann dies dazu führen, dass Fehler übersehen werden, weil man sich zu sehr auf die automatisierte (KI-)Auswertung verlässt [15] [16].

- Ankereffekt (engl. *anchoring bias*) [9]: Erste Informationen (s.g. „Anker") beeinflussen nachfolgende Urteile unverhältnismäßig stark. So kann ein erstes KI-Urteil (z. B. „kein Defekt") die weitere menschliche Bewertung prägen – auch wenn später Hinweise auf das Gegenteil auftreten.
- Bestätigungsfehler (engl. *confirmation bias*) [9]: Neigung, Informationen so auszuwählen oder zu interpretieren, dass sie die eigenen Erwartungen oder Vorannahmen bestätigen. Beispielsweise, wenn ein Prüfer bereits eine Hypothese hat (z. B. „kein Fehler"), werden KI-Ergebnisse, die das bestätigen, eher akzeptiert – und abweichende Hinweise möglicherweise ignoriert.

Solche und viele andere kognitive Verzerrungen sind normale menschliche Denkmuster und Abkürzungen, die uns allen helfen, im Alltag effizient zu handeln – doch gerade in der ZfP, wo sich die Hauptaufgabe zunehmend in Richtung Bewertung und Entscheidungsfindung verlagert, müssen wir uns dieser Tendenzen bewusster denn je sein.

2.3 Risikoeinschätzung in der Praxis: Ein Vergleich mit dem Glücksspiel

Möchte man bewerten, ob eine aktuelle Situation risikofreudiges Verhalten begünstigt, kann man dies mit der Prospekttheorie und einem Glücksspiel vergleichen. Ein Spieler hat die Wahl zwischen zwei Verlustszenarien: a) Sicherer Verlust von 500 € und b) Risiko eines höheren Verlusts von 550 €, jedoch mit der Möglichkeit, weniger zu verlieren oder sogar zu gewinnen. Rein rational betrachtet wäre Szenario a) die bessere Wahl, da es weniger Verlust bedeutet. In der Realität entscheiden sich Menschen jedoch häufig für b) – ein Verhalten, das durch die Prospekttheorie erklärt wird [9].

Übertragen auf die ZfP heißt das: Wenn z. B. der Einsatz eines Prüfverfahrens als teuer oder aufwendig wahrgenommen wird, neigen Entscheidungstragende möglicherweise dazu, auf die Prüfung zu verzichten – selbst wenn dadurch ein höheres Ausfallrisiko entsteht. Die Entscheidung wirkt situative rational, ist aber strategisch riskant.

2.4 Zwei große Unsicherheiten in der ZfP

Neben der menschlichen Komponente gibt es zwei wesentliche systemische Unsicherheiten, die jede Entscheidung in der ZfP beeinflussen:

- Unklare Folgekosten: Die langfristigen Konsequenzen eines Schadens sind oft nicht vollständig absehbar. Besonders in Bereichen wie Brückenbau oder Infrastruktur ist es schwierig, den Wert einer geprüften Komponente exakt zu bestimmen.
- Indirekte Verbindung zwischen Prüfverfahren und tatsächlicher Ausfallrate: Selbst wenn ein Prüfverfahren eine hohe Defekterkennungsrate aufweist, bedeutet dies nicht zwangsläufig, dass sich dadurch die Ausfallrate in der Realität linear reduzieren lässt.

Diese Unsicherheiten zeigen: Objektive Entscheidungsfindung ist nicht nur eine Frage der Technik – sie bleibt ein komplexer Abwägungsprozess, der Erfahrung, Kontextverständnis und Systemkenntnis erfordert. Eines ist jedoch klar: Ohne Kenntnis der Grenzen eines Prüfverfahrens ist weder die Ausfallwahrscheinlichkeit abschätzbar noch eine fundierte Entscheidungsfindung möglich.

3. Unterstützung der Entscheidungsfindung in der KI-gestützten ZfP

Ein weit verbreiteter Trugschluss ist es, die Lösung *allein* in der Ausbildung der Prüfenden zu sehen. Natürlich ist eine gezielte Qualifizierung unerlässlich – beispielsweise im Rahmen des international vorgeschlagenen „*NDT Engineer*" [21] oder der nach DIN 19293 [22] geforderten "*fähigen Person*", die für die Bewertung der Zuverlässigkeit von Prüfverfahren und deren Einsatzentscheidungen verantwortlich ist. Gerade bei sicherheitskritischen Anwendungen mit geringen Ausfallwahrscheinlichkeiten ist eine fundierte Entscheidungsfindung besonders wichtig.

Eine entscheidende Frage bleibt jedoch unbeantwortet: *Wo* sollen diese Fähigkeiten erworben werden und *wie* können sie systematisch nachgewiesen werden? Derzeit liegt die Verantwortung oft bei den jeweiligen Unternehmen, die die entsprechenden Kompetenzen in ihren Qualitätshandbüchern definieren. Dieser Ansatz hat einige deutliche Vorteile: a) er schlägt eine Brücke zum Qualitätsmanagement und verankert die zerstörungsfreie Prüfung stärker in den betrieblichen Entscheidungsprozessen; und b) er fördert die Eigenverantwortung und bietet den Freiraum, eine zuverlässige, kontextgerechte Anwendung der Prüfverfahren zu gestalten.

Gleichzeitig zeigen Studien wie [15][16][17] und Erfahrungen aus der Praxis, dass Qualifikation allein nicht ausreicht. Die Entscheidungsfindung kann und sollte u.a. auch durch folgende Faktoren unterstützt werden:

- Transparente und erklärbare Systeme (Stichwort: erklärbare KI), die eine nachvollziehbare Entscheidungsgrundlage bieten,
- Intuitive, benutzerfreundliche und vertrauenswürdige Schnittstellen, die eine Informationsüberflutung vermeiden,
- realistische Erwartungen an KI-gestützte Systeme, um sowohl übermäßigen als auch unzureichenden Einsatz zu vermeiden,
- organisatorische Unterstützung, die eine offene Fehlerkommunikation ermöglicht, Entscheidungsdruck reduziert und Verantwortlichkeiten klar definiert.

Entscheidungsfindung ist somit kein rein individueller Akt, sondern das Ergebnis einer Interaktion zwischen Mensch, Technik und Organisation. Systeme müssen so gestaltet sein, dass sie intuitiv bedienbar, aber auch kritisch hinterfragbar sind. Prüfer benötigen nicht nur Fachwissen, sondern auch kognitive Unterstützung, um ihrer Rolle gerecht werden zu können.

Zusammenfassend lässt sich sagen, dass fundierte Entscheidungen in der ZfP nicht nur das Ergebnis von Schulungen sind, sondern auch von einem Arbeitsumfeld, das systematisch menschliche Faktoren, Benutzerfreundlichkeit und Verantwortung berücksichtigt. Nur so können Sicherheits- und Qualitätsziele auch in hochautomatisierten Prozessen zuverlässig erreicht werden.

Im Folgenden werden neun Kernpunkte vorgestellt, die eine fundierte und verantwortungsvolle Entscheidungsfindung in der KI-gestützten ZfP systematisch unterstützen.

3.1 Kernpunkte einer fundierten Entscheidungsfindung in der KI-gestützten ZfP

Gute Entscheidungen in der ZfP entstehen nicht allein durch Fachwissen. In einer zunehmend automatisierten Prüfwelt sind sie das Ergebnis eines Zusammenspiels aus menschlichen Fähigkeiten, verständlicher Technik und klaren organisatorischen Strukturen. Neun zentrale Aspekte tragen dazu bei, Prüfende gezielt zu unterstützen:

1. **Menschliche Faktoren berücksichtigen:** Entscheidungen in der ZfP werden maßgeblich durch menschliche Faktoren beeinflusst – z. B. durch Intuition, Erfahrung, Aufmerksamkeit oder kognitive Verzerrungen wie den Automation Bias. Diese Faktoren müssen bei der Gestaltung von Systemen, Prozessen und Schulungskonzepten berücksichtigt werden [7].
2. **Systemverständnis für KI und Automatisierung (Ausbildung):** Nur wer versteht, wie automatisierte Systeme funktionieren, wo ihre Stärken und wo ihre Grenzen liegen, kann ihre Ergebnisse richtig interpretieren. Die Arbeit mit KI erfordert neue Fähigkeiten – sowohl im technischen als auch im kognitiven Bereich [7].
3. **Nutzerzentriertes Systemdesign (User-Centred Design):** Intuitive, transparente und erklärbare Schnittstellen reduzieren kognitive Überlastung, verbessern die Informationsverarbeitung und erhöhen die Systemakzeptanz. Nutzer sollten frühzeitig in die Entwicklung eingebunden werden [7][23][24][25]
4. **Erklärbare KI (Explainable AI):** KI-Systeme müssen nachvollziehbar sein. Visualisierungen, Confidence-Level und transparente Entscheidungspfade ermöglichen Vertrauen, Verständnis und Kontrolle über automatisierte Vorschläge.
5. **Realistische Bewertung von Prüfverfahren:** ZfP-Verfahren liefern keine absoluten Wahrheiten. Kennwerte wie die Probability of Detection (POD) müssen im Kontext interpretiert und Unsicherheiten aktiv kommuniziert werden, um Fehleinschätzungen zu vermeiden.
6. **Organisatorische Verankerung der ZfP:** ZfP-Entscheidungen müssen klar in Management- und Qualitätsprozesse eingebunden sein – etwa nach DIN 19293 [22]. Verantwortlichkeiten und Entscheidungswege sind eindeutig zu definieren und zu dokumentieren.
7. **Ökonomisch sinnvoller Prüfeinsatz**: Der Einsatz von ZfP muss wirtschaftlich begründet erfolgen – mit Blick auf Aufwand, Nutzen und Risiken bei potenziellen Ausfällen [26].
8. **Wirtschaftliche und rechtliche Rahmenbedingungen:** Prüfentscheidungen betreffen auch Haftung, Kosten und Verantwortung. Normen, Richtlinien und gesetzliche Vorgaben (z. B. Produkthaftung, AI Act) müssen bekannt und in Entscheidungen integriert sein. Diese Faktoren müssen in Schulung, Systemauslegung und Entscheidungsprozessen von Anfang an mitgedacht werden [26][27].
9. **Aufgabenverteilung nach Stärken:** In der Mensch-Technik-Interaktion sollte die Aufgabenverteilung nicht ausschließlich danach erfolgen, was technologisch machbar ist, sondern was sinnvoll ist. KI-Systeme sind leistungsfähig bei der Verarbeitung großer Datenmengen, Mustererkennung und Konsistenz – Menschen hingegen bringen Kontextverständnis, ethische Abwägung und Verantwortungsbewusstsein mit. Systeme sollten so gestaltet sein, dass sie diese komplementären Stärken unterstützen – etwa durch interaktive Funktionen, die zu einer aktiven Auseinandersetzung mit den Systemvorschlägen anregen, statt diese unhinterfragt zu übernehmen [7].

Fazit: Entscheidungen lassen sich nicht vollständig objektivieren – aber durch gezielte System- und Umfeldgestaltung unterstützen. So bleibt die ZfP auch im Zeitalter der KI ein verlässliches, verantwortungsvoll nutzbares Instrument der Qualitätssicherung.

4. Schlussfolgerung

Die Menschen werden auch in Zukunft ein zentraler Bestandteil des Qualitätsmanagements und der Prüfung von Bauteilen bleiben. Allerdings ist absehbar, dass sich ihre Rolle grundlegend verändern wird: vom prüfenden Personal hin zu der entscheidenden Instanz. Diese neue Verantwortung bringt sowohl neue Aufgaben als auch neue Herausforderungen mit sich. Entscheidungen müssen systematisch und fundiert getroffen werden – denn Fehlentscheidungen können erhebliche Folgekosten verursachen. Ein wesentlicher Bestandteil dieser neuen Rolle ist die statistische Aufklärung sowie der Einsatz von Hilfsmitteln aus der statistischen Heuristik und Psychologie, um komplexe Zusammenhänge verständlich und nachvollziehbar zu machen.

Gleichzeitig entstehen neue Rahmenbedingungen: Aktuelle Normen und Richtlinien – wie z. B. DIN 19293, der AI Act oder die Deutsche Normungsroadmap KI – helfen dabei, die neue Rolle der Prüfenden zu definieren und systematisch zu unterstützen. Um dieses Wissen in der Breite verfügbar zu machen, braucht es gezielte Dissemination: über Schulungen, interdisziplinäre Zusammenarbeit, normungsnahe Gremienarbeit und praxisorientierte Kommunikation. Nur so kann die ZfP auch im Zeitalter der KI ein verlässliches, menschlich verantwortetes Instrument zur Qualitätssicherung bleiben.

Referenzen

[1] Deutsche Telekom Stiftung, "Was motiviert zum MINT-Lernen? – Eine qualitativ-quantitative SINUS-Studie unter 10- bis 16-jährigen," Nov. 2024. [Online]. Verfügbar: https://www.telekom-stiftung.de/sites/default/files/files/dts-sinus-studie-mint-motivation-zusammenfassung.pdf

[2] J. Taylor, "Australian lawyer caught using ChatGPT filed court documents referencing 'non-existent' cases," *The Guardian*, Feb. 1, 2025. [Online]. Verfügbar: https://www.theguardian.com/australia-news/2025/feb/01/australian-lawyer-caught-using-chatgpt-filed-court-documents-referencing-non-existent-cases

[3] t-online, "Tesla verwechselt im Test Bild mit Straße – und rast durch Wand," *t-online*, Mar. 19, 2025. [Online]. Verfügbar: https://www.t-online.de/mobilitaet/aktuelles/id_100642496/lidar-gegen-kameras-tesla-versagt-im-test-und-verwechselt-wand-mit-strasse.html

[4] Europäische Union, "Verordnung (EU) 2024/1689 ... (Verordnung über künstliche Intelligenz)," *Amtsblatt der Europäischen Union*, Jun. 13, 2024. [Online]. Verfügbar: https://eur-lex.europa.eu/legal-content/DE/TXT/?uri=OJ%3AL_202401689

[5] M. Bastian, "Forscher liefern weiteren Grund zur Skepsis bei KI-Benchmarks," *The Decoder*, Jan. 6, 2025. [Online]. Verfügbar: https://the-decoder.de/forscher-liefern-weiteren-grund-zur-skepsis-bei-ki-benchmarks/

[6] G. Gigerenzer and W. Gaissmaier, *Intuition und Führung: Wie gute Entscheidungen entstehen*. Wiesbaden: Springer Fachmedien, 2015.

[7] M. Bertovic and I. Virkkunen, "NDE 4.0: New Paradigm for the NDE Inspection Personnel," in *Handbook of Nondestructive Evaluation 4.0*, N. Meyendorf, I. Ida, R. Singh, and J. Vrana, Eds. Cham: Springer, 2021, ch. 9-1, pp. 1–31.

[8] S. Löffler, "Regeln, was sind denn schon Regeln," *Frankfurter Allgemeine Zeitung*, Jan. 1, 2025. [Online]. Verfügbar: https://www.faz.net/aktuell/sport/mehr-sport/anarchie-bei-blitzschach-wm-eklat-um-dresscode-und-geteilten-weltmeistertitel-110205349.html

[9] D. Kahneman, *Thinking, Fast and Slow*. New York: Macmillan, 2013.

[10] D. Kahneman, O. Sibony, and C. R. Sunstein, *Noise – A Flaw in Human Judgment*. New York: HarperCollins, 2021.

[11] D. Kahneman and A. Tversky, "Prospect Theory: An analysis of decision under risk" Econometrica, Vol.47, S.263-291, 1979;

[12] K. Kort, "Sauerstoff der Titan-Crew womöglich aufgebraucht," *Handelsblatt*, Jun. 22, 2023. [Online]. Verfügbar: https://www.handelsblatt.com/unternehmen/dienstleister/titanic-tauchboot-sauerstoff-der-titan-crew-womoeglich-aufgebraucht/29217964.html

[13] NTV, "Cybertrucks von Tesla verliert Karosserieteile," *n-tv*, Mar. 20, 2025. [Online]. Verfügbar: https://www.n-tv.de/wirtschaft/Cybertrucks-von-Tesla-verlieren-Karosserieteile-article25643859.html

[14] DIN EN 16991:2018, "Risikobasierte Inspektion (RBIF)," Deutsche Fassung, DIN e.V., 2018.

[15] M. Bertovic, *Human Factors in Non-Destructive Testing (NDT): Risks and Challenges of Mechanised NDT*, Doctoral dissertation, Technische Universität Berlin. BAM-Dissertationsreihe, Band 145. Berlin: Bundesanstalt für Materialforschung und -prüfung (BAM), 2016.
[16] M. Bertovic, "A human factors perspective on the use of automated aids in the evaluation of NDT data," *AIP Conf. Proc.*, vol. 1706, p. 020003, 2016. doi: 10.1063/1.4940449.
[17] M. Bertovic, M. Gaal, C. Müller, and B. Fahlbruch, "Investigating human factors in manual ultrasonic testing: Testing the human factor model," *Insight – Non-Destructive Testing and Condition Monitoring*, vol. 53, no. 12, pp. 673–676, 2011. doi: 10.1784/insi.2011.53.12.673.
[18] *Tagesspiegel*, "Einsturz der Carolabrücke in Dresden: Erste Vermutung zu Ursache bestätigt," Oct. 2, 2024. [Online]. Verfügbar: https://www.tagesspiegel.de/gesellschaft/panorama/einsturz-der-carolabrucke-in-dresden-erste-vermutung-zu-ursache-bestatigt-12475886.html
[19] K. R. Feinberg, *What is Life Worth? The Unprecedented Effort to Compensate the Victims of 9/11*. New York: Public Affairs, 2005.
[20] K. R. Feinberg, *Who Gets What: Fair Compensation after Tragedy and Financial Upheaval*. New York: Public Affairs, 2012.
[21] P. Trampus, V. Krstelj, and G. Nardoni, "NDT integrity engineering – A new discipline," presented at *The 3rd International Conference on Structural Integrity (ICSI)*, 2019.
[22] DIN 19293:2025, "Zerstörungsfreie Prüfung – Auswertung von Prüfdaten – Leitfaden zur Einschätzung der Zuverlässigkeit von Anwendungen," DIN, e.V., 2025.
[23] DIN EN ISO 9241-110:2020, "Ergonomics of human-system interaction – Part 110: Interaction principles," DIN e.V., 2020.
[24] DIN EN ISO 9241-210:2020, "Ergonomics of human-system interaction – Part 210: Human-centred design for interactive systems," DIN e.V., 2020.
[25] DIN EN ISO 9241-11:1999, "Ergonomic requirements for office work with visual display terminals (VDTs). Part 11: Guidance on usability," DIN e.V., 1999.
[26] D. Kanzler, "How Good is the Collaboration with AI in NDT?," presented at *3rd International Conference on NDE 4.0*, Mar. 3, 2025, Bengaluru, India.
[27] Deutsches Institut für Normung, *Deutsche Normungsroadmap Künstliche IntelligenzI*, 2nd ed. 2022. [Online]. Verfügbar: https://www.din.de/resource/blob/916792/20bf33d405710a703aa26f81362493bb/nrm-ki-deutsch-2022-final-web-250-data.pdf

Hinweis zum Einsatz von KI-gestützter Sprachassistenz: Beim sprachlichen Feinschliff des Manuskripts wurde ChatGPT von OpenAI unterstützend eingesetzt. Sämtliche inhaltlichen Beiträge, Analysen und Quellen wurden von den Autor*innen selbst eingebracht und überprüft

DGZfP-Jahrestagung 2025

Vorstellung der aktuellen Arbeit der Normungsgremien für den Bereich Ultraschallprüfung

AUTOR*INNEN
Michael SCHMITT [1], Thomas HECKEL [2]
[1] DIN Deutsches Institut für Normung, Berlin, Deutschland, michael.schmitt@din.de
[2] Bundesanstalt für Materialforschung und -prüfung (BAM), Berlin, Deutschland, thomas.heckel@bam.de

Kurzfassung. Normung und Standardisierung in der Ultraschallprüfung finden auf nationaler, europäischer und internationaler Ebene statt. Es findet Normung zu den Grundlagen der Ultraschallprüfung, zur Ultraschallprüfausrüstung und zur Prüfung von Schweißverbindungen statt.

Deutschland nimmt hier eine führende Rolle in diesem Bereich ein, was durch Leitungsfunktionen in europäischen und internationalen Gremien sowie durch Projektleitungen bei der Überarbeitung bzw. Aktualisierung von Normen sowie bei der Entwicklung von neuen Normen gewährleistet wird.

Damit diese Normungsaktivitäten die deutsche Wirtschaft unterstützen, benötigen diese Projektleitungen fachliche Zuarbeit und inhaltliche Steuerung durch in den Normungsgremien vertretene Fachexperten der deutschen Wirtschaft.

Keywords: Normung, Qualifizierung, Qualitätssicherung, Ultraschall

Einführung

Normen und Standards sind ein wichtiger Teil des technischen Regelwerks im Bereich der Ultraschallprüfung. Normung und Standardisierung finden auf nationaler, europäischer und internationaler Ebene in Gremien der Normungsinstitute statt. Während die Organisation durch die Normungsinstitute (in Deutschland DIN) geschieht, liegen die Inhalte der erarbeiteten Normen und Standards in der Verantwortung der beteiligten Fachexpert*innen. Die Gestaltungshoheit liegt daher allein bei der interessierten Fachöffentlichkeit, woraus sich eine starke Einflussmöglichkeit ergibt. Durch die Übernahme von Leitungsfunktionen in Gremien, in denen Normen erarbeitet und verantwortet werden, kann dieser Einfluss auf Themen und Inhalte der Normung im Bereich der Ultraschallprüfung noch weiter gesteigert werden. Im Umkehrschluss bedeutet dies, dass Firmen und Institute sich entweder im Standardisierungsprozess selbst engagieren oder die Gestaltungshoheit ihren Wettbewerbern überlassen müssen. Denn, wer sich an dem System nicht beteiligt, muss letztlich Normen anwenden, die andere nach ihren eigenen Interessen gestalten.

https://doi.org/10.58286/31271

1. Organisation

Normung und Standardisierung in der Ultraschallprüfung findet auf internationaler Ebene im ISO/TC 135 "Non-destructive testing", Unterkomitee 3 „Ultrasonic testing" statt, das von Deutschland geleitet wird. Die fachliche Arbeit erfolgt zurzeit in den Arbeitsgruppen WG 5 „Ultrasonic test equipment" und WG 7 „Basic principles", die beide ebenfalls unter deutscher Leitung stehen.

Zusätzlich gibt es bei ISO noch eine Arbeitsgruppe, die sich im ISO/TC 44/SC 5 "Testing and inspection of welds" mit der zerstörungsfreien Prüfung von Schweißnähten beschäftigt (Tabelle 2). ISO/TC 44/SC 5/WG 2 „Ultraschallprüfung" wurde in der Vergangenheit von Deutschland geleitet. Mangels Fachexpertise im nationalen Gremium wurde die Leitung niedergelegt und ist derzeit vakant.

Neben dem ISO/TC 44/SC 5 hat auch das International Institute of Welding IIW das Recht, ISO-Normen in dem Bereich der Schweißnahtprüfung zu erstellen. Dieses Recht nimmt das IIW auch wahr, wobei darauf geachtet werden muss, dass keine Doppelnormung entsteht.

Auf europäischer Ebene findet die Normung der zerstörungsfreien Prüfung von Schweißnähten in der CEN/TC 121/WG 21 "Prüfung von Schweißungen" statt. Die Arbeitsgruppe beschränkt sich aber auf die Übernahme von internationalen Normen, ist also nicht aktiv.

Der Gegenpart zum ISO/TC 135/SC 3 auf europäischer Ebene ist die Arbeitsgruppe WG 2 „Ultraschallprüfung" des CEN/TC 138 "Zerstörungsfreie Prüfverfahren". Die Arbeitsgruppe steht unter deutscher Leitung.

Abbildung 1 zeigt den Zusammenhang zwischen den Normungsgremien für Ultraschall auf den 3 Ebenen.

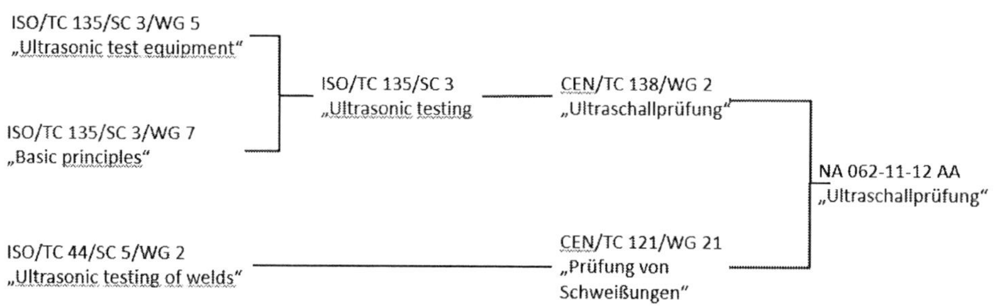

Abb. 1. Zusammenhang der nationalen, europäischen und internationalen Normungsgremien der Ultraschallprüfung.

2. In den Jahren 2024 und 2025 veröffentlichte Normen

In den Jahren 2024 und 2025 wurden die in Tabelle 1 aufgeführten Normen veröffentlicht. Fast alle Normen wurden in den Arbeitsgruppen des ISO/TC 135/SC 3 und damit unter deutscher Gremienleitung erarbeitet.

Tabelle 1. In den Jahren 2024 und 2025 veröffentlichte Normen aus dem Bereich der Ultraschallprüfung

Nummer	**Titel**	**Ausgabedatum**	**Datum der Ausgabe, die mit Veröffentlichung der Norm zurückgezogen wird**
DIN EN ISO 16946	Zerstörungsfreie Prüfung — Ultraschallprüfung — Beschreibung des Stufenkeil-Standardkörpers	2024-09	2015-07
DIN EN ISO 16810	Zerstörungsfreie Prüfung — Ultraschallprüfung — Allgemeine Grundsätze	2025-01	2014-07
DIN EN ISO 16823	Zerstörungsfreie Prüfung — Ultraschallprüfung — Durchschallungstechnik	2025-04	2014-07
DIN EN ISO 18563-2	Zerstörungsfreie Prüfung — Charakterisierung und Verifizierung der Ultraschall-Prüfausrüstung mit Phased-Array — Teil 2: Array-Prüfköpfe	2024-12	2017-12
DIN EN ISO 18563-3	Zerstörungsfreie Prüfung — Charakterisierung und Verifizierung der Ultraschall-Prüfausrüstung mit Phased-Array — Teil 3: Vollständige Prüfsysteme	2024-09	2016-06
DIN EN ISO 19675	Zerstörungsfreie Prüfung — Ultraschallprüfung — Festlegungen für einen Kalibrierkörper für die Phased-Array-Prüfung (PAUT)	2024-10	---

DIN EN ISO 16946 unterscheidet sich von der Vorgängerausgabe dahingehend, dass der Begriff "Kalibrierkörper" durch den Begriff "Standardkörper" ersetzt wurde und die Struktur des Abschnitts über die Herstellung des Stufenkeil-Standardkörpers an DIN EN ISO 2400 und DIN EN ISO 7963 angepasst wurde. Außerdem wurde der Norm ein Link zu einer CAD-Datei des Stufenkeil-Standardkörpers hinzugefügt.

DIN EN ISO 16810 unterscheidet sich von der Vorgängerausgabe hauptsächlich dadurch, dass die Norm redaktionell korrigiert wurde und die Lesbarkeit und Verständlichkeit des Dokuments verbessert wurde, z. B. durch die Harmonisierung der Begriffe.

DIN EN ISO 16823 unterscheidet sich von der Vorgängerausgabe dadurch, dass einige Bilder zum besseren Verständnis überarbeitet wurden und die Begriffe im gesamten Dokument vereinheitlicht wurden. Außerdem wurde das gesamte Dokument redaktionell überarbeitet.

Die neueste Ausgabe der DIN EN ISO 18563-2 unterscheidet sich von seiner Vorgänger-Ausgabe im Wesentlichen dadurch, dass die Begriffe und Arten der Array-Prüfköpfe nach DIN EN ISO 23423 geändert wurden. Die Norm wurde außerdem redaktionell verbessert, so wurde z. B. die Benennung „Messung“ wie jeweils anwendbar durch „Bestimmung“ oder „Bewertung“ ersetzt.

DIN EN ISO 18563-3 wiederum unterscheidet sich von der Vorgängerausgabe dadurch, dass Matrix-Array-Prüfköpfe in die Norm integriert wurden und der Anwendungsbereich der Norm um Signalverarbeitungstechniken mit Arrays (z. B. TFM) erweitert wurde. Außerdem wurden die Prüfungen von Gruppe 1 und Gruppe 2 ersatzlos gestrichen. Zusätzlich wurde zur Vereinfachung für eine funktionellere Norm ein Abschnitt zur Anwendung bildgebender Verfahren für die Verifizierung des vollständigen Prüfsystems aufgenommen. Die Informationen bezüglich der Charakterisierung von Schallbündeln wurde in Anhang A verschoben.

Bei DIN EN ISO 19675 handelt es sich um eine neue Europäische Norm, die durch die unveränderte Übernahme von ISO 19675:2017 als europäische Norm entstanden ist. Die Norm legt Anforderungen an die Maße, den Werkstoff und die Herstellung eines Stahlblocks für die Kalibrierung von Ultraschallprüfgeräten fest, die bei der Ultraschallprüfung mit der Phased-Array-Technik verwendet werden. Aktuelle Arbeiten an der Terminologie im Rahmen der ISO 5577 haben das Ziel, den Begriff der Kalibrieung wo immer anwendbar durch die Begriffe Einstellung und Justierung zu ersetzen.

3. Aktuelle Normungsprojekte

Die in Tabelle 2 aufgeführten Normen befanden sich im Jahr 2025 in Erarbeitung. Bei allen Projekten handelt es sich um Überarbeitungen bestehender Normen, so dass gleichzeitig mit ihrer Veröffentlichung die entsprechende Vorgängerausgabe zurückgezogen wird. Alle Normen werden in den Arbeitsgruppen des ISO/TC 135/SC 3 und damit unter deutscher Gremienleitung erarbeitet. Bei allen Überarbeitungen ist mit einer Veröffentlichung der Normen im Jahr 2025 zu rechnen.

Tabelle 2. Im Jahr 2025 in Überarbeitung befindliche Normen aus dem Bereich der Ultraschallprüfung

Nummer	**Titel**	**Ausgabe-datum Norm-Entwurf**	**voraussicht-liches Ausgabe-datum**	**Ausgabedatum der Ausgabe, die mit Veröffentlich-ung der Norm zurückgezogen wird**
DIN EN ISO 2400	Zerstörungsfreie Prüfung — Ultraschallprüfung — Beschreibung des Standardkörpers Nr. 1	2024-09	2025-09	2013-01
DIN EN ISO 5577	Zerstörungsfreie Prüfung — Ultraschallprüfung — Begriffe	2024-09	2025-12	2017-05
DIN EN ISO 16809	Zerstörungsfreie Prüfung — Dickenbestimmung mit Ultraschall	2024-07	2025-10	2020-02
DIN EN ISO 16811	Zerstörungsfreie Prüfung — Ultraschallprüfung — Empfindlichkeits- und Entfernungsjustierung	2024-02	2025-07	2014-06
DIN EN ISO 16826	Zerstörungsfreie Prüfung — Ultraschallprüfung — Prüfung auf Inhomogenitäten senkrecht zur Oberfläche	2024-03	2025-07	2014-06
DIN EN ISO 16827	Zerstörungsfreie Prüfung — Ultraschallprüfung — Beschreibung und Größenbestimmung von Inhomogenitäten	2024-03	2025-10	2014-06
DIN EN ISO 16828	Zerstörungsfreie Prüfung — Ultraschallprüfung — Beugungslaufzeittechnik zum Auffinden und Ausmessen von Inhomogenitäten	2024-07	2025-11	2014-06

4. Neues Thema Prüfung mit luftgekoppeltem Ultraschall

Es ist geplant, sich zukünftig dem Thema der Luftultraschallprüfung zu widmen. Auf Betreiben von Deutschland wird als erste Initiative die Norm ISO 25222-2, *Zerstörungsfreie Prüfung — Charakterisierung und Verifizierung von Prüfausrüstung für luftgekoppelten Ultraschall — Teil 2: Prüfköpfe* entwickelt. Der Normungsantrag basiert auf der DGZfP US 08, *Charakterisierung und Verifizierung der luftgekoppelten Ultraschallprüfköpfe*. Interessierte sind herzlich eingeladen, im DIN-Arbeitsausschuss NA 062-11-12 AA an der Entwicklung der Norm mitzuarbeiten. Eine zusätzliche Mitarbeit auf internationaler Ebene ist ebenfalls möglich und wünschenswert.

5. Gefährdete Normung: Ultraschallprüfung von Schweißnähten

Obwohl sich die deutsche Fachöffentlichkeit in den letzten Jahren nur in unzureichendem Ausmaß an der Normung zur Schweißnahtprüfung mit Ultraschall beteiligt hat, befand sich die zuständige internationale Arbeitsgruppe ISO/TC 44/SC 5/WG 2 "Ultraschallprüfung von Schweißverbindungen" unter deutscher Leitung. Nachdem der Arbeitsgruppenleiter im Jahr 2024 überraschend verstarb, wurde in der deutschen Community nach einer Nachfolge gesucht. Dies war leider erfolglos, so dass die Leitung der Arbeitsgruppe abgegeben werden musste. Dies hat zu einem temporären Stillstand in der Normung zur Schweißnahtprüfung geführt. Außerdem wird die Normung in diesem Bereich zukünftig von anderen Ländern an- und durchgeführt, sofern keine deutsche Leitung gestellt werden kann. Die anstehenden Überarbeitungen von acht wichtigen Normen in diesem Bereich werden nach aktuellem Stand nur mit geringerer deutscher Beteiligung durchgeführt werden. Interessenten der deutschen Fachöffentlichkeit sind daher dringend eingeladen, sich an der Überarbeitung der Normen zu beteiligen.

6. Zusammenfassung

Mit großer Initiative einzelner Fachexpert*innen konnte im Jahr 2024 die Normung im Bereich der Ultraschallprüfung auf internationaler Ebene im ISO/TC 135/SC 3 „Ultraschallprüfung" vorangebracht werden. Es konnten 6 Normen veröffentlicht werden und 7 Normen konnten so weit entwickelt werden, dass sie voraussichtlich im Jahr 2025 veröffentlicht werden können. Diese Normen wurden/werden in letzter Konsequenz auch unverändert als deutsche Normen (DIN EN ISO) veröffentlicht. Es handelte sich dabei im Wesentlichen um Überarbeitungen bestehender Normen. Da dieses Arbeitspaket vor dem Abschluss steht, wird nun mit der Entwicklung von Normen zur Prüfung mit luftgekoppeltem Ultraschall begonnen. Als erstes diesbezügliches Projekt wurde die Charakterisierung und Verifizierung der entsprechenden Prüfköpfe in das Arbeitsprogramm aufgenommen. Die Normung im Bereich der Schweißnahtprüfung mit Ultraschall findet zurzeit ohne deutsche Beteiligung statt. Daher ergeht ein Aufruf an die deutsche Fachöffentlichkeit, sich zu an der entsprechenden Normung zu beteiligen.

Normung in der zerstörungsfreien Prüfung im Umbruch

AUTOR*INNEN
Michael SCHMITT [1], Friedrich BAKE [2], Thomas SCHWENDER [3], Franziska BAENSCH [4], Daniel MUELLER [5]
[1] DIN Deutsches Institut für Normung, Berlin, Deutschland, michael.schmitt@din.de
[2] Bundesanstalt für Materialforschung und -prüfung, Berlin, Deutschland, friedrich.bake@bam.de
[3] Fraunhofer Institut für zerstörungsfreie Prüfung IZFP, Saarbrücken, Deutschland, thomas.schwender@izfp.fraunhofer.de
[4] DIN Deutsches Institut für Normung, Berlin, Deutschland, franziska.baensch@din.de
[5] DIN Deutsches Institut für Normung, Berlin, Deutschland, daniel.mueller@din.de

Kurzfassung. Normung und Standardisierung in der zerstörungsfreien Prüfung finden auf nationaler, europäischer und internationaler Ebene statt. Durch die Übernahme von Leitungsfunktionen in diesen Gremien kann ein deutlicher steuernder Einfluss auf zukünftige Themen und Schwerpunkte sowie die Inhalte der Normen genommen werden. Unabdingbar für den Erfolg ist zusätzlich das persönliche Engagement der Fachexpert*innen in den nationalen Gremien und den europäischen und internationalen Arbeitsgruppen. Denn wer nicht normt, wird genormt und muss letztlich Regeln verwenden, die andere entsprechend ihren eigenen Interessen gestalten.

Eine wichtige Voraussetzung für die Teilnahme in europäischen und internationalen Normungsgremien ist die Mitwirkung in den entsprechenden nationalen Spiegelgremien der nationalen Normungsinstitute, welche die europäischen (CEN) und internationalen (ISO) Aktivitäten spiegeln und die Fachexperten für die Mitarbeit autorisieren. Diese nationalen Normungsgremien wurden nun einer Reorganisation unterworfen, um den aktuellen Aufgaben und zukünftige Herausforderungen optimal entsprechen zu können. Dies hat zu einer Umorganisation der bestehenden Gremien aber auch zur Gründung neuer Gremien zur Abdeckung neuer Aufgabenfelder geführt. So wurden die nationalen Arbeitsausschüsse zur Normung der zerstörungsfreien Prüfung in einem neuen eigenen Fachbereich (NA 062-11 FB) innerhalb des DIN-Normenausschusses Materialprüfung (NMP) zusammengefasst.

Auf internationaler Ebene findet die Normungsarbeit in den technischen Komitees ISO/TC 135 "Non-destructive testing" und ISO/TC 44/SC 5 "Testing and inspection of welds" und ihren Arbeitsgruppen statt. Das europäische Pendant ist das CEN/TC 138 "Zerstörungsfreie Prüfverfahren" mit seinen Arbeitsgremien. Die DIN-Spiegelgremien sind für die deutsche Mitarbeit in diesen Gremien und die inhaltliche Bewertung der Norm-Entwürfe zuständig.

Keywords: Normung, Qualifizierung, Qualitätssicherung

Einführung

Normen und Standards sind ein wichtiger Teil des technischen Regelwerks. Normung und Standardisierung in der zerstörungsfreien Prüfung finden auf nationaler, europäischer und internationaler Ebene in Gremien der Normungsinstitute statt. Während die Organisation durch die Normungsinstitute geschieht, liegen die Inhalte der erarbeiteten Normen und Standards in der Verantwortung der beteiligten Fachexpert*innen. Die Gestaltungshoheit liegt daher allein bei der interessierten Fachöffentlichkeit, woraus sich eine starke Einflussmöglichkeit ergibt. Durch die Übernahme von Leitungsfunktionen in Gremien, in denen Normen erarbeitet und verantwortet werden, kann dieser Einfluss auf Themen und Inhalte der Normen noch weiter gesteigert werden. Im Umkehrschluss bedeutet dies, dass Firmen und Institute sich entweder im Standardisierungsprozess selbst engagieren oder die Gestaltungshoheit ihren Wettbewerbern überlassen müssen. Denn, wer sich an dem System nicht beteiligt, muss letztlich Normen anwenden, die andere entsprechend ihrer eigenen Interessen gestalten. Dies können nationale Wettbewerber mit zuwiderlaufenden Interessen sein, aber auch internationale Wettbewerber, deren Vorstellungen von Normungsinhalten sich von denen deutscher oder europäischer Fachexpert*innen fundamental unterscheiden können. Um den deutschen Einfluss im europäischen und internationalen Wettbewerb optimal gestalten zu können, wurde die nationale Gremienstruktur einer Reorganisation unterzogen.

1. Bedeutung

Die Platzierung deutscher Produkte auf dem Weltmarkt und der damit verbundene wirtschaftliche Erfolg gehen seit vielen Jahren Hand in Hand mit der Platzierung der diese Produkte unterstützenden deutschen Standards in den internationalen Normungsgremien. Der gute Ruf deutscher Produkte auf dem Weltmarkt fußt auf einem Qualitätsgedanken, der sich in dem hohen Standard der zerstörungsfreien Prüfung in Deutschland widerspiegelt, die sich auch aufgrund des jahrelangen erfolgreichen Engagements deutscher Fachexpert*innen in den jeweiligen Normungsgremien in dem europäischen und internationalen technischen Regelwerk niederschlägt. Die deutschen Fachexpert*innen haben sich in der andauernden Auseinandersetzung mit den Fachexpert*innen der anderen beteiligten Länder einen herausragenden Status erarbeitet, der die deutsche Normung im Vergleich in die Lage versetzt, die Themen und die Inhalte der europäischen und internationalen ZfP-Normen maßgeblich zu beeinflussen.

Normung spiegelt den aktuellen Stand der Technik wider und damit auch die Erwartung an die Leistung eines Produkts oder die Durchführung einer Tätigkeit. Dem ist im Sinne der deutschen Wirtschaft am besten gedient, wenn die in der Norm festgeschriebene Erwartung den deutschen Stand der Technik widerspiegelt. Eine zu hohe Erwartung könnte eine Leistung unnötig verteuern. Eine zu geringe Erwartung könnte einen technischen Vorsprung verspielen oder ein Qualitäts- oder Sicherheitsniveau unzulässig senken. Die Anpassung des technischen Inhalts einer Norm an die eigene Erwartungshaltung ist eine Leistung, die allen Fachexpert*innen und Marktteilnehmer*innen selbst obliegt. Aktuell betreiben manche Unternehmen einen großen Aufwand, um das Abweichen von Normungsinhalten von ihrem eigenen Leistungsstandard aufzufangen. Dieser Aufwand ist signifikant höher als der Aufwand, der für eine aktive Normungsarbeit notwendig ist.

2. Individuelle Einflussmöglichkeiten

Es gibt zwei Möglichkeiten, an der Normung zu partizipieren. Die einfachste Möglichkeit besteht darin, Norm-Entwürfe innerhalb ihrer Einspruchsfrist zu kommentieren. Alle nationalen und europäischen Normen, die bei dem deutschen Normungsinstitut DIN und/oder dem europäischen Normungsinstitut CEN erarbeitet werden, werden vor ihrer Veröffentlichung der Fachöffentlichkeit als Norm-Entwurf zur Kommentierung vorgelegt und entsprechend veröffentlicht. Diese Entwürfe kann jeder kommentieren und seine Kommentare an DIN senden. Dies ist online im Norm-Entwurfsportal möglich [1] oder per E-Mail an die Geschäftsstelle des zuständigen Normenausschusses. Diese Kommentare werden dem zuständigen DIN-Gremium vorgelegt und dort unter Beteiligung der kommentierenden Person diskutiert. Einsprüche zu Norm-Entwürfen der internationalen Normungsorganisation ISO können auf diese Art und Weise nur abgegeben werden, wenn diese Norm-Entwürfe gleichzeitig nationale oder europäische Norm-Entwürfe (DIN ISO oder DIN EN ISO) sind. Rein internationale Norm-Entwürfe können nur vom national zuständigen Gremium bei DIN bewertet und kommentiert werden.

Die zweite Möglichkeit, die zusätzlich einen Einfluss auf rein internationale Normen ermöglicht, besteht in der aktiven Mitarbeit in dem zuständigen DIN-Gremium. Dadurch stehen die vollständigen Möglichkeiten offen, wie z.B.:

- Möglichkeit, Themen zu platzieren
- Möglichkeit, Normungsinhalte zu beeinflussen
- Kostenreduzierung durch Durchsetzung der eigenen Technologie
- Möglichkeit nachteilige Festlegungen zu verhindern
- Wissen über neue Technologien zu erlangen
- Wissen über neue Anforderungen zu erlangen
- Netzwerkpflege

Durch die aktive Mitarbeit in dem nationalen Gremium bei DIN steht auch die Möglichkeit der aktiven Teilnahme in europäischen (CEN) und internationalen (ISO) Normungsgremien offen. Eine Teilnahme in diesen Gremien erhöht den Grad der Einflussnahme nochmal deutlich, insbesondere bei der Übernahme einer Leitungsfunktion. Diesen Vorteilen steht der Nachteil gegenüber, dass eine aktive Beteiligung an der Normung personelle Ressourcen bindet. Eine entsprechende Kosten/Nutzen-Abwägung, die jedem Marktteilnehmer selbst obliegt, sollte im vollen Bewusstsein der Konsequenzen durchgeführt werden.

3. Stand der Normung bei CEN und ISO

Normung und Standardisierung in der zerstörungsfreien Prüfung ist auf europäischer und internationaler Ebene organisatorisch untergliedert nach den einzelnen Prüfverfahren.

Zuständig ist auf internationaler Ebene das von Japan geleitete ISO/TC 135 "Non-destructive testing". Unter der Verantwortung des ISO/TC 135 wurden bisher 101 Normen und Standards veröffentlicht, die aktuell gültig sind. 37 Normen und Standards befinden sich derzeit in der Entwicklung. Dies sind neue Normen und Standards, aber auch Überarbeitungen von Normen und Standards, die nicht mehr den aktuellen Stand der Technik widerspiegeln. Organisatorisch gliedert sich das ISO/TC 135 in 8 methodenspezifische Unterkomitees (Tabelle 1).

Tabelle 1. Gremien des ISO/TC 135 „Zerstörungsfreie Prüfung" und Anzahl der zugehörigen Normen und Projekte

Bezeichnung	Titel	Anzahl veröffentlichte Normen	Anzahl aktuelle Projekte
ISO/TC 135	Zerstörungsfreie Prüfung	1	1
ISO/TC 135/SC 2	Oberflächenverfahren	14	5
ISO/TC 135/SC 3	Ultraschallprüfung	22	10
ISO/TC 135/SC 4	Wirbelstromprüfung	7	1
ISO/TC 135/SC 5	Durchstrahlungsprüfung	28	15
ISO/TC 135/SC 6	Dichtheitsprüfung	4	1
ISO/TC 135/SC 7	Personalqualifizierung	7	2
ISO/TC 135/SC 8	Thermografieprüfung	5	1
ISO/TC 135/SC 9	Akustische Emissionsprüfung	13	1

Die Unterkomitees ISO/TC 135/SC 3 "Ultraschallprüfung" und ISO/TC 135/SC 5 "Durchstrahlungsprüfung" werden von Deutschland geleitet.

Zusätzlich gibt es bei ISO noch 2 Arbeitsgruppen, die sich im ISO/TC 44/SC 5 "Testing and inspection of welds" mit der zerstörungsfreien Prüfung von Schweißnähten beschäftigen (Tabelle 2). ISO/TC 44/SC 5/WG 2 „Ultraschallprüfung" wurde in der Vergangenheit von Deutschland geleitet. Mangels Fachexpertise im nationalen Gremium wurde die Leitung niedergelegt. ISO/TC 44/SC 5/WG 1 „Durchstrahlungsprüfung" ist zurzeit inaktiv.

Tabelle 2. ZfP-Gremien des ISO/TC 44/SC 5 „Prüfung von Schweißungen" und Anzahl der zugehörigen Normen und Projekte

Bezeichnung	Titel	Anzahl veröffentlichte Normen	Anzahl aktuelle Projekte
ISO/TC 44/SC 5/WG 1	Durchstrahlungsprüfung	4	0
ISO/TC 44/SC 5/WG 2	Ultraschallprüfung	12	9

Neben dem ISO/TC 44/SC 5 hat auch das International Institute of Welding IIW das Recht, ISO-Normen in dem Bereich der Schweißnahtprüfung zu erstellen. Dieses Recht nimmt das IIW auch wahr, wobei darauf geachtet werden muss, dass keine Doppelnormung entsteht.

Auf europäischer Ebene findet die Normung der zerstörungsfreien Prüfung von Schweißnähten in der CEN/TC 121/WG 21 "Prüfung von Schweißungen" statt. Die Arbeitsgruppe beschränkt sich aber auf die Übernahme von Normen des ISO/TC 44/SC 5, ist also nicht aktiv.

Der Gegenpart zum ISO/TC 135 auf europäischer Ebene ist das CEN/TC 138 "Zerstörungsfreie Prüfverfahren". Das von Frankreich geleitete Technische Komitee (TC) untergliedert sich in 9 methodenspezifische Arbeitsgruppen (WG, siehe Tabelle 3). Die Arbeitsgruppen zu Durchstrahlungsprüfung, Ultraschallprüfung, Eindringprüfung, Dichtheitsprüfung, sowie Infrarot- und thermografische Prüfung stehen unter deutscher Leitung.

Tabelle 3. Aktive Gremien des CEN/TC 138 „Zerstörungsfreie Prüfung" und Anzahl der zugehörigen Normen und Projekte

Bezeichnung	Titel	Anzahl veröffentlichte Normen	Anzahl aktuelle Projekte
CEN/TC 138/WG 1	Durchstrahlungsprüfung	31	9
CEN/TC 138/WG 2	Ultraschallprüfung	19	9
CEN/TC 138/WG 3	Wirbelstromprüfung	6	1
CEN/TC 138/WG 4	Eindringprüfung	10	1
CEN/TC 138/WG 5	Magnetpulverprüfung	4	0
CEN/TC 138/WG 6	Dichtheitsprüfung	8	2
CEN/TC 138/WG 7	Akustische Emissionsprüfung	11	1
CEN/TC 138/WG 10	Röntgenstrahlenbeugung	2	1
CEN/TC 138/WG 11	Infrarot- und thermografische Prüfung	5	0

4. Stand der Normung bei DIN

4.1 Neuer Fachbereich „Zerstörungsfreie Prüfung"

Die Basis für eine erfolgreiche Wahrnehmung deutscher Interessen auf europäischer und internationaler Ebene bilden die nationalen deutschen Gremien. Die grundlegenden Arbeitsgremien zur zerstörungsfreien Prüfung sind im DIN-Normenausschuss Materialprüfung (NMP) angesiedelt. Die Gremien wurden 2024 einer umfassenden Reorganisation unterzogen. Es wurde ein neuer Fachbereich „Zerstörungsfreie Prüfung" gegründet, der von dem ebenfalls neu gegründeten Fachbereichsbeirat gesteuert wird. Der Fachbereich ist untergliedert in die beiden Sachgebiete „Querschnittsthemen" und „Prüfverfahren", denen jeweils die entsprechenden Arbeitsausschüsse (AA) zugeordnet sind (Abb. 1).

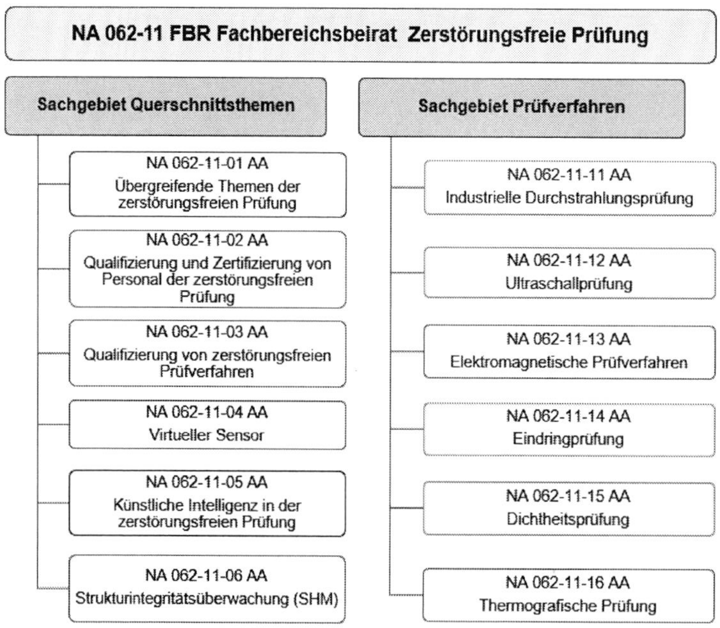

Abb. 1. Der neu gegründete Fachbereich "Zerstörungsfreie Prüfung" bei DIN

Tabelle 4. ZfP-Gremien des DIN-Fachbereichs „Zerstörungsfreie Prüfung" und Anzahl der zugehörigen nationalen Normen und Projekte

Bezeichnung	**Titel**	**Anzahl veröffentlichte Normen**	**Anzahl aktuelle Projekte**
NA 062-11 FBR	Fachbereichsbeirat Zerstörungsfreie Prüfung	-	-
NA 062-11-01 AA	Übergreifende Themen der zerstörungsfreien Prüfung	0	0
NA 062-11-02 AA	Qualifizierung und Zertifizierung von Personal der zerstörungsfreien Prüfung	7	2
NA 062-11-03 AA	Qualifizierung von zerstörungsfreien Prüfverfahren	1	1
NA 062-11-04 AA	Virtueller Sensor	0	2
NA 062-11-05 AA	Künstliche Intelligenz in der zerstörungsfreien Prüfung	0	0
NA 062-11-11 AA	Industrielle Durchstrahlungsprüfung	52	15
NA 062-11-12 AA	Ultraschallprüfung	55	21
NA 062-11-13 AA	Elektromagnetische Prüfverfahren	20	2
NA 062-11-14 AA	Eindringprüfung	11	2
NA 062-11-15 AA	Dichtheitsprüfung	10	2
NA 062-08-27 AA	Thermografische Prüfung	13	0

Die nationalen Gremien sind für alle nationalen Normen und für die nationale Meinungsbildung hinsichtlich der europäischen und internationalen Normen zuständig und geben bei Abstimmungen ein konsolidiertes Meinungsbild an die europäischen und internationalen Gremien ab. Bei Uneinigkeit im Gremium oder fehlender Expertise enthält sich Deutschland bei Abstimmungen. Die nationalen Gremien benennen außerdem alle nationalen Expert*innen zur Mitarbeit in den europäischen und internationalen Gremien.

Die Abhängigkeit der nationalen, europäischen und internationalen Gremien voneinander ist in Abbildung 2 dargestellt.

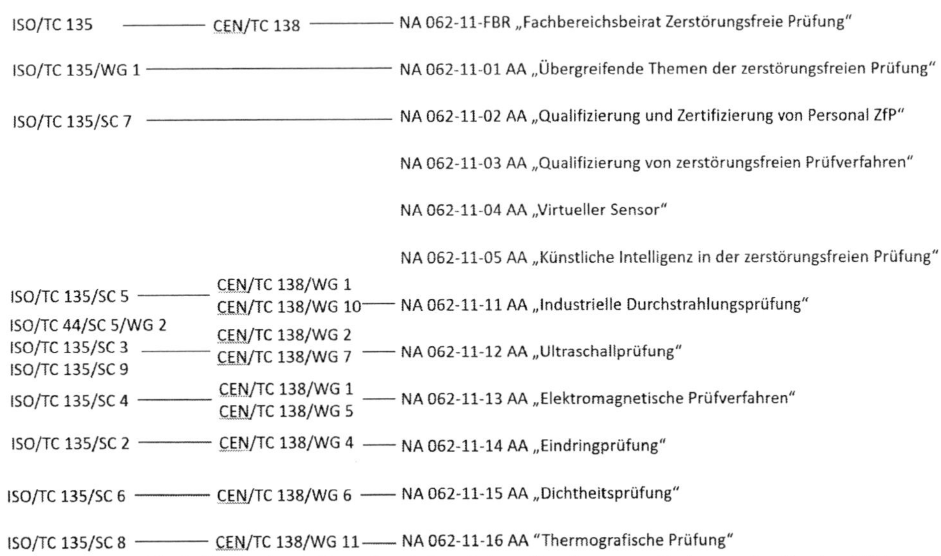

Abb. 2. Gegenüberstellung der aktiven internationalen, europäischen und nationalen ZfP-Normungsgremien

4.2 Neuer Arbeitsausschuss „Übergreifende Themen der zerstörungsfreien Prüfung"

Im November 2024 wurde der neue Arbeitsausschuss NA 062-11-01 AA "Übergreifende Themen der zerstörungsfreien Prüfung" gegründet. Er beschäftigt sich mit Normungsprojekten, die für mehrere Prüfverfahren gelten werden. Dies sind zum einen traditionelle Themen wie eine verfahrensübergreifende Terminologie. Er wird sich nach eigenem Wunsch aber auch mit Anforderungen für die Auswahl, Charakterisierung und den Betrieb von Einrichtungen (z. B. Geräte und Software) und deren Anwendungen beschäftigen, sofern es sich um Geräte und Software handelt, die für mehrere Prüfverfahren verwendet werden kann. Außerdem werden Normen für Datenformate, Schnittstellen und digitale Zertifikate bei der Zerstörungsfreien Prüfung entwickelt. Das genaue Arbeitsprogramm wird noch auf den ersten Sitzungen festgelegt. Interessierte sind herzlich eingeladen, in dem Gremium mitzuarbeiten.

4.3 Neuer Arbeitsausschuss „Künstliche Intelligenz in der zerstörungsfreien Prüfung“

Im Oktober 2024 wurde der neue Arbeitsausschuss NA 062-11-05 AA “Künstliche Intelligenz in der zerstörungsfreien Prüfung” gegründet. Er beschäftigt sich mit Normungsprojekten, die Anforderungen für den Einsatz von künstlicher Intelligenz für die zerstörungsfreie Prüfung regeln. Darunter fallen Themen wie Maschinelles Lernen, Neuronale Netze, Deep Learning und verwandte Verfahren. Auch in diesem Arbeitsausschuss wird das genaue Arbeitsprogramm noch in den ersten Sitzungen festgelegt. Interessierte sind herzlich eingeladen, in dem Gremium mitzuarbeiten.

4.4 Geplanter Arbeitsausschuss „Strukturintegritätsüberwachung (SHM)“

Die Überwachung von Strukturen (Structural Health Monitoring) wird als wichtiges Anwendungsfeld der zerstörungsfreien Prüfung angesehen, das sich von der Zustandsüberwachung vom Maschinen unterscheidet. Daher wird diesem Themengebiet im Fachbereich “Zerstörungsfreie Prüfung” zukünftig ein Arbeitsausschuss zur Verfügung gestellt, in dem eine entsprechende Normung im passenden Umfeld erfolgen kann. Der Termin der Gründungssitzung ist der 6. Mai 2025. Interessierte sind weiterhin herzlich eingeladen, in dem Gremium mitzuarbeiten.

4.4 Weiterhin kein Normungsgremium für die visuelle Prüfung

Neben den oben genannten positiven Entwicklungen hat die Normung in der zerstörungsfreien Prüfung auch mit Problemen zu kämpfen, die sich insbesondere aus dem Desinteresse der Fachöffentlichkeit an der Mitarbeit an bestimmten Normungsthemen ergeben. Darunter hat aktuell insbesondere ein Themenbereich zu leiden: die visuelle Prüfung.

Derzeit wirken innerhalb des DIN-Fachbereichs „Zerstörungsfreie Prüfung“ keine Personen mit, die über eine ausreichende Fachkenntnis auf dem Gebiet der visuellen Prüfung verfügen, um die Entwicklung von Normen in diesem Themenkreis zu rechtfertigen. Daher wurden die nationalen Normungsaktivitäten zur visuellen Prüfung im Jahr 2024 eingestellt. Die europäischen und internationalen Normungsaktivitäten finden seither ohne deutsche Beteiligung statt. Unabhängig vom Ausmaß der deutschen Beteiligung müssen Europäische Normen unverändert in das deutsche Normenwerk übernommen werden. In Deutschland gültige Normen zur visuellen Prüfung, an die sich die deutsche Fachöffentlichkeit halten muss, werden somit zurzeit ohne deutsche Beteiligung erstellt. Eine Wiederaufnahme der Arbeiten in Deutschland wäre wünschenswert. Dazu müsste ein neuer Arbeitsausschuss gegründet werden. Interessenten an der Normung und Standardisierung von Verfahren zur visuellen Prüfung sind eingeladen, sich bei den Autor*innen zu melden.

5. Zusammenfassung

Die Normung und Standardisierung im Bereich der zerstörungsfreien Prüfung findet auf nationaler, europäischer und internationaler Ebene statt, wobei die drei Ebenen eng verzahnt sind. Zurzeit kann Deutschland aufgrund des großen Engagements einiger weniger Fachexpert*innen einen großen Einfluss auf den Inhalt der europäischen und internationalen Normen ausüben, was letztendlich der deutschen Allgemeinheit und damit der deutschen Industrie zugutekommt. Es ist absehbar, dass dieser Einfluss in naher Zukunft abnehmen wird, wenn sich die deutsche Industrie weiterhin nur auf dem aktuellen Niveau in die Normung der zerstörungsfreien Prüfung einbringt. Um dem schwindenden Einfluss organisatorisch entgegenzuwirken, wurde bei DIN der neue Fachbereich „Zerstörungsfreie Prüfung" gegründet, in dem alle Gremien organisiert sind, die sich im Schwerpunkt mit der zerstörungsfreien Prüfung auseinandersetzen. Die Steuerung erfolgt durch den ebenfalls neu gegründeten Fachbereichsbeirat. In dieser Funktion hat der Fachbereichsbeirat die Gründung eines Arbeitsausschusses zur künstlichen Intelligenz und eines Arbeitsausschusses für übergeordnetete Themen, wie z. B. Datentransfer und Datenkopplung, initiiert. Zum Thema Structural Health Monitoring befindet sich zurzeit ein weiterer Arbeitsausschuss in Gründung.

Referenzen

[1] www.entwuerfe.din.de

Einblick in gängige Normen der ZfP – offene Fragen in Durchführungs- und Bewertungsnormen der ZfP – Teil 2

Bernd HEUTLING[1], Günter MOß[2]
[1] GSI mbH, NL SLV Hannover, Hannover, Deutschland, heutling@slv-hannover.de
[2] GSI mbH, NL SLV Hannover, Hannover, Deutschland, moss@slv-hannover.de

Kurzfassung. Spätestens in den Schulungen für ZfP-Prüfer der Stufen 2 und 3 sind nationale und internationale Normen und Standards für die Durchführung von ZfP-Prüfungen und die Bewertung festgestellter Unregelmäßigkeiten bzw. Anzeigen wichtiger Bestandteil der Kurse.
In der Anwendung dieser (und sicherlich auch anderer) Regelwerke tauchen allerdings immer wieder nicht-eindeutige Festlegungen und andere Unsicherheiten auf, so dass die hinsichtlich Durchführung und Bewertung genannten Angaben oftmals lückenhaft oder gar widersprüchlich sind.
Diese Tatsachen wurden bereits 2018 [2] und 2024 [1] dargestellt und sollen hier nun wieder mit Bezug auf Details in aktuellen Normen in die heutige Situation von Prüfen und Bewerten eingeordnet werden.
Darüber hinaus soll auch herausgearbeitet werden, dass es durch objektive und subjektive Aspekte in der Interpretation und Anwendung von Regelwerken durchaus völlig unterschiedliche Ergebnisse geben kann. Dazu sollen konkrete Beispiele aus z.B. Sichtprüfung, Eindringprüfung, Magnetpulverprüfung u.a. dargestellt und zur Diskussion gestellt werden.

Keywords: ZfP, Normen, lückenhaft, Beispiele

Einführung

Normen sind nach allgemeinem Verständnis freiwillige Regeln, die (im Gegensatz zu Standards) durch anerkannte Organisationen oder Gremien von Personen mit Expertenkenntnissen in einem Konsensverfahren festgelegt werden. Sie sind im Idealfall Sammlungen des Know-How von Menschen, die auf ihrem Fachgebiet umfangreiches Expertenwissen haben und die Bedürfnisse der Unternehmen und Kunden kennen, die sie repräsentieren. Normen bestehen also aus Fachwissen und bilden den Stand von anerkannter Wissenschaft und Technik ab. Sie sind im besten Fall leistungsstarke Werkzeuge, die helfen, Betriebsrisiken zu reduzieren, Produktivität zu steigern und Nachhaltigkeit zu gewährleisten. [1] [2]

Daher ist die Denkweise anerkannt, dass bei Einhaltung der Vorgaben und Rahmenbedingungen der verwendeten Norm(en) das Produkt qualitativ gut, belastbar, sicher und nachhaltig sein wird. [1] [2]

Um die Normvorgaben befolgen zu können, erwartet man von Normen lückenlose Definitionen, vollständige Angaben und schlüssige Vorgehensweisen. Allerdings werden Normen von Menschen gemacht und allen Anstrengungen zum Trotz, die Qualität von Normen zu maximieren, schleichen sich immer mal Ungänzen in Normen ein.

Dieser Beitrag betrachtet eine Auswahl typischer und oft genutzter Normen aus der zerstörungsfreien Prüfung in Hinsicht auf unklare Sachverhalte, offene Fragen und hinterfragbare Elemente. [1] [2]

1. Anmerkungen zu DIN EN 10228-1:2016

1.1 Die fünffache Länge

In DIN EN 10228-1:2016 [3] zur Durchführung und Bewertung von Magnetpulver-/MT-Prüfungen an Schmiedestücken aus Stahl werden Klasseneinteilungen vorgegeben und festgelegt, wann Anzeigen Einzelanzeigen und wann sie zusammenhängende Anzeigen sind (im Folgenden verkürzt zitiert aus Abschnitt "13 Klasseneinteilung der Anzeigen"):

a) *Eine Anzeige ist als Einzelanzeige zu betrachten, wenn sie wie diese [andere Anzeigen] ausgerichtet, aber von ihnen* ***mehr als das 5fache*** *der Länge der längeren der beiden betrachteten Anzeigen entfernt ist.*

b) *Zusammenhängende Anzeigen sind 2 (oder mehrere) Anzeigen [..], wenn ihr Abstand* ***weniger als das 5fache*** *der Länge der längeren der beiden betrachteten Anzeigen beträgt. [..]*

Sollte der sicherlich eher theoretische Fall auftreten, dass die Anzeigen genau das 5-fache der Länge [..] entfernt liegen, kann man hiernach keine Entscheidung treffen.

Im Pendant dieser Norm für die Durchführung und Bewertung von Eindring-/PT-Prüfungen an Schmiedestücken aus Stahl DIN EN 10228-2:2016 [4] ist diese Unterscheidung sauber definiert mit *"weniger oder gleich als das 5fache [..]"*.

1.2 Beispiel in der Norm für die „Zusammenfassungsregel"

Darüber hinaus wird in beiden Normen (Teil 1 [3] und Teil 2 [4]) im Abschnitt 13 zu der „*Klasseneinteilung der Anzeigen*" das folgende Beispiel gezeigt (hier zum leichteren Nachvollziehen bereits mit den in der Legende genannten Maßen [rot] versehen).

Auszug aus der Legende zur Abbildung 1 nach [3, 4]:

1) einzelne Anzeigen *2) zusammenhängende Anzeigen* *3) Bezugsfläche*

Bezugsfläche: 148 mm x 105 mm (d. h. A6-Format)

d_1 = 15,2 mm, d_2 = 11,6 mm, d_3 = 63,5 mm, L_1 = 5,8 mm, L_2 = 3,6 mm, L_3 = 10,1 mm,

L_4 = 7,9 mm, L_5 = 5,1mm, L_6 = 10,1 mm || $d_1 < 5L_1$; $d_2 < 5L_3$; $d_3 > 5L_3$

L_1, L_2 und L_3 = *Einzellängen von zusammenhängenden Anzeigen*

In beiden Normteilen wird das in der Abbildung 1 genannte Beispiel zur Zusammenfassungsregel aufgeführt. In den Legenden werden die gleichen Maße genannt und es wird in beiden Fällen geschlossen, dass die Längen L_1, L_2 und L_3 Einzellängen einer zusammenhängenden Anzeige sind gemäß der in 13 b genannten Regel.

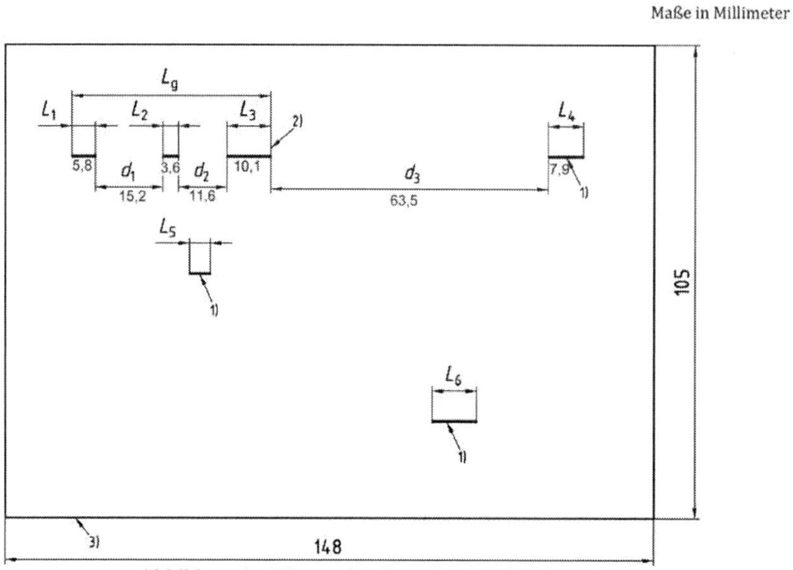

Abbildung 1: „Klasseneinteilung der Anzeigen" nach [3, 4] mit ergänzten Maßangaben gemäß Legende

1.3 Fehlen eines Bezugs auf die Registriergrenzen und Konsequenzen derer Anwendung

Hier ist zu beachten, dass diese Beispiele [3, 4] keine Qualitätsklassen nennen; Qualitätsklassen werden erst zu einem späteren Zeitpunkt (Abschnitt 14 in beiden Normteilen) (Abbildung 2 und Abbildung 3) eingeführt. Daher liegt der Schluss nahe, dass die Qualitätsklassen und die (erst) dort eingeführten Registriergrenzen (jeweils Abschnitt 14 [3, 4]) für die Betrachtung *„zusammenhängend oder nicht"* (jeweils Abschnitt 13 [3, 4]) keine Rolle spielen.

Tabelle 2 – Qualitätsklassen, Registriergrenzen und Zulässigkeitskriterien[c]

Parameter	Qualitätsklasse			
	1	2	3[a]	4[b]
Registriergrenze: Anzeigenlänge, (mm)	≥ 5	≥ 2	≥ 2	≥ 1
Zulässige Länge von Einzelanzeigen L und zulässige Länge zusammenhängender Anzeigen L_g, (mm)	≤ 20	≤ 8	≤ 4	≤ 2
Zulässige kumulative Länge von Anzeigen auf der Bezugsfläche, (mm)	≤ 75	≤ 36	≤ 24	≤ 5
Größte zulässige Anzahl von Anzeigen auf der Bezugsfläche	15	10	7	5

a Qualitätsklasse nicht anwendbar für die Prüfung von Flächen mit einer Bearbeitungszugabe größer als 3 mm je Fläche.

b Qualitätsklasse nicht anwendbar für die Prüfung von Flächen mit einer Bearbeitungszugabe größer als 1 mm je Fläche.

c Bezugsfläche: 148 mm x 105 mm (d. h. A6-Format). Für Teile kleiner als die Bezugsfläche, oder falls die zu prüfende Fläche kleiner als die Bezugsfläche ist, dann sollte die größte zulässige Anzahl von Anzeigen auf der Bezugsfläche ermittelt und vereinbart werden (die Zulässigkeitskriterien für die Länge bleiben allerdings bestehen).

Abbildung 2: *„Tabelle 2 - Qualitätsklassen, Registriergrenzen und Zulässigkeitskriterien"* aus *„14 Registrier- und Zulässigkeitskriterien"* [3]

Dennoch gibt es eine nicht endende Diskussion, ob nicht doch zuerst die Einzelanzeigen zu registrieren sind und nur die registrierpflichtigen Einzelanzeigen hinsichtlich der Zusammenfassungsregel ([3] 13 b) zu betrachten sind.

Gedankenspiel zu [3] mit vorheriger Festlegung(!) der Qualitätsklasse 1:

1. vorherige Festlegung: Qualitätsklasse 1 [3]
2. ⇒ L_2 (3,6 mm) < Registriergrenze (5 mm) ⇒ darf nicht berücksichtigt werden
3. ⇒ Aussage der Legende „*L_1, L_2 und L_3 - Einzellängen zusammenhängender Anzeigen*" ist dann **falsch** (weil *L_2* inkludiert ist)
   (Tatsächlich bleibt die Gesamtlänge der zusammenhängenden Anzeige gleich; die L_2 inkludierende Aussage 3 ist allerdings falsch.)

1.4 Ausblick auf die baugleiche Norm für die Eindringprüfung [4]

DIN EN 10228-2 [4] (für das ZfP-Verfahren PT) enthält nur hinsichtlich der Registriergrenzen abweichende Angaben (Abbildung 3).

Tabelle 2 – Qualitätsklassen, Registriergrenzen und Zulässigkeitskriterien

Parameter	Qualitätsklasse			
	1	2	3	4[a]
Registriergrenze (mm)[b]	≥ 7	≥ 3	≥ 3	≥ 1
Zulässige Länge L linearer Einzelanzeigen bzw. größte zulässige Länge L_g von zusammenhängenden Anzeigen (mm)[b]	≤ 20	≤ 8	≤ 4	≤ 2
Zulässige kumulative Länge linearer Anzeigen auf der Bezugsfläche (mm)[b]	≤ 75	≤ 36	≤ 24	≤ 5
Zulässige Größe einzelner runder Anzeigen (mm)[b]	≤ 30	≤ 12	≤ 8	≤ 3
Größte zulässige Anzahl nachweisbarer Anzeigen auf der Bezugsfläche[c]	15	10	7	5

[a] Qualitätsklasse 4 ist nicht anwendbar für die Prüfung von Flächen mit einer Bearbeitungszugabe ≥ 0,5 mm je Fläche.

[b] Die tabellierten Werte gelten für die Anzeigengröße, nicht aber für die wahre Fehlergröße an der Oberfläche.

[c] Bezugsfläche: 148 mm x 105 mm (d.h. A6-Format). Für Teile kleiner als die Bezugsfläche, oder falls die zu prüfende Fläche kleiner als die Bezugsfläche ist, dann sollte die größte zulässige Anzahl von Anzeigen auf der Bezugsfläche ermittelt und vereinbart werden (die Zulässigkeitskriterien für die Länge bleiben allerdings bestehen).

Abbildung 3: „*Tabelle 2 - Qualitätsklassen, Registriergrenzen und Zulässigkeitskriterien*" aus „*14 Registrier- und Zulässigkeitskriterien*" [4]

Gedankenspiel zu [4] mit vorheriger Festlegung(!) der Qualitätsklasse 1:

1. vorherige Festlegung(!): Qualitätsklasse 1 [4] ⇒
2. L_1 (5,8 mm) < Registriergrenze (7 mm) ⇒ darf nicht berücksichtigt werden
3. L_2 (3,6 mm) < Registriergrenze (7 mm) ⇒ darf nicht berücksichtigt werden
4. L_3 (10,1 mm) > Registriergrenze (7 mm) ⇒ muss berücksichtigt werden
5. Aussage in der Legende „*L_1, L_2 und L_3 - Einzellängen zusammenhängender Anzeigen*" ist komplett **falsch**, denn es gibt nichts mehr zusammenzufassen, da nur eine Einzelanzeige (L_3) übrigbleibt

1.5 Erstes Beispiel zur unterschiedlichen Wirkung der Registrierung vor oder nach der Anwendung der Zusammenfassungsregel

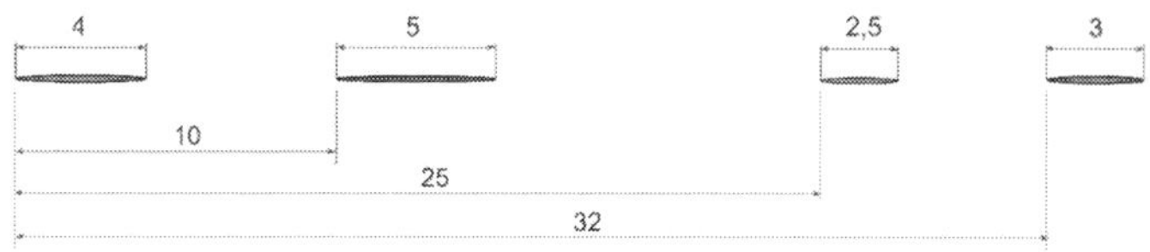

Abbildung 4: Beispiel 1

Für das Beispiel 1 aus Abbildung 4, bei dem die Abstände erkennbar unter dem Verhältnis $d:l < 5:1$ liegen, die Zusammenfassung also obligatorisch ist, ergeben sich je nach Anwendung der Regel „*Registrierung vor bzw. nach Zusammenfassung*" folgende Ergebnisse:

	DIN EN 10228-1 [3], Qualitätsklasse 1		DIN EN 10228-2 [4], Qualitätsklasse 1	
	Registrierung **nach** Zusammenfassung	Registrierung **vor** Zusammenfassung	Registrierung **nach** Zusammenfassung	Registrierung **vor** Zusammenfassung
Reg.-grenze	5		7	
l_{max}	$l_{zusam} = 35$	$l_{Einzel} = 5$	$l_{zusam} = 35$	keine Anzeige!
Bewertung	unzulässig	zulässig	unzulässig	zulässig

1.6 Zweites Beispiel zur unterschiedlichen Wirkung der Registrierung vor oder nach der Anwendung der Zusammenfassungsregel

Im Beispiel 2 einer geschmiedeten Welle (Abbildung 5) enthält diese Welle drei sichtbare Ungänzen (L_I=6 mm, L_{II}=3 mm, L_{IV}=7 mm) sowie eine verdeckte Ungänze (L_{III}=6 mm). Eine Betrachtung analog zu der für Abbildung 4 führt zu dem im Folgenden aufgeführten Ergebnis nach [3]:

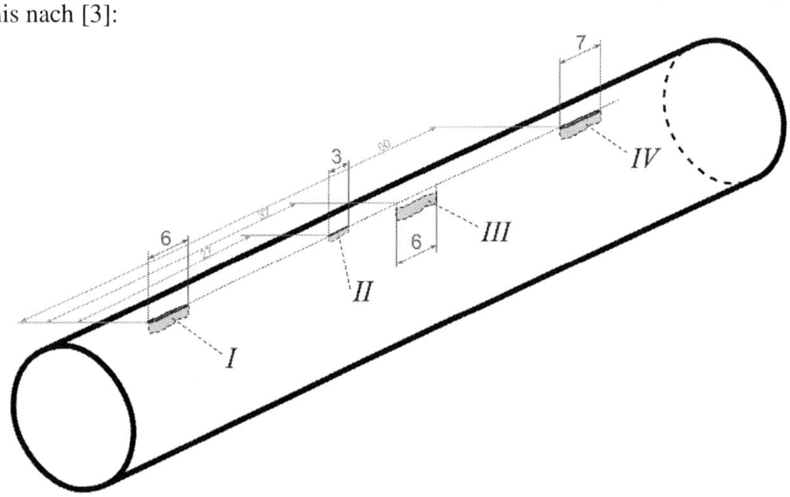

Abbildung 5: Beispiel 2

	DIN EN 10228-1 [3], Qualitätsklasse 1		DIN EN 10228-2 [4], Qualitätsklasse 1	
	Registrierung **nach** Zusammenfassung	Registrierung **vor** Zusammenfassung	Registrierung **nach** Zusammenfassung	Registrierung **vor** Zusammenfassung
Reg.-grenze	5		7	
l_{max} [mm]	$l_{zus} = 80$	$l_{Einzel,IV} = 7$ [1)] $l_{Einzel,II} = 6$	$l_{zus} = 80$	$l_{Einzel,IV} = 7$ [1) 2)]
Bewertung	unzulässig	zulässig	unzulässig	zulässig

[1)] *L_{II}=3 mm ist kleiner als die Registriergrenze [3], also wie „nicht da“*
[2)] *L_{I}=6 mm ist kleiner als die Registriergrenze [4], also wie „nicht da“*

Besonders beachtenswert ist, dass die Schutzfunktion, die die Zusammenfassung von Anzeigen hinsichtlich der Inkludierung von verdeckten Anzeigen bietet, durch die vorzeitige Registrierung und damit Nicht-Beachtung von Anzeigen ausgehebelt wird. In beiden Fällen der „*Registrierung vor der Zusammenfassung*“ wird der Bereich der verdeckten Anzeige *III* durch den Wegfall der Zusammenfassung nicht als unzulässig deklariert!

1.7 Schlussfolgerungen

Die o.g. Beispiele sind für die Autoren hinreichender Anlass, die Zusammenfassungsregel unabhängig von der später anzuwendenden Qualitätsklasse / Registriergrenze anzuwenden („*Registrierung* ***nach*** *Zusammenfassung*“); daraus lässt sich die folgende Grundregel formulieren:

1.) *Zusammenfassungsregel anwenden unabhängig von Qualitätsklassen / Registriergrenzen,*

2.) *dann erst die resultierenden Einzel- und zusammenhängenden Anzeigen gemäß der gewählten Qualitätsklasse / Registriergrenze der Bewertung zuführen.*

Da dies nicht schriftlich in der Norm verankert ist, ist dies nur die Regel, nach der die Autoren eine sicherheitsbewusste Bewertung durchführen.

2. Gedanken zur DIN EN ISO 23278:2015

2.1 Einführung und erste Wirrungen

Aufbauend auf den Grundgedanken des vorigen Abschnitts werden diese Strukturen beispielhaft in der DIN EN ISO 23278 [5] betrachtet:

> „{I} *Zulässigkeitsgrenzen für linienartige Anzeigen sind diese, die der Beobachtungsschwelle entsprechen.* {II} *Die Anzeigen mit geringen Abmessungen dürfen nicht berücksichtigt werden.* {III} *Annehmbare Anzeigen müssen nicht aufgezeichnet werden.*
>
> {IV} *Benachbarte Anzeigen sind als eine einzige, kontinuierliche Anzeige anzusehen, wenn ihr Abstand kleiner als die Hauptabmessung der kleineren Anzeige ist.*“ [5]

Als erstes Problem entpuppt sich in [5]{II} die Formulierung „*mit geringen Abmessungen*“; was ist „gering“? Tatsächlich ist dies offenbar eine ungenaue Übersetzung, denn in der englischen Variante heißt es „*indications lower than this*“, so dass es besser heißen sollte „*mit Abmessungen geringer als die Zulässigkeitsgrenzen*“.

Der erste Absatz lässt sich dann so zusammenfassen: „*Anzeigen mit Abmessungen kleiner als die Zulässigkeitsgrenzen dürfen nicht beachtet oder protokolliert werden.*"

2.2 Zulässigkeitsgrenzen und die Anwendung der Zusammenfassungsregel – Beispiel Zulässigkeitsgrenze 2

Tabelle 1 — Zulässigkeitsgrenzen für Anzeigen von Unregelmäßigkeiten

Maße in Millimeter

Anzeigentyp	**Zulässigkeitsgrenzen**[a] 1	2	3
Linienartige Anzeige l = Länge der Anzeige	$l \leq 1{,}5$	$l \leq 3$	$l \leq 6$
Nichtlinienartige Anzeige d = größter Achsendurchmesser	$d \leq 2$	$d \leq 3$	$d \leq 4$

[a] Zulässigkeitsgrenzen 2 und 3 dürfen mit einer Endung „X" versehen werden, wenn die nachgewiesenen linienartigen Anzeigen nach der Zulässigkeitsgrenze 1 beurteilt werden müssen. Im Vergleich zu den ursprünglichen Zulässigkeitsgrenzen kann die Nachweiswahrscheinlichkeit jedoch niedrig sein.

Abbildung 6: Zulässigkeitsgrenzen gemäß DIN EN ISO 23278 [5]

Es wird festgelegt, dass Zulässigkeitsgrenze 2 gelten soll. Betrachtet wird Beispiel 3 aus Abbildung 7:

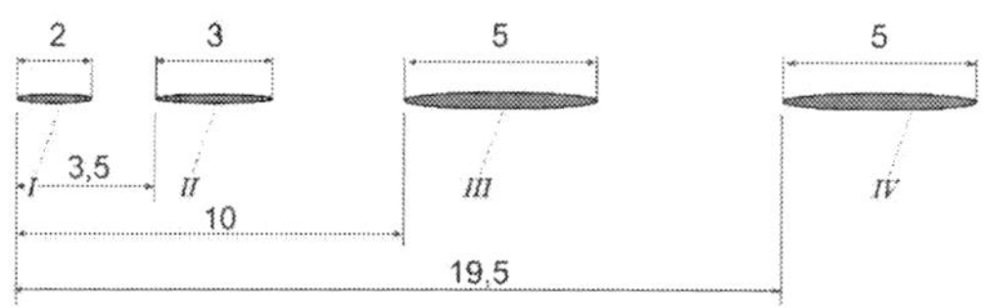

Abbildung 7: Beispiel 3

In [5]{I} bis [5]{III} wird festgelegt, dass Anzeigen kleiner als die Zulässigkeitsgrenze (hier $ZG_{lin.} \leq 3$ mm) nicht zu beachten sind.

Wird die Registrierung zuerst durchgeführt, ergibt sich folgendes Bild: Anzeige I ist „nicht da". Der Abstand zwischen II und III ($d_{II\text{-}III} = 7$ mm) ist größer als die Länge der kleineren Anzeige II ($l_{II} = 3$ mm); die beiden Anzeigen sind also nicht zusammenzufassen. Der Abstand zwischen III und IV ($d_{III\text{-}IV} = 4{,}5$ mm) ist größer als die Länge von III bzw. IV ($l_{III} = 5$ mm); die beiden Anzeigen III und IV sind also zusammenzufassen zu einer Anzeige mit der Gesamtlänge $l_{ges} = 14{,}5$ mm. Auf der Fläche sind also zwei Anzeigen: eine 3 mm lange Anzeige (II) und eine zusammengefasste Anzeige III+IV. Anzeige II ist zulässig, Anzeige III+IV ist unzulässig.

Wird zuerst die Zusammenfassungsregel [5]{IV} angewendet, ergibt sich gemäß dem obigen Vorgehen folgendes Bild: die Anzeigen I und II sind zusammenzufassen ($l_{ges,I+II} = 6{,}5$ mm), die Anzeigen II und III sind nicht zusammenzufassen, die Anzeigen III und IV sind zusammenzufassen ($l_{ges,I+II} = 14{,}5$ mm). Anzeige I+II und Anzeige III+IV sind unzulässig.

2.3 Zulässigkeitsgrenzen und die Anwendung der Zusammenfassungsregel – Beispiel Zulässigkeitsgrenze 3

Noch auffälliger wird die Diskrepanz für Zulässigkeitsgrenze 3:

Fall 1: *Registrierung zuerst durchführen*: es sind **keine** Anzeigen mehr vorhanden, da alle Anzeigen kleiner als die Zulässigkeitsgrenze (hier $ZG_{lin.} \leq 6$ mm) sind und nicht zu berücksichtigen sind.

Fall 2: *Zusammenfassungsregel zuerst durchführen*: Ergebnis wie oben: Anzeige I+II und Anzeige III+IV sind unzulässig.

2.4 Schlussfolgerungen

Die Norm [5] enthält keine ausdrücklichen Hinweise darauf, welches Vorgehen das richtige ist; ob es ratsam ist, sich allein auf die Abfolge der Textblöcke ({I} → {IV}) zurückzuziehen, ist neben der Berücksichtigung von Sicherheitsaspekten natürlich auch die Frage der unterschiedlichen Interessenfragen von Hersteller und Kunde.

Die fast wortgleiche Norm für die Eindringprüfung an Schweißnähten [6] beinhaltet diese Problematik genauso wie die Gussteil-Normen [7], [8] und [9], auf die hier aus Platzgründen nicht eingegangen wird.

3. Zusammenfassung und Ausblick

Es geht auch hier nicht darum, Normgesellschaften, -ausschüsse oder -ausschussmitglieder anzugreifen! Wer schon mal in einem Normausschuss mitgearbeitet hat, weiß, wie schwer die Konsensfindung dort sein kann.

In diesem Beitrag geht es vielmehr darum, die Nutzer von Normen zur Achtsamkeit und Vorsicht aufzurufen. Denn in vielen Normen und Standards sind kleinere oder größere Inkonsistenzen enthalten, die bei Nichtbeachtung früher oder später zu Diskussionen zwischen den Vertragspartnern führen können („so gesehen zulässig, anders gesehen unzulässig“).

Als Behelf bis zur Bereinigung von unklaren Stellen in Normen kann es hilfreich sein, diese Inkonsistenzen in Verfahrensbeschreibungen oder Prüfanweisungen aufzugreifen und eine Klärung herbeizuführen, die von Kunden und Auftraggeber akzeptiert werden kann.

Ziel aller Bemühungen, angefangen vom Normentwurf bis zur Prüfanweisung, sollte und muss es sein, ZfP-Prüfungen verständlich, verlässlich und nachvollziehbar zu machen, so dass mehrere Prüfer bei Einhaltung von Normen und Prüfanweisungen zu den gleichen Ergebnissen (Anzeigen, ihre Klassifizierung und ihre Bewertungen) kommen.

Referenzen

[1] Heutling, B., Moß, G.: Einblick in gängige Normen der ZfP – offene Fragen in Durchführungs- und Bewertungsnormen der ZfP; DGZfP-Jahrestagung 2024; Osnabrück; 2024

[2] Weinert. T., Keitel, S.: Zulässig oder unzulässig – die Probleme des Prüfers bei der Entscheidungsfindung; DGZfP-Jahrestagung 2018; Leipzig; 2018

[3] DIN EN 10228-1:2016 Zerstörungsfreie Prüfung von Schmiedestücken aus Stahl; Teil 1: Magnetpulverprüfung; DIN Media, Berlin

[4] DIN EN 10228-2:2016 Zerstörungsfreie Prüfung von Schmiedestücken aus Stahl; Teil 2: Eindringprüfung; DIN Media, Berlin

[5] DIN EN ISO 23278:2015 Zerstörungsfreie Prüfung von Schweißverbindungen – Magnetpulverprüfung von Schweißverbindungen – Zulässigkeitsgrenzen; DIN Media, Berlin

[6] DIN EN ISO 23277:2015 Zerstörungsfreie Prüfung von Schweißverbindungen – Eindringprüfung von Schweißverbindungen – Zulässigkeitsgrenzen; DIN Media, Berlin

[7] DIN EN 1369:2013 Gießereiwesen – Magnetpulverprüfung; DIN Media, Berlin

[8] DIN EN 1371-1:2012 Gießereiwesen - Eindringprüfung - Teil 1: Sand-, Schwerkraftkokillen- und Niederdruckkokillengussstücke;

[9] DIN EN 1371-2:2015 Gießereiwesen - Eindringprüfung - Teil 2: Feingussstücke; DIN Media, Berlin

Hinweis für die Suche: der Beuth Verlag hat sich 2025 in *DIN Media* umbenannt

DGZfP-Jahrestagung 2025

DGZfP-Richtlinie AZfP 01
-
fachliche Unterstützung bei der internen und externen Begutachtung von ZfP-Prüflaboren

Steffen BESSERT[1]
[1] DGZfP Ausbildung und Training GmbH, Berlin, Deutschland, bs@dgzfp.de

Kurzfassung

Diese Richtlinie hat das Ziel, ZfP-Prüflabore bei der Umsetzung normativer Forderungen in der täglichen Arbeit zu unterstützen. Die Umsetzung der normativen Forderungen werden bei internen und externen Begutachtungen des Prüflabors überprüft und sollen eine Konformität zu den angewendeten Normen sicherstellen. Somit ist diese Richtlinie auch ein fachliche Anleitung für beauftragte ZfP-Gutachter bzw. ZfP-Auditoren.

Die Richtlinie zeigt dabei verfahrensunabhängige Anforderungen auf und formuliert hierzu Beispiele. Es geht dabei z. B. um das Schreiben von Prüfanweisungen, die Umsetzung der ZfP-Aufsicht im Prüflabor, um die Kompetenz des Prüfpersonals, um das Thema „Metrologische Rückführung/Messunsicherheit“ in der ZfP.

Im verfahrensspezifischen Teil werden für jedes ZfP-Verfahren und für die darin enthaltenen Prüfmethoden auf verfahrenstypische Merkmale hingewiesen. Man findet hierzu auch z.B. bei den Verfahren UT und ET Hinweise, die sowohl die manuelle als auch auf die mechanisierte Prüfung betreffen.

In den Anhängen zu der Richtlinie finden sich zu den ZfP-Verfahren Informationen zu den Überprüfungen bzw. Kalibrierungen von Geräten und Referenzkörpern. Die hierzu notwendigen Regelwerke sind benannt und Hinweise zu den Überprüfungs- bzw. Kalibrierfristen sind beigefügt.

Weiterhin sind in der Richtlinie Checklisten für die ZfP-Prüfverfahren aufgeführt, welche die Vorbereitung und Durchführung der Begutachtung unterstützen sollen.

Keywords: Akkreditierung, Zertifizierung, Audit, Begutachtung, Messunsicherheit

FÜR SICHERHEIT.
JEDEN TAG.

DGZfP-Richtlinie AZfP 01 – fachliche Unterstützung bei der internen und externen Begutachtung von ZfP - Prüflaboren

Steffen Bessert, DGZfP Training und Ausbildung GmbH

dgzfp

DGZfP-Richtlinie AZfP 01

Gliederung

1. Motivation
2. Inhalt der Richtlinie AZfP 01
3. Verfahrensunabhängige Kriterien
4. Verfahrensspezifische Kriterien
5. Anhang 1: Beispiele für Bestimmung bzw. Abschätzung der Messunsicherheit
6. Anhang 3: Liste ZfP -Geräte für Überprüfung nach Norm
7. Checkliste zur Vorbereitung

dgzfp

DGZfP-Richtlinie AZfP 01

1. Motivation

- Notwendigkeit im Rahmen von Zulassungen Begutachtungen bzw. Audits in der Firma durchzuführen bzw. Durchführungen zu gestatten
- Regelwerke (Normen, Richtlinien, Anweisungen u.a.) bilden die Rahmenbedingung für die Audits bzw. Begutachtungen, eine Konformität in den Tätigkeiten hierzu sollte das Ergebnis darstellen
- Unterschiedliche Herangehensweisen an Regelwerke und unterschiedliche Auslegungen erschweren die fachliche Ausrichtung von Prüflaboren

DGZfP-Richtlinie AZfP 01

2. Inhalt der Richtlinie AZfP 01

1 Geltungsbereich	**4 Verfahrensspezifische Kriterien**
2 Begriffe	4.1 Durchstrahlungsprüfung (RT)
3 Beschreibung	4.2 Ultraschallprüfung (UT)
3.1 Verfahrensunabhängige Kriterien	4.3 Magnetpulverprüfung (MT)
3.1.1 Allgemein	4.4 Eindringprüfung (PT)
3.1.2 Personal	4.5 Wirbelstromprüfung (ET)
3.1.3 Metrologische Rückführbarkeit	4.6 Dichtheitsprüfung (LT)
3.1.4 Technische Aufzeichnungen und Bericht	4.7 Sichtprüfung (VT)
	4.8 Infrarotthermografieprüfung (TT)
3.1.5 Sicherung der Validität von Prüfergebnissen	4.9 Schallemissionsprüfung (AT)
	4.10 Mobile Härteprüfung (HT)

DGZfP-Richtlinie AZfP 01

Anhang 1: Beispiele für die Bestimmung bzw. Abschätzung der Messunsicherheit
Anhang 2: Muster einer Sehfähigkeitsbescheinigung
Anhang 3: Liste ZfP-Prüfgeräte für Überprüfung nach Norm
Anhang 4: Liste ZfP-Prüfgeräte für Kalibrierung nach Norm
Anhang 5: Checkliste in den Verfahren der Zerstörungsfreien Prüfung (Checkliste ZfP)
Anhang 6: Hinweise zu verwendeten Regelwerken

dgzfp

DGZfP-Richtlinie AZfP 01

3. Verfahrensunabhängige Kriterien (Beispiel)

- Metrologische Rückführung

Unterscheidung zwischen qualitativer und quantitativer ZfP:

Bereich 1. Rein qualitative ZfP:

Im Allgemeinen sind die klassischen genormten Prüfverfahren in der ZfP rein qualitative Verfahren, es werden registrierpflichtige Anzeigen berichtet. Angaben von Abständen und/oder Maßen zur Lage der Anzeige dienen hier lediglich der Orientierung (z. B. für die Reparatur) und gelten nicht als rückgeführter Wert.

Bereich 2. Rein quantitative ZfP:

Einige genormte Verfahren der ZfP sehen vor, dass mit ihnen quantitative Ergebnisse ermittelt werden (z. B. EN ISO 16809 Dickenmessung mit Ultraschall, EN ISO 2360 Messen der Schichtdicken–Wirbelstromverfahren).

Bereich 3. Qualitative genormte ZfP-Methoden, die zur Ermittlung von quantitativen Ergebnissen außerhalb des genormten Anwendungsbereichs verwendet werden:

Zu den Bereichen 1 und 2 existiert derzeit ein dritter Bereich, in dem die klassischen qualitativen ZfP-Methoden zur Ermittlung von quantitativen Ergebnissen (z. B. Dickenmessung, Risstiefe, Risslage, Rissvolumen) verwendet werden. Diese Methoden nutzen i. d. R. normativ erzeugte Prüfergebnisse und gewinnen durch softwarebasierte Ansätze bzw. Methoden weitere Prüfinformationen.

dgzfp

DGZfP-Richtlinie AZfP 01

4. Verfahrensspezifische Kriterien (Beispiel)

- Spezifische Anforderungen in den ZfP -Verfahren: RT/CT, UT, MT, PT, ET, LT, VT, TT, AT, HT
- Geräte
  - Spezielle Vorgaben nach Norm zu Überprüfung und Überwachung
  - Kennzeichnung von Unterlagen (z.B. RT-Aufnahmen)
  - Datenspeicherung
- Zusätzliche Anforderungen an Prüftechniken
  - Zusatzqualifikation von Prüfpersonal
  - Erstellen von Prüfanweisungen
  - Gerätesystemtechnik und deren Überprüfung
- Anforderungen an mechanisierte bzw. automatisierte Prüfung
  - Gerätesystemtechnik und deren Überprüfung

dgzfp

DGZfP-Richtlinie AZfP 01

5. Anhang 1: Beispiele für Bestimmung bzw. Abschätzung der Messunsicherheit

- Beispiel 1: Beleuchtungsstärke bei Tageslicht nach Kalibrierzertifikat
  - EN ISO 3059: ≥ 500 lx; lt. Kalibrierzertifikat Messunsicherheit 10% (u) vom Messwert
  - Erweiterte Messunsicherheit 2u (somit 20 %) und Wahrscheinlichkeit des Messwertes > 95%
  - Ergebnis: 500 lx * 1,25 = 625 lx, d.h. für eine ausreichende Beleuchtungsstärke unter Berücksichtigung der Messunsicherheit müssen mindestens 625 lx am Betrachtungsort gemessen werden
- Beispiel 2: Interne/Externe Kalibrierung von UT -Kalibrierkörper K1
  - EN ISO 2400: Abweichung der Hauptabmessungen 0,1 mm
  - Ziel: Justierung im Justierbereich s_B = 0-100 mm
  - Kalibrierzertifikat z.B. 100,0 mm± 0,19 µm; Abweichung mit rückgeführtem Messschieber und interner Kalibrierung z.B. 100,0 mm± 0,01 mm
  - Weitere Einflüsse: Ablesefehler am UT-Gerät: 1% (1 mm), Vollwellengleichrichtung mit Messart „Spitze": λ/4 (0,8 mm mit 4 MHz und 3255 m/s)
  - Ergebnis: 2u ergibt gleichen Wert von ± 2,04 mm, d.h. Messung intern ausreichend

dgzfp

DGZfP-Richtlinie AZfP 01

6. Anhang 3: Liste ZfP -Prüfgeräte für Überprüfung nach Norm (z.B. RT)

Verfahren	Geräte	Regelwerk	Abschnitte der Regelwerke	Fristen
RT-F RT-D RT-S	Röntgenröhre	Laboreigene Verfahrensanweisung		Brennfleck; Ursprungszeugnis des Herstellers
	Isotop	Laboreigene Verfahrensanweisung		Strahlergröße; Ursprungszeugnis des Herstellers
	Betrachtungsgerät (nur RT-F)	ISO 5580	4.6; 5.3	jährlich
	Draht-Typ-Bildgüteprüfkörper Stufe/Loch-Typ-Bildgüteprüfkörper Doppeldraht-Typ-Bildgüteprüfkörper (RT-D, RT-S)	EN ISO 19232-1 EN ISO 19232-2 EN ISO 19232-5	4.2; 4.3; 4.4 4.2; 4.3; 4. 4.2; 4.3; 4.	Zustandskontrolle vor jeder Anwendung, jährlich dokumentierte Überprüfung
	Entwicklereinheit (nur RT-F: Kontrolle der Filmverarbeitung)	Filmverarbeitungskontrolle nach EN ISO 11699-2	7.	Tägliche Zustandskontrolle vor Entwicklungsbeginn
	Densitometer (nur RT-F)	Laboreigene Verfahrensanweisung		Täglich oder vor jeder Filmbewertung
	Auswertung von digitalen Bilddaten am Monitor	EN ISO 17636-2	7.10	jährlich
RT-CT	Bildschirm	EN ISO 17636-2	7.10	jährlich
	Bildgütenachweise und räumliche Auflösung: - zylinderförmiger Probestab	EN ISO 15708-3	5.	Zustandskontrolle vor jeder Prüfung

dgzfp

DGZfP-Richtlinie AZfP 01

7. Checklisten für ZfP -Verfahren:

1.1	Ultraschallprüfung -manuell-	Ja	Nein	Entfällt
1.01	Ausreichende Personal-Qualifikation nach EN ISO 9712 Bei externer UT-Stufe 3 durchgeführte Aktivitäten dokumentiert?			
	Bei Prüftätigkeiten nach ASME: Ausreichende Personal-Qualifikation nach ASME-Code (Written Practice, Arbeitgeberzertifizierung, gültiger Nachweis der Sehfähigkeit)			
1.02	Gültiger Nachweis der Sehfähigkeit			
1.03	Geräte und Prüfköpfe entsprechend dem Stand der Technik und regelmäßig (mind. jährlich) auf Eignung überprüft (Anweisung zur Überprüfung vorhanden und Protokoll, siehe EN ISO 22232-1 Abs. 7, Tab. 2 mit Verweis auf EN ISO 22232-3 Abs. 4, Tab. 1)			
1.04	Regeln, Normen und Vorschriften aktuell			
1.05	Prüfanweisungen vorhanden – allgemeine/objektspezifische			
1.06	Kalibrierkörper - dokumentiert (Herstellerkonformitätsbestätigung) und Vermessung/Überprüfung mit rückgeführten Messmitteln erfolgt - jährlich 1 x auf Beschädigungen überprüft, dokumentiert und Anweisung zur Überprüfung mit Angabe der Ausschlusskriterien vorhanden			
1.07	Testkörper (soweit vorhanden und genutzt) - dokumentiert (Zeichnung mit Bemaßung der Testfehler -> Angabe zur Toleranz/ Messunsicherheit, Vermessung mit rückgeführtem Messmittel erfolgt) - jährlich 1 x auf Beschädigungen überprüft, dokumentiert und Anweisung zur Überprüfung mit Angabe der Ausschlusskriterien vorhanden			
1.08	Werden Wanddickenmessungen durchgeführt? - Messunsicherheitsbetrachtung vorhanden - Kalibrierkörper (Stufenkeil) mit rückgeführtem Messmittel überprüft und dokumentiert - Geräteüberprüfung unter Beachtung der EN 15317 Konformitätsbestätigung des Herstellers Tab. 2 (Gruppe1) einmalig, ansonsten Gruppe 2 und Gruppe 3			
1.09	Prüfkopfdaten vorhanden			
1.10	Prüfberichte/Aufzeichnungen/Dokumentation eingesehen, - Normenangaben mit Ausgabestand, - verwendete Prüfmittel, - Kennzeichnung Kundenvorgaben, - Hinweis auf Akkreditierung nach EA, - Vervielfältigungshinweis, - Ergebnisse beziehen sich auf den Prüfgegenstand, dokumentiert Rückverfolgbarkeit gegeben und Prüfberichte normenkonform Angaben bei Wanddickenmessungen angemessen			
1.11	Wurde eine Eignungsprüfung/Vergleichsprüfung innerhalb eines festgelegten Überwachungszyklus durchgeführt			

dgzfp

DGZfP-Richtlinie AZfP 01

Mitglieder des Fachausschusses „Anforderungen an Zerstörungsfreie Prüfverfahren"

- Dirk Furtmann; KARL DEUTSCH Prüf - und Messgerätebau GmbH + Co KG
- Thorsten Arndt; Delta Test GmbH
- Andreas Eck; MBQ Qualitätssicherungs -GmbH
- Ralf Ettlich; DB Systemtechnik GmbH
- Gerhard Heck; Österreichische Gesellschaft für Zerstörungsfreie Prüfung
- Martin Kurras; Fraunhofer -Institut für Zerstörungsfreie Prüfverfahren IZFP
- Hans-Joachim Malitte; privat (vormals BAM Berlin)
- Thomas Mädel; TÜV Nord MPA GmbH & Co. KG
- Jürgen Müller; Ing.-Büro Jürgen Müller
- Christian Stapf; Christian Stapf Elektronik
- Burkhard Wyen; DEKRA Incos GmbH

und die Vorsitzenden der verfahrensbezogenen FA der DGZfP

dgzfp

FÜR SICHERHEIT.
JEDEN TAG.

Vielen Dank für die Aufmerksamkeit!

Steffen Bessert: bs@dgzfp.de

dgzfp

DGZfP-Jahrestagung 2025

Mikroferoskop im Fokus: Anwendungs-potenziale der digitalen Speckle-Holografie in der Serienprüfung von Kleinstbauteilen

Jessica PLASSMANN[1,2], Tim EIS[1], Ann-Kathrin MOLITOR[1], Valentin BASTGEN[1,2], Michael SCHUTH[1]
[1]Technikum OGKB, Hochschule Trier, Deutschland
[2] Rheinland-Pfälzische Technische Universität Kaiserslautern Landau, Deutschland

Kontakt: plassmaj@hochschule-trier.de

Kurzfassung. Die vollständige Kontrolle von Kleinstbauteilen ist in Anwendungsbereichen wie der Elektroindustrie und der Medizintechnik von zentraler Bedeutung. Das patentierte Verfahren des Technikums OGKB Trier begegnet dieser Herausforderung durch die Kombination des interferometrischen Messverfahrens der digitalen Speckle-Holografie mit der Mikroskopie. Das sogenannte Mikroferoskop ermöglicht die präzise Detektion von Fehlstellen in einer Größenordnung von wenigen Quadratmillimetern.

Das Messprinzip basiert auf der optischen Erfassung von Verformungen im Mikrometerbereich, die durch definierte Belastungen, wie beispielsweise thermische Ausdehnung, induziert werden. Anhand charakteristischer Verformungsanomalien können Fehlstellen identifiziert werden. Für die Analyse von Verformungen senkrecht zur Bauteilebene, also out-of-plane (oop), wurde ein Modul entwickelt, das durch orthogonale Beleuchtung optimale Messbedingungen schafft. Ergänzend kommt ein 3D-Modul zum Einsatz, das mithilfe dreier Laserdioden unterschiedlicher Wellenlängen sowohl Verformungen in der Bauteilebene, in plane (ip), als auch senkrecht dazu erfasst.

Die hohe Sensitivität der oop-Holografie birgt in der Anwendung das Risiko, dass Störfaktoren, die unter Serienbedingungen auftreten, das Ergebnis beeinträchtigen. Es ist jedoch zu hinterfragen, ob die hohe Sensitivität in der Anwendung tatsächlich erforderlich ist und ob eine Senkung derselben eine Steigerung der Robustheit des Verfahrens bewirken könnte. Aufgrund der geneigten Beleuchtung der Prüfbauteile weist das 3D-Modul eine geringere Sensitivität auf als das speziell für oop-Messungen optimierte Modul. Im Rahmen eines Vergleichs werden daher mit beiden Modulen Messungen an Kleinstbauteilen durchgeführt, um deren Einsatzmöglichkeiten zu evaluieren.

Keywords: Mikroferoskop, MFK, Speckle -Holografie, ESPI, Serienprüfung

1. Einführung

Die Qualitätssicherung in der industriellen Serienprüfung erfordert Prüfverfahren, die sowohl zuverlässig als auch effizient kleinste Fehlstellen detektieren können. Etablierte Methoden wie Thermografie, Ultraschall und Röntgenprüfung haben sich in vielen Bereichen bewährt, bringen jedoch jeweils spezifische Einschränkungen mit sich. Während die Thermografie thermische Eigenschaften des Materials ausnutzt, um Defekte sichtbar zu machen, erfasst die Ultraschallprüfung interne Strukturveränderungen über akustische Wellen. Die Röntgen-prüfung wiederum ermöglicht eine Durchstrahlungsanalyse, ist jedoch mit hohen Sicherheitsanforderungen verbunden.

https://doi.org/10.58286/31257

Ein alternatives, aber noch wenig verbreitetes Verfahren, ist die Shearografie: Ein interferometrisches Messverfahren, das Verformungsänderungen zwischen zwei Belastungszuständen erfasst. Aufgrund ihrer Robustheit gegenüber Umgebungsstörungen hat sie das Potenzial, sich in der Serienprüfung weiter zu etablieren. Allerdings erfordert das Verfahren für eine hohe Genauigkeit einen ausreichend großen Shearbetrag, wodurch es sich nur eingeschränkt für die Detektion kleinster Fehlstellen eignet. Wenn der Shearbetrag, also der Betrag der Verkippung der Aufnahme, zu groß ist, kann es passieren, dass sich Defekte auf den Differenzbildern nicht mehr überlagern und nicht mehr eindeutig identifizierbar sind. Dies stellt insbesondere bei sicherheitskritischen Kleinstbauteilen – etwa in der Medizintechnik oder Mikroelektronik – eine Herausforderung dar. In Anwendungen, bei denen das Versagen eines Bauteils schwerwiegende Sach-, Umwelt- oder Personenschäden verursachen kann, sind daher noch präzisere Prüfverfahren erforderlich. [1] [2]

Eine vielversprechende Alternative stellt die digitale Speckle-Holografie (engl.: Electronic Speckle Pattern Interferometry, ESPI) dar, die im Gegensatz zur Shearografie nicht nur die Verformungsänderung, sondern die tatsächliche Verformung eines Bauteils misst. Somit können selbst kleinste Fehlstellen sichtbar gemacht werden. Allerdings bringt die hohe Sensitivität der Holografie auch eine erhöhte Störanfälligkeit mit sich, was ihre Anwendung unter Serienbedingungen erschwert. [2]

In dieser Arbeit wird das Mikroferoskop (MFK) vorgestellt – ein patentiertes Messsystem [3], das die digitale Speckle-Holografie mit der Mikroskopie kombiniert und speziell für die hochpräzise Prüfung von Kleinstbauteilen entwickelt wurde. Um die Herausforderungen der Sensitivität zu adressieren, wurden zwei Messmodule entwickelt: Ein oop-Modul, das hochsensitive Messungen senkrecht zur Bauteilebene (out-of-plane – oop) ermöglicht, sowie ein 3D-Modul, das durch den Einsatz mehrerer Laserwellenlängen eine robustere Erfassung sowohl in-plane (ip) als auch out-of-plane erlaubt. Ziel dieser Arbeit ist es, die Sensitivität dieser beiden Module anhand eines Beispielbauteils zu analysieren.

2. Grundlagen

2.1 Digitale Speckle-Holografie: Aufbau und Funktionsweise

Die optische Interferometrie bietet hochpräzise Messtechniken für eine Vielzahl technischer Anwendungen. Die ESPI nutzt dabei das Prinzip der Interferenz, um kleinste Verformungen eines Objekts mit einer Genauigkeit im Submikrometerbereich zu erfassen. Diese Messungen basieren auf der Überlagerung einer Referenzwellenfront E_r mit einer von der Objektoberfläche reflektierten Wellenfront E_o. Die resultierende Phasenverschiebung $\Delta\varphi$ liefert Informationen über irreguläre Verformungen oder Defekte im Messobjekt.

Der Messaufbau ([4], vgl. S. 77) umfasst eine kohärente Lichtquelle, üblicherweise einen Laser, der durch einen Strahlteiler in zwei Teilstrahlen aufgeteilt wird. Ein Teilstrahl wird über einen piezoelektrisch gesteuerten Spiegel auf die Kamera geleitet und dient als Referenzwellenfront. Der Spiegel erzeugt gezielte Phasenverschiebungen $\Delta\varphi$ des Referenzstrahls. Der zweite Teilstrahl wird auf das Messobjekt gerichtet, von dessen Oberfläche reflektiert und trifft als Objektwellenfront auf die Kamera. Diese interferiert mit der Referenzwellenfront und erzeugt Interferenzmuster, die die Verformungen des Objekts widerspiegeln.

Die Intensität der Interferenzmuster wird nach [5] durch die Überlagerung der Referenzwellenfront E_r und der Objektwellenfront E_o bestimmt. die Intensität I ergibt sich dabei aus dem Betragsquadrat der beiden Wellenfronten:

$$I = | E_r + E_o |^2 = | E_r |^2 + | E_o |^2 + 2\Re(E_r E_o^*)$$

Der Interferenzterm $2\Re\,(E_r.E_o^*)$, der die komplex konjugierte Objektwellenfront E_o^* enthält, liefert die Phaseninformation, die zur Bestimmung der Verformung des Objekts notwendig ist. Zur Extraktion dieser Phaseninformation wird unter anderem häufig die Technik des zeitlichen Phasenschiebens (engl.: Temporal Phase Shifting, TPS) eingesetzt. Dabei werden mehrere Messbilder aufgenommen, die jeweils eine kleine Phasenverschiebung $\Delta\varphi_i$ aufweisen $(\Delta\varphi_1, \Delta\varphi_2, ...)$. Durch die Analyse dieser Phasenverschiebungen können die Phaseninformation präzise extrahiert und somit die Verformungen des Objekts exakt berechnet werden.

Abbildung 2.1 zeigt den schematischen Aufbau des ESPI-Systems für eine out-of-plane Messung. Bei vielen holografisch arbeitenden Interferometern wird die Referenzfront koaxial zur Beobachtungsachse eingekoppelt ([6], vgl. Kap. 21.3). In diesem spezifischen Fall ist dieser Schritt jedoch nicht explizit dargestellt, da er für die Messung nicht von Bedeutung ist. Der dargestellte Aufbau entspricht den wesentlichen Komponenten und dem Prinzip, wie die Interferenzmuster zur Berechnung der Oberflächenverformungen genutzt werden.

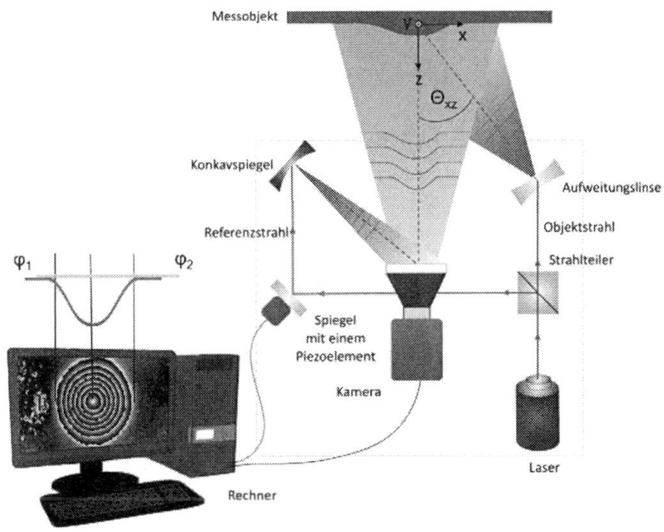

Abb. 2.1. Typischer Messaufbau der out-of-plane Holografie mit TPS nach [7], vgl. S. 32

Im Gegensatz dazu erfasst die in-plane Holografie (ip-Holografie) Verformungen, die in der Ebene der Messoberfläche liegen. Typischerweise wird hierzu eine dual-beam-illumination Methode eingesetzt, bei der das Objekt mit zwei kohärenten Lichtstrahlen symmetrisch beleuchtet wird ([2] vgl. Kap. 3.2.5).

2.2 Mathematische Grundlagen: Sensitivität der Holografie

Die Holografie misst nicht die absolute Höhe eines Punktes auf der Oberfläche, sondern die relativen Deformationen durch die Phasendifferenz zwischen zwei Belastungszuständen. Für die in Abb. 2.1 dargestellte allgemeine Beleuchtungsgeometrie mit einem Einfallswinkel Θ_{xz} ergibt sich die Beziehung zwischen der gemessenen Phasenverschiebung und den Verformungskomponenten des Messobjekts nach [7], vgl. Kap. 2.4.a. zu:

$$N = \frac{1}{\lambda}[D_u \sin(\Theta_{xz}) + D_w(1 + \cos(\Theta_{xz}))]$$

In dieser Gleichung repräsentiert N die gemessene Streifenordnung, die durch die Überlagerung der Referenz- und Objektwellenfront entsteht. Der Begriff Streifenordnung bezeichnet dabei die Anzahl der Interferenzstreifen, die auf der Kamera als Resultat der Phasenverschiebung sichtbar werden. D_u stellt die in-plane-Verformung in x-Richtung dar, also die Deformation, die parallel zur Oberfläche auftritt. D_w beschreibt die out-of-plane-Verformung, die sich senkrecht zur Oberfläche des Objekts vollzieht und typischerweise als Höhenänderung in der z-Richtung wahrgenommen wird. λ beschreibt die Wellenlänge des verwendeten Lasers. Diese ist für die Messung entscheidend, da sie die Auflösung der Messungen beeinflusst.

Die Formel zeigt, dass sich die gemessene Phasenverschiebung aus zwei Teilen zusammensetzt: Einem in-plane Anteil D_u sin (Θ_{xz}), der sich aufgrund der Deformation in der xy-Ebene ergibt, und einem out-of-plane Anteil D_w $(1 + \cos(\Theta_{xz}))$, der durch die Veränderung in der z-Richtung verstärkt wird. Der Faktor $(1 + \cos(\Theta_{xz}))$ erhöht die Sensitivität der Messung für Formänderungen, die sich senkrecht zur Oberfläche des Objekts abspielen. Das bedeutet, dass Formänderungen, die in der z-Richtung (out-of-plane) auftreten, im Vergleich zu in-plane-Änderungen stärker gewichtet werden, was die Messung besonders für diese Art der Deformation empfindlicher macht.

2.3 Funktionsprinzip des Mikroferoskops

Das Mikroferoskop (MFK) stellt eine einzigartige Kombination aus einem handelsüblichen Mikroskop und einem speziell entwickelten Zusatzmodul dar. Es wurde 2010 beim deutschen Patent- und Markenamt eingereicht und erhielt 2020 das Patent [3]. Der entscheidende Vorteil des MFK liegt in der Flexibilität des Zusatzmoduls, das als Steckbauteil in das Mikroskop integriert ist. Diese modulare Struktur ermöglicht eine einfache Umrüstung des Mikroskops, sodass neben der visuellen Betrachtung auch mikroskopische ESPI-Messungen durchgeführt werden können.

Der ursprüngliche Prototyp des MFK wurde entwickelt, um die oop- und ip-Verformungen von Objekten zu messen. Der Interferometer-Aufbau des MFK orientierte sich ursprünglich am Mach-Zehnder-Interferometer, wurde jedoch weiterentwickelt, um den spezifischen Anforderungen der Mikroskopie gerecht zu werden. Die Mikroskopbrennweite des Geräts lässt sich in 13 Stufen von 35,5 mm bis 575 mm verstellen, sodass Messflächen von 40 mm x 25 mm bis zu 2,5 mm x 2 mm erfasst werden können.

Einen zentralen Aspekt der MFK-Entwicklung stellte die Integration des Laserlichts über Glasfaser dar, das durch den optischen Pfad im Messgerät geführt wird. Abhängig von der Position interner Abschattungsvorrichtungen wird entweder ein Referenzstrahl oder ein Objektstrahl erzeugt, um Messungen in der oop- oder ip- Richtung durchzuführen. Diese ersten Versionen des MFK ermöglichten hochqualitative Messungen mit statischer Anregung des Messobjekts und legten den Grundstein für die späteren Weiterentwicklungen.

Der ursprünglich entwickelte Prototyp des MFK wurde kontinuierlich weiterentwickelt, um Schwächen zu beseitigen und das Gerät für eine breitere Anwendung zu optimieren. Die aktuelle Version des Prototyps, das MFK 2, setzt anstelle der bisherigen Technik des zeitlichen Phasenschiebens auf das räumliche Phasenschieben (engl.: Spacial Phase Shifting, SPS). Dabei wird eine speziell entwickelte Blende verwendet, um eine Trägerfrequenz zu erzeugen, die in Kombination mit einer Fast Fourier Transformation (FFT) die Phaseninformation aus nur einer Messaufnahme pro Zustand extrahiert. Dieser Ansatz ermöglicht es, dynamische Belastungen zu messen, bei denen traditionelle Methoden des zeitlichen Phasenschiebens aufgrund der prinzipbedingt nacheinander aufgenommenen

Bilder fehleranfällig wären. Allerdings führt die Verwendung der Blende zu einer Begrenzung der Lichtausbeute, was wiederum die Genauigkeit des räumlichen Phasenschiebens im Vergleich zum zeitlichen Phasenschieben leicht reduziert. Die beiden zentralen Module des MFK 2 sind das oop-Modul und das 3D-Modul, die jeweils auf spezifische Messanforderungen optimiert wurden und in Abbildung 2.2 dargestellt werden.

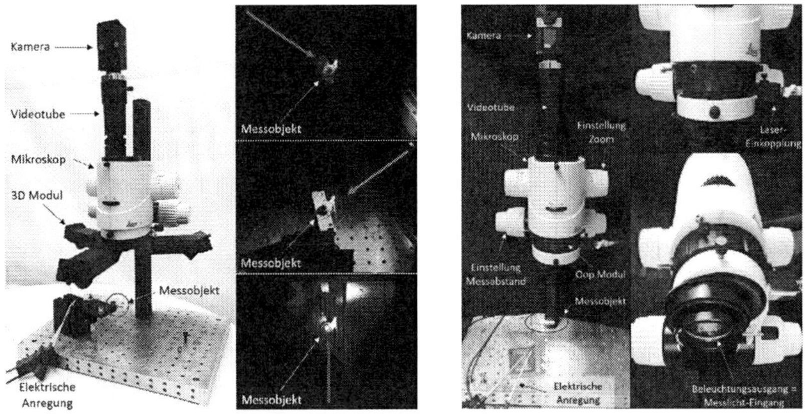

Abb. 2.2. 3D-Modul (links) und oop-Modul (rechts) nach [7], vgl. S. 147 ff

Das oop-Modul wurde kompakter und leichter gestaltet als sein Vorgänger und wird direkt am unteren Teil des Mikroskops montiert, was den Aufbau lediglich um 35 mm erhöht. Diese Verbesserung macht das MFK 2 benutzerfreundlicher und handlicher. Ein wesentlicher Vorteil des oop-Moduls ist der 0° Beleuchtungswinkel, der präzise Messungen der oop-Verformungen ohne Bildabschattungen ermöglicht. Der vom Laser beleuchtete Bereich befindet sich unterhalb des Mikroskops, wodurch Sicherheitsanforderungen verbessert werden. Erste Messungen, wie die Prüfung von Gummi-Dichtringen unter transienter Anregung mit einer blauen Laserdiode, lieferten bereits vielversprechende Ergebnisse ([8], [9], [10]).

Das 3D-Modul ermöglicht die Messung der gesamten Verformung eines Objekts, einschließlich der oop- und ip-Anteile. Im Vergleich zum ursprünglichen MFK wurde dieses Modul mit einer geometrischen Anordnung von drei RGB-Laserdioden entwickelt, die eine hochsensitive 3D-ESPI-Messung ermöglichen. Diese Messanordnung ist besonders vorteilhaft bei komplexen Verformungen, bei denen mehrere Ebenen erfasst werden müssen.

Die Verwendung von RGB-Laserdioden mit unterschiedlichen Wellenlängen hat einen direkten Einfluss auf die Sensitivität der Messungen. Laut der mathematischen Herleitung der Streifenordnung (siehe Kapitel 2.2) ist diese invers zur Wellenlänge des verwendeten Lasers. Das bedeutet, dass bei kleineren Wellenlängen die Streifenordnung steigt, was zu einer höheren Sensitivität und besseren Auflösung führt. Umgekehrt führt die Verwendung von größeren Wellenlängen zu einer geringeren Streifenordnung, was die Empfindlichkeit für kleine Verformungen reduziert. Diese Eigenschaft ermöglicht es, durch die Wahl der Laserwellenlängen gezielt die Auflösung und Sensitivität der Messung anzupassen, je nachdem, welche Verformungsarten der zu untersuchenden Bauteile zu erwarten sind. Gleichsam soll durch die Messung von oop- sowie ip-Anteilen in x und y eine 3D Messung möglich werden. Das Modul ist mit einer Armlänge von 180 mm statt der ursprünglichen 300 mm ausgelegt, wodurch die Sensitivität bei der Messung von oop-Verformungen erhöht und eine höhere Präzision insbesondere bei kleinen Objekten erreicht wird. Zudem wurde das Gewicht auf lediglich 400 Gramm reduziert.

3. Sensitivität der Messmodule – Vergleich und Diskussion

Das Prüfobjekt besteht aus einem zylindrischen Aluminiumkörper mit einer zentralen Flachbodenbohrung von 5 mm Durchmesser und einer Restwandstärke von 0,5 mm an der Oberfläche. Der Außendurchmesser des Zylinders wurde so gewählt, dass die strukturelle Steifigkeit ausreichend hoch ist, um Verformungen außerhalb der geschwächten Zone zu minimieren. Dadurch kann die Deformation in der Umgebung der Flachbodenbohrung isoliert betrachtet werden, ohne dass globale Verformungseffekte die Messergebnisse verfälschen. Abb. 3.1, mittig zeigt einen Querschnitt des Prüfobjekts.

Das Bauteil wurde während der Messungen unterhalb des MFKs fixiert, um die Reproduzierbarkeit der Ergebnisse sicherzustellen. Die Belastung erfolgte durch rückseitige, zentrische Einleitung von Überdruck in den Stufen 1 bar und 3 bar unmittelbar in die Bohrung. In Abb. 3.1 rechts ist die simulierte Verformung des Bauteils bei 3 bar dargestellt. Die Simulationsergebnisse wurden zur Visualisierung mit dem Faktor 500 skaliert.

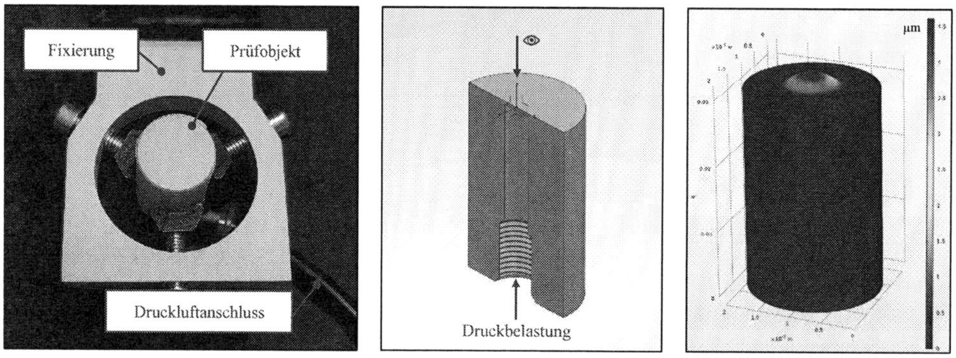

Abb. 3.1. links: Fixiertes Prüfobjekt, mittig: CAD-Darstellung des Prüfobjekts im Schnitt durch die eingebrachte Fehlstelle, rechts: Finite-Elemente-Simulation des Prüfobjekts mit einer Belastung von 3 bar

Tabelle 3.1. Gegenüberstellung der ungefilterten Messergebnisse des Prüfkörpers für eine Druckbelastung von je 1 bar und 3 bar, Messlichtwellenlänge 532 nm

	λ	1 bar	3 bar
oop-Modul	532 nm (grün)	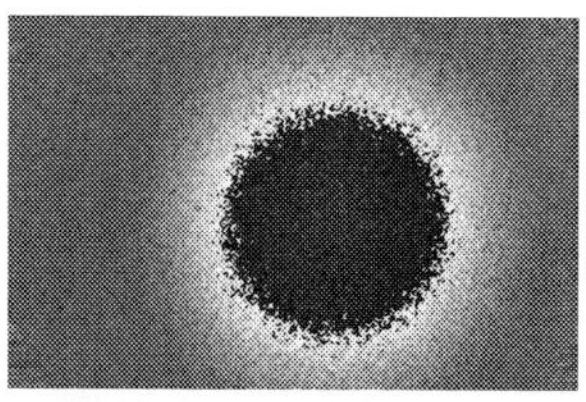	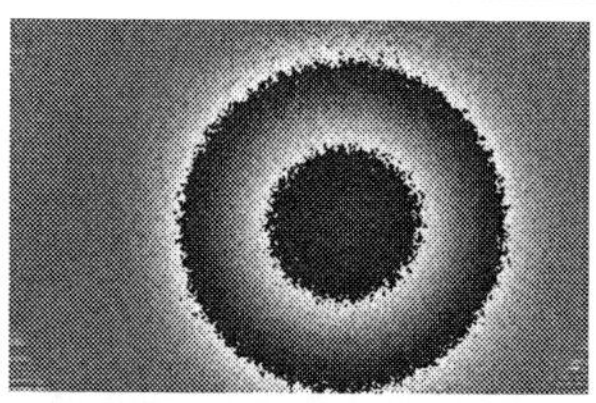
3D-Modul (45°)	532 nm (grün)	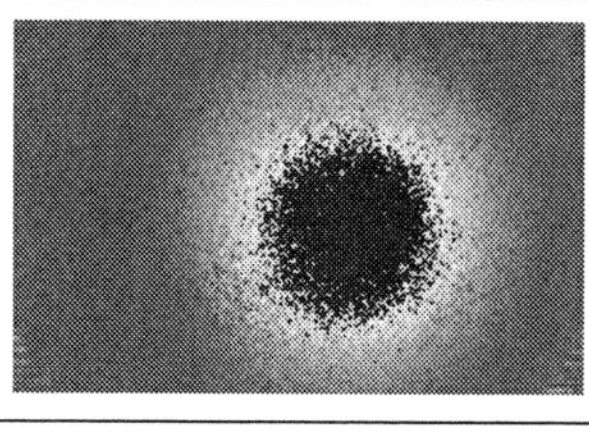	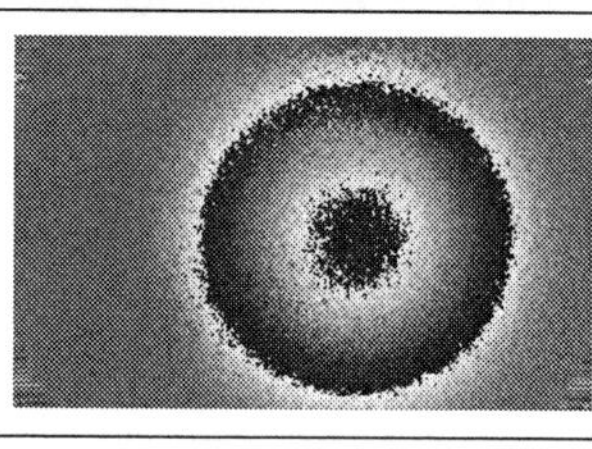

Zur quantitativen Analyse der Verformung wurde die Messung mit beiden Modulen unter Verwendung eines aufgeweiteten Neodym-dotierten Yttrium-Aluminium-Granat-Laser (Nd:YAG-Laser) mit einer Wellenlänge von 532 nm durchgeführt. Die Messungen hinsichtlich der erfassten Phasenstreifenanzahl sowie der Qualität der Streifenbilder verglichen. Die Referenzaufnahme erfolgte im Grundzustand (0 bar). Die Ergebnisse sind in Tabelle 3.1 gegenübergestellt. Bei einem Überdruck von 1 bar zeigen sowohl das oop- als auch das RGB-Modul jeweils einen Phasenstreifen. Die Messung mit dem RGB-Modul weist dabei eine stärkere Rauschüberlagerung auf. Diese Unschärfe kann zu erhöhten Unsicherheiten bei der quantitativen Auswertung führen. Bei einem Überdruck von 3 bar zeigt sich eine insgesamt stärkere Verformung des Prüfkörpers, was sich in der Ausbildung von zwei Phasenstreifen in beiden Messsystemen widerspiegelt.

In Tabelle 3.2 wird das 3D-Modul nochmals spezifisch betrachtet, indem die Messungen für verschiedene Wellenlängen miteinander verglichen werden. Für die Messungen mit dem RGB-Modul wurden verschiedene Laserlichtquellen eingesetzt: Während die grüne Wellenlänge (532 nm) mit dem Nd:YAG-Laser realisiert wurde, kamen für die roten (657 nm) und blauen (448 nm) Messungen Laserdioden zum Einsatz. Während mit grünem und blauem Laserlicht bereits bei 1 bar ein Phasenstreifen detektiert wird, bleibt der Messwert für das rote Licht unter dieser Belastung ohne Phasenübergang. Dies deutet darauf hin, dass die Sensitivität für die rote Wellenlänge geringer ist und kleinere Verformungen in diesem Bereich nicht mit ausreichend hoher Genauigkeit erfasst werden können. Bei einer Erhöhung des Drucks auf 3 bar zeigen die Messungen mit blauem und grünem Licht jeweils zwei Phasenstreifen, während für das rote Laserlicht lediglich ein Phasenstreifen registriert wird.

Tabelle 3.2. Gegenüberstellung der ungefilterten Messergebnisse des Prüfkörpers bei Messungen mit dem RGB-Modul und unterschiedlichen Messlichtwellenlängen für eine Druckbelastung von je 1 bar und 3 bar,

	3D-Modul (Beleuchtungswinkel 45°)		
	448 nm (blau)	532 nm (grün)	657 nm (rot)
1 bar	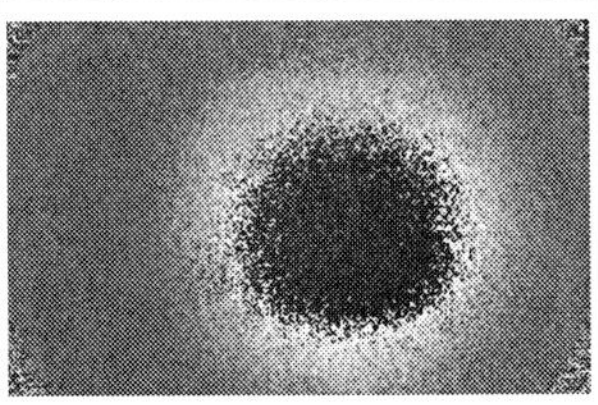	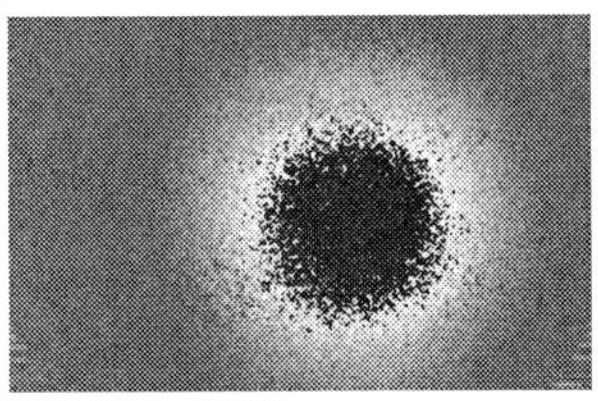	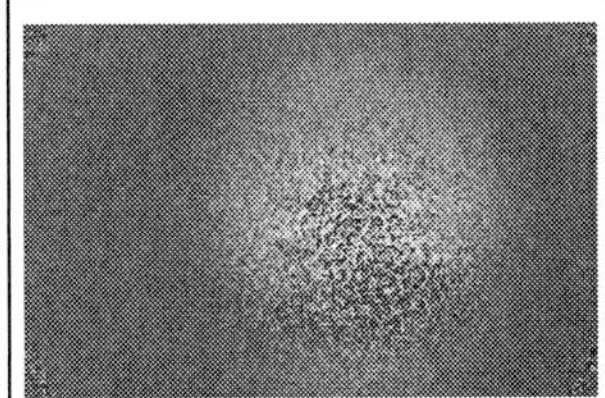
3 bar	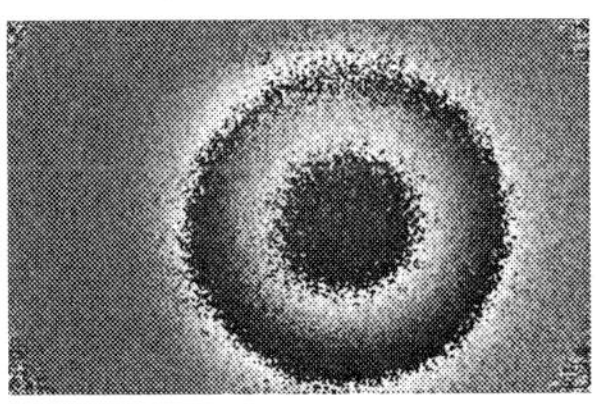	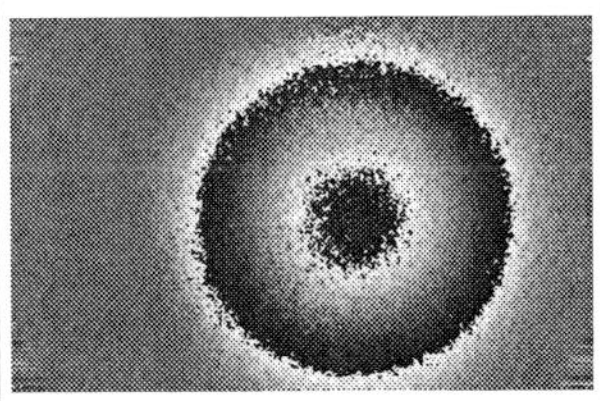	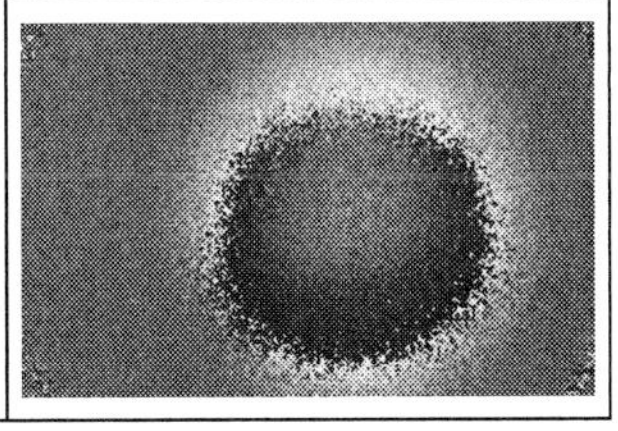

Die Messergebnisse zeigen darüber hinaus, dass der Phasenkontrast bei der Messung mit grünem Laserlicht deutlich ausgeprägter ist als bei den roten und blauen Laserdioden. Die Phasenstreifen sind schärfer abgegrenzt und weniger verrauscht, was eine genauere Auswertung ermöglicht. Dies ist vermutlich auf die jeweils verwendete Laserlichtquelle zurückzuführen: Während für die grüne Wellenlänge ein Nd:YAG-Laser mit hoher spektraler Reinheit eingesetzt wurde, stammen die roten und blauen Wellenlängen aus Laserdioden. Zum einen weisen die Laserdioden eine geringere Kohärenzlänge auf, welche das Messergebnis beeinflusst und durch Abbildungsoptiken, Gläser oder Linsen zur Strahlformung weiter vermindert wird ([7], Vgl. S. 14 ff). Zum anderen weisen Laserdioden aufgrund ihrer Emissionscharakteristik ein breiteres Spektrum auf. Besonders beim räumlichen Phasenschieben stellt das breitere Spektrum der Laserdioden eine zusätzliche Fehlerquelle dar. Diese Methode basiert auf der gezielten Brechung des Laserlichts an einer Blende, um eine Trägerfrequenz zu erzeugen, aus der die Phaseninformation berechnet wird.

Da die Phasenverschiebung direkt von der Wellenlänge abhängt, führt ein breites Emissionsspektrum zu Überlagerungseffekten unterschiedlicher Wellenlängenanteile, die das Interferenzmuster verfälschen und die Messgenauigkeit verringern. Die Ergebnisse verdeutlichen somit, dass neben der Wellenlänge auch die spektrale Reinheit und Kohärenz der Lichtquelle einen wesentlichen Einfluss auf die Qualität der Phasenmessung hat.

4. Fazit und Ausblick

Die durchgeführten Messungen zeigen, dass das oop-Modul aufgrund seiner höheren Sensitivität und der optimierten Messbedingungen für die Detektion von Verformungen exaktere Phasenbilder liefert. Im Vergleich dazu wies das 3D-Modul, das mit Laserdioden unterschiedlicher Wellenlängen arbeitet, eine geringere Sensitivität auf. Dies führte zu stärkeren Störungen und verrauschten Phasenstreifen, insbesondere bei der Messung mit rotem Licht. Der Unterschied in der Leistung der beiden Module wurde unter anderem durch die Art der Lichtquelle beeinflusst, wobei das grüne Laserlicht im oop-Modul präzisere Auswertung ermöglichte. Die Messungen mit dem 3D-Modul, welches für die in-plane Messung unter 45° ausgelegt wurde, zeigten jedoch ebenfalls valide Ergebnisse, insbesondere bei der Verwendung von grünem und blauem Licht und einer ausreichend reinen Laserquelle.

Die Ergebnisse verdeutlichen, dass jedes Messmodul je nach Anwendung und spezifischer Messanforderung seine Stärken und Schwächen hat. Derzeit wird im Rahmen eines Forschungsprojekts, das durch das Zentrale Innovationsprogramm Mittelstand (ZIM) gefördert wird, daran gearbeitet, die Vorteile beider Module zu kombinieren und ein integriertes System zu entwickeln. Ziel ist es, ein noch robusteres und flexibleres Messsystem zu schaffen, das die Anforderungen an Genauigkeit und Effizienz für die industrielle Serienprüfung von Kleinstbauteilen besser erfüllt. Ein vielversprechender Ansatz ist auch die Untersuchung steilerer Beleuchtungswinkel, da diese zu einer höheren Reinheit und Schärfe der Phasenbilder führen. In der aktuellen Arbeit wurde primär die oop-Messung betrachtet, jedoch wird angestrebt, einen Winkel zu definieren, der sowohl für oop- als auch ip-Messungen gute Ergebnisse liefert. Dieser integrative Ansatz könnte die Effizienz des Systems steigern und die Flexibilität für unterschiedliche Anwendungen verbessern.

Referenzen

[1] P. Jacquot und J.-M. Fournier, Interferometry in Speckle Light: Theory and Applications, H. Springer Berlin, Hrsg., 2000, pp. XV, 659.

[2] W. Buerakov, Interferometrische Bauteilprüfung, Schadenserkennung und Ermittlung der modalen Parameter mittels dynamischer Belastung, Shaker Verlag Aachen, 2016.

[3] M. Schuth, D. Robert und D. Nägel, „Mikroferoskop“. Deutschland Patent DE 102010020860 A1, 30 Januar 2020.

[4] U. Schnars und W. Jueptner, Digital Hologram Recording, Numerical Reconstruction, and Related Techniques, Springer Berlin, Heidelberg, 2010.

[5] P. Hariharan, Optical Holography: Principles, Techniques and Applications, Cambridge University Press, 1996.

[6] H. Naumann, G. Schröder und M. Löffler-Mang, Handbuch: Bauelemente der Optik, vollständig überarbeitete und erweiterte Auflage, München: Hanser Verlag, 2014.

[7] C. Petry, Weiterentwicklung der Shearografie mit räumlichem Phasenschieben als zerstörungsfreies Prüfverfahren für die automatisierte Serienprüfung, Saarbrücken: Universität des Saarlandes, 2021.

[8] C. Petry, L. Petry und J. Petry, „Mikroferoskop - Verformungsanalyse in Echtzeit bei der Dichtringkontrolle,“ *inVision 3,* Juni 2019.

[9] C. P. L. Petry, J. Petry und M. Schuth, „Dichtring-Qualitätskontrolle mit einem industrietauglichen Mikroferoskop,“ *Laser Magazin,* pp. 21-23, September 2019.

[10] C. Petry, L. P. J. Petry und M. Schuth, „Simpler Aufbau, hohe Messgenauigkeit - optische Kontrolle von Gummidichtringen mit dem Mikroferoskop,“ *QZ Zeitschrift,* Nr. 1, August 2020.

Roboterbasierte Eindringprüfung (PT) großer Gussbauteile

Christopher WITZKE, Philipp HETTICH
Laubinger + Rickmann GmbH & Co. KG, Nordwalde, Deutschland,
mail@laubinger-rickmann.de

Kurzfassung

Der Einsatz von Aluminium-Gussbauteilen im Fahrzeug nimmt aufgrund der allgemein bekannten Notwendigkeit zur Gewichts- und Emissionsreduktion seit Jahren weltweit zu. Damit einher geht ebenso die Funktionsintegration und letzten Endes die Größe der Bauteile, die im aktuellen Trend zu Mega- oder Gigcastings gipfelt.

Bei der (Hoch-) Serienprüfung von großen Gussbauteilen (bspw. Querträger, Hinterachsträger) im Fahrzeugbau treffen nun gleich mehrere Herausforderungen aufeinander. Zum einen ist ein manuelles Handling bei Bauteilgewichten jenseits der 25kg sowohl bei der Bestückung als auch der Inspektion zu vermeiden oder bestenfalls gänzlich auszuschließen. Zum anderen geht die Kombination aus hohen Stückzahlen und großformatigen Bauteilen beim Einsatz klassischer Fördertechnik mit einem drastisch erhöhten Flächenbedarf einher, denn die Einhaltung der Prozesszeiten wird in der Regel über die Länge der Förderstrecke bei der zu erzielenden Taktung definiert. Zu guter Letzt wird dies um bauteilspezifische Herausforderungen (hohler) Sandgussbauteile und komplexer Geometrien ergänzt.

Es bedarf neuer Anlagen-Konzepte, um diesen anspruchsvollen Aufgabenstellungen gerecht zu werden.

Der Beitrag stellt ein umgesetztes Anlagenkonzept zur Prüfung großer Aluminium-Gussteile, hergestellt im Sandguss, vor. Dabei wird nicht nur auf die zuvor genannten Herausforderungen eingegangen, sondern auch die individuelle, kundenspezifische Umsetzung in Nordamerika beleuchtet – von der ersten Konzeptidee bis zur Serienprüfung ohne Anlaufkurve.

Keywords:
Eindringprüfung, PT, FPI, Rissprüfung, Gussbauteile, roboterbasierte Prüfung, Automatisierung

TECHNOLOGY FOR QUALITY

DGZFP JAHRESTAGUNG 2025

Laubinger Rickmann

Technology for Quality

1

DAS UNTERNEHMEN

DGZFP JAHRESTAGUNG 2025

Laubinger Rickmann

Technology for Quality

2

DAS UNTERNEHMEN | ÜBERSICHT

- Ein Team von **150 Personen**, deren Fokus Ihre Lösung ist.
- Seit **52 Jahren** entwickeln wir Sondermaschinen…
- …für Anwendungen bei denen **Qualität und Präzision** im Fokus stehen.
- Alle Komponenten sind durch unsere Teams geplant, entwickelt und implementiert. **Wir kennen jede Schraube – und ihren Platz.**
- **Wir fangen dort an, wo es keinen Standard mehr gibt.** Und werden eine Lösung speziell für Ihre Anforderungen schaffen.

Mai 2025 Seite 3 DGZfP Jahrestagung 2025 | Maschinen und Konzepte sind geistiges Eigentum von L+R Laubinger Rickmann

3

DAS UNTERNEHMEN | GESCHÄFTSFELDER

Ultraschall Grundsysteme

Eindringprüfanlagen

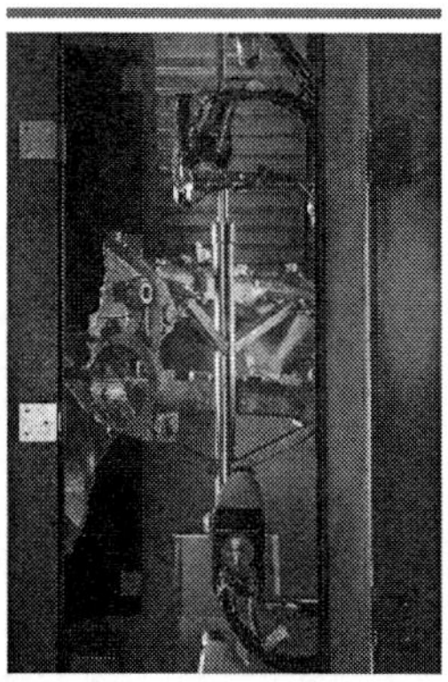

Richtanlagen

Automatisierung

Mai 2025 Seite 4 DGZfP Jahrestagung 2025 | Maschinen und Konzepte sind geistiges Eigentum von L+R Laubinger Rickmann

4

DAS UNTERNEHMEN | STANDORTE

Deutschland

Laubinger & Rickmann
GmbH & Co. KG
Nordwalde
+49 2573 93 88 – 0
mail@laubinger-rickmann.de

USA (Service)

LR Q-Systems Inc.

Chattanooga, TN
+1 423 4855888
mail@lr-q.com

China (Teilfertigung)

LR Special Machines Co. Ltd.

Taicang, Jiangsu
+86 171 62129278
asia@laubinger-rickmann.de

Mai 2025 Seite 5 | DGZfP Jahrestagung 2025 | Maschinen und Konzepte sind geistiges Eigentum von L+R | Laubinger Rickmann

5

DAS UNTERNEHMEN | KAPAZITÄTEN

SOFTWARE

15 Personen

- NC
- SPS
- Virtuelle Inbetriebnahme
- Robotik div. Hersteller
- Hochsprachen

ELEKTRO

25 Personen

- 3D Schaltschrank-Konstruktion inkl. Wiring
- Vollautomatische Schaltschrankfertigung
- Vollautomatische Kabelkonfektionierung

KONSTRUKTION

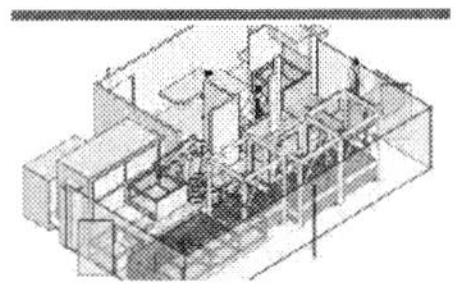

28 Personen

- 3D Konstruktion
- FEM-Simulation
- Ablaufsimulation & Taktzeitanalyse
- Digitaler Zwilling

SERVICE

25 Personen

- Eigener Service in Deutschland, China und den USA
- Umfassende Remote-Unterstützung
- Service-Glasses inklusive Live-Translator

Mai 2025 Seite 6 | DGZfP Jahrestagung 2025 | Maschinen und Konzepte sind geistiges Eigentum von L+R | Laubinger Rickmann

6

DAS UNTERNEHMEN | KAPAZITÄTEN

BEARBEITUNG

15 Personen
+ 5-Achs-Fräsmaschinen
+ CAD CAM Programmierung
+ Automatischer Plattenwechsler
+ Drehmaschinen mit Doppelspindel

SCHWEIßTECHNIK

7 Personen
+ Zertifizierter Schweißfachbetrieb
+ MIG/MAG-Schweißen
+ WIG-Schweißen
+ Laser-Schweißen
+ PT, Rot-Weiß-Prüfung

BLECH-VERARBEITUNG

10 Personen
+ Laserschneider
+ Kantbänke bis 4,2 m und 320 t Druckkraft
+ Eigene Gesellschafts-Einheit inkl. Lohnfertigung

MESSTECHNIK

3 Personen
+ Mobile Lasertracker
+ Mobile Messarme inkl. Scanner
+ Polyworks-Anbindung
+ Umfassende Auftragsvermessungen

35 weitere Personen in Verwaltung, Projektmanagement und mechanischer Montage tätig

Laubinger Rickmann

7

ROBOTERBASIERTE EINDRINGPRÜFUNG (PT) GROßER GUSSBAUTEILE

Laubinger Rickmann
Technology for Quality

8

DIE AUSLEGUNG EINER PT-ANLAGE

DGZFP JAHRESTAGUNG 2025

Laubinger Rickmann

Technology for Quality

9

DIE AUSLEGUNG

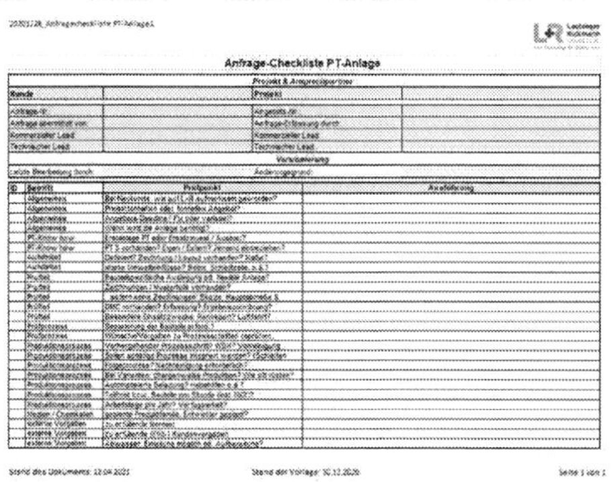

Anfrage-Checkliste

Erfassen

- + Produktionsprozess
- + Kundenanforderungen
- + Beteiligte & Erfahrungen
- + Endkunden-Spezifikation
- + Lokale Gegebenheiten
- + Schnittstellen
- + Limitationen

Verstehen

- + Ziele des Kunden
- + Anforderungen Dritter

Mai 2025 Seite 10

DGZfP Jahrestagung 2025 | Maschinen und Konzepte sind geistiges Eigentum von L+R

Laubinger Rickmann

10

DIE AUSLEGUNG

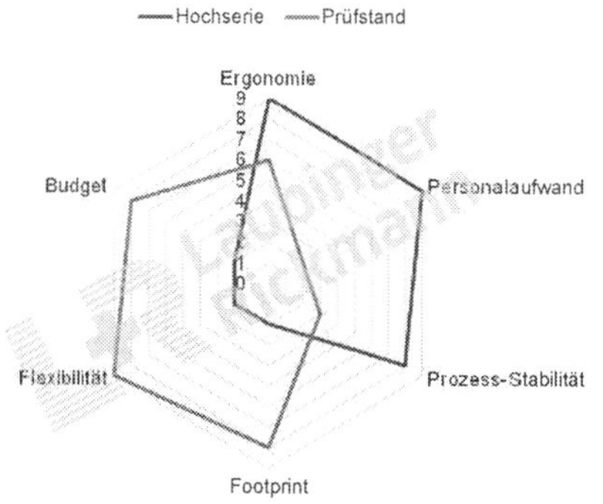

Ziele & Priorisierung

Prozess-Schritt	Umsetzung					
Fördertechnik	Schubstangen	Hängeförderer, getaktet: getaktet	power&free	kombiniert	Schüttgut	Roboter
Bauteileingabe	manuell		Pick & Place		Roboter	
Vorreinigung	Chemisch		Mechanisch		Keine	
Vor-Trocknung	Ja			Nein		
Bauteil-Kühlung	Ja			[illegible]		
Eindringmittel-Typ	I (fluoreszent)			II (Rot)		
Eindringmittel-Empfindlichkeit	0,5	1	2	3	4	–
Eindringmittel-Applikation	Sprühen, Luftdruck	Sprühen, Airless	[illegible]	Tauchen	Streichen	Übergießen
Zwischenreinigung	Wasser, Tauchen	[illegible]	Kombiniert	nachemulgierbar, lipophil	nachemulgierbar, hydrophil	–
Wasseraufbereitung	[illegible]		AK-Filter		Sonstige Verfahren	
Entwicklereinsatz	Trockenpulver, E-Statik	Wb, suspendiert, Tauchen	Wb, gelöst, Tauchen	Lösemittelbasiert, Airless	Lösemittelbasiert, Dose	kein Einsatz
Handling, Inspektion	Im Warenträger	Manipulationshilfe	Bauteilpräsentation	Ablagesysteme	–	
Datenzuordnung	manuell (Papier)		Anzeigenerfassung		Anzeigen- und Parameter-Erfassung	
Bauteil-Ausförderung (NiO)	manuell			Förderband		
Bauteil-Ausförderung (IO)	manuell	Förderband	Pick & Place		Roboter	

Prozess-Matrix

Mai 2025 Seite 11 DGZfP Jahrestagung 2025 | Maschinen und Konzepte sind geistiges Eigentum von L+R Laubinger Rickmann

11

DIE AUSLEGUNG | BEISPIELE

Parallele Förderstränge

+ Ziel:
  + Engpass Inspektionsdauer auflösen
  + Prüferabhängigkeit minimieren
+ Umsetzung:
  + Parallele Versorgung der Prüfplätze
  + Individuelle Fördertechnik
+ Ergebnis:
  + Optimale Versorgung der Prüfer
  + Prüfung entspricht Produktionsalltag

Mai 2025 Seite 12 DGZfP Jahrestagung 2025 | Maschinen und Konzepte sind geistiges Eigentum von L+R Laubinger Rickmann

12

DIE AUSLEGUNG | BEISPIELE

Berührungsloses Bauteilhandling

+ Ziel:
  + Erfassung der vollständigen Oberfläche
  + Minimale Belastung des Prüfers
+ Umsetzung:
  + Adaption Prüfkopfmanipulator UT
  + Zentrales spreizen zur Aufnahme
+ Ergebnis:
  + Vollständige Prüfung im Gehänge
  + Einfaches Bauteilhandling via Joystick

13

DIE AUSLEGUNG | BEISPIELE

Gehänge mit Sperrluft

+ Ziel:
  + Automatisierter Prozess
  + Vermeidung der Medien-Ansammlung
+ Umsetzung:
  + Entwicklung Spezial-Gehänge
  + Einsatz von Sperrluft
+ Ergebnis:
  + Keine Flüssigkeit im Bauteilinneren
  + Vollautomatisierte Hochserienanlage

14

DIE AUSLEGUNG | BEISPIELE

Verkettung mit Strahlanlage

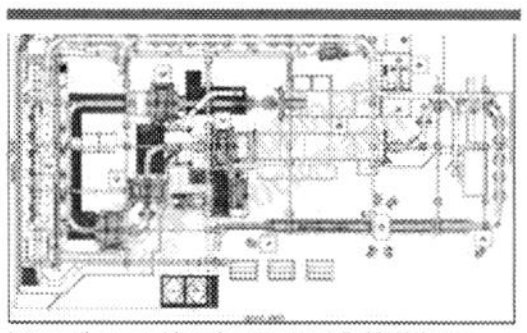

Integration von Vorreinigung und Schleifplätzen

+ Ziel:
  + Verkettung mit Vor- und nachgelagerten Prozessen
+ Umsetzung:
  + Integration von
    + Röntgen-
    + Vorreinigungs- und/oder
    + Strahlanlagen sowie
    + Nacharbeitsplätzen / Schleifstationen
  + Prozessübergreifendes Bauteilhandling
+ Ergebnis:
  + Kein manueller Eingriff vor/nach PT

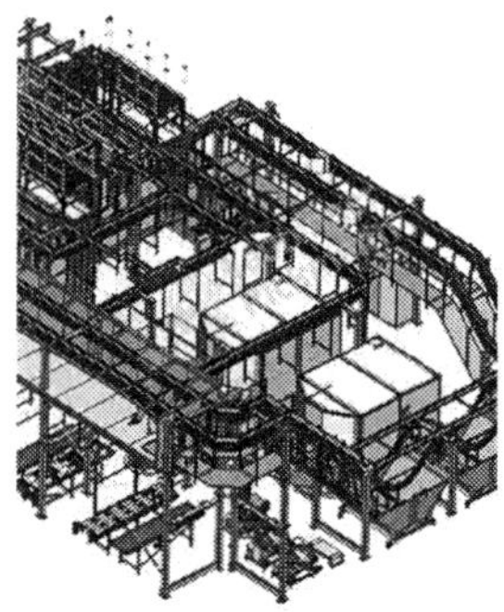

Verkettung mit RT-S

15

DIE AUTOMOTIVE TRENDS
DGZFP JAHRESTAGUNG 2025

Laubinger Rickmann

Technology for Quality

16

DIE TRENDS

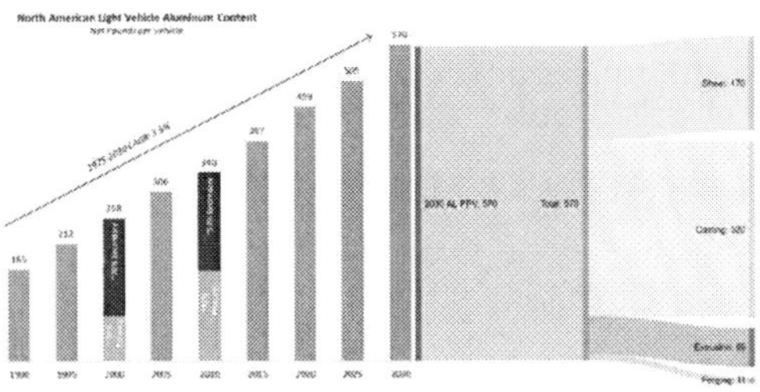

Source: Ducker April 2020 data, Aluminum Association 2020 Report

Al Penetration*	2018 (kgpv, %)		2025 (kgpv, %)		2030 (kgpv, %)	
Chassis & Suspension	11.6	26%	39.5	70%	59.4	94%
Wheels & Brakes	24.6	66%	44.7	96%	49.7	96%
Battery System	42.2	100%	53.3	100%	59.3	100%
Driveline	9.8	66%	17.4	93%	19.4	93%
Heat Transfer	11.1	90%	14.9	96%	16.6	96%
Body Structures	9.9	8%	18.5	11%	25.3	14%
Body Closures	8.0	12%	23.3	28%	36.3	39%
Crash Management Systems	6.9	66%	9.4	71%	11.1	75%
Other Component	4.4	93%	5.8	97%	6.4	97%
Total	128.4	31%	226.8	50%	283.5	56%

Source: CM Group

17

DIE TRENDS

Various Online Sources

Funktionsintegration:

+ Grenzen zwischen Struktur- und (prüfpflichtigen) Chassis-Bauteilen verschwimmen
+ Welche Bauteilfunktionen und damit Prüfpflichten kommen auf welche Bauteile zu?

18

DIE TRENDS

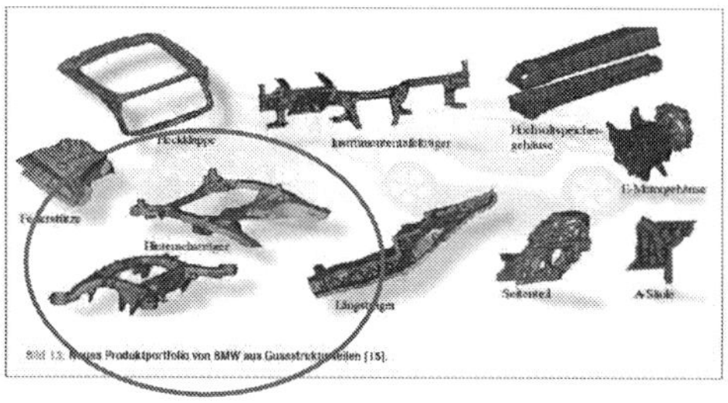

Funktionsintegration:

- Verschiedene Produktionsverfahren im Wettbewerb
  (Druckguss greift nach Sand- bzw. Kokillenguss)
- Materialmix vs. Maximale Integration

Source: Fent, D. A.: Der Wandel einer Druckgießerei von Motor- zu Strukturbauteilen, Aalener Gießerei-Kolloquium, BMW Group, 2015

19

DIE HERAUSFORDERUNGEN
DGZFP JAHRESTAGUNG 2025

Laubinger Rickmann
Technology for Quality

20

DIE HERAUSFORDERUNGEN

Hilfsrahmen / Subframes

+ Anfragen mit Koffermaßen 1.300 x 1.000 x 600 mm³ und mehr
+ Bauteilgewichte >35kg
+ Abrufzahlen von bis zu 200.000 Einheiten p.a.
  → Taktzeit ca. 45s
  (bei 5% Ausschuss und OEE des Gesamtprozesses von 85%)

Mai 2025 Seite 21

DGZfP Jahrestagung 2025 | Maschinen und Konzepte sind geistiges Eigentum von L+R

Laubinger Rickmann

21

DIE HERAUSFORDERUNGEN

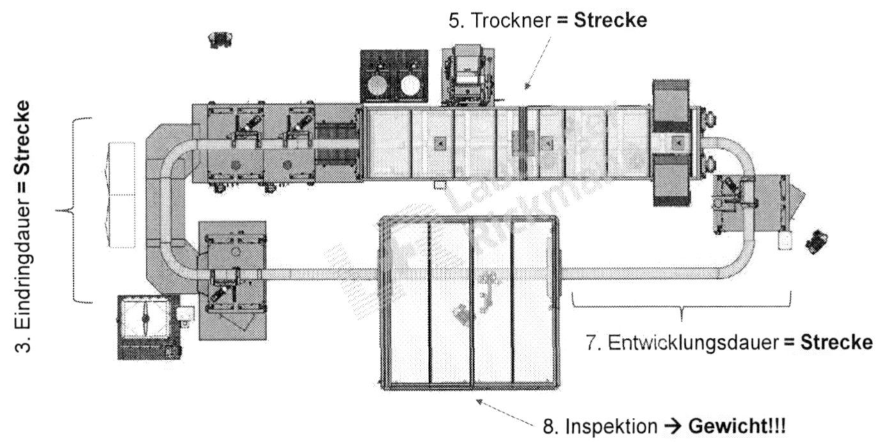

Mai 2025 Seite 22

DGZfP Jahrestagung 2025 | Maschinen und Konzepte sind geistiges Eigentum von L+R

Laubinger Rickmann

22

DIE HERAUSFORDERUNGEN

Gehänge-Abstand, in mm	1750
Takt, in s	45
Eindringdauer, in Min.	8
Eindringdauer, in Takten	11
Eindringdauer, in m	**19.25**
Trocknungsdauer, in Min.	15
Trocknungsdauer, in Takten	20
Trocknungsdauer, in m	**35**
Entwicklungsdauer, in Min.	16
Entwicklungsdauer, in Takten	22
Entwicklungsdauer, in m	**39**
Summe, in m (zzgl. Prüfkabinen etc.)	**93.25**

Mai 2025 Seite 23 DGZfP Jahrestagung 2025 | Maschinen und Konzepte sind geistiges Eigentum von L+R Laubinger Rickmann

23

DIE HERAUSFORDERUNGEN

Ergonomie???

Mai 2025 Seite 24 DGZfP Jahrestagung 2025 | Maschinen und Konzepte sind geistiges Eigentum von L+R Laubinger Rickmann

24

DAS KONZEPT
DGZFP JAHRESTAGUNG 2025

Laubinger Rickmann
Technology for Quality

25

DAS KONZEPT

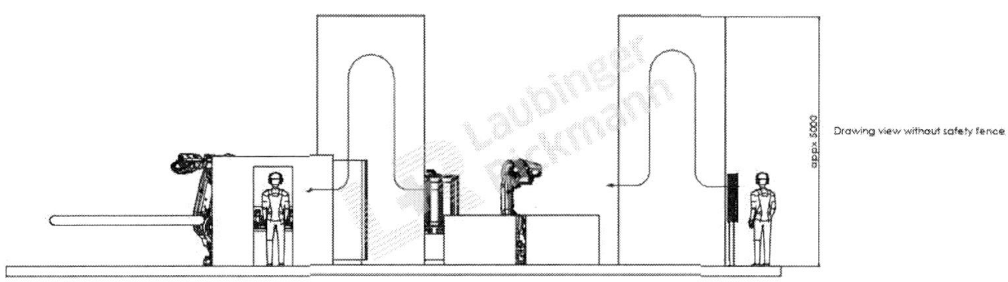

Nutzung der verfügbaren Höhe

Mai 2025 Seite 26 DGZfP Jahrestagung 2025 | Maschinen und Konzepte sind geistiges Eigentum von L+R
Laubinger Rickmann

26

DAS KONZEPT

2.5 Inspection station

After the dryer a second robot transports the part to the inspection turning table. Here it is positioned and can be turned around the vertical in order to inspect it without human handling (shown in Figure 1).

Figure 1 Example of part position on the inspection table

Ergonomische Lösung für den Prüfer

DAS KONZEPT

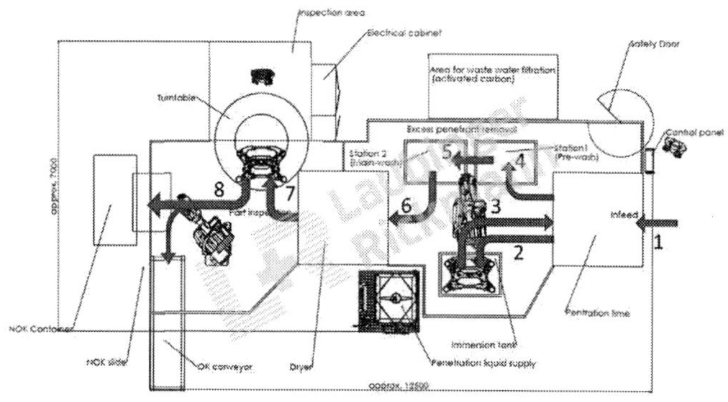

Prüfablauf

DAS KONZEPT

+ Bessere **Raumnutzung** als Hängeförderer, da vertikale Lagerung f. Eindringdauer und Trocknung
+ **Zuverlässige Benetzung** in Tauchtechnik
+ Kein **Ex-Schutz** notwendig
+ Auf andere Bauteile / **Bauteilvarianten** rüstbar
+ Einbringung von **Puffer-Zonen** in Regalsystem realisierbar (vgl. getakteter Hängeförderer)
+ Prozesse und Prozesszeiten können **unabhängig voneinander justiert** werden
+ **Ergonomische** Prüfung schwerer Bauteile ermöglicht
+ Verlässliche **Bauteilseparierung**, keine Verwechslungs- oder Fehlausschleusungs-Gefahr

Mai 2025 Seite 29 DGZfP Jahrestagung 2025 | Maschinen und Konzepte sind geistiges Eigentum von L+R Laubinger Rickmann

29

DIE (ERSTE) UMSETZUNG

DGZFP JAHRESTAGUNG 2025

Laubinger Rickmann
Technology for Quality

30

DIE UMSETZUNG

Theorie Vs. Praxis

+ Offensichtliche Herausforderungen:
  + Abschirmung des Roboters,
  + Schnittstellen zwischen den einzelnen Geräten
  + Positioniergenauigkeit der Bauteilablagen
  + Manuelle Bauteileingabe
+ In der Praxis:
  + Keine Zukauf-Regalsysteme, die den Anforderungen gerecht wurden (Temperatur, Empfindlichkeit, Produktionsumgebung)
  + Roboter-Lieferzeit >50 Wochen (COVID)
  + Aufbau und Inbetriebnahme des Grundsystems deutlich Schneller als erwartet
  + Priorisierung der Arbeitsschritte

Mai 2025 Seite 31

DGZfP Jahrestagung 2025 | Maschinen und Konzepte
sind geistiges Eigentum von L+R

Laubinger Rickmann

31

DIE UMSETZUNG

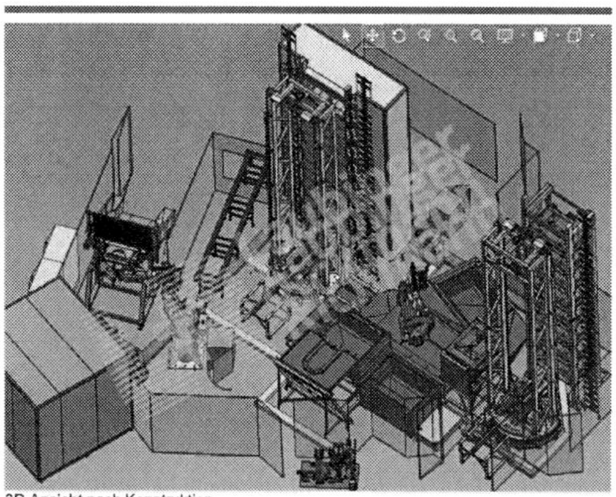

3D Ansicht nach Konstruktion

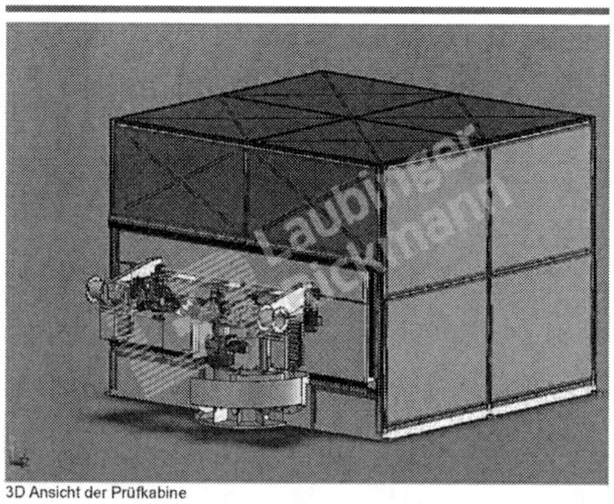

3D Ansicht der Prüfkabine

Mai 2025 Seite 32

DGZfP Jahrestagung 2025 | Maschinen und Konzepte
sind geistiges Eigentum von L+R

Laubinger Rickmann

32

DIE UMSETZUNG

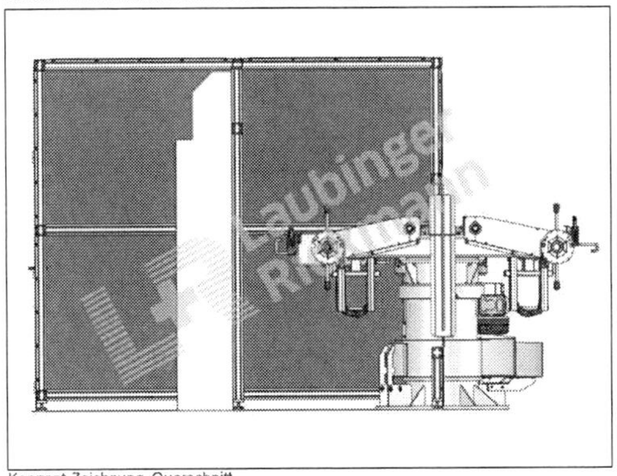

Konzept-Zeichnung, Querschnitt

Drehtisch in Produktion

Mai 2025 Seite 33

DGZfP Jahrestagung 2025 | Maschinen und Konzepte sind geistiges Eigentum von L+R

Laubinger Rickmann

33

WEITER GEDACHT…

DGZFP JAHRESTAGUNG 2025

Laubinger Rickmann

Technology for Quality

34

WEITER GEDACHT… | SKALIERUNG

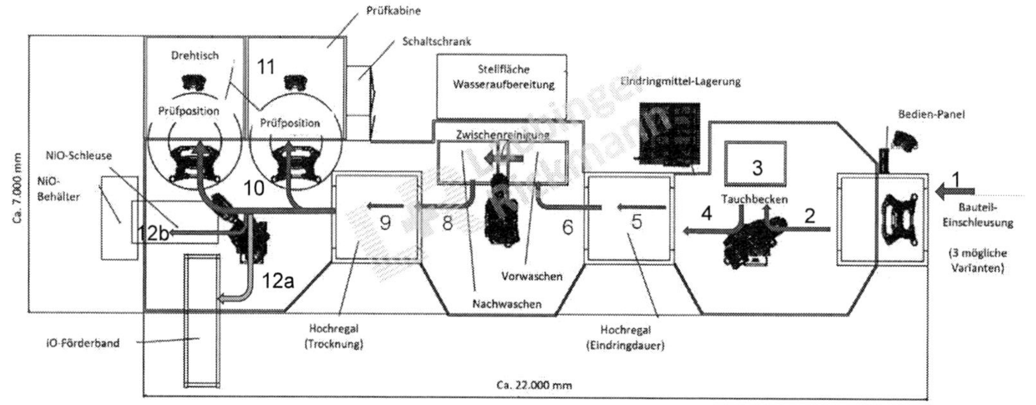

Mai 2025 Seite 35

DGZfP Jahrestagung 2025 | Maschinen und Konzepte sind geistiges Eigentum von L+R

Laubinger Rickmann

35

WEITER GEDACHT… | APPLIKATION

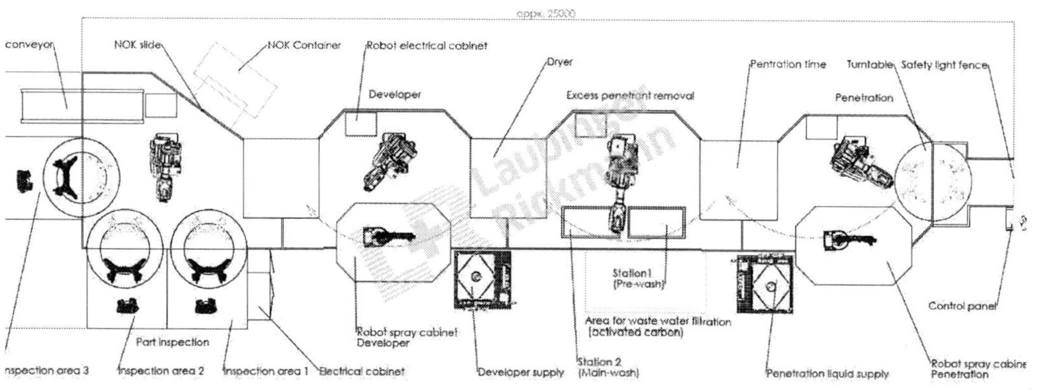

Mai 2025 Seite 36

DGZfP Jahrestagung 2025 | Maschinen und Konzepte sind geistiges Eigentum von L+R

Laubinger Rickmann

36

UND NOCH WEITER...
DGZFP JAHRESTAGUNG 2025

Laubinger Rickmann
Technology for Quality

37

NOCH WEITER... | FLEXIBILITÄT & REZEPTE

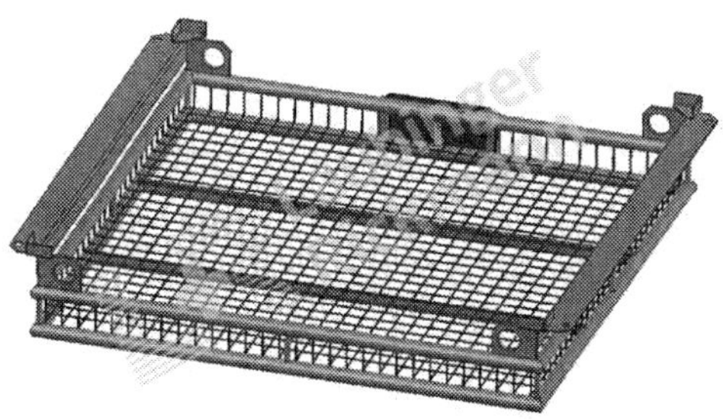

Korb-Träger | Beispiel

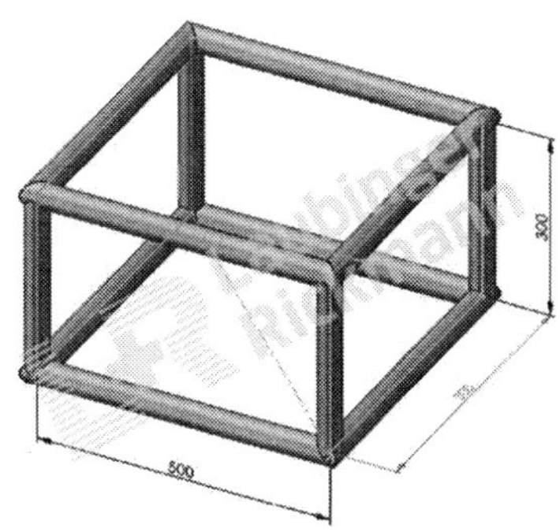

Korb-Träger | Beispiel

Laubinger Rickmann

38

NOCH WEITER... | FLEXIBILITÄT & REZEPTE

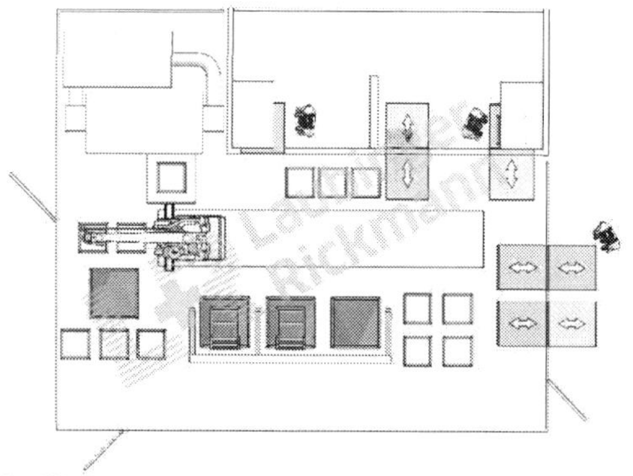

Draufsicht Gesamtanlage (ohne Inspektionskabinen)

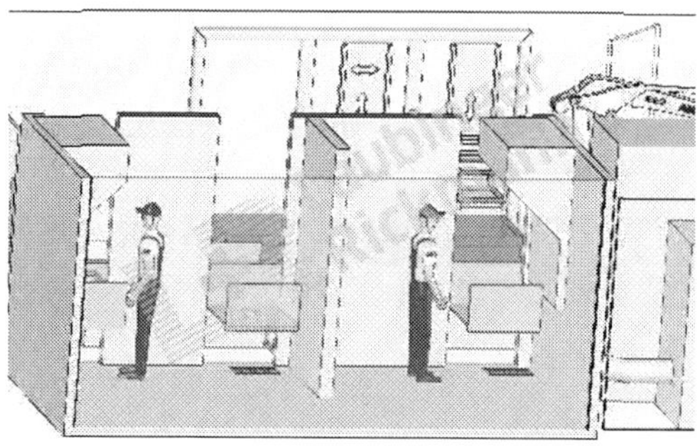

Ansicht des manuellen Arbeitsbereichs

Mai 2025 Seite 39

DGZfP Jahrestagung 2025 | Maschinen und Konzepte sind geistiges Eigentum von L+R

Laubinger Rickmann

39

NOCH WEITER... | FLEXIBILITÄT & REZEPTE

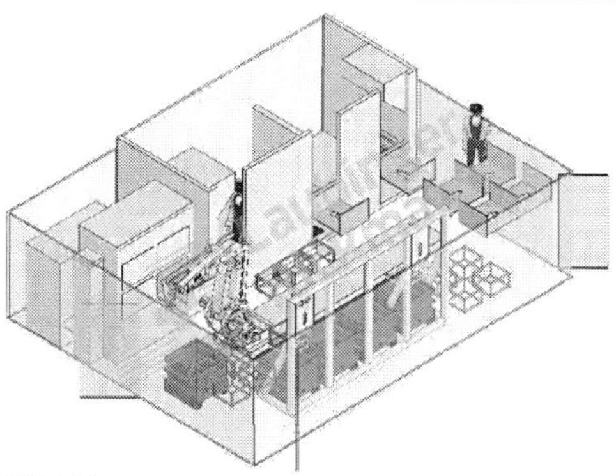

3D-Ansicht 1

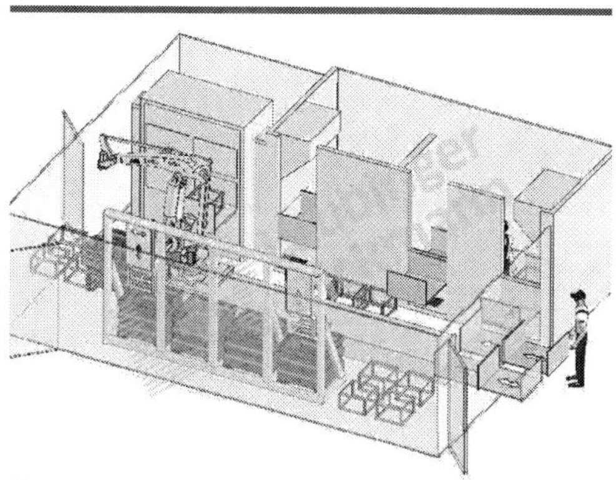

3D-Ansicht 2

Mai 2025 Seite 40

DGZfP Jahrestagung 2025 | Maschinen und Konzepte sind geistiges Eigentum von L+R

Laubinger Rickmann

40

NOCH WEITER… | FLEXIBILITÄT & REZEPTE

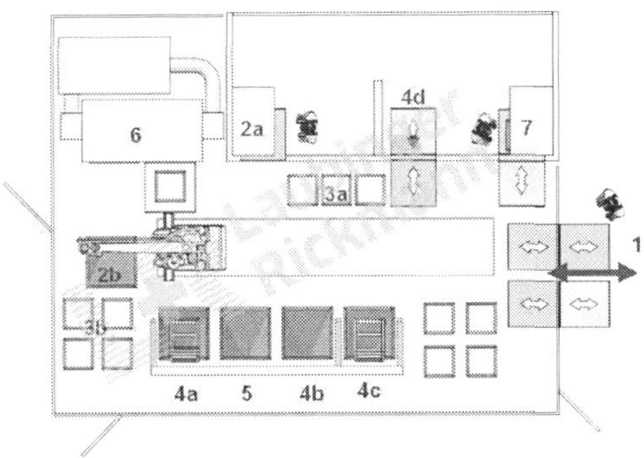

1. Warenträger-Eingabe/-Ausgabe
2. Eindringmittel-Auftrag
   a) Manuell, Sprühen (inkl. Absaugwand)
   b) Automatisiert, Tauchen
3. Eindringdauer
   a) Für manuellen Auftrag
   b) Für automatisierten Auftrag
4. Zwischen-Reinigung
   a) Tauch- und Sprühtechnik (Nachemulgierbar)
   b) Tauch- und Sprühtechnik (Emulgator-Stopp)
   c) Tauch- und Sprühtechnik (Wasser-Abwaschbar)
   d) Handwaschplatz
5. Emulgator
6. Trockner
7. Manuelle Entwickler-Station

Mai 2025 Seite 41

DGZfP Jahrestagung 2025 | Maschinen und Konzepte sind geistiges Eigentum von L+R

L+R Laubinger Rickmann

41

Vielen Dank!

L+R Laubinger Rickmann

Technology for Quality

42

PHILIPP HETTICH

Leitung weltweiter Vertrieb

Mail hettich@laubinger-rickmann.de

Phone +49 2573 9388 77

Mobile +49 160 109 4 209

Contact Me

DGZfP-Jahrestagung 2025

DEFECTOVIEW HS - Hochgeschwindigkeitsbildgebung in Drahtwalzwerken

Frank EIBOFNER [1], Patrick HAIST [1], Sven KOHN [1], Sargon YOUSSEF [1],
Ilja HEINZ [2], Christoph STEIMER [2], Daniel STICHER [2]
[1] Institut Dr. Foerster GmbH & Co. KG, Reutlingen, Deutschland,
frank.eibofner@foerstergroup.com
[2] Saarstahl AG, Völklingen, Deutschland

Kurzfassung. Der Einsatz der Wirbelstromprüfung zur Qualitätskontrolle und Prozessüberwachung in Drahtwalzwerken ist seit Jahren etabliert. Bei der Wirbelstromprüfung werden Fehler als Amplituden- und Phasensignal dargestellt, ein Rückschluss auf den Fehlertyp ist dabei nahezu unmöglich. Eine visuelle Beurteilung der Fehler am Ende des Walzprozesses ist unwirtschaftlich, weil der Draht aufgewickelt wird und nur ein kleiner Teil der Drahtoberfläche sichtbar ist. Zur Prozessüberwachung und zur Vermeidung von Ausschuss ist es aber erforderlich, dass die Ursache von Fehlern schnell gefunden wird.

Zu diesem Zweck wurde das DEFECTOVIEW HS entwickelt, ein optisches Dokumentationssystem, das den gesamten Drahtumfang abbildet. In der Walzlinie wird es hinter dem Wirbelstromprüfgerät positioniert und zeichnet zu den detektierten Fehlern aus der Wirbelstromprüfung entsprechende Bildsequenzen auf. Die Bildbereiche vor und nach dem Fehler werden zur Beurteilung der Fehlerursache herangezogen. Zusätzlich lassen sich auch weitere Bilder von Drahtbereichen aufnehmen, um beispielsweise festzustellen, ob für das Wirbelstromprüfgerät nicht auffindbare Längsfehler auf dem Draht vorhanden sind.

Das DEFECTOVIEW HS kann durch seine kompakte Bauform einfach in Walzlinien integriert werden. Für den Betrieb ist lediglich ein Gebläse notwendig, das zur Kühlung der Komponenten, sowie zur Reinigung von Zunder- und Wasserrückständen aus dem Glaszylinder dient. Der Wartungsaufwand des Systems beschränkt sich auf eine wöchentliche Reinigung des Glaszylinders.

Die ersten DEFECTOVIEW HS sind seit über einem Jahr bei der Saarstahl AG in Betrieb.

Keywords: optische Verfahren, Wirbelstromprüfung, Drahtwalzwerk, Heißdraht

1. Ausgangssituation:

1.1 Drahtwalzwerk:

In einem Drahtwalzwerk (siehe Abb. 1) werden aus quadratischen Knüppeln mit Kantenlängen von z.B. 150 mm x 150 mm runde Drähte mit Durchmessern von 4,5 mm – 17 mm hergestellt. Der erste Schritt im Herstellungsprozess besteht darin, den Knüppel mit einer qualitätsabhängigen Erwärmungsstrategie in einem Ofen zu erhitzen. Nachdem der Knüppel die vorgegebene Walzanfangstemperatur erreicht hat, wird er in mehreren Umformschritten auf den gewünschten Drahtdurchmesser heruntergewalzt. Bei jedem Umformschritt verringert sich die Querschnittsfläche des Drahtes, wodurch sich aufgrund des konstanten Volumens

https://doi.org/10.58286/31279

dessen Geschwindigkeit erhöht. Die letzten Umformschritte werden im sogenannten Fertigblock durchgeführt. Während des gesamten Walzprozesses kann die Qualität des Drahtes überprüft werden, z.B. mit einer Profilvermessung oder Oberflächenprüfung. Danach wird der nun bis zu 100 m/s schnelle Draht in einem Windungsleger gefahren und anschließend auf einem Windungs- / Kühltransportband abgelegt, bevor er seine finale Coilform in der nachgelagerten Bundbildekammer erhält. Das abgekühlte Drahtcoil wird in einer Kettentransportbahn zur Drahtprüfstation transportiert und dort mit verschiedenen Methoden, je nach Qualitätsvorgabe, geprüft.

Abb. 1 Vieradrige Walzstraße in Saarbrücken Burbach.

1.2 Wirbelstromprüfung:

Die Prüfung auf Oberflächenfehler wird bei der Drahtherstellung normalerweise mit dem Wirbelstromverfahren durchgeführt. Das Wirbelstromprüfgerät wird dabei direkt nach dem Fertigblock installiert. Bei der aufgrund der hohen Drahttemperaturen mit Wasser gekühlten Sensorik (siehe DEFECTOTHERM Spule in Abb. 2) handelt es sich um eine sogenannte transformatorische Durchlaufspulenanordnung, d.h. die Erreger- und Messspulen umfassen den gesamten Draht. Dabei wird von der Erregerspule ein auf den Draht wirkendes Wechselfeld erzeugt, welches bei Anwesenheit eines Oberflächenfehlers eine Spannung in der Empfangsspulenanordnung, die in Differenz verschaltet ist, induziert. Diese Signale werden digitalisiert und anschließend mit Hilfe einer sogenannten kohärenten Demodulation in der komplexen Impedanzebene ausgewertet. Überschreitet die komplexe Amplitude des Fehlersignals eine vordefinierte Schwelle, wird diese Messung als ein Fehler klassifiziert und zur Dokumentation ein Fehlerereignis in eine Datenbank geschrieben. Durch unterschiedliche Schwellen können die Fehler nach Größen sortiert werden. Nach Abschluss des Walzvorgangs wird für jeden Draht mittels einer trivialen Logik eine Qualitätszahl aus Fehleranzahl und Fehlergröße errechnet. Diese Qualitätszahl repräsentiert eine statistische Größe, die es jedoch nicht erlaubt, die Art des Fehlers und die zugrunde liegende Ursache zu beurteilen.

Abb. 2 DEFECTOTHERM Spule im Einsatz

1.3 Problemstellung:

Die Drahtherstellung ist ein komplexer Prozess, bei dem in unterschiedlichen Prozessschritten typische Restfehler auf der Drahtoberfläche auftreten, die je nach Einsatzzweck qualitätsrelevant werden können. Das Wirbelstromsignal wird als komplexe Impedanz dargestellt, aus dem die Auswerteelektronik die Amplituden- und Phaseninformation berechnet (Abb. 3). Aus der komplexen Impedanz lässt sich jedoch in den wenigsten Fällen ein Rückschluss auf die Ursache des Fehlers ziehen. Diese Fehler mit dem Wirbelstromverfahren im Sinne eines schnellen Screenings zu detektieren ist nur der erste Schritt. Viel wichtiger ist es im Anschluss die Ursache der Fehler zu finden, damit möglichst schnell geeignete Maßnahmen zur weiteren Reduzierung der walzdrahttypischen Restfehler getroffen werden können. Zu klären ist, ob die Fehlerursache im Walzprozess liegt oder ob bereits das Ausgangsmaterial, d.h. der Knüppel, Mängel aufgewiesen hat.

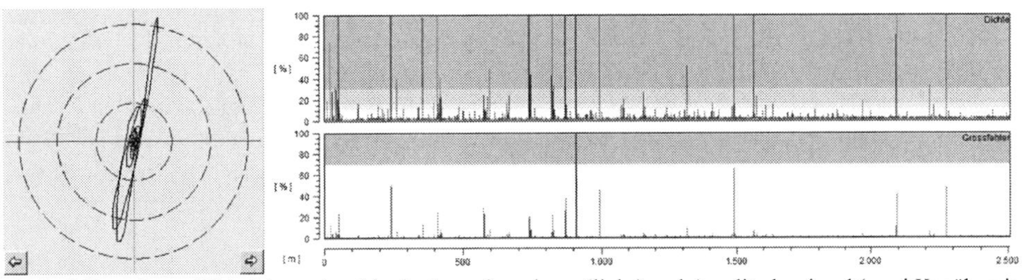

Abb. 3 Komplexes Wirbelstromsignal in der Impedanzebene (links) und Amplitudensignal (zwei Kanäle mit unterschiedlicher Verstärkung) in der Anzeige auf dem Prüfgerät (rechts).

Anhand der Form eines Oberflächenfehlers können erfahrene Walzwerker die prozessbedingte Ursache annähernd bestimmen. Allerdings ist die visuelle Überprüfung der aufgewickelten Drähte nach Abschluss des Walzprozesses nicht immer erfolgreich, vor allem wenn es sich bei dem zu suchenden Fehler um ein Einzelereignis handelt. Der gewickelte Draht (Abb. 4) lässt eine visuelle Inspektion nur an seiner äußeren Oberfläche zu und auch dort ist es schwierig einen Einzelfehler in kurzer Zeit zu finden.

Abb. 4 Drahtcoil, der nur die Betrachtung eines kleinen Teils der Oberfläche zulässt

Die Wirbelstromprüfung in Kombination mit einer visuellen Inspektion ist also prinzipiell geeignet, die Fehlerursache dem Walzprozess zuzuordnen, eine 100 % Zuordnung ist jedoch mit einem sehr hohen und unwirtschaftlichen Aufwand verbunden.

2. Lösungsansatz:

2.1 Optische Fehlerdokumentation:

Das Wirbelstromsignal allein ist selten geeignet eine Aussage über die Fehlerart und die Ursache des Fehlers zu treffen. Deshalb ist es naheliegend die Wirbelstromprüfung mit einem System zu kombinieren, das Bilder der Fehler im Walzprozess aufnimmt. Die Anforderungen an ein solches hüttentaugliches System sind, dass es die rauen Bedingungen im Drahtwalzwerk, beispielsweise hohe Drahttemperaturen von bis zu 1250 °C, Zunder und Wasser, ohne Beschädigung übersteht und gleichzeitig scharfe Bilder des bis zu 150 m/s schnellen Drahtes aufnehmen kann. Außerdem müssen die Wirbelstromdaten mit den dazugehörigen Bilddaten direkt nach der Detektion zugänglich sein, so dass zeitnah in den Prozess eingegriffen werden kann.

2.2 Kooperation SAG und Entwicklungsprototyp:

Das DEFECTOVIEW HS für die optische Fehlerdokumentation wurde in einer Kooperation mit der Firma Saarstahl AG entwickelt. Zu Beginn wurden erste Versuche mit einem Laboraufbau durchgeführt, um vor allem die Leistungsfähigkeit der bildgebenden Komponenten zu testen. Danach wurde ein Prototyp direkt in der Walzlinie aufgebaut, der bereits in Kombination mit dem Wirbelstromprüfgerät arbeitete. Dieser wurde dann im letzten Schritt durch ein überarbeitetes Seriensystem ersetzt (Abb. 5). Durch diese Versuche und Installationen in die Walzlinie konnte das System dank der Unterstützung des Kooperationspartners auf die rauen Bedingungen im Walzwerk erprobt und optimiert werden.

Seit August 2023 sind bei der Saarstahl AG am Standort Burbach vier DEFECTOVIEW HS dauerhaft installiert und im Einsatz.

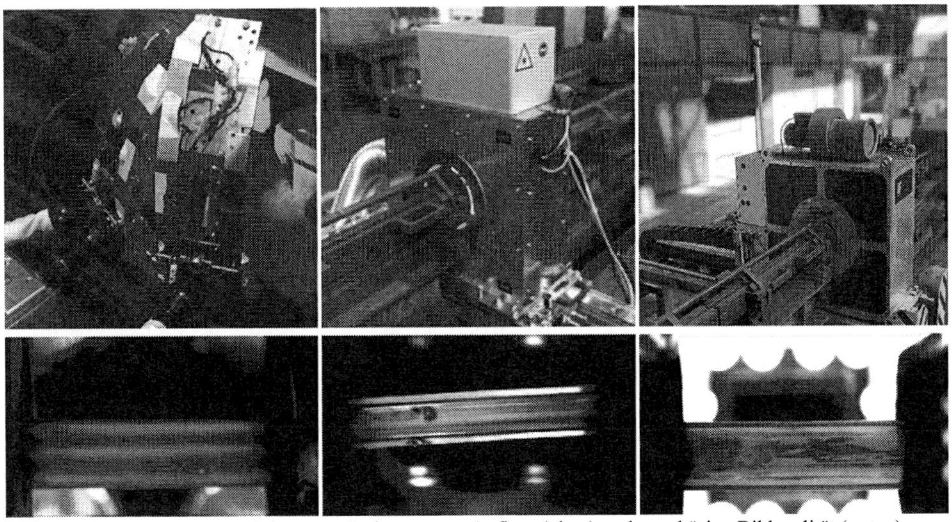

Abb. 5 Entwicklungsschritte zum Seriensystem, Aufbau (oben) und zugehörige Bildqualität (unten) von links nach rechts: erste Aufnahmen im Walzwerk, Prototyp, Seriensystem

2.3 Komponenten:

Das DEFECTOVIEW HS besteht aus einem Kamerasystem (Abb. 6), das in die Walzstraße integriert ist und aus einer Auswerteeinheit, die in einem Elektronikschrank verbaut ist.

Abb. 6 DEFECTOVIEW HS Kamerasystem

2.3.1 Kamerasystem:

Der heiße Draht wird mit zwei Schutzdüsen durch das Kamerasystem geführt. Ein Abstand zwischen den Schutzdüsen erlaubt mit Hilfe von vier Flächenkameras die gesamte Drahtoberfläche zu erfassen. Durch den Einsatz von Flächenkameras werden im Vergleich zu Zeilenkameras Bildartefakte, die durch das Schwärmen des Drahtes in der Materialführung und durch Toleranzen der Geschwindigkeitsinformation entstehen, vermieden und es werden scharfe und verzerrungsfreie Bilder erzeugt. Mit einer nominalen Bildauflösung von 10 Pixel pro mm können auch kleine Defekte auf den Bildern erkannt werden. Der Abstand der beiden Schutzdüsen, d.h. der ungeführte Bereich des Drahtes von ungefähr 40 mm entspricht dabei der Drahtlänge, die pro Bild erfasst werden kann. Beleuchtet wird der Draht von LEDs, die in vier Einheiten zusammengefasst sind und durch Blitzcontroller mit Strompulsen versorgt werden. Die LED-Einheiten sind so angeordnet, dass die Oberfläche des Drahtes homogen ausgeleuchtet wird. Durch die Verwendung von Blitzcontrollern wird vermieden, dass sich die LEDs erhitzen, da sie nur für eine kurze Zeit angeschaltet werden. Die Ansteuerung der Blitzcontroller erfolgt direkt über die Kameras, so dass die Belichtungszeiten der Kameras und die Beleuchtungen synchronisiert sind. Zur Erkennung des Drahtes ist im Kamerasystem eine Lichtschranke verbaut.

Das Kamerasystem ist in zwei Bereiche unterteilt: je ein Bereich, in dem sich der Draht und die elektronischen Komponenten befinden. Um die elektronischen Komponenten vor Verschmutzungen zu schützen, sind die beiden Bereiche durch einen Glaszylinder getrennt. Zur Kühlung der elektronischen Komponenten und zum Schutz des Glaszylinders vor Verschmutzung ist ein Luftstrom erforderlich. Dieser kann durch ein Gebläse erzeugt werden, das direkt auf dem Kamerasystem verbaut ist oder von einem entfernten Gebläse mit einem Schlauch in das Kamerasystem geleitet werden. Das Gebläse sorgt im Bereich mit den elektronischen Komponenten für einen leichten Überdruck, der verhindert, dass Schmutz eindringen kann. Durch einen kleinen Spalt kann der Luftstrom in den Glaszylinder eindringen und dort im Prozess vorhandenes Wasser, Zunder und Schmutz direkt aus dem System transportieren. Der konstante Luftstrom im System reicht aus, um die Temperatur des glühenden Drahtes abzuführen und die elektronischen Komponenten zu kühlen.

Durch den verbauten Glaszylinder, der mit wenigen Handgriffen entnommen werden kann, wird der Aufwand für die notwendigen Wartungsarbeiten auf ein Minimum reduziert. Mit den im Kamerasystem vorhandenen Sensoren (Temperatur, Druck, Endschalter) kann der Zustand des Systems über eine SPS überwacht und auf der Bedienoberfläche des DEFECTOVIEW HS dargestellt werden.

2.3.2 Auswerteeinheit:

Die Auswerteeinheit des DEFECTOVIEW HS ist in einem Elektronikschrank untergebracht, der nicht direkt an der Walzlinie stehen muss. Hauptbestandteil der Auswerteeinheit ist ein Rechner mit einem großen Arbeitsspeicher (z.B. 128 GB RAM) und zwei Bildeinzugskarten, sogenannten Framegrabbern. Die Framegrabber sind mit der Walzstraße, dem Wirbelstromprüfgerät und den Kameras verbunden. Zum einen werden die Kameras über den Framegrabber konfiguriert und zum anderen werden die von den Kameras bereitgestellten Bilder vom Framegrabber eingezogen, mit Triggersignalen und Positionsdaten versehen und dann in einen Ringspeicher verschoben.

Auf dem Rechner läuft außerdem die Software, mit der das DEFECTOVIEW HS konfiguriert wird. Über diese können die Blitzcontroller und die im Kamerasystem verbaute SPS konfiguriert werden und die Statusinformationen der Sensoren abgefragt werden. Die Konfigurationssoftware ist als Webapplikation realisiert, wodurch dessen Konfiguration und auch die Abfrage von relevanten Statusinformationen sowohl am Elektronikschrank wie auch an anderen Orten über das Netzwerk durchgeführt werden kann.

2.4 Systemaufbau:

In der Regel findet die Wirbelstromprüfung im Drahtwalzwerk nach dem Fertigblock statt. Hinter dem Fertigblock und nach der Einbauposition der DEFECTOTHERM Spule folgt dann ein Kühlstrecke. Hinter dieser befindet sich das Kamerasystem des DEFECTOVIEW HS. Zwischen der DEFECTOTHERM Spule und dem Kamerasystem liegt dadurch ein Abstand von ungefähr 10 m. Abb. 7 zeigt beispielhaft die schematische Einbausituation der DEFETOTHERM Spule und des Kamerasystems in einer Walzlinie und die Verbindungen der einzelnen Komponenten bis hin zu den Datenbanken und der kombinierten Anzeige im ImageViewer.

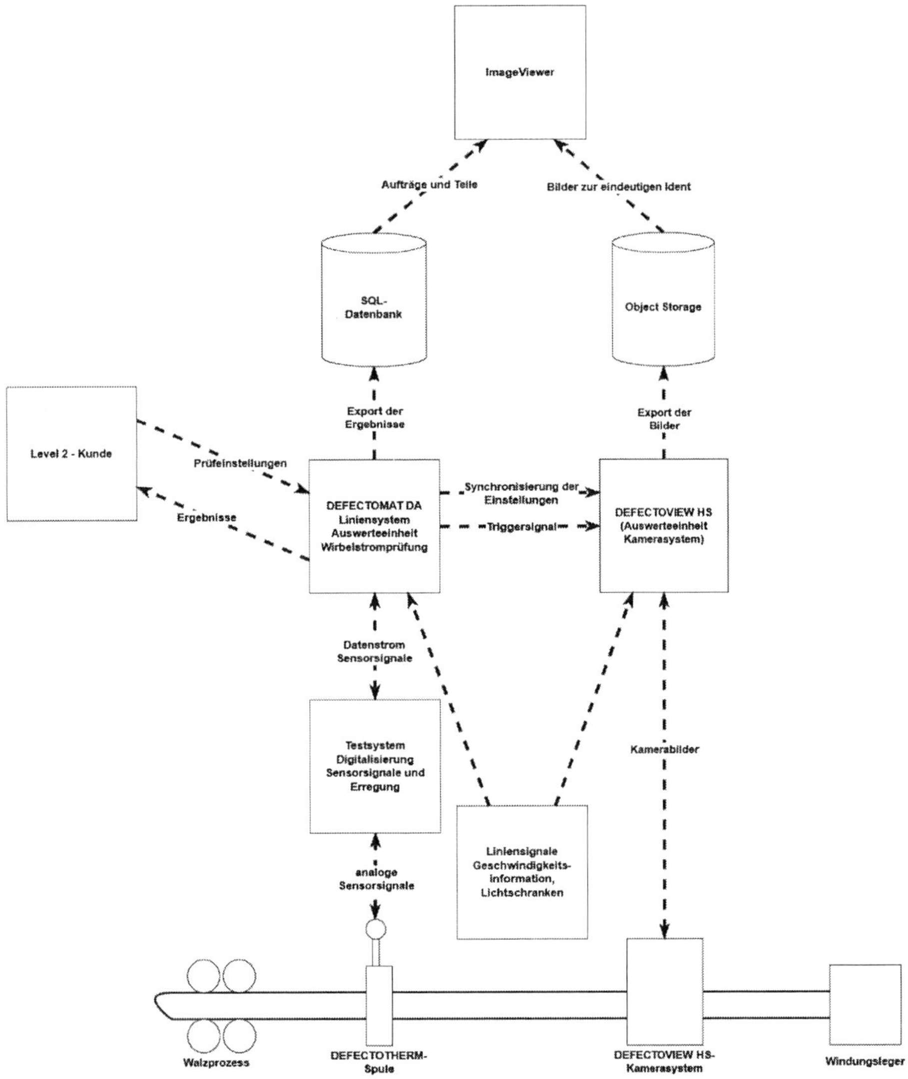

Abb. 7 Systemüberblick Verbindungen DEFECTOMAT DA und DEFECTOVIEW HS, sowie Speicher

Der Walzprozess in einem Drahtwerk ist normalerweise derart hoch automatisiert, dass das Wirbelstromprüfsystem DEFECTOMAT DA automatisch über eine Level 2 Schnittstelle parametriert wird. Die relevanten Einstellungen (z.B. Materialdurchmesser, maximale Geschwindigkeit, usw.) werden dann durch ein Synchronisierungsprogramm vom DEFECTOMAT DA direkt auf die Auswerteeinheit des DEFECTOVIEW HS übertragen. Der Zugriff auf das DEFECTOVIEW HS erfolgt über eine Webapplikation, wodurch eine Parametrierung oder eine Überprüfung der Einstellungen über Fernzugriff erfolgen kann.
Liniensignale, wie die Lichtschranken und die Geschwindigkeitsinformationen, sind jeweils fest an die Hardware der Auswerteeinheiten des Wirbelstromprüfgeräts und des DEFECTOVIEW HS angeschlossen.

Die Wirbelstromsignale werden hardwarenah an der Spule im Testsystem digitalisiert und dann störungsfrei an das Liniensystem in der Auswerteelektronik übertragen. Das Liniensystem wertet dann die Daten entsprechend den Prüfeinstellungen aus und sendet die

Ergebnisse über die Level 2 Schnittstelle an das Leitsystem. Zusätzlich werden die Prüfergebnisse in eine SQL-Datenbank exportiert, so dass einerseits eine Ergebnisansicht und andererseits eine Archivierung möglich ist. Erreicht ein Wirbelstromsignal eine voreingestellte Fehlerschwelle, sendet der DEFECTOMAT DA ein Triggersignal an das DEFECTOVIEW HS.

Das DEFECTOVIEW HS Kamerasystem liefert permanent Kamerabilder an die Auswerteeinheit, ohne diese langfristig zu speichern. Wenn ein Triggersignal vom Wirbelstromprüfgerät die Auswerteeinheit erreicht, werden die zu dem auslösenden Fehler gehörenden Bilder in einen Object Storage abspeichert. Die Bilder können in jedem S3 kompatiblen Object Storage abgespeichert werden, wodurch eine Cloudanbindung möglich ist. Es ist optional auch möglich, Bilder abzuspeichern, ohne dass ein Triggersignal vorhanden ist, um beispielsweise Informationen über den aktuellen Zustand des Materials und des Prozesses zu erhalten. Solche Samplebilder können über einen Parameter einstellbar an verschiedenen Positionen des Drahtes automatisch oder manuell aufgenommen werden. Durch die Verwendung eines Object Storage, können die Samplebilder mit anderen Markierungen, sogenannten Tags, als die Fehlerbilder versehen werden. Um die Datenmenge des Langzeitspeichers zu reduzieren, ist es möglich einen zweiten, lokalen Object Storage in die Auswerteeinheit des DEFECTOVIEW HS zu integrieren. In den lokalen Object Storage werden alle Bilder mit Fehler, sowie alle Samplebilder geschrieben. Diese Bilder können über den ImageViewer direkt ausgewertet werden. Beim Übertragen der Bilder aus dem lokalen Object Storage in den Langzeitspeicher (siehe Abb. 8) werden Samplebilder, die nur während des Prozesses relevant sind, nicht übertragen.

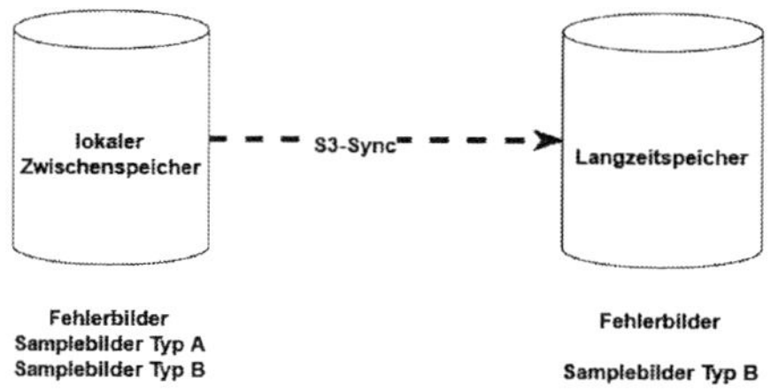

Abb. 8 Speicherung

Mit dem ImageViewer können die Wirbelstromdaten aus der SQL-Datenbank des DEFECTOMAT DA geladen werden, inklusive der Auftragsdaten. Über die Auftragsdaten findet der ImageViewer die zu den jeweiligen Prüfteilen gehörenden Bilder im Object Storage und zeigt diese gemeinsam mit den Wirbelstromsignalen an. Der ImageViewer ist eine Software, die auf jedem Rechner installiert werden kann, der Zugriff auf die SQL-Datenbank und den Object Storage hat.

2.5 Funktionsweise Rücksprung:

Der DEFECTOMAT DA benötigt vom Empfang der Wirbelstromsignale bis zum Aussenden des Triggersignals eine gewisse Zeit, in der sich der fehlerhafte Drahtabschnitt in der Walzlinie weiter nach hinten bewegt. Als Markierposition wird die Strecke bezeichnet, die vor Auslösen des Triggersignals und nach Passieren der DEFECTOTHERM Spule zurückgelegt wird. Entsprechend den vorliegenden Drahtgeschwindigkeiten beträgt die Markierposition ca. 30 m. In vielen Drahtwalzwerken ist es aber aus Platzgründen nicht möglich das Kamerasystem 30 m hinter dem Wirbelstromprüfgerät einzubauen.

Dieses Problem kann durch Zwischenspeicherung der Bilder im DEFECTOVIEW HS umgangen werden. Das Kamerasystem kann dadurch unabhängig von der Markierposition in die Walzlinie integriert werden. Der Rücksprung (Abb. 9) entspricht dem Abstand vom Kamerasystem zur Markierposition.

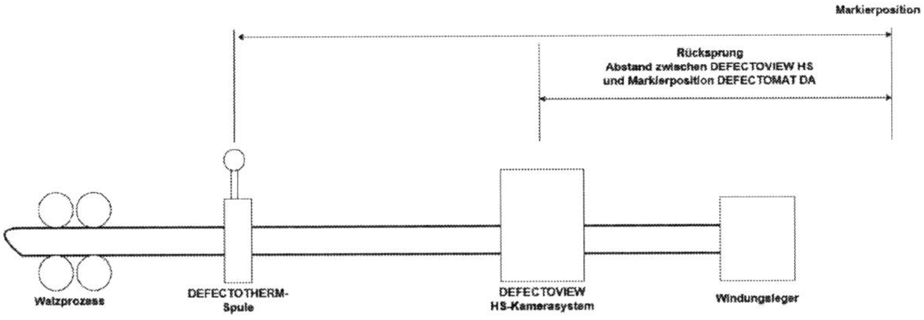

Abb. 9 Abstandswerte für Rücksprung

Sobald von der Lichtschranke des Wirbelstromprüfgerätes ein neuer Draht erkannt wird, startet im DEFECTOVIEW HS ein Wegtaktzähler. Dieser hört auf zu zählen, wenn die Lichtschranke im Kamerasystem belegt ist. Die Anzahl der gezählten Wegtakte entspricht dann dem Abstand zwischen DEFECTOTHERM Spule und Kamerasystem. Durch den gemessenen Abstand zwischen den beiden Systemen und der eingestellten Markierposition, kann der Rücksprung berechnet werden. Die Bilddaten, die der Länge des Rücksprungs entsprechen, müssen zwischengespeichert werden, zum Beispiel in Form eines Ringspeichers, der im Arbeitsspeicher (RAM) des Rechners umgesetzt ist. Bei einem Rücksprung von 10 m müssen somit pro Kamera mindestens 250 Bilder im RAM zwischengespeichert werden. Im Ringspeicher werden genauso viele Bilder vorgehalten, wie für den Rücksprung notwendig sind. Kommen weitere Bilder in den Ringspeicher, so werden nicht mehr notwendige Bilder, wie in Abb. 10 zu erkennen, entfernt.

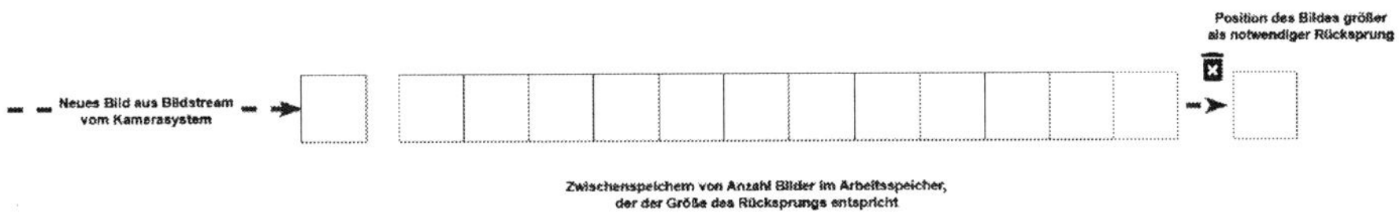

Abb. 10 Bild wird aus dem Ringspeicher entfernt

Kommt ein Triggersignal vom Prüfgerät, so wird im Ringspeicher das entsprechende Bild (Abb. 11) ausgewählt und abgespeichert.

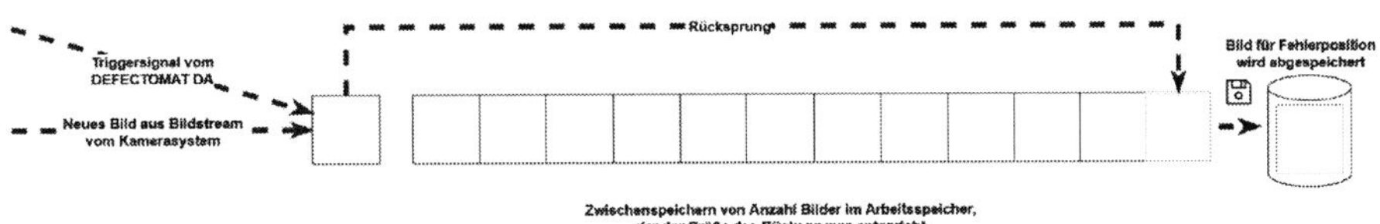

Abb. 11 Bild wird aus dem Ringspeicher abgespeichert

Bei jedem Triggersignal wird ein Bildbereich von ungefähr 1 m abgespeichert. Dadurch lassen sich z.B. Toleranzen des Wegtaktes kompensieren, die sich besonders bei großen Abständen zwischen DEFECTOTHERM Spule und DEFCTOVIEW HS auf die Treffergenauigkeit auswirken.

2.6 Einsatz bei der Saarstahl AG:

Im Drahtwalzwerk Burbach sind vier DEFECOTIEW HS im Durchmesserbereich von 4.5 mm – 17 mm im Einsatz. Bei diesen Durchmessern liegt die Geschwindigkeit des Drahtes bei bis zu 100 m/s. Die DEFECTOVIEW HS liefern über den gesamten Durchmesserbereich, bei einer konstanten Materialführung von 20 mm, scharfe und kontrastreiche Bilder. Für jedes Fehlersignal wird eine Bildsequenz mit einem Längenabschnitt von ca. 1 m Draht abgespeichert. So kann nicht nur der eigentliche Fehler beurteilt werden, sondern bei längeren Fehlern auch auf deren Entstehung geschlossen werden.

Unterschiedlich konfigurierte Instanzen des ImageViewer werden in der Revision und in der Qualitätsstelle verwendet. In der Revision zeigt der ImageViewer die letzten Drähte an, wobei die Bildsequenzen automatisch abgespielt werden. In der Qualitätsstelle (Beispiele siehe Abb. 12 und Abb. 13) werden mit dem ImageViewer gezielt Drähte mit Fehlersignalen analysiert und die Bilder zur Unterstützung der finalen Beurteilung der Qualität herangezogen.

Die zusätzlich zum Wirbelstromsignal vorhandenen Bilder haben die Zeiten, die zur Beurteilung der Fehlertypen und ob ein Eingriff in den Prozess notwendig ist, deutlich reduziert. Falls ein Eingriff in den Prozess notwendig ist, bieten die Bilder eine wertvolle Unterstützung zur Ermittlung möglicher Ursachen.

Die Wartung des Kamerasystems beschränkt sich aktuell auf eine geplante Reinigung des Glaszylinders am Wochenende und den Austausch des Gebläsefilters einmal pro Monat. Der Zeitaufwand pro Kamerasystem beträgt dabei weniger als 10 Minuten.

Seit Einbau der vier DEFCOVIEW HS ist es zu keinem nennenswerten Schaden oder Ausfall der Komponenten gekommen.

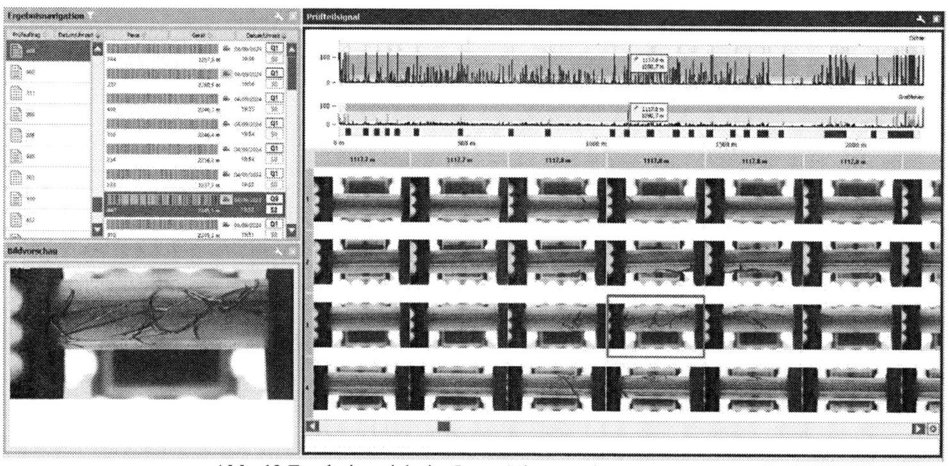

Abb. 12 Ergebnisansicht im ImageViewer mit Einwalzungen

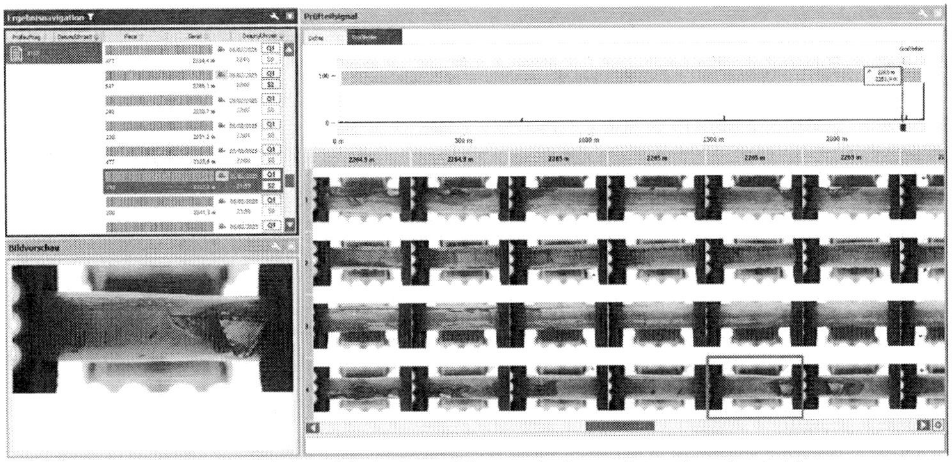

Abb. 13 Ergebnisansicht im ImageViewer mit eingewalztem Fremdmaterial

3. Fazit:

Die Kombination des Wirbelstromprüfgeräts mit dem optischen Dokumentationssystem DEFECTOVIEW HS erlaubt es unter Verwendung von Flächenkameras, vom Wirbelstromsystem gefundene Fehler direkt nach Abschluss des Drahtes zu identifizieren und in realitätsgetreuer Darstellung zu visualisieren.

Im ImageViewer werden die Bilder dem Amplitudensignal zugeordnet und je nach Fehlertyp und -größe kann der Walzwerker entscheiden, ob in den Walzprozess eingegriffen werden muss. Die Verkürzung der Zeit, die bis zu dieser Entscheidung vergeht, führt zur signifikanten Reduktion bis hin zur vollständigen Vermeidung von Ausschuss.

Durch die langjährige Erprobung mit Hilfe eines Prototyps, konnte das DEFECTOVIEW HS Seriensystem für den Einsatz im Drahtwalzwerk optimiert werden. Statt Druckluft und Kühlgerät genügt ein einfaches Gebläse, welches nicht nur Verschmutzungen aus dem Kamerasystem entfernt, sondern auch die elektronischen Komponenten kühlt. Dadurch verringert sich auch der wöchentliche Wartungsaufwand zur Reinigung des Systems auf ein Minimum. Die Verwendung von LEDs in Kombination mit Blitzcontroller sorgt für lange Standzeiten der Beleuchtungseinheiten.

Fliegende Roboter für ZfP an schwer erreichbaren Stellen in der Höhe

Felix STADLER [1],
[1] Voliro AG, Zürich, Schweiz, fjs@voliro.com

Kurzfassung

Neuartige Drohnen ermöglichen ZfP in der Höhe und helfen so Betriebsausfälle und Risiken für Betreiber industrieller Anlagen zu minimieren.

Die zerstörungsfreie Prüfung ist eine etablierte Methode zur Überprüfung der Integrität industrieller Anlagen. Sie umfasst kontaktbasierte Tests, die Korrosion, Erosion oder Materialfehler in Stahl und anderen Materialien aufdecken. Traditionell werden diese Tests mit Instrumenten durchgeführt, welche von Hand mit der Oberfläche in Kontakt gebracht werden. Für Messungen in der Höhe sind jedoch Gerüste, Hebebühnen oder Industriekletterer erforderlich – zeitaufwändige und kostenintensive Methoden mit beträchtlichem Risikomanagement.

Die Nutzung von Drohnen für ZfP-Messungen in der Höhe eliminiert die Notwendigkeit von Gerüsten oder Industrieklettern. Dabei kommen neuartige Drohnendesigns zum Einsatz, die erstmals 2022 eingeführt wurden und speziell für kontaktbasierte Inspektionen entwickelt wurden. Im Gegensatz zu herkömmlichen Multikoptern, die beim Ausüben von Druck instabil werden, ermöglichen diese Designs eine präzise und stabile Krafteinwirkung auf Oberflächen.

Drohnen-gestützte Inspektionen sind bis zu 10-mal schneller und reduzieren Sicherheitsrisiken, indem sie Arbeiten in der Höhe ersetzen. Zudem können Anlagen im laufenden Betrieb inspiziert werden, wodurch Ausfälle vermieden werden.

Keywords: ZfP, Drohne, Robotik

PRÄSENTATOR

Felix Stadler

Head of Customer Success

Bei Voliro seit 4 Jahren

Verantwortlich für Bestandskundenbetreuung

VOLIRO

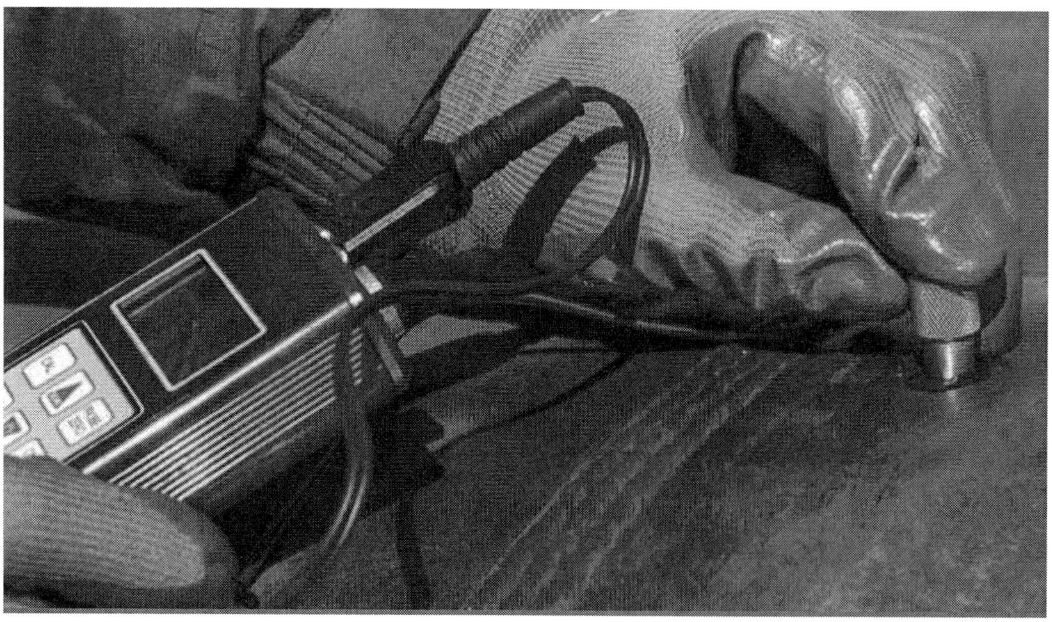

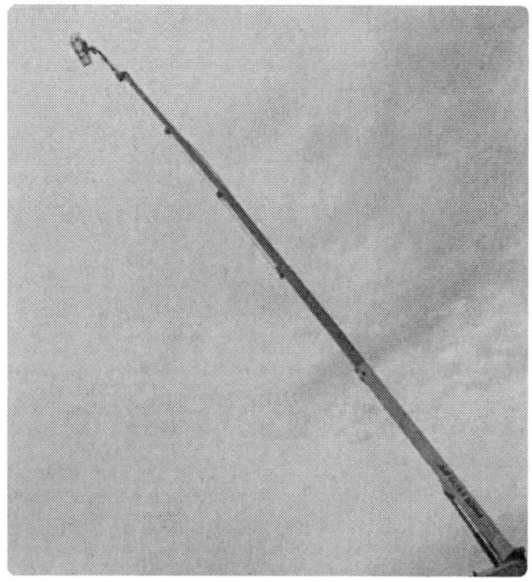

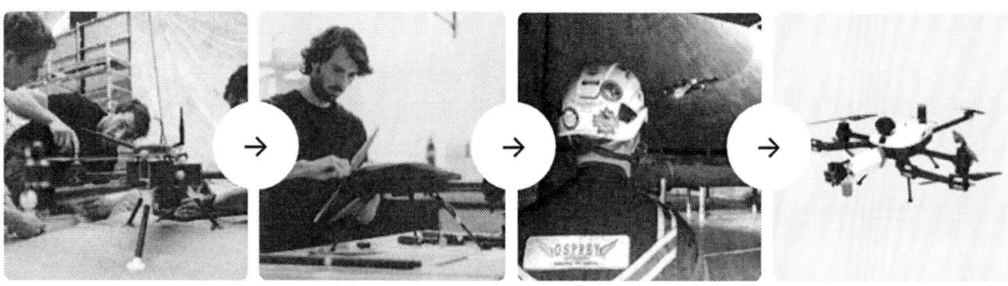

2016: Forschung

Forschungsprojekt am Autonomous Systems Lab der ETH Zurich

2019: Gründung

Gründung der Voliro AG

2020/21: Produktentwicklung

Validierung der Anwendungen in der Industrie

2022: Produkteinführung

Lancierung der Voliro T Drohne als Produkt

TECHNISCHE MERKMALE

Kontakt an Oberflächen

Mit einer Kraft von bis zu 30 N

In jeglicher Richtung

360-Grad-Orientierung der Drohne

Stabil und robust

Hält starkem Wind und Turbulenzen stand

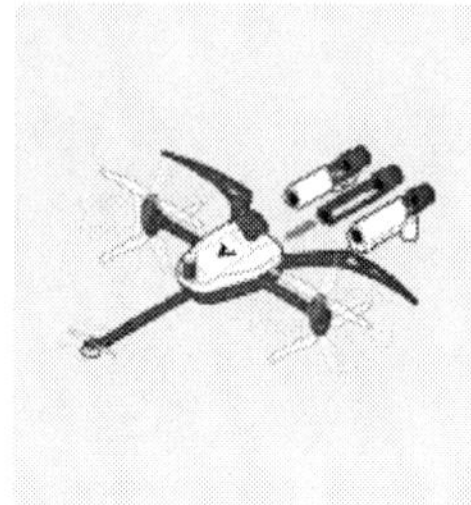

Plattform-Design

Modulare Sensoren

ZFP SENSOREN

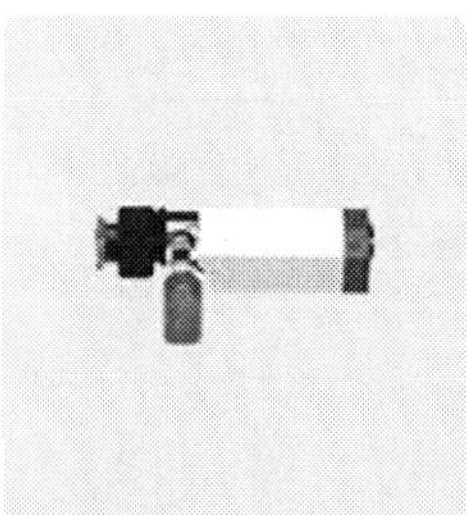

UT (Ultrasonic Transducer)

Echo-to-Echo and Pulsed Echo test mode with live A-Scan. Thickness Range 2-150 mm, resolution 0.06 mm.

EMAT (Electromagnetic acoustic transducer)

Ideal for corroded and dirty surfaces with a maximal lift-off of 4 mm.

DFT (Dry Film Thickness)

Magnetic induction and eddy current measurement mode with a range of 0 to 1.5 mm (F) / 0.7 mm (N).

PEC (Pulsed Eddy Current)

Maximum lift-off distance of 100 mm. Wall thickness range (low carbon steel): 3 - 18 mm.

FALLSTUDIE — BERGBAU

Zwischenbunker für Erz bei Suncor Energy in Kanada

Abschaltung stoppt Produktion

Hohe Risiken für Arbeiten im inneren des Behälters

Sehr hohe Sicherheitsanforderungen

OSPREY INTEGRITY

FALLSTUDIE — BERGBAU

Zwischenbunker für Erz bei Suncor Energy in Kanada

Prüfung der Kacheln

Planmäßig sollten 100% proaktiv ersetzt werden

40 Minuten Zeitfenster ZfP mit der Drohne

Kein Unterbruch der Produktion (kurz genug)

Resultat: Nur 60% mussten ersetzt werden, daher kein Gerüstbau, sonder nur Kletterer

Einsparung von 1.200.000 CAD

OSPREY INTEGRITY

ZFP IN DER HÖHE — WENIGER RISIKEN, MEHR EFFIZIENZ

Tanks

Kamine

Produktionsstätten

Windkraftanlagen

VIELEN DANK

Felix Stadler — fjs@voliro.com

Innovative Ansätze der automatisierten Laserthermografie in industrienaher Anwendung

Philipp Daniel HIRSCH [1], Sruthi Krishna KUNJI PURAYIL [1], Julien LECOMPAGNON [1], Mathias ZIEGLER [1]
[1] Bundesanstalt für Materialforschung und –prüfung (BAM), 12205 Berlin, Deutschland, philipp-daniel.hirsch@bam.de

Kurzfassung

Die Kombination aus aktiver Thermografie und Robotik eröffnet neue Möglichkeiten für die automatisierte zerstörungsfreie Prüfung (ZfP) komplexer Bauteilgeometrien. In einem modular aufgebauten Prüfsystem kommen ein Faserlaser zur gezielten thermischen Anregung, ein Industrieroboter zur präzisen Positionierung sowie eine gekühlte IR-Kamera und weitere Sensorik zur Datenerfassung zum Einsatz.

Im Zentrum der entwickelten Methode stehen vier wesentliche Komponenten: Erstens ermöglicht der Robotereinsatz eine hochpräzise und wiederholbare thermografische Inspektion entlang frei definierbarer Prüfpfade. Zweitens wird mithilfe einer Ulbricht-Kugel die ortsaufgelöste Emissionsgradkorrektur nicht-homogener Oberflächen durchgeführt – ein entscheidender Schritt für die quantitative Auswertung thermischer Signale. Drittens erhöht die Kombination von Infrarot- und VIS-Bilddaten die Sensitivität der Defekterkennung, insbesondere bei der Erkennung von Kanten und Strukturen. Viertens erlaubt die Integration eines 3D-Scanners die automatisierte Bahnplanung auf Basis von Punktwolken. Dadurch lassen sich komplexe Prüfteile flexibel erfassen, ohne aufwendige Vorrichtungen oder manuelle Vorpositionierung. Zusätzlich ermöglicht der Vergleich mit der Sollgeometrie eine Detektion geometrischer Abweichungen oder Verformungen.

Das Gesamtsystem erlaubt eine vollständig automatisierte, anpassbare und hochsensitive Prüfung verschiedenster Bauteile. Die vorgestellte Kombination aus robotergestützter Thermografie, Emissionskorrektur, multispektraler Bildverarbeitung und intelligenter Bahnplanung stellt eine vielversprechende Lösung für ZfP-Anwendungen im industriellen Umfeld dar. Die Ergebnisse unterstreichen das Potenzial dieser Technologie als vielseitig einsetzbares, skalierbares Prüfverfahren.

Keywords: Robotik, Laserthermografie, Emissionskorrektur, Multispektrale Bildverarbeitung, 3D-Bahnplanung, Automatisierte ZfP

Innovative Ansätze der automatisierten Laserthermografie in industrienaher Anwendung

BAM
Bundesanstalt für Materialforschung und -prüfung

Philipp Daniel HIRSCH, Sruthi Krishna KUNJI PURAYIL, Julien LECOMPAGNON und Mathias ZIEGLER

Fachbereich 8.3: Thermografische Verfahren

Bundesanstalt für Materialforschung und –prüfung (BAM), 12205 Berlin

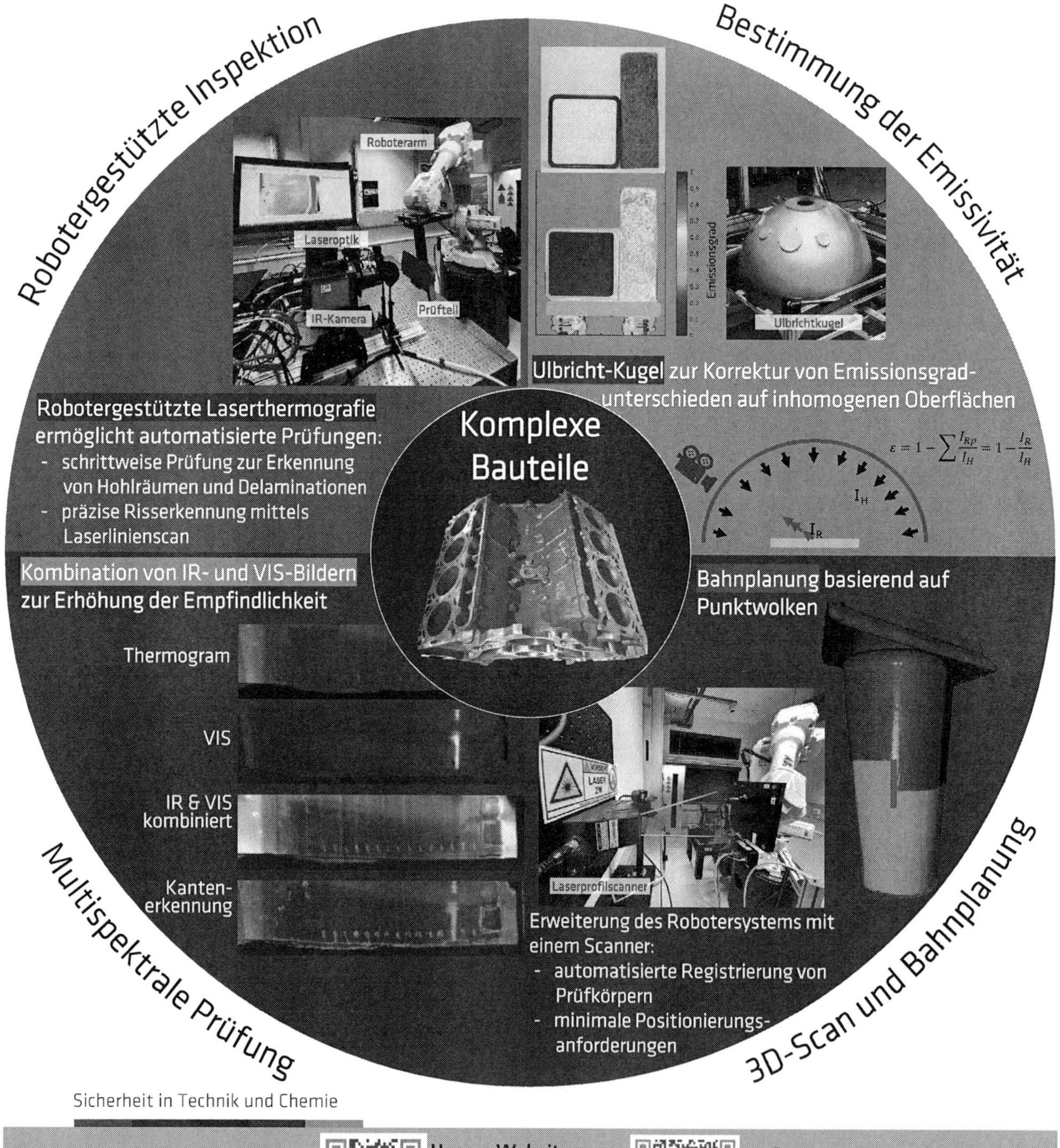

Sicherheit in Technik und Chemie

www.bam.de

Unsere Website

LinkedIn

FOCUS AREA INFRASTRUCTURE

Anwendung der Lock-in-Thermographie zur Defekterkennung in additiv gefertigten AMC Schweißnähten

Jihed BEN MAJED [1], Ralph SCHACHT [1], Tim FISCHER [1], Marek ZAJACZKOWSKI [2], Daniel MAY [3], Kaushal Arun Pareek PAREEK [3], Syed Ali ZAFAR [4]
[1] Brandenburgische Technische Universität Cottbus – Senftenberg, Brandenburg, Deutschland, JI.BENM@OUTLOOK.COM
[2] Berliner Nanotest und Design GmbH, Berlin, Deutschland, ZAJACZKOWSKI@NANOTEST.EU
[3] Technische Universität Chemnitz, Sachsen, Deutschland, DANIEL.MAY@ETIT.TU-CHEMNITZ.DE
[4] Access e.V. Cottbus, Sachsen, Deutschland, A.ZAFAR@ACCESS-TECHNOLOGY.DE

Kurzfassung. Die Lock-in-Thermographie (LIT) ermöglicht eine präzise, zerstörungsfreie Defekterkennung in additiv gefertigten Aluminium-Matrix-Verbundwerkstoff (AMC)-Schweißnähten. In dieser Arbeit wird ein speziell entwickelter experimenteller Aufbau verwendet, der die Analyse der thermischen Reaktion auf periodische Wärmeanregung optimiert. Durch die Kombination experimenteller Untersuchungen mit transienten thermischen Simulationen und Fourier-Analysen konnten Defekte ab 2 mm Größe bis in eine Tiefe von 1,5 mm zuverlässig identifiziert werden. Zudem wurde ein Algorithmus zur automatisierten Bilddifferenzanalyse entwickelt, der die Detektion erleichtert und beschleunigt. Die Ergebnisse zeigen, dass LIT eine hohe Sensitivität für innere Materialunregelmäßigkeiten aufweist und sich für industrielle Anwendungen eignet. Die Messdauer konnte durch eine optimierte Erwärmung der Probe auf eine Minute reduziert werden, was die Effizienz des Verfahrens steigert.

Keywords: Lock-in-Thermographie, Defekterkennung, Aluminium-Matrix-Verbundwerkstoffe (AMC)-Schweißnaht, Zerstörungsfreie Prüfung, Numerische Simulation

Einführung

Die zerstörungsfreie Prüfung (ZfP) hat sich als essenzielles Werkzeug in der Qualitätssicherung moderner Fertigungsprozesse etabliert, insbesondere bei der Herstellung von (AMC) Werkstoffen. Diese innovativen Materialien, die sich durch ihr geringes Gewicht und ihre hohe Festigkeit auszeichnen, finden Anwendung in der Luft- und Raumfahrt, der Automobilindustrie und anderen sicherheitskritischen Bereichen. Jedoch stellt die während des Schweißprozesses auftretende Porenbildung eine signifikante Herausforderung dar, da sie die mechanischen Eigenschaften der Bauteile erheblich beeinträchtigen kann. [1]
Die Lock-in-Thermographie (LIT) hat sich als fortschrittliche Methode zur Detektion von Materialdefekten etabliert. Im Gegensatz zu herkömmlichen thermografischen Verfahren ermöglicht LIT durch den Einsatz periodisch modulierter Wärmequellen eine präzisere

https://doi.org/10.58286/31221

Erkennung von Defekten unterhalb der Oberfläche. Frühere Arbeiten, wie die von Maldague (2001) [2] sowie Breitenstein und Langenkamp (2003) [3], haben die Effektivität der LIT bei der Untersuchung unterschiedlicher Materialien und Defektarten bestätigt. Obwohl thermografische Verfahren in bestehenden Schweißnahtprüfnormen wie der EN ISO 10042 (2018) [4] nicht berücksichtigt wurden, gibt es mittlerweile eine eigene Norm für die thermografische Prüfung von Schweißverbindungen. Die DIN 54186 (2022) [5] legt Verfahren zur Prüfung von Laserstrahlschweißverbindungen mithilfe aktiver Thermografie fest und definiert Anforderungen an die Anlagentechnik, das Prüfverfahren sowie die Dokumentation. Aktuelle Studien, darunter die Untersuchungen von Mund et al. (2021) [6] und Srajbr et al. (2022) [7], haben die Anwendungen der LIT weiterentwickelt. Während Mund et al. (2021) die Qualifizierung der aktiven Thermografie zur Detektion von Unregelmäßigkeiten in Schweißnähten durch realistische Prüfkörper und systematische Validierungen untersuchten, entwickelten Srajbr et al. (2022) ein Prüfkonzept, das spezifische Anforderungen der aktiven Thermografie für Schweißverbindungen adressiert. Sie setzten dabei innovative Validierungsmethoden ein, um die praktische Anwendung dieser Verfahren weiter voranzutreiben.

Ein zentraler wissenschaftlicher Beitrag dieser Arbeit ist die Anwendung der LIT auf additiv gefertigte AMC-Schweißnähte, wodurch nachgewiesen wurde, dass Defekte ab einem Durchmesser von 2 mm bis zu einer Tiefe von 1,5 mm zuverlässig identifiziert werden können. Darüber hinaus wurde die Messzeit optimiert, was den industriellen Einsatz der Methode praktikabler macht. Zur Absicherung der experimentellen Ergebnisse wurden transiente thermische Simulationen in ANSYS durchgeführt und mit Fourier-Analysen in MATLAB ausgewertet. Erstmals kamen hierbei 3D-gescannte Proben in der Simulation zum Einsatz, um realistische Material- und Oberflächeneigenschaften zu berücksichtigen. Ein weiterer innovativer Aspekt der Studie ist die Entwicklung eines Algorithmus zur Bilddifferenzanalyse, der eine schnellere und präzisere Identifikation von Defekten ermöglicht. Diese Methode kann zukünftig als Grundlage für KI-gestützte Inspektionen dienen und die Anwendung der LIT optimieren.

1. Methodik

Die Methodik beschreibt die Herangehensweise zur Untersuchung der Eignung der Lock-in-Thermographie (LIT) für die Detektion von Defekten in additiv gefertigten AMC-Schweißnähten. Der Schwerpunkt liegt auf experimentellen Untersuchungen an realen Proben und deren Validierung durch FEM-Simulationen. Das Ziel ist die Entwicklung und Optimierung einer zuverlässigen Methode zur Detektion von Defekten in Schweißnähten. Im Folgenden werden die verwendeten Verfahren detailliert beschrieben.

1.1 Experimentelle Untersuchung

Die experimentelle Untersuchung der LIT basiert auf fünf zentralen Komponenten (siehe Abbildung 1): Eine Infrarotkamera zur Erfassung thermischer Signale, ein Laser-Linse-Diffusor-System zur gezielten Anregung der Schweißnahtproben, eine Heizplatte zur präzisen Temperaturkontrolle, ein PC mit Lock-in-Software zur Datenverarbeitung sowie ein Synchronisationsmodul zur synchronisierten Steuerung von Kamera und Laser. Die Lock-in-Software ermöglicht die Analyse der Messergebnisse, darunter Amplituden- und Phasenbilder sowie 0°- und -90°-Bilder. Diese Komponenten arbeiten zusammen, um Defekte unter der Schweißnaht zuverlässig zu identifizieren. Für die experimentellen Untersuchungen wurde die Infrarotkamera (ImageIR® 8300 InfraTec) verwendet, während die Datenverarbeitung und Analyse mit der Software (IRBIS® 3 InfraTec) erfolgte. Als modulierte Wärmequelle kam ein infrarotes Einzelemitter-Laserdioden-System (element®

e06 nLIGHT) zum Einsatz. Dieses System liefert eine optische Ausgangsleistung von bis zu 45 W über eine 200-µm-Faser bei einer Wellenlänge von 885 ± 5 nm. Der Laserstrahl wurde mithilfe einer plankonvexen Linse (LA1951-AB Thorlabs) aus N-BK7-Glas geformt. Diese Linse ist mit einer Antireflexbeschichtung versehen, um Reflexionen zu minimieren und die Effizienz des optischen Systems zu maximieren. Plankonvexe Linsen sind ideal zum Fokussieren eines kollimierten Strahls sowie zur Kollimation von Licht aus einer Punktquelle. Dabei sollte eine kollimierte Lichtquelle die gekrümmte Oberfläche der Linse treffen, während eine Punktlichtquelle auf die planare Oberfläche gerichtet sein sollte, um sphärische Aberrationen zu reduzieren. [8]

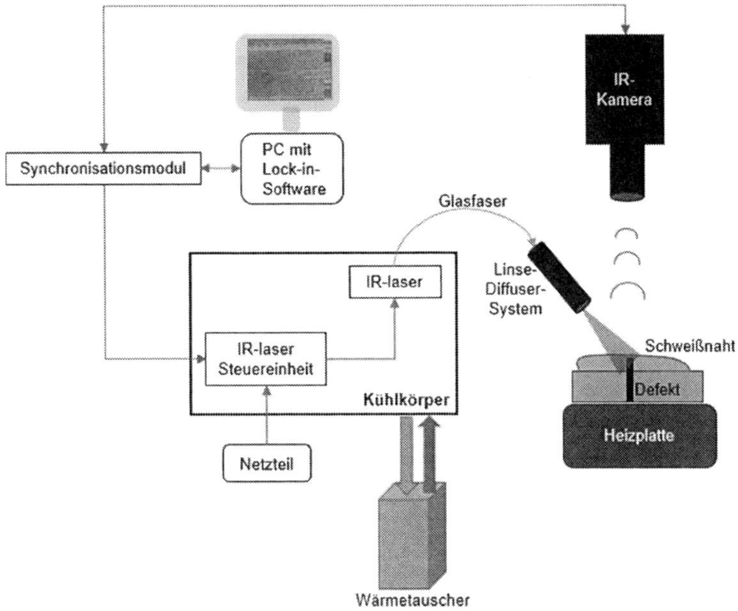

Abb. 1. Aufbau zur Lock-in-Thermographie für die Defekterkennung in Schweißnähten

Eine effiziente Kühlung wurde implementiert, da die Steuereinheit und der Laser während des Betriebs erhebliche Wärme erzeugt. Diese erfolgt durch eine Kombination aus einer Kühlplatte und einem Wärmetauscher. Die Kühlplatte leitet die Wärme direkt von der Steuereinheit ab, während der Wärmetauscher die überschüssige Wärme an ein externes Kühlmedium abgibt. Durch diese optimierte Temperaturregulierung wird sichergestellt, dass die experimentellen Bedingungen konstant bleiben, und Messungen nicht durch thermische Schwankungen beeinflusst werden. Die untersuchten Proben (siehe Abbildung 2) bestanden aus der Aluminiumlegierung AlSi10Mg und wurden mit spezifischen geometrischen Abmessungen (siehe Tabelle 1) gefertigt und enthielten gezielt eingebrachte Defekte, um realitätsnahe Untersuchungen durchzuführen.

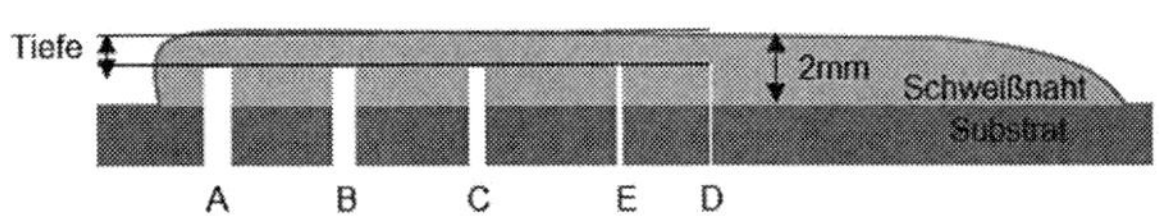

Abb. 2. Schweißnaht Proben Geometrie

1.2 Optische Strahlführung und Anregung der Probe

Nachdem der Laserstrahl die Glasfaser verlässt, trifft dieser zunächst auf die plane Seite der plankonvexen Linse, wo er so umgeformt wird, dass divergierende Lichtstrahlen parallelisiert werden. Dies ermöglicht eine kontrollierte und effiziente Ausbreitung des Strahls. Anschließend trifft der Strahl auf einen Diffusor aus ZEONOR-Polymer (ED1-C20-MD, Thorlabs). Dieser Diffusor sorgt für eine gleichmäßige Streuung des Laserstrahls, wodurch eine homogene, quadratische Energieverteilung auf der Oberfläche der Probe erzeugt wird. Die gezielte Strahlformung spielt eine entscheidende Rolle für die thermische Anregung, deren Grundlagen und Richtlinien in verschiedenen Normen, darunter DIN 54185 [9] und DIN EN 17501 [10], definiert sind. Ein wichtiger Parameter in diesem Aufbau ist der Abstand zwischen dem Linsen-Diffusor-System und der Probe. Eine Vergrößerung dieses Abstands führt zwar zu einer größeren bestrahlten Fläche, verringert jedoch gleichzeitig die auf die Schweißnaht auftreffende Laserleistung. Sobald der Laserstrahl auf die Schweißnaht trifft, wird er entweder absorbiert oder reflektiert. Die Absorption und Reflexion des Laserstrahls an der Schweißnahtoberfläche hängen von mehreren Faktoren ab, darunter die Wellenlänge, die Polarisation, der Einfallswinkel des Strahls sowie die optischen Eigenschaften des Materials, in diesem Fall der Aluminiumlegierung AlSi10Mg. Darüber hinaus kann die Oberflächenrauheit der Schweißnaht die Lichtausbreitung beeinflussen, indem sie zu einer ungleichmäßigen Absorption führt und lokale Intensitätsschwankungen verursacht. [11]

1.3 Modulation und Detektion

Der Laser wird über die Steuereinheit (LDP-C 50-05, PicoLAS) moduliert und auf eine definierte Frequenz eingestellt. Durch die Modulation entsteht ein periodischer Laserstrahl, der die Probe gezielt erwärmt. Dabei breiten sich thermische Wellen aus, deren Verhalten von der Materialstruktur sowie möglichen Defekten beeinflusst wird. Die Infrarotkamera mit einer thermischen Empfindlichkeit von 20 mK und einer Bildauflösung von 640 × 512 Pixeln erfasst die daraus resultierenden Temperaturveränderungen mit einer Bildrate von 120 Hz. Die erfassten Messdaten wurden anschließend weiterverarbeitet und in verschiedene Bildformate umgewandelt, darunter das 0°-Bild, das -90°-Bild sowie Phasen- und Amplitudenbilder. Diese Bildanalyse ermöglicht eine Präzise Auswertung der thermischen Signale und erlaubt eine Exakte Lokalisierung von Defekten innerhalb der Probe.

1.4 Defekterkennung durch Bilddifferenzanalyse in der Lock-in-Thermographie

In dieser Studie wird ein Python-Programm zur Analyse und Visualisierung von Differenzen zwischen zwei experimentellen Bilddatensätzen aus Lock-in-Thermographie-Experimenten vorgestellt (siehe Abbildung 3). Die Rohdaten bestehen aus pixelweise gespeicherten Messwerten, die in Excel-Dateien abgelegt sind und aus Experimenten mit und ohne Laseranregung stammen. Ziel des Programms ist es, Differenzbilder zu erzeugen, um thermische Anomalien sichtbar zu machen und Defektpositionen präzise zu identifizieren. Das Verfahren basiert auf der Subtraktion der beiden Bilddatensätze, wodurch thermische Unterschiede hervorgehoben werden. Zur verbesserten Visualisierung wird das resultierende Bild mit einer -90°-Phasenverschiebung zur Anregung rekonstruiert. Ein zentraler Bestandteil des Algorithmus ist die Identifikation der Pixelpositionen mit den größten Differenzen, die rot markiert werden, da sie in der Regel mit Defekten korrelieren. Das Programm gruppiert zusammenhängende markierte Pixel und identifiziert sie als Defektbereiche, während vereinzelte Markierungen entfernt werden, um Fehlinterpretationen zu vermeiden. In der LIT führen Defektstellen zu signifikanten

Temperaturdifferenzen, da strukturelle Unregelmäßigkeiten wie Risse, Einschlüsse oder Poren die Wärmeleitung lokal beeinflussen. Diese Störungen verändern die Phasenantwort des Materials, wodurch Defekte im Differenzbild als kontrastreiche Bereiche erscheinen. Die vorgestellte Methode ermöglicht somit eine zuverlässige und objektive Identifikation von Materialdefekten.

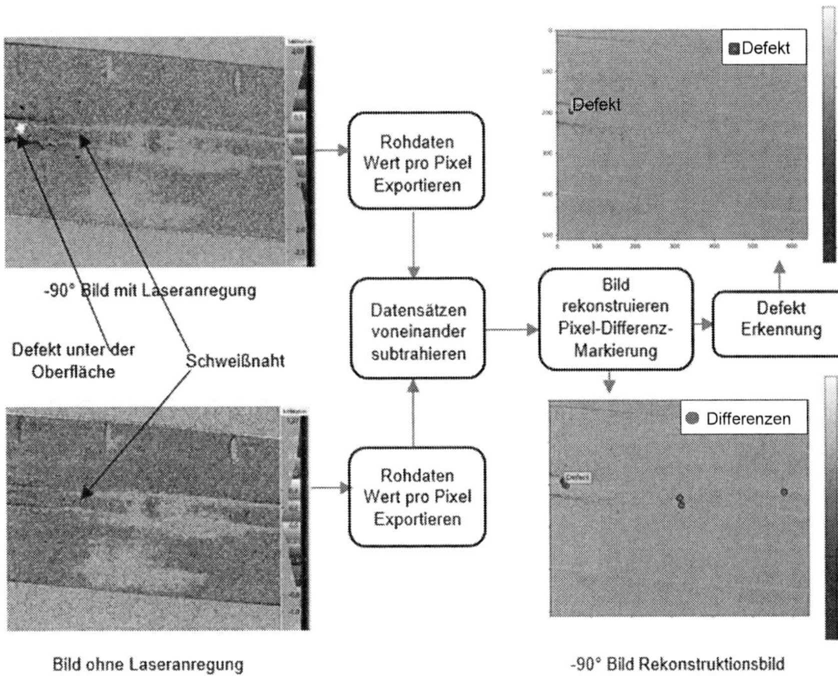

Abb. 3. Schematische Darstellung der Automatisierte Defekterkennung mittels Bilddifferenzanalyse in der Lock-in-Thermographie

1.5 Validierung durch FEM Simulation

Zur Validierung der experimentellen Ergebnisse sowie zur Optimierung der Anregungsbedingungen wurden numerische Simulationen durchgeführt. Basierend auf der Dissertation von May [12] kam ein transienter thermischer Ansatz in ANSYS zum Einsatz, um die Wärmeausbreitung in additiv gefertigten AMC-Schweißnähten realitätsnah zu modellieren. Die Probengeometrie wurde mithilfe eines 3D-Scanners digitalisiert, um Oberflächenunregelmäßigkeiten exakt zu erfassen. Die Simulation berücksichtigte materialtypische Eigenschaften wie Dichte, Wärmeleitfähigkeit und Diffusivität, die aus der ANSYS-Datenbank für AlSi10Mg entnommen wurden. Die Laseranregung wurde durch definierte Parameter wie Leistung und Modulationsfrequenz präzise nachgebildet. Defekte wurden als Volumendiskontinuitäten mit variabler Größe, Form und Tiefe integriert, um ihren Einfluss auf die thermischen Wellen zu analysieren. Die Simulation der Wärmeausbreitung erfolgte unter transienten Bedingungen, um die zeitabhängige Entwicklung der thermischen Signale realistisch darzustellen. Nach der Durchführung der transienten thermischen Simulation in ANSYS wurde ein Simulationsvideo erstellt, das die zeitliche Entwicklung der Wärmeausbreitung visualisiert. Dieses Video wurde anschließend in MATLAB integriert, um eine detaillierte Analyse der thermischen Eigenschaften durchzuführen. Im MATLAB-Skript begann die Verarbeitung des Videos mit der Definition des Dateipfads sowie der Festlegung zentraler Parameter. Dazu zählten die Anzahl der zu analysierenden Frames, die Minimal- und Maximaltemperaturen sowie die thermische Empfindlichkeit der verwendeten Kamera. Die Frames des Videos wurden extrahiert, in Graustufen konvertiert und auf die einheitliche Größe von 640 × 512 Pixeln wie bei der Realen Aufnahme skaliert. Um die thermischen Eigenschaften genauer zu untersuchen, wurde eine Fourier-Transformation entlang der Zeitachse angewendet, wie in DIN EN 17119 (2018) [13] beschrieben. Dabei wurden das In-Phase-Bild (0°) und das Quadratur-Bild (-90°) berechnet. Basierend auf diesen Ergebnissen wurden ein phasenunabhängiges Amplitudenbild sowie ein Phasenbild generiert. Zur Optimierung der Analyse wurde eine Phasenkorrektur durchgeführt, und die Amplitudenbilder wurden entsprechend den definierten Temperaturgrenzen skaliert. Dieser umfassende Verarbeitungsprozess ermöglichte eine präzise Analyse der thermischen Eigenschaften, die in den Simulationsdaten enthalten waren.

2. Ergebnisse

Die experimentellen Untersuchungen zeigen, dass die Lock-in-Thermographie (LIT) ein vielversprechendes Verfahren zur Detektion von Defekten in additiv gefertigten AMC-Schweißnähten darstellt. Durch die Kombination aus experimenteller Analyse und numerischer Simulation konnte eine detaillierte Charakterisierung der thermischen Reaktion auf Defekte unterschiedlicher Größe und Tiefe erreicht werden. Die erzielten Ergebnisse unterstreichen das Potenzial der Methode, eine präzise und zuverlässige Defekterkennung in industriellen Anwendungen zu ermöglichen.

2.1 Experimentelle Ergebnisse

Zur Untersuchung der Detektionsgrenzen der LIT wurden drei Proben mit künstlich eingebrachten zylindrischen Defekten analysiert. Die Defekte wiesen Durchmesser von 2,5 mm bis 0,5 mm auf und befanden sich in Tiefen von 1,0 mm bis 2,0 mm (siehe Tabelle 1).

Tabelle 1. Proben Abmessungen

Probe	Tiefe	Bohrung A	Bohrung B	Bohrung C	Bohrung D	Bohrung E
Probe 1	1mm	2,5mm	2mm	1,5mm	1mm	0,5mm
Probe 2	1,5mm	2,5mm	2mm	1,5mm	1mm	0,5mm
Probe 3	2mm	2,5mm	2mm	1,5mm	1mm	0,5mm

Defekt A (Siehe Abbildung 4) in Probe 1 konnte sowohl in den Phasenbildern als auch in den -90°-Bildern klar identifiziert werden. Die experimentellen Ergebnisse wurden durch eine Kombination aus thermischer Anregung und hochauflösender Bildgebung gewonnen.

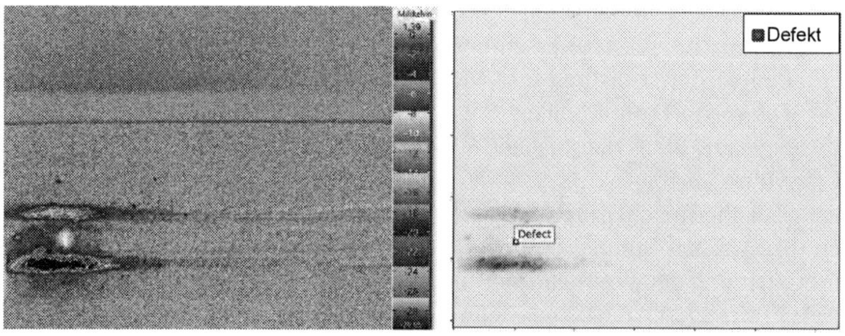

Abb.4. Lock-in-Thermographie-Ergebnisse Probe 1: -90°-Bild der Defektposition A (links) und deren automatische Defekterkennung (rechts)

Die thermische Anregung erfolgte mittels einer sinusförmig modulierten Laserquelle, während die resultierenden Temperaturverteilungen mit der Infrarotkamera erfasst wurden. Für die Analyse wurden sowohl Phasenbilder als auch -90°-Bilder ausgewertet, da diese eine hohe Sensitivität gegenüber Defekten in unterschiedlichen Tiefen aufweisen. Zusätzlich wurde ein weiteres Bild mit der automatischen Defekterkennung durch Bilddifferenzanalyse erstellt. Dieses Bild hebt den Defekt deutlich hervor und zeigt die Effektivität der Bilddifferenzanalyse bei der Identifikation von Defekten. Auch Defekt B konnte in den Phasenbildern und -90°-Bildern sehr deutlich erkannt werden. Die Bilddifferenzanalyse durch das Python-Programm bestätigte die Ergebnisse und zeigte eine ebenso klare Identifikation des Defekts wie bei Defekt A. Im Fall von Defekt C (siehe Abbildung 5) war die Sichtbarkeit in den Phasenbildern und -90°-Bildern weniger klar im Vergleich zu Defekt A und B. Dennoch konnte der Defekt durch die Bilddifferenzanalyse zuverlässig erkannt werden. Dies zeigt, dass der Algorithmus auch bei weniger deutlichen Defekten effektiv arbeitet und eine automatische Identifikation ermöglicht. Defekt E und D waren nicht mehr zu erkennen.

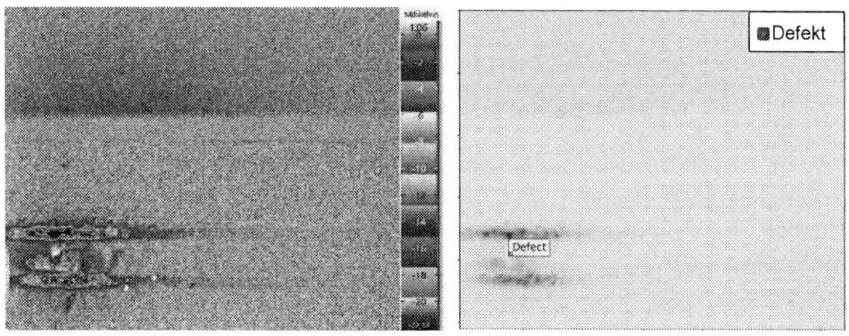

Abb. 5. Lock-in-Thermographie-Ergebnisse Probe 1: -90°-Bild der Defektposition C (links) und deren automatische Defekterkennung (rechts)

In Probe 2 konnten nur die Defekte A und B detektiert werden, jedoch mit einer reduzierten Signalintensität im Vergleich zu Probe 1. Für die Analyse wurden zwei Phasenbilder herangezogen, da diese die höchste Sensitivität für Defekte in dieser Tiefe aufwiesen. Die thermischen Signale zeigten eine abgeschwächte Amplitudenmodulation, was auf die zunehmende Dämpfung der Wärmewellen hinweist. Die -90°-Bilder lieferten in diesem Fall keinen ausreichenden Kontrast zur Defektidentifikation. Defekt C, E und D waren nicht mehr zu erkennen. In Probe 3 konnten keine signifikanten thermischen Anomalien detektiert werden. Die in dieser Tiefe stark abgeschwächten thermischen Wellen führten zu keinem ausreichenden Kontrast zwischen Defekt und Grundmaterial. Dieses Ergebnis bestätigt die erwartete Detektionsgrenze der LIT unter den gewählten experimentellen Bedingungen.

2.1 Bewertung der Messergebnisse

Die experimentellen Ergebnisse zeigen, dass mit einer Anregungsfrequenz von 10 Hz Defekte bis zu einer Tiefe von 1,5 mm ab einem Durchmesser von 2 mm zuverlässig detektiert werden können. Besonders die Phasenbilder erwiesen sich als geeignet zur Identifikation tieferliegender Defekte, während die -90°-Bilder eine zusätzliche Kontrastierung ermöglichten. Defekte in 2,0 mm Tiefe waren mit der gewählten Anregungsfrequenz und Signalverarbeitung nicht mehr eindeutig nachweisbar. Dies legt nahe, dass für tiefere Defekte eine Optimierung der Messparameter erforderlich ist, beispielsweise durch eine Reduzierung der Anregungsfrequenz und eine Intensivierung der Wärmequelle. Darüber hinaus zeigte sich, dass eine Vorwärmung der Schweißnahtproben auf 60 °C die Defekterkennung signifikant verbesserte und die Messzeit um die Hälfte reduzierte. Die erhöhte Ausgangstemperatur der Probe beeinflusst die thermische Diffusivität des Materials positiv, wodurch sich der Temperaturkontrast zwischen intakten und defektbehafteten Bereichen verstärkt. Besonders in den -90°-Bildern wurden Defekte durch die verstärkte Wärmeleitung deutlicher hervorgehoben, was eine präzisere Identifikation ermöglichte. Ein weiterer entscheidender Vorteil der Methode liegt in der kurzen Messdauer von nur einer Minute pro 2 cm^2, wodurch sich die LIT für den industriellen Einsatz besonders eignet. Zudem zeigte die automatische Bilddifferenzanalyse eine hohe Präzision bei der Identifikation der Defektpositionen, was den gesamten Inspektionsprozess weiter optimieren kann. Diese Erkenntnisse liefern eine wesentliche Grundlage für die Weiterentwicklung der LIT in der zerstörungsfreien Prüfung von AMC-Schweißnähten.

2.2 Numerische Simulationen Ergebnisse

Die Modellierung der Wärmeausbreitung unter Berücksichtigung der realen Materialeigenschaften zeigte, dass Defekte durch charakteristische Phasenverschiebungen und Amplitudenänderungen in der thermischen Signatur erkennbar sind. Insbesondere bei der Analyse der -90°-Bilder (Siehe Abbildung 6) konnten die Defekte klar identifiziert werden, da die Phasenverschiebung die thermischen Anomalien deutlich hervorhob. Für Probe 1 wurden die Defekte A, B und C simuliert und analysiert. Die Simulation zeigte, dass Defekt A sowie Defekt B sowohl in den Phasenbildern als auch in den -90°-Bildern deutlich erkennbar waren. Die thermische Signatur des Defekts wies eine klare Phasenverschiebung auf, die mit den experimentellen Ergebnissen übereinstimmte. Defekt C war in der Simulation etwas weniger deutlich, konnte jedoch durch die Phasenkorrektur zuverlässig identifiziert werden.

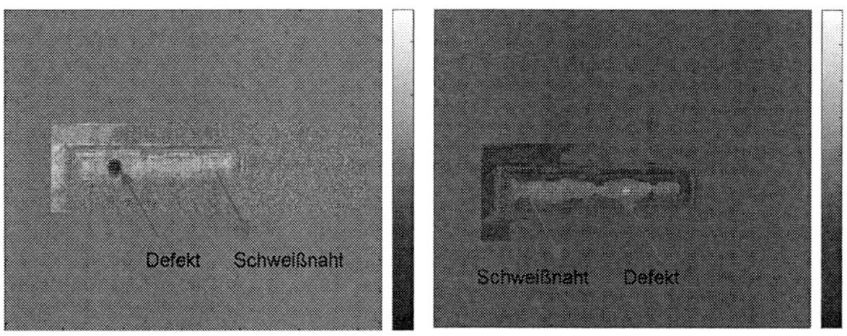

Abb. 6. Lock-in-Thermographie-Simulationsergebnisse Probe 1: -90°-Bild der Defektposition A (links) - -90°-Bild der Defektposition C (rechts)

Für Probe 2 wurden die Defekte A und B simuliert. Die Ergebnisse der Simulation zeigten, dass die thermischen Signale in dieser Tiefe abgeschwächt waren, was den experimentellen Beobachtungen entsprach. Defekt A konnte in den Phasenbildern noch identifiziert werden, während Defekt B ein ähnliches Verhalten zeigte.

2.3 Validierung der Simulationsergebnisse

Die hohe Übereinstimmung zwischen dem simulierten und dem experimentellen Phasenbild (siehe Abbildung 7) bestätigt die Zuverlässigkeit der angewandten Methodik. Die numerische Simulation konnte nicht nur die Position und Größe der Defekte korrekt abbilden, sondern auch die thermischen Eigenschaften der Materialien realistisch nachbilden.

Besonders die Phasenbilder zeigen, dass die Simulation die Wärmeströme und deren Verzögerung präzise erfasst, was essenziell für die Defekterkennung ist.

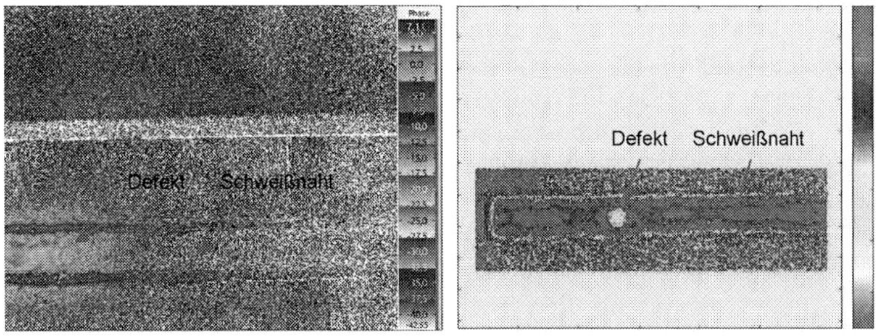

Abb. 7. Lock-in-Thermographie-Vergleichsergebnisse Probe 2: Phasenbild der Defektposition B (links) und deren Simulationsergebnis (rechts)

Diese Ergebnisse verdeutlichen, dass numerische Simulationen ein wertvolles Werkzeug zur Validierung und Optimierung der LIT darstellen. Durch die genaue Nachbildung der thermischen Wellenausbreitung und die Berücksichtigung realer Materialeigenschaften lassen sich Defekte in verschiedenen Tiefen und Größen zuverlässig identifizieren.

4. Zusammenfassung und Ausblick

Die vorliegende Arbeit hebt die Bedeutung der Lock-in-Thermographie (LIT) als ein effektives und zerstörungsfreies Prüfverfahren zur Detektion von Defekten in additiv gefertigten AMC-Schweißnähten hervor. Durch die Kombination eines innovativen Prüfaufbaus mit experimentellen und numerischen Methoden wurde eine präzise Methodik entwickelt, die sowohl oberflächennahe als auch tiefere Defekte zuverlässig identifiziert. Die kurze Prüfzeit von nur einer Minute macht die LIT besonders attraktiv für industrielle Anwendungen, da sie eine schnelle und effiziente Qualitätskontrolle ermöglicht.

Trotz der vielversprechenden Ergebnisse gibt es noch Optimierungspotenzial. So könnten weitere Studien zur Anwendung der LIT auf andere Materialklassen oder komplexere Geometrien durchgeführt werden. Um die Flexibilität der Methode zu erweitern und noch bessere Ergebnisse zu erzielen, könnte der Einsatz eines stärkeren Hochleistungslasers in Betracht gezogen werden. Ein solcher Laser würde eine intensivere Anregung ermöglichen, insbesondere bei Proben mit Oberflächenanomalien, die derzeit die Zuverlässigkeit der Methode beeinträchtigen können. Ebenso könnte der Einsatz einer Infrarotkamera mit höherer thermischer Empfindlichkeit die Detektion von Defekten verbessern, da feinere Temperaturunterschiede präziser erfasst werden könnten. Diese Verbesserungen würden dazu beitragen, die Methode robuster und vielseitiger zu machen. Auch die Integration von künstlicher Intelligenz (KI) zur automatisierten Auswertung und Klassifizierung von Defekten bietet großes Potenzial für weitergehende Entwicklungen. Insbesondere in Kombination mit maschinellem Lernen könnte die Effizienz und Präzision der Defekterkennung weiter gesteigert werden.

Obwohl diese Studie erfolgversprechend ist, gibt es einige Limitationen und mögliche Fehlerquellen. Ein wichtiger Aspekt ist die räumliche Homogenität der Wärmequelle. Kleine Abweichungen in der Strahlverteilung oder ungleichmäßige Absorption auf der Probenoberfläche können zu einer ungleichmäßigen Erwärmung führen, insbesondere bei komplexen Geometrien. Dies erschwert die Interpretation der Messergebnisse. Auch die Wahl der Modulationsfrequenz ist entscheidend. Während niedrige Frequenzen tiefere Defekte sichtbar machen, ermöglichen hohe Frequenzen eine bessere Auflösung oberflächennaher Defekte. Die optimale Abstimmung stellt daher eine Herausforderung dar, insbesondere bei Defekten in unterschiedlichen Tiefen. Zusätzlich sind die experimentellen Grenzen der Infrarotkamera zu berücksichtigen. Die thermische Empfindlichkeit und räumliche Auflösung setzen eine Grenze für die Detektion sehr kleiner oder schwach ausgeprägter Defekte. Ein weiterer Punkt ist die Validierung durch numerische Simulationen. Diese beruhen auf idealisierten Annahmen zu Materialeigenschaften und Randbedingungen, was zu Abweichungen zwischen Simulation und Experiment führen kann. Zukünftige Arbeiten sollten diese Limitationen weiter untersuchen, unter anderem durch die Verfeinerung der Modellierung und den systematischen Einsatz von einer Multi-Modulationsfrequenzen Analyse. Zudem könnte der Anwendung KI-gestützter Algorithmen die automatisierte Auswertung der Thermografie Daten weiter verbessern und die Defekterkennung optimieren.

Danksagung

Wir danken die Berliner Nanotest und Design GmbH, die Brandenburgische Technische Universität Cottbus-Senftenberg und der Förderinitiative „WIR! – Wandel durch Innovation in der Region" des Bundesministeriums für Bildung und Forschung (BMBF), die es uns ermöglicht haben, im Rahmen des Teilvorhabens „Inline-Prozesskettenentwicklung zur Reparatur von AMC-Werkstoffen sowie Porendetektion mittels aktiver IR-Thermographie zur Prozessüberwachung" eine Machbarkeitsstudie durchzuführen.

Referenz

[1] C. Gollnow (2010). Beeinflussung des metallurgischen Verhaltens strahl geschweißter Verbindungen an Stählen durch prozess spezifische Einflussgrößen. Institut für Werkstoff- und Fügetechnik, Otto-von-Guericke-Universität, Magdeburg.
[2] Maldague, X. P. (2001). Theory and Practice of Infrared Technology for Nondestructive Testing. Wiley-Interscience.
[3] Breitenstein, O., & Langenkamp, M. (2003). Lock-in Thermography: Basics and Use for Evaluating Electronic Devices and Materials. Springer.
[4] Deutsche Norm EN ISO 10042 (2019). Schweißen Lichtbogenschweißverbindungen an Aluminium und seinen Legierungen Bewertungsgruppen von Unregelmäßigkeiten. Beuth Verlag GmbH.
[5] Deutsche Norm DIN 54186 (2022). Zerstörungsfreie Prüfung–Prüfung von Laserstrahl-schweißverbindungen mit aktiver Thermografie. Beuth Verlag GmbH.
[6] Mund, M. et al. (2021). Additive Manufacturing Defect Detection Using Lock-in Thermography. Journal of Nondestructive Evaluation.
[7] Srajbr et al. (2022). Qualifizierung der aktiven Thermografie zur Prüfung von Schweißverbindungen. DGZfP Thermographie-Kolloquium 2022.
[8] Bahaa E. A. Saleh und Malvin Carl Teich (2020). Optik und Photonik Dritte Auflage. WILEY-VCH.
[9] Deutsche Norm DIN 54185 (2021) Zerstörungsfreie Prüfung–Thermografische Prüfung–Lock-in-Thermografie mit optischer Anregung. Beuth Verlag GmbH.
[10] Deutsche Norm DIN EN 17501 (2022) Zerstörungsfreie Prüfung Thermografische Prüfung Aktive Thermografie mit Laser-Anregung. Beuth Verlag GmbH.
[11] Jens bliedtner, Hartmut Müller, Andrea Barz. Lasermaterialbearbeitung Grundlagen-Verfahren Anwendungen-Beispiele. Fachbuchverlag Leipzig im Carl Hanser verlag
[12] Daniel May (2015). Dissertation „Transiente Methoden der Infrarot-Thermografie zur zerstörungsfreien Fehleranalytik in der mikroelektronischen Aufbau- und Verbindungstechnik". Technischen Universität Chemnitz
[13] Deutsche Norm DIN EN 17119 (2018). Zerstörungsfreie Prüfung Thermografische Prüfung Aktive Thermografie. Beuth Verlag GmbH.

Effiziente photothermische Defektrekonstruktion für zeitlich nicht-äquidistante Puls-Thermographiedaten in Compositebauteilen

Lukas GAHLEITNER [1], Günther MAYR [1], Umut CAKMAK[2], Peter BURGHOLZER[3]
[1] University of Applied Sciences Upper Austria, Wels, Österreich, lukas.gahleitner@fh-wels.at
[2] Johannes Kepler University Linz, Linz, Österreich, umut.cakmak@jku.at
[3] Research Center for Non Destructive Testing GmbH, Linz, Österreich, peter.burgholzer@recendt.at

Kurzfassung

Für die Rekonstruktion von Defekten werden in der Aktiven Thermografie üblicherweise Temperaturdaten herangezogen, die zeitlich äquidistant abgetastet wurden. Allerdings nimmt mit zunehmender Zeit, während des Wärmediffusionsvorgangs, auch der für die Rekonstruktion nutzbare Informationsgehalt aus dem Temperatursignal ab. Aus diesem Grund wird in dieser Studie eine physikalisch-basierte Abtaststrategie für Puls-Thermographiedaten demonstriert.

Um die photothermische Rekonstruktion aus zeitlich nicht-äquidistanten Puls-Thermographiedaten zu bewerkstelligen, wird eine Erweiterung des Konzepts der Virtuellen Welle präsentiert. Um dies zu erreichen, wird ein verallgemeinerter diskreter Transformationskern eingeführt, der es erlaubt, beliebige zeitliche Abtaststrategien zu berücksichtigen. Zunächst wird die Evidenz der vorgeschlagenen Strategie für analytische Temperatursignale gezeigt. Darüber hinaus werden die Vorteile für simulierte Temperatursignale, die von einer orthotropen Probe mit Defektschnittstellen in unterschiedlichen Tiefenpositionen stammen, demonstriert.

Zur experimentellen Verifizierung wird die gepulste Thermographie für eine Compositeprobe mit verschiedenen realen Fremdmaterialeinschlüssen aus der Luftfahrtindustrie angewandt. Es kann gezeigt werden, dass eine effiziente Zeitabtastung im Rahmen des Konzepts der Virtuellen Welle eine erhebliche Reduktion der Anzahl der Datenpunkte im Vergleich zu einer äquidistanten zeitlichen Abtastung ermöglicht, ohne dabei die Qualität der Rekonstruktionsergebnisse zu verschlechtern.

Keywords:

Aktive Thermographie; Faserverbundwerkstoffe; Photothermische Rekonstruktion; Virtuelles Wellenkonzept; Sampling

Effiziente photothermische Defektrekonstruktion für zeitlich nicht-äquidistante Puls-Thermographiedaten in Compositebauteilen

Lukas Gahleitner[1], Günther Mayr[1], Umut Cakmak[2], Peter Burgholzer[3]

[1] University of Applied Sciences Upper Austria, Research Group of Thermography & NDT

[2] Johannes Kepler University Linz, Institute of Polymer Product Engineering (IPPE)

[3] Research Center for Non Destructive Testing GmbH

Motivation der Studie

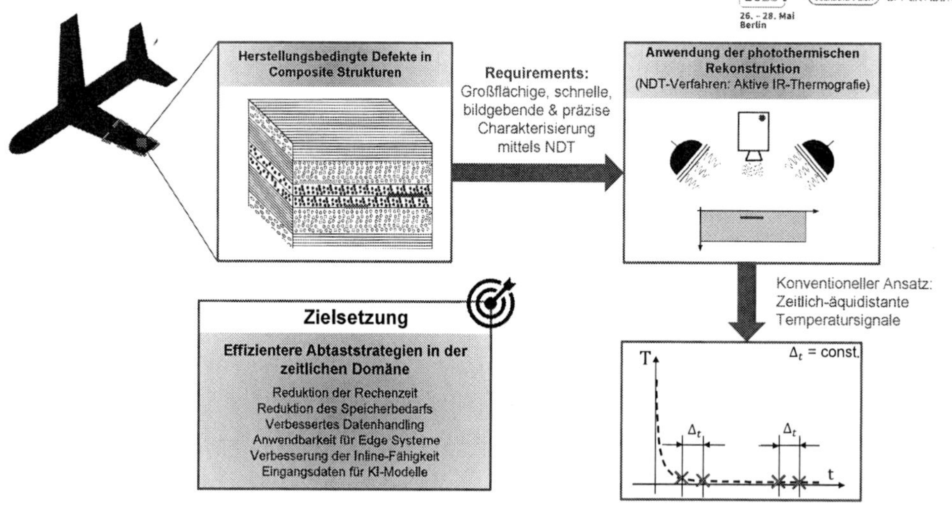

Virtuelles Wellen Konzept (VWC) [1]

dgzfp JAHRESTAGUNG 2025 26. - 28. Mai Berlin | fh OBERÖSTERREICH UNIVERSITY OF APPLIED SCIENCES UPPER AUSTRIA

1. Rekonstruktionsschritt:

Beziehung zwischen Temperatur- und Virtuellem Wellensignal:

$$T(\mathbf{r},t) = \int K(t,t') T_{\text{virt}}(\mathbf{r},t') \mathrm{d}t' \quad \text{Fredholm Integralgleichung}$$

$$\text{mit} \quad K(t,t') = \frac{c_{33}}{\sqrt{\pi \alpha_{33} t}} \exp\left(-\frac{c_{33}^2 t'^2}{4\alpha_{33} t}\right) \quad \text{für } t > 0$$

Transformationskernel (kontinuierlich)

- 1D-Transformation ⟶ Lokale Berechnung von $\mathbf{T}_{\text{virt}}$ für individuellen Pixelpunkt **r** = *(x,y,z)*
- Regularisierung notwendig ⟶ Schlecht gestelltes inverses Wärmeleitungsproblem (IHTP)
- Virtuelles Wellensignal $\mathbf{T}_{\text{virt}}$ folgt mathematisch der fundamentalen Wellengleichung

2. Rekonstruktionsschritt:

Anwendung von Verfahren aus der Ultraschalltechnik für die Rekonstruktion von Defekten (Örtliche Mittelung – Erhöhung SNR | Informationen über Wärmeleitungstensor – Defekt sizing)

Nicht-äquidistante Zeitschritte im VWC [2]

dgzfp JAHRESTAGUNG 2025 26. - 28. Mai Berlin | fh OBERÖSTERREICH UNIVERSITY OF APPLIED SCIENCES UPPER AUSTRIA

Diskretisierung (1)

- Diskretisierung der Fredholm Integralgleichung:

$$T(t) \approx \int_{t'_0}^{t'_{\text{end}}} \underbrace{K(t,t') T_{\text{virt}}(t')}_{g(t,t')} \, \mathrm{d}t' = \sum_{j=0}^{N_{\text{tv}}-1} \Delta t'_j \boxed{K(t,t'_j)} T_{\text{virt}}(t'_j).$$

- Einsetzen der Definition des kontinuierlichen Transformationskernels:

$$T(t) \approx \sum_{j=0}^{N_{\text{tv}}-1} \left\{ \Delta t'_j \boxed{\frac{c}{\sqrt{\pi \alpha t}} \exp\left(-\frac{c^2 t'^2_j}{4\alpha t}\right)} T_{\text{virt}}(t'_j) \right\}$$

- Evaluierung für beliebigen Zeitschritt k des Temperatursignals:

$$T(t_k) \approx \sum_{j=0}^{N_{\text{tv}}-1} \left\{ \Delta t'_j \frac{c}{\sqrt{\pi \alpha t_k}} \exp\left(-\frac{c^2 t'^2_j}{4\alpha t_k}\right) T_{\text{virt}}(t'_j) \right\} \xrightarrow{c \,=\, \tilde{c}\, \Delta z_j / \Delta t'_j} T(t_k) \approx \sum_{j=0}^{N_{\text{tv}}-1} \left\{ \underbrace{\frac{\tilde{c} \Delta z_j}{\sqrt{\pi \alpha t_k}} \exp\left(-\frac{\tilde{c}^2 \Delta z_j^2 \, t'^2_j}{\Delta t'^2_j \, 4\alpha t_k}\right)}_{\tilde{K}(k,j)} T_{\text{virt}}(t'_j) \right\}$$

Nicht-äquidistante Zeitschritte im VWC [2]

Diskretisierung (2)

Lokale Transformation: $\mathbf{T} = \widetilde{\mathbf{K}}\,\mathbf{T}_{\text{virt}}$ Kernel matrix $\widetilde{\mathbf{K}}$ beinhaltet Informationen über nicht-äquidistante Diskretisierung

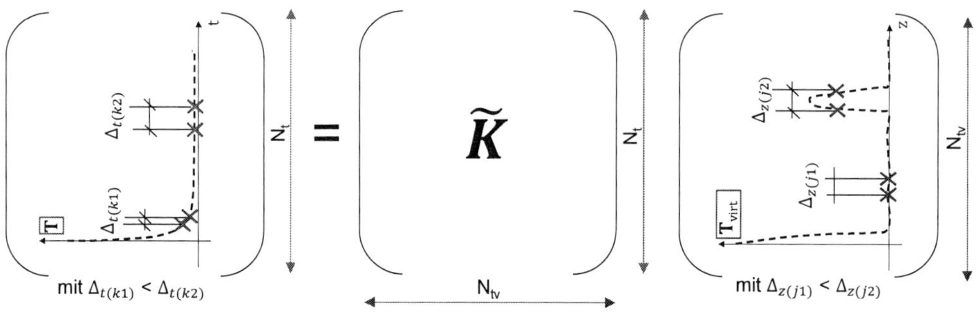

→ Regularisierung mit **ADMM** (alternating direction method of multipliers)!

Nicht-äquidistante Strategie für Temperatursignal [2]

Betrachtete Bedingungen/Aspekte:

- Nicht-äquidistanter Ansatz für gesamtes Signal (globale Betrachtung) von t_0 bis t_{end}
- Diskretisierung auf Basis der Fundamentallösung des halb-unendlichen Körpers $T_{z20}(t)$
- Einbeziehung der maximalen Frame-Rate der IR-Kamera
- Annahme von gleicher Signal-Energie in jedem „Einzelsegment“

Oberflächentemperatursignal für pixel p über Messzeit t

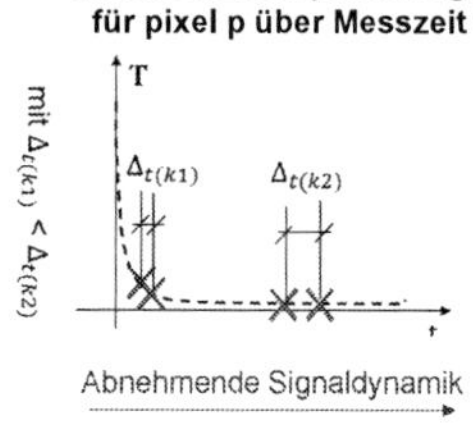

$$t_k^{(nq)} = \tau\left[\left(\frac{t_{end}}{\tau}+1\right)^{\frac{k}{N_t-1}}-1\right]$$

Details siehe Ref. [2]

Nebenbedingung: $(t_1 - t_0) = \tau\left[\left(\frac{t_{end}}{\tau}+1\right)^{\frac{1}{N_{t,ref}-1}}-1\right] \geq \Delta t_{min}$

Maximale Aufzeichnungsrate $f_{max} = 1/\Delta t_{min}$ der IR-Kamera einbeziehen!

Nicht-äquidistante Strategie für Virtuelles Wellensignal (Tiefendomäne) [3]

dgzfp JAHRESTAGUNG 2025 26. – 28. Mai Berlin | fh OBERÖSTERREICH UNIVERSITY OF APPLIED SCIENCES UPPER AUSTRIA

Betrachtete Bedingungen/Aspekte:

- Nicht-äquidistanter Ansatz für gesamtes Signal (globale Betrachtung) von z_0 bis z_{end}
- Diskretisierung auf Basis des Tiefen-Auflösungsvermögen der „thermischen Wellen" (frequenzbasierte Cut-Off Bedingung [4])
- Einbeziehung der maximalen Frame-Rate der IR-Kamera

Virtuelles Wellensignal für pixel p über Tiefe z

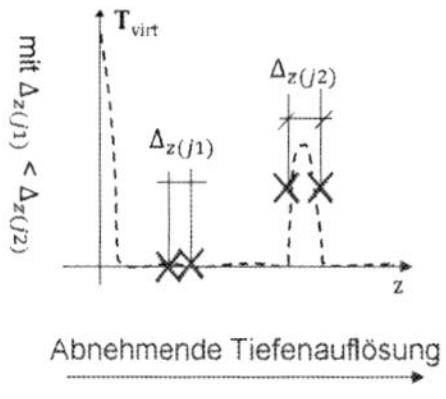

$$t_j^{\prime(\mathrm{nq})} = \left(\frac{j}{N_{\mathrm{tv}} - 1} + \tau' \right)^2 - \tau'^2$$

Details siehe Ref. [3]

Nebenbedingung: $\Delta t'_{\min} = (t'_1 - t'_0) = \left(\frac{1}{N_{\mathrm{tv,ref}} - 1} + \tau' \right)^2 - \tau'^2 \geq \Delta t_{\exp}$

Verifizierung für analytische Temperatursignale [2]

dgzfp JAHRESTAGUNG 2025 26. – 28. Mai Berlin | fh OBERÖSTERREICH UNIVERSITY OF APPLIED SCIENCES UPPER AUSTRIA

1D Wärmeübertragung (Stab):

$$T_{Z22}(z = 0, t) = \frac{2\hat{T}_0}{\sqrt{4\pi\alpha t}} \sum_{n=-\infty}^{\infty} \exp\left(-\frac{(2nL)^2}{4\alpha t}\right)$$

- Zeitlich nicht-äquidistant: $N_t = 1000$, $\tau = 0.91$ s, $\Delta t_{min} = 1/350$ s
- Zeitlich äquidistant: $N_t = 1000$, $\Delta t =$ const.

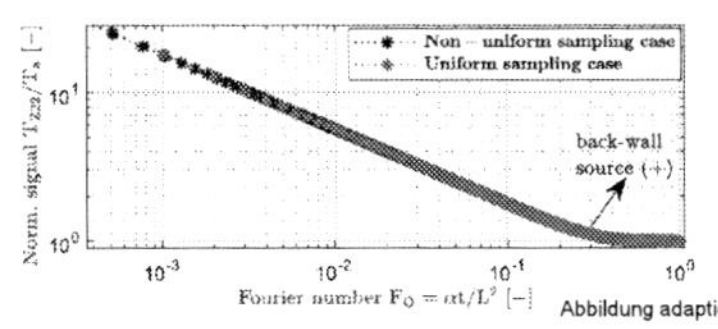

Abbildung adaptiert aus [2]

$t'_j = t_j$

Virtuelle Wellensignale (a) theor., (b) T-SVD, (c) ADMM:

a) $T_{\mathrm{virt},Z22}(z = 0, t') = \frac{\hat{T}_0}{c} \sum_{n=-\infty}^{\infty} \left[\delta\left(t' + \frac{2nL}{c}\right) + \delta\left(t' - \frac{2nL}{c}\right)\right]$

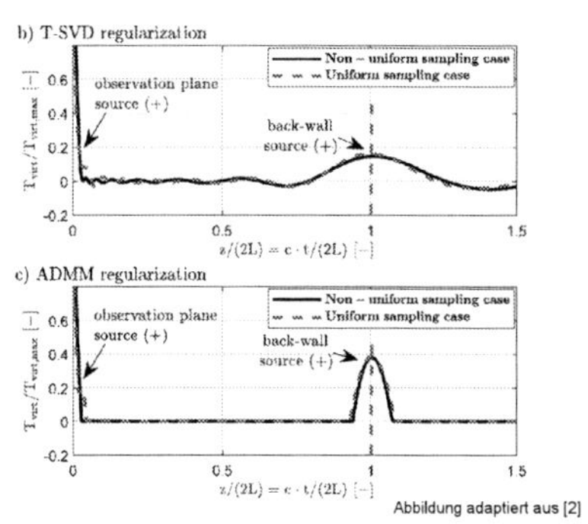

Abbildung adaptiert aus [2]

Verifizierung für numerisch-generierte Temperatursignale [2]

dgzfp JAHRESTAGUNG 2025 26. – 28. Mai Berlin | fh OBERÖSTERREICH UNIVERSITY OF APPLIED SCIENCES UPPER AUSTRIA

Querschnitt (und Randbedinungen) der betrachteten **Probe mit Teflon-Fremdmaterialeinschlüsse**:

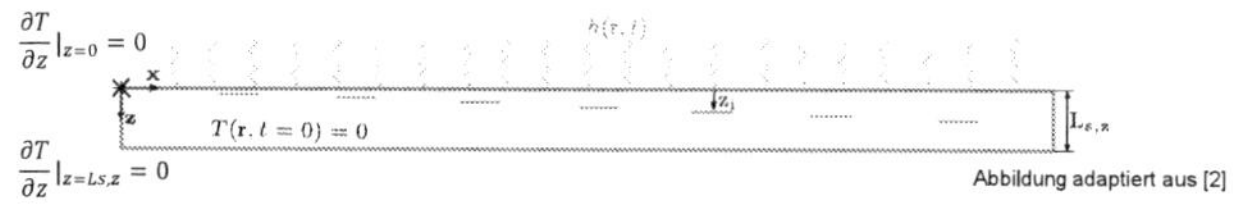

Abbildung adaptiert aus [2]

B-scan Darstellung (äquidistanter und nicht-äquidistanter Fall):

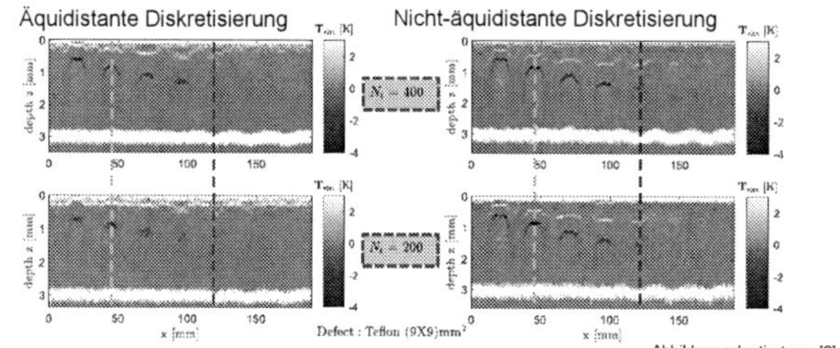

Abbildung adaptiert aus [2]

Untersuchung der Positionsgenauigkeit der auftretenden Features

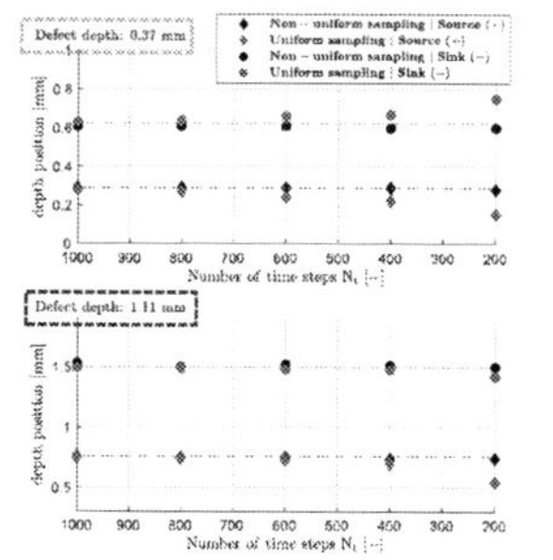

Abbildung adaptiert aus [2]

Experimentelle 3D Defektrekonstruktion in UD-CFK [3]

dgzfp JAHRESTAGUNG 2025 26. – 28. Mai Berlin | fh OBERÖSTERREICH UNIVERSITY OF APPLIED SCIENCES UPPER AUSTRIA

2. Rekonstruktionsschritt (auf Basis nicht-äquidistanter Daten):

Anwendung von Verfahren aus der Ultraschalltechnik für die Rekonstruktion von Defekten: Örtliche Mittelung – Erhöhung SNR | Informationen über Wärmeleitungstensor – Defekt sizing, Kompensation von lateralen Wärmeflüssen

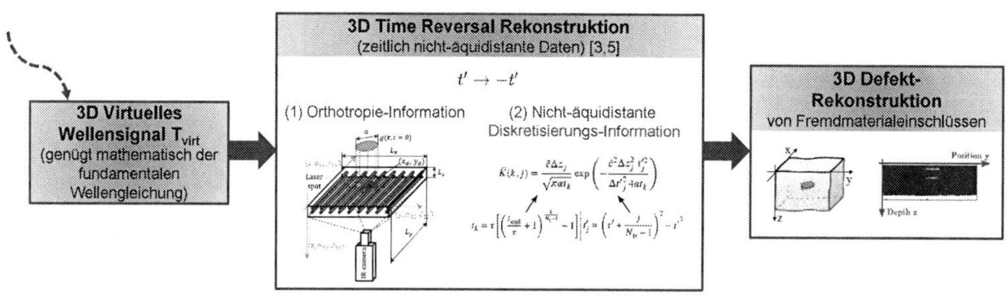

Experimentelle 3D Defektrekonstruktion in UD-CFK [3]

dgzfp JAHRESTAGUNG 2025 26. – 28. Mai Berlin

UNIVERSITY OF APPLIED SCIENCES UPPER AUSTRIA

B-Scan Darstellung (äquidistanter und nicht-äquidistanter Fall):

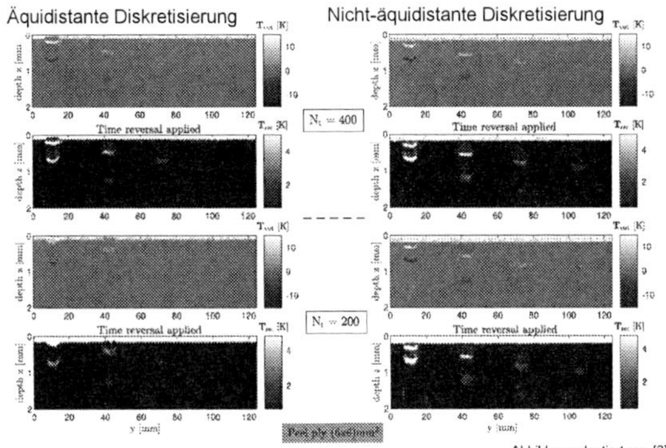

Abbildung adaptiert aus [3]

3D-Defektrekonstruktion (nicht-äquidistant):

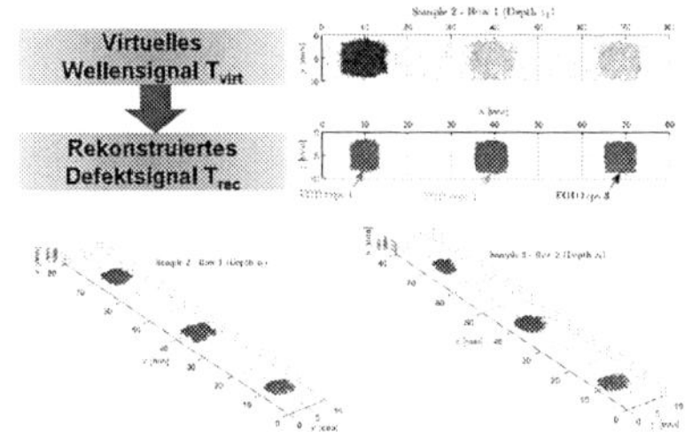

Abbildung adaptiert aus [3]

Reduktion der Berechnungsdauer (≈ Faktor 5)
- Äquidistanter Standard: $N_t = 1000$, $N_{tv} = 850$
- Nicht-äquidistanter Fall: $N_t = 200$, $N_{tv} = 200$

Reduktion des Speicherbedarfs, Einfaches Handling der Daten

Zusammenfassung

dgzfp JAHRESTAGUNG 2025 26. – 28. Mai Berlin

UNIVERSITY OF APPLIED SCIENCES UPPER AUSTRIA

→ Erweiterung des Virtuellen Wellenkonzepts (VWC) in der Thermografie für nicht-äquidistante Diskretisierung

→ Einführung eines generalisieren Transformationskernels für beliebige Abtaststrategien im VWC

→ Diskussion von physikalisch-basierten Strategien zur sinnvollen Abtastung im Temperatur- sowie Virtuellem Wellensignal

→ Verifizierung anhand von analytischen und simulierten Temperatursignalen mit Defekten in orthotropem Material

→ Anwendung von Time Reversal für die experimentelle 3D Defektrekonstruktion in orthotropem CFK und nicht äquidistanten Eingangssignalen

→ Berechnungstechnische Effizienzsteigerung und Reduktion des Speicherbedarfs bei gleichbleibender Rekonstruktionsqualität

Förderung

dgzfp JAHRESTAGUNG 2025 26. – 28. Mai Berlin
fh OBERÖSTERREICH UNIVERSITY OF APPLIED SCIENCES UPPER AUSTRIA

→ Austrian Research Funding Association (FFG) within the dissertation funding program of the University of Applied Sciences Upper Austria, within the project “Interface” (Contract No. 905121).

→ Project “HyperMAT” by the federal government of Upper Austria and the European Regional Development Fund (EFRE) in the framework of the EU-program IBW/EFRE & JTF 2021–2027.

Wissenschaftliche Kooperation

fh OBERÖSTERREICH UNIVERSITY OF APPLIED SCIENCES UPPER AUSTRIA
JKU JOHANNES KEPLER UNIVERSITÄT LINZ
iPPE
RECENDT RESEARCH CENTER NON DESTRUCTIVE TESTING

Referenzen

dgzfp JAHRESTAGUNG 2025 26. – 28. Mai Berlin
fh OBERÖSTERREICH UNIVERSITY OF APPLIED SCIENCES UPPER AUSTRIA

→ [1] P. Burgholzer, M. Thor, J. Gruber, and G. Mayr. Three-dimensional thermographic imaging using a virtual wave concept. J. Appl. Phys., 121(10):105102. doi: 10.1063/1.4978010 (2017).

→ [2] L. Gahleitner, G. Mayr, G. Mayr, P. Burgholzer, U. Cakmak. Efficient defect reconstruction from temporal non-uniform pulsed thermography data using the virtual wave concept. NDT & E Int. 147:103200. doi: 10.1016/j.ndteint.2024.103200 (2024).

→ [3] L. Gahleitner, G. Mayr, P. Burgholzer, U. Cakmak. Three-dimensional defect reconstruction in carbon fiber-reinforced composites with temporally non-uniform pulsed thermography data. NDT & E Int. 154:103363. doi:10.1016/j.ndteint.2025.103363 (2025).

→ [4] P. Burgholzer, G. Mayr, G. Thummerer, M. Haltmeier. Linking information theory and thermodynamics to spatial resolution in photothermal and photoacoustic imaging. J Appl Phys. 0021-8979, 128 (17), doi: 10.1063/5.0023986 (2020).

→ [5] P. Burgholzer, G. J. Matt, M. Haltmeier, G. Paltauf. Exact and approximative imaging methods for photoacoustic tomography using an arbitrary detection surface. Phys Rev E, Stat Nonlinear Soft Matter Phys. 1539-3755, 75: 046706, doi: 10.1103/PhysRevE.75.046706 (2007).

Erkennung von Innenkorrosion in Fässern für radioaktive Abfälle mit thermographischen Methoden

Anton AVERIN[1], Philipp Daniel HIRSCH [1] und Julien LECOMPAGNON [1]
[1] Bundesanstalt für Materialforschung und -prüfung (BAM), Berlin, Deutschland, anton.averin@bam.de

Kurzfassung

In Deutschland lagern über 130.000 Kubikmeter schwach- und mittelradioaktiver Abfälle, davon etwa 90 % des gesamten radioaktiven Abfalls in 200-Liter-Fässern. Diese Fässer werden aktuell in Zwischenlagern aufbewahrt und dort, um die Sicherheit zu gewährleisten, manuell auf äußere Korrosion überprüft, während das vorgesehene Endlager, das Bergwerk Konrad, voraussichtlich erst bis 2030 fertiggestellt wird. Manuelle Inspektionen bergen jedoch das Risiko menschlicher Fehler, was die Zuverlässigkeit und Sicherheit des Prozesses beeinträchtigt und sind auf die Erkennung äußerlich sichtbarer Unregelmäßigkeiten beschränkt.

Ziel des FORKA-geförderten Projektes „ZIKA" ist es, diese Inspektionen durch eine vollautomatisierte zerstörungsfreie Prüfung (NDT) zu automatisieren. Ein Schwerpunkt liegt dabei auf der Anwendung von aktiver Infrarotthermografie zur Erkennung innerer Defekte in Metallfässern, die mit visuellen Inspektionen nicht entdeckt werden können. Die Genauigkeit thermografischer Prüfung wird dabei durch Oberflächenkontaminationen (Kratzer, Schmutz, Aufkleber), Farbschichten mit geringer Wärmeleitfähigkeit und die Krümmung der Fässer, welche die Wärmeverteilung beeinflussen, beeinträchtigt.

Die hier präsentierte Studie analysiert die Eignung von aktiver thermografischer Prüfung zur Detektion innenliegender Korrosion unter Zuhilfenahme verschiedener Wärmequellen – Blitzlampen und Laser –, sowie verschiedener Anregetechniken, wie Impulsthermografie und Lock-in-Thermografie. Zudem werden auch Methoden zur Datenverarbeitung, darunter Hauptkomponententhermografie (PCT), Puls-Phasen-Thermografie (PPT) und thermische Signalrekonstruktion (TSR) evaluiert. Um Messrauschen und Artefakte in den thermografischen Bildern zu filtern, wurden maschinelle Lernmodelle entwickelt. Die Ergebnisse zeigen, dass die Kombination fortschrittlicher Thermografietechniken mit maschinellem Lernen die Defekterkennung verbessert und automatisierte, zuverlässigere Inspektionsprozesse für die Lagerung radioaktiver Abfälle ermöglicht.

Keywords: Thermografie, Zerstörungsfreie Prüfung, Radioaktive Abfälle, Automatisierte Inspektion, Thermische Datenverarbeitung, Maschinelles Lernen, Defekterkennung

Sicherheit in Technik und Chemie

BAM

Bundesanstalt für Materialforschung und -prüfung

FEDERAL INSTITUTE FOR MATERIALS RESEARCH AND TESTING

Automated thermographic inspection of radioactive waste drums.

Averin Anton, Philip Hirsch, Julien Lecomagnon

www.bam.de

1

Problem description

200 L Storage Drum

Internal corrosion from outer

Drums are typically intended for **short-term storage**, making them vulnerable to **aging, corrosion, and structural degradation**.

Current inspection methods rely on visual evaluation by human inspectors, which are subjective and do not offer early detection or prediction of leakage.

2

State of the art and goals

The project inherits another- EMOS (2020-22)

We aimed to use thermography for inner damage detection

EMOS from inside

EMOS was focused on detecting external corrosion on radioactive waste barrels using VIS cameras and laser scanners

Drum shell

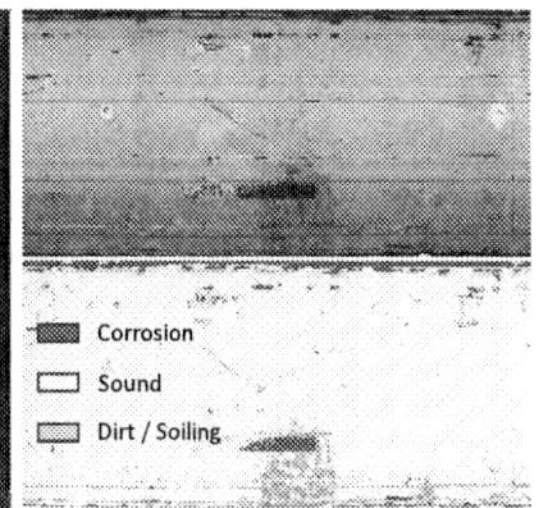

Drum lid

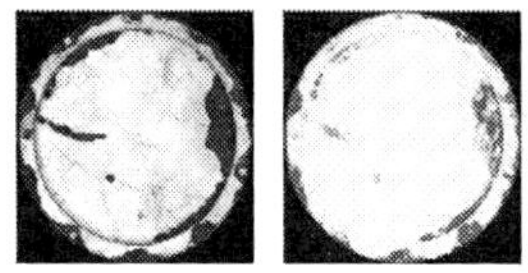

3

Thermography challenge

Factors that make defect detection with thermography ambiguous:

Complex shape

Surface artifacts – scratches, dirt, paintings, dents, paint layer thickness inhomogeneities and its delamination

Investigated barrel cut #1
(thickness 2.5 mm, material – ferritic steel)

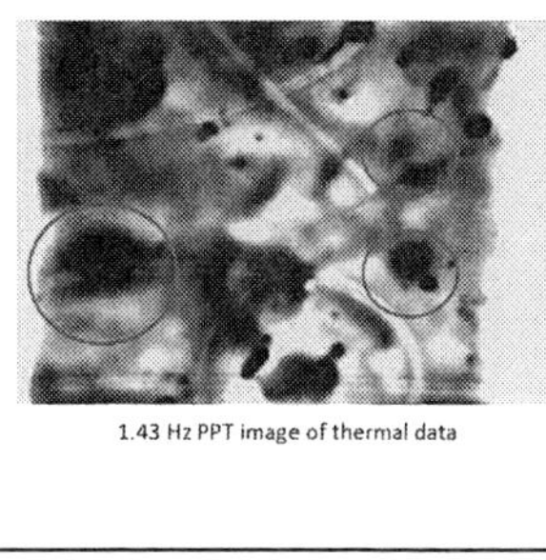

1.43 Hz PPT image of thermal data

Outer surface

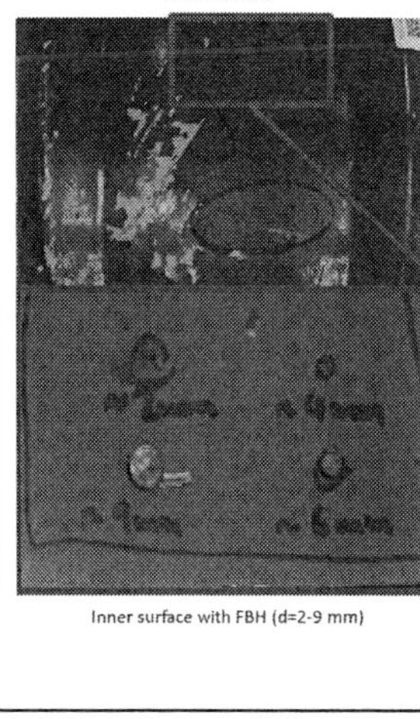

Inner surface with FBH (d=2-9 mm)

4

Setups.

We have tested 2 kinds of setups:

Flash-thermography

low budget, broad wavelength range, high energy

Key parameters:

- $\lambda = 250 - 1500$ nm
- Power density 0.3 W/cm^2
- Sensing area 16x16 cm^2
- Pulse time 200 ms

Laser-thermography + Robot arm assistance

narrow bandwidth, different excitation schemes [Pulse therm. (PT) and Lock-in therm. (LIT)].

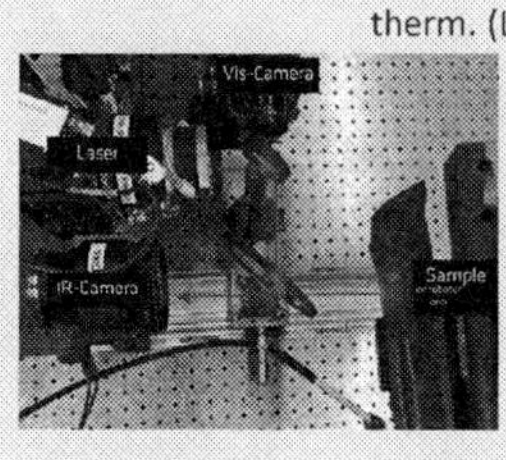

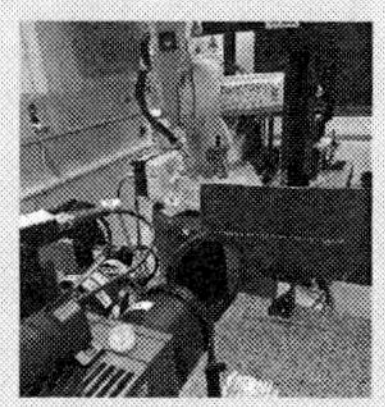

Key parameters:

- $\lambda = 940 \pm 5$ nm
- Power density 11 W/cm^2
- Sensing area 7x7 cm^2
- Pulse time 500 ms (PT)

5

Laser thermography results

PT results:

PCT 2

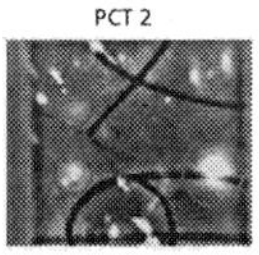

Principal Component Thermography (PCT)

PPT 1.143 Hz

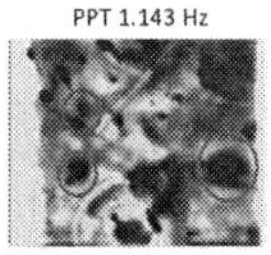

Pulse Phase Thermography (PPT)

TSR 4

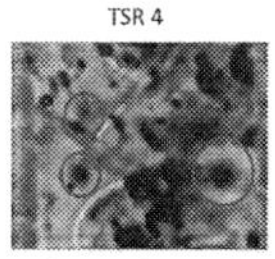

Thermal Signal Reconstruction (TSR)

LIT results:

1 Hz. Lock-in. Phase 1.27 Hz

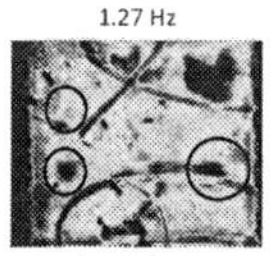

1.5 Hz. Lock-in. Phase 0.33 Hz

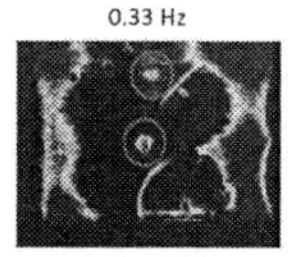

1.5 Hz. Lock-in. Phase 0.33 Hz

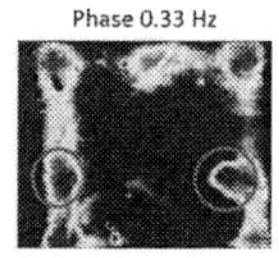

Optimal frequencies were defined from 1D model of periodic heat propagation in slab

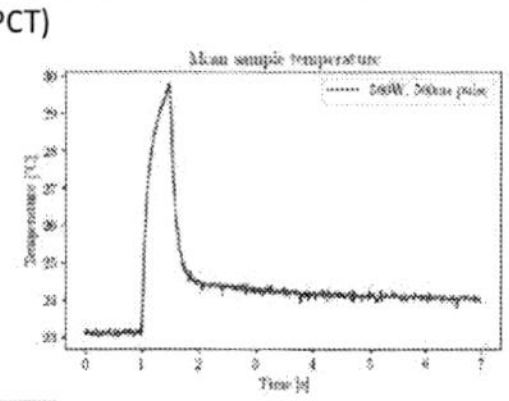

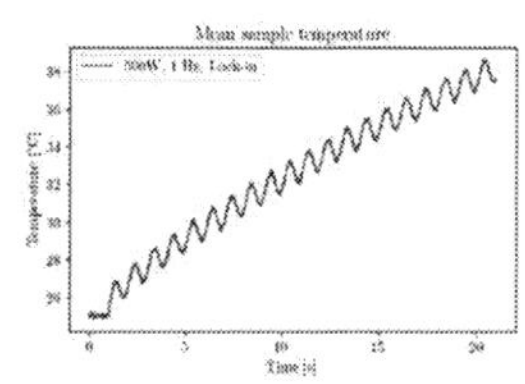

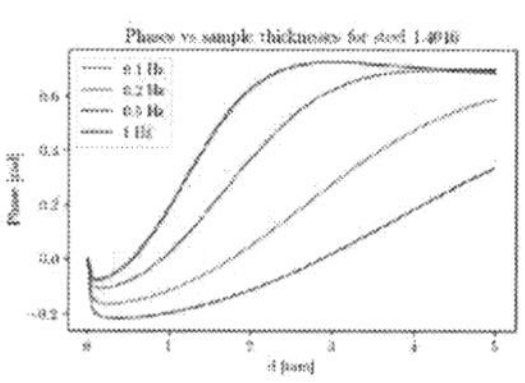

6

Flash thermography results

For Flash experiment data we tested TSR, PPT, PCT

Problem — In both laser and flash FBHs cannot be distinguished for sure from surface artifacts

Proposal — Combine postprocessing methods

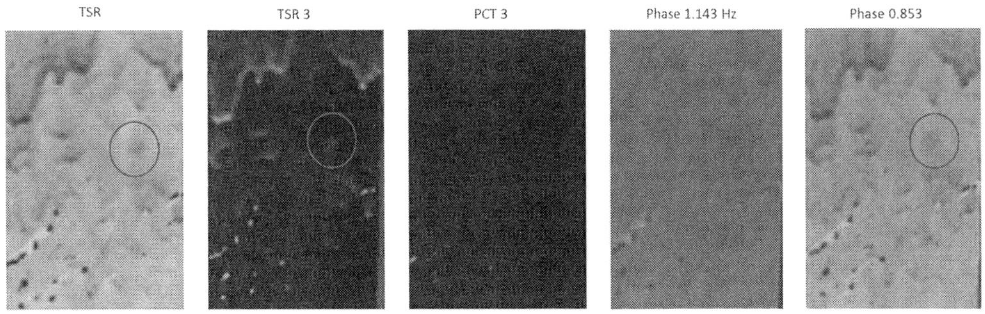

Setup underperforms laser with only 1 of 4 defects is visible

7

Preparation for statistical evaluation

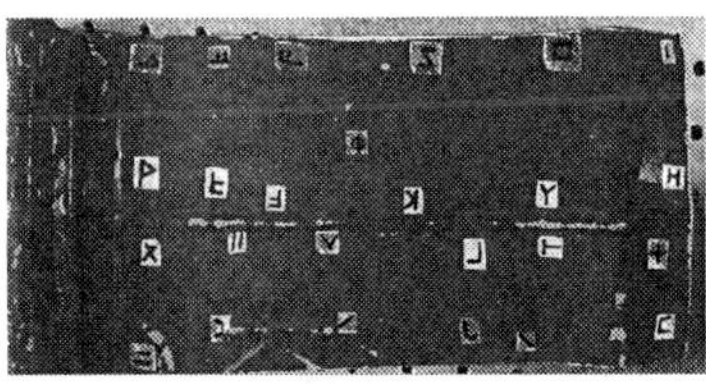

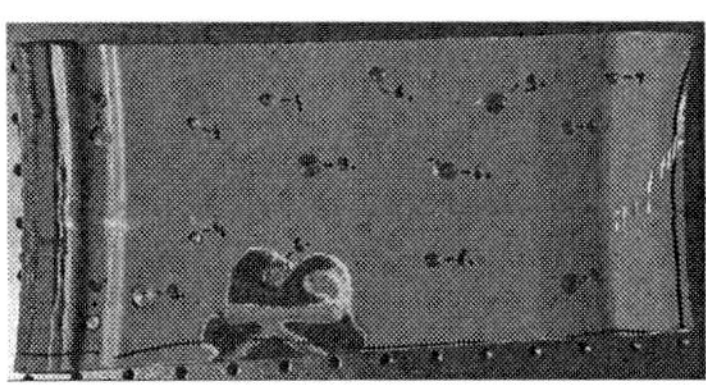

- For statistical evaluation we used sample with 15 FBHs (4, 6, 9 mm)
- "Ground truth" for each thermogram were generated based on VIS images overlay (shown here)
- Labels are used for better overlay between VIS and thermograms

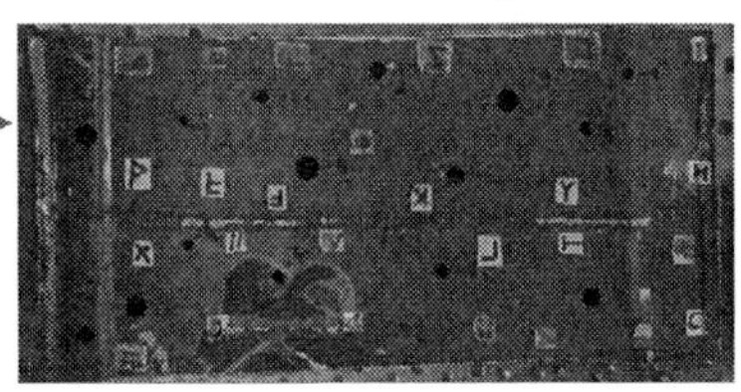

8

Preparation for statistical evaluation

Thermograms were merged with overlays using labels and surface artifacts visible in both images

From mentioned overlays we generated a "ground truth" binary masks

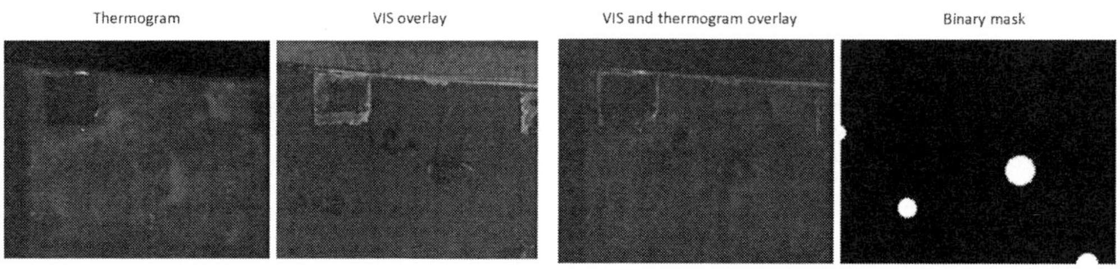

9

Advanced postprocessing

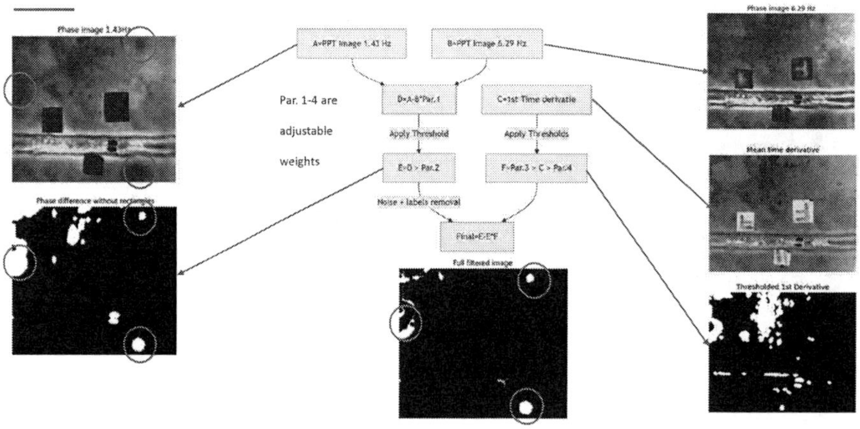

10

Advanced postprocessing

- Differential evolution method was used to find best weights
- Method optimizes F1 score (balancing false negative and false positives)
- $F1 = 2 * \frac{PRECISION * RECALL}{PRECISION + REC}$
- $PRECISION = \frac{TP}{TP+}$
- $RECALL = \frac{TP}{TP}$
- TP: Number of True positive predictions
- FP: Number of False positive predictions
- FN: Number of False negative predictions

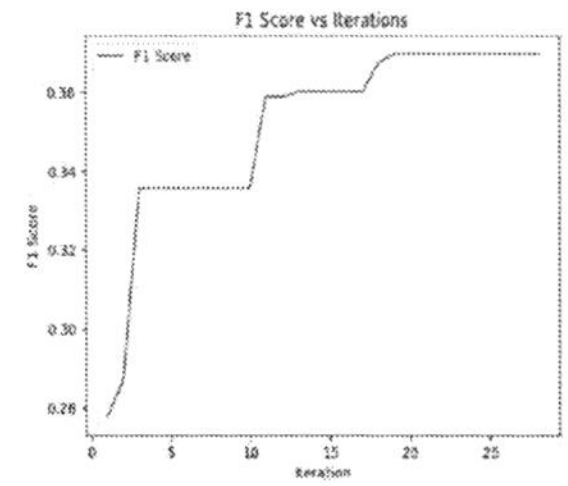

Manual parameter search	
Precision	0.297
Recall	0.204
F1	0.2422
Differential evolution	
Precision	0.382
Recall	0.379
F1	0.369

- It is a pixel statistic
- On defect scale:
  - 18 of 23 were correctly found
  - Only 3 false predictions

11

Postprocessing with NNs

Architectures:

- Multi Layer Perception (MLP) – pixel classification
- Convolutional Neural Network (CNN) – image classification

Data structure

- 15 Images – CNN
- $5*10^6$ pixels – MLP

Learning data:

- postprocessed (TSR, PPT,...)
- raw (T)

MLP

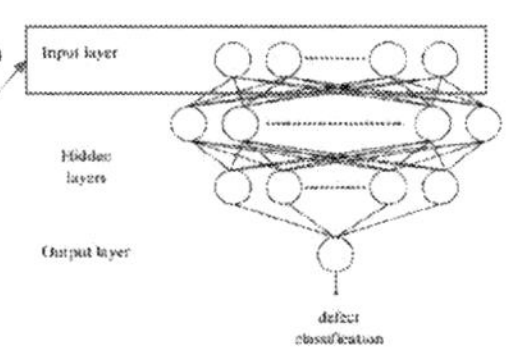

CNN

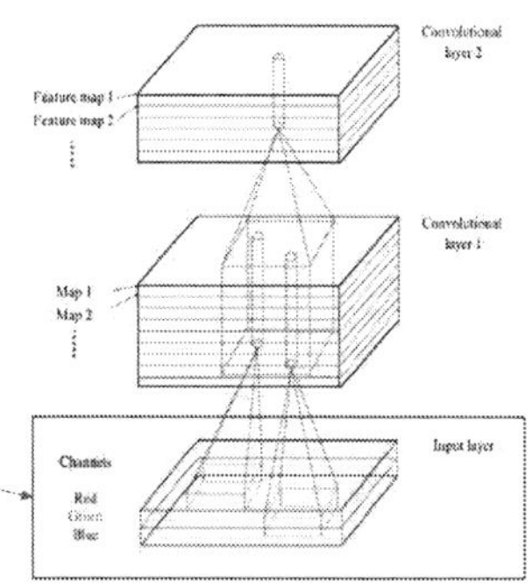

12

Postprocessing with NNs

Best model CNN –

2D_32(3x3)+2D_32(3x3)+Dropout(0.3)+2D_1(1x1)

Small data amount – overfitting prevention:

- simple architectures (low levels)
- data augmentation (image rotation, displacement)
- cross-validation (5 folds, 80/20 train/validation)

Dataset – TSR (0-5), PPT (0-25 Phase, 0-25 Amp),

PCT (0-5), time deriv. (1st, 2nd).

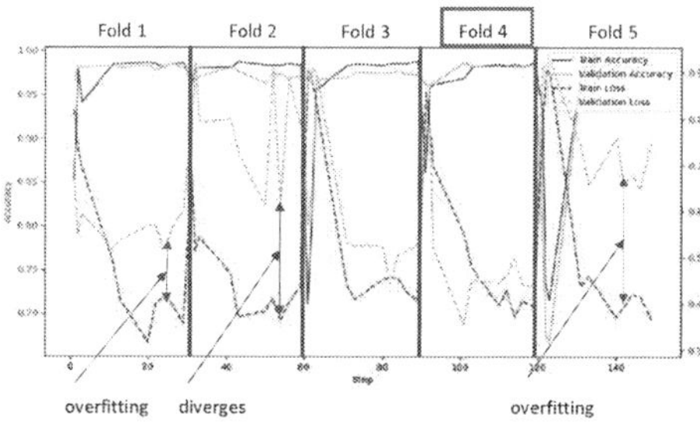

Model	F1		Precision		Recall	
Differential evolution	0.369		0.382		0.379	
MLP	0.315		0.19		0.932	
CNN	Val+Train 0.542	Test: 0.55	Val+Train 0.556	Test: 0.546	Val+Train 0.53	Test: 0.556

13

Automated detection

The defect-scale results:

- 19 of 23 correctly classified
- 4 is false positive

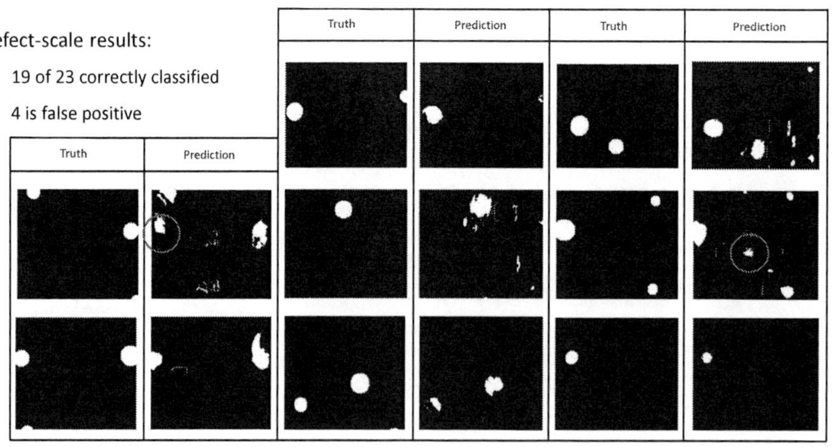

14

Next steps

- Improve NNs (improve loss function, improve metrics, MLP+CNN combination, LSTMs, Encoders,..)
- Full barrel scan (More robust setup, more data)
- Using different filaments for barrel

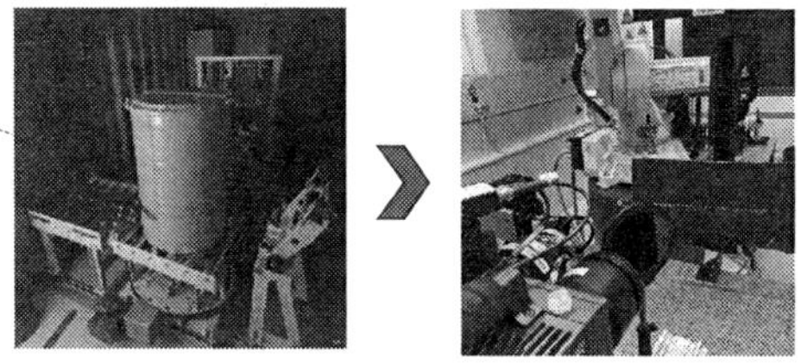

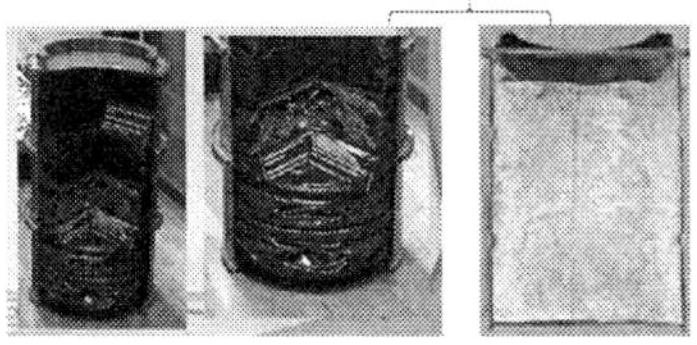

15

Sicherheit in Technik und Chemie

BAM
Bundesanstalt für Materialforschung und -prüfung

FEDERAL INSTITUTE FOR MATERIALS RESEARCH AND TESTING

Automated thermographic inspection of radioactive waste drums.

Averin Anton, Philip Hirsch, Julien Lecomagnon

www.bam.de

16

DGZfP-Jahrestagung 2025

Untersuchung und Kompensation des Size-of-Source Effektes der Thermografie zur Prozessüberwachung in der metallischen additiven Fertigung

Alec HILBERG[1], Simon J. ALTENBURG[2] Andreas KROLL [3]
[1] Bundesanstalt für Materialforschung und -prüfung, Berlin, Deutschland, alec.hilberg@bam.de
[2] Bundesanstalt für Materialforschung und -prüfung, Berlin, Deutschland, simon.altenburg@bam.de
[3] Universität Kassel - Fachgebiet Mess- und Regelungstechnik, Kassel, Deutschland, andreas.kroll@mrt.uni-kassel.de

Kurzfassung

Der Einsatz der Thermografie zur Prozessüberwachung in der additiven Fertigung von Metallen ist aktueller Schwerpunkt diverser Forschungsprojekte. Prozessbedingt können im Bauteil Risse, innere Spannungen und sonstige interne Defekte auftreten, deren Entstehung durch wichtige thermische Größen wie Phasenumwandlungstemperaturen, Abkühlgeschwindigkeiten und thermische Gradienten beeinflusst wird. Um diese Einflüsse zu untersuchen, ist eine möglichst genaue Messung dieser physikalischen Größen notwendig.

Neben den häufig betrachteten Einflüssen des Emissionsgrades, der Umgebungstemperatur und der atmosphärischen Absorption besteht ein weiterer Einfluss, hervorgerufen durch das optische System. Der Kameraaufbau, das verwendete Objektiv und Teilchen im Strahlengang können zur Beugung, Streuung, Reflektion und Aberration der elektromagnetischen Strahlung führen, wodurch die thermografische Messung verfälscht wird. Dies ist ein in der Pyrometrie intensiv erforschter Effekt und wird, aufgrund der Abhängigkeit zur betrachteten Objektgröße, als Size-of-Source Effekt bezeichnet und in dieser Arbeit untersucht.

Im Laser-Pulver-Auftragsschweißen (LPA) wird mittels eines Lasers auf einer Substratplatte ein Schmelzbad erzeugt und durch Einblasen von Metallpulver die lagenweise Fertigung von komplexen Bauteilen ermöglicht. Dabei können, je nach geometrischer Auflösung der Kamera, Größen des Schmelzbades im Bereich weniger Prozente der gesamten Bildfläche auftreten. Aus durchgeführten Referenzmessungen an Schwarzkörperstrahlern kann in der gleichen Größenordnung eine Abweichung der Strahlungswerte im zweistelligen Prozentbereich festgestellt werden. Daher ist in diesem Anwendungsfall der dadurch verursachte Fehler nicht zu vernachlässigen, weshalb auf Basis der Referenzmessungen eine Methode entwickelt wurde, um in Abhängigkeit der Größe des Objektes einen Faktor zur Kompensation des Size-of-Source Effektes zu berechnen.

In diesem Beitrag werden erste Ergebnisse dieser im Rahmen des DFG Projektes QT-LPA (Nr. 516965606) durchgeführten Arbeiten präsentiert.

Keywords: Additive Manufacturing, Thermography, Monitoring, Size of Source Effect

BAM
Bundesanstalt für Materialforschung und -prüfung

Sicherheit in Technik und Chemie

05.2025

Untersuchung und Kompensation des Size-of-Source Effektes der Thermografie zur Prozessüberwachung in der metallischen additiven Fertigung

Alec Hilberg*, Simon J. Altenburg*, Andreas Kroll**

www.bam.de

* Bundesanstalt für Materialforschung und –prüfung (BAM)
** Universität Kassel, Fachgebiet Mess- und Regelungstechnik

Laser-Pulver-Auftragschweißen (LPA)

BAM

- Einsatz eines fokussierten Lasers erzeugt Schmelzbad
- Zufuhr von Metallpulver zur lagenweisen Erzeugung additiver Strukturen
- Lokale Intergasatmosphäre zur Reduktion von Oxidation

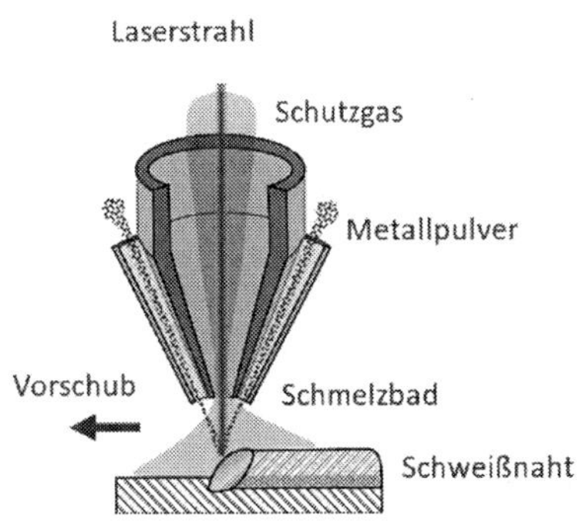

Schematische Darstellung des LPA Prozesses

Laser-Pulver-Auftragschweißen (LPA)

BAM

- ➢ Thermografie als Methode zur in-situ Prozessüberwachung
- Zusammenhang zwischen thermografischen Merkmalen und Bauteilfehlern [3,4,5]
- Möglichst genaue Messung der elektromagnetischen Strahlung notwendig
- ➢ Viele Einflüsse auf Messung (Emissionsgrad, Winkel, Size-of-source Effekt, ...)

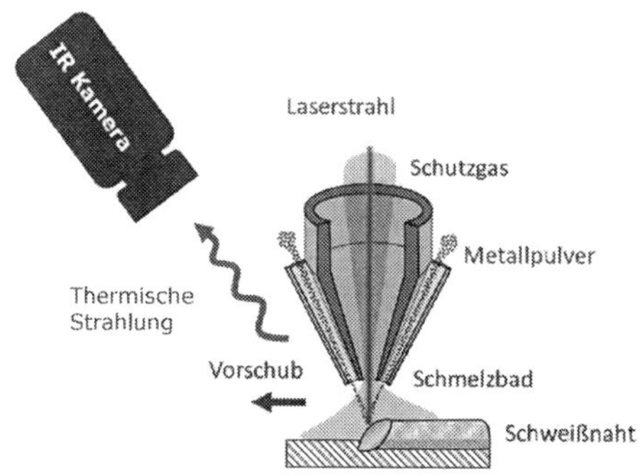

Schematische Darstellung des LPA Prozesses

Size of Source Effekt (SSE)

BAM

Einfluss der Größe eines Objektes auf das thermografische Signal

➢ Scheinbare Temperatur sinkt für kleinere Objekte

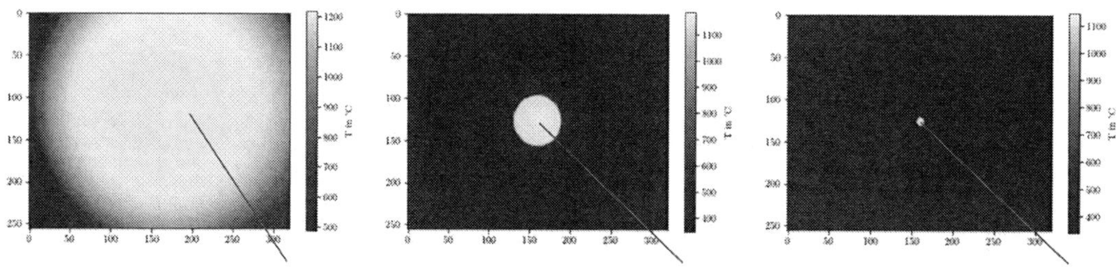

1198 °C
Signal = 100 %

1180 °C
Signal = 93 %

1140 °C
Signal = 84,6 %

Thermogramme eines Schwarzkörperstrahlers mit 1200 °C, Verwendung einer verstellbaren Blende zur Variierung der Größe

Size of Source Effekt (SSE)

BAM

Einfluss der Größe eines Objektes auf das thermografische Signal

➢ Scheinbare Temperatur sinkt für kleinere Objekte

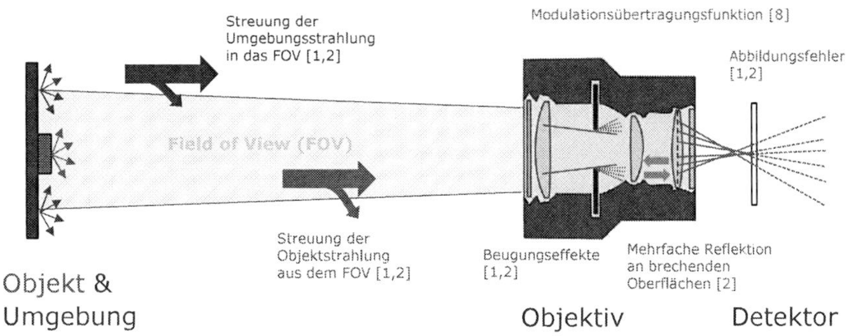

Prozessdaten

BAM

- Mittelwellen Infrarot (MWIR) Thermografie während LPA Prozess der Fertigung einer liegenden Wand
- In Abhängigkeit des optischen Setups kann Schmelzbadgröße anteilig wenige Prozente im Sichtfeld einnehmen

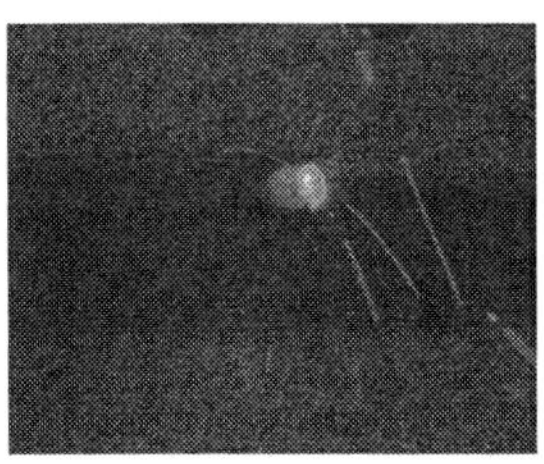

Visuelle HDR Aufnahme des LPA Prozesses, Aufnahme von Lars Sommerlade

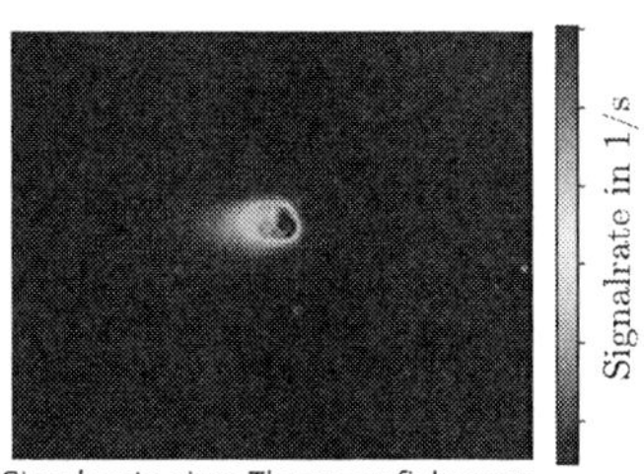

Signalwerte einer Thermografiekamera im MWIR Bereich während LPA Prozess

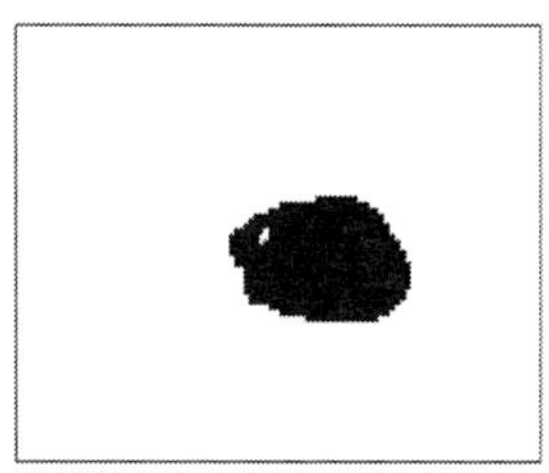

Identifikation und Markierung Schmelzbad, aktuelle Größe = 0,76%

Datenerhebung

- Variation der Temperatur durch Schwarzkörperstrahler (SKS)
  - HGH ECN100:30 °C – 600 °C
- Variation der Objektgröße durch verstellbare Blende
- Messung mit MWIR Thermografiekamera im Bereich 1,5 µm – 5,5 µm
  - Telops MS M3*k*

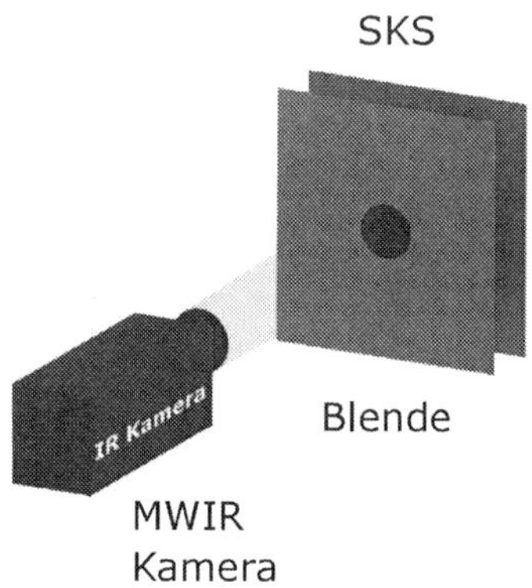

Schematische Darstellung des Messaufbaus

Datenerhebung

- Keine direkte Auswertung der Temperatur
- Verwendung des digitalen Signals, normiert mit der Belichtungszeit (Signalrate)
- Als Referenzpunkt für Signalraten bei jeweiliger Temperatur wird eine Blendengröße verwendet, die größer als FOV ist

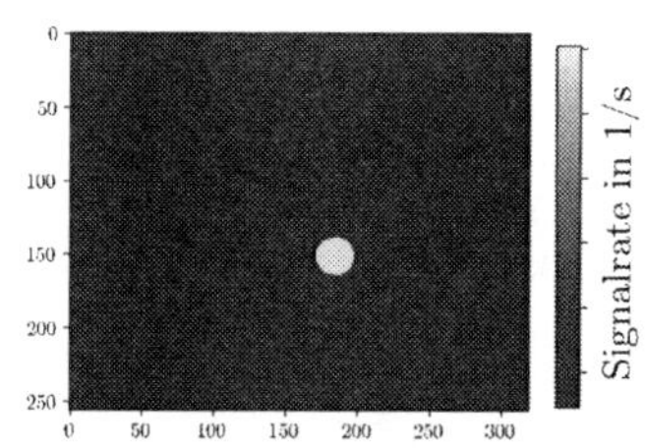

Signalrate für verringerte Blendengrößen bei der Untersuchung des SSE

Auswertung - Parameter

BAM

Aktuelle Signalrate	S	Aktuelle gemittelte Signalrate der Quelle
Referenz Signalrate	S_{ref}	Referenzsignalrate bei der Temperatur des Schwarzkörperstrahlers
Hintergrund Signalrate	S_{bg}	Mittlere Signalrate des aktuellen Hintergrundes
Quellgröße	***size***	*Prozentuale Größe der Quelle im Bild*
SSE Parameter	***sse***	Prozentualer Wert der aktuellen Signalrate der Quelle zwischen Referenzsignalrate und Hintergrundsignalrate

Auswertung - Parameter

BAM

$$size = \frac{Anzahl\ Pixel\ der\ Quelle}{Anzahl\ Pixel\ Gesamt}$$

$$sse(size) = \frac{S(size) - S_{bg}(size)}{S_{ref} - S_{bg}(size)}$$

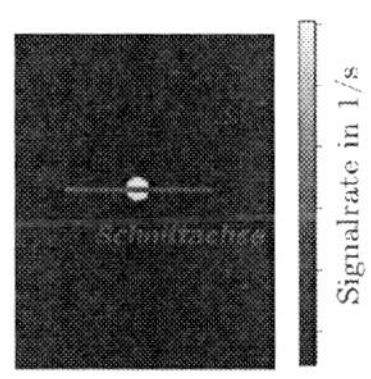

Signalrate eines BB-Strahlers bei 500 °C mit verkleinerter Blende

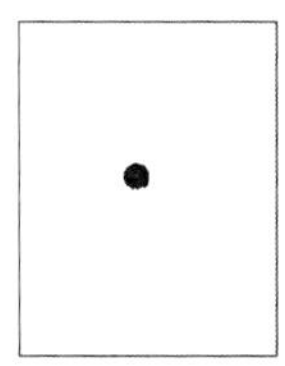

Zugehörige Pixel des BB-Strahlers zur Berechnung der size

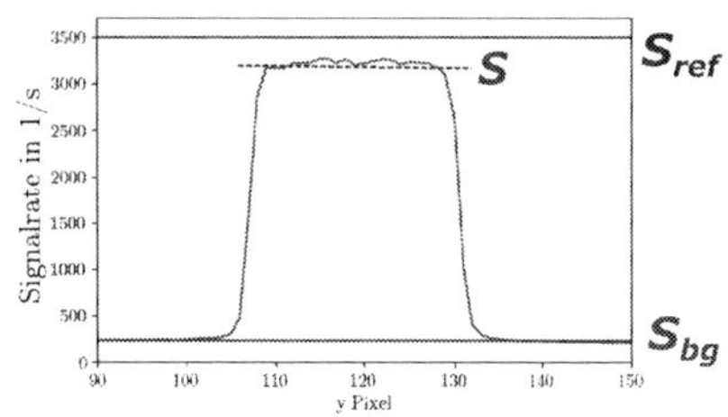

Darstellung der Signalrate des Schnittes durch den Blendenmittelpunkt, Markierung der Referenz- und Hintergrundsignalrate

Auswertung

BAM

- ***sse*** Werte verändern sich in Abhängigkeit der Größe
- ***sse***-Kurven scheinen sich zu überlagern, unabhängig von der eingestellten Temperatur
- Überlagerte Kurve kann approximiert werden

➢ Berechnung der ursprünglichen Referenz

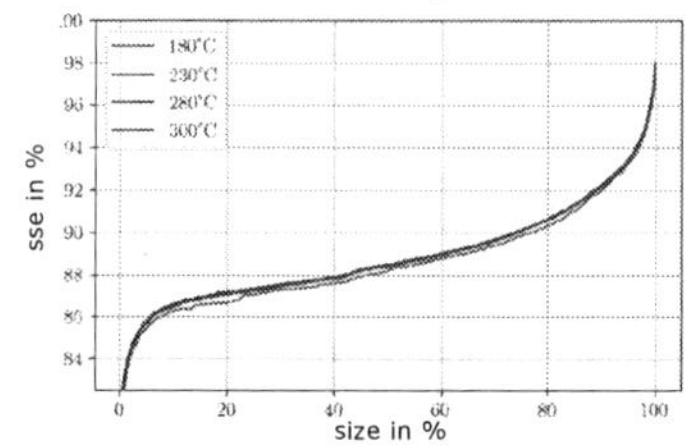

sse Kurven in Abhängigkeit der ***size*** für unterschiedliche Temperaturen

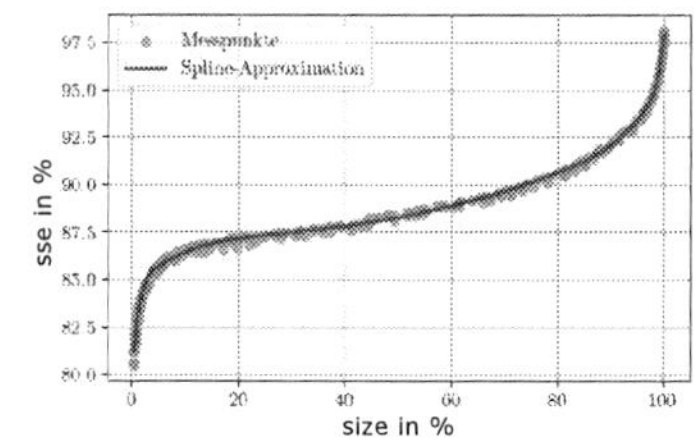

sse Messpunkte in Abhängigkeit der ***size*** aller Messungen und approximierte Kurve

Korrekturfaktor

BAM

- Korrekturfaktor k

$$k \cdot S = S_{ref}$$

- S_{ref} über ***sse***-Kurve, S und S_{bg} bestimmen

$$S_{ref} = \frac{S - S_{bg}}{sse(size)} + S_{bg}$$

➢ $$k = \frac{S - S_{bg}}{sse(size) \cdot S} + \frac{S_{bg}}{S}$$

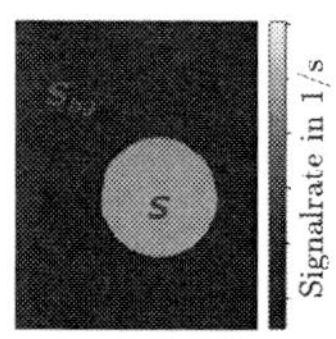

MWIR Thermogramm SK-Strahler, T = 350°C

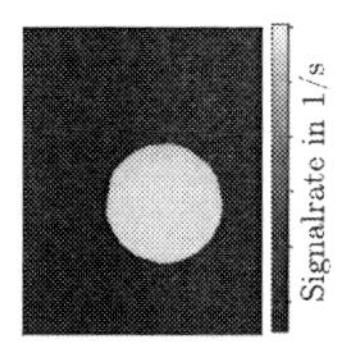

MWIR Thermogramm SK-Strahler mit durchgeführter SSE Korrektur, T = 350°C

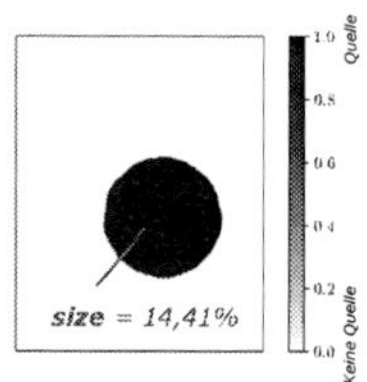

*Identifizierte Quelle bei 350°C mit einer **size** von 14,41%*

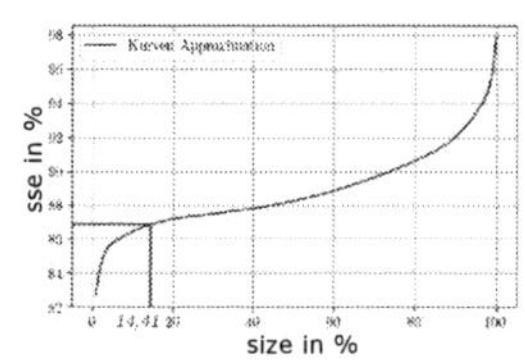

Approximierte Kurve des Zusammenhanges zwischen **sse** und ***size***

Korrekturbewertung

BAM

- ➢ Bewertung anhand neuer Messung bei 350° C
- Korrektur wird für gemittelte Werte der Quelle durchgeführt
- Verlauf des relativen Fehlers ohne Korrektur führt zu Werten über 15%
- Für korrigierte Bilder liegt relativer Fehler für alle gemessenen Größen unter 0,4%

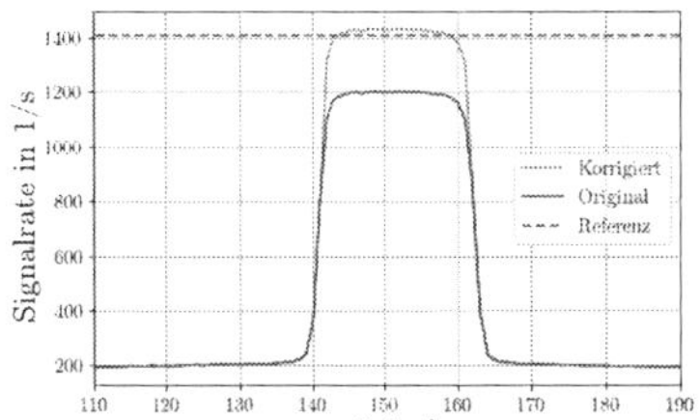

Signalrate des Schnittes durch Blende bei T = 350°C für originales und korrigiertes Bild

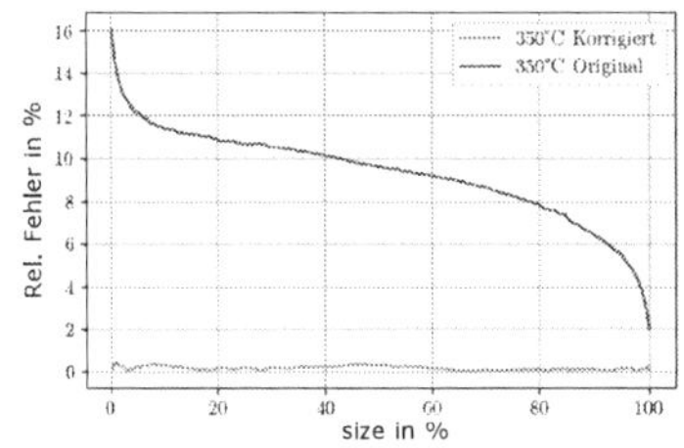

Relativer Fehler der Signalrate zur Referenz bei T = 350°C für originales und korrigiertes Bild

05.2025 A. Hilberg, alec.hilberg@bam.de 13

Ausblick - Prozessübertragbarkeit

BAM

- Korrektur an Schwarzkörperstrahlern bringt gute Ergebnisse
- Im LPA-Prozess graduelle Änderung der Signalrate, insbesondere hinter Schmelzbad, Überlagerung von Temperaturbereichen
- ➢ Explizite Untersuchungen notwendig, um Vorgehen zur Übertragbarkeit auf Prozess zu entwickeln

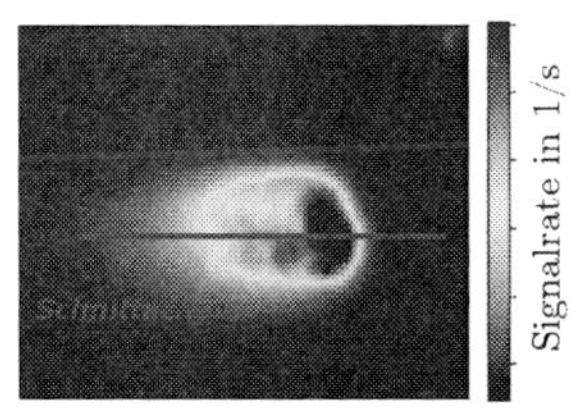

Signalwerte einer Thermografiekamera im MWIR Bereich während LPA Prozess, Fokus Schmelzbad

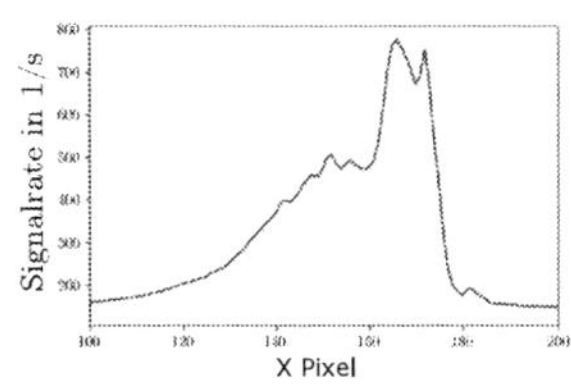

Signalwerte einer Thermografiekamera im MWIR Bereich während LPA Prozess, Verlauf des Schnittes durch Schmelzbad

05.2025 A. Hilberg, alec.hilberg@bam.de 14

Vielen Dank für die Aufmerksamkeit

BAM

Gefördert durch die Deutsche Forschungsgemeinschaft (DFG) - 516965606,

DFG Deutsche Forschungsgemeinschaft

Dankeschön an die
Thermografie Abteilung 8.3 der Bundesanstalt für Materialforschung und -prüfung,
die Universität Kassel,
Robert Schmoll,
Lars Sommerlade &
Miguel Mendez des Fachbereichs Mess- und Regelungstechnik

UNIKASSEL VERSITÄT

mrt

05.2025 A. Hilberg, alec.hilberg@bam.de 15

Literatur

BAM

1. Peter Saunders & Hamish Edgar (2008). *On the characterization and correction of the size-of-source effect in radiation thermometers.* BIPM and IOP Publishing Ltd.
2. B. Gutschwager, J. Hollandt, T. Jankowski & R. Gärtner (2008). *A Vacuum Infrared Standard Radiation Thermometer at the PTB.* Springer Science+Business Media.
3. G. Mohr, S. J. Altenburg, A. Ulbricht, P. Heinrich, D. Baum, C. Maierhofer & K. Hilgenberg (2020). *In-Situ Defect Detection in Laser Powder Bed Fusion by Using Thermography and Optical Tomography—Comparison to Computed Tomography.* Metals, 10, 103.
4. A. Ulbricht et al. (2020). *Separation of the Formation Mechanisms of Residual Stresses in LPBF 316L*, Metals, 10.
5. G. Mohr, S. J. Altenburg, and K. Hilgenberg (2020). *Effects of inter layer time and build height on resulting proper-ties of 316L stainless steel processed by laser powder bed fusion*. Addit. Manuf., 32, 101080.
6. Schwarzkörperstrahler HGH ECN 100, https://hgh-infrared.com/ecn-blackbody/, abgerufen am 26.03.2025
7. Kalibrierstrahler PYROTHERM CS 1200, https://dias-infrared.de/produkte/kalibrierstrahler-und-emitter/, abgerufen am 26.03.2025
8. Helmut Budzier & Gerald Gerlach (2001). *The size-of-source effect in thermography*. J. Sens. Sens. Syst., 10, 179–184, 2021

05.2025 A. Hilberg, alec.hilberg@bam.de 16

Einsatz neuartiger Erwärmungsstrategien mit zeitlicher Modulation zur Steigerung der Detektionsgrenzen in der aktiven Laserthermografie

Julien LECOMPAGNON [1], Philipp Daniel HIRSCH [1] und Mathias ZIEGLER [1]
[1] Bundesanstalt für Materialforschung und -prüfung (BAM), Berlin, Deutschland,
julien.lecompagnon@bam.de

Kurzfassung

Moderne Lasersysteme haben sich als vielseitige Wärmequellen für aktive thermografische Prüfungen erwiesen. Im Vergleich zu herkömmlichen Lichtquellen wie Blitz- oder Halogenlampen können ihre Ausgangsleistungen leicht moduliert werden, was eine Vielzahl komplexer Anregungen ermöglicht. Diese Studie untersucht die Wirksamkeit verschiedener zeitlich strukturierter Erwärmungsmethoden zur Detektion und Tiefenbestimmung von Defekten.

Insgesamt werden verschiedene komplexe zeitlich strukturierte Erwärmungsmuster verglichen, darunter Einzelpuls und kodierte Anregung mit Barker-Codes und Legendre-Sequenzen, sowie klassische Lock-in-Modulation und neuartige Multifrequenz-Lock-in Anregung. Die Untersuchung konzentriert sich auf dabei auf die Erreichung einer maximalen Temperaturerhöhung zur Steigerung des Signal-Rauschverhältnisses auch bei Anregungen mit großen Frequenzinhalt und die erreichte Qualität der Defekterkennung bei unterschiedlichen Anregungsschemata.

Die experimentellen Ergebnisse zeigen, dass kodierte Anregungen wie Barker- und Legendre-Sequenzen im Vergleich zur Einzelpulsheizung nicht nur höhere Temperaturanstiege, sondern bei gleichem Frequenzinhalt auch eine bessere Defekterkennung ermöglichen. Insbesondere die Mehrton-Modulation zeigte den höchsten Temperaturanstieg und ermöglicht die Bewertung in verschiedenen Tiefen ohne Verlängerung der Messzeit. Die Verwendung kodierter Anregungen bietet signifikante Vorteile bei der thermografischen Erkennung interner Defekte. Die Ergebnisse dieser Studie tragen dazu bei, die Effizienz und Genauigkeit der thermografischen Prüfmethoden zu verbessern, insbesondere bei der Untersuchung von Materialien mit hoher Wärmeleitfähigkeit und tief verborgenen Defekten.

Besonders an den untersuchten Modulationsmethoden profitieren alle Anregungsmethoden, die aktuell technologie- oder prinzipbedingt unter geringen realisierbaren Ausgangsleistungen leiden. Besonders hervorzuheben ist hier die Nutzung von Laserprojektoren als Wärmequelle, die vielseitig einsetzbar eine räumlich strukturierte Erwärmung für thermografische Prüfungen ermöglichen, jedoch bisher stark in ihrer optischen Ausgangsleistung limitiert sind.

Keywords: Thermografie, Laser, Kodierte Anregung, Pulskompression

Sicherheit in Technik und Chemie

BAM
Bundesanstalt für Materialforschung und -prüfung

28.05.2025

Einsatz neuartiger Erwärmungsstrategien mit zeitlicher Modulation zur Steigerung der Detektionsgrenzen in der aktiven Laserthermografie

Julien Lecompagnon[1], Philipp Daniel Hirsch[1] und Mathias Ziegler[1]

[1] Fachbereich 8.3: Thermografische Verfahren, BAM

www.bam.de

Rekonstruktion tiefliegender Defekte
Problembeschreibung

BAM

Die Herausforderung

Wärme DGL: $\frac{1}{\alpha}\cdot\frac{\delta T}{\delta t}-\frac{\delta^2 T}{\delta r^2}=0$ diffusiv!

Tiefe: $L = 0$ mm, $L = 0.5$ mm, $L = 1$ mm

Auflösungs-verschlechterung

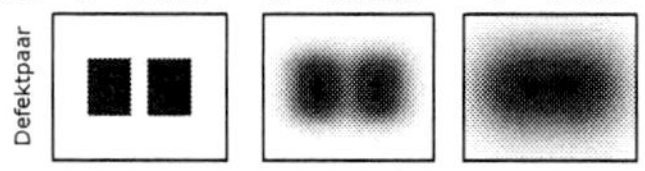

Unsere Lösung

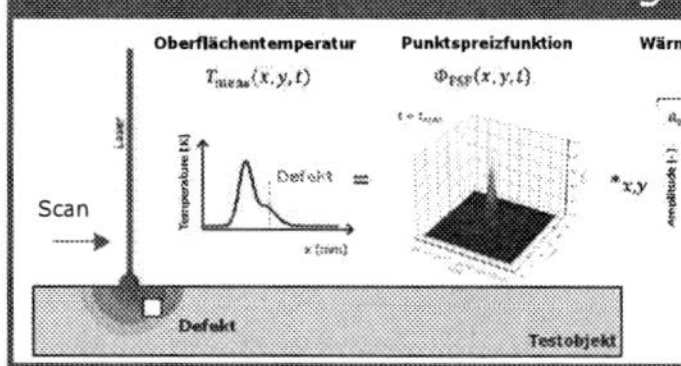

Strukturierte Beleuchtung

Das Werkzeug

Leistungsstark aber langsam!

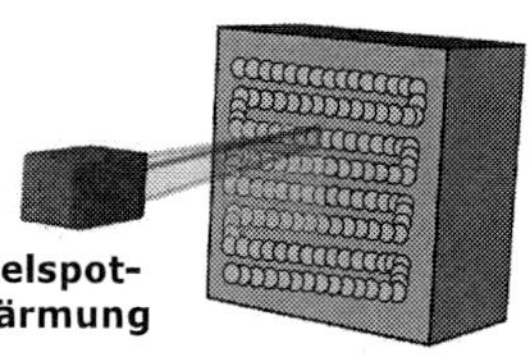

Einzelspot-Erwärmung

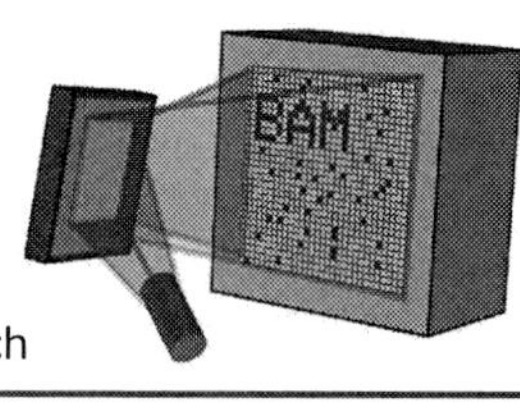

Vollflächige Modulation

Schnell aber leistungsschwach

Vollflächig strukturierte Beleuchtung
Limitationen aktueller Gerätetechnologie

- 1280×800 pixel; $10.8\ \mu m$ Pixelgröße
- $86\ \mathrm{W}$ max. optische Ausgangsleistung @ $270\ \mathrm{W}$ optische Eingangsleistung (!)
- Nur digitale Anregung (an/aus) => Füllfaktor bestimmt Erwärmung

➔ Kaum leistungsstark genug für die Untersuchung metallischer Objekte

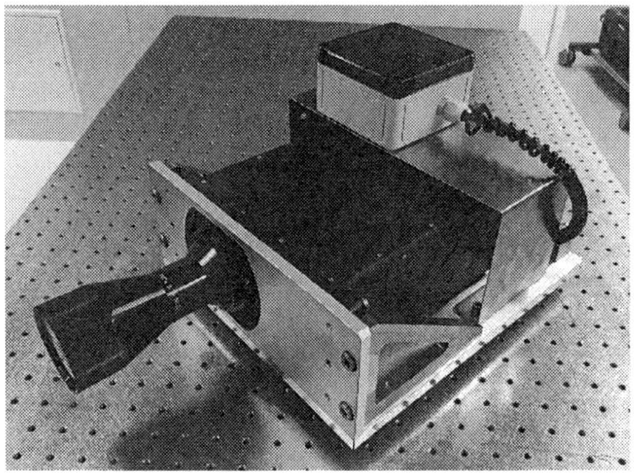

Beispiel für Anregungsmuster

Muster: 128 x 80 Muster: 32 x 20 Muster: 8 x 6

Mehrere solcher Anregungsmuster werden zur Rekonstruktion der inneren Objektstruktur benötigt

28.05.2025 DGZfP-Jahrestagung 2025 | Dr.-Ing. Julien Lecompagnon 3

Vollflächig strukturierte Beleuchtung
Limitationen aktueller Gerätetechnologie

- 1280×800 pixel; $10.8\ \mu m$ Pixelgröße
- $86\ \mathrm{W}$ max. optische Ausgangsleistung @ $270\ \mathrm{W}$ optische Eingangsleistung (!)
- Nur digitale Anregung (an/aus) => Füllfaktor bestimmt Erwärmung

➔ Kaum leistungsstark genug für die Untersu

Unser Ziel für diese Studie:

- Finden einer Lösung zum Umgang mit der Leistungslimitation bestehender Projektortechnologie
- Qualitativer Vergleich und Exploration möglicher Methoden in einem praktischen Anwendungsszenario

zur ötigt

28.05.2025 DGZfP-Jahrestagung 2025 | Dr.-Ing. Julien Lecompagnon 3

Experimenteller Aufbau
Laserthermografieaufbau und Probekörper

BAM

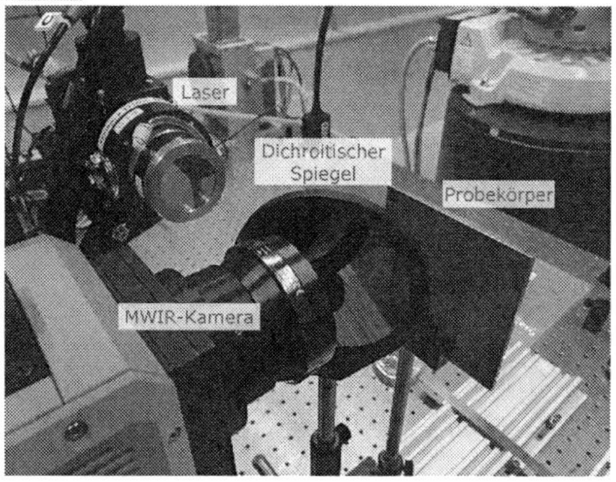

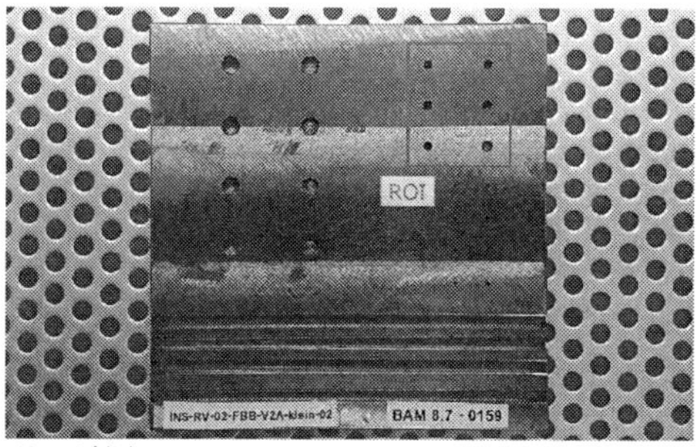

Materialdicke = 4 mm
FBH-Durchmesser = 2 mm
FBH-Lochtiefen = {0.4, 0.76, 1.23, 1.66, 2.15, 2.66} mm

- 500 W, 940 nm Laserquelle; fokussiert auf eine Fläche von 68x68mm²
- Unbeschichteter Edelstahlprobekörper (1.4301) mit verschiedenen Flachlochbohrungen (FBHs)

28.05.2025 DGZfP-Jahrestagung 2025 | Dr.-Ing. Julien Lecompagnon 4

Einzelpulsanregung (Referenz)
Anregung mit hoher Bandbreite

BAM

Kurzer Puls => mehr Bandbreite ☺ & weniger Energieinhalt ☹
Langer Puls => weniger Bandbreite ☹ & mehr Energieinhalt ☺

Referenzmessung

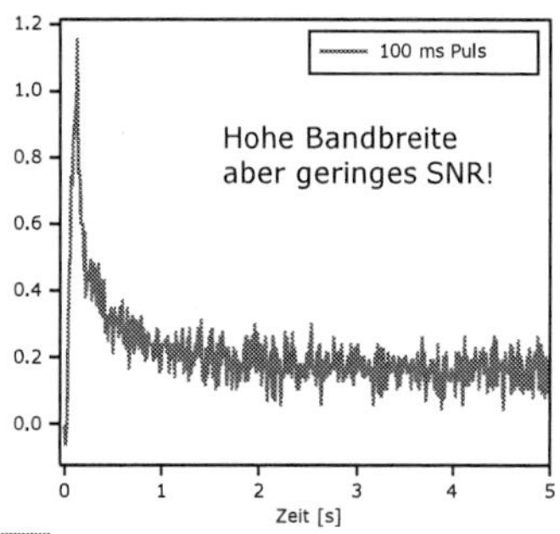

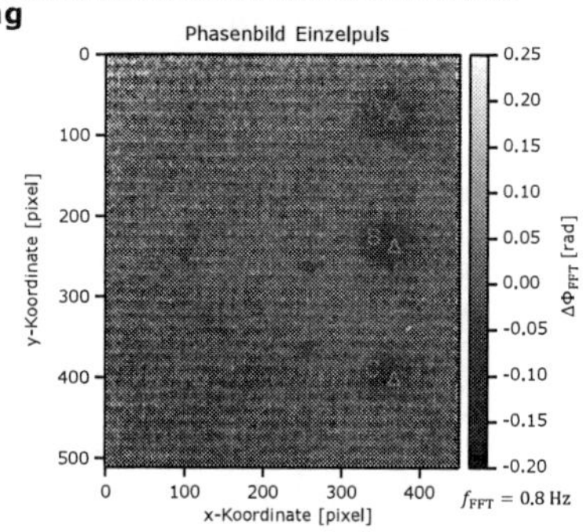

28.05.2025 DGZfP-Jahrestagung 2025 | Dr.-Ing. Julien Lecompagnon 5

BAM

Ansatz 1:
Pulskompression

www.bam.de

BAM

Pulskompression
Grundlegendes Prinzip

$T_{\text{meas}}(x,y,t) = G_{\text{heat}}(x,y,t) \quad *_t \quad I_t(t)$

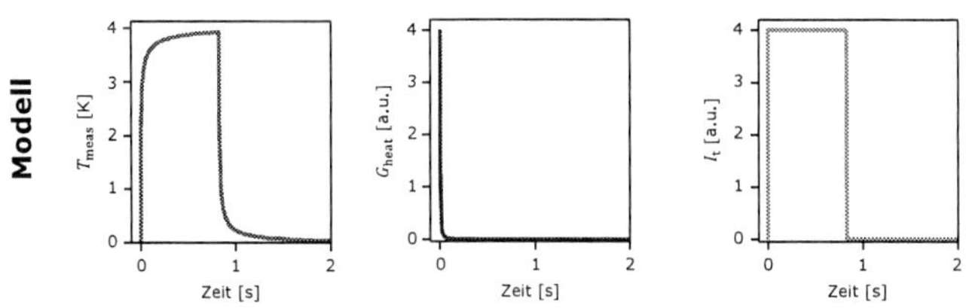

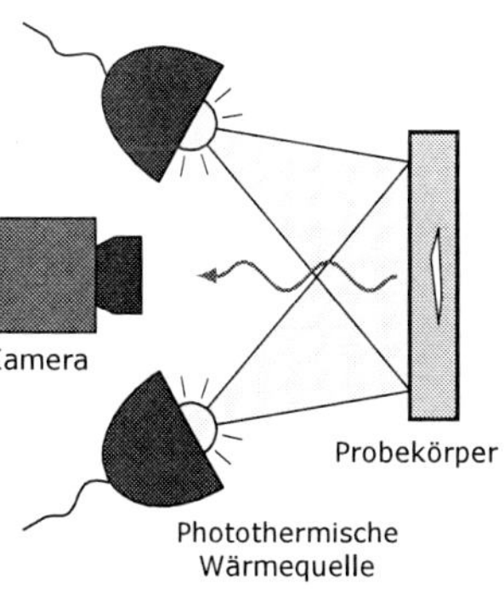

Pulskompression:

$$T_{\text{meas}}(x,y,t) = G_{\text{heat}}(x,y,t) *_t \underbrace{\overbrace{I_t(t) \quad *_t \, S(t)}^{\text{Vergrößerter Energieinhalt!}}}_{\text{Verlängert das Anregungssignal}} \quad \underbrace{*_t \, S_{inverse}(t)}_{\text{Revidiert den Einfluss der Verlängerung}}$$

Wie? → Funktionen mit kleiner Autokorrelation

$$S(t) \star S(t) \approx \delta$$

$$\Rightarrow S_{inverse}(t) \approx S(-t)$$

Pulskompression
Barker-Code

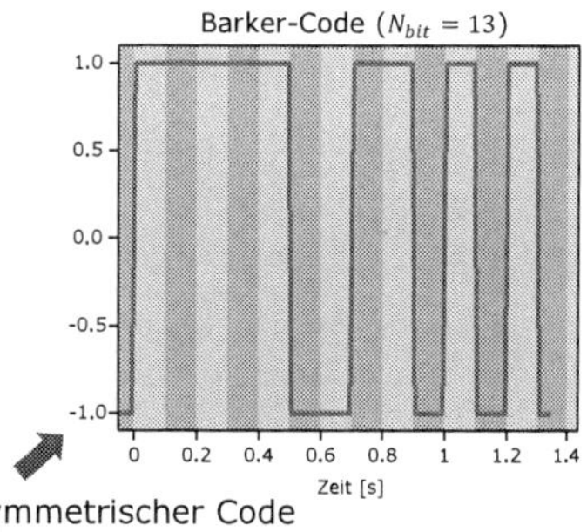

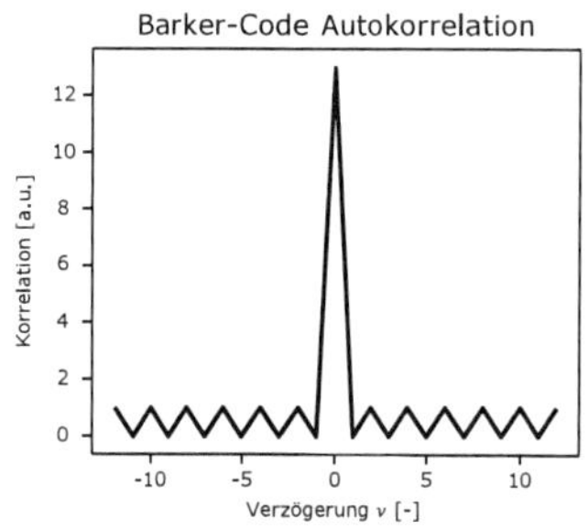

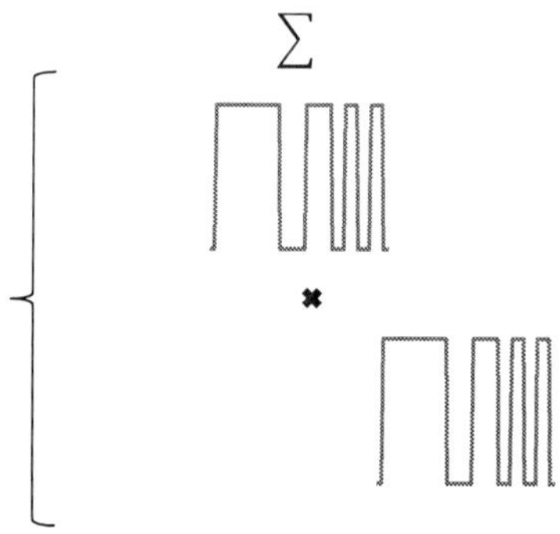

Symmetrischer Code
→ DC-Offset notwendig!

- Binäre Sequenz deren Autokorrelation nur Nebenmaxima ≤ 1 aufweist
- Barker-Codes nur bekannt für Längen: 2, 3, 4, 5, 7, 11 und 13

Barker-Code Erwärmung
Ergebnisse

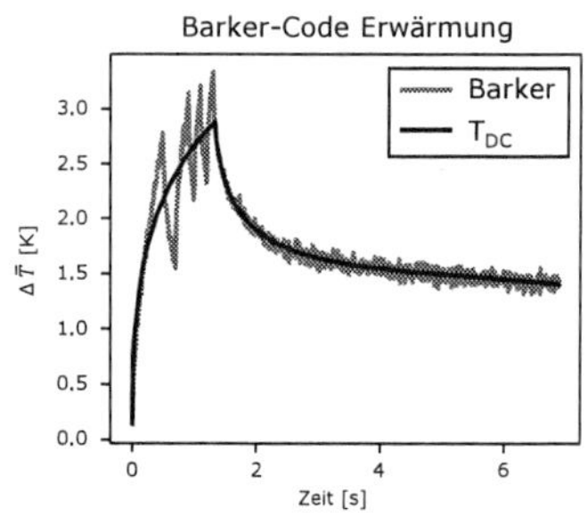

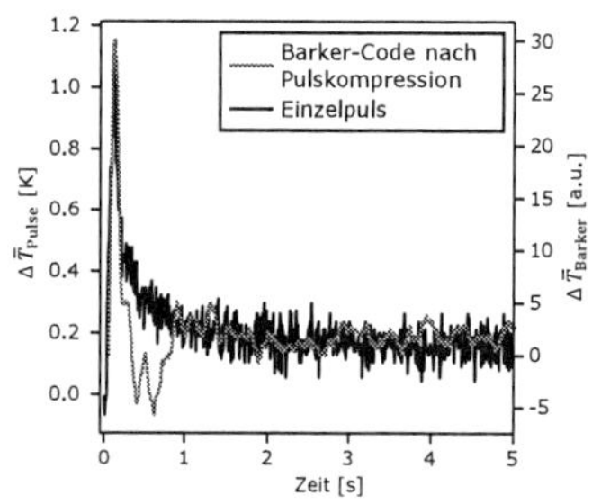

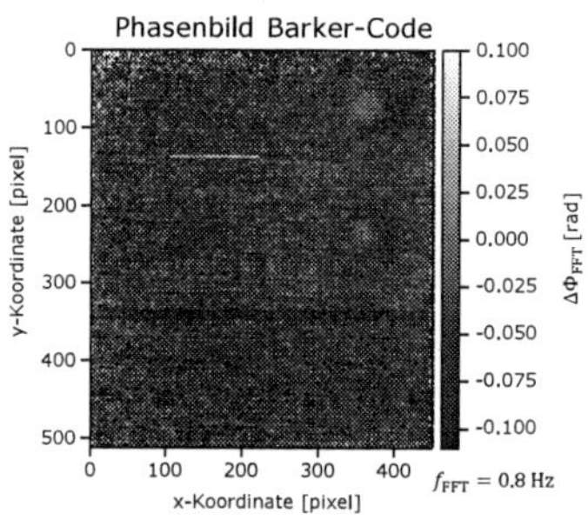

- DC-Offset muss durch Polynomialfit entfernt werden (Erwärmung und Abkühlung!)
- Ergebnisse zeigen Artefakte (Nebenmaxima!)
- Phasenbild liefert nur schlechte Ergebnisse!

Pulskompression
Legendre-Sequenzen

- Können für alle primzahligen Längen generiert werden
- Weisen eine sehr geringe Autokorrelation auf

Jedoch,

- einmal wiederholt, ist ihre zyklische Autokorrelation nahezu perfekt (kleiner DC-offset)

Wegweisende Arbeit:
M. Ricci, R. Zito and S. Laureti: „Pseudo-noise pulse-compression thermography: a powerful tool for time-domain thermography analysis". NDT&E International. June 2024, 10.1016/j.ndteint.2024.103218

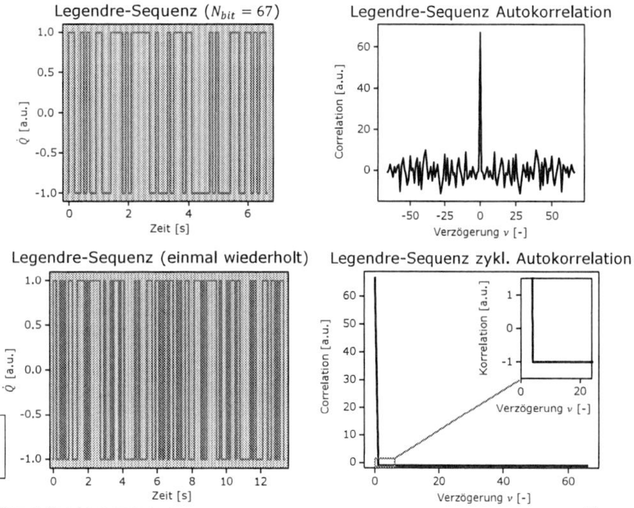

Legendre-Sequenz Erwärmung
Ergebnisse

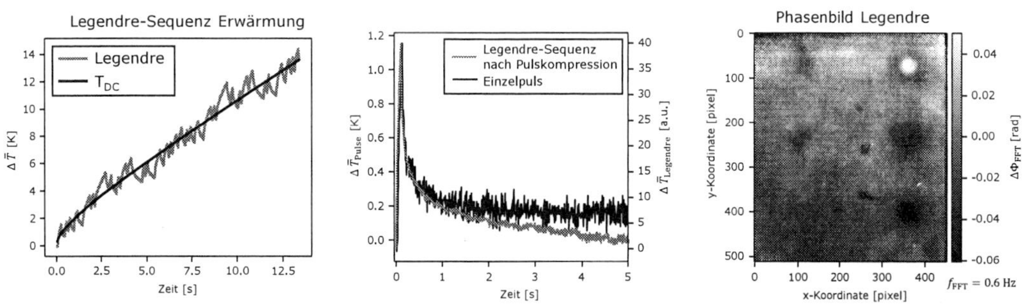

- DC-Offsetkorrektur durch Polynomialfit (nur Erwärmung!)
- Pulsekompression zeigt keine Artefakte
- Qualität des Phasenbilds drastisch verbessert!

Gleiche Bandbreite wie 100 ms Einzelpuls bei drastisch verssertem SNR.

Einfach in der Anwendung.

BAM

Ansatz 2:
Lock-In Thermografie

www.bam.de

28.05.2025 DGZfP-Jahrestagung 2025 | Dr.-Ing. Julien Lecompagnon 12

Einzelton Lock-In Thermografie
Referenz

BAM

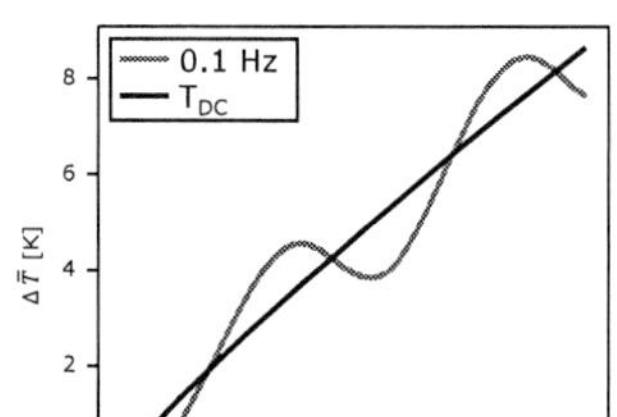

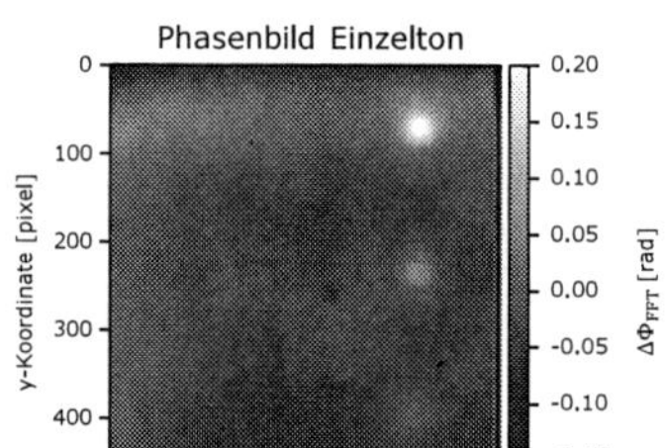

- Anregung mit fester Frequenz: $S_{ST}(t) = A_{ST}\cos\left(2\pi f t - \frac{\pi}{2}\right)$
- Frequenz bestimmt Eindringtiefe: $\mu = \sqrt{\frac{\alpha}{\pi f}}$
- Typische Frequenzen $\ll 1\,\mathrm{Hz}$ ➔ lange Messzeiten und geringe Bandbreite!

28.05.2025 DGZfP-Jahrestagung 2025 | Dr.-Ing. Julien Lecompagnon 13

Multiton Lock-in Thermografie
Überblick

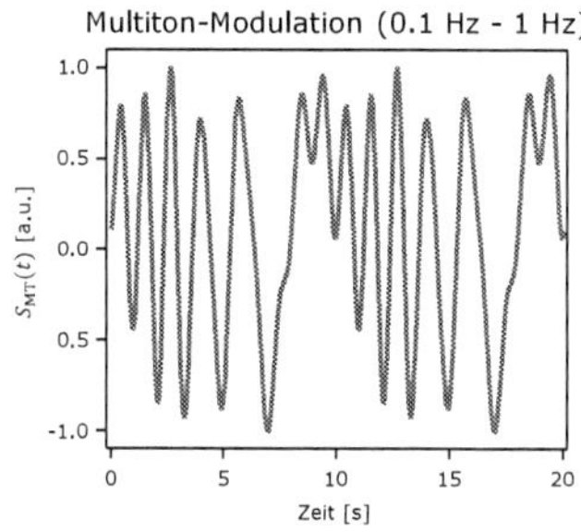

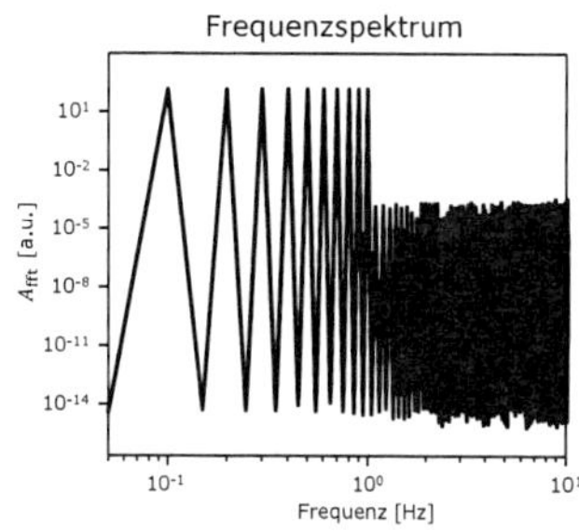

- Vielfache einer Grundfrequenz können zu einer einzelnen Anregung kombiniert werden
- Scheitelfaktor der resultierenden Anregung muss minimiert werden
- Generierende Funktion: $S_{MT}(t) = A_{MT} \sum_{k=1}^{N_f} cos\left(2\pi k f_{base} t - \frac{\pi k^2}{N_f}\right)$ ← Scheitelfaktor-minimierung

Wegweisende Arbeit:
S. Laureti, P. Bison, G. Ferrarini, R. Zito and M. Ricci: "Simultaneous Multi-frequency lock-in Thermography: A new flexible and effective Active Thermography scheme". NDT&E International. June 2024, 10.1016/j.ndteint.2024.103144.

28.05.2025 DGZfP-Jahrestagung 2025 | Dr.-Ing. Julien Le 14

Multiton Lock-in Thermografie
Ergebnisse

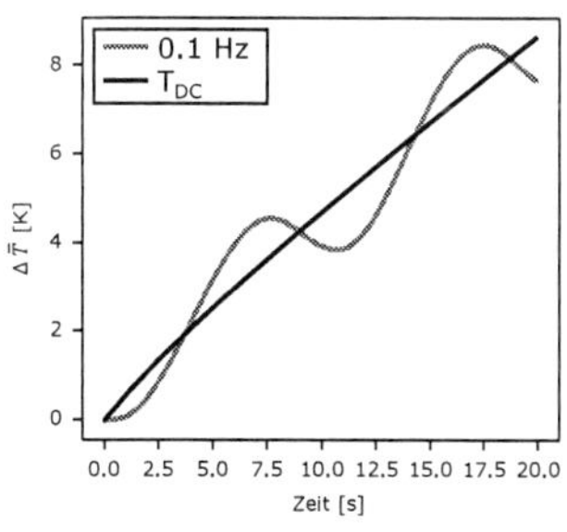

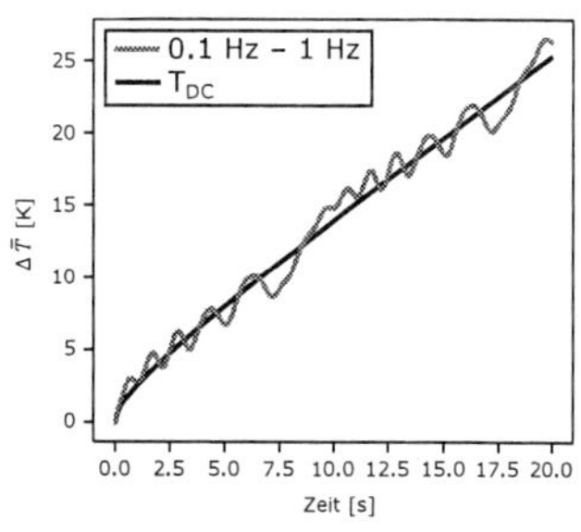

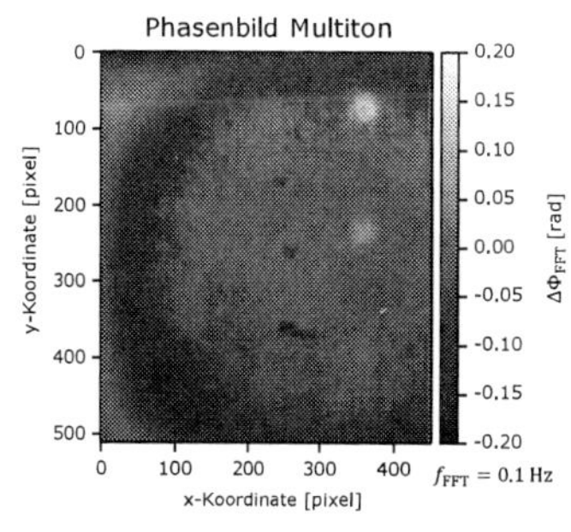

- Anregung mit 10 unterschiedlichen Frequenzen (keine zusätzliche Messzeit notwendig)
- Höheres SNR als klassische Einzeltonanregung!
- Qualitativ keine Degradation der Defektdetektion sichtbar (Ringförmige Struktur im Bild?)

28.05.2025 DGZfP-Jahrestagung 2025 | Dr.-Ing. Julien Lecompagnon 15

Zusammenfassung
Was haben wir gelernt?

Pulskompression

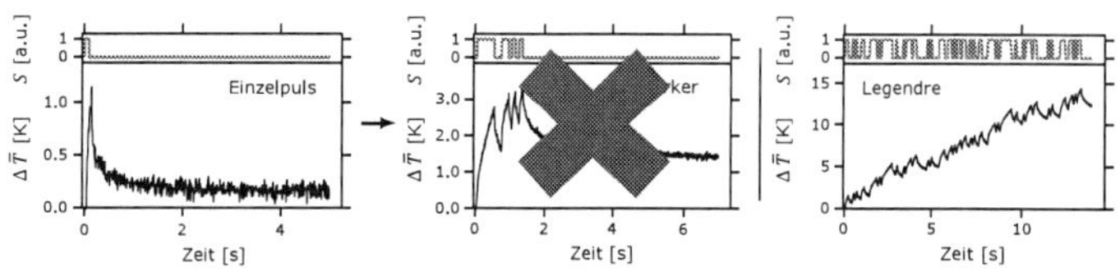

- Die Kurzpulsanregung wird durch Pulskompression gut ergänzt.
- Legendre-Sequenzen bieten einen perfekten Kompromiss zwischen Leistung und Messzeit

Lock-In Varianten

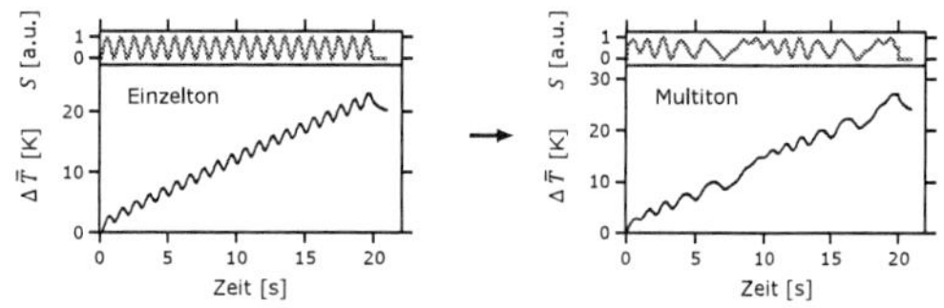

- Barker-Codes bieten nicht die notwendige Flexibilität und sind schwer zu handhaben
- Lock-In Thermografie ist ein mächtiges Werkzeug, aber eine Erweiterung auf Mehrtonanregung lohnt sich.
- Die Ergebnisse sind auf alle ausreichend schnellen optischen Anregungsvarianten (Laser, LED, etc.) übertragbar.

Ihr Interesse geweckt?
Nehmen Sie Kontakt mit uns auf!

Julien Lecompagnon
julien.lecompagnon@bam.de

Unsere Website

In Kontakt bleiben

www.bam.de

Bessere räumliche Auflösung für Thermographie, Ultraschall und andere Methoden der zerstörungsfreie Bildgebung durch Virtuelle Wellen

Peter BURGHOLZER[1], Günther MAYR[2], Lukas GAHLEITNER[2], Markus HALTMEIER[3]
[1] Research Center for Non Destructive Testing, Linz, Österreich, peter.burgholzer@recendt.at
[2] University of Applied Sciences Upper Austria, Wels, Österreich, guenther.mayr@fh-wels.at
[3] Universität Innsbruck, Innsbruck, Österreich, markus.haltmeier@uibk.ac.at

Kurzfassung. In der zerstörungsfreien und biomedizinischen Bildgebung werden räumliche Muster im Inneren einer Probe abgebildet, ohne die Probe z. B. durch Schneiden zu zerstören. Dazu werden sich ausbreitende Wellen, z.B. elektromagnetische oder Ultraschall-Signale, oder sogar diffuse Wärme durch diese inneren Muster erzeugt oder modifiziert und übertragen diese Strukturinformationen an die Probenoberfläche. Dort können die Signale detektiert und aus den gemessenen Signalen ein Bild der inneren Struktur rekonstruiert werden. Die Menge an Informationen über das Innere der Probe, die aus den detektierten Signalen an der Probenoberfläche gewonnen werden kann, wird maßgeblich durch die Ausbreitung von der internen Struktur zur Oberfläche beeinflusst. In der realen Welt ist jede Signalausbreitung mehr oder weniger irreversibel. Die bei der Ausbreitung erzeugte Entropie entspricht dem Informationsverlust.

In einem idealisierten Modell werden solche sich ausbreitenden Wellen, die als virtuelle Wellen bezeichnet werden, durch die Wellengleichung beschrieben. Sie bleiben gültige Lösungen der Wellengleichung, wenn die Richtung der Zeit umgekehrt wird, weisen also Reversibilität auf und zeigen keine Entropieproduktion und daher keinen Informationsverlust während der Ausbreitung.

Eine Fredholmsche Integralgleichung der ersten Art setzt die Diffusionsfelder lokal für jeden Punkt im Raum in Beziehung zu den virtuellen Wellen als deren sich ausbreitende Gegenstück. Diese virtuellen Wellen können aus dem gemessenen Diffusionsfeld an jedem Detektionspunkt berechnet werden, indem diese Integralgleichung invertiert wird. In der Vergangenheit wurde dieses Fredholm-Integral mit Hilfe sogenannter „thermischer Wellen" für die Thermografie abgeleitet. Dies war oft verwirrend, da die thermische Diffusion nicht durch die Wellengleichung beschrieben werden kann. Hier haben wir diese Beziehung im realen Raum hergeleitet, ohne den Fourier-Frequenzraum verwenden zu müssen.

Wir haben die lokal berechneten virtuellen Wellen aus den gemessenen diffusiven Oberflächensignalen zur Bildrekonstruktion unter Verwendung etablierter Time-of-Flight-Methoden aus der Ultraschall- oder RADAR-Bildgebung verwendet. Dies verbessert die räumliche Auflösung in der Thermografie und kompensiert die Dispersion von Quantenwellenpaketen in der Atomsonden-Tomographie.

Keywords: Aktive Thermographie, Virtuelles Wellenkonzept, Photothermische Rekonstruktion, Atomsonden-Tomographie

https://doi.org/10.58286/31276

1. Einführung

In der Bildgebung breiten sich Wellen von der abzubildenden Struktur zu einem Detektor aus, wo ihr Signal zur Rekonstruktion des Bildes verwendet wird. Wie bereits Ernst Abbe 1873 zeigte, ist die Ortsauflösung in optischen Mikroskopen beugungsbegrenzt, etwa um die Wellenlänge[1]. In der Akustik, z.B. in der photoakustischen Bildgebung, wird die detektierbare Wellenlänge nicht nur durch technische Einschränkungen wie die Bandbreite des akustischen Detektors begrenzt, sondern auch durch die frequenzabhängige akustische Dämpfung während der Ausbreitung von der internen Struktur zum Detektor auf der Probenoberfläche[2]. In Flüssigkeiten nimmt der akustische Dämpfungskoeffizient typischerweise mit dem Quadrat der Frequenz zu[3]. Dies führt zu einer Verbreiterung der Wellenpakete durch Dissipation, in diesem Fall sogar ohne Dispersion, da bei der quadratischen Abhängigkeit der Dämpfung von der Frequenz die Schallgeschwindigkeit für alle Frequenzen konstant bleibt.

In der photoakustischen Bildgebung wird in der Regel ein Kurzpulslaser verwendet, um eine halbtransparente Probe, wie z. B. biomedizinisches Gewebe, zu beleuchten[4]. Das Licht, das lokal unterschiedlich stark absorbiert wird, bewirkt einen plötzlichen Temperaturanstieg und erzeugt durch unmittelbare Ausdehnung ein Ultraschall-Drucksignal. Die photoakustische Rekonstruktion besteht in der Berechnung der räumlichen Anfangsdruckverteilung aus den Ultraschallsignalen, die an mehreren Detektionspunkten außerhalb der abgebildeten Probe gemessen werden. Dieser Anfangsdruck ist lokal proportional zur absorbierten optischen Energie. Mit idealen akustischen Detektoren und ohne akustische Dämpfung konnte das Problem der inversen Rekonstruktion exakt gelöst werden, z.B. durch Zeitumkehr[5]. Ein solches zeitinvariantes Signal wird als virtuelles Wellensignal bezeichnet, da in der Realität eine akustische Dämpfung immer zu einem höheren Verlust der höherfrequenten Komponenten bei der Wellenausbreitung führt. Seit fast 20 Jahren gibt es mehrere Versuche, diese Abschwächung zu kompensieren, z.B. von La Rivière et al.[6,7], von Ammari[8], von Dean-Ben et al.[2], oder von Burgholzer et al.[9]. In einer zeitdiskretisierten Version, bei der der Druck als Funktion der Zeit ein Vektor ist, wird dies erreicht, indem eine lineare Beziehung zwischen dem gemessenen Drucksignalvektor $\mathbf{p}_{meas}$ und dem virtuellen Wellensignal $\mathbf{p}_{virt}$ an derselben Detektorstelle eingeführt wird ("lokale Transformation"), was eine Lösung der Wellengleichung wäre:

$$\mathbf{p}_{meas} = \mathbf{M}\mathbf{p}_{virt}. \tag{1}$$

Die Matrix **M** kann nur in wenigen Spezialfällen analytisch berechnet werden, z.B. für eine quadratische frequenzabhängige akustische Dämpfung[9,10]. Die Bestimmung von $\mathbf{p}_{virt}$ aus dem gemessenen Signal $\mathbf{p}_{meas}$ in Gl. (1) stellt ein schlecht gestelltes oder schlecht konditioniertes inverses Problem dar. Die direkte Umkehrung dieser Gleichung würde zu einer starken Rauschverstärkung führen, die zu unbrauchbaren Lösungen führt. Daher werden für die Inversion Regularisierungsmethoden verwendet. Dies ist eine Folge des Informationsverlustes durch akustische Dämpfung höherfrequenter Komponenten bei der Druckausbreitung auf die Oberfläche.

Wir konnten das in Gl. (1) beschriebene Konzept der virtuellen Welle von abgeschwächten Schallwellen mit normalerweise eher geringer Dämpfung während der Ausbreitung einer Wellenlänge auf die thermische Diffusion[11–13] übertragen. Die sogenannten "thermischen Wellen", also die Frequenzanteile im Fourier-Frequenzraum, werden deutlich um den Faktor $1/e \approx 0.37$ innerhalb einer Wellenlänge abgeschwächt. Das Konzept der teilweisen Kompensation der Wellenpaketverbreiterung scheint für die zerstörungsfreie und biomedizinische Bildgebung im allgemeinen nützlich zu sein. Zum Beispiel wurden Anwendungen der Umwandlung der diffusiven elektromagnetischen Welle in ein Wellenfeld von Lee et al. gezeigt[14,15] und für geophysikalische inverse Probleme von Gershenson[16], der erstmals feststellte, dass das Konzept der virtuellen Welle auch auf die Interpretation des zeitabhängigen Wärmeflusses anwendbar ist[17]. Er erwähnte, dass Gl. (1) für die Lokalisierung von entfernten zeitabhängigen Wärmequellen, wie z.B. aktiven Vulkanen in der Geophysik, verwendet werden könnte, aber er verwendete es nicht für die thermographische Bildgebung durch Ultraschallrekonstruktion aus den virtuellen Wellensignalen. Die Wärmediffusion ermöglicht die Abbildung von Strukturen im Inneren der Probe mittels Thermografie in der photothermischen Bildgebung[13]. Die Wellenlänge der thermischen Welle bestimmt die räumliche Auflösung. Höherfrequente Komponenten mit kleineren Wellenlängen ermöglichen eine bessere räumliche Auflösung. Die im Vergleich zur Akustik höhere Dämpfung von Wärmewellen führt jedoch zu einer starken Unschärfe und einer geringen räumlichen Auflösung bei der Abbildung tieferer Strukturen[18]. Mandelis hat einen vereinheitlichenden Rahmen für die Behandlung verschiedener diffusionsbezogener periodischer Phänomene unter der globalen mathematischen Bezeichnung Diffusionswellenfelder entwickelt, wie z.B. thermische Wellen, Ladungsträgerdichtewellen, diffuse Photonendichtewellen, aber auch modulierte Wirbelströme, Neutronenwellen oder harmonische Massentransport-Diffusionswellen[19].

Hier stellen wir das Konzept der virtuellen Welle ohne Verwendung der sogenannten "thermischen Wellen" vor, in Anlehnung an die bereits 1986 eingeführte Ableitung von Romanov[20]. Bei der Untersuchung inverser Probleme für parabolische und elliptische Differentialgleichungen bewies er die Möglichkeit, zur Untersuchung eines äquivalenten Problems für Gleichungen hyperbolischen Typs überzugehen. Wir werden dies für die thermografische Tomographie und zur Kompensation der Dispersion von Quantenwellenpaketen verwenden.

Für die Ausbreitung von Quantenteilchen wird die zeitliche Entwicklung der Wellenfunktion durch die Schrödinger-Gleichung beschrieben. Die Breite eines Gaußschen Wellenpakets, das die Ausbreitung eines freien oder homogen beschleunigten Quantenteilchens beschreibt, nimmt proportional zur Ausbreitungszeit zu[21,22]. Der physikalische Grund dafür ist die Heisenbergsche Unschärferelation: da das Wellenpaket das Quantenteilchen lokalisiert, kann der Impuls, der proportional zur Geschwindigkeit des Quantenteilchens ist, keinen bestimmten Wert annehmen, sondern verteilt sich um einen Mittelwert. Die entsprechende virtuelle Welle verwendet die gewöhnliche Wellengleichung anstelle der Schrödinger-Gleichung und weist daher keine Dispersion auf. Ähnlich wie bei der akustischen Dämpfung und der thermischen Diffusion kann die Verbreiterung des Quantenpakets durch Dissipation kompensiert werden, indem für jeden Detektorstandort eine virtuelle Welle berechnet wird.

2. Virtuelles Wellenkonzept

In diesem Abschnitt setzen wir das diffusive Wellenfeld mit virtuellen Wellen als ihren Ausbreitungsdualen in Beziehung. Die resultierende Beziehung ist die gleiche, die wir aus der thermischen Wellentheorie unter Verwendung des Fourier-Frequenzraums[11] erhalten haben. Die Wärmediffusionsgleichung ist ein einfaches Beispiel für eine parabolische partielle Differentialgleichung[23]

$$\frac{\partial}{\partial t}T(\mathbf{r},t) = \alpha \nabla^2 T(\mathbf{r},t), \tag{2}$$

wobei $T(r,t)$ die Temperatur als Funktion von Raum und Zeit ist, ∇^2 die Laplace-Ableitung (zweite Ableitung im Raum) und α die Wärmeleitfähigkeit ist, von der angenommen wird, dass sie in der Probe homogen ist. Diese Gleichung beschreibt die Temperaturauswertung aus der Wärmediffusion für eine Zeit $t > 0$ mit der Anfangstemperatur $T0(\mathrm{r}) \equiv T(r, t = 0)$. Die entsprechende virtuelle Welle $Tvirt(r,t)$ als ihr propagatives Gegenstück löst die

Wellengleichung eines hyperbolischen Typs und hat die gleiche Anfangstemperatur. Da sie zeitumkehrbar ist, ist sie gerade in t und ihre erste Zeitableitung bei $t = 0$ muss Null sein:

$$\frac{\partial^2}{\partial t^2} T_{virt}(\mathbf{r}, t) = c^2 \nabla^2 T_{virt}(\mathbf{r}, t) \qquad \text{with initial conditions}$$
$$T_{virt}(\mathbf{r}, t = 0) = T_0(\mathbf{r}) \tag{3}$$
$$\frac{\partial}{\partial t} T_{virt}(\mathbf{r}, t = 0) = 0,$$

wo die Ausbreitungsgeschwindigkeit c der virtuellen Wellen beliebig gewählt werden kann. Die gemessene Oberflächentemperatur und die virtuelle Welle werden durch eine lokale Transformation verbunden, d.h. für den gleichen Ort r:

$$\begin{aligned} T(\mathbf{r}, t) \quad &= \int_{-\infty}^{\infty} T_{\text{virt}}(\mathbf{r}, t') K(t, t') \mathrm{d}t', \\ &\qquad \text{with} \\ K(t, t') \quad &\equiv \frac{c}{\sqrt{\pi \alpha t}} \exp\left(-\frac{c^2 t'^2}{4 \alpha t}\right) X \text{for} X t > 0. \end{aligned} \tag{4}$$

Dies kann nach Romanov[20] bewiesen werden durch:

$$\begin{aligned} \frac{\partial}{\partial t} K(t, t') &= \left(-\frac{1}{2t} + \frac{c^2 t'^2}{4 \alpha t^2}\right) K(t, t') \\ \frac{\partial}{\partial t'} K(t, t') &= -\frac{c^2 t'}{2 \alpha t} K(t, t') \\ \frac{\partial^2}{\partial t'^2} K(t, t') &= \left(-\frac{c^2}{2 \alpha t} + \left(\frac{c^2 t'}{2 \alpha t}\right)^2\right) K(t, t') \\ &= \frac{c^2}{\alpha} \frac{\partial}{\partial t} K(t, t') \end{aligned} \tag{5}$$

Wenn wir diese letzte Beziehung für $K(t,t')$ verwenden, können wir aus Gl. (4):

$$\frac{\partial}{\partial t} T(\mathbf{r}, t) = \int_{-\infty}^{\infty} T_{\text{virt}}(\mathbf{r}, t') \frac{\alpha}{c^2} \frac{\partial^2}{\partial t'^2} K(t, t') \mathrm{d}t' \tag{6}$$

Durch zweimalige partielle Integration, wobei $K(t,t')$ und $Tvirt(r,t')$ für $t' = \pm\infty$ gegen Null gehen, und aus Gl. (3) erhält man:

$$\begin{aligned} \frac{\partial}{\partial t} T(\mathbf{r}, t) &= \int_{-\infty}^{\infty} \frac{\alpha}{c^2} K(t, t') \frac{\partial^2}{\partial t'^2} T_{\text{virt}}(\mathbf{r}, t') \mathrm{d}t' \\ &= \int_{-\infty}^{\infty} \alpha\, K(t, t') \nabla^2 T_{\text{virt}}(\mathbf{r}, t') \mathrm{d}t' \\ &= \alpha \nabla^2 \int_{-\infty}^{\infty} K(t, t') T_{\text{virt}}(\mathbf{r}, t') \mathrm{d}t' \\ &= \alpha \nabla^2 T(\mathbf{r}, t) \end{aligned} \tag{7}$$

Für die letzte Gleichung wurde Gl. (4) verwendet, so dass die durch die Beziehung in Gl. (4) definierte Temperatur die Wärmediffusionsgleichung (2) erfüllt. Diese Fredholm-Integralgleichung der ersten Art kann invertiert werden, um aus dem gemessenen Diffusivfeld T den Ausbreitungsdual $Tvirt$ zu berechnen.

Akustische, Thermografische und Atomsonden-Bildgebung

RECENDT

Akustische Bildgebung

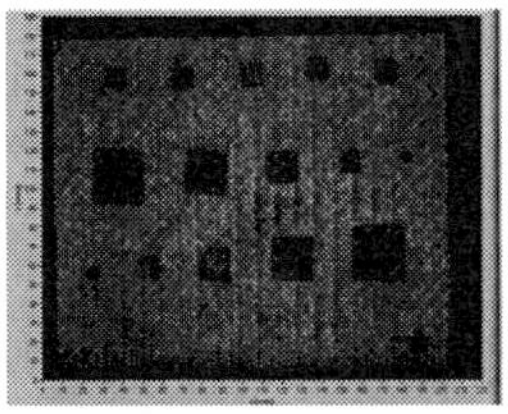

Thermografische Bildgebung

Atomsonden- Tomographie

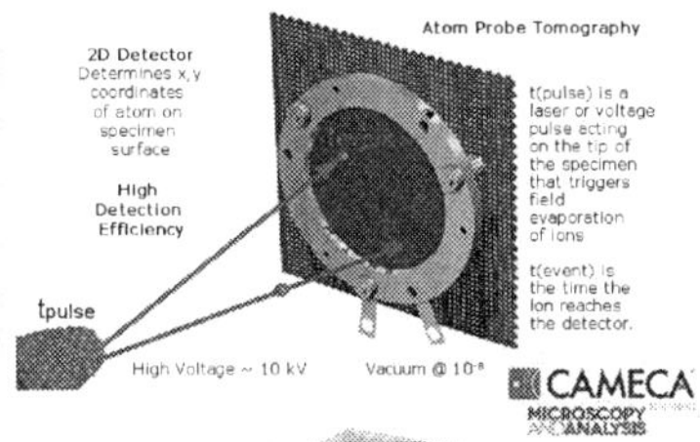

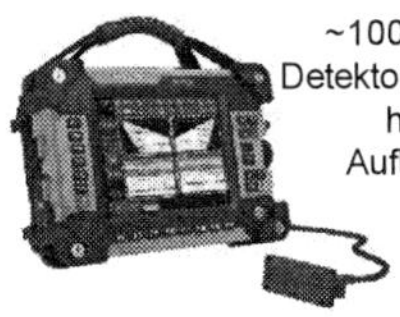

~100 – 1000 Detektorelemente, hohe Auflösung

... mehr als 1 Million Detektor Elemente , berührungslos

APT kann die Element Konzentration mit nm Auflösung in 3D bestimmen

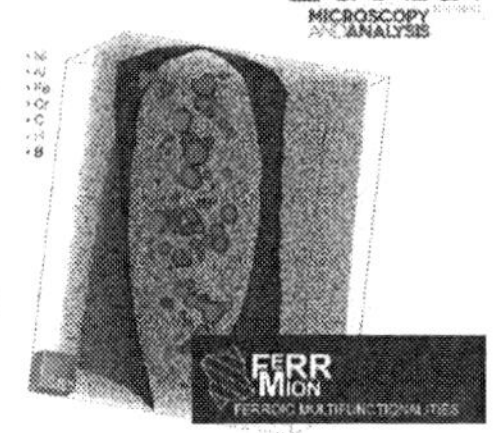

Jupiter durch ein 250-mm -Teleskop in IR, Rot und Blau

RECENDT

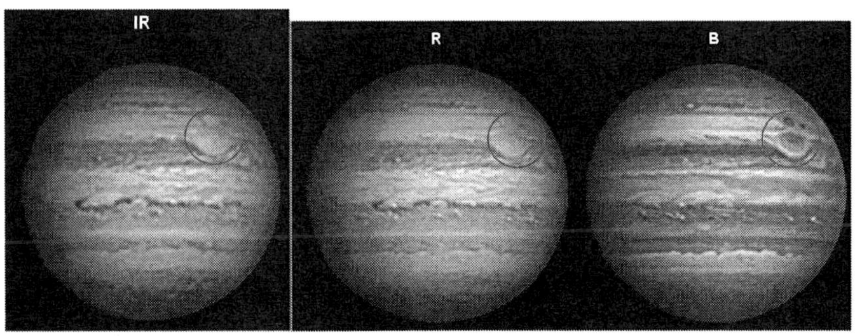

The optical resolution and the wavelength

VWC: lokale Transformation zwischengemessener Temperatur und virtuellem Wellensignal (2D-Simulation)

RECENDT

$T_{virt}(\boldsymbol{r},t) := \frac{p(\boldsymbol{r},t)}{\beta c^2 \rho_0}$ $\quad T(\boldsymbol{r},t) = \int_{-\infty}^{\infty} T_{virt}(\boldsymbol{r},t') K(t,t') dt',$

$$with\ K(t,t') \equiv \frac{c}{\sqrt{\pi \alpha t}} exp\left(-\frac{c^2 t'^2}{4 \alpha t}\right)\ for\ t > 0$$

Diskretisiert, um eine Matrixgleichung zu erzeugen:

$$\boldsymbol{T} = \mathbf{K}\, \boldsymbol{T}_{virt}$$

Truncated Singular Value Decomposition (SVD)-Methode: Rekonstruiertes Signal $\boldsymbol{T}_{rec}$ als Schätzung für $\boldsymbol{T}_{virt}$ vom thermografische Signal $\boldsymbol{T}$:

$\mathbf{K} = \mathbf{USV}^t$ mit unitären Matrizen $\mathbf{U}$ und $\mathbf{V}$

In der Matrix $\mathbf{S}^{+}$ werden für alle singulären Werte in $\mathbf{S}$ ihre Kehrwerte genommen oder Null für alle singulären Werte, die kleiner sind als $1/SNR$:

$\boldsymbol{T}_{rec} = \mathbf{K}^{+}\boldsymbol{T}$ mit der Pseudoinversen $\mathbf{K}^{+} := \mathbf{VS}^{+}\mathbf{U}^{t}$

Oder Minimierung $\|\mathbf{K}\, \boldsymbol{T}_{virt} - \boldsymbol{T}\|_2^2 + \lambda \|\boldsymbol{T}\|_1$ mit DR oder ADMM

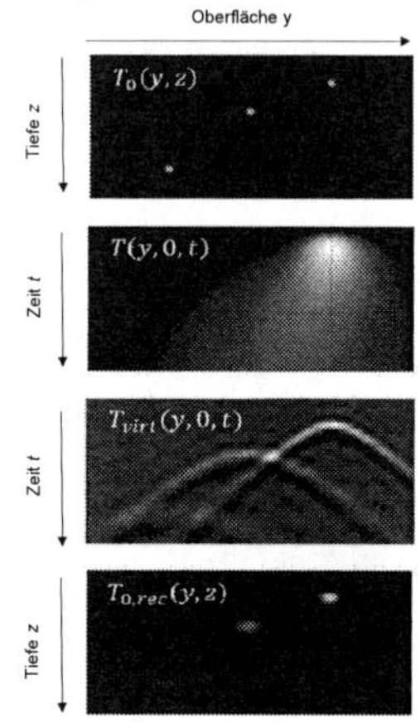

Graphitstäbe eingebettet in Epoxidharz in einer Tiefe von 1,6/ 2,6 und 3,6 mm

RECENDT

G. Thummerer, G. Mayr, M. Haltmeier, and P. Burgholzer, "Photoacoustic reconstruction from photothermal measurements including prior information," Photoacoustics 19, 10017 (2020).

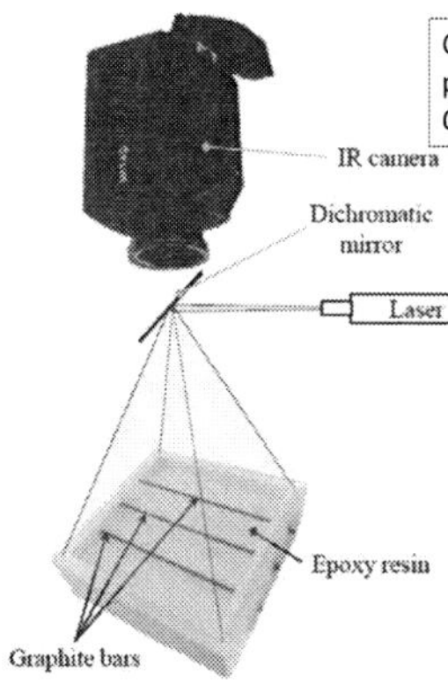

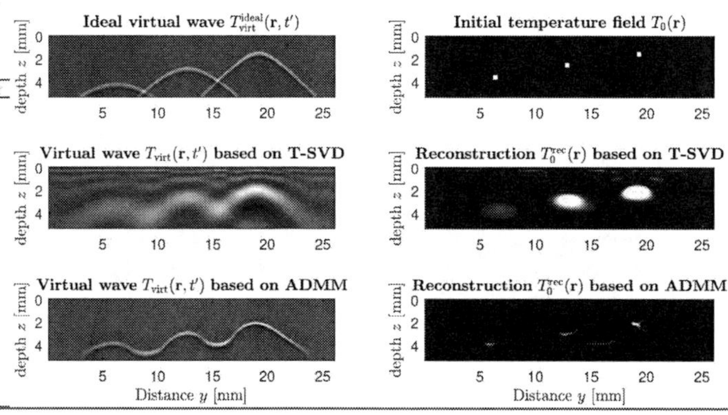

Dispersion von Quanten-Wellenpaketen (1D)

RECENDT

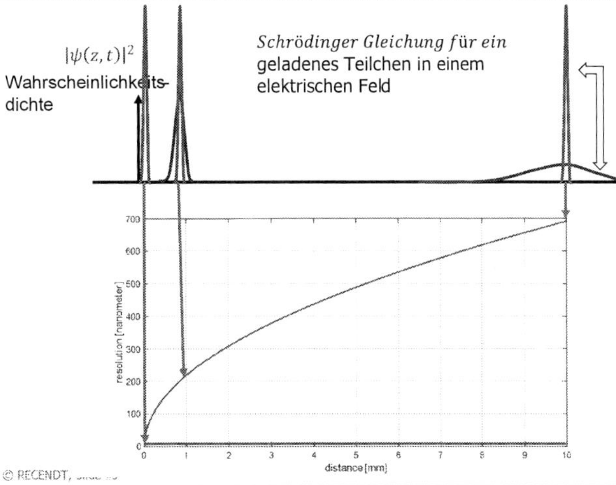

$$|\psi_{\text{virt}}(z,t)|^2 = \frac{1}{a}\sqrt{\frac{2}{\pi}}\exp\left(-\frac{2(z-\eta)^2}{a^2}\right)$$

$$\psi = \mathbf{K}_Q\, \psi_{\text{virt}}$$

$$|\psi(z,t)|^2 = \frac{1}{\varepsilon}\sqrt{\frac{2}{\pi}}\exp\left(-\frac{2(z-\eta)^2}{\varepsilon^2}\right)$$

$$with\ \varepsilon^2 = a^2 + \frac{h^2t^2}{\pi^2m^2a^2}$$

$$and\ \eta = \frac{qEt^2}{2m}$$

$$K_Q(t,t') \equiv \frac{c\sqrt{2m}}{\sqrt{i\hbar\pi t}}exp\left(i\frac{mc^2t'^2}{2\hbar t}\right)$$

Zusammenfassung

RECENDT

- Das VWC verbindet lokal die Lösung der Diffusionsgleichung (Temperatur T) und der Wellengleichung (virtuelle Temperatur T_virt oder Wellenfunktion ψ) für dieselbe Anregung.
- Bei Wärmediffusion werden laterale Wärmeflüsse werden mitbetrachtet (echte 2D oder 3D Rekonstruktion möglich)
- Hauptvorteile von VWC:
  - Besseres SNR: N Pixel, sqrt(N) Erhöhung des SNR
  - Zusätzliche Informationen wie Sparsity oder Positivity können problemlos integriert werden

Referenzen

[1]E. Abbe, Archiv für mikroskopische Anatomie 9, 413 (1873).
[2]X. L. Deán-Ben, D. Razansky, and V. Ntziachristos, Physics in medicine and biology 56, 6129 (2011).
[3]S. K. Patch and A. Greenleaf, in *Photons Plus Ultrasound: Imaging and Sensing 2007: The Eighth Conference on Biomedical Thermoacoustics, Optoacoustics, and Acousto-optics*, SPIE Proceedings, edited by A. A. Oraevsky and L. V. Wang (SPIE, 2007) p. 643726.
[4]L. V. Wang, *Photoacoustic imaging and spectroscopy*, Optical science and engineering, Vol. 144 (CRC Press, Boca Raton, 2009).
[5]P. Burgholzer, G. J. Matt, M. Haltmeier, and G. Paltauf, Physical review. E, Statistical, nonlinear, and soft matter physics 75, 046706 (2007).
[6]P. J. La Riviere, J. Zhang, and M. A. Anastasio, in *IEEE Nuclear Science Symposium Conference Record, 2005* (IEEE, October 23 - 29, 2005) pp. 1841–1845.
[7]P. J. La Rivière, J. Zhang, and M. A. Anastasio, Optics letters 31, 781 (2006).
[8]H. Ammari, ed., *Mathematical modeling in biomedical imaging II: Optical, ultrasound, and opto-acoustic tomographies / Habib Ammari, editor*, Lecture notes in mathematics, 0075-8434, Vol. 2035. Mathematical biosciences subseries (Springer, Heidelberg, 2012).
[9]P. Burgholzer, J. Bauer-Marschallinger, and M. Haltmeier, Photoacoustics 19, 100191 (2020).
[10]P. Burgholzer, J. Bauer-Marschallinger, B. Reitinger, and T. Berer, Journal of Imaging 5, 13 (2019).
[11]P. Burgholzer, M. Thor, J. Gruber, and G. Mayr, Journal of Applied Physics 121, 105102 (2017).
[12]P. Burgholzer, G. Mayr, G. Thummerer, and M. Haltmeier, Journal of Applied Physics 128, 171102 (2020).
[13]P. Burgholzer, G. Mayr, G. Thummerer, and M. Haltmeier, Journal of Applied Physics 131 (2022), 10.1063/5.0093315.
[14]K. H. Lee, G. Liu, and H. F. Morrison, GEOPHYSICS 54, 1180 (1989).
[15]K. H. Lee and G. Xie, GEOPHYSICS 58, 780 (1993).
[16]M. Gershenson, GEOPHYSICS 62, 763 (1997).
[17]M. Gershenson, in *1993 SEG Annual Meeting* (1993).
[18]P. Burgholzer, T. Berer, J. Gruber, and G. Mayr, Applied Physics Letters 111, 031908 (2017).
[19]A. Mandelis, *Diffusion-Wave Fields* (Springer New York, New York, NY, 2001).
[20]V. G. Romanov, *Inverse Problems of Mathematical Physics* (De Gruyter, Berlin, Boston, 1986).
[21]A. Messiah, *Quantum Mechanics*, Vol. 1, p.222 (North-Holland, New York, NY, 1961).
[22]A. Shukla and K. Bhattacharya, American Journal of Physics 78 (2010), 10.1119/1.3290866.
[23]H. S. Carslaw and J. C. Jaeger, *Conduction of heat in solids*, 2nd ed., Oxford science publications (Clarendon, Oxford, 1959 (1986 printing)).

Innovative Infrarot-Kameratechnologie für die Aktiv-Thermografie in der ZfP

AUTOR
Torsten KÖNIG
InfraTec GmbH Infrarotsensorik und Messtechnik, Dresden, Deutschland,
t.koenig@InfraTec.de

Kurzfassung

Zerstörungsfreie Prüfverfahren mittels aktiver und passiver Thermografie haben sich mehr und mehr in der Praxis etabliert. Ein großer Vorteil dieser Technologie ist die schnelle bildgebende und berührungslose Erfassung von Defekten und Fehlstellen. Die Kombination von geeigneten Anregungsquellen und passenden Thermografiesystemen sowie den zugehörigen Auswertealgorithmen (u. a. Lock-In- und Puls-Phasen-Thermografie, Quotienten-Methode) haben die Prüfmöglichkeiten in der ZfP deutlich erweitert.

Die Infrarot-Kameratechnologie hat sich in den vergangenen Jahren sehr stark weiterentwickelt. Kamerasysteme mit gekühlten und ungekühlten Detektoren verfügen über bessere thermische Empfindlichkeiten, ebenso sind die Pixelanzahl sowie die Bildwiederholraten stark angestiegen. Daraus ergeben sich deutlich leistungsfähigere Systeme für die zerstörungsfreien Prüfverfahren.

Die Präsentation gibt einen Einblick in die Technik und die Technologie der Thermografie. Welche Grenzen, welche Möglichkeiten bietet diese Technologie? Wie ist der aktuelle Stand der Technik? Es werden die messtechnischen Besonderheiten der passiven sowie der aktiven Thermografie vorgestellt. Hierbei wird der Fokus auf die technischen Spezifikationen der aktuellen Thermografiesysteme gelegt sowie Applikationsbeispiele der passiven und aktiven Thermografie präsentiert.

Keywords: Thermografie, zerstörungsfreie Prüfung, zerstörungsfreie Werkstoffprüfung, Infrarotkamera, Thermografiekamera, Wärmebildkamera, Messgerät, aktive Thermografie

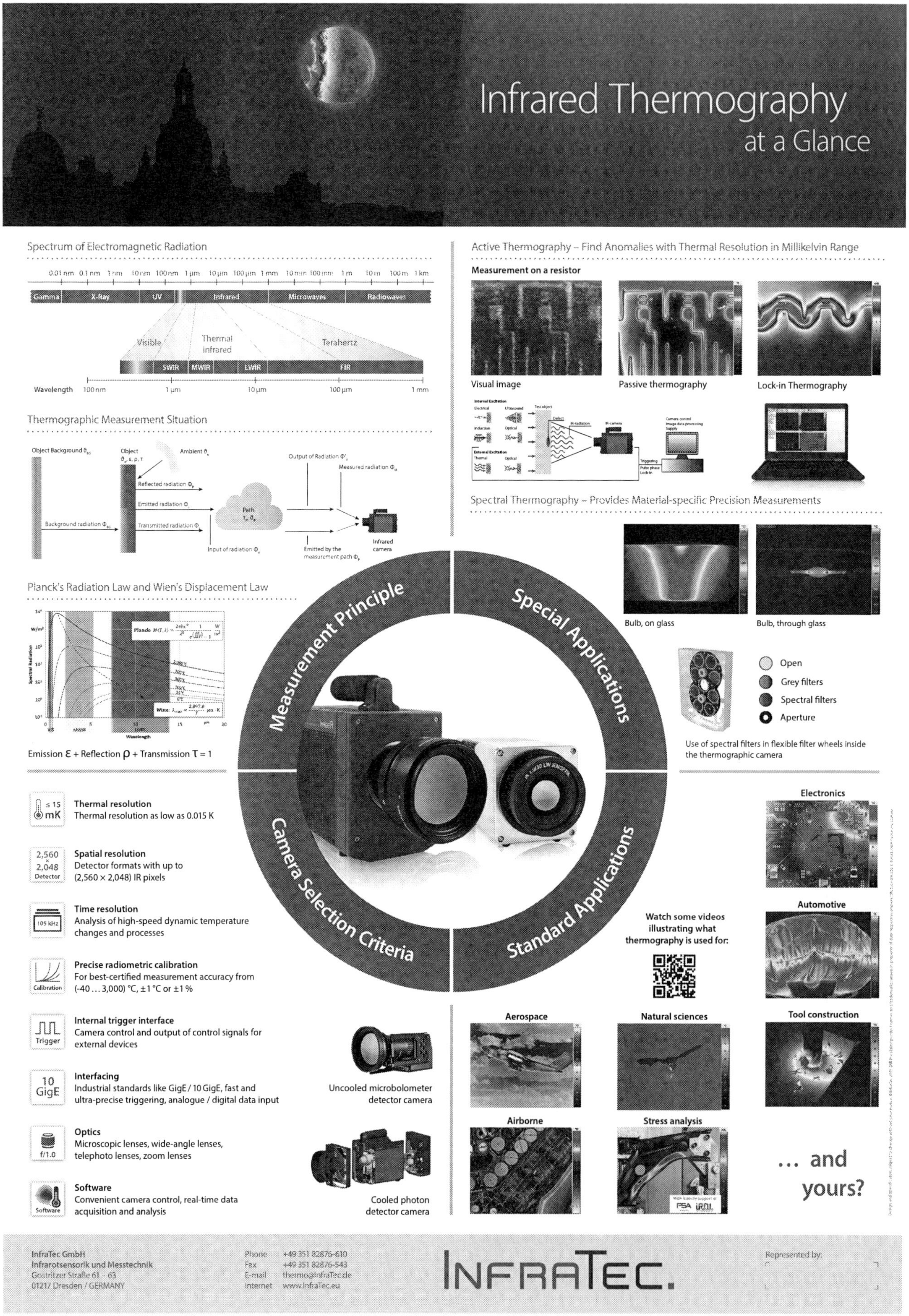

Infrared Thermography
at a Glance
Spectrum of Electromagnetic Radiation
Gamma
X-Ray
UV
Infrared
Microwaves
Radiowaves
Visible
Thermal infrared
Terahertz
SWIR
MWIR
LWIR
FIR
Wavelength
Thermographic Measurement Situation
Planck's Radiation Law and Wien's Displacement Law
Emission ε + Reflection ρ + Transmission τ = 1
Active Thermography – Find Anomalies with Thermal Resolution in Millikelvin Range
Measurement on a resistor
Visual image
Passive thermography
Lock-in Thermography
Spectral Thermography – Provides Material-specific Precision Measurements
Bulb, on glass
Bulb, through glass
Open
Grey filters
Spectral filters
Aperture
Use of spectral filters in flexible filter wheels inside the thermographic camera
Measurement Principle
Special Applications
Camera Selection Criteria
Standard Applications
Thermal resolution
Thermal resolution as low as 0.015 K
Spatial resolution
Detector formats with up to (2,560 × 2,048) IR pixels
Time resolution
Analysis of high-speed dynamic temperature changes and processes
Precise radiometric calibration
For best-certified measurement accuracy from (-40 … 3,000) °C, ±1 °C or ±1 %
Internal trigger interface
Camera control and output of control signals for external devices
Interfacing
Industrial standards like GigE / 10 GigE, fast and ultra-precise triggering, analogue / digital data input
Optics
Microscopic lenses, wide-angle lenses, telephoto lenses, zoom lenses
Software
Convenient camera control, real-time data acquisition and analysis
Uncooled microbolometer detector camera
Cooled photon detector camera
Watch some videos illustrating what thermography is used for:
Electronics
Automotive
Aerospace
Natural sciences
Tool construction
Airborne
Stress analysis
… and yours?
InfraTec GmbH
Infrarotsensorik und Messtechnik
Gostritzer Straße 61 – 63
01217 Dresden / GERMANY
Phone +49 351 82876-610
Fax +49 351 82876-543
E-mail thermo@InfraTec.de
Internet www.InfraTec.eu
INFRATEC.
Represented by:

Anforderungen an die Scan-Genauigkeit für die Ultraschallprüfung mit SAFT

Hubert MOOSHOFER [1], Karsten SCHÖRNER [1], Christian BORGMANN [2]
[1] Siemens AG, München (Garching), Deutschland
hubert.mooshofer@siemens.com
karsten.schoerner@siemens.com
[2] Siemens Energy Global GmbH & Co. KG, Mülheim a. d. Ruhr, Deutschland
borgmann@siemens-energy.com

Kurzfassung

Durch SAFT (Synthetic Aperture Focusing Technique) lässt sich die Ultraschallprüfung wesentlich verbessern: Defekte können (bei gleicher Prüffrequenz) besser aufgelöst werden, was oftmals eine Trennung von Gruppenanzeigen erlaubt. Außerdem wird die Amplitude von Gefügerauschen reduziert, und damit die Nachweisgrenze verbessert. Da im Rahmen der SAFT-Analyse die HF-Signale vieler Prüfpositionen phasenrichtig miteinander verrechnet werden, spielt die Positionsgenauigkeit bei dem Verfahren eine wichtige Rolle. Aus diesem Grund muss eine ausreichende Positionsgenauigkeit sichergestellt sein, um die Vorteile von SAFT voll zu nutzen.

Im Rahmen dieses Beitrages wird dargestellt, wie Positionsfehler zu Phasenfehlern im HF-Signal und letztendlich zu Amplitudenfehlern im SAFT-Ergebnis führen. Durch Vorgabe des maximal zulässigen Amplitudenfehlers lassen sich Anforderungen an die Positionsgenauigkeit ableiten. Es wird gezeigt, dass es zweckmäßig ist verschiedene Arten von Positionsfehlern zu unterscheiden, da beispielsweise ein globaler Positionsversatz oder ein Skalierungsfehler weniger schwer wiegen als ein statistisch zufällig verteilter Positionsjitter. Der Zusammenhang zwischen Positionsfehler und Amplitudenfehler wird quantitativ angegeben. Außerdem wird dargestellt, dass die Anforderungen an die Positionsgenauigkeit bei Prüfung mit Longitudinalwellen im Allgemeinen weniger anspruchsvoll sind als bei Prüfung mit Transversalwellen.

Im Hinblick auf den Praxiseinsatz wird exemplarisch gezeigt, wie die Anforderungen an die Positionsgenauigkeit bei der Prüfung von zylindrischen Bauteilen festgelegt werden, und durch welche Maßnahmen die Positionsgenauigkeit nachgewiesen und verbessert werden kann. Damit werden die Voraussetzungen für die SAFT-Prüfung großer Bauteile insbesondere im Hinblick auf die quantitative Bewertung der SAFT-Amplitude (QSAFT) geschaffen.

Keywords:
Synthetic Aperture Focusing Technique (SAFT), Scan-Genauigkeit, Phasenfehler Amplitudenfehler, Genauigkeitsanforderungen

Anforderungen an die Scan-Genauigkeit bei der Ultraschall-Prüfung mit SAFT (Synthetic Aperture Focusing Technique)

Hubert Mooshofer[1], Karsten Schörner[1], Christian Borgmann[2]
[1]Siemens AG, München (Garching); [2] Siemens Energy Global GmbH & Co. KG, Mülheim an der Ruhr

SAFT-Prüfung

- Verbessert Signal-Rausch-Verhältnis und Defektlokalisierung
- Defektgrößenbewertung: Quantitative SAFT (QSAFT bzw. DGS-SAFT)
- Anwendung z.B. zur Prüfung von Turbinenscheiben

$$SAFT(\vec{x}_{rec}) = \sum_{i=1}^{N} \mathrm{UT}_i\left(t = \left[\frac{\|\vec{x}_{rec} - \vec{x}_i\|}{c/2}\right]\right)$$

Amplitudenfehler

Phasenfehler

Positionsfehler

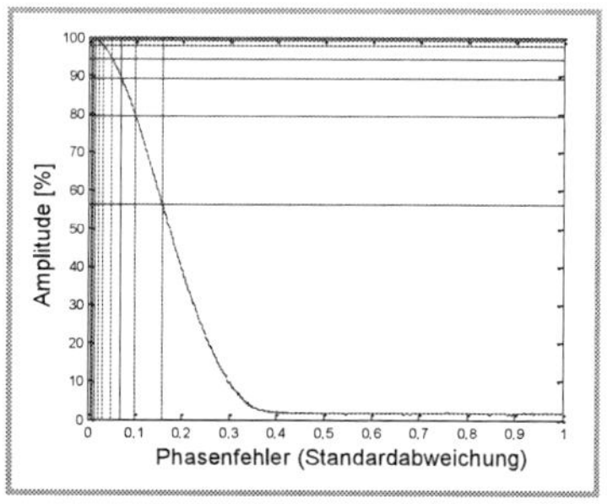

Genauigkeitsanforderungen

- Bei klassischer UT reicht die Sicherstellung der Prüfabdeckung
- Bei SAFT werden Echosignale (HF) phasenrichtig verrechnet, d.h. ein Phasenfehler (=Signal-Verschiebung) ändert das Ergebnis
- Reduktion der konstruktiven Interferenz bei Defektanzeigen bedeutet Unterbewertung (bei Amplitudenbewertung)

→ Notwendigkeit von Genauigkeitsanforderungen

Definition Fehlertypen

- Positionsfehler: Maß für die Abweichungen zwischen den tatsächlichen Prüfpositionen und dem nominellen Prüfraster
- Phasenfehler: Zeitliche Verschiebung der Echosignale (A-Scans) Ein Positionsfehler beim Scan führt zu einem Phasenfehler
- Amplitudenfehler: Abweichung des SAFT-Ergebnis

Positionsfehler ds auf ebener Oberfläche

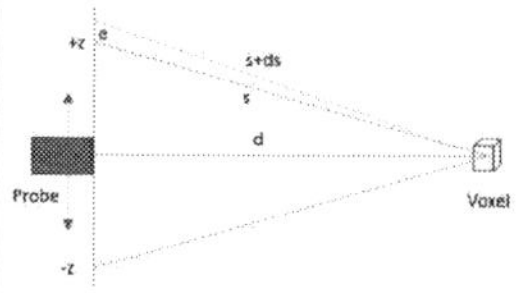

$$ds = \sqrt{d^2 + (z+e)^2} - \sqrt{d^2 + z^2}$$

$$ds \approx e \cdot \frac{z}{s} = e \cdot \sin(\vartheta)$$

Positionsfehler ds auf Zylinder-Oberfläche

(x,y)→(x',y')

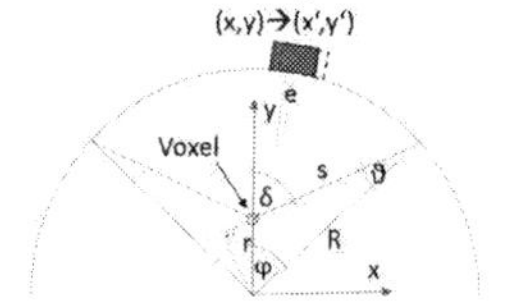

$$ds = \sqrt{x^2 + (y-r)^2} - \sqrt{x'^2 + (y'-r)^2}$$

$$ds \approx e \cdot \frac{r}{s} \cdot \sin(\varphi) = e \cdot \sin(\vartheta)$$

Zusammenhang und Bewertung

Statistische Simulation des Zusammenhangs zwischen Phasen- und Amplitudengenauigkeit (siehe Abbildung rechts)

- Vorgabe des max. Amplitudenfehlers im SAFT-Ergebnis

→ Ableitung der erforderlichen Phasengenauigkeit dp

→ Ableitung der Anforderungen an die Scan-Genauigkeit e

Formel für Zusammenhang zwischen Phasengenauigkeit und Scan-Genauigkeit:

- Umso kleiner die Wellenlänge ist bzw. um so größer der Einschallwinkel desto schärfere Anforderungen ergeben sich für die Scan-Genauigkeit e

Max. Einschallwinkel Randstrahl:
Senkrechteinschallung: ϑ
Schrägeinschallung: $\beta+\vartheta$

→ $ds = e \cdot \sin(\beta + \vartheta)$

$ds = (\lambda/2) \cdot dp$

$e = (\lambda/2) \cdot dp / \sin(\beta + \vartheta)$

β = Einschallwinkel
ϑ = Divergenzwinkel
λ = Wellenlänge
e = Positionsfehler
ds = Schallwegfehler
dp = Phasenfehler

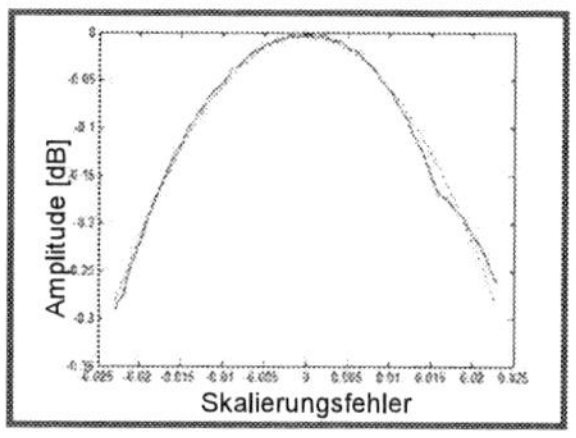

Je nach Art des Positionsfehlers ergeben sich unterschiedliche Auswirkungen:

- Globaler Positionsversatz → Anzeigenverschiebung gleich Versatz
- Skalierungsfehler → Anzeigenverschiebung > Versatz / moderate Amplitudenreduktion
- Statistisch zufällig verteilter Positionsjitter → Amplitudenreduktion

→ Der Positionsjitter hat den größten Einfluss im Hinblick auf eine Amplitudenreduktion

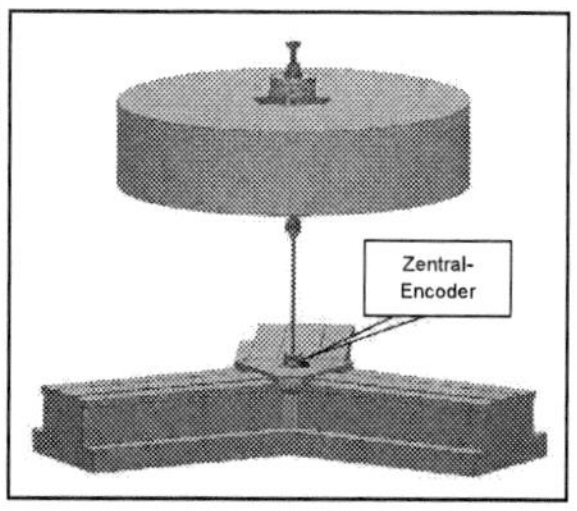

Steigerung der Scan-Genauigkeit

- Hohe Steifigkeit des Manipulatorarms
- Verwendung eines Zentralencoder statt Encoder-Rad
- Präzise Bestimmung der Scheibenparameter

Bestimmung der Scan-Genauigkeit

- Messung der Deformation des Manipulatorarms z.B. mittels optischem Sensor
- Auswertung der Anzeigenwanderung bei Schrägeinschallung

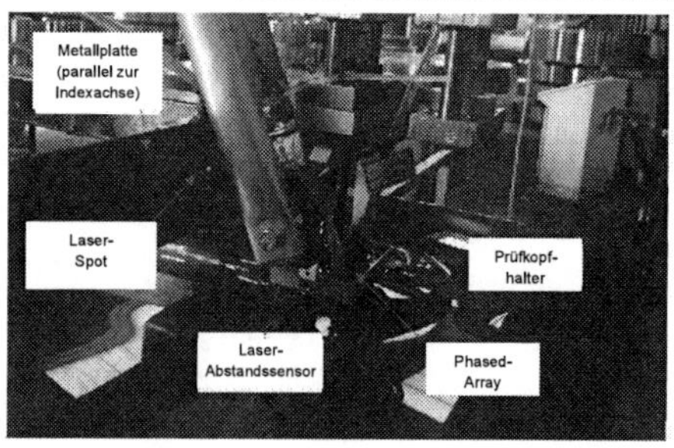

Ergebnisse

- Erforderliche Scan-Genauigkeit für die SAFT-Prüfung von Siemens Energy Turbinenscheiben erreicht
- Die Transversalwellen-Prüfung ist grundsätzlich anspruchsvoller als die Senkrechtprüfung mit Longitudinalwellen
- Positionsgenauigkeit in Index- und Scan-Richtung 0.4 mm bzw. 0.11mm

dgzfp Jahrestagung 2025 Berlin – Poster

SIEMENS energy SIEMENS

Die Bedeutung der Amplitudenstabilität bei der Ultraschallprüfung mit TFM und SAFT

Christian BORGMANN [1], Hubert MOOSHOFER [2], Karsten SCHÖRNER [2]
[1] Siemens Energy Global GmbH & Co. KG, Mülheim a. d. Ruhr, Deutschland
borgmann@siemens-energy.com
[2] Siemens AG, Garching, Deutschland
hubert.mooshofer@siemens.com
karsten.schoerner@siemens.com

Kurzfassung

Die Rekonstruktionsverfahren TFM/FMC und SAFT erfreuen sich zunehmender Beliebtheit. Neben einer besonders hohen Defektauflösung verbessern beide Verfahren auch die Nachweisgrenze der Ultraschallprüfung. Dabei handelt es sich um inverse Verfahren, die eine genaue Rekonstruktion von Defekten, deren Ausdehnung und -Form ermöglichen. Bei diesen Verfahren wird das Prüfvolumen in Pixel bzw. Voxel aufgeteilt und für diese jeweils eine Defektamplitude berechnet. Diese Aufteilung wird üblicherweise in Form eines Rasters vorgenommen, wobei dessen geeignete Wahl entscheidend ist. Ein zu feines Raster führt zu unnötigem Rechenaufwand, der überproportional wächst, und dazu führt, dass die Frame-Rate sinkt (bei TFM) bzw. die Prüfauswertung länger dauert (bei SAFT). Ein zu grobes Raster führt dazu, dass kleine Defekte übersehen oder unterbewertet werden. ISO 23865 definiert die Amplitudenstabilität als Verhältnis der Signalamplitude bei einem diskreten Raster und einem unendlich feinen Raster. Besonders anspruchsvoll ist dabei QSAFT, bei dem die Amplitudenwerte im SAFT-Ergebnis quantitativ ausgewertet werden.

Im Rahmen dieses Beitrages wird untersucht, wie das Rekonstruktionsraster optimal gewählt werden kann. Dabei macht es einen entscheidenden Unterschied, ob die reelle oder komplexe SAFT- bzw. TFM-Rekonstruktion verwendet wird, d.h. ob das HF-Signal oder das daraus resultierende analytische Signal phasenrichtig superponiert werden. Dadurch verschwindet die Lambda/2-Welligkeit von Anzeigen und es verbleibt die Einhüllende. Mit Hilfe des Konzeptes der Amplitudenstabilität wird gezeigt, dass sich die Anforderungen an das Rekonstruktionsraster erheblich reduzieren lassen. Hinausgehend über ISO 23865 - wird gezeigt, dass sich die Amplitudenstabilität rechnerisch bestimmen lässt. Damit ergibt sich ein quantitativer Zusammenhang zwischen Anzeigengröße und Rekonstruktionsraster. Dies ermöglicht eine sichere Prüfaussage bei optimaler Rekonstruktionsgeschwindigkeit.

Keywords:
Synthetic Aperture Focusing Technique (SAFT), Total Focusing Method (TFM), Amplitudenstabilität, Rekonstruktionsraster

SIEMENS energy

Die Bedeutung der Amplitudenstabilität bei der Ultraschallprüfung mit TFM und SAFT

DGZfP Jahrestagung 2025

SIEMENS energy SIEMENS

Amplitudenstabilität ISO 23865

- ISO 23865 definiert die **Amplitudenstabilität** im Hinblick auf das **Rekonstruktionsraster**
- ISO 23865 fordert maximal **2dB Abweichung** falls die Signalamplitude zum *‚Sizen'* genutzt wird
- ISO 23865 nennt eine stabile Amplitude bei einem Rekonstruktionsraster von $\boldsymbol{\lambda/5}$

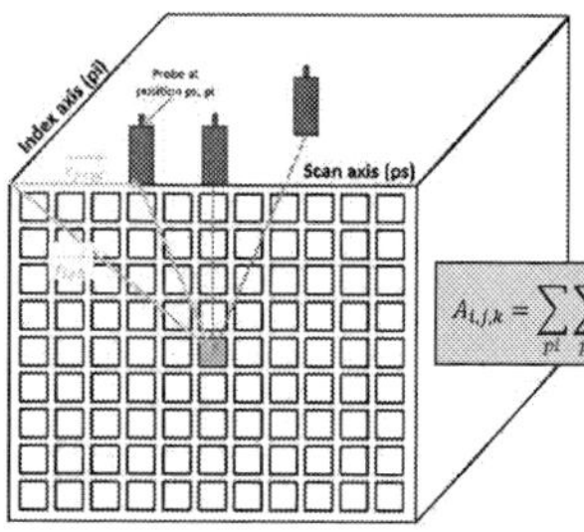

$$A_{i,j,k} = \sum_{pi}\sum_{ps} A_{ps,pi}\left(t = \frac{2\left|\vec{r}_{i,j,k} - \vec{r}_{ps,pi}\right|}{c}\right)$$

INTERNATIONAL STANDARD ISO 23865

First edition 2021-01

Non-destructive testing — Ultrasonic testing — General use of full matrix capture/total focusing technique (FMC/TFM) and related technologies

Essais non destructifs — Contrôle par ultrasons — Utilisation générale de l'acquisition de la matrice intégrale/technique de focalisation en tous points (FMC/TTF) et de techniques associées

C.1.1 General

This annex gives guidance on checking the FMC/TFM setup, ROI and grid. Due to FMC/TFM being an imaging method consisting of a data acquisition and an imaging step, which both involve sampling, special care needs to be taken for the grid point spacing. Coverage and resolution are implicitly tested

$$\beta = 20.\log\left(\frac{\beta_H}{\beta_L}\right)$$

where

β is the amplitude stability in dB;

β_H is the highest amplitude value as a percentage of FSH;

β_L is the lowest amplitude value as a percentage of FSH.

Reference number ISO 23865:2021(E)

Borgmann, Mooshofer, Schörner

Amplitudenstabilität SAFT

SAFT-Signal einer Fehlstelle im unendlich feinen Raster (**Ortskoordinate x kontinuierlich**)

$$\mathrm{A}(x) = \cos\left(x \cdot 2 \cdot \frac{2\pi}{\lambda}\right) \cdot e^{-\left(\frac{2x\sqrt{\ln 2}}{R}\right)^2}$$

$\mathrm{A}(x_{-2dB}) = 0{,}8$ für $R > x$

$$\boldsymbol{x_{-2dB}} \approx \pm\frac{\lambda}{4\pi}\cos^{-1}0{,}8 = \pm 0{,}05 \cdot \lambda = \pm\frac{\lambda}{20}$$

Rasterabstand für -2dB: $G = 2x_{-2dB} \approx \frac{\lambda}{10}$

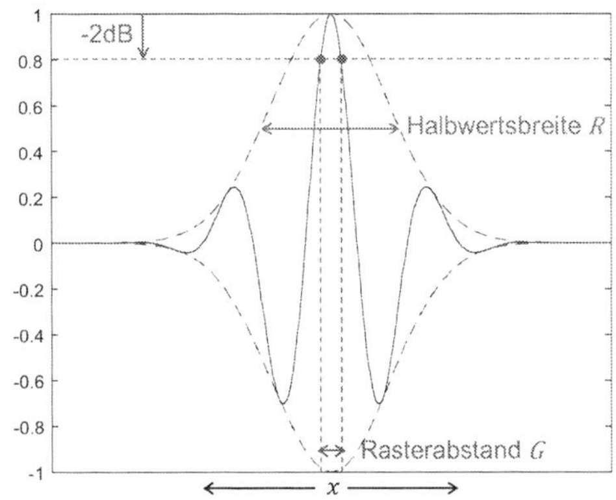

DGZfP Jahrestagung 2025

Borgmann, Mooshofer, Schörner
Unrestricted © Siemens Energy, 2025 3

Komplexe SAFT-Berechnung

SAFT-Implementierung bei **Siemens Energy**

- SAFT aus **Superposition** der analytischen Signale
- Analytische Signal = **Einhüllende mit Phaseninformation**

$$A_{ps,analytical}(t) = A_{ps}(t) + i\mathcal{H}\{A_{ps}(t)\}$$

Vorteile dieser SAFT-Implementierung

- Automatische Gleichrichtung
- Bestimmung der maximalen Aanzeigenamplitude
- Interpretation der Ergebnisse
- **Größerer Rasterabstand**

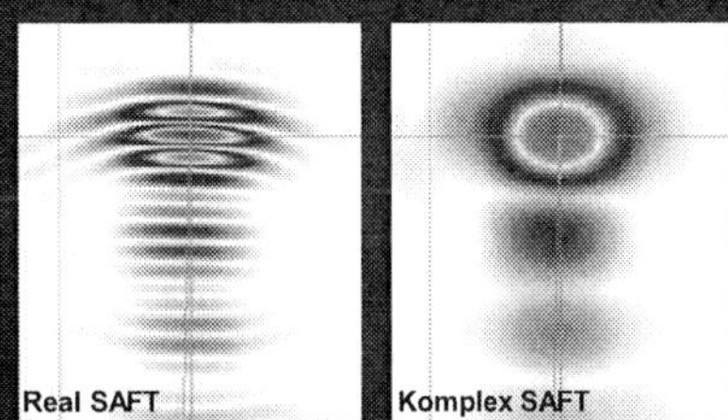

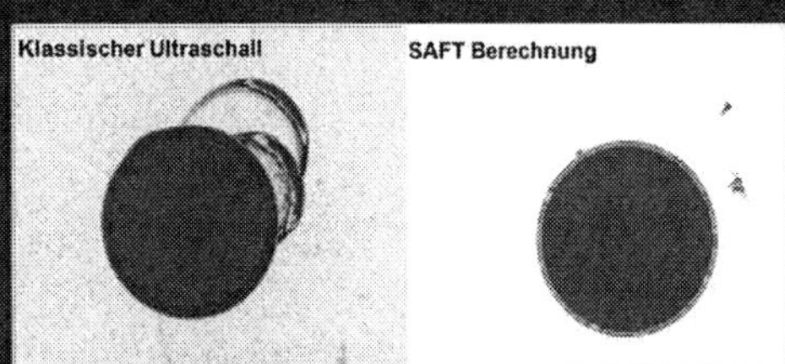

DGZfP Jahrestagung 2025

Borgmann, Mooshofer, Schörner
Unrestricted © Siemens Energy, 2025 4

Amplitudenstabilität SAFT

SAFT-Einhüllende einer Fehlstelle im unendlich feinen Raster (**Ortskoordinate x kontinuierlich**)

$$\mathrm{A}(x) = e^{-\left(\frac{2x\sqrt{\ln 2}}{R}\right)^2}$$

$$\mathrm{A}(x_{-2dB}) = 0{,}8$$

$$\boldsymbol{x_{-2dB}} = \pm 0{,}5 \cdot R \cdot \sqrt{-\frac{\ln 0{,}8}{\ln 2}} = \pm 0{,}28 \cdot R$$

Rasterabstand für -2dB: $G = 2x_{-2dB} \approx 0{,}56 \cdot R$

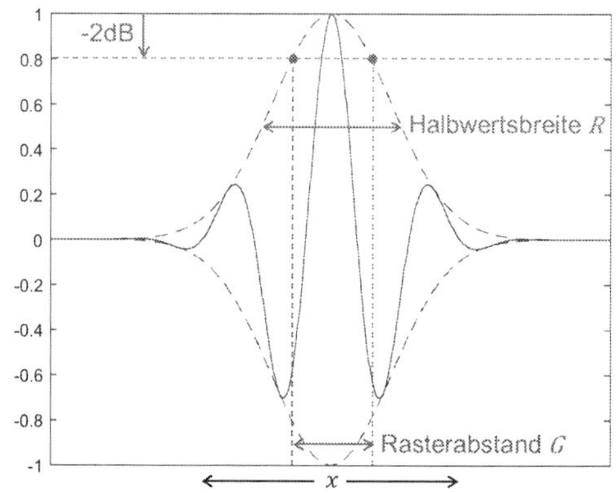

Amplitudenstabilität SAFT

SAFT-Einhüllende einer Fehlstelle im unendlich feinen Raster (**Ortskoordinate x kontinuierlich**)

$$\mathrm{A}(x) = e^{-\left(\frac{2x\sqrt{\ln 2}}{R}\right)^2}$$

Amplitudenstabilität bei gegebenen Raster G

$$\beta_H = \mathrm{A}(x = 0) = 1$$

$$\beta_L = \mathrm{A}(x = G/2) = e^{-\left(\frac{G\sqrt{\ln 2}}{R}\right)^2}$$

$$\boldsymbol{\beta} = 20\mathrm{dB} \cdot \lg e \cdot \ln e \cdot \left(\frac{G}{R}\right)^2 = \mathbf{6dB} \cdot \left(\frac{\boldsymbol{G}}{\boldsymbol{R}}\right)^2$$

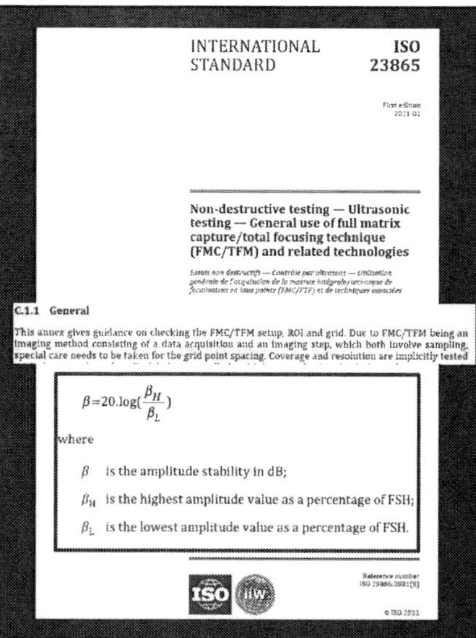
INTERNATIONAL STANDARD

ISO 23865

First edition 2021-01

Non-destructive testing — Ultrasonic testing — General use of full matrix capture/total focusing technique (FMC/TFM) and related technologies

Essais non destructifs — Contrôle par ultrasons — Utilisation générale de l'acquisition de la matrice intégrale/technique de focalisation en tous points (FMC/TFP) et de techniques associées

C.1.1 General

This annex gives guidance on checking the FMC/TFM setup, ROI and grid. Due to FMC/TFM being an imaging method consisting of a data acquisition and an imaging step, which both involve sampling, special care needs to be taken for the grid point spacing. Coverage and resolution are implicitly tested

$$\beta = 20.\log\left(\frac{\beta_H}{\beta_L}\right)$$

where

β is the amplitude stability in dB;

β_H is the highest amplitude value as a percentage of FSH;

β_L is the lowest amplitude value as a percentage of FSH.

Reference number ISO 23865:2021(E)

© ISO 2021

Amplitudenstabilität SAFT

$$\beta = -6\mathrm{dB}\cdot\left(\left(\frac{G_1}{R_1}\right)^2+\left(\frac{G_2}{R_2}\right)^2+\left(\frac{G_3}{R_3}\right)^2\right)$$

G: Grid point spacing (Rekonstruktionsraster)
R: SAFT-Fokusgröße (Halbwertbreite)
1/2/3: Raumrichtung (z.B. x/y/z oder r/φ/z)

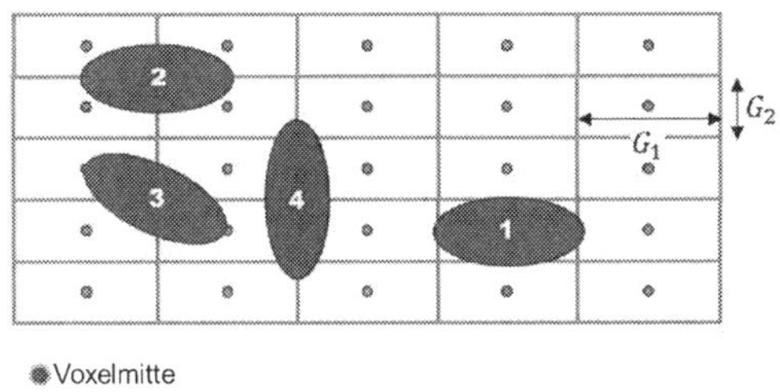

DGZfP Jahrestagung 2025

Borgmann, Mooshofer, Schörner 7
Unrestricted © Siemens Energy, 2025

SAFT an Gasturbinenscheiben

Testscheibe mit Bohrungen

- D_a = 1530 mm
- D_i = 300 mm
- c_L = 5954 m/s
- f = 2 MHz
- λ ≈ 3 mm
- Radiale Einschallung
- Prüfkopfverschiebung in Umfangsrichtung
- SAFT-Fokussierung in Umfangsrichtung
- R = 1.4 mm

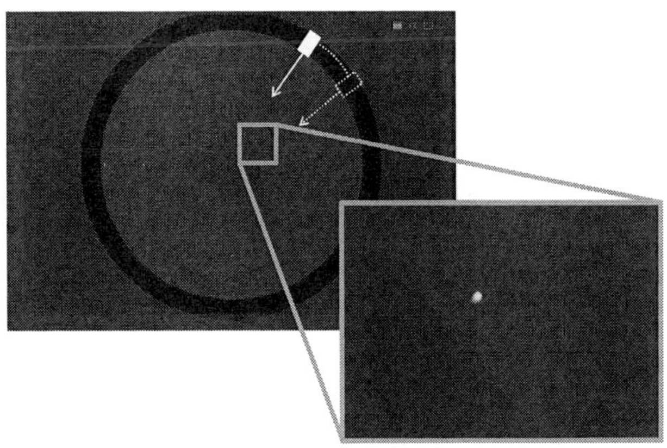

DGZfP Jahrestagung 2025

Borgmann, Mooshofer, Schörner 8
Unrestricted © Siemens Energy, 2025

Amplitudenstabilität -2dB komplexe SAFT-Rekonstruktion

Rasterabstand für -2dB: $G \approx 0,56 \cdot R$

R = 1.4 mm

G = 0.8 mm

Amplitudenstabilität -2dB reelle SAFT-Rekonstruktion

Rasterabstand für -2dB: $G \approx \frac{\lambda}{10}$

λ = 3 mm

G = 0.3 mm

Gröberes **Raster** bei komplexer SAFT-Rekonstruktion: **Faktor 2,7**

Reduzierter **Rechenaufwand** bei komplexer SAFT-Rekonstruktion: **Faktor 0,14**

DGZfP Jahrestagung 2025

Borgmann, Mooshofer, Schörner 9
Unrestricted © Siemens Energy, 2025

Zusammenfassung

- Basierend auf der Definition der Amplitudenstabilität von ISO 23685 wurden theoretische Formeln für die reelle und komplexe SAFT-Rekonstruktionsraster G bei β = -2dB hergeleitet
  - Reelle SAFT-Rekonstruktion $G \approx \frac{\lambda}{10}$
  - Komplexe SAFT-Rekonstruktion $G \approx 0,56 \cdot R$
- Für gegebenes Raster G wurde für die komplexe SAFT-Rekonstruktion ein funktionaler Zusammenhang zwischen SAFT-Auflösung R und Amplitudenstabilität β hergeleitet

$$\beta = -6\text{dB} \cdot \left(\left(\frac{G_1}{R_1}\right)^2 + \left(\frac{G_2}{R_2}\right)^2 + \left(\frac{G_3}{R_3}\right)^2 \right)$$

- Die komplexe SAFT-Rekonstruktion erlaubt ein deutlich gröberes Prüfraster bei gleicher Amplitudenstabilität (geringerer Rechenaufwand)
- Die Vorgabe eines Rekonstruktionsraster in ISO 23685 für β = -2dB ist zu überprüfen $G = \frac{\lambda}{5}$ vs. $G = \frac{\lambda}{10}$

DGZfP Jahrestagung 2025

Borgmann, Mooshofer, Schörner 10
Unrestricted © Siemens Energy, 2025

Kontakt

Dr. Christian Borgmann
Siemens Energy Global GmbH & Co. KG
borgmann@siemens-energy.com

Dr. Hubert Mooshofer
Siemens AG
hubert.mooshofer@siemens.com

Dr. Karsten Schörner
Siemens AG
karsten.schoerner@siemens.com

DGZfP Jahrestagung 2025

Borgmann, Mooshofer, Schörner

12

Simulation für die Schienenprüfung unter Einbezug realer Rauschsignaturen

Tianyun Zhang[1], Thomas Heckel[2]
[1] Bundesanstalt für Materialforschung und -prüfung (BAM), Berlin, Deutschland, tianyun.zhang@bam.de
[2] Bundesanstalt für Materialforschung und -prüfung (BAM), Berlin, Deutschland, thomas.heckel@bam.de

Kurzfassung

Zur Erkennung von Schädigungen an der Eisenbahnschiene und zur Sicherstellung eines störungsfreien Bahnbetriebs werden u. a. Prüftechniken eingesetzt, die eine Kombination aus Ultraschall- und Wirbelstromprüfverfahren verwenden. Zur Optimierung der Auflösung und Qualität der Prüfanzeigen in Schienen werden Fehlererkennungsalgorithmen auf Basis von künstlicher Intelligenz (KI) aufgebaut. Im Rahmen des Forschungsprojekts „Artificial Intelligence for Rail Inspection" (AIFRI) werden entsprechende Datensätze zum Trainieren eines KI-Algorithmus durch Simulation erstellt.

Hierbei werden reale Degradationen und Artefakte in Schienen als skalierbare Modelle in der Simulation nachgebildet. Dabei wird die Verarbeitung der Simulationsdaten analog zum Datenverarbeitungsprozess durchgeführt, wie er bei der Datenaufnahme im Feld erfolgt. Ein Fokus der Simulationen liegt auf den spezifischen Anzeigemustern, die durch verschiedene Schienentypen (z. B. Formanzeigen), Umwelteinflüsse (z. B. Ankopplung) und Störsignale (z. B. Rauschen) entstehen. Das akustische und elektrische Rauschen wirkt sich signifikant auf die Leistung und Präzision der Prüfung aus. Die Einbindung realistischer Rauschsignaturen in die Simulationsdaten wird zu einem wichtigen Bestandteil der Modellentwicklung und stellt eine Herausforderung bei der Erstellung der realitätsnahen Trainingsdaten dar. Für unsere Zielsetzung werden die Felddaten nach verschiedenen Kriterien statistisch ausgewertet und repräsentative Daten für die Simulationsdaten herangezogen.

Das Projekt AIFRI wird im Rahmen der Innovationsinitiative mFUND unter dem Förderkennzeichen 01FV2014C durch das Bundesministerium für Digitales und Verkehr gefördert.

Keywords: Ultraschallprüfung, Eisenbahnschienen

Simulation für die Schienenprüfung unter Einbezug realer Rauschsignaturen

BAM
Bundesanstalt für Materialforschung und -prüfung

Tianyun Zhang, Thomas Heckel
Bundesanstalt für Materialforschung und –prüfung (BAM)

Forschungsprojekt „Artificial Intelligence for Rail Inspection" (AIFRI)
Im Rahmen des Forschungsprojekts wurden Datensätze zum Trainieren eines KI-Algorithmus durch Simulation erstellt. Hierbei wurden reale Fehlertypen und Artefakte in Schienen als skalierbare Modelle in der Simulation nachgebildet. Dabei wurde die Verarbeitung der Simulationsdaten analog zum Datenverarbeitungsprozess durchgeführt, wie er bei der Datenaufnahme in Feld erfolgt. Das akustische und elektrische Rauschen sowie weitere Störsignale stellten in ihrer Nachbildung in der Simulation eine Herausforderung dar. Die Einbindung realistischer Signalsignaturen in die Simulationsdaten war ein wichtiger Bestandteil für die Erstellung realitätsnaher Trainingsdaten. Für diese Zielsetzung wurden die Felddaten nach verschiedenen Kriterien statistisch ausgewertet und mehrere repräsentative Datensätze entwickelt, die entsprechende Signalsignaturen für die Simulation bereitstellten.

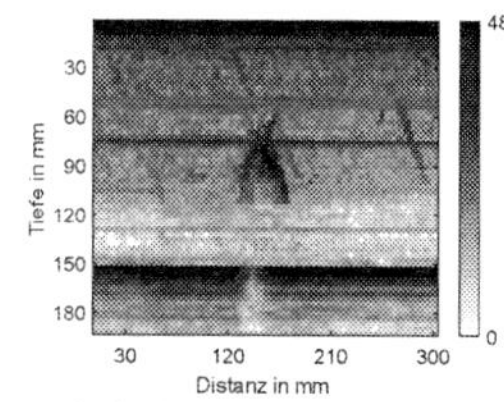

Amplitudendaten einer Bohrung im Feld

Vergleich der Rauschsignaturen

Die Analyse der Felddaten erfolgte mit Datensätzen aus neun unabhängigen Prüffahrten mit einer Länge von insgesamt 189,5 km im Offenen Digitalen Testfeld des DZSF, verteilt auf drei Prüfbereiche und drei Prüfsysteme. Für die hier vorgestellte Auswertung wurden die anzeigenfreien Abschnitte mit dem Schienenprofil 54E4 des jeweiligen Datensatzes zusammengestellt und für die Analyse verwendet. Dies entspricht einer Länge von insgesamt 93,6 km bzw. 49,4 % der zur Verfügung stehenden Felddaten. Um die Ähnlichkeit der Signalsignaturen der erzeugten Signaturdatensätze zu überprüfen, wurden diese qualitativ und quantitativ verglichen. Die Vergleiche basierten auf der Analyse der Amplitudendaten aus überlagerten B-Bildern, aufgenommen mit einer Pixelgröße von 3 mm × 3 mm.

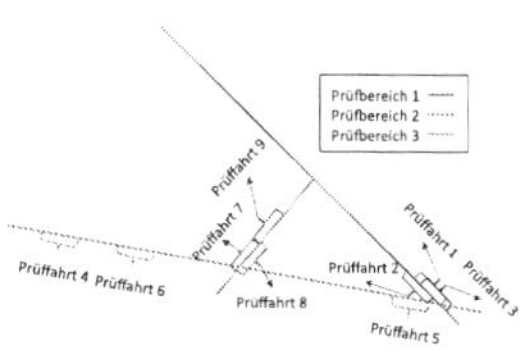

Auswahl der Prüffahrten aus dem Offenen Digitalen Testfeld

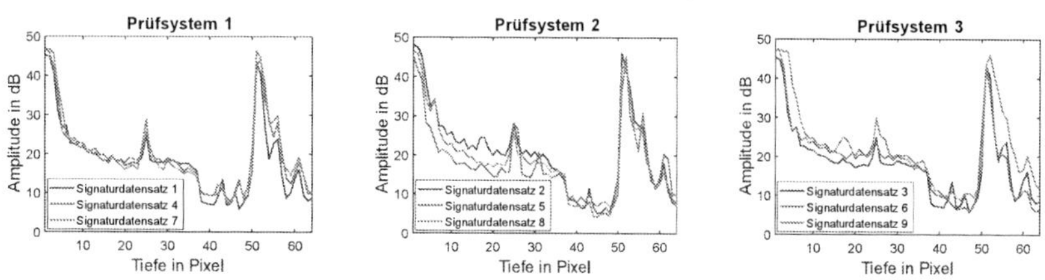

Amplitudenverlauf der neun Prüffahrten unterteilt in die Prüfsysteme 1, 2 und 3

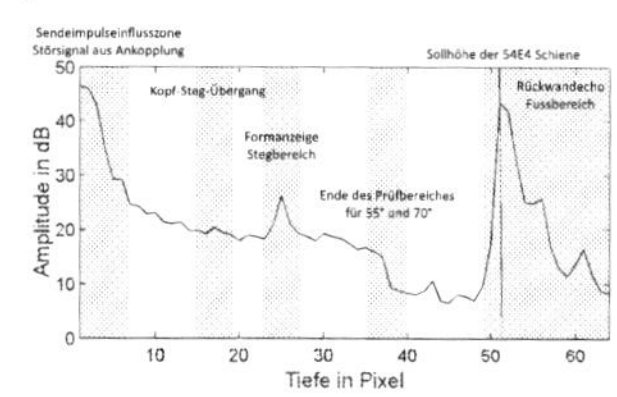

Mittelwerte der Amplituden über die Tiefe

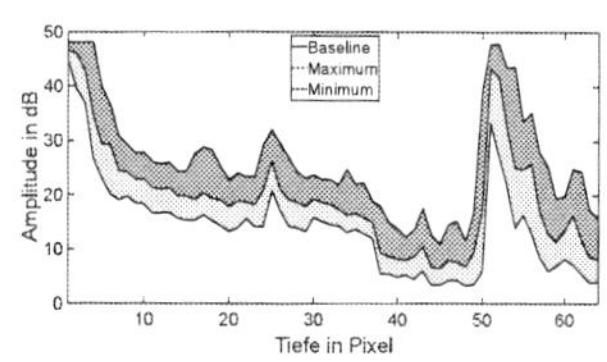

Amplitudenverteilung über die Tiefe

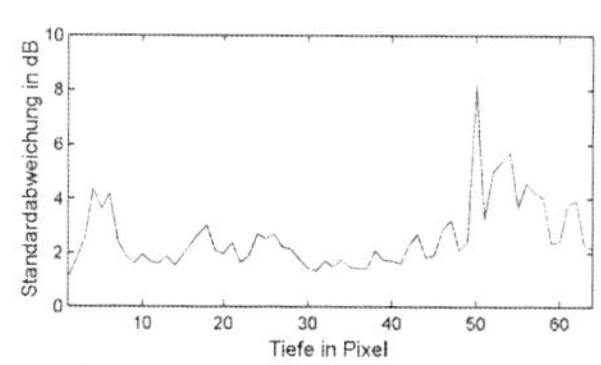

Standardabweichung über die Tiefe

Qualitativer Vergleich

Für den qualitativen Vergleich der Signaturdatensätze wurden für jedes Pixel entlang der Ortskoordinate zeilenweise die Mittelwerte sowie Maxima und Minima bestimmt und über die Tiefenkoordinate aufgetragen. Die Mittelwerte aller neun Signaturdatensätze wurden dabei als Basislinie verwendet. Maximal- und Minimalwerte weisen eine Spanne von 12,3 dB auf. Die Standardabweichung variiert dabei je Zeile zwischen 1,1 dB und 8,2 dB. Im Bereich des Fahrspiegels und im Bereich der Rückwand zeigt sich eine stärker ausgeprägte Variation.

Fazit

Die neun generierten Signaturdatensätze weisen unterschiedliche Empfindlichkeiten auf und unterscheiden sich deutlich voneinander. Einflüsse wie z. B. Gleisgeometrie, Ankoppelbedingungen und Prüfgeschwindigkeit führen zu wechselnden Signaturen im Feld, die nur mit erheblichem Aufwand nachgebildet werden können. Um eine entsprechende Variation bei der Prüfung im Feld darzustellen und für das KI-Training abzubilden, konnten die erzeugten Datensätze für die Simulation des Schienentyps 54E4 angewendet werden.

Quantitativer Vergleich

Um die Ähnlichkeit der Signalsignaturen quantitativ zu bewerten wurde ein zeilenweiser Vergleich der Daten mittels des Mann-Whitney-U-Test durchgeführt. Insgesamt wurden 36 Tests für jeweils 64 Zeilen berechnet. Das Ergebnis zeigt, dass sich die Signalsignaturen insgesamt zu 77 % voneinander unterschieden und nur zu 23 % ähnlich sind. Große Werte für die Ähnlichkeit von 40 % bis 50 % treten z. B. bevorzugt im Bereich des Rückwandechos auf.

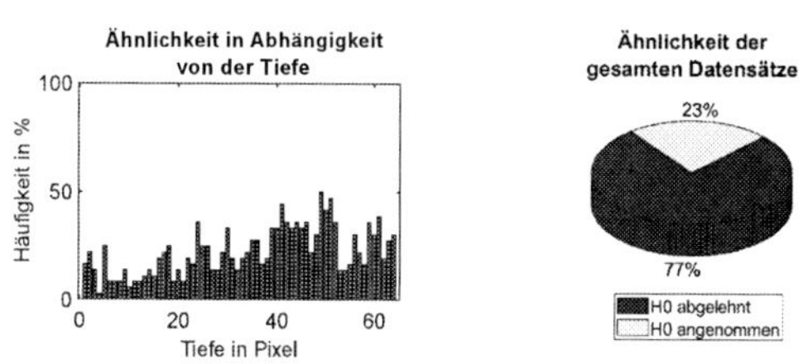

Ähnlichkeitsbetrachtung der Amplitudendaten mittels Mann-Whitney-U-Test

Sicherheit in Technik und Chemie

www.bam.de

THEMENFELD INFRASTRUKTUR

Qualitätssicherung an additiv gefertigten Großbauteilen mit Ultraschall Phased-Array-Verfahren und digitaler Durchstrahlung

Antje ZÖSCH [1], Christopher SEIDEL [2]
imq - Ingenieurbetrieb GmbH, Crimmitschau / Deutschland
[1] a.zoesch@imq-gmbh.com; [2] ch.seidel@imq-gmbh.com

Kurzfassung. Im Vortrag werden Untersuchungen an einem additiv gefertigten Demonstrator-Bauteil mit verschiedenen Inhomogenitäten vorgestellt. Die Werkstoffeigenschaften wurden ermittelt und mit den Eigenschaften eines adäquaten Gussbauteils verglichen. Zur Erkennung und Bewertung von Unregelmäßigkeiten kamen zerstörungsfreie Prüfverfahren zum Einsatz, deren Ergebnisse durch Referenzverfahren verifiziert wurden. Schwerpunkt bildete das Ultraschall Phased Array-Verfahren. Konzeption und Prüfkonfiguration wurden durch vorherige Schallfeldsimulationen erarbeitet. Die Prüfergebnisse zeigen die Möglichkeiten und Grenzen des Verfahrens auf und liefern Ansätze für die weitere Verbesserung des Einsatzes des UT-PA Verfahrens zur Bewertung der Bauteilqualität.
Das Vorhaben wurde bearbeitet innerhalb des Verbundprojektes „WIR! – Wandel durch Innovation in der Region“, welches das vom BMBF gefördert wurde.

Keywords: Ultraschall Phased-Array, Total-Focusing-Method, Additive Fertigung, Walzenbrecher, digitale Durchstrahlung, Verschleißsimulation

1. Einführung

Additive Fertigung wird für immer mehr Unternehmen eine attraktiv Alternative zu etablierten Fertigungsverfahren. Besonders für komplexe Bauteile und maßgeschneiderte Produkte lässt sich der Einsatz wirtschaftlich darstellen. Besonders bei Reparaturarbeiten bietet die Wiederherstellung des Bauteils gegenüber der Neufertigung deutliche Vorteile. Die vorliegende Anwendung richtet das Hauptaugenmerk auf eine ressourcenschonende Verlängerung der Lebensdauer von Großbauteilen mit hohem Materialeinsatz und komplexen Geometrien durch die Reparatur der Verschleißschäden mittels additiver Fertigung.

In der Industrie müssen Komponenten hohe Qualitäts- und Sicherheitsstandards erfüllen. Derzeit erfordert der Mangel an produkt- bzw. verfahrensspezifischen Normen und Standards, dass Prozesse für die Fertigung von Bauteilen für jeden Einzelfall von Grund auf definiert werden müssen. Die Norm DIN EN ISOASTM 529202023-12 (Ersatz für die DIN SPEC 17071:2019-12) stellt die notwendigen Anforderungen an die Prozessgestaltung für hochwertige additiv gefertigte Bauteile ganzheitlich dar [1, 2]. Wichtige Aktivitäten sind dabei eine gezielte Vorbereitung der additiven Fertigung mit Hilfe von digitalen Werkzeugen wie z. B. Simulationen und die Sicherstellung der geforderten Bauteilqualität über die gesamte Prozesskette hinweg.

NDTnet
Medien- und Publikationspartner

https://doi.org/10.58286/31255

Im konkreten Anwendungsfall sollte eine Reparatur von abgetragenen Becherzähnen von Walzenbrechern (Sizern) mittels additiver Fertigung realisiert werden. Im Vorfeld wurden dazu Versuchspläne erarbeitet, um die werkstoffliche und technische Eignung des additiven Werkstoffs zu überprüfen, und ein Vergleich mit dem Bauteil aus konventioneller Fertigung angestrebt. Zur Beurteilung der Bauteilqualität kamen zerstörende und zerstörungsfreie Werkstoffprüfverfahren zum Einsatz.

2. Versuchsplan und Durchführung

2.1. Herstellung des Probekörpers

Die Versuchsquader wurden an der BTU Cottbus-Senftenberg und bei TAKRAF mittels Lichtbogenauftragsschweißens (Wire Powder Arc Additive Manufacturing, WPAAM) gefertigt. Für den 3D-Druck sind Brechzähne ganz (Neudruck) oder teilweise (Reparaturdruck) auf ein Substratblech aufzutragen. Die Brechzähne können in erster Näherung durch Quader vergleichbarer Abmessung angenähert werden, um somit den Druckprozess hinsichtlich Auftragsrate und Materialgüte zu optimieren (Abbildung 1). Als Werkstoffe wurden die Stähle S355J2+N und AM70 (S700) gewählt.

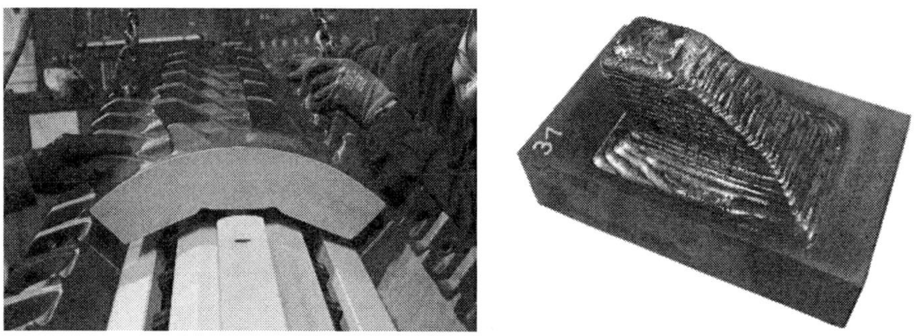

Abb. 1. Gussbauteil Brechersegment (links) / Brecherzahn Neudruck (rechts)

2.2. Untersuchungsprogramm

Die gefertigten Quader wurden im ersten Schritt zerstörungsfrei mit Ultraschall Phased-Array-Techniken geprüft. Für die Untersuchungen wurde ein Omniscan MX (32/128 PR) der Firma Evident Scientific verwendet. Die Einschallung wurden mit hochfrequenten Prüfköpfen wie dem 5L64-A12 und 7,5L64- realisiert.

Im Anschluss wurden Referenzproben anhand von erstellten Probeentnahmeplänen entnommen. Zur Beurteilung der werkstofflichen Eignung wurden Zugversuche, Kerbschlagbiegeversuche, Härtemessungen und Gefügeanalysen durchgeführt (Abb. 2). Die Referenzproben wurden hierbei mittels digitaler Durchstrahlung gegen geprüft und bildgebend dokumentiert. Für die Einstrahlung wurde eine Röntgenröhre vom Typ CP200 genutzt. Die belichteten Speicherfolien wurden mit einem Carestream-Scanner Model HPX-1 digitalisiert.

Im zweiten Schritt wurden Dauerschwingproben in Anlehnung an die DIN EN 6072 konstruiert und gefertigt.

3. Ergebnisse

3.1. Referenzuntersuchungen

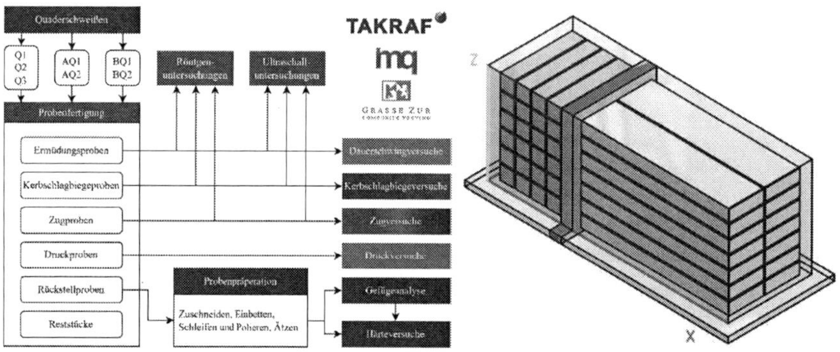

Abb. 2. Übersicht über die durchgeführten Untersuchungen aller Projektpartner (links) und ein Probenentnahmeplan für mech.-technologisch Kennwertermittlung (rechts)

Die geforderten werkstofflichen Parameter wurden auf Basis des Ausgangsmaterials und Verschleiß-Simulationen ermittelt. TAKRAF hat dazu gemeinsam mit ANSYS ROCKY ein Bruchmodell (TAKRAF Breakage Model TBM) entwickelt und damit DEM-Simulationen durchgeführt [5]. Auf der Basis der durchgeführten Verschleiß- und Bruchsimulationen können die aufgelisteten Mindestanforderungen an die Brechzähne für die Zerkleinerung des als „mittelhart" geltenden Bauxit bestätigt und wie folgt konkretisiert werden:

- Rockwell-Härte $\geq$ 55HRC
- Streckgrenze $\geq$ 600N/mm^2
- Kerbschlagarbeit $\geq$ 27J bei -40°C
- Scherkraft $\geq$ 600kN

Die durchgeführten Referenzuntersuchungen konnten umfassend zeigen, dass die werkstofflichen Parameter wie Zugfestigkeit, Kerbschlagarbeit und Härte die Anforderungen erfüllen konnten [4] [5].

Additiv gefertigte Materialen können jedoch nicht als homogen angesehen werden. Es entstehen Schwankungen in den mechanisch-technologischen Eigenschaften. Diese sind in der Regel abhängig von der Aufbauhöhe. Metallographische Untersuchungen haben gezeigt, dass die Porosität mit zunehmender Aufbauhöhe leicht ansteigt (von 0,6% auf 1,2%). Gleichzeigt steigt ebenfalls die Härte leicht an. Aufgrund der gefundenen Zusammenhänge wurde die digitale Durchstrahlung von Referenzproben angestrebt. Dieses Verfahren bietet eine schnelle, digitale Bildgebung und liefert verlässliche Aussagen über Volumenfehler innerhalb der Referenzproben (Abb. 3).

Die Bilder der digitalen Durchstrahlung wurden mit einer Scan-Auflösung von 25 µm digitalisiert und zeigen auch kleinste Poren und Materialtrennungen. Die Interpretation der Ergebnisse der mechanisch-technologischen Prüfung wird deutlich verbessert und der Einfluss von Porositäten auf die Materialeigenschaften herausgearbeitet. Eine erhöhte Porenanzahl führte zu einer deutlichen Verringerung der Kerbschlagarbeit, diese Verringerung kann bis zu 50% des Ausgangswertes betragen [4]. Proben mit größeren Inhomogenitäten wurden nicht in die Auswertung der Ergebnisse einbezogen.

Verfahrensbedingt ist die maximale Wandstärke bei der Durchstrahlung mit der hohen Auflösung stark begrenzt, so dass für die Überprüfung der additiv gefertigten Brecherzähne ein weiteres Volumenverfahren zur Anwendung kommen musste.

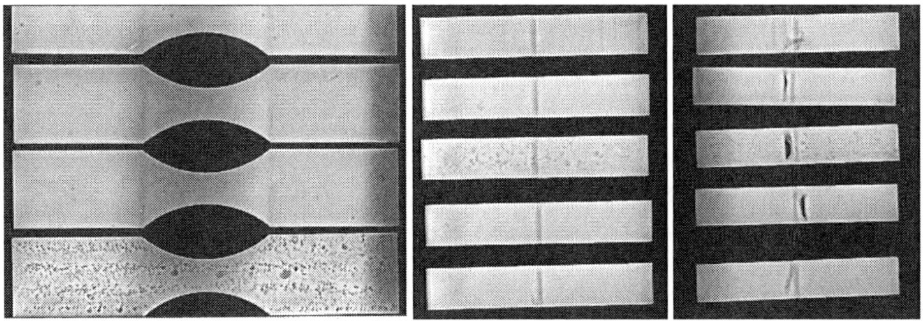

Abb. 3. Zugproben und Kerbschlagproben mit ausfälligen Poren und Fehlstellen im Volumen

3.2. Ultraschall - Vorbetrachtungen

Das Ultraschall Phased-Array-Verfahren bietet eine Reihe von Vorteilen gegenüber konventionellen Methoden. Dazu gehören die bildgebende Darstellung und der vollständig digitale Prüfablauf. Im Gegensatz zu konventionellen Methoden können Phased-Array-Systeme nicht nur A-Bilder, sondern auch hochauflösende Querschnittsbilder des Materials erzeugen (C-Scan, B-Bild). Diese erleichtern die Identifikation und Analyse von Fehlern innerhalb des Prüfvolumens. Die Fähigkeit, den Einschallwinkel und die Fokustiefe elektronisch zu ändern, ermöglicht eine bessere Anpassung an die Prüfanforderungen. Die Reduzierung der Prüfmechanik erleichtert die Prüfbarkeit bei Bauteilen mit komplexer Außenkontur. Es werden weniger Prüfköpfe benötigt, was Prüfablauf und Datenerfassung erleichtert und beschleunigt.

Für die vorliegende Prüfaufgabe wurden Prüfköpfe mit großer, aktiver Apertur genutzt (5L64 und 7,5L64). Große Aperturen geben aufgrund des großen resultierenden Nahfelds mehr Spielraum bei der Ausrichtung und Fokussierung der Schallfelder. Im ersten Schritt wurden zwei Parametersätze für die Prüfung auf Anbindungsfehler und für die Prüfung von Volumenfehlern realisiert (Abb. 4). Anbindungsfehler sind für additive Strukturen typisch, welche auf ein Substrat gedruckt werden. Im vorliegenden Fall bestand das Substrat aus S355J2+N und hatte eine Wandstärke von 40 mm. Es ist entscheidend, dass die Anbindung der gedruckten Brecherzahns möglichst fehlerfrei ist, da es während des Einsatzes zu erheblichen Scher- und Druckkräften kommt [5].

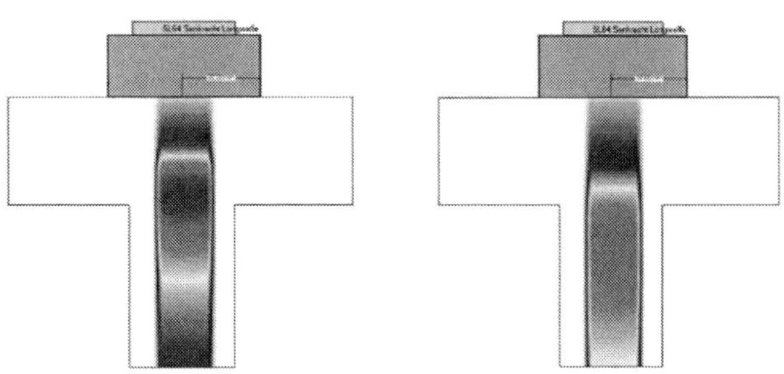

Abb. 4. Schallfeldsimulation für verschiedene Fokusparameter für Anbindungsfehler (links) und Volumenfehler (rechts)

3.3. Phased-Array Ultraschall

Die Prüfung der Versuchsbauteile wurde mit zwei Parametersätzen durchgeführt und zeigte Anbindungsfehler und Volumenfehler vor allem in den Randbereichen der Versuchsquader (Abb. 5). In den Randbereichen treten die größten Temperaturgradienten beim Schweißvorgang auf, so dass die Wahrscheinlichkeit für die Ausbildung von Materialtrennungen und Gaseinschlüssen steigt.

Abb. 5. Ultraschallscans eines Versuchsquaders mit Fehleranzeigen im Randbereich

Das Prüfsystem wurde unter Verwendung eines individuellen Vorlaufkeils und eines Vergleichskörpers aus einem artverwandten Werkstoff weiterentwickelt (Auftragswerkstoff AM70 - 8MnNiMoCrSi7-6-5). Der resultierende Prüfbereich wurde so auf die gesamte Aufbauhöhe erweitert (circa 100 mm). Die Justierung wurde über den gesamten Tiefenbereich mit Querbohrungen von 3,0 mm in definierten Tiefenlagen erstellt. Aus den resultierenden Echohöhen der Querbohrungen wurden tiefenabhängige Verstärkungskurven (TCG – time corrected gain) erstellt. Das Verfahren wurde für alle Prüfkopfelemente und Fokusgesetzte durchgeführt, um eine konstante Verstärkung und Empfindlichkeit im gesamten Prüfvolumen zu gewährleisten.

Die Divergenz des Schallfeldes ist außerhalb des Fokusbereichs größer und wirkt sich negativ auf die erreichbare Ortsauflösung aus. Die Phased-Array-Fokussierung führt zu einer deutlichen Verbesserung der Empfindlichkeit im Fokusbereich, wodurch die Erkennungsleistung im fraglichen Bereich erhöht wird. Diese erhöhte Empfindlichkeit ist jedoch auf eine programmierbare, festgelegte Tiefe im Prüfteil beschränkt. Reflektoren außerhalb des Fokusbereichs erscheinen weniger scharf und etwas größer als der gleiche Reflektor im Fokusbereich.

Mit Hilfe des optimierten Prüfsystems wurde alle gefertigten Demonstrator-Bauteile geprüft und vollständig dokumentiert (Beispiel Abb. 6). Im Vergleich zu den Bauteilen aus den Vorversuchen konnte die Fehleranzahl deutlich reduziert werden. Die Fehleranzeigen der Demonstrator-Bauteile können in zwei Gruppen eingeteilt werden. Fehleranzeigen zwischen dem Substratblech und aufgebautem Material sind als Anbindungsfehler klassifiziert und machen 60% aller Fehleranzeigen aus. Die verbleibenden 40% setzten sich aus Fehleranzeigen im Volumen des aufgebauten Materials zusammen.

Vergleichsuntersuchungen im Zuge der Kennwertermittlung und der werkstofflichen Analysen konnten die gefundenen Fehleranzeigen mit hoher Übereinstimmung bestätigen. Es zeigte sich jedoch, dass Porositäten innerhalb des aufgebauten Materials auftraten, welche bei der PAUT-Prüfung mit der eingestellten Empfindlichkeit entsprechend einer Querbohrung 3,0 mm nicht zur Anzeige gekommen waren. Die Poren und Porennester konnten in verschiedenen Aufbauhöhen lichtmikroskopisch dokumentiert werden. Anhand der Vermessungen der mikroskopischen Aufnahmen wurden mehrheitlich Poren kleiner

1,5 mm ermittelt. Für eine sichere Detektion dieser Fehlerbilder, wurde die Gesamtempfindlichkeit im Prüfbereich angehoben. Die Empfindlichkeit wurde auf eine äquivalente Querbohrung von 1,5 mm angehoben und die Ultraschallscans der Demonstrator-Bauteile erneut bewertet. Die Darstellung erfolgt über C-Bilder und B-Bilder (Abb. 7) und ermöglicht somit eine detaillierte Betrachtung der interessanten Tiefenbereiche innerhalb des Prüfvolumens. Es konnten Einzelporen und Porennester mit ausreichender Amplitude nachgewiesen werden, das Signal-Rausch-Verhältnis lag bei allen Fehleranzeigen zwischen 15 dB bis 21 dB und damit deutlich oberhalb der Anforderungen gängiger Regelwerke.

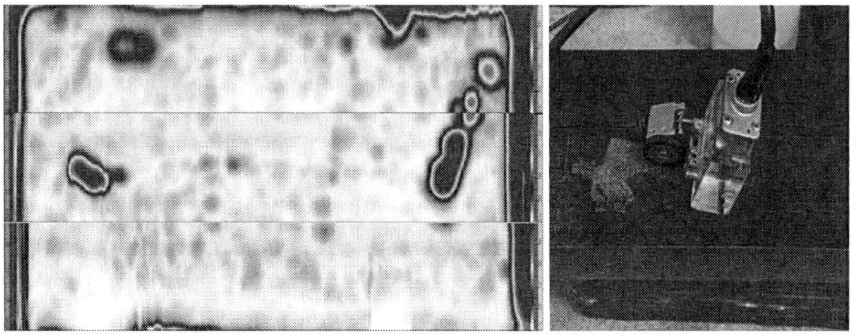

Abb. 6. Ultraschallscans eines Demonstrator-Bauteils mit vereinzelten Fehleranzeigen

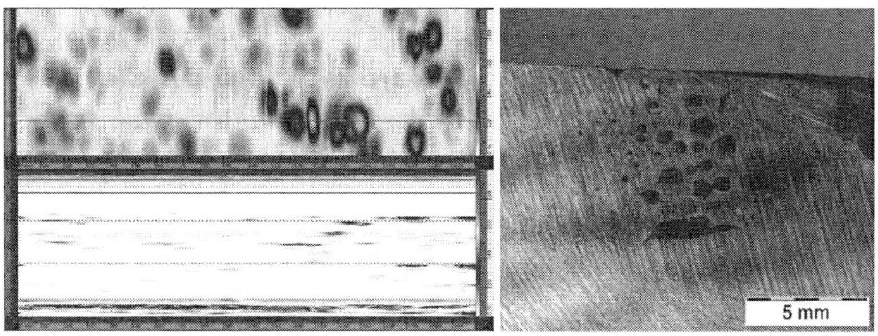

Abb. 7. Ultraschallscans eines Versuchsquaders mit Fehleranzeigen im Randbereich

3.4. Ultraschall - Total-Focusing-Method TFM

Als alternativer Ansatz wurden auch synthetisch fokussierte Schallfelder für die zerstörungsfreie Prüfung der Demonstrator-Bauteile in Betracht gezogen. Diese modernen Rekonstruktionsverfahren (Total-Focusing-Method TFM) verfolgen bei der Einschallung und Auswertung der reflektierten Schallwellen einen anderen Ansatz. Die Grundlage bildet die Erfassungsstrategie in Form des Full Matrix Capture (FMC). Der FMC-Datensatz enthält alle Informationen hinsichtlich der Schallausbreitung zwischen allen Sensorelementen, einschließlich Reflexionen an Grenzflächen und Schallstreuung durch Fehler. Dabei kann jede Art der PAUT-Prüfdatenerfassung kann mittels FMC rekonstruiert werden. Derzeit ist TFM eine sehr effektive Methode, um ein Bild des Prüfbereichs zu generieren, welches überall und in jeder Tiefe fokussiert ist.

Die Messungen an Versuchsbauteilen zeigen eine klare Darstellung des Prüfvolumens (Abb. 8). Der Empfindlichkeitsabfall ist im gesamten Tiefenbereich kleiner als 4 dB, dieser Wert deckt sich auch mit der Simulation. Durch die synthetische

Fokussierung des Schallfeldes können alle Fehlerarten wie Anbindungsfehler und Porositäten im Volumen scharf abgebildet werden. Die geometrische Ausdehnung und die Flächenhaftigkeit der Fehleranzeigen können dadurch deutlich besser beurteilt werden. Eine Aufweitung der reflektierten Anzeigen wie bei der Phased-Array-Prüfung ist nicht zu beobachten (Abb. 8). Das TFM-Verfahren lieferte im vorliegenden Anwendungsfall deutliche und messbare Vorteile was Ortsauflösung und Darstellbarkeit von Fehleranzeigen betrifft. Im Rahmen der Anwendbarkeit von Rekonstruktionsverfahren gibt es großes Potential zur Weiterentwicklung und Optimierung. Diese Untersuchungen sind Bestandteil weiterer Arbeiten, diese werden im Rahmen eines Folgeprojekts realisiert.

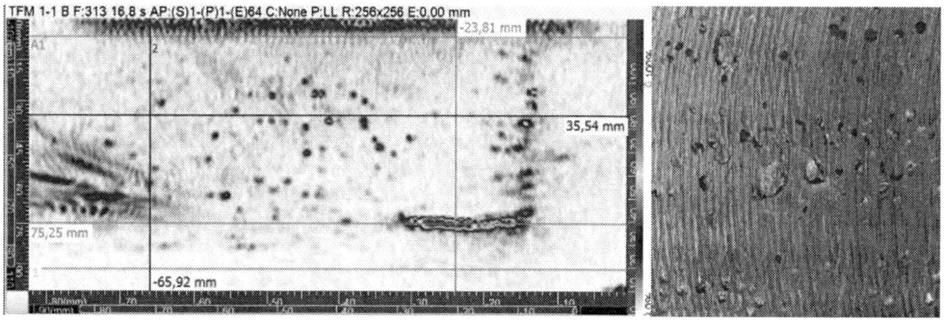

Abb. 8. TFM-Scans Demonstrator-Bauteil mit Fehlerstellen im Volumen und einem großflächigen Anbindungsfehler / Porositäten im Volumen im Vergleich mit einem Schnitt durch das Bauteil (Abbildung nicht maßstäblich)

Die Anwendbarkeit der zerstörungsfreien Prüfmethoden konnte an einer Reihe von Versuchsbauteilen und Demonstrator-Bauteilen nachgewiesen werden, die unter realen Fertigungsbedingungen hergestellt wurden und den gleichen strukturellen Aufbau wie die tatsächlich herzustellenden Brecherzähne aufweisen. Es konnte ein Großteil der typischen Fertigungsfehler bzw. Schweißfehler sicher detektiert, katalogisiert und mit den Herstellungsparametern korreliert werden. Die Auswertung der Ultraschalldaten erfolgte bei zunächst von Hand, für eine fertigungsbegleitende Prüfung musste eine automatisierte Lösung geschaffen werden.

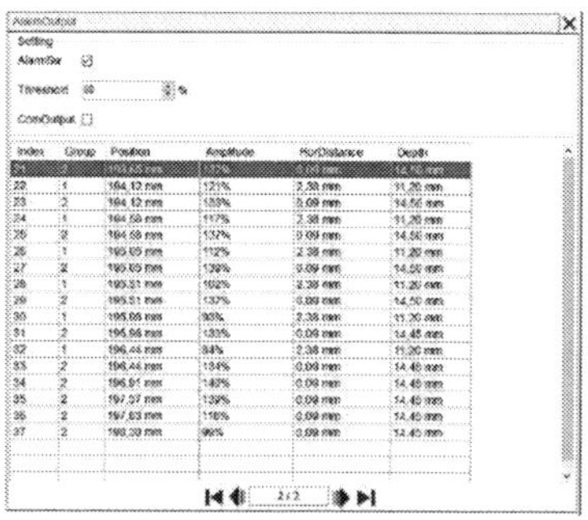

Abb. 9. Automatisch generierte Fehlertabelle mit Anzeigen aus einem Scanabschnitt

Für eine automatische Auswertung und Erfassung von Fehleranzeigen wurde eine automatische Blendendiskriminierung herangezogen. Die Blendendiskriminierung wurde anhand des Anforderungsprofils während der Projektarbeiten bei imq GmbH in Zusammenarbeit mit dem Gerätehersteller entwickelt. Bei diesem Verfahren werden die Amplituden in einem zweidimensionalen Blendenbereich (Wegkoordinate, Tiefenkoordinate) aufgezeichnet und zur Auswertung gespeichert. Die Registrierschwelle kann dabei variabel festgelegt werden, im vorliegenden Fall wurden 80% BSH gewählt. Die

Blendendiskriminierung ermöglicht so die Überwachung eines definierten Tiefenbereichs über die gesamte Anbindungsfläche eines Brecherzahns oder mehrerer Brecherzähne. Die erfassten Fehlerbereiche werden mit ihrer Länge entlang der Scankoordinate und den zugehörigen Amplituden tabellarisch dokumentiert (siehe Abbildung 9). Zusätzlich werden die Fehlerkoordinaten und die Tiefenlagen der Anzeigen gespeichert. Die Prüfergebnisse zu jedem Bauteil können als csv.-Dateien exportiert und so in andere Datenbanksysteme integriert werden.

4. Zusammenfassung:

Innerhalb der Projektarbeiten konnte gezeigt werden, dass der gewählte Werkstoff im Zusammenspiel mit dem Substratmaterial für die additive Fertigung der Brechzähne geeignet ist. Die werkstofflichen Kennwerte und die ermittelte Festigkeit entsprechen den Anforderungen hinsichtlich Einsatzzweck und Haltbarkeit. Im Laufe der Untersuchungen wurden eine Reihe von Besonderheiten, welche die additive Fertigungstechnik mit sich bringt, aufgedeckt und gelöst. Das betrifft vor allem die kritische Analyse von Referenzproben mittels digitaler Durchstrahlung und die Detektion von Fertigungsfehlern nach dem Schweißprozess. Die Ultraschall-Phased-Array Prüfung hat sich als stabiles und sicheres Verfahren zur Detektion von Anbindungsfehlern und Volumenfehlern etabliert. Die detektierten Fehleranzeigen wurden an Musterbauteilen zerstörend referenziert und mit großer Übereinstimmung bestätigt. Es konnten flächige Anbindungsfehler, Risse im aufgebauten Material und Volumenfehler bis hin zu Makroporen nachgewiesen werden. Die Grenzen des Verfahrens hinsichtlich der lokalen Fehlerauflösung und der verfahrensbedingten Fokussierung der Schallfelder wurden klar herausgestellt. Um die kritischen Punkte zu verbessern und alternative Technologien zu beleuchten, wurden Untersuchungen mit weiteren Verfahren wie TFM (Total-Focusing-Method) realisiert. Es wurde ein Konzept für eine teilautomatische Prüfung mittels 2D-Flächenscannern erarbeitet und eine automatische Auswerteroutine mit einer vollständigen digitalen Dokumentation der Prüfdaten und der Prüfergebnisse.

5. Referenzen

[1] DIN EN ISO/ASTM 52920:2023-12: Additive Fertigung - Grundsätze der Qualifizierung - Anforderungen an industrielle additive Fertigungsverfahren und Produktionsstätten (ISO/ASTM 52920:2023); Norm; Deutsche Fassung EN ISO/ASTM 52920:2023

[2] DIN SPEC 17071: 2019-09 Additive Fertigung – Anforderungen an qualitätsgesicherte Prozesse für additive Fertigungszentren; Text in Deutsch und Englisch

[3] Raiko Lorenz Bachelorarbeit: Erhöhung der Auftragsrate des WAAM-Prozesses und Analyse des Einflusses der Fertigungsparameter auf die Werkstoffstoffeigenschaften 2024

[4] Dr.-Ing. Marko Schmidt Abschlussbericht: WEARINGGREEN: Additive Fertigung und Aufarbeitung von Verschleißteilen Teilprojekt TP3-1.2.5 von: DIGMATEST: Digitalisierung in der Materialprüfung 2024

[5] Salzgitter Mannesmann Forschung GmbH, Duisburg MTU Aero Engines AG, München Kalibrierung des TFM-Verfahrens mittels additiv gefertigter Kalibrier-Reflektoren und Vergleich mit Simulationen DACH-Jahrestagung 2019

6. Danksagung

Die Autoren danken dem BMBF für die finanzielle Unterstützung des Projektes „DIGMATEST: Digitalisierung in der Materialprüfung“ innerhalb des Förderprogramms „Wandel durch Innovation in der Region (WIR!)“. Besonderer Dank gilt den Projektpartnern der TAKRAF GmbH für die konstruktive und erfolgreiche Zusammenarbeit.

Neue Ansätze zur Inline-Prozessüberwachung der Klebschichtaushärtung mittels transversaler Ultraschallwellen

Christian MATHISZIK [1], Franz HIELSCHER [1], Johannes KOAL[1]
[1] Technische Universität Dresden, Institut für Fertigungstechnik, Professur für Fügetechnik und Montage, Dresden, Germany
christian.mathiszik@tu-dresden.de

Kurzfassung. Die industrielle Verarbeitung von chemisch reagierenden Klebstoffen auf ein- und zweikomponentiger Basis (1K, 2K) wird branchenübergreifend in verschiedenen Bereichen des Handwerks sowie der Großindustrie angewendet. Für Serien- und Einzelanwendungen (z. B. Baustellenanwendungen) ist ein stabiler und verlässlicher Klebprozess die Grundvoraussetzung, um die geforderten Verbindungsqualitäten zu erfüllen. Entscheidende Prozessphasen sind die Klebstoffdosierung, -mischung sowie die Verarbeitung inklusive des Klebstoffauftragens. Die Topfzeit definiert den Zeitraum, innerhalb dessen die zulässige Verarbeitbarkeit des Klebstoffs nach dem Anmischen liegt. Mit dem Ende der Topfzeit beginnt der Klebstoff relevant auszuhärten und stellt damit eine äußerst qualitätsrelevante Prozessgröße dar. Die Länge der Topfzeit ist abhängig von verschiedenen Faktoren, wie der chemischen Zusammensetzung und dem Mischungsverhältnis des Klebstoffs, der Verarbeitungstemperatur und z. T. der Luftfeuchtigkeit. Bisherige Prozessüberwachungen in industrielle Anwendungen betrachten den Aushärteprozess und somit auch den Zeitpunkt bis zur Topfzeit kaum. Hier wird meist auf die oft konservativen Herstellerangaben aus dem Datenblatt zurückgegriffen. Für eine bedarfsgerechte und ressourcenschonende Weiterverarbeitung ist jedoch die exakte Kenntnis über den Aushärtegrad von entscheidender Bedeutung. So können im Idealfall die geklebten Bauteile nach durchlaufener Weiterverarbeitungszeit, jedoch bereits vor vollständiger Aushärtung für weitere Fertigungsschritte in Betracht gezogen werden.

In diesem Beitrag erfolgt eine Bewertung des Aushärtegrads der Klebverbindung mittels transversaler Ultraschallwellen im Puls-Echo-Verfahren. Im Rahmen der vorliegenden Untersuchung werden die Schalllaufzeiten und das Dämpfungsverhalten der Ultraschallwellen an verschiedenen Klebstoffdicken in Abhängigkeit des Aushärteverlaufs analysiert. Die Ergebnisse zeigen, dass der Übergang von der flüssigen in die feste Phase des Klebstoffs beobachtet werden kann. Dies eröffnet potenzielle Anwendungsmöglichkeiten für eine Inline-Prozessüberwachung der Klebschichtaushärtung.

Keywords: Klebverbindungen, Topfzeit, in-situ-Aushärtegradmessung, transversaler Ultraschall, Ultraschallmessung

https://doi.org/10.58286/31256

Einführung

Die Zerstörungsfreie Prüfung von Klebverbindungen ist für den Klebprozess essenziell, weil es anders als andere Fügeverfahren nur unzureichend über normierte Prozesszustandsgrößen überwacht werden kann. Für die zerstörungsfreie Prüfung von Klebverbindung gibt es unterschiedliche Verfahren zur Bestimmung von Fehlstellen, Einschlüssen oder Kissing Bonds [1, 2]. Die Prüfverfahren werden nicht, wie sonst üblich, nach ihren physikalischen Wirkprinzipien eingeteilt, sondern nach der Einteilung des National Materials Advisory Board [1]. So ist die Unterteilung in Mechanische, Elektrisch, Thermisch, Visuell, Chemische und in durchdringende Strahlung dabei werden zum aktuellen Zeitpunkt keine Chemisch Verfahren zur zerstörungsfreien Prüfung von Klebverbindungen verwendet [1]. Innerhalb der Gruppe der Strahlungsverfahren wird die Röntgenstrahlung als ein Beispiel angeführt. Dieses Verfahren ermöglicht die Erfassung von Tiefeninformationen bezüglich der Klebverbindung sowie potenzieller Fehler [3]. Allerdings ist der Prozess im Vergleich zu alternativen Verfahren deutlich zeitintensiver. Für die Gruppe der thermischen Verfahren wird die Thermografie als Vertreter gewählt, für die visuellen Verfahren das Verfahren der Sherografie [4] oder der Zeitbereichsreflektometrie (TDR) [2]. Alle Verfahren können jedoch keine Daten aus der Mitte der Klebschicht darstellen. Die Thermografie nutzt die Temperaturänderung infolge der exothermen Reaktion des Klebstoffs während der Aushärtung. Die Shearografie und TDR hingegen untersuchen die Oberfläche nach Unregelmäßigkeiten. Ein Beispiel für Mechanische Verfahren ist Ultraschall. Ultraschall bietet wie die Röntgenstrahlung auch Tiefeninformationen der Klebverbindung, hat aber den Vorteil einer kürzeren Messzeit. Die Bewertung des Aushärtegrades ist ein so eigenes Thema, dass es dafür eigene Messmethoden gibt. Die zerstörungsfreie Prüfung des Aushärtegrads kann zur Taktzeitoptimierungen, einer effizienteren Fertigung klebtechnisch hergestellter Produkte führen und zur Überwachung der Klebqualität dienen. Bisher kann nur mit wenigen Verfahren eine prozessbegleitende Bewertung durchgeführt werden, da die meisten Verfahren nur eine Bewertung der Oberfläche der Klebverbindung ermöglichen. Unter Laborbedingungen kann die Aushärtung von Klebstoffen mit verschiedenen Verfahren gemessen werden:

- Dynamisch-mechanische Analyse (DMA) [5]
- dynamische Differenzkalometrie (DSC) [6]
- Dielektrische Analyse (DEA)
- Terahertz Strahlung [7]
- Unilateraler Kernspinresonanz (uNMR) [8]
- Ultraschallmessung (US) [9, 10]

Die DMA benötigt direkten Zugang zum Klebstoff über einen Tastkopf, damit eine Messung möglich ist. Gleiches gilt für die DSC, bei der die Wärmedifferenz einer Referenzprobe mit der untersuchten Probe verglichen wird. Bei der DEA wird die Probe direkt zwischen zwei Elektroden eingespannt, um das temperatur-frequenzabhängige dielektrische Verhalten zu bewerten. Dabei können Phasenübergänge bestimmt werden, wodurch auf das Aushärteverhalten rückgeschlossen werden kann. Die Verfahren, mit denen ein Test außerhalb von Laborbedingungen möglich ist, sind die Terahertz Strahlung, die uNMR und der Ultraschall. Das Terahertz-Verfahren wurde bislang nicht an Proben mit realen Bedingungen getestet wie eine Klebschicht zwischen zwei Metallen [7]. Die Messung der uNMR wird mit den Referenzverfahren, wie beispielsweise der DSC oder der DEA, abgeglichen und korreliert. Da sich alle zuvor aufgeführten Verfahren nicht für den Einsatz in der realen Produktion eignen, ist die Entwicklung eines alternativen Verfahrens zur Bestimmung des Aushärtegrades erforderlich. Die Ultraschallverfahren bieten derzeit das größte Potenzial. Gegenwärtig existieren zwei Ansätze zur Bestimmung der Aushärtung: die Verwendung von luftgekoppeltem Ultraschall und die berührende Ultraschallprüfung. Beide nutzen longitudinalen Wellen [9, 10]. Bei der kontaktierten Ultraschallprüfung werden die Proben als Scherzugproben untersucht. Die zwei überlappenden Metallstreifen werden über eine 25 mm x 25 mm große Klebfläche miteinander verbunden. Bei den experimentelle Untersuchungen in [10] wurde der luftgekoppelte Ultraschall an Acrylplatten als Deck- und Grundplatten mit einer Klebschicht von 2 mm getestet.

Im vorliegenden Beitrag werden experimentelle Untersuchungen mit einem adaptierten Ultraschallsystem vorgestellt. Es werden Transversalwellen im Puls-Echo-Verfahren verwendet, da Transversalwellen in Flüssigkeiten stark gedämpft werden und somit keine Möglichkeit zur Ausbreitung haben. Die Einkopplung erfolgt im direkten Kontakt mit den Klebproben. Dabei wird die Aushärtung des Klebstoffs untersucht. Der Übergang von der flüssigen in die feste Phase des Klebstoffs beeinflusst die Ausbreitung der Transversalwellen u. a. in ihrem Dämpfungsverhalten und ihrer Ausbreitungsgeschwindigkeit. Dies kann für den Ansatz der Puls-Echo-Technik genutzt werden. Aufgrund des mit fortschreitender Klebstoffaushärtung zunehmenden Schubmoduls G nimmt die Schallgeschwindigkeit zu. Dies führt zu einer kontinuierlichen Zunahme der Schallimpedanz [11]. Die Folge ist eine Einkopplung der Schallwellen in die Klebschicht und eine Abnahme der Echo-Schallamplitude am Schallempfänger.

1. Methodik

1.1 Versuchsaufbau, Versuchswerkstoffe und Versuchsumfang

Für die klebtechnische Anwendung ist eine Adaption eines in-situ-Prüfsystems für die Ultraschallprüfung von Punktschweißverbindungen während des Schweißprozesses vorgesehen. Das System nutzt die Ultraschallsender und -empfänger des Bosch Rexroth PSQ 6000 Ultraschallsystems. Es wurde mit eigener Steuerung und Auswertesoftware in [12, 13] entwickelt. Der Ultraschallsender (US-Emitter) wird über einen Hochspannungsimpuls angeregt, wodurch ein Ultraschallsignal im Bereich von 50 bis 500 kHz ausgesendet wird [13]. Die ausgesendete horizontal polarisierte Transversalwelle wird vom Ultraschallempfänger (US-Receiver) nach dem Impuls-Echo-Prinzip empfangen. Die Transversalwelle durchläuft die Probe, die aus einem Deckblech, einer Klebschicht und einem Grundblech besteht. Der Versuchsaufbau ist in Abbildung 1, links, zu sehen. An den unterschiedlichen Grenzschichten kommt es zu Schallreflexionen. Diese führen zu mehreren Schalllaufzeiten (TOF0 – TOF4, TOF: time off flight), was in Abbildung 1, rechtes, schematisch dargestellt ist.

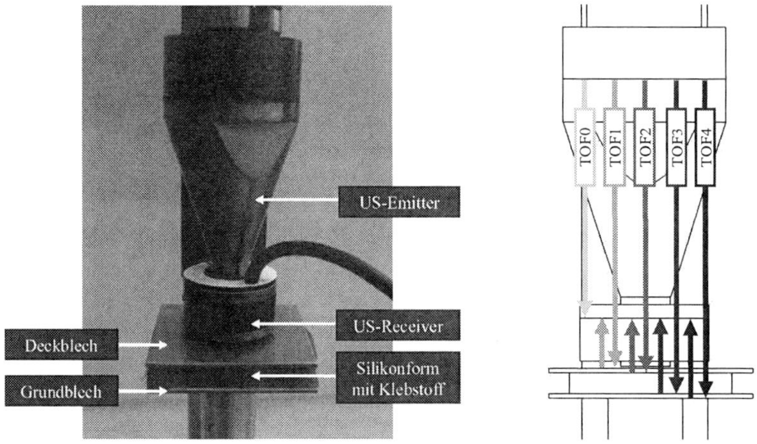

Abbildung 1: links: Versuchsaufbau zur zerstörungsfreien Bewertung der Klebstoffaushärtung mittels transversaler Ultraschallwellen im Puls-Echo-Verfahren; rechts: schematische Darstellung der verschiedenen Schalllaufzeiten (TOF0 – TOF4) infolge der Grenzschichtreflexionen

Die Klebschicht wird durch eine Silikonform in einer definierten Größe gehalten. Die Versuche werden mit dem Klebstoff Scotch-Weld DP460 der Firma 3M durchgeführt. Es handelt sich um einen Zweikomponenten-Konstruktionsklebstoff (2K-Klebstoff). Die Topfzeit liegt lt. Datenblatt bei 60 Minuten. Nach technischem Datenblatt ist der Klebstoff nach vier bis sechs Stunden ausgehärtet[14]. In Abbildung 2 ist der Aufbau der Klebstoffprobe ohne Deckblech während der manuellen Klebstoffapplikation dargestellt. Als Grund- und Deckblech wurde ein mikrolegierte, duktiler Tiefziehstahl CR300LA+GI50/50 mit verzinkter Oberfläche verwendet. Aufgenommen werden die Ultraschallmessungen bei drei unterschiedlichen Blechdicken t_s (Versuchs-ID: SAS (Sheet-Adhesive-Sheet)). Für Plausibilitätsmessungen werden zusätzlich Messungen an den drei Blechdicken ohne Klebstoff innerhalb der Silikonform durchgeführt (Versuchs-ID: SNA (Sheet-no-Adhesive)). Die verwendeten Dicken können aus Tabelle 1 entnommen werden. Jede Versuchskombination wurde dreifach experimentell durchgeführt.

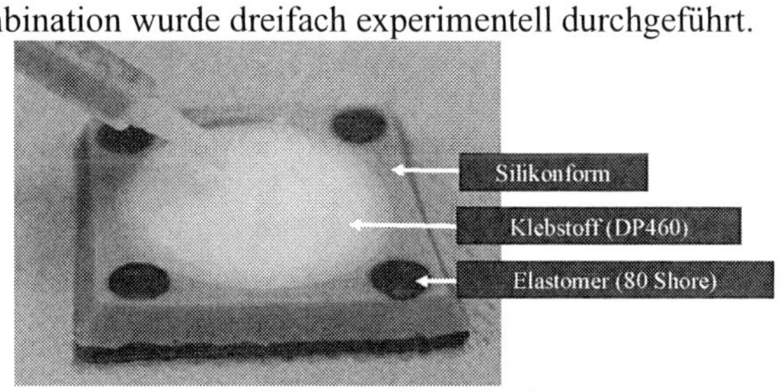

Abbildung 2: Klebstoffprobe

Tabelle 1. Versuchsmatrix

Versuchs-ID	**Probenart, Beschreibung**	**Blechdicke t_s**	**Klebstoff**
SNA-10/15/20	Einzelblech, ohne Klebstoff	1,0 mm; 1,5 mm; 2,0 mm	ohne
SAS-10-i*	Blech – Klebstoff – Blech	1,0 mm	DP460
SAS-15-i*	Blech – Klebstoff – Blech	1,5 mm	DP460
SAS-20-i*	Blech – Klebstoff – Blech	2,0 mm	DP460

i: Probennummer (1…3)

1.3 Versuchsablauf

Vor Beginn der Messungen werden sowohl das Grundblech als auch das Deckblech sorgfältig wie im Klebprozess üblich mit Isopropanol gereinigt, um Verunreinigungen zu entfernen. Die Silikonform wurde präpariert, indem die vier Elastomer-Zylinder eingesetzt werden, um die Druckkräfte auf das Silikon zu minimieren und einen definierten Klebspalt zu gewährleisten. Für die Applikation des Klebstoffes wurde für jede Probe ein separates Mischrohr verwendet, wobei vor dem Auftragen eine kleine Menge Klebstoff verworfen werden musste, um eine definierten, homogenen Mischzustand während der Klebstoffapplikation zu gewährleisten. Im Anschluss wurde die präparierte Silikonform auf das Grundblech aufgesetzt und bis zur Oberkante mit Klebstoff gefüllt. Abschließend wurde das Deckblech aufgesetzt und die Probe in den Versuchsaufbau eingelegt. Für eine bessere Einkopplung der Schallwellen, wird der Sender mit 114 N auf das Deckblech gepresst. Die Messung wurde mit dem Einlegen der Probe gestartet und erfasste über einen Zeitraum von mehr als sechs Stunden kontinuierlich Daten innerhalb einer Pulssequenz. Eine Pulssequenz ist 200 ms lang und beinhalten mindestens acht einzelne US-Pulse. Die US-Pulse werden alle 10 ms im US-Emitter angeregt. Die Pulssequenzen erfolgen in einem Intervall von 30 Sekunden. Abbildung 3 zeigt den schematischen Messablauf der Ultraschallmessungen während der Klebstoffaushärtung. Die Messfrequenz beträgt 20 MHz. Als Messsystem wird ein PicoScope 5444D MSO eingesetzt.

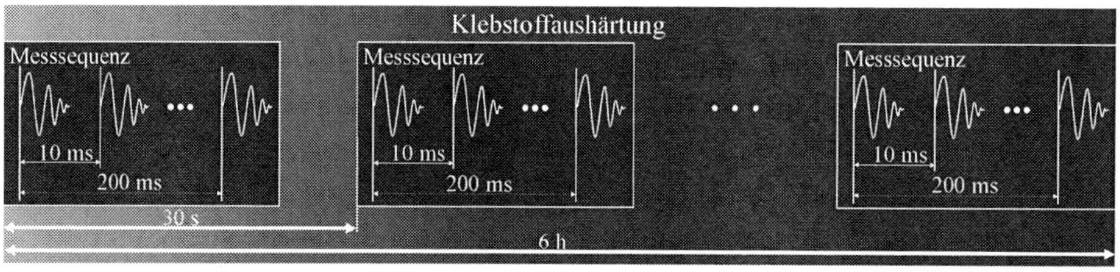

Abbildung 3: Schematischer Messablauf für Ultraschallmessungen während der Klebstoffaushärtung

1.4 Versuchsauswertung

Die US-Pulse jeder Messsequenz werden synchronisiert übereinandergelegt. Anschließend wird der mittlere Pulsverlauf berechnet. Anschließend werden die gemittelten US-Pulse über die Aushärtezeit und Pulsdauer hinweg analysiert. Diese Analyse erfolgt in einem Intensitätsdiagramm. Für die Analyse wird die Laufzeit TOF2 der Ultraschallwelle verwendet. Sie wird an der Grenzfläche zwischen dem oberen Deckblech und der Klebstoffschicht reflektiert. TOF2 liegt je nach Blechdicke im Bereich von 36 µs bis 38 µs, was nach Gleichung 1 berechnet werden kann. Die Schallwellen werden daher in einem Zeitfenster von 30 µs bis 45 µs näher betrachtet. Es werden jeweils die Effektivwerte von neun gemessenen Messwert berechnet. Abbildung 4 zeigt den schematischen Ablauf der Datenaufbereitung und Datenauswertung für die Ultraschallmessungen während der Klebstoffaushärtung.

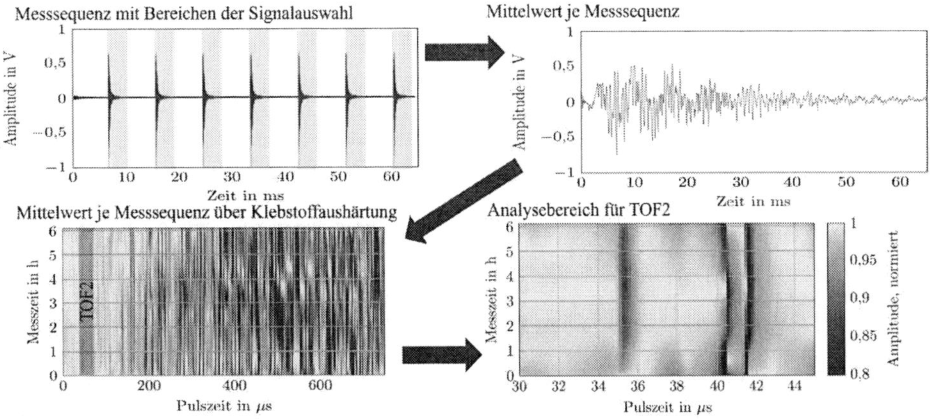

Abbildung 4: Schematischer Ablauf der Datenaufbereitung und Datenauswertung für die Ultraschallmessungen während der Klebstoffaushärtung (dargestellt an Probe SAS-10-1)

Es ist von einer Abnahme der Schallamplitude in dem Zeitbereich auszugehen, da die Schallgeschwindigkeit im Klebstoff durch die fortschreitende Aushärtung steigt. Dies bewirkt eine Steigerung der Klebstoffschallimpedanz Z_K. Folglich kommt es unter der Annahme eines senkrechten Auftreffens der Wellen zu einer abnehmenden Reflexion R und gleichzeitigen zunehmenden Durchlässigkeit D der Schallwellen an der Grenzfläche Klebstoff – oberes Stahlblech (vgl. Gleichungen 2, 3) [11]. Die Schallimpedanz von Stahl Z_{St} wird während der Versuche als konstant angenommen.

$$\mathrm{TOF2} = \frac{l_1+l_2}{c_{T.Cu}} + \frac{2t_s}{c_{T.St}} \qquad (1)$$

mit:

l_1= 70,45 mm, Länge Sender-Piezo – Kontakt US-Sender – oberes Deckblech
l_2= 12,6 mm, Länge Empfänger-Piezo – Kontakt US-Sender – oberes Deckblech
t_s: Blechdicke
$c_{T.Cu}$= 2310 m/s, Schallgeschwindigkeit Transversalwelle Kupfer
$c_{T.St}$= 3200 m/s, Schallgeschwindigkeit Transversalwelle Stahl

$$R=\frac{Z_K-Z_{St}}{Z_K+Z_{St}} \text{ und } D=\frac{2Z_K}{Z_K+Z_{St}}, \text{ mit } Z=\varrho c \qquad (2, 3)$$

mit:

ϱ: Dichte
c: Schallgeschwindigkeit

2. Ergebnisse

Die Ergebnisse der Proben mit und ohne Klebstoff innerhalb der Silikonform zwischen den Blechen sind in Abbildung 5 zu sehen. Es ist zu erkennen, dass die Proben ohne Klebstoff (SNA) einen anderen Verlauf zeigen als die Proben mit Klebstoff (SAS). Dies bestätigt die Annahme, dass der Klebstoff im Analysebereich um die Schalllaufzeit TOF2 einen Einfluss auf das reflektierte Ultraschallsignal hat.

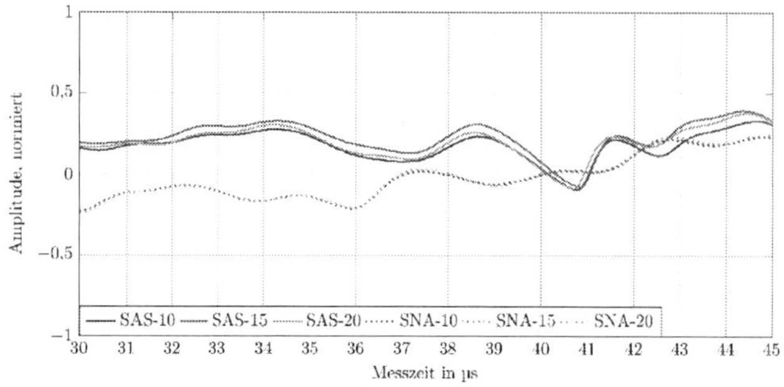

Abbildung 5: Signalvergleich zwischen Proben mit Klebstoff (SAS) und ohne Klebstoff (SNA) im Analysebereiche der Schalllaufzeit TOF2

Die berechneten Effektivwerte der Schallverläufe ausgewählter Proben über die gesamte Messzeit von 6 Stunden im Analysebereich von 30 µs bis 45 µs um TOF2 sind in Abbildung 6 dargestellt. Die abgebildete Amplitude über die Pulszeit wurde normiert. Die Verläufe zeigen Ähnlichkeiten. In den Zeitbereichen zwischen 35 µs und 40 µs sind Bereiche abnehmender Amplituden zu erkennen, was auf die Aushärtung des Klebstoffs hindeutet. Insbesondere bei SAS-10-3 und SAS-20-3 sind diese deutlicher ausgeprägt als bei SAS-15-2. Zur genaueren Analyse werden die Verläufe über die Messzeit einzeln im Nahbereich von TOF2 mit ±3 µs betrachtet. Dabei werden fallende Verläufe jeder Probe identifiziert. Die Verläufe sind in Abbildung 7 dargestellt. Für die Proben SAS-10 liegen die TOF2 im Bereich von 35,3 µs bis 37,2 µs und schwanken mit ±1 µs um die berechnete TOF2 für die Blechdicke t_s = 1,0 mm. Die Proben SAS-15 mit Blechdicken t_s = 1,5 mm zeigen die kleinste Schwankung der TOF2 mit 0,3 µs, wobei diese vor der berechneten blechdickenspezifischen TOF2 (t_s = 1,5 mm) = 36,9 µs liegen. Die Messdaten der Proben SAS-20 weisen die größten Schwankungen bezüglich der identifizierten TOF2 von 6,2 µs auf. Bei allen Proben kann eine fallende Tendenz der gemessenen Schallamplitude über die Messzeit und somit über die Aushärtedauer des Klebstoffs beobachtet werden. Die Proben SAS10-1 und SAS15-1 zeigen dabei zu Beginn der Messungen zunächst einen Anstieg der Schallamplitude, welche anschließend in einen fallenden Verlauf übergeht. SAS-10-3 und SAS-20-3 zeigen den deutlichsten Trend einer Abnahme der Schallamplitude.

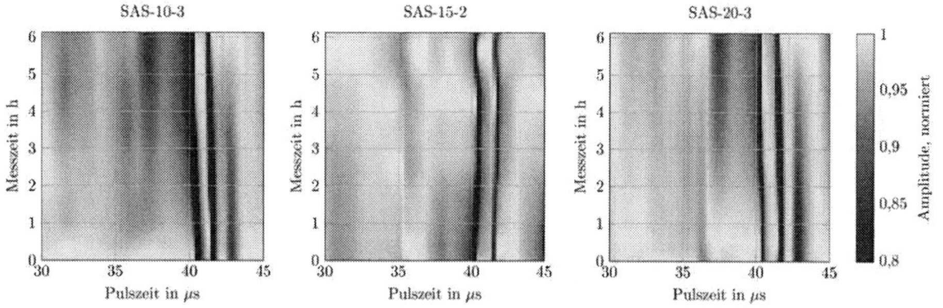

Abbildung 6: Schallverläufe über die gesamte Messzeit von 6 Stunden im Analysebereich von 30 µs bis 45 µs um TOF2 für die Proben SAS-10-3, SAS-15-2 und SAS-20-3

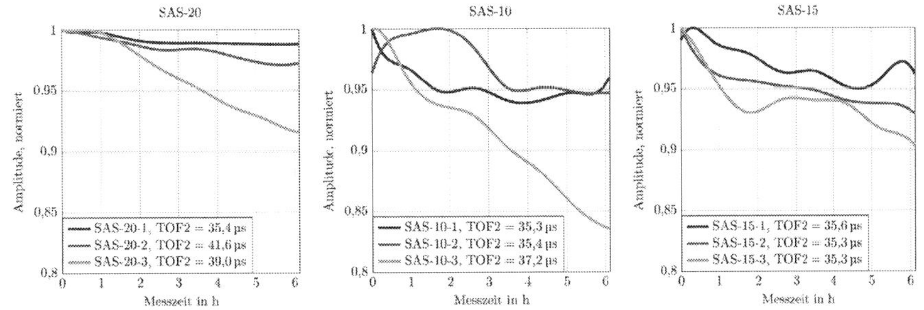

Abbildung 7: Schallverläufe über die gesamte Messzeit von 6 Stunden für die probenspezifische Schalllaufzeiten TOF2 aller Versuchsproben mit Klebstoff

3. Diskussion

Die Analyse der reflektierten Ultraschallsignale während des Aushärtungsprozesses des betrachteten Klebstoffs gezeigt, dass die Schallamplitude Rückschlüsse auf Materialveränderungen ermöglicht. Allgemein lässt sich beobachten, dass die gemessenen Amplituden der Signale während des Aushärtungsprozesses abnehmen. Die Streuungen in den identifizierten Schalllaufzeiten TOF2 der reflektierten Schallwellen kann mit material- und prozessbedingten Schwankungen während der Messzeit zusammenhängen. Diese Schwankungen können auch Indizien für eine Inhomogenität der Klebstoffschicht und den spezifischen Aushärtungsbedingungen sein und bedürfen weiteren Analysen.
Die gezielte Auswertung der Schalllaufzeiten ermöglicht die Identifizierung charakteristischer Zeitbereiche, in denen signifikante Änderungen während der Klebstoffaushärtung (z. B. Topfzeit) stattfinden. Dies kann zu einer verbesserten Prozesssteuerung führen und ermöglicht Rückschlüsse auf die mechanischen Eigenschaften des Aushärteprozesses des Klebstoffs.

4. Zusammenfassung

Zur zerstörungsfreien Bewertung des Aushärtevorgangs werden am Beispiel eines ausgewählten zweikomponentigen Klebstoffs Ultraschallmessungen durchgeführt. Dazu wurde ein vorhandenes Ultraschallsystem, welches für die in-situ-Punktschweißprüfung konzipiert wurde, an Klebproben angewendet. Die transversalen Ultraschallwellen werden nach dem Puls-Echo-Verfahren gemessen und analysiert. Es zeigt sich, dass der Aushärtevorgang des untersuchten Klebstoffs beobachtet und bewertet werden kann. Die Messungen unterliegen Schwankungen. Diese können wichtige Indizien spezifischer Aushärtephänomene oder Rückschlüsse auf eine schlechte Durchmischung der Klebstoffkomponenten liefern. Die vorliegende Analyse unterstreicht die Bedeutung akustischer Prüfmethoden für die Beurteilung der Klebstoffhärtung und zeigt ihr Potenzial für industrielle Anwendungen in der Produktionsüberwachung auf. In weiteren Versuchen wird der Versuchsaufbau um die Transmission Messung erweitert, zusätzlich sollen Versuche zu unterschiedlichen Klebstoffdicken, Stöchiometrien, Klebstoffen und mit Aushärtetemperaturen.

Referenzen

[1] Gerd Habenicht, *Kleben: Grundlagen, Technologien, Anwendungen.* 6., aktualisierte Auflage, 6th ed. Berlin Heidelberg: Springer-Verlag Berlin Heidelberg, 2009.

[2] P. J. Steinbild, R. Höhne, R. Füßel, and N. Modler, "A sensor detecting kissing bonds in adhesively bonded joints using electric time domain reflectometry," *NDT & E International*, vol. 102, pp. 114–119, 2019, doi: 10.1016/j.ndteint.2018.11.013.

[3] J. E. Jakes *et al.*, "X-ray methods to observe and quantify adhesive penetration into wood," *J Mater Sci*, vol. 54, no. 1, pp. 705–718, 2019, doi: 10.1007/s10853-018-2783-5.

[4] DGZfP, Ed., *Vergleich der induktiv angeregten Shearografie und Thermographie zur Fehlerdetektion an hochfesten Strukturklebungen und elastischen Dickschichtklebungen*, 2017.

[5] Peter Guttmann, "Dynamisch mechanische Analyse von Epoxidharzen," Masterarbeit, Institut für Werkstoffkunde und Prüfung der Kunststoffe, Montanumuniversität Leoben, Leoben, 2009. Accessed: Mar. 20 2025. [Online]. Available: https://pureadmin.unileoben.ac.at/ws/portalfiles/portal/1903684/AC07968850n01vt.pdf

[6] W. Krenkel, Ed., *Verbundwerkstoffe*: Wiley, 2009.

[7] Carina Scheller, "Überwachung der Klebstoffaushärtung mittels Terahertz," *ZfP-Zeitung : Zeitschrift der DACH-Gesellschaften DGZfP, ÖGfZP und SGZP*, no. 175, pp. 45–51, 2022. [Online]. Available: https://www.ndt.net/search/docs.php3?id=26516

[8] Norbert Halmen, "Zerstörungsfreie Prüfung von Klebeverbindungen durch den Einsatz unilateraler NMR," *ZfP-Zeitung : Zeitschrift der DACH-Gesellschaften DGZfP, ÖGfZP und SGZP*, pp. 34–38, 2020. [Online]. Available: https://www.ndt.net/search/docs.php3?id=25482

[9] E. Y. Maeva, I. Severina, and G. B. Chapman, "Ultrasonic Analysis of the Degree of Cure and Cohesive Properties of the Adhesive in a Bond Joint," *Research in Nondestructive Evaluation*, vol. 18, no. 2, pp. 121–138, 2007, doi: 10.1080/09349840601128747.

[10] Giovanni SCHOBER, "Überwachung der Klebstoffaushärtung mittels luftgekoppelten Ultraschalls," in *DGZfP-Jahrestagung 2018.*

[11] J. Krautkrämer and H. Krautkrämer, *Werkstoffprüfung mit Ultraschall.* Berlin, Heidelberg: Springer Berlin Heidelberg, 1986.

[12] C. Mathiszik, "Verifizieren eines ausgewählten Ultraschallmoduls zur Linsendurchmesserbestimmung beim Widerstandspunktschweißen und Entwicklung eines erkennenden Algorithmus," Großer Beleg, Technische Universität Dresden, Professur Fügetechnik und Montage, Dresden, 2012.

[13] Marcus Heide, "Aufbau einer Einrichtung zur Bestimmung des Linsendurchmessers für das Widerstandspunktschweißen," Oberseminar, Technische Universität Dresden, Professur Fügetechnik und Montage, Dresden, 2013.

[14] 3M Deutschland GmbH, *Technisches Datenblatt: Scotch-WeldTM DP 460.*

Ultraschall Phased Array als Screening Tool zur Inspektion auf Wasserstoff induzierte Risse

Ulrich BAUMGARTEN
DEKRA Incos GmbH, Großmehring, ulrich.baumgarten@dekra.com

Kurzfassung: In der chemischen und petrochemischen Industrie ist der Schädigungsmechanismus der Wasserstoff induzierten Risse (HIC) eine häufige Schädigung.

Als ein geeignetes und insbesondere schnelles Screening Tool erweis sich dabei die Ultraschall Phased Array Prüfung von Schweißnaht T-Stoßen, die bei Bedarf um ergänzende Prüfungen erweitert wurde.

Keywords: UT-Phased Array; HIC

Einführung

Wasserstoffschädigung ist ein häufig auftretender Schaden im Bereich der petrochemischen Industrie. Mit Ultraschall Phased Array in Senktrechteinschallung hat sich ein schnelles, geeignetes Screening Tool ergeben. Bei geringen Anforderungen an die Oberfläche und einem Prüfbereich ausschließlich im Bereich von Schweißnaht T-Stößen ist somit eine Prüfung von bis zu 40 Behältern im Rahmen eines Anlagenstillstandes möglich.

https://doi.org/10.58286/31261

1. Schadensmachanismus

Wasserstoff induzierte Risse (eng: Hydrogen induced cracks - HIC) sind typische Schäden im Bereich der petrochemischen Industrie. Aus zumeist flüssigen Kohlen-wasserstoffen diffundiert elementarer Wasserstoff in ferritisch-perlitische Stähle. Die Anlagerung des elementaren Wasserstoffes erfolgt nach den in Abbildung 1 dargestellten Reaktionen. Schwefelsaure Medien begünstigen die Anlagerung.

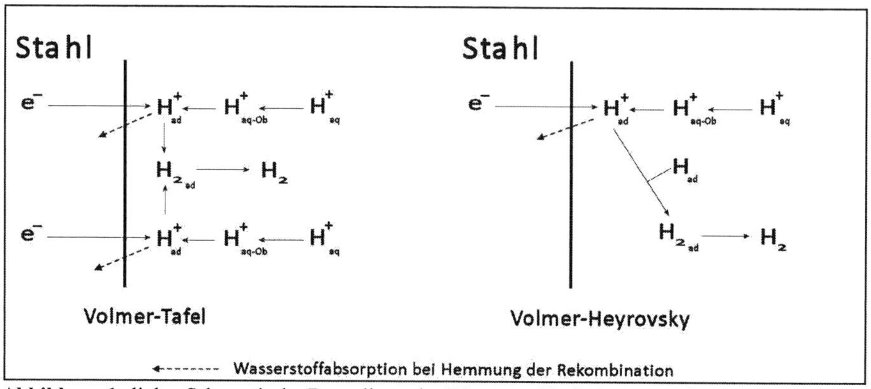

Abbildung 1: links: Schematische Darstellung der Wasserstoffaufnahme aus einer Lösung in Stahl nach der Tafel-Reaktion. In Wasser gelöste Protonen adsorbieren an Stahl und rekombinieren unter Aufnahme von Elektronen zu molekularem Wasserstoff, der wieder in Lösung geht. - rechts: Schematische Darstellung der Wasserstoffaufnahme aus einer Lösung in Stahl nach der Heyrovsky-Reaktion. Ein in Wasser gelöstes Proton adsorbiert an Stahl und rekombiniert unter Aufnahme von einem Elektron mit einem Wasserstoffatom zu molekularem Wasserstoff, der wieder in Lösung geht. Bei Hemmung der Rekombination diffundieren die Protonen in den Stahl – beide Reaktionen.

Die temperaturabhängige Löslichkeit von Wasserstoff ist in Stählen mit einer kubisch-raumzentrierten (krz) Elementarzelle (ferritisch-perlitische Stähle) gering. Aus diesem Grund ist die Wasserstoffschädigung ausschließlich in ferritisch-perlitischen Stählen relevant – dargestellt in Abbildung 2.

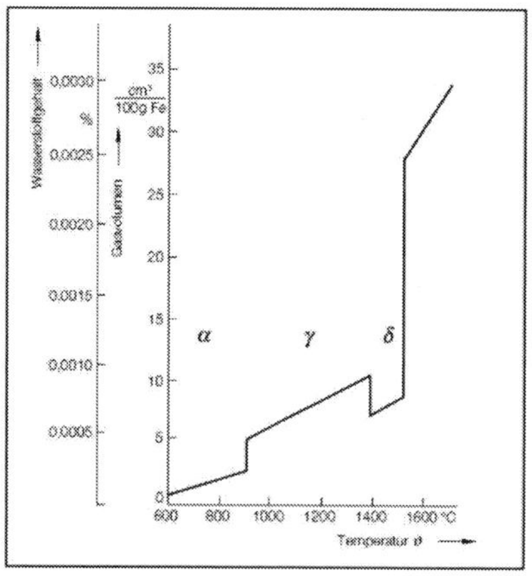

Abbildung 2: Wasserstofflöslichkeit in reinem Eisen in Abhängigkeit von der Temperatur. Der maximale atomare Wasserstoffgehalt, ist über der Temperatur aufgetragen. Bei Erhöhung der Temperatur steigt die Löslichkeit von Wasserstoff in Eisen. Im kubisch flächenzentrierten γ-Eisen ist die Wasserstofflöslichkeit erheblich höher als im kubisch flächenzentrierten Gitter des α-Eisens und des δ-Eisens.

Das Wasserstoffatom belegt auf Grund seines geringen Durchmessers (ca. 10^{-10} m) bevorzugt Zwischengitterplätze. In einem krz-Gitter sind dies Tetraederplätze, in einem kubisch-flächenzentrierten (kfz) Gitter Oktaederplätze.

Die Rekombination zu molekularem Wasserstoff im Stahl erfolgt bevorzugt in so genannten Wasserstofffallen. Wasserstofffallen sind Bereiche mit erhöhter atomarer wie molekularer Wasserstoffkonzentration im Stahl. Zu Wasserstofffallen können Gitterfehler, innere Phasengrenzen, Poren, Lunker, Einschlüsse, Ausscheidungen, mikroskopische Anrisse oder Bereiche mit einer erhöhten Anzahl an Versetzungen.

In Wasserstofffallen sammelt sich zunächst atomarer Wasserstoff und rekombiniert zu molekularem Wasserstoff. Dabei werden Drücke von bis zu 1000 bar erreicht. Durch die hohen Drücke von molekularem Wasserstoff sowie die erhebliche Volumenzunahme entstehen Materialtrennungen im Bauteil, die Wasserstoffschädigungen.

Abhängig vom Ort des Auftretens der Wasserstoffschädigung im Bauteil bzw. ob das Bauteil unter äußeren Zugspannungen steht, wird die Wasserstoffschädigung in mehrere Formen unterschieden.

HIC tritt zumeist in Walzrichtung im Bauteil auf, kann aber auch senkrecht zu angelegten Zugspannungen verlaufen. Außerdem ist eine Mischform aus waagrecht und senkrecht im Bauteil orientierten Rissen möglich. Oberflächennahe Wasserstoffschädigung tritt in Form von Blasen auf, wenn das Bauteil nicht unter äußeren Zugspannungen steht. Die Schädigung reicht dabei in eine Tiefe von ca. 1 mm. Blasen treten hauptsächlich in weichen Stählen niedriger Festigkeit auf. Die Blasenbildung erfolgt dabei hauptsächlich in Bereichen nichtmetallischer Einschlüsse.

Im Schweißgut und in der Wärmeeinflusszone (WEZ) können Zugspannungen durch Temperatureinwirkung oder äußere Belastungen auftreten. Unter diesen Voraussetzungen tritt die Wasserstoffschädigung in Form von SSC (Sulfid Stress Cracks, dt: Sulfid-Spannungsrisse) und SOHIC (Stress Oriented Hydrogen Induced Cracks, dt: spannungs-orientierte wasserstoffinduzierte Risse) auf. SOHIC sind kurze, (bis zu 1 mm) in Walzrichtung waagrecht übereinander orientierte Risse im Bauteil, die mitunter die Werkstoffoberfläche erreichen. SSC sind senkrecht zu den angelegten Zugspannungen orientierte Risse. Das Wachstum der SSC beginnt an der Bauteiloberfläche

Im Inneren von spannungslosen Bauteilen tritt die Wasserstoffversprödung in Form von HIC auf, welche die häufigste Form der Wasserstoffversprödung ist. HIC, auch als SWC (Stepwise Cracks, dt: stufenförmige Risse) bezeichnet, sind im Inneren eines Bauteils in beliebiger Tiefe in Walzrichtung waagrecht orientierte Risse. Der Verlauf der HIC ist zumeist stufenweise versetzt. Dargestellt sind verschiedenen Formen der Wasserstoffschädigung in den Abbildung 3 und 4. Ein beispielhafter metallographischer Schliff ist in Abbildung 5 dargestellt.

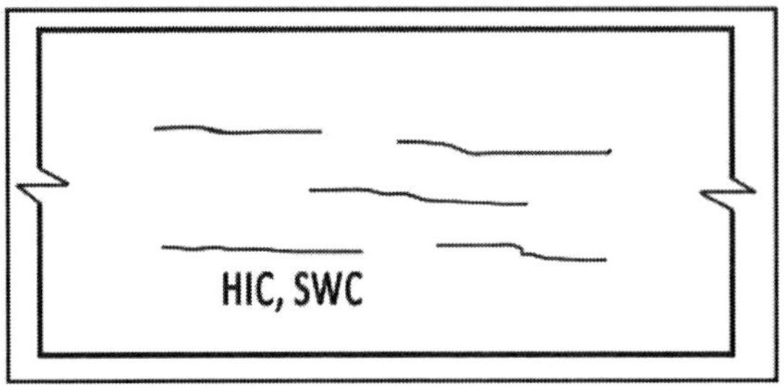

Abbildung 3: Skizze eines spannungslosen Bauteils mit mehreren einzelnen HIC bzw. SWC. Die verschiedenen Risse sind in Walzrichtung orientiert und zeigen einen mitunter stufenförmigen Verlauf ohne Kontakt zur Bauteiloberfläche.

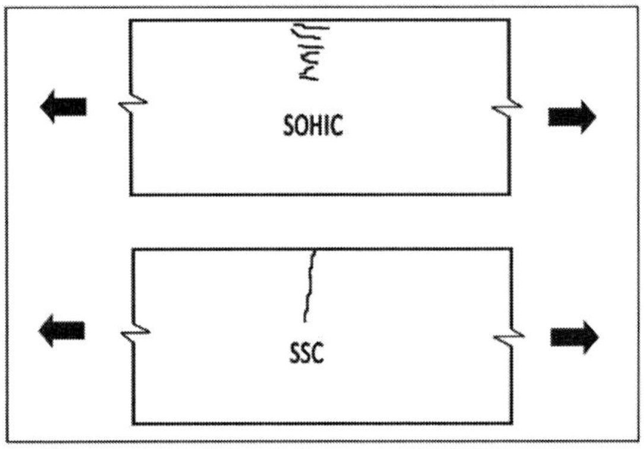

Abbildung 4: oben: Skizzierte Erscheinungsform von SOHIC in einem unter Zugspannungen stehendem Bauteil. Dabei bilden sich mehrere parallel in Walzrichtung verlaufende Risse - unten: Skizzierte Erscheinungsform von SSC in einem unter Zugspannungen stehenden Bauteil. Hierbei erfolgt die Rissbildung beginnend an der Bauteiloberfläche senkrecht zur angelegten Spannung.

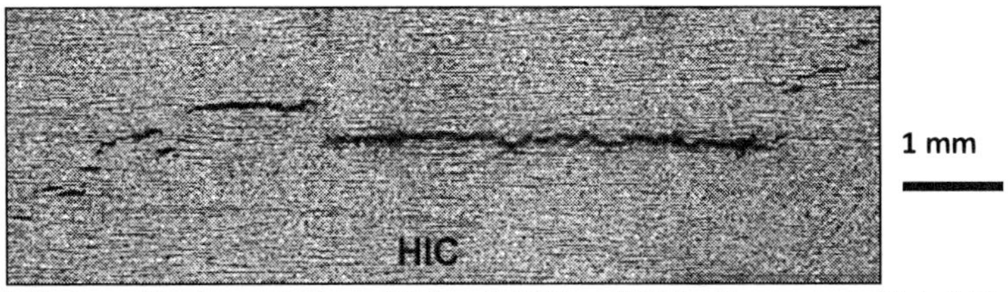

Abbildung 5: Photographie eines metallographischen Schliffes mit Makroätzung von in Walzrichtung orientierter HIC mit unterschiedlichen Fortschritt. Im linken Bereich sind vereinzelte nicht verbundene HIC gegeben, die in einem stufenförmigen Verlauf resultieren würden. Im rechten Bereich ist eine HIC im fortgeschrittenen Stadium dargestellt

2. Prüfaufgabe

Mit dem Hauptaugenmerk auf HIC war das Grundmaterial im Bereich von T-Stößen von Schweißnähten zu prüfen. Beispielhaft ist ein T-Stoß in Abbildung 6 dargestellt.

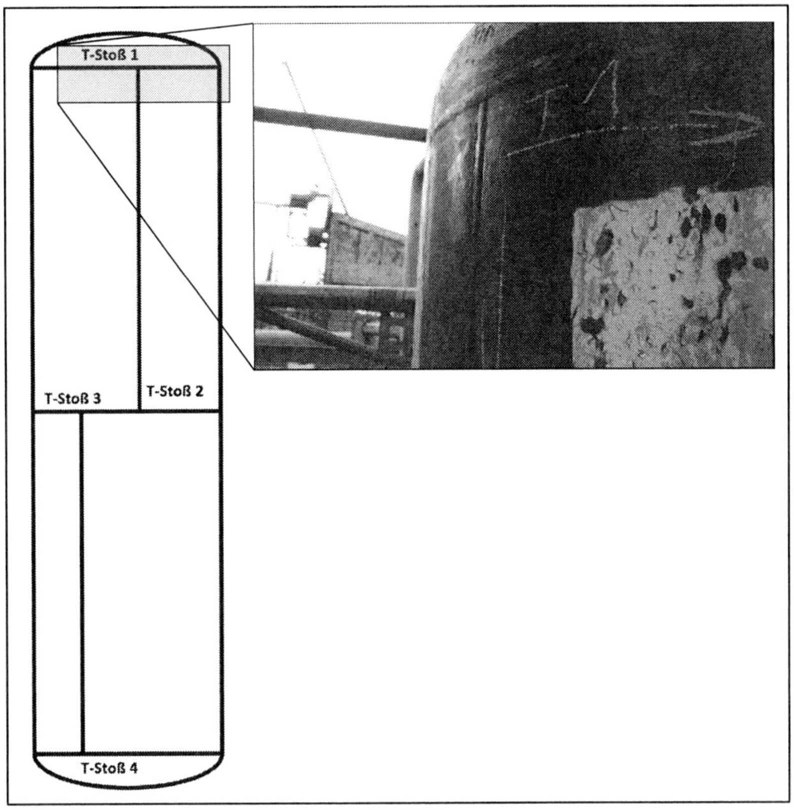

Abbildung 6: Beispielhafte Darstellung eines Behälters in der Petrochemischen Industrie. Der Bereich des zu prüfenden T-Stoßes Nr. 1 ist als Foto abgebildet.

Die Anforderung an die Oberfläche sind gering. Festanhaftende Farbschichten und Rost müssen nicht entfernt werden. Dargestellt ist dies in Abbildung 7.

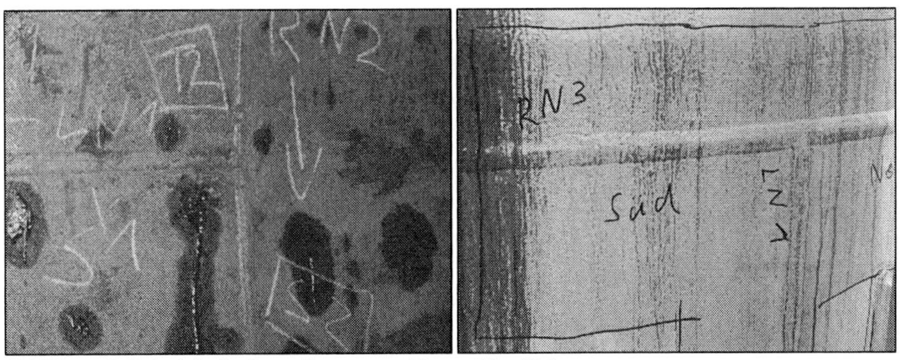

Abbildung 7: Beispielhafte Darstellung von prüfbaren Oberflächen.

3. Durchgeführte Prüfungen und Ergebnisse

Die Prüfung erfolgte in Senkechteinschallung mit einem 5L64-Prüfkopf. Die Vorlaufstrecke bestand aus Wasser, und eine Tiefenausgleichskurve wurde verwendet.

Die von Hand aufgenommenen Scans wurden beidseitig der Schweißnähte geführt. Die Auswertung erfolgte hierbei im Ultraschall C-Bild – in Draufsicht unter farblicher Angabe der Signalamplitude (Abbildung 8) und unter farblicher Angabe der Signallaufzeit (Abbildung 9).

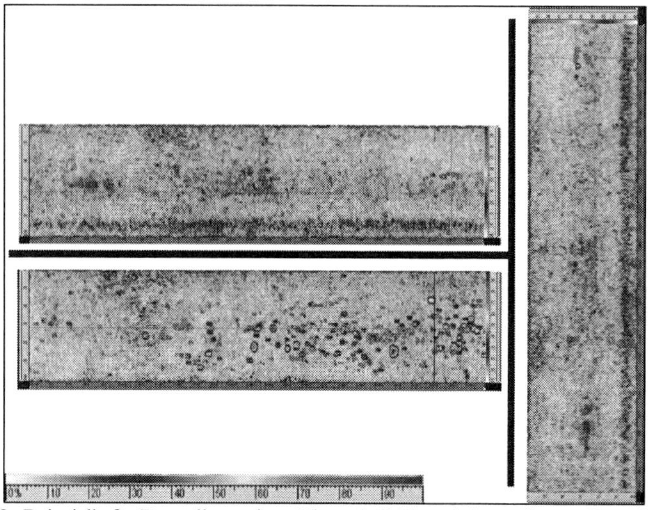

Abbildung 8: Beispielhafte Darstellung eines Ultraschall Phased Array Scans. Geprüft wurden die Bereiche eines T-Stoßes (die Schweißnähte sind schwarz skizziert). Dargestellt sind die Signalamplituden mit der Legende links unten.

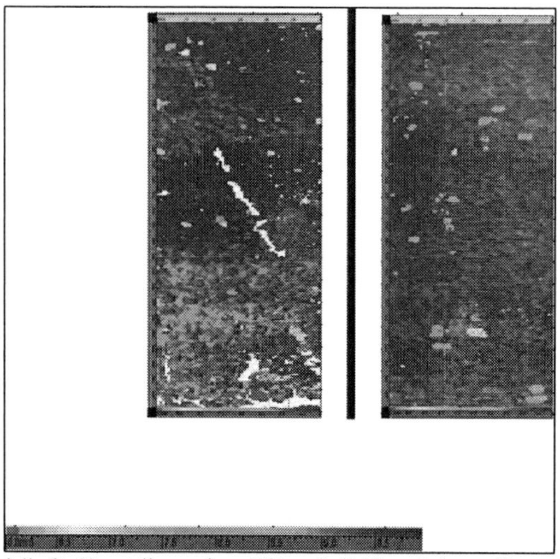

Abbildung 9: Beispielhafte Darstellung eines Ultraschall Phased Array Scans. Geprüft wurden die Bereiche eines T-Stoßes (die Längsnaht ist schwarz skizziert, die Rundnaht war nicht zugänglich). Dargestellt ist die Signallaufzeit mit der Legende links unten.

In den in Abbildung 8 und in Abbildung 9 dargestellten Scans sind Wasserstoffschädigungen als rote und gelbe Bereiche erkennbar. Auffällige Bereiche wurden weitergehenden Ultraschallprüfungen unterzogen.

Durchführung und quantitative Auswertung der Kalibrierung von kundenspezifischen Phased-Array-Prüfgeräten des Ultraschall-Prüfsystems zur Prüfung von Blankstahl „LOCIS-PAUT“ nach DIN EN ISO 18563-1

Sanjeevareddy KOLKOORI [1], Claudia JASPERNEITE [2], Cord ASCHE [2], Daniel KOERS [2], Josef MAIER [3]
[1] Rosenxt Creation Center GmbH, Alzenau, Deutschland, skolkoori@rosen-nxt.com
[2] IDS GmbH, Lingen/Stutensee, Deutschland
[3]Voestalpine BÖHLER Edelstahl GmbH Co & KG, Kapfenberg, Österreich

Kurzfassung

Zur Durchführung der wiederkehrenden Prüfungen und der regelmäßigen Zertifizierung der Ultraschall-Phased-Array-Prüfgeräten in automatisierten Prüfanlagen nach der internationalen Norm DIN EN ISO 18563-1 ist eine vollständige Verifizierung der Funktionsfähigkeit der eingesetzten Prüfelektronik in der Produktion beim Kunden erforderlich. Im Rahmen eines kundenspezifischen Projektes von voestalpine-Böhler und Rosenxt wurde ein neuartiges PAUT-Prüfsystem “LOCIS-PAUT” für die Stangenprüfung entwickelt und beim Kunden eingesetzt. Die vollständige und genaue Kalibrierung des Prüfgerätes der automatisierten PAUT-Prüfanlage dient nicht nur zur Bestimmung der Leistungsfähigkeit des Prüfgerätes, sondern auch zur 100%igen Sicherstellung der Qualität von hochwertigen Luftfahrtstählen in der Produktion.

In diesem Beitrag werden die vollständigen Kalibriermessungen und deren quantitative Auswertung des integrierten 32/128 gemultiplexten Phased-Array-Prüfgerätes des LOCIS-PAUT-Prüfsystems nach DIN EN ISO 18563-1: Gruppe 2 vorgestellt. Die Kalibriermessungen wurden mit einem im Haus Entwickelten automatisierten PAUT-Kalibriersystem durchgeführt. Die normgerecht gemessenen Eigenschaften der Sende- und Empfangskanäle der 32/128-PAUT-Prüfelektroniken werden vorgestellt. Insgesamt wurden ca. 12000 Messungen an den 4x 32/128 integrierten Prüfelektroniken durchgeführt. Die Datenerfassung, Auswertung, Visualisierung und die automatische Erstellung des Kalibrierzertifikats werden normgerecht durchgeführt. Anschließend wird die Bedeutung der Messungen für die Zertifizierung des kundenspezifischen Prüfsystems „LOCIS-PAUT“ zur Prüfung von Blankstahl für Luftfahrtanwendungen in der ZfP-Praxis erläutert.

Keywords: Ultraschall-Phased-Array, PAUT-Prüfelektronik, automatisierte Prüfanlage, Kalibrierzertfikat, DIN EN ISO 18563-1, Blankstahl, LOCIS-PAUT

Durchführung und quantitative Auswertung der Kalibrierung von kundenspezifischen Phased-Array-Prüfgeräten des Ultraschall Prüfsystems zur Prüfung von Blankstahl "LOCIS-PAUT" nach DIN EN ISO 18563-1

rosenxt

Sanjeevareddy Kolkoori[1], Claudia Jasperneite[2], Cord Asche[2], Daniel Koers[2], Josef Maier[3]
[1] Rosenxt Creation Center GmbH, Alzenau / [2] IDS GmbH, Stutensee/Lingen / [3]Voestalpine BÖHLER Edelstahl GmbH Co & KG, Kapfenberg, Österreich

Motivation

- Überprüfung der Leistungsfähigkeit des automatisierten PAUT-Prüfsystems in regelmäßigen Abständen gemäß DIN EN ISO 18563-3
- Um eine 100%ige Qualitätskontrolle im Feld durch den Einsatz der präzise kalibrierten PAUT-Prüfelektronik nach DIN EN ISO 18563-1 zu erreichen.
- Eine gültiges Zertifikat nach DIN EN ISO 18563-1 für die PA-Prüfung von hochwertigen Luftfahrtwerkstoffen ist erforderlich

Ziele

- Vollständige Charakterisierung applikationsspezifischer PAUT–Prüfelektroniken gemäß DIN EN ISO 18563-1: Gruppe2
- Reduzierung der Stillstandszeiten in der Produktionslinie durch Einsatz eines automatisierten und zuverlässigen Kalibrierprüfsystems
- Quantitative Auswertung und Visualisierung von großen Datenmengen
- Automatisierte Erstellung vom Kalibrierzertifikat vor Ort

Abb1. Automatisierte Ultraschall-Phased-Array Prüfanlage für die Prüfung von Blankstahl „LOCIS-PAUT" mit integrierter, im Hausentwickelter applikationsspezifischer PAUT-Prüfelektronik

Messergebnisse

Quantitative Auswertung nach DIN EN ISO 18563-1

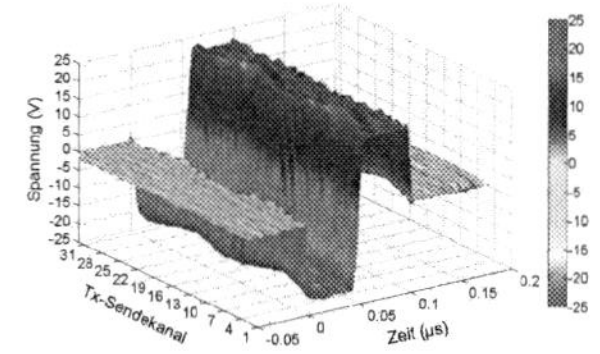

Abb2. Gemessener bipolarer Sendeimpuls der PAUT-Prüfelektronik 1x32/128

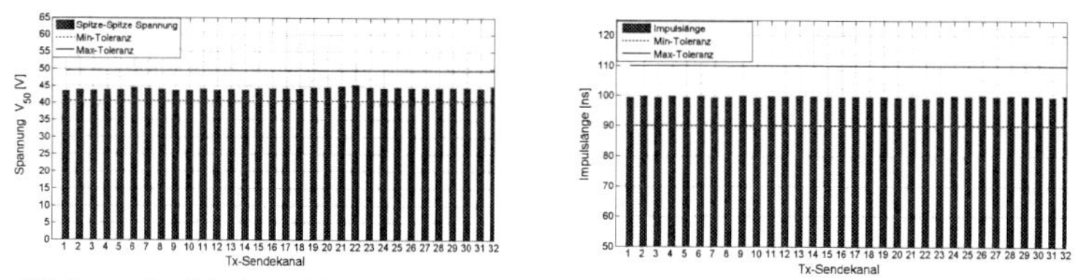

Abb3. Gemessene Eigenschaften der Tx-Sendekanäle der gemultiplexten PAUT-Prüfelktronik 1x32/128 (links) Spitze-Spitze Spannung und (rechts) Impulslänge.

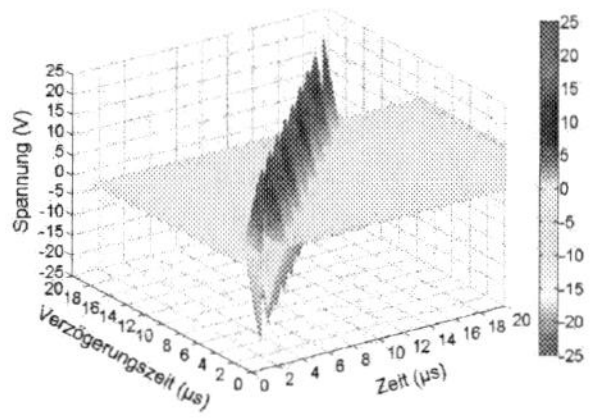

Abb4. Linearität der Zeitverzögerungen der Tx-Sendekanäle

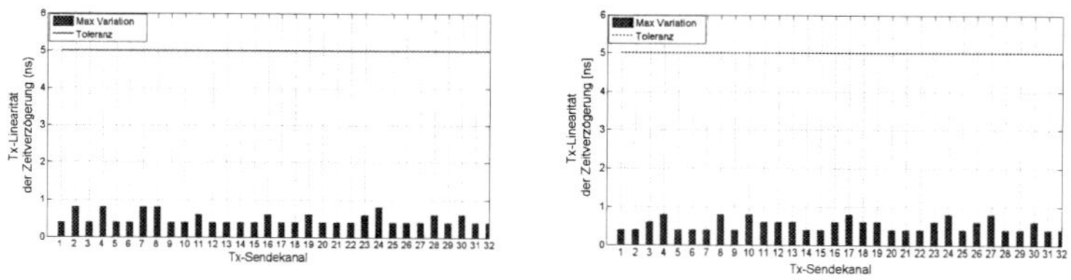

Abb5. Gemessene Linearität der Zeitverzögerungen der Tx-Sendekanäle der zwei PAUT-Prüfelktroniken 2x32/128 (links) Prüfelektronik 1 und (rechts) Prüfelektronik 2

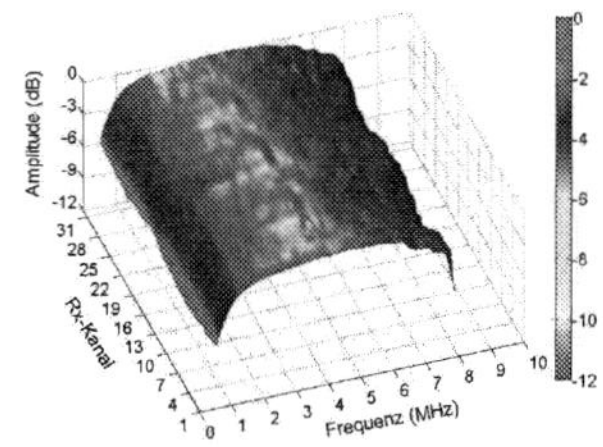

Abb6. Frequenzeigenschaften der Rx-Empfängerkanäle

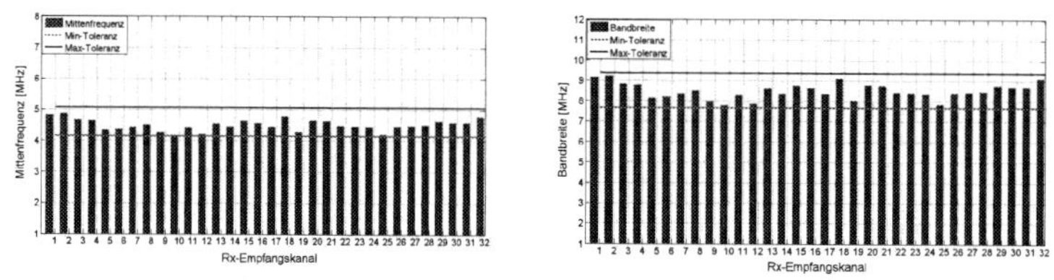

Abb7. Gemessene Frequenzeigenschaften der Rx-Empfängerkanäle der gemultiplexten PAUT-Prüfelktronik 1x32/128 (links) Mittenfrequenz und (rechts) Bandbreite

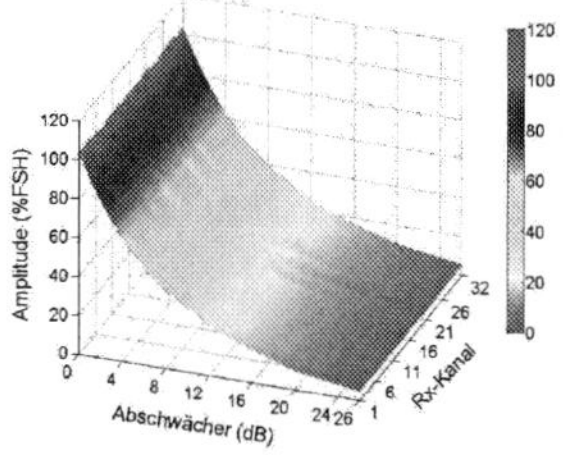

Abb8. Linearität der vertikalen Anzeige der PAUT Prüfelektronik 1x32/128

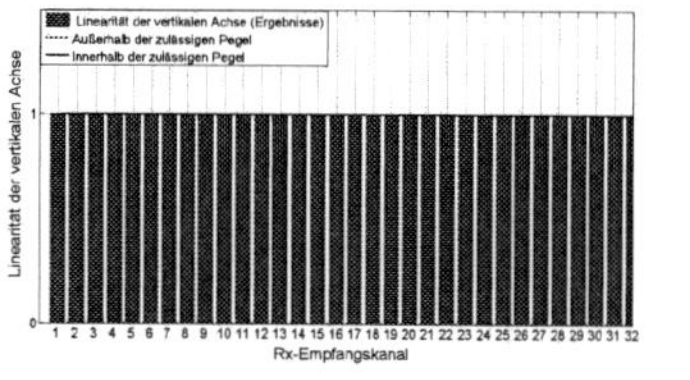

Abb9. Automatische Auswertung der Linearität der vertikalen Achse der PAUT-1x32/128

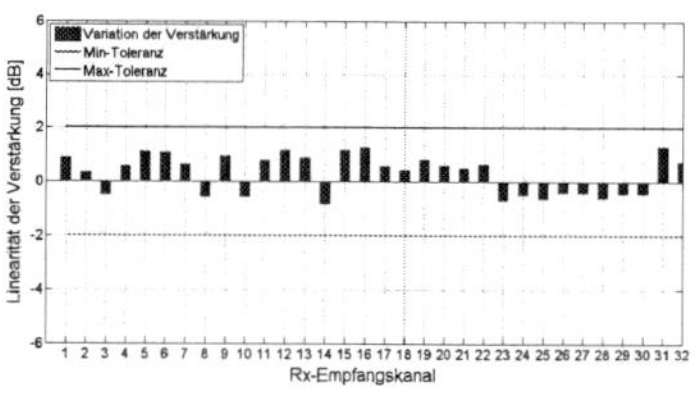

Abb10. Gemessene Linearität der Verstärkung der PAUT-Prüfelektronik 1x32/128

Kontakt: Dr.-Ing. Sanjeevareddy Kolkoori
UT Level 3
Tel.: +49-60239199906, E-Mail: skolkoori@rosen-nxt.com
ids@rosen-nxt.com | www.rosen-nxt.com/ids
Poster mit Kurzpräsentation, Mo.4.A-Poster Slam, DGZfP-Jahrestagung in Berlin, Mai 2025

DGZfP-Jahrestagung 2025

IWEX PORTABLE
Der Traum eines jeden ZfP-Prüfers

Thomas Hud [2]
[1] Applus RTD Deutschland Inspektionsgesellschaft mbH
, **44894 Bochum Deutschland**
www.applus.com

Kurzfassung

Wenn ein Ultraschallprüfer von „Bildgebung" spricht, meint er meist Darstellungen wie B-Scans, C-Scans, P-Scans, ToFD-Bilder oder Phased-Array-Sektorscans. Diese Bilder vermitteln zwar einen Eindruck realer Strukturen – zum Beispiel von einem Schweißfehler, sie zeigen jedoch keine exakte physikalische Abbildung.
Eine echte Darstellung müsste die genaue Position, Orientierung, Form und Größe des Fehlers in Bezug zum Werkstück zeigen – in 2D oder sogar 3D. Und das möglichst ohne Verzerrungen durch technische Einschränkungen des Prüfverfahrens.
Ein gutes Beispiel für eine solche realitätsnahe Abbildung ist die Computertomographie. Auch 3D-Bilder ungeborener Kinder im Mutterleib zeigen, was heute medizinisch möglich ist. Übertragen auf die industrielle Prüfung, wäre das Ziel, solche Bildqualität mit der Geschwindigkeit moderner Ultraschallanlagen zu kombinieren.
Das ist der Traum vieler ZfP-Prüfer.
Applus+ hat diesen Traum mit dem IWEX 3D Imager Wirklichkeit werden lassen. Dieses neue Verfahren ermöglicht eine besonders realitätsnahe Darstellung von Schweißfehlern. Es erlaubt eine eindeutige Charakterisierung und präzise Größenbestimmung von Anzeigen.
Die einzige Grenze liegt in der physikalischen Wellenlänge des Ultraschalls. Damit erreicht die zerstörungsfreie Prüfung eine neue Stufe – näher an der Realität als je zuvor.

Keywords:

- Inverse Wave field Extrapolation (IWEX)
- 3D-Darstellung
- Total Focusing Method (TFM) mit bis zu 20 Signalpfade
- Schweißnahtprüfung
- Inspektion von Guss- und Schmiedeteilen
- Prüfung sicherheitskritischer Komponenten

Drei Jahrzehnte Generalisierte Punktquellensynthese – Rück- und Ausblick

Martin SPIES [1]
[1] Baker Hughes Digital Solutions GmbH, Frankfurt am Main
E-Mail: martin.spies@bakerhughes.com

Kurzfassung. Simulationsverfahren im Bereich der Ultraschallprüfung helfen beim Verständnis komplexer Wellenausbreitungsphänomene, bei der Optimierung von Prüfparametern und bei der Interpretation experimenteller Ergebnisse. Um rechenintensive Simulationen mittels numerischer Verfahren zu umgehen, wurde das halb-analytische Verfahren der Generalisierten Punktquellensynthese implementiert. Dieser Beitrag gibt eine kurze Einführung in die theoretischen Grundlagen des Verfahrens und präsentiert ausgewählte Anwendungen der Methode im Bereich der Prüfkopf-Optimierung und der Simulation von Prüfvorgängen.

Keywords: Modellierung, Simulation, Ultraschall, Schallfeld, Prüfkopf, Phased-Array, Fehlerprüfung

Einführung

Die Simulation zerstörungsfreier Prüfverfahren, lange Jahre den Forschungsinstitutionen vorbehalten, hat inzwischen eine beträchtliche Bedeutung für praktische Anwendungen erlangt. Dies betrifft verschiedene Aspekte, wie etwa die Optimierung von Sensoren und die Durchführung von virtuellen Inspektionen zur Prüfvorbereitung und -optimierung. Auf der Basis geeigneter physikalischer Modelle helfen Simulationsverfahren im Bereich der Ultraschallprüfung beim Verständnis komplexer Wellenausbreitungsphänomene, bei der Optimierung von Prüfparametern und bei der Interpretation experimenteller Ergebnisse.

Um rechenintensive Simulationen mittels numerischer Verfahren zu umgehen, wurde das halb-analytische Verfahren der Generalisierten Punktquellensynthese implementiert und über die Jahre weiterentwickelt. Wesentliche Anwendungen betreffen die Wellenausbreitung in isotropen und anisotropen Medien und den Einsatz von Ein- und Mehrelement-Prüfköpfen. Zur Simulation der Fehlerprüfung wurde ein Ultraschall-Fehler-Wechselwirkungsmodell auf Basis der Kirchhoff-Näherung implementiert.

Dieser Beitrag gibt eine kurze Einführung in die theoretischen Grundlagen des Verfahrens und präsentiert ausgewählte Anwendungen der Methode im Bereich der Prüfkopf-Optimierung und der Simulation von Prüfvorgängen. Im Hinblick auf die quantitative Simulation der Fehlerprüfung an Bauteilen mit ebenen sowie gekrümmten Oberflächen werden Ergebnisse präsentiert, die im Rahmen eines internationalen Benchmarks erarbeitet wurden. Da es heutzutage sehr populär geworden ist, sich auf kommerzielle Simulationswerkzeuge zu verlassen, soll dieser Beitrag nicht zuletzt auf die Notwendigkeit aufmerksam machen, die Kenntnis und die Verfügbarkeit der theoretischen Grundlagen der Ultraschallsimulation in der ZfP-Community aufrechtzuerhalten.

https://doi.org/10.58286/31232

1. Theoretische Grundlagen des Verfahrens

Eine ganze Reihe von Verfahren stehen im Bereich der Ultraschallsimulation zur Verfügung, die in Abbildung 1 schematisch aufgelistet sind. Diese können gruppiert werden in Strahlverfolgungsverfahren (‚Ray-Tracing') sowie (halb-)analytische und numerische Verfahren. Durch die Anwendung von Näherungslösungen und anschließende Validierung lässt sich die Effektivität semi-analytischer Verfahren zusätzlich steigern. So wurde die Generalisierte Punktquellensynthese (GPSS) für isotrope und anisotrope Medien zur Schallfeldberechnung und Prüfkopfoptimierung [1,2] sowie zur Simulation der Ultraschall-Fehlerprüfung [3] entwickelt.

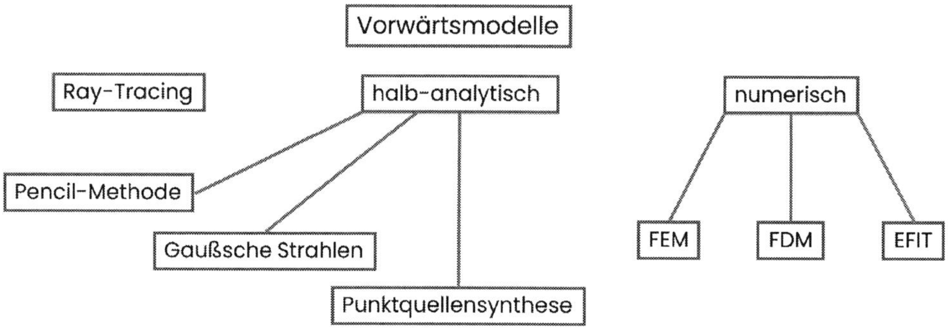

Abb. 1. Verschiedene Verfahren der Vorwärtssimulation - schematischer Überblick. Die numerischen Verfahren umfassen die Finite-Elemente- und die Finite-Differenzen-Methode sowie die Elastodynamische Finite Integrationstechnik.

Den physikalischen Hintergrund der GPSS-Methode bildet das Huygenssche Prinzip: Jeder Punkt einer Wellenfront ist Startpunkt einer Elementarwelle, die neue Wellenfront entsteht durch Überlagerung aller Elementarwellen. Deren Ausbreitungsfunktionen werden abhängig von den jeweiligen physikalischen Randbedingungen mit den entsprechenden Punktrichtwirkungen versehen. Die verschiedenen physikalischen Prozesse, die mit dem gestellten Prüfproblem verbunden sind, müssen entsprechend modelliert werden. Bei der Schallfeldberechnung sind dies (i) die Abstrahlung der Ultraschallwellen durch Ein- oder Mehrelement-Prüfköpfe unterschiedlicher Geometrien sowie die Ausbreitung der Wellen im jeweiligen Medium und (ii) der Reflexions- und Transmissionsprozess an der Grenzfläche zwischen Vorlaufmedium und Bauteil beim Einsatz von Winkelprüfköpfen oder bei Anwendung der Immersionstechnik.

Abbildung 2 zeigt den formalen Weg zur Herleitung der Oberflächenintegrale für die Schallfeldberechnung einerseits und für die Berechnung von Fehlersignalen andererseits. Ausgangspunkt ist die mathematische Formulierung des Huygensschen Prinzips unter Berücksichtigung der dyadischen und triadischen Greenschen Funktionen $\mathbf{G}$ und Σ, welche die Übertragungseigenschaften des jeweiligen Mediums charakterisieren [4]. Durch Berücksichtigung der Spannungsfreiheit an der Material- bzw. der Fehleroberfläche ergeben sich die entsprechenden Oberflächenintegrale, welche die gleiche Struktur aufweisen, lediglich die eingehenden physikalischen Größen sind unterschiedlich. Weitere Details sind in den Referenzen [1-3] zu finden.

Die numerische Berechnung dieser Oberflächenintegrale erfolgt unter Beachtung des Sampling-Theorems sukzessiv auf der Basis äquidistanter Gitter, mit denen die Prüfkopfapertur, die Grenzflächen zwischen zwei Medien sowie die Fehleroberfläche diskretisiert werden. Die transiente Modellierung von Zeitsignalen erfolgt durch Berechnung der harmonischen Wellenfelder für viele Frequenzen und deren anschließende numerische

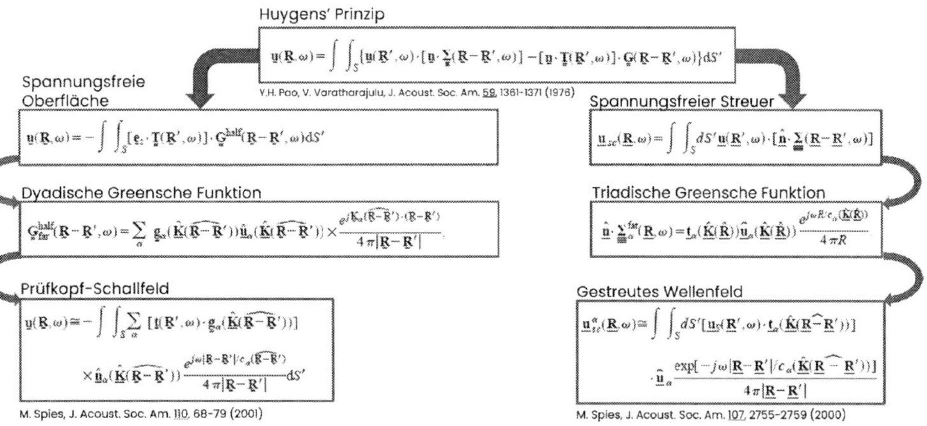

Abb. 2. Der Weg zur Generalisierten Punktquellensynthese ausgehend von der mathematischen Formulierung des Huygensschen Prinzips.

Fourier-Transformation in den Zeitbereich. Dazu wird eine geeignete Spektralfunktion zur Charakterisierung des Prüfkopfsignals herangezogen. Alternativ kann die Berechnung auch im Zeitbereich erfolgen, dazu wird eine entsprechende Impuls-Funktion verwendet.

2. Ein ausgewähltes Beispiel der Validierung

Die Validierung der Schallfeldberechnung sowie der Berechnung von Fehlersignalen auf Basis der Kirchhoff-Näherung erfolgte für zahlreiche Anwendungsfälle sowohl für isotrope als auch anisotrope Materialien. Als repräsentatives Beispiel betrachten wir hier die Validierung anhand des internationalen UT-Benchmarks 2007 [5]. Abbildung 3 illustriert das gestellte Prüfproblem. Unter Einsatz eines sphärisch fokussierenden Prüfkopfes (Mittenfrequenz 5.4 MHz, Durchmesser 19.05 mm) werden mit einem Wasservorlauf von 150 mm Longitudinalwellen in einen Testblock eingeschallt. Experimentell wurden 5 Testblöcke mit ebener sowie konvexer und konkaver Oberfläche unterschiedlicher Krümmungsradien (2 und 8 Zoll) untersucht, in denen in Tiefen von 3.2 mm (1/8 Zoll) bis 102.4 mm (4 Zoll) Flachbodenbohrungen (FBB) von 0.8 mm Durchmesser eingebracht worden waren[1].

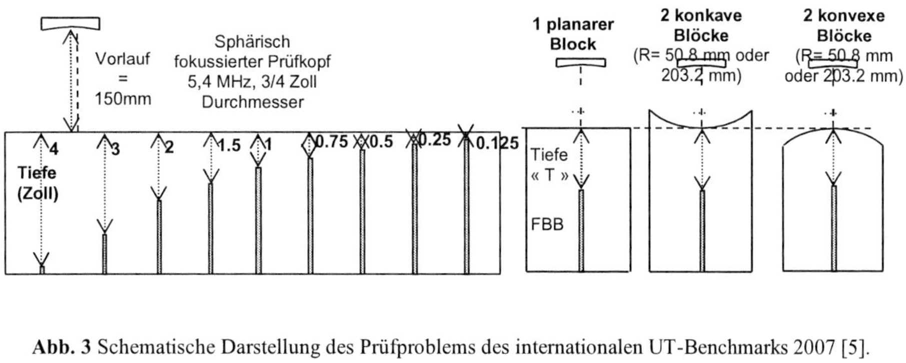

Abb. 3 Schematische Darstellung des Prüfproblems des internationalen UT-Benchmarks 2007 [5].

[1] Die Experimente wurden vom Laboratoire d'Intégration des Systèmes et des Technologies LIST der französischen Atomenergiebehörde CEA an entsprechenden Testkörpern aus Aluminium (Dichte 2.75 g/cm^3, V_{long} = 6416 m/s, V_{trans} = 3163 m/s) durchgeführt.

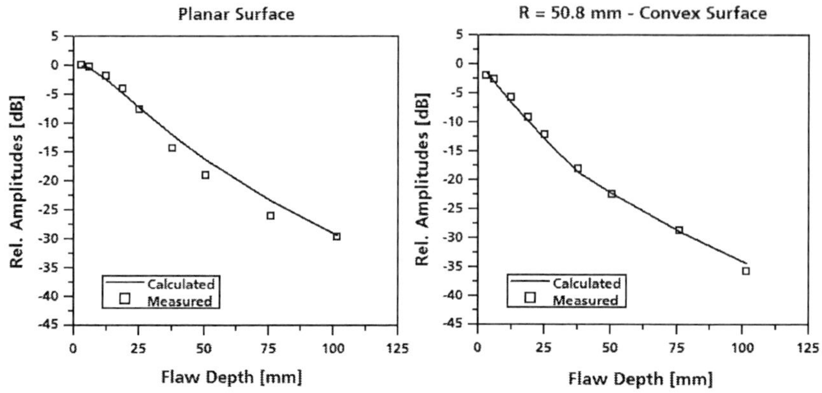

Abb. 4 Simulierte vs. gemessene relative Maximalamplituden als Funktion der Fehlertiefe im ebenen (links) und konvexen (rechts) Testblock (normiert auf die FBB in 3.2 mm Tiefe im ebenen Testblock).

Abbildung 4 zeigt exemplarisch die mittels Generalisierter Punktquellensynthese berechneten Fehleramplituden im Vergleich zu den experimentell ermittelten Maximalamplituden für die Testkörper mit ebener Oberfläche sowie der stark gekrümmten konvexen Oberfläche. Die Normierung erfolgte jeweils auf die FBB in 3.2 mm Tiefe im planaren Testblock. Die Übereinstimmung ist hervorragend, die maximale Abweichung von etwa 2 dB (FBB in 76 mm Tiefe, Abb. 4 links) liegt in dem üblicherweise zu erwartenden Bereich von 1 bis 2 dB, vorausgesetzt die der Berechnung zugrundeliegenden Näherungsannahmen liegen im Gültigkeitsbereich der jeweiligen Prüfapplikation. Für die beiden Testkörper mit konkaver Oberfläche ergaben sich für die beiden Fehler in 76 und 102 mm Tiefe größere Abweichungen, die in [5] näher kommentiert sind.

3. Optimierung von Sensoren und Verfahren für spezielle Prüfanwendungen

3.1 Optimierung von Phased-Array-Prüfköpfen

Im Vordergrund der Anwendungen der GPSS-Methode stand von jeher das Thema Sensor-Optimierung, insbesondere die Auslegung und Optimierung von Phased-Array-Prüfköpfen. So wurden erste Optimierungen speziell für den Einsatz bei der kerntechnischen Inspektion im Hinblick auf eine verbesserte Abdeckung des Prüfbereiches durchgeführt [6]. In [7] wird über das Design eines multifunktionalen Array-Prüfkopfes berichtet, welches die Anregung dreier Wellenarten - Longitudinal-, Transversal- und Oberflächenwelle (‚Sub-Surface Longitudinal Wave') - in unterschiedlichen Tiefenbereichen von Rohrleitungen ermöglicht. Ein als Array betriebener elektromagnetisch-akustischer Prüfkopf (‚EMAT') wurde in [8] hinsichtlich der Anzahl der anzusteuernden Segmente und deren Unterelemente mit dem Ziel der Nebenkeulenunterdrückung optimiert. Die gezielte Optimierung eines Phased-Array-Sensors für die Pipeline-Inspektion adressierte insbesondere die Unterdrückung der Nebenkeulen höherer Ordnung (‚Grating Lobes'), um das damit verbundene Problem der Aufzeichnung von Scheinanzeigen zu beheben [9,10].

Im Folgenden betrachten wir unkonventionelle Applikationsbeispiele aus dem Bereich der Inspektion von Triebwerksbauteilen, bei denen der Einsatz der Generalisierten Punktquellensynthese zu einer deutlichen Verbesserung der Prüfempfindlichkeit geführt hat.

3.2 Optimierung von Prüfköpfen für stark gekrümmte Bauteile

Das Bauteil und die Prüfanordnung, anhand derer der Optimierungsprozess illustriert wird, sind in Abbildung 5 schematisch dargestellt. Das zylindrische Bauteil hat einen Bohrungsinnendurchmesser von 34 mm. Die Prüfung soll mittels Immersionstechnik von innen erfolgen, mit der höchsten Empfindlichkeit in einem Bereich von 4 bis 8 mm Tiefe. Durch den Einfluss der starken Innenkrümmung ergibt sich bei Einsatz eines konventionellen Prüfkopfes neben dem natürlichen ein zweiter Fokusbereich und damit eine inhomogene Verteilung der Schallenergie über einen unerwünscht großen Bereich.

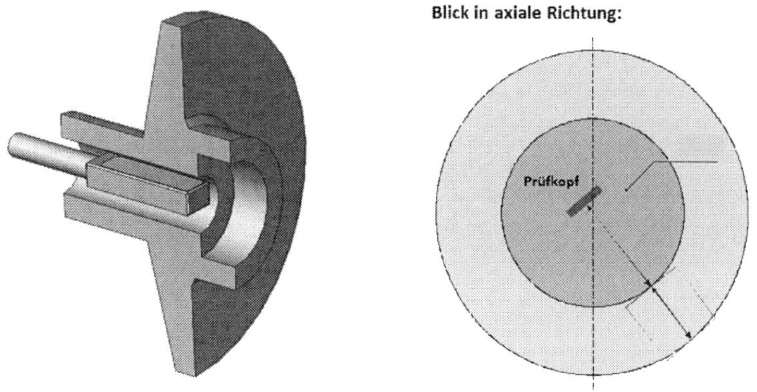

Abb. 5 Schematische Darstellung der zu prüfenden Triebwerksscheibe in Vordrehkontur mit Prüfkopf in der Innenbohrung (links); der zu bewertende Prüfbereich ist rechts gezeigt (blaues Quadrat).

Um die Schallenergie im spezifizierten Prüfbereich zu fokussieren und diesen somit optimal abzudecken, wurde folgende Prozedur zur Ermittlung einer speziellen Schwingergeometrie angewandt [11]:

- Eine zirkulare oder rechteckige, ebene Prüfkopfapertur wird als zweidimensionales Array von Punktquellen betrachtet, die äquidistant angeordnet sind.
- Unter Berücksichtigung der speziellen Prüfkonfiguration, d.h. der Eigenschaften des Vorlaufmediums und des Bauteils sowie der geometrischen Parameter (z.B. Wasservorlaufstrecke, Bauteilabmessungen und -geometrie) werden die Verzögerungszeiten für jede Punktquelle ermittelt, die erforderlich sind, um das Schallfeld auf den spezifizierten Tiefenbereich zu fokussieren.
- Die auf Basis des Fermat-Prinzips ermittelten Verzögerungszeiten werden dann in Weglängen konvertiert. Diese charakterisieren die Form des Piezoelements, welches das Schallfeld in der spezifizierten Tiefe fokussiert.

Mittels dieser Vorgehensweise und durch Variation der Frequenz und der Dimensionierung des Piezoelements wurde ein Prüfkopf erarbeitet, der auf eine Tiefe von 6 mm bei einer Wasservorlaustrecke von 15 mm fokussiert. Die Mittenfrequenz beträgt 10 MHz und die Abmessung 7 x 7 mm^2. Die Rechteckapertur besitzt eine konvexe Krümmung mit Radius 80 mm in der Radialebene und eine konkave Krümmung mit Radius 40 mm in der Axialebene, der Schwinger ist also hyperboloidisch geformt.

Abbildung 6 zeigt das für diese Konfiguration berechnete Schallfeld in der Axial- und der Radialebene im Vergleich mit dem Schallfeld des bis dato eingesetzten zylindrisch gekrümmten, bi-fokussierenden Prüfkopfes (Frequenz 15 MHz, Durchmesser 9 mm). Das Schallfeld des optimierten Sensors zeigt eine Amplitudenvariation von 0 bis -6 dB im avisierten Fokusbereich von 4 bis 8 mm, die Maximalamplitude beträgt etwa das Fünffache des ursprünglich eingesetzten Prüfkopfes, womit eine wesentliche Steigerung der Prüfempfindlichkeit erzielt wurde.

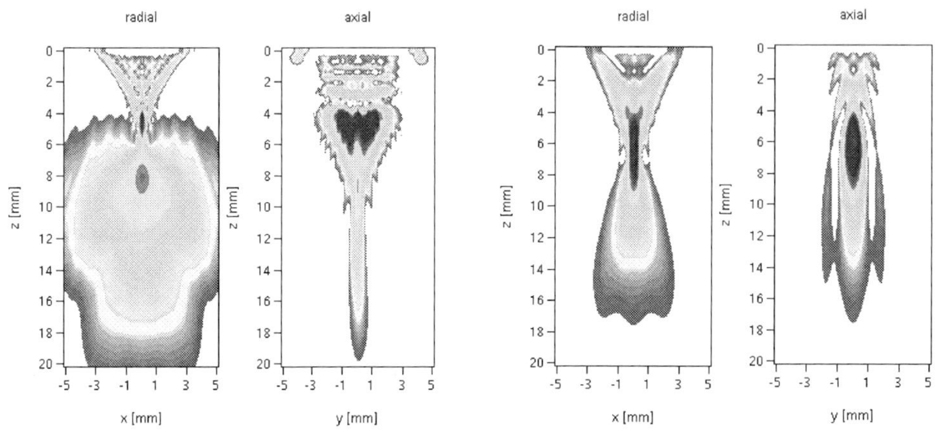

Abb. 6 Schallfelder des zylindrisch gekrümmten bi-fokussierenden Prüfkopfes und des optimierten fokussierenden Prüfkopfes in der Axial- und der Radialebene.

3.3 Optimierung von Umlenkspiegeln für die Ultraschallprüfung

Bei der Ultraschall-Inspektion von Bauteilen mit komplexer Geometrie besteht die Möglichkeit, dass der interessierende Prüfbereich nur mit Einschränkungen zugänglich ist. In solchen Fällen hat sich der Einsatz von Spiegeln bewährt, die eine Umlenkung des Ultraschallfeldes auf die Bauteiloberfläche erlauben. In diesem Beispiel betrachten wir die Optimierung eines Umlenkspiegels mit dem Ziel, eine verbesserte Prüfung des Fügezonenbereiches eines sogenannten Blisks (‚Bladed Disk') mittels Fokussierung des Schallfeldes durch die Geometrie des Spiegels zu ermöglichen [12]. Bei dem betrachteten Bauteil kommt das Fügeverfahren des linearen Reibschweißens zum Einsatz, um die Schaufeln an die Triebwerksscheibe anzufügen. Abbildung 7 zeigt den Niederdruckkompressor eines Triebwerks mit der Anordnung dreier Blisks sowie eine Illustration des Fügeprozesses.

Abb. 7 Niederdruckkompressor (links) und Illustration des linearen Reibschweißprozesses.

Der für die Ultraschallprüfung interessante Bereich ist in Abbildung 8 dargestellt, die einen Querschnitt der Fügezone zeigt sowie einen Ausschnitt mit dem Koordinatensystem, das für die nachfolgenden Erwägungen genutzt wurde. Die Prüfung wird in Immersionstechnik durchgeführt unter Einsatz eines Spiegels, um das einfallende Schallfeld auf die Bauteiloberfläche (x-y-Ebene) umzulenken. Der Spiegel wird unter einem Winkel von 45° getroffen, sodass eine Longitudinalwelle unter 41° im Fügebereich erzeugt wird.

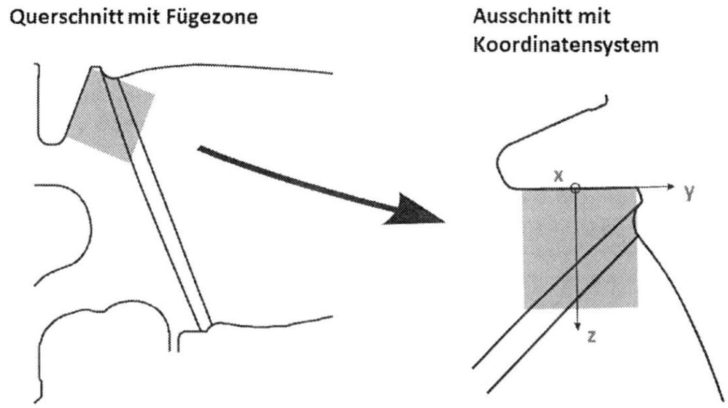

Abb. 8 Darstellung der Bauteilgeometrie im interessierenden Prüfbereich.

Die Berechnung des Schallfeldes erfolgt in zwei Schritten: zunächst wird die von einem ebenen Prüfkopf auf der Spiegeloberfläche erzeugte Teilchenverschiebungsverteilung berechnet, im zweiten Schritt wird dann der Spiegel wie ein Prüfkopf behandelt, der in das Bauteil einschallt. Um eine Fokussierung in einer Tiefe von 20 mm durch einen entsprechend geformten Spiegel zu erreichen, wird die in Abschnitt 3.2 beschriebene Prozedur angewandt, wobei in diesem Fall der Spiegel als Array von Punktquellen betrachtet wird. Die Form des Spiegels ergibt sich dann durch Konvertierung der zur Fokussierung notwendigen Verzögerungszeiten in Weglängen. Die entsprechende geometrische Projektion, die von der Orientierung des Spiegels abhängt, wurde für einen Abstand des Prüfkopfes zum Spiegel und des Spiegels zur Bauteiloberfläche von jeweils 30 mm berechnet.

Folgende Parameter wurden für den Rechteckspiegel ermittelt: (i) Abmessung15 x 15 mm^2, (ii) konkave Krümmung in der x-z-Ebene, Radius 1000 mm, (iii) konkave Krümmung in der y-z-Ebene, Radius 650 mm. Da die ermittelte Krümmung in der x-z-Ebene sehr gering ist (was auf die natürliche Fokussierung durch die Bauteilkrümmung zurückzuführen ist), wurde diese bei der Fertigung des Spiegels vernachlässigt. In der Tat zeigten die Simulationsergebnisse, dass lediglich vernachlässigbare Verbesserungen sowohl bei der Maximalamplitude als auch der Schallfeldbreite durch diese zusätzliche Krümmung erzielt werden. Daher wurde ein zylindrisch gekrümmter Spiegel mit einem Radius von 650 mm in der y-z-Ebene gefertigt und getestet. Der Schallfeldberechnung sagte eine Erhöhung der Maximalamplitude von ca. 3 dB und eine Verringerung der Schallfeldbreite im Fokusbereich von etwa 70 % im Vergleich zum Einsatz eines ebenen Spiegels voraus.

Das ermittelte zylindrische Profil des Spiegels wurde mittels Präzisionsdrahterodieren einer flachen rechteckigen Messingplatte gefertigt. Abbildung 9 zeigt den eingesetzten ebenen Standardsensor (Frequenz 25 MHz, Durchmesser 12.7 mm) mit Halterung und dem gefertigten Spiegel. Um den fokussierenden mit dem bisher eingesetzten ebenen Spiegel zu vergleichen, wurde ein Testblock aus Titan-6-4 mit einer 0.2 mm Flachbodenbohrung in einer Tiefe von 10 mm unter der Oberfläche untersucht. Die A-Scans zeigten eine Erhöhung der Echoamplitude der 0.2 mm FBB um ca. 3 dB bei Einsatz des modifizierten Spiegels. Die durch die Fokussierung erzielte schärfere Fehlerabbildung ist in der C-Scan-Darstellung zu erkennen (Abb. 10).

Schließlich wurde die Prüfkopf-Spiegel-Konfiguration an einem Blisk-Dummy mit erodierten Nuten in der Fügezone getestet, um die Auflösungsgrenzen dieser Prüftechnik zu untersuchen. Abbildung 11 zeigt auf der linken Seite den Blisk-Dummy im Wassertank, zusammen mit der Sensoreinheit an der interessierenden Prüfposition; auf der rechten Seite sind die Prüfergebnisse dargestellt. Im vergrößert dargestellten C-Scan-Bereich sind klare Anzeigen der erodierten Nuten der Größe 0.4 x 0.4 mm^2 zu sehen, die mit einem Signal-

Rausch-Verhältnis von größer 6 dB detektiert wurden. Diese hohe Auflösung in einem Prüfbereich mit eingeschränkter Zugänglichkeit an einem Bauteil mit komplexer Geometrie unter Einsatz eines Standardprüfkopfes wurde durch die simulationsbasierte Optimierung eines problemlos zu fertigenden Umlenkspiegels erreicht.

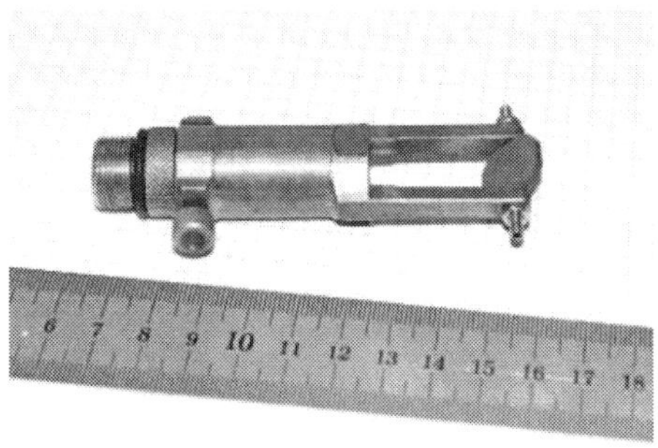

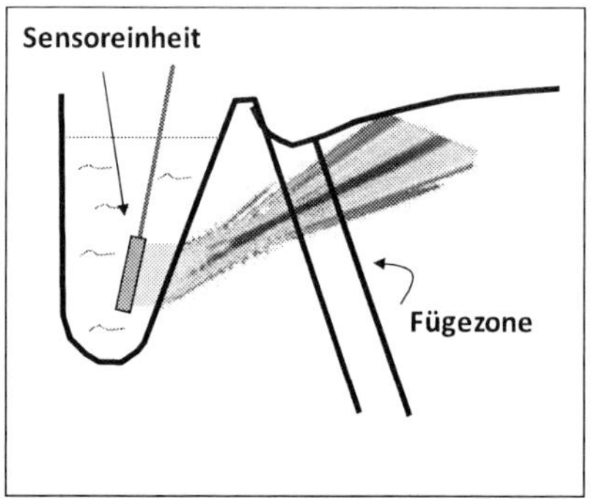

Abb. 9 Sensoreinheit mit Prüfkopf, Halterung und Umlenkspiegel (links), schematische Darstellung der Prüfanordnung (rechts).

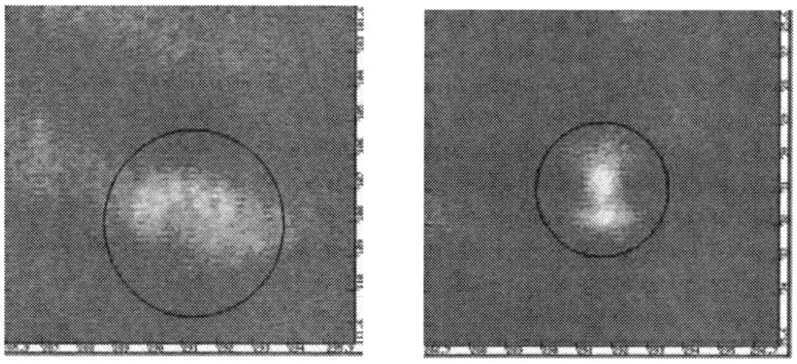

Abb. 10 C-Scan-Darstellung der Anzeige der 0.2 mm FBB unter Einsatz des ebenen (links) und des optimierten Umlenkspiegels (rechts).

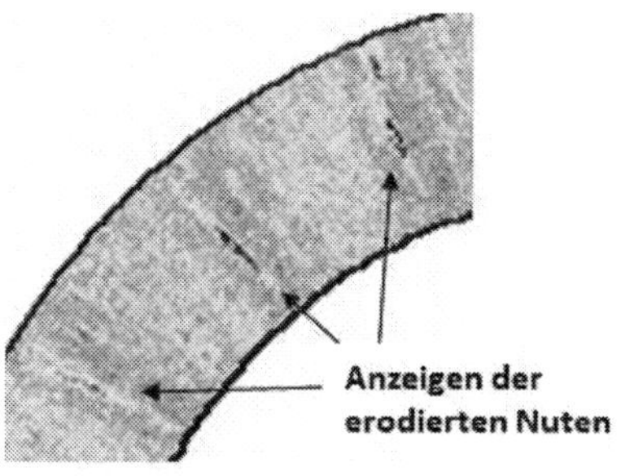

Abb. 11 Prüfanordnung mit Blisk-Dummy im Wasserbad (links); C-Scan mit Nutanzeigen (rechts).

4. Rück- und Ausblick

Ausschlaggebend für die Implementierung der Generalisierten Punktquellensynthese vor mehr als 3 Jahrzehnten waren der verstärkte Einsatz von anisotropen Faserverbund-werkstoffen und die bei deren Prüfung zu berücksichtigenden komplexen Ausbreitungseigenschaften des Ultraschalls einerseits sowie generell die zunehmend anspruchsvollen Prüfanwendungen mit der einhergehenden Notwendigkeit der gezielten Prüfkopfoptimierung. Zu etwa der gleichen Zeit, Mitte der 1990er Jahre, wurden kommerzielle Simulationsverfahren für den Bereich der Ultraschallprüfung verfügbar. Heutzutage wird der Einsatz kommerzieller Simulationssoftware immer beliebter, allerdings in der Regel, ohne sich mit den in diesen Tools implementierten Modellen und deren Gültigkeitsbereich näher zu befassen. Ein zuverlässiger Einsatz dieser Software erfordert jedoch die Kenntnis der zugrundeliegenden Ultraschallphysik sowie der Grenzen der jeweiligen implementierten (Näherungs-) Verfahren. Vor diesem Hintergrund gilt es, die Kenntnis und die Verfügbarkeit der theoretischen Grundlagen der Ultraschallsimulation in der ZfP-Community aufrechtzuerhalten.

Danksagung

Die beiden Optimierungsbeispiele aus dem Bereich der Inspektion von Triebwerksbauteilen wurden im Rahmen meiner Tätigkeit am Fraunhofer-Institut für Zerstörungsfreie Prüfverfahren IZFP in Kooperation mit der MTU Aero Engines AG, München, erarbeitet. Mein besonderer Dank gilt Herrn Dr. Joachim Bamberg, inzwischen im Ruhestand, für die hervorragende, freundschaftliche Zusammenarbeit und die Gelegenheit, innovative Prüfansätze unter Einsatz des GPSS-Verfahrens zu erarbeiten.

Referenzen

[1] M. Spies; Transducer-Modeling in General Transversely Isotropic Media via Point-Source-Synthesis. Theory; J. Nondestr. Eval., Vol. 13 (1994), pp. 85-99

[2] M. Spies; Semi-Analytical Elastic Wavefield Modeling Applied to Arbitrarily Oriented Orthotropic Media; J. Acoust. Soc. Am., Vol. 110 (2001), pp. 68-79

[3] M. Spies; Kirchhoff Evaluation of Scattered Elastic Wavefields in Anisotropic Media; J. Acoust. Soc. Am., Vol. 107 (2000), pp. 2755-2759

[4] Y.H. Pao, V. Varatharajulu; Huygens' Principle, Radiation Conditions and Integral Formulas for the Scattering of Elastic Waves; J. Acoust. Soc. Am., Vol. 59 (1976), pp. 1361-1371

[5] M. Spies; Validierung semi-analytischer Simulation der Fehlerprüfung an ebenen und gekrümmten Bauteilen anhand des internationalen UT-Benchmarks 2007; DGZfP Berichtsband BB 113-CD, DACH-Jahrestagung 2008, P35

[6] M. Spies, M. Kröning; 3D-Modellierung der Abstrahlung, Ausbreitung und Beugung elastischer Wellen mittels Generalisierter Punktquellensynthese (GPSS); in: Fortschrittliche ZfP - ein Instrument für Ökologie und Ökonomie, DGZfP Jahrestagung 1995, DGZfP-Berichtsband 43, Berlin, pp. 443-450

[7] M. Spies, W. Gebhardt, M. Kröning, M. Dalichow; Design and Optimization of a Multifunctional Phased Array Search Unit; in: Review of Progress in Quantitative NDE, Vol. 15A, Plenum Press, New York (1996), pp. 1035-1042

[8] M. Spies, F. Walte, G. Hübschen; Einsatz der Generalisierten Punktquellensynthese zur Vorbereitung automatisierter Ultraschallprüfungen; DGZfP-Berichtsband 57, Zweites Deutsch-Französisches Seminar 'Automatisierung und Modellierung für die ZfP', Aachen (1997), pp. 39-48

[9] M. Spies, H. Rieder, I. Lachtchouk, M. Tschuch; Applikationsspezifische Optimierung eines Phased-Array Sensors am Beispiel der Pipeline-Inspektion; Berichtsband BB-171 DACH-Jahrestagung (2019), DGZfP, Berlin, Di.1.A.2, ISBN 978-3-947971-02-2

[10] M. Spies, O. Müller, I. Lachtchouk, M. Tschuch; Untersuchung und quantitative Bewertung der Leistungsfähigkeit eines neuen Phased-Array-Sensors für die Pipeline-Inspektion; Berichtsband BB-176 DGZfP Jahrestagung (2021), DGZfP, Berlin, Manuskript 1, ISBN 978-3-947971-18-3

[11] M. Spies; Efficient Optimization of Single and Multiple Element Transducers for the Inspection of Complex-Shaped Components; NDT&E International, Vol. 37 (2004), pp. 455-459

[12] M. Spies, J. Bamberg; Advanced Ultrasonic NDT of Aero Engine Components Using Standard Transducers and Optimized Mirrors; DGZfP-Berichtsband 103-CD, 9th European Conference on NDT, 25.-29. September 2006, Berlin

Ultraschallprüfung von Buckelschweißmuttern: Ein Vergleich zwischen labor- und praxisnaher Anwendung

Christian MATHISZIK [1], Martin WEBER [2], Franz HIELSCHER [1], Johannes, KOAL [1], Uwe FÜSSEL [1], Hans Christian SCHMALE [1], Jere HYVÖNEN [2], Ari SALMI [2]
[1] Technische Universität Dresden, Institut für Fertigungstechnik, Professur für Fügetechnik und Montage, Dresden, Germany, christian.mathiszik@tu-dresden.de
[2] University of Helsinki, Electronics Research Laboratory, Helsinki, Finland, martin.weber@helsinki.fi

Kurzfassung. Das Buckelschweißen ist ein sehr effizientes Schweißverfahren, welches mit einem hohen Automatisierungsgrad angewendet werden kann. Es wird in verschiedenen Branchen zur Herstellung u. a. von Automobilkarosserien, Schienenfahrzeugen, Küchengeräten sowie in der Elektroindustrie eingesetzt. Die Industriezweige sehen sich mit einer Reihe von Herausforderungen konfrontiert, darunter steigende Sicherheitsanforderungen, zunehmende Materialvielfalt, sowie eine fortschreitend regulierte Ressourceneffizienz im Sinne der Kreislaufwirtschaft. Aus diesen Gründen wird eine zuverlässige Qualitätssicherung der Buckelschweißverbindungen immer wichtiger. Eine Sichtprüfung ist für die Qualitätssicherung meist nicht anwendbar, da die Schweißverbindungen von außen nicht sichtbar sind. Die Überwachung der Prozessparameter wird neben regelmäßigen zerstörenden Prüfungen zur Bewertung der Verbindungsqualität angewendet. Letzteres steht einer guten Ressourceneffizienz entgegen. Im Bereich der zerstörungsfreien Prüfverfahren (ZfP) wurde noch kein Industriestandard festgelegt. Naheliegende Prüfmethoden aus dem Bereich des Widerstandspunktschweißens, wie die manuelle Ultraschallprüfung, konnten sich bisher nicht etablieren. Die Gründe dafür liegen in der Vielfalt möglicher Buckelschweißverbindungen und der daraus hohen geometrischen Variabilität der zu prüfenden Bauteile.

Im Beitrag werden Untersuchungsergebnisse präsentiert, welche in labor- und praxisnahen Ultraschallprüfungen an Buckelschweißmuttern gewonnen wurden. Dabei werden mittels coded excitation scanning acoustic microscopy (CESAM) gewonnene Prüfergebnisse denen der manuellen Ultraschallprüfung gegenübergestellt und diskutiert. Mögliche Weiterentwicklungsschritte werden abschließend durch einen Vergleich mit den entsprechenden Ergebnissen der zerstörenden Prüfungen der Schweißverbindungen abgeleitet.

Keywords: Buckelschweißen, Widerstandsschweißen, ZfP, Ultraschall, CESAM

Einführung

Buckelschweißverbindungen werden in der blechverarbeitenden Industrie breit gefächert eingesetzt. Hierzu zählen der Fahrzeugbau, der Metallbau, der Gerätebau, der Maschinenbau, die Möbelindustrie und die Elektrotechnik. Die stoffschlüssigen Verbindungen werden durch Widerstandspressschweißverfahren hergestellt. Vorteilhaft sind die niedrigen Kosten und die

 https://doi.org/10.58286/31254

hohe Produktivität. Verfahrensbedingt entsteht die Schweißverbindung zwischen den Fügepartnern und ist von außen nicht sichtbar. Die visuelle Prüfung, die Magnetpulverprüfung oder die Farbeindringprüfung können daher nicht für die Qualitätsbewertung genutzt werden. Somit kommen nur Methoden der zerstörungsfreien Prüfung (ZfP) in Frage, die Informationen aus dem Inneren der Fügeverbindungen liefern. Dazu zählen u. a. die Durchstrahlungsprüfung, die Thermographie und die Ultraschallprüfung. Die Durchstrahlungsprüfung erfordert aufgrund der Strahlenbelastung einen hohen apparativen Aufwand und ist dadurch mit hohen Kosten verbunden. Für die Anwendung in der Serienfertigung von Buckelschweißverbindungen ist sie daher ungeeignet und wird nur bei sehr hochpreisigen und sicherheitsrelevanten Einzelteilen sowie in der Forschung und Entwicklung eingesetzt. Die Thermographie wurde für die Prüfung von Punktschweißverbindungen untersucht, hat sich aber wegen der unterschiedlichen Emissivität der Oberflächen, des Einflusses von Materialinhomogenitäten und der langen Messzeiten in der Serienfertigung nur vereinzelt durchgesetzt [1–3]. Die Ultraschallprüfung mit Bildgebung ist eine weit verbreitete Methode zur Prüfung von Punktschweißverbindungen [4–7]. Für die Auswertung wird u. a. die Laufzeit durch Falschfarbendarstellung ortsabhängig abgebildet [4, 6, 8–11]. Dieses Vorgehen kann zur Bestimmung der Anbindungsfläche beim Buckelschweißen eingesetzt werden, ist aber bisher nicht üblich. Als Gründe werden das fehlende Rückwandecho, Bohrungen in den Blechteilen (z. B. bei Schweißmuttern) und das aufzubringende Koppelmittel genannt. In [12] wird die Entwicklung und Anwendung eines mobil einsetzbaren Ultraschallverfahrens beschrieben, das auch für Buckelschweißverbindungen geeignet ist. Die Ergebnisse erscheinen vielversprechend und es wird über eine erfolgreiche Erprobung in der Serienfertigung berichtet, ohne dass hierzu weitere Veröffentlichungen vorliegen.

Zusammenfassend zeigt sich, dass sich bisher keine der genannten ZfP-Methoden für die Qualitätsbewertung von Buckelschweißverbindungen etablieren konnte. Es fehlen Grundlagenuntersuchungen.

Der vorliegende Artikel greift diese Wissenslücke auf und vergleicht Ergebnisse aus Laborversuchen mit praxisnahen Messungen mittels Ultraschallprüfung an Buckelschweißverbindungen. Dabei werden die gewonnen Prüfergebnisse der coded excitation scanning acoustic microscopy (CESAM) der manuellen Ultraschallprüfung gegenübergestellt und diskutiert. Mögliche Weiterentwicklungsschritte werden abschließend durch einen Vergleich mit den entsprechenden Ergebnissen der zerstörenden Prüfungen der Schweißverbindungen abgeleitet.

1 Werkstoffe und Versuchsmethodik

1.1 Werkstoffkombinationen

Die Untersuchungen werden an M6-Vierkantmuttern nach DIN 928-M6-ST und M8-Massebolzen mit drei massiven Segmentbuckeln durchgeführt (Abb. 1). Diese werden auf quadratische, zinkbeschichtete Bleche aus CR300LA+GI50/50 und 22MnB5+AS150 mit einer Blechdicken von jeweils 1,5 mm geschweißt. Die Bleche für die Vierkantmuttern besitzen ein zentrisches Loch mit einem Durchmessen von 7 mm. Die Kantenlänge aller Bleche beträgt 48 mm. Die Proben werden in einer Portalschweißanlage hergestellt, die über Stromquellen für das Gleichstrom-Mittelfrequenz- (MFDC) und für das Kondensatorentladungsschweißen (KE) verfügt. Beide Arten der Schweißstromquellen werden eingesetzt. Die Anlage hat den Vorteil, dass die unterschiedlichen Schweißsysteme MFDC und KE den gleichen mechanischen Aufbau zum Schweißen der Proben nutzen. Dies ermöglicht Vergleiche zwischen beiden Schweißsystemen. Es werden Elektroden nach ISO 5821-C0-20-22 der Legierungszusammensetzung CuCr1Zr eingesetzt. Für die M6-Vierkantmuttern hat die untere

Elektrode eine Bohrung von 7,0 mm für den Zentrierdorn zum sicheren und reproduzierbaren Positionieren der Schweißmuttern. Bei den Massebolzen besitzt die untere Elektrode eine Sacklochbohrung mit einem Durchmesser von 17 mm und einer Bohrungstiefe von 14 mm. Das Loch dient der sicheren und reproduzierbaren Positionierung der M8-Massebolzen und zur Aufnahme des Gewindebolzens. Es fließt kein Schweißstrom durch den Gewindebolzen, wodurch dieser vom Schweißprozess unbeeinflusst bleibt. Die Portalschweißanlage und der experimentelle Aufbau zum Herstellen der Buckelschweißproben ist in Abb. 2 zu sehen.

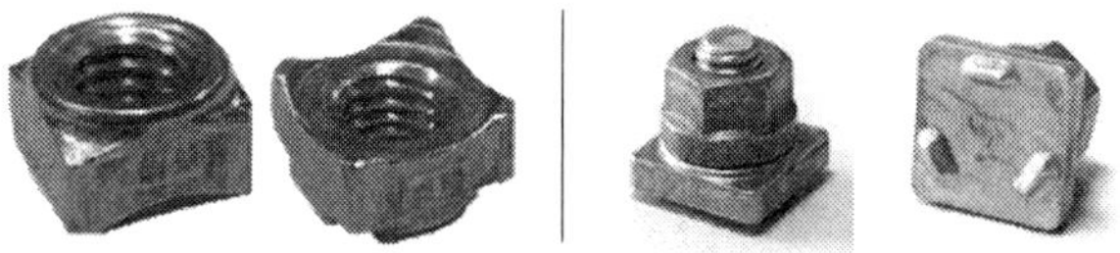

Abb. 1. M6-Vierkantmutter (DIN 928-M6-ST, links) und M8-Massebolzen (rechts)

Abb. 2. Schweißanlage zum KE- und MFDC-Schweißen als Portalsystem an der Technischen Universität Dresden (links) und detaillierter Darstellung des Arbeitsraums mit eingelegter M6-Vierkantmutter nach DIN 928-M6-ST und Blech (rechts)

Zum Herstellen unterschiedlicher Qualitätsstufen werden je Verbindungskombination zwei Parametersätze (PS1, PS2) gewählt. In Tabelle 1 sind die Versuchskombination mit den jeweiligen Schweißparametern zum Herstellen der Schweißproben dargestellt. Diese Festlegungen basieren auf zuvor ermittelten Schweißbereichen in [13]. Für die MFDC-Schweißungen wird eine Schweißzeit von 20 ms und eine jeweilige Vor- und Nachhaltezeit von 400 ms gewählt. Diese Parameter orientieren sich am IGF-Vorhaben Nr. 19.878 BR der Otto-von-Guericke-Universität Magdeburg [14]. Die KE-Schweißungen werden mit einer Kondensatorkapazität von 7,26 mF bei einem Übersetzungsverhältnis 1:20 mit Einzelpuls hergestellt. Die Schweißproben werden in randomisierter Reihenfolge der verschiedenen Parametersätze geschweißt. Dadurch werden äußere Einflüsse, wie Elektrodenverschleiß, oder Variation der Kühlwassertemperatur, geringgehalten. Es werden je Parametersatz 30 Proben hergestellt, aus denen je zwei Proben für die weiteren Untersuchungen dieses Beitrags zufällig ausgewählt werden.

Tabelle 1: Materialkombination mit Blech und aufgeschweißtem Fügeelement der einzelnen Versuchsreihen (VR) und Schweißparameter der Parameterstufen (PS)

R	Fügeelement (Festigkeit)	Blech (Blechdick, ZugfestigkeitR_m)	Kr aft	Sc hweißzeit	MF: Schweißstrom/ KE: Ladeenergie	
					P S1	P S2
KM1	M6- Vierkantmutter DIN 928-M6-ST (460 MPA)	CR300LA+GL50 /50 (1,5 mm, 447,6 MPa)	3,0 kN	20 ms	1 1 kA	1 7 kA
MK2	M6- Vierkantmutter DIN 928-M6-ST (460 MPA)	22MnB5+AS150 (1,5 mm, 447,6 MPa)	3,0 kN	20 ms	1 2 kA	1 8 kA
KM3	M6- Vierkantmutter DIN 928-M6-ST (460 MPA)	22MnB5+AS150 (1,5 mm, ~1500 MPa)	10, 0 kN	Ein zelpuls	1 ,10 kJ (550 V)	2 ,04 kJ (750 V)
B1	M8- Massebolzen (460 MPA)	CR300LA+GI50/ 50 (1,5 mm, 447,6 MPa)	6,0 kN	20 ms	1 2 kA	2 2 kA

1.2 Ultraschallprüfung

1.2.1 Laborbasierte Ultraschallprüfung mittels Ultraschallmikroskopie

Die Proben werden mittels Ultraschallmikroskopie zerstörungsfrei untersucht. Dazu wird ein coded excitation scanning acoustic microscope (CESAM) eingesetzt. Das Messsystem wird an der Universität von Helsinki entwickelt und betrieben. Detaillierte Information zu dem verwendeten Messsystem können der Literatur entnommen werden [15, 16]. Ein 30 MHz Schallwandler (Panametrics V375 30/.25 F=.75"OLF) wird im Aufbau verwendet. Das Anregungssignal ist ein Chirpsignal mit einer Gaußförmigen Hüllkurve (Startfrequenz: 10 MHz, Endfrequenz: 40 MHz, Dauer: 20 µs, linearer Frequenzanstieg). Die mittlere Wellenlänge liegt bei 60 µm. Mittels Pulskompression wird das empfangene reflektierte elektrische Signal in einen Impuls überführt, um eine hohe Zeitauflösung zu erhalten [15]. Während der Messung sind die Proben in deionisiertes Wasser eingetaucht, welches zur Schallübertragung notwendig ist. Die Fläche wird mit einer Schrittweite von 25 x 25 μm^2 erfasst. Die Messungen erfolgten blechseitig durch die 1,5 mm dicken Bleche hindurch. Die Fokusebene wird auf die Rückseite des Blechs gelegt, da sich in dieser Ebene auch die Kontaktzone der Schweißverbindung befindet. Die Amplitude des reflektierten Signals innerhalb der Fokuszone wird als C-Bild dargestellt. Beim Übergang vom Blech in das Wasser wird ein Teil des Ultraschallsignals reflektiert, was in den Diagrammen als grauer Hintergrund zu sehen ist. Lufteinschlüsse und andere Inhomogenitäten zeigen eine starke Reflektivität und sind als helle Stellen in den Diagrammen zu erkennen. Eine optimale Schweißverbindung zweier gleichartiger Materialien zeichnet sich durch eine geringe Reflektivität aus. Dementsprechend sind diese Bereiche dunkel dargestellt. Die Ermittlung der Anbindungsfläche erfolgt durch Schwellwertanalysen der erzeugten C-Bilder. Der Bereich des Lochs im Blech wir dabei ausgeblendet. Abb. 3 zeigt den Versuchsaufbau und den schematischen Ablauf zur Steuerung des CESAM.

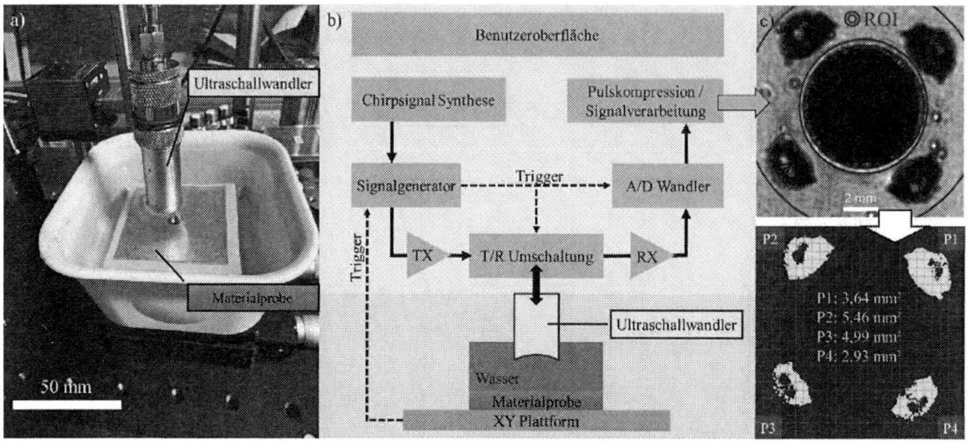

Abb. 3. In a) ist der Aufbau einer Messung dargestellt. Die Materialprobe ist in Wasser eingetaucht und der Schallwandler wird über die Probe geführt. In b) ist der schematische Aufbau des Mikroskops gezeigt. Die Steuerung erfolgt über die Benutzeroberfläche. Das gewünschte Chirpsignal wird von dem Signalgenerator erzeugt und elektrisch verstärkt. Anschließend wird es über einen Sende-/Empfangsumschalter zu dem Schallwandler geführt. Dort wird es in eine Ultraschallsignal umgesetzt, welche an der Probe teilweise reflektiert wird und anschließend wieder zum Schallwandler gelangt. Von dort gelangt es zum Empfangsverstärker und wird anschließend digitalisiert. Die Pulskompression wird mittels GPU-Beschleunigung durchgeführt. Aus der Positionsinformation der XY-Plattform wird ein Triggersignal abgeleitet, um eine synchrone Datenerfassung zu gewährleisten. c) Darstellung zur Ermittlung der Anbindungsfläche innerhalb des Analysebereichs (ROI) mittels Schwellwertanalyse

1.2.2 Praxisnahe Ultraschallprüfung mittels industriellen mobilen Ultraschallprüfgerät

Als Ultraschallprüfsystem wird das bildgebenden Ultraschallprüfsystem PHAsisBLU der Firma VOGT Ultrasonics GmbH eingesetzt. Der verwendete Prüfkopf ist mit einem US-Phased-Array-Wandler mit 121 Einzelschwingern, 1,0 mm Pitch, ausgestattet. Die Messungen erfolgen in einer Messvorrichtung. Durch diese Messvorrichtung wird der sonst manuell geführte Prüfkopf fixiert und mittels Linearführungen senkrecht auf die Schweißproben abgesenkt. Es wird eine Vorlaufstrecke aus Kunststoff verwendet. Die Schweißproben werden in einem Wasserbad (Koppelmittel: Wasser) positioniert. Das Wasserbad kann manuell verschoben werden. Dies ist notwendig, da die Fläche des Prüfkopfs nicht die gesamte Fläche der Fügeelemente/Verbindungen der Versuchsreihen mit einer Messung abdecken. Nach erfolgter Positionierung der Proben und Absenken des Prüfkopfs auf die Proben wird die Messung manuell ausgelöst. Der eigentliche Messprozess wird automatisiert durchgeführt. Damit ist es nicht von dem Anwender abhängig und liefert reproduzierbare und objektive Ergebnisse [17]. Jeder Schweißbuckel wird mit drei Wiederholungen von der Blechseite gemessen. Die Blechdicke bzw. die fügeelementseitige Ebene wird im System als Blende eingestellt. Reflexionen weisen daher auf nicht verbundene Bereiche hin. In verschweißten Bereichen kommt es nicht zu Reflexionen aus der Ebene. Dies betrifft auch den Bereich des Lochs im Blech. Dies muss daher ausgeblendet werden. Das Prüfsystem und die zugehörige Auswertesoftware ist speziell für Punktschweißverbindungen entwickelt. Sie kann daher nur bedingt für die Bestimmung der Anbindungsflächen genutzt werden. Daher werden die vom Prüfsystem erzeugten D-Bilder mit einer selbstentwickelten Auswertung bewertet. Abb. 4 zeigt links die Prüfvorrichtung und rechts eine beispielhafte Darstellung der Flächenbestimmung. Die Flächen werden nur im Analysebereich (ROI) ermittelt. Dafür erfolgt eine Kantendetektion der D-Bilder. Die innenliegende Fläche kann somit ermittelt werden.

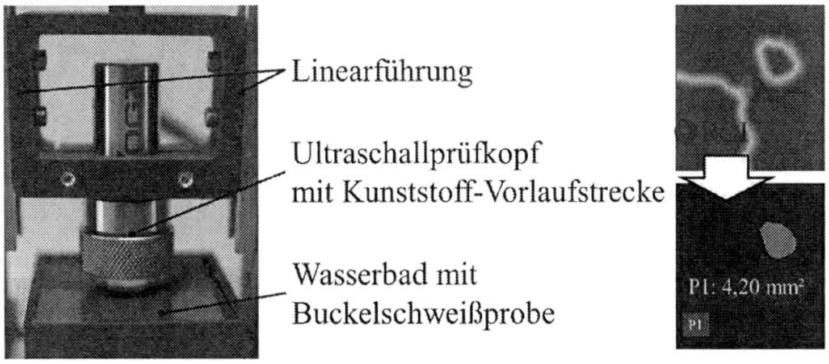

Abb. 4. Ultraschallprüfvorrichtung mit einem Phased-Array- Prüfkopf mit 121 Einzelschwingern (links) und Darstellung zur Ermittlung der Anbindungsfläche innerhalb des Analysebereichs (ROI) mittels Kantendetektion

1.2.3 Anlagentechnik für die zerstörende Prüfung

Die ZfP-Ergebnisse werden mit den Ergebnissen der quasistatischen zerstörenden Prüfung der Schweißproben verglichen. Die Proben der Versuchsreihen mit Schweißmuttern werden auf einen Gewindebolzen in einer Kopfzugvorrichtung geschraubt (Abb. 5, links). Die Proben mit Massebolzen werden über den Gewindebolzen in die Zugvorrichtung eingeschraubt. Die Zugvorrichtung zieht in Achsrichtung am Gewindebolzen, was einer Kopfzugprüfung entspricht. Die Bleche werden durch eine Matrize fixiert, die die Fügeelemente mit einem Umfangsabstand von 1 mm umschließen. Die Matrizengeometrie ist beispielhaft für die M6-Vierkantmuttern (DIN 928-M6-ST) in Abb. 5, rechts, dargestellt. Es wir an den Fügelementen gezogen, bis diese vom Blech abreißen. Die Kraft F wird über den Traversenweg s der Zugprüfmaschine aufgezeichnet. Die Prüfgeschwindigkeit für alle Proben wird mit 10 mm/min festgelegt.

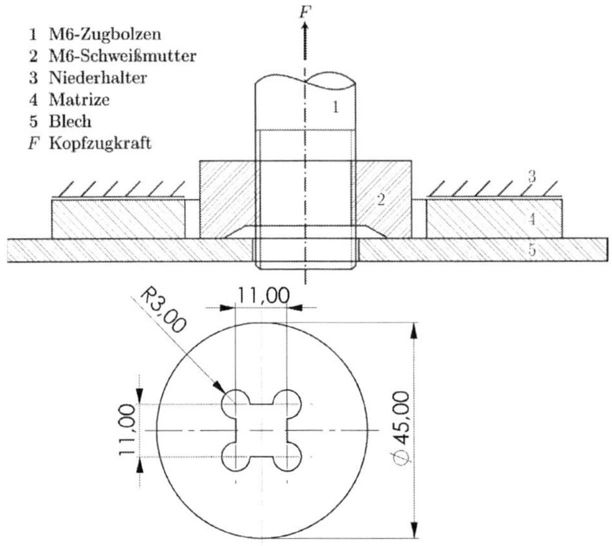

Abb. 5. Schematische Darstellung der Vorrichtung für die zerstörende Prüfung der Schweißproben mittels Kopfzugversuch (links) und technische Darstellung der verwendeten Matrizengeometrie für die M6-Vierkantmutter (DIN 928-M6-ST) (rechts)

2 Ergebnisse

Die laborbasierten und praxisnahen Ultraschallmessungen zeigen Unterschiede bei den einzelnen Versuchsreihen und den verschiedenen Schweißparametersätzen. In Abb. 6 sind die jeweiligen Bilder am Beispiel von Probe VKM1-MF_PS1 zu sehen, repräsentativ für die Proben mit M6-Vierkantmutter. In den dunklen Bereichen des C-Bilds der CESAM-Messung werden keine reflektierenden Ultraschallwellen gemessen. Es liegt eine Schweißverbindung zwischen den Fügepartnern vor. Der innenliegende runde dunkle Bereich ist die Bohrung im Blech. Dieser Bereich wird bei der Flächenermittlung herausgefiltert. Es ist zu erkennen, dass innerhalb der Buckelflächen Reflexionen detektiert werden. Dies kann auf eine schlechte Schweißverbindung im Inneren hinweisen. Es ist zu erwarten, dass die Buckel nur am Rand miteinander verschweißt sind. Die entsprechende Messung mittels des Phased-Array-Prüfsystems ist in Abb. 6, mittig, zu sehen. Die Buckel sind in den jeweiligen Bildquadranten abgebildet, alle Schweißverbindungen und das Loch sind hellgrau dargestellt. Dunkelgrau sind Reflexionen aus der Blechebene. Der jeweils größere hellgraue Bereich ist das Loch des Blechs. In Abb. 6, rechts ist die lichtmikroskopische Aufnahme dargestellt. Die Umrisse der Bruchflächen gleichen denen der CESAM-Messungen. Unterschiede sind beim Vergleich zwischen praxisnaher Ultraschallmessung und den Bruchflächen zu sehen. Es kann festgestellt werden, dass die Umrisse nicht übereinstimmen. Weiter sind zusammenlaufende Bereiche zwischen Loch und Buckelschweißverbindung in Q4 in Abb. 6, mittig, zu erkennen. Durch die Anpassung der ROI kann eine Trennung der Bereiche erzielt werden. Dabei wird der Innenradius der ROI auf konstant 3,75 mm eingestellt, um die Bereiche des Lochs gleichermaßen bei allen Proben herauszufiltern.

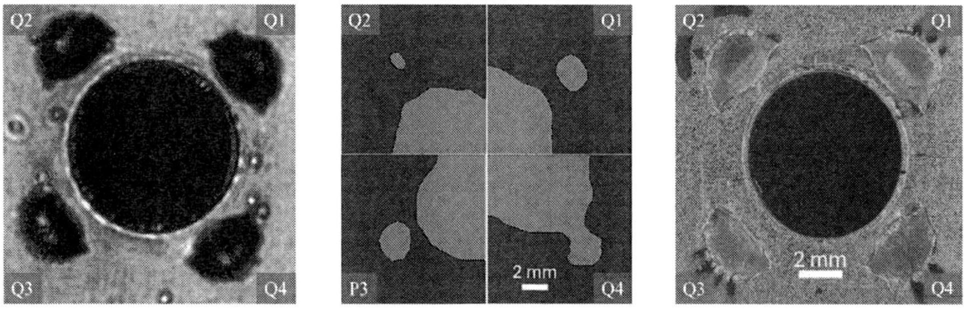

Abb. 6. Messungen mittels CESAM (links), Phased-Array-Prüfsystem (mittig) und der zerstörenden Prüfung mittels Kopfzugversuch (rechts) am Beispiel der Versuchsreihe VKM1-MF Probe PS1-1 mit M6-Vierkantmutter (DIN 928-M6-ST)

Abb. 7 zeigt die Bilder der Ultraschallmessungen und der am Mikroskop aufgenommen Buckelschweißverbindungen mit M8-Massbolzen am Beispiel von Probe MB1-PS2-1. Die Schweißbuckel sind bei beiden Ultraschallmessungen deutlich zu erkennen. Geringere geometrische Abweichungen zwischen den Außenkonturen sind zu beobachten. Es ist zu beachten, dass bei der zerstörenden Prüfung die Schweißbuckel aus dem Blech ausgeknöpft sind. Die entsprechenden Zugkräfte beziehen sich daher bei der Festigkeitsbetrachtung auf andere Flächen als die gemessenen Flächen der Ultraschallsysteme.

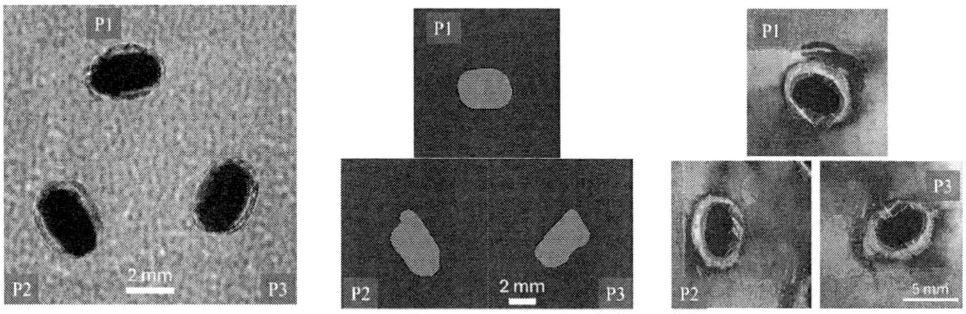

Abb. 7. Messungen mittels CESAM (links), Phased-Array-Prüfsystem (mittig) und der zerstörenden Prüfung mittels Kopfzugversuch (rechts) am Beispiel der Versuchsreihe MB1-MF Probe PS2-1 mit M8-Massebolzen

Die einzelnen ermittelten ZfP-Flächen der jeder Versuchsreihe sind in einer Gegenüberstellung zu den gemessenen Flächen nach erfolgter zerstörender Prüfung in Abb. 8 zu sehen. Innerhalb der Versuchsreihen sind die Flächen den jeweiligen Parametersätzen PS1 und PS2 zugeordnet. Es sind bei allen Messungen gute Korrelationen zu beobachten. Die Determinationskoeffizienten R^2 werden hierfür berechnet. Sie liegen bei den CESAM-Messungen zwischen 0,85 und 0,97. Für die praxisnahe Ultraschallprüfung liegen diese im Mittelwert höher zwischen 0,90 und 0,97. Zur besseren Veranschaulichung ist der Ideale 1:1-Zusammenhang als schwarze gestrichelte Linie dargestellt. Es ist zu erkennen, dass die Geraden der linearen Regression für die CESAM-Messungen im Anstieg dem idealen Zusammenhang ähnlicher sind als die der praxisnahen Ultraschallmessungen, welche einen wesentlich größeren Anstieg aufweisen. Dabei ist festzustellen, dass die praxisnahe Ultraschallmessung kleine Flächen in PS1 unterbewertet und größere Flächen aus PS2 überbewertet. Für die CESAM-Messungen sind die Flächen meist kleiner als die der zerstörenden Prüfung.

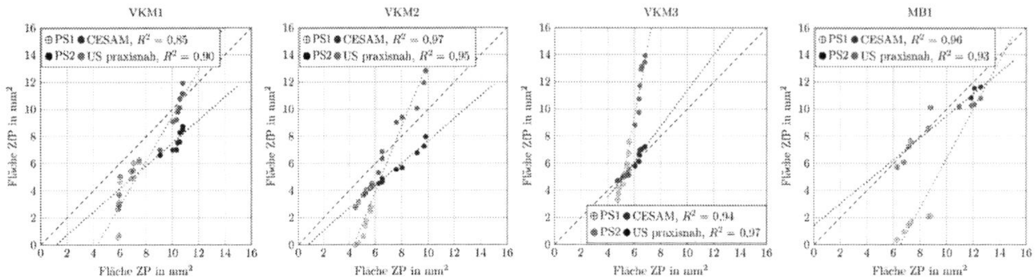

Abb. 8. Gegenüberstellung der ermittelten ZfP-Flächen aus CESAM und Phased-Array-Prüfsystem (US, praxisnah) gegenüber den Flächen nach zerstörender Prüfung (ZP) mittels Kopfzugversuch für alle Versuchsreihen mit Angabe der jeweiligen Determinationskoeffizienten $\boldsymbol{R^2}$ durch Linearregression

3 Diskussion

Die Zusammenhänge der Ergebnisse lassen sich in der statistischen Auswertung in Abb. 9 bestätigen und hervorheben. Die Mittelwerte der CESAM-Flächen liegen bis auf VKM2-PS2 und VKM3-PS2 unterhalb der ZP-Flächen. Für VKM2-PS2 und VKM3-PS2 sind die Unterschiede nicht signifikant. Die gemittelten Flächen für die praxisnahe Ultraschallmessung weichen hingegen deutlich von denen der zerstörenden Prüfung ab. Sie Streuen mehr als die CESAM- und ZP-Messungen. Falschbewertung mit größer bewerteten Flächen stellen für die Anwendung ein Risiko dar. Es lässt sich daher der Bedarf an höher auflösenden praxistauglichen Ultraschallsystemen für die Bewertung kleiner Flächen, wie z. B. Buckelschweißverbindungen dieses Artikels, ableiten. Für die Bewertung einer notwendigen Auflösung eines praxisnahen Ultraschallmesssystems werden die CESAM-Messungen in ihrer Auflösung zunächst reduziert. Anschließend wird die ursprüngliche Auflösung auf 0,025 mm rekonstruiert. Beide Schritte erfolgten mittels quadratischer Interpolation. Abb. 10 zeigt die absoluten und prozentualen Abweichungen der ermittelten Gesamtfläche aller Schweißbuckel über die Auflösung. Es ist zu erkennen, dass mit abnehmender Auflösung die Abweichungen zunehmen. Es ist bei einer Auflösung größer 1 mm mit bis zu 15 % Abweichung zu rechnen. Für Abweichungen kleiner 1 % über alle Versuchsreihen ist eine maximale Auflösung von 0,25 mm zulässig.

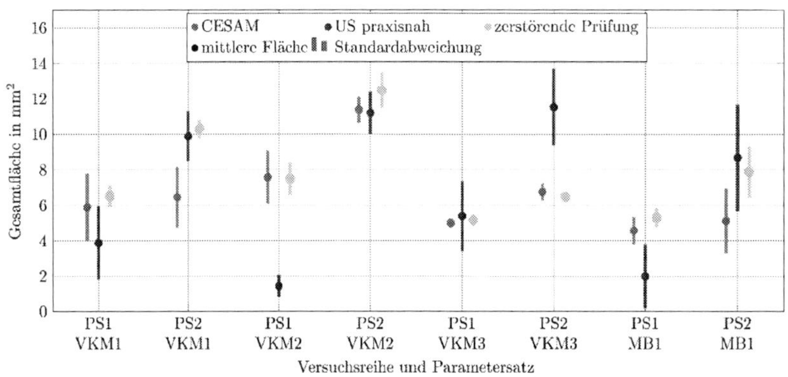

Abb. 9. Statistische Auswertung der ermittelten Flächen aus CESAM, Phased-Array-Prüfsystem und den Flächen nach zerstörender Prüfung und der zerstörenden Prüfung mittels Kopfzugversuch für alle Versuchsreihen

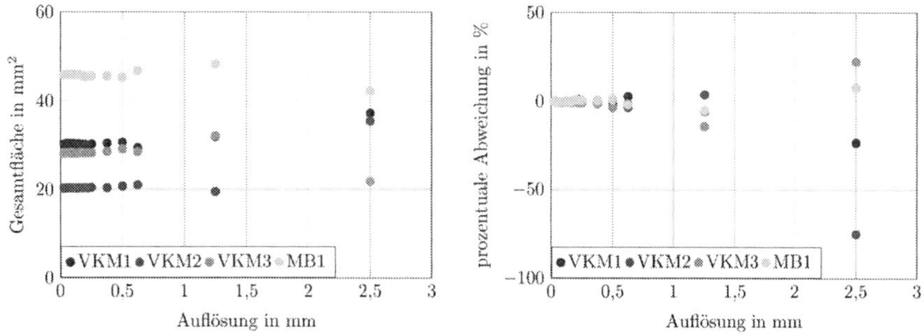

Abb. 10. Abweichung der ermittelten Flächen nach reduzierter und rekonstruierter Auflösung der CESAM-Messungen für die verschiedenen Versuchsreihen mit der Flächenabweichung (links) und der prozentualen Abweichung (rechts)

4 Zusammenfassung

Für Buckelschweißverbindungen kleiner Querschnitte existieren aktuell keine serientauglichen ZfP-Systeme. Bildgebende Ultraschallsysteme, die für die Punktschweißprüfung eingesetzt werden, scheinen die erste Wahl für diesen Zweck zu sein. Die Systeme sind jedoch bisher nicht ausreichend für die vielfältigen geometrischen Variationen im Bereich von Buckelschweißverbindungen vorbereitet. Komplexe Geometrien der Schweißbuckel verhindern oftmals eine reproduzierbare Qualitätsbewertung. Weiter sind die Auflösung der verfügbaren Systeme zu gering, um die kleinen Anbindungsflächen der Schweißbuckel zuverlässig zu detektieren. Die durchgeführten Untersuchungen bestätigen diesen Aspekt. Den Vergleich zwischen hochauflösenden CESAM-Messungen und praxisnaher Ultraschallmessungen konnte gezeigt werden, welche weiteren Entwicklungsschritte für zuverlässige Qualitätsbewertungen mindestens notwendig sind. Hierzu zählt insbesondere einer Verbesserung der Auflösung auf mindestens 0,25 x 0,25 mm^2.

Referenzen

[1] J. Chen, Z. Feng, and D. C. Warren, "On-Line Weld NDE with IR Thermography," Oak Ridge, Jun. 16 2014. Accessed: Nov. 23 2022.

[2] J. Schlichting, S. Brauser, L. A. Pepke, C. Maierhofer, M. Rethmeier, and M. Kreutzbruck, "Thermographic testing of spot welds," *NDT&E International*, vol. 48, no. 1, pp. 23–29, 2012.

[3] B. Zubert, F. Hönl, C. Gühmann, and M. Rethmeier, "Information capacity of the spot weld emissivity," *Materials Testing*, vol. 56, no. 2, pp. 96–103, 2014.

[4] R. G. Maev, R. Kitzmann, and Y. Oberdörfer, "Fortschritte in der Anwendung von künstlicher Intelligenz bei der zerstörungsfreien Prüfung im Karosseriebau," *e-Journal of Nondestructive Testing*, vol. 27, no. 07, 2022. [Online]. Available: https://www.ndt.net/search/docs.php3?id=27050

[5] C. Köhler, G. Vogt, and S. Berg, "Robot-based spot weld inspection - almost couplant-free, imaging phased array based inspection with PHAsis, integrated and automated by ABB Robotics," *eJNDT*, vol. 28, no. 8, 2023, doi: 10.58286/28194.

[6] P. Poltersdorf, J. Mußmann, and G. Vogt, "Easy to go and innovative validation process using the spot weld inspection system PHAsis and related software," *eJNDT*, vol. 28, no. 8, 2023, doi: 10.58286/28096.

[7] L. Yang *et al.*, "Ultrasonic Non-Destructive Testing and Evaluation of Stainless-Steel Resistance Spot Welding Based on Spiral C-Scan Technique," *Sensors (Basel, Switzerland)*, vol. 24, no. 15, 2024, doi: 10.3390/s24154771.
[8] J. K. Na, "Nondestructive inspection of resistance spot welds using matrix phased array ultrasonic technology," in *Advanced Materials & Processes*, pp. 22–24.
[9] J. K. Na and S. T. Gleeson, "Portable matrix phased array spot weld inspection system," US 2014/0165730 A1, USA 14/183,643, Jun 19, 2014.
[10] OLYMPUS CORPORATION, *Einführung in die Gruppenstrahlertechnik.* [Online]. Available: https://www.olympus-ims.com/de/ndt-tutorials/phased-array/ (accessed: Feb. 4 2019).
[11] G. Vogt, J. Mußmann, B. Vogt, and W.-K. Stiller, "Imaging spot weld inspection using Phased Array technology – new features and correlation to destructive testing," in *12th ECNDT, Gothenburg, Sweden, 2018: 12th European Conference on Non-Destructive Testing, Gothenburg, Sweden, 2018 : proceedings*, 2018.
[12] R. Augner, K. Loth, K. Harsch, H. Krauter, and G. Maier, "Schweißverbindungsprüfung im Automobilbau - Entwicklung, Erprobung und Serienanwendung eines mobilen Ultraschallminiscanners," in *DACH-Jahrestagung 2008*, St. Gallen, 2008.
[13] H. C. Schmale, C. Mathiszik, J. Zschetzsche, and U. Füssel, "Zerstörungsfreie Prüfung von Buckelschweißverbindungen an Blechstrukturen: Schlussbericht IGF 21.542 BR (non-destructive characterization of spot welds by imaging analysis of the residual magnetic flux density)," Technische Universität Dresden, Professur für Fügetechnik und Montage, Dresden, 2023.
[14] S. Jüttner, *Entwicklungen und Untersuchungen von Qualitätskriterien beim Kurzzeitwiderstandsschweißen mit hoher Wärmestromdichte: Schlussbericht zu IGF-Vorhaben Nr. 19878 BR*. Magdeburg: Universitätsbibliothek, 2020.
[15] A. Meriläinen, J. Hyvönen, A. Salmi, and E. Hæggström, "CESAM-Coded excitation scanning acoustic microscope," *The Review of scientific instruments*, vol. 92, no. 7, p. 74901, 2021, doi: 10.1063/5.0047351.
[16] J. Hyvonen, A. Merilainen, A. Salmi, L. Hupa, N. Lindfors, and E. Haggstrom, "Three Megapixel Ultrasonic Microscope Imaging," in *2019 IEEE International Ultrasonics Symposium (IUS)*, Glasgow, United Kingdom, 2019, pp. 1886–1889.
[17] C. Mathiszik, J. Koal, J. Zschetzsche, U. Füssel, and H. C. Schmale, "Non-destructive characterization of resistance projection welded joints by ultrasonic and passive magnetic flux density testing," *Weld World*, vol. 68, no. 10, pp. 2671–2682, 2024, doi: 10.1007/s40194-024-01808-3.

DGZfP-Jahrestagung 2025

Charakterisierung der Porosität und Porengeometrie mittels akustischer Resonanzanalyse

Linus LITTNER*, Igor SOLODOV und Marc KREUTZBRUCK

Institut für Kunststofftechnik, Universität Stuttgart; Stuttgart, Deutschland
* E-Mail: linus.littner@ikt.uni-stuttgart.de; Telefon: +49 711 685 62897

Kurzfassung. Die Fähigkeit, Poren exakt zu detektieren, ist insbesondere bei Faserverbundstrukturen in der Luft- und Raumfahrt, aber auch bei additiv gefertigten metallischen Komponenten und im Metallguss ausschlaggebend für die Beurteilung der Bauteilsteifigkeit und -festigkeit. Dabei ist die Ultraschallprüfung eine etablierte Methode zur Bestimmung der Porosität, bei der eine direkte Korrelation zwischen frequenzabhängiger Schallabschwächung und Porengehalt vorliegt. Voraussetzung für die Zuverlässigkeit dieser Methode ist eine gleichmäßige Form und Größe der Poren. Da dies in realen Bauteilen selten der Fall ist, können Abweichungen des gemessenen Porengehalts von über 25 % auftreten. Ein akustisches Verfahren, welches die exakte Form und Lage der Poren bestimmen könnte, würde daher die Prüfzuverlässigkeit deutlich steigern.

Als Schritt in diese Richtung untersucht diese Arbeit die Möglichkeit, die Resonanzfrequenzen einer einzelnen Luftblase in festem Epoxidharz als Indikator für dessen Geometrie zu verwenden. Dazu wurden die theoretischen Grundlagen der Wechselwirkung akustischer Wellen mit Luftblasen analysiert und eine allgemeine Abhängigkeit zwischen Resonanzfrequenz und Porenradius erarbeitet. Zur experimentellen Validierung wurden Prüfkörper mit definierten Poren im Millimeterbereich hergestellt und deren Resonanzfrequenzen mittels Piezo-Keramik-Aktoren und einem Laser-Doppler-Vibrometer gemessen.

Die Ergebnisse zeigen, dass die gemessenen Porenresonanzen mit den Vorhersagen übereinstimmen und eine lineare reziproke Abhängigkeit vom Porenradius aufweisen. Die entwickelte Theorie erlaubt somit Rückschlüsse auf die Porengröße und kann zur Optimierung bestehender Methoden zur Bestimmung der Porosität beitragen.

Keywords: Ultraschallprüfung, Lokale Defektresonanz, Porosität, Poren

1. Einführung

Luftblasen in Flüssigkeiten sind nicht nur wegen der Neugier von Kaffeeliebhabern auf das Geräusch eines kochenden Wasserkochers in den Blickpunkt der Akustikforschung geraten, sondern auch wegen deren wichtigen Rolle in einigen technischen angewandten Anwendungen. So geht in hydro- und aerodynamischen Anlagen die Blasenbildung und ihr anschließender Kollaps mit sehr hohen lokalen Temperaturen einher und kann zu Kavitationsschäden an Bauteilen führen [1].

In jüngster Zeit haben sich neue Anwendungsaufgaben ergeben, die sich mit der zerstörungsfreien Prüfung (ZfP) von Porosität in Verbundwerkstoffen [2], der additiven Fertigung [3] und Metallgussbauteilen [4] befassen. Es handelt sich um ein unvermeidliches Herstellungsproblem bei laminierten Verbundwerkstoffstrukturen wie kohlenstofffaserverstärkten Kunststoffen (CFK), dass durch eine unzureichende Infiltration oder durch während des Aushärtungsprozesses von Epoxidharz freigesetzte Luft Poren entstehen. Diese mit eingeschlossener Luft gefüllten Hohlräume vermindern die interlaminare Festigkeit und

 https://doi.org/10.58286/31273

beeinträchtigen andere mechanische Eigenschaften so stark, dass ein Porositätsgrad von 2 % für viele Verbundwerkstoffbauteile zur Akzeptanzschwelle geworden ist [2]. Eine herkömmliche Methode zur Bestimmung der Porosität ist die Ultraschallprüfung, bei der die frequenzabhängige Schallschwächung bestimmt wird, die eine direkte Korrelation mit der Porosität ermöglicht [5]. Voraussetzung für die Zuverlässigkeit dieser Methode ist eine konstante Form und Größe der Poren sowie eine statistische Verteilung derselben. Da dies bei realen Bauteilen nicht gewährleistet werden kann, führt dies zu einer Abweichung des gemessenen Porengehalts von bis zu 25 % [6, 7].

Neben der Bestimmung der frequenzabhängigen Schallschwächung kann mit der Impulsthermografie die Porosität in CFK mit modellbasierten Analysemethoden bestimmt werden [8]. Im Vergleich zu scannenden Ultraschallverfahren ermöglicht diese Methode die direkte Abbildung von großflächigen Bauteilen und es konnte gezeigt werden, dass die Messabweichung auf ca. 10 % reduziert werden kann. Allerdings ist auch hier zu berücksichtigen, dass die Genauigkeit der Thermografie mit zunehmender Probendicke aufgrund der lateralen Wärmeströme abnimmt. Diese Methode ist daher auf dünne Proben mit einer Dicke von höchstens wenigen Millimetern beschränkt. In [8] wurde daher nur für dünne Prüfkörper mit einer Dicke von weniger als 2,16 mm eine Verbesserung der Ergebnisse im Vergleich zur Standard-Ultraschalldämpfungsmethode erzielt. Zudem hängt die Genauigkeit beider Methoden stark von der Form, Größe und Verteilung der Poren ab. Um den Zustand eines Bauteils bestmöglich zu analysieren, ist eine Methode erforderlich, mit der die Geometrie und Lage einzelner Poren präzise bestimmt werden kann. Verschiedene aufwendige ZfP-Techniken wie die Röntgen-Computertomographie sind derzeit die Kandidaten für die Inspektion von Poren in Verbundwerkstoffen [9]. Ein einfaches ZfP-Verfahren, das unabhängig von der Porenmorphologie und den Faser- und Matrixwerkstoffen Aufschluss über den Grad der Porosität in einem Verbundwerkstofflaminat geben könnte, muss jedoch erst noch entwickelt werden.

Als einen Schritt in diese Richtung untersucht die vorliegende Arbeit die Möglichkeit der lokalen Defektresonanz (LDR) [10, 11] durch die Auswertung der Frequenzantwort (FR) einer einzelnen Luftblase in festem Epoxidharz als Indikator für die Porosität. Es werden die theoretischen Grundlagen der Wechselwirkung von akustischen Wellen mit einer Pore dargelegt, gefolgt von experimentellen Ergebnissen mit Poren unterschiedlicher Größe.

2. Theoretische Grundlagen der Porenresonanz

Wir gehen von einem kugelförmigen Koordinatensystem im Zentrum einer Luftblase aus, um deren Schwingung im Medium zu betrachten. Die Wellenlänge der zur Anregung genutzten akustischen Welle ist größer als die Größe der Pore (Radius R). Die in der Pore zu erwartende stehende kugelförmige Schwingung erfordert die folgenden Beziehungen für den Druck p und die Schwingungsgeschwindigkeit v im Inneren der Blase[12]:

$$p = \frac{\sin(kr)}{r}; \qquad v = \frac{1}{i\rho ckr}\frac{\sin(kr) - kr\cos(kr)}{r} \tag{1}$$

Durch Kombination von (1) erhält man für die akustische Impedanz der Schwingung $Z_b = p/v$ in der Blase:

$$\frac{kr\tan(kr)}{\tan(kr) - kr} = Z_b \tag{2}$$

Die Werte von k und die zulässigen Porenresonanzfrequenzen ergeben sich aus der Randbedingung bei $r = R$:

$$Z_b(r = R) = Z_{in} \tag{3}$$

wobei Z_{in} die akustische Eingangsimpedanz des Wellenfeldes im umgebenden Medium ist.

Um den Ansatz einfach zu halten, sind die beiden spezifischen Fälle für das umgebende Medium recht einfach zu betrachten: frei schwingende Blase $Z_{in} = 0$ (schallweiches Medium) und fest eingespannte Blase $Z_{in} = \infty$ (schallhartes Medium). Diese beiden Fälle decken den Bereich der möglichen Medien (und der Tiefe der Blase) ab und ermöglichen es, die Erwartungen für verschiedene experimentelle Situationen zu bewerten.

Im ersten Fall $Z_{in} = 0$ und Gl. (2) ergibt die Resonanzfrequenzen:

$$kr = \pi n \text{ oder } f_n = \frac{c}{2R} n \tag{4}$$

wobei n eine ganze Zahl ist. Das Ergebnis für den Satz von LDR-Frequenzen (4) ist leicht zu interpretieren: ein ganzzahliges Vielfaches der Wellenlängen in Luft passt zum Durchmesser der Blase, damit die Resonanz auftritt. Die erste LDR-Frequenz $f_1 = \frac{c}{2R}$ entspricht der sphärisch-symmetrischen Schwingung der Pore (Breathing Mode), deren radiale Auslenkungen im experimentellen Versuchen leicht zu beobachten ist. Für eine Pore mit einem Radius von 1 mm und die Schallgeschwindigkeit in Luft c ≈ 340 m/s liegt dieser Wert nach (4) schätzungsweise im Bereich von ≈ 170 kHz. Für die Bewertung von LDR mit beliebigem Blasenradius ist es zweckmäßig, das Produkt $(f_1 R)$ im Auge zu behalten, dessen Wert aus (4) ≈ 0,5c ist.

Für den zweiten Fall $Z_{in} = \infty$ wird Gleichung (2) reduziert auf:

$$\tan(kR) = kR \tag{5}$$

Die Lösung von (5) lässt sich veranschaulichen, indem man die beiden Seiten als Funktionen von kR aufträgt (siehe Abb. 1).

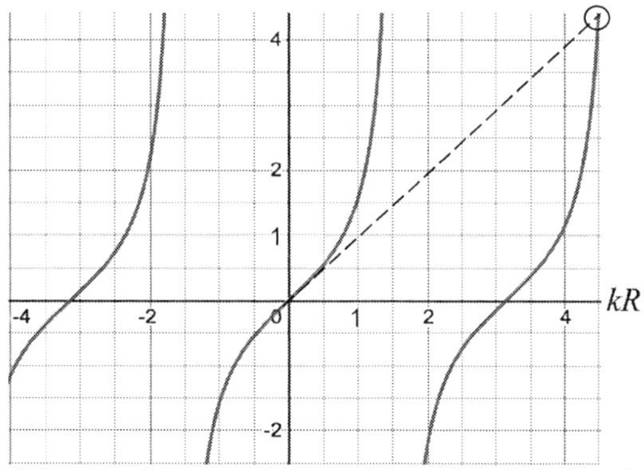

Abb. 1. Diagramme der linken (durchgezogene Linie) und der rechten Seite (gestrichelte Linie) von Gl. 2.

Aus Abb. 1 geht hervor, dass der erste Schnittpunkt der Kurven $kR \approx 4{,}5$ entspricht, was für die Resonanzfrequenz des ersten LDR $f_1 \approx 244$ kHz für eine Blase von 1 mm Radius ergibt. Die LDR-Frequenz für andere Radien kann in diesem Fall leicht durch Verwendung des Produkts $(f_1 R) \approx 0{,}72c$ geschätzt werden. Unter der Berücksichtigung von beiden Grenzwerte ($Z_{in} = 0$ und $Z_{in} = \infty$) muss das Produkt $(f_1 R)$ einer Blase in einem beliebigen festen Material zwischen 0,5c und 0,72c liegen.

Da $(fR)/c$ die Anzahl der Wellenlängen angibt, die zum Radius passen (R/λ), deutet der Wert für $Z_{in} = \infty$ auf einen gewissen Phasenverlust in der Reflexion hin, der eine einfache Interpretation der im ersten Fall gegebenen Resonanz erschwert. Es zeigt auch, dass alle Zwischenfälle von Z_{in} eine detaillierte Bewertung des Phasenverlusts der Reflexion erfordern, um eine LDR-Frequenz zu erhalten. Der obige vereinfachte Ansatz zeigt jedoch erstens, dass der Bereich möglicher Werte von $(f_1 R)$ für Poren von mm-Größe im Hochfrequenzbereich liegt, und zweitens könnte er zur schnellen Überprüfung der mittels ZfP-Verfahren ermittelten Werte für LDR-Frequenzen verwendet werden.

3. Methodik der Versuche

3.1 Herstellung von definierten Poren in Epoxidharz

Zur Herstellung der porösen Probekörper wurde in einem ersten Schritt eine zylindrische Form mit etwa 20 ml Harz/Härter-Gemisch gefüllt. Es wurde eine Epoxidharz/Härter-Mischung mit einer durchschnittlichen Topfzeit von 40 Minuten verwendet. Der harzgefüllte Hohlraum wurde dann für 20 Minuten einem Vakuum von 0,1 bar ausgesetzt, um eine unkontrollierte Porenbildung beim anschließenden Aushärtungsprozess zu verhindern. Um eine definierte Pore zu erzeugen, muss das reagierende Epoxidharz eine hohe Viskosität aufweisen, damit gezielt eingebrachten Lufteinschlüsse nicht aufsteigen können. Diese Bedingung war nach einer Reaktionszeit von etwa 65 Minuten erfüllt. Anschließend wurde mit einer Spritze und einer Kanüle eine definierte Luftmenge (~0,005 ml) in das reagierende Harz injiziert. Diese Luftmenge erzeugt Poren mit einem Durchmesser von ≈ 3-4 mm.

Vier Proben mit Poren wurden nach dem oben beschriebenen Verfahren hergestellt. Eine genaue Charakterisierung der hergestellten Poren wurde mit Hilfe einer Röntgen-CT-Anlage (FF20, Yxlon International, Deutschland) durchgeführt (siehe Abb. 2). Aus der mit Avizo (Thermo Fisher Scientific, U.S.A.) durchgeführten Analyse der rekonstruierten Proben wurden der Porendurchmesser, die Probendicke und das Porenvolumen bestimmt. Die Sphärizität der Pore wurde als Qualitätsmerkmal berechnet, um die Abweichung von einer idealen Kugelform zu quantifizieren. Die Daten sind in Tabelle 1 aufgeführt.

Tabelle 1. Gemessene Parameter der Poren

Pore	Radius (mm)	Volumen (mm^3)	Sphärizität	Probendicke (mm)
N1	1,62	17,93	0,979	6,61
N2	1,73	21,77	0,980	8,56
N3	1,84	26,30	0,973	8,48
N4	1,88	27,95	0,949	24,8

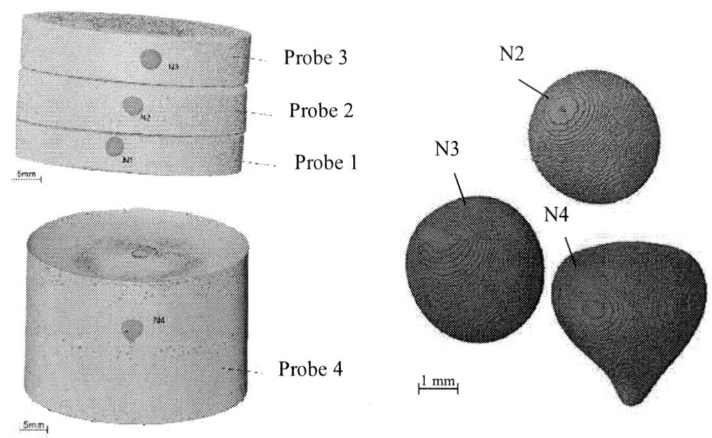

Abb. 2. CT-Scans der Proben mit den rekonstruierten Poren N1 - N4 (links) und Vergrößerung der Poren N2 - N4 (rechts)

3.2 Experimentelle Bestimmung der Porenresonanz

Zur Schwingungsanregung der Poren wurden piezokeramische Aktoren für Longitudinalwellen an flachen Oberflächen der zylindrischen Proben unterschiedlicher Dicke angebracht. Die Grundfrequenzen der Aktoren liegen im Bereich (7-15 kHz), sie bieten jedoch eine zuverlässige breitbandige Anregung im Frequenzbereich bis 200 kHz. Jenseits von 300 kHz konnte kein Frequenzgang gemessen werden.

Die Aktoren wurden mit einem Arbiträr-Funktionsgenerator (HP 33120A, Keysight, USA) durch breitbandige Spannungssignale (1kHz-1MHz, periodische Chirp-Signale) betrieben, die von einem High-Amplitude-Verstärker (HVA B100, isi-sys, Deutschland) verstärkt wurden. Dadurch wurde eine Eingangsamplitude von 30-50 V für den Piezo-Aktor erreichen. Zur Messung und Analyse der Frequenzantwort (FR) der Poren wurde ein Scanning-Laservibrometer mit 1 MHz Bandbreite (PSV 300, Polytec, Deutschland) verwendet. Um die Schwingungen zu untersuchen, konnte der Strahl des Messlasers entweder durch den flachen Boden oder über die gekrümmte Seitenfläche des transparenten Epoxidzylinders ausgerichtet werden. Die erste Messanordnung (Abb. 4, a) bot einen weiten Sichtwinkel und ermöglichte eine Messung sowohl innerhalb als auch außerhalb der Blase, während bei der zweiten Messanordnung (Abb. 4, b) das Messfeld aufgrund der Oberflächenkrümmung verringert war und hauptsächlich zur Überprüfung der mit dem ersten Aufbau erzielten Ergebnisse verwendet wurde.

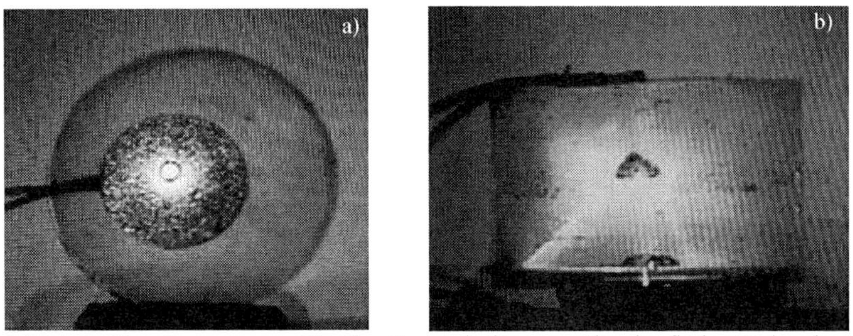

Abb. 4. Laserstrahlsondierung von sphärischen Porenschwingungen über den flachen Boden einer transparenten Probe (a); seitlicher Laseraufbau für die tropfenförmige Pore (b).

4. Messergebnisse der Porenresonanz

Die in diesem Abschnitt für die Pore N1 (siehe Tabelle 1) gemessene Frequenzantwort ist in Abb. 5 dargestellt. Beim Auftreffen des Laserstrahls auf die Pore ist eine deutliche LDR bei $f_1 \approx 138{,}8$ kHz zu beobachten (Abb. 4, a). Für den Blasenradius R ≈ 1,62 mm (Tabelle 1) beträgt $(f_1 R)$ daher ≈ 225 m/s, d. h. $0{,}66c$. Der gemessene Wert liegt damit zwischen den theoretischen Erwartungen aus Abschnitt 2: .$0{,}5c \leq (f_1 R) \leq 0{,}72c$.

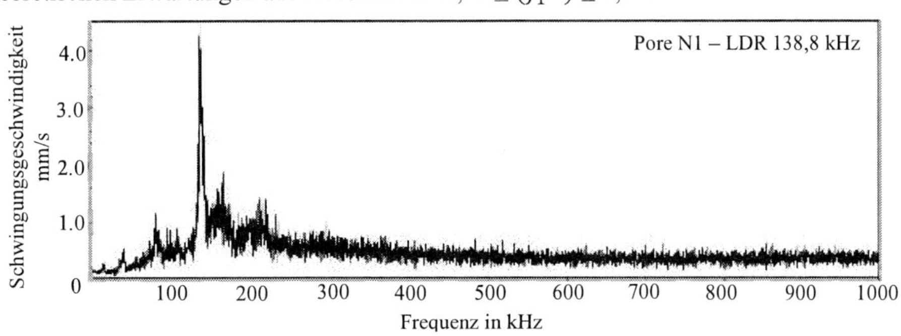

Abb. 5. Linearer FR gemessen für Pore N1.

Eine ähnliche Schlussfolgerung kann aus den FR-Messungen für die anderen kugelförmigen Poren, z. B. N3, gezogen werden, die in Abb. 6 dargestellt sind. Die entsprechenden gemessenen LDR-Frequenzen und die Werte des Parameters $(f_1 R)$ sind in Tabelle 2 zusammengefasst. Alle erhaltenen Werte liegen um 0,6c, was ebenfalls in den Bereich passt, der aus der theoretischen Analyse in Abschnitt 2 erwartet wurde.

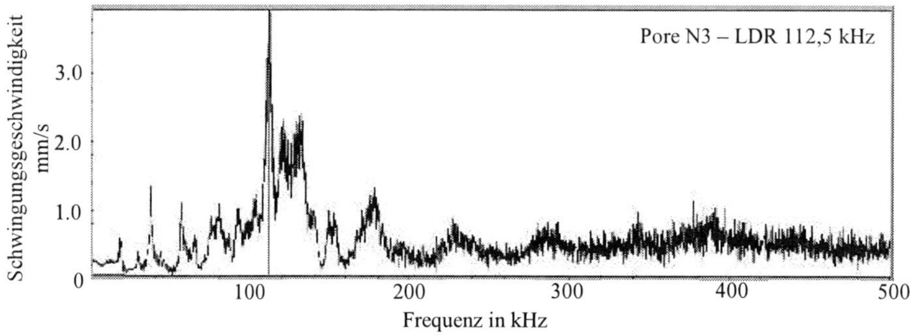

Abb. 6. Linearer FR gemessen für Pore N3.

Tabelle 2. LDR-Messergebnisse für kugelförmige Poren N 1-4

Pore	f_1 (kHz)	$f_1 R$	f_t (kHz)	f_{pore} (kHz)
N1	138.8	0.66c	190.6	398.1
N2	124.06	0.63c	147.2	290.8
N3	112.5	0.61c	148.6	297.4
N4	108-141.7	0.60c	50.8	105.7

Zur weiteren Einordnung der gemessenen Resonanzfrequenzen wurden weitere aus der Literatur bekannte Resonanzfälle betrachtet. Hierzu wurde zum einen die Probengrundresonanz (Dickenresonanzfrequenz) herangezogen. Zur Berechnung wurde die Longitudinalwellengeschwindigkeit in der Epoxidmatrix gemessen (c_m= 2,52 10^3 m/s) und mit der Probendicke d kombiniert, um die Dickenresonanz jeder Probe mit $f_t = c_m/(2d)$ zu ermitteln. Diese Werte sind auch in Tabelle 2 aufgeführt und zeigen, dass die Resonanzfrequenzen der Proben um ca. (20-50) kHz von der gemessenen FR entfernt sind und ebenfalls in der Frequenzantwort in Abb. 5 und Abb. 6 zu finden sind.

Außerdem wurde der in [4] hergeleitete analytische Ansatz a zur Berechnung der verwendet Porenresonanz f_{pore} (siehe Tabelle 2) berechnet. Die Ergebnisse für f_{pore} sind allerdings signifikant höher für die Poren N1 - N3, womit sie außerhalb des Frequenzbereichs der verwendeten Piezo-Aktoren liegen und experimentell nicht nachgewiesen werden konnten. Die Pore N4 zeigt hingegen eine bessere Übereinstimmung mit nur geringen Abweichungen. Allerdings wurde eine Korrelation zwischen der Porenresonanz und der Dickenresonanz $f_{pore} \approx 2f_t$ für die untersuchten Geometrien identifiziert, die es in weiteren Arbeiten näher zu untersuchen gilt.

5. Fazit

Die untersuchten Poren in festem Material im mm-Bereich zeigen scharfe Resonanzen der (LDR) im hohen kHz-Frequenzbereich, die auf die sphärisch-symmetrischen Schwingung (Breathing Mode) zurückzuführen sind. Diese Schlussfolgerung wird durch theoretische Grundlagen untermauert, wobei eine Analogie zwischen Blasenschwingungen in Flüssigkeiten und Festkörpern verwendet wird, die eine klare Welleninterpretation der Blasenresonanz ermöglicht. Nach der entwickelten Analyse ist die Resonanzfrequenz f_1 für die Breathing Mode reziprok zum Porenradius R, so dass sie durch den Wert des Parameters

$(f_1 R)$ für eine bestimmte Pore abgeschätzt werden kann. Dieses Produkt soll im Bereich zwischen 0,5c und 0,72c liegen, wobei c die Schallgeschwindigkeit in Luft ist.

Die experimentelle Überprüfung für mm-große Luftblasen in festem Epoxidharz zeigt, dass der Wert dieses Parameters $\geq 0{,}6c$ beträgt, d. h. gut zu den theoretischen Erwartungen passt. Der Wert des gemessenen Produkts ermöglicht auch die Bewertung des Porenradius und kann so zur Verbesserung von bestehenden ZfP-Methoden zur Bestimmung der Porosität durch Erweiterung der Berechnungsmodelle um die tatsächliche Porengrößenverteilung beitragen.

6. Danksagung

Teile der hier vorgestellten Ergebnisse entstanden im Rahmen des von der Deutschen Forschungsgemeinschaft e.V. (DFG) geförderten Forschungsprojekts 549528608.

Ein besonderer Dank geht an meinen Co-Autor und Mentor Igor Solodov. Er war mir in wissenschaftlicher wie menschlicher Hinsicht ein Vorbild und hat mich stets unterstützt – seine Impulse wirken über diese Arbeit hinaus.

Referenzen

1. Flynn, H.G. Physics of acoustic cavitation in liquids. In: Physical Acoustics, 1964, (Vol. 1, Part B), 57–172.
2. PARK, S.-J. and M.-K. SEO. Composite Characterization. In: *Interface Science and Composites:* Elsevier, 2011, 631–738. ISBN 9780123750495.
3. SLOTWINSKI, J.A., E.J. GARBOCZI and K.M. HEBENSTREIT. Porosity Measurements and Analysis for Metal Additive Manufacturing Process Control. In: *Journal of research of the National Institute of Standards and Technology,* 2014, **119**, 494–528. Doi:10.6028/jres.119.019
4. TANMOY, B. and R. SUBHANKAR. Defect resonance frequency of spherical pore within isotropic elastic solid. In: *Ultrasonics,* 2022, **124**, 106759. Doi:10.1016/j.ultras.2022.106759
5. DANIEL, I.M., S.C. WOOH and I. KOMSKY. Quantitative porosity characterization of composite materials by means of ultrasonic attenuation measurements. In: *Journal of Nondestructive Evaluation,* 1992, **11**(1), 1–8. Doi:10.1007/BF00566012
6. BIRT, E.A. and R.A. SMITH. A review of NDE methods for porosity measurement in fibre-reinforced polymer composites. In: *Insight - Non-Destructive Testing and Condition Monitoring,* 2004, **46**(11), 681–686. Doi:10.1784/insi.46.11.681.52280
7. LIN, L., M. LUO, H.T. TIAN, X.M. LI und G.P. GUO. *Experimental investigation on porosity of carbon fiber-reinforced composite using ultrasonic attenuation coefficient.* Shanghai, 2008.
8. MAYR, G. *Characterization of porous carbon fiber reinforced polymer components using optical excited pulse thermography.* Doctoral dissertation. Johannes Kepler Universität Linz. Institut für Polymer Product Engineering. Linz, 2016.
9. STAUDE, A., M. BARTSCHER, K. EHRIG, J. GOEBBELS, M. KOCH, U. NEUSCHAEFER-RUBE and J. NÖTEL. Quantification of the capability of micro-CT to detect defects in castings using a new test piece and a voxel-based comparison method. In: *NDT & E International,* 2011, **44**(6), 531–536. Doi:10.1016/j.ndteint.2011.05.006
10. SOLODOV, I., J. BAI, S. BEKGULYAN and G. BUSSE. A local defect resonance to enhance acoustic wave-defect interaction in ultrasonic nondestructive evaluation. In: *Applied Physics Letters,* 2011, **99**(21). Doi:10.1063/1.3663872
11. SOLODOV, I., M. RAHAMMER and M. KREUTZBRUCK. Analytical evaluation of resonance frequencies for planar defects: Effect of a defect shape. In: *NDT & E International,* 2019, **102**, 274–280. Doi:10.1016/j.ndteint.2018.12.008
12. ISAKOVICH, M.A. *Obschaya akustika [General acoustics].* Moscow: Nauka Publ., 1973.

Zuverlässigkeitsbasierter Technologietransfer am Beispiel Risslumineszenz

Von der ZfP im Labor zum KI-basierten SHM-System für die Industrie – Statusbericht

Paul GERARDS-WÜNSCHE, Falk HILLE
Bundesanstalt für Materialforschung und -prüfung, Berlin, Deutschland,
paul.gerards-wuensche@bam.de

Kurzfassung

Der Übergang neuer zerstörungsfreien Prüfmethoden (ZfP) aus dem Labor in die industrielle Anwendung stellt erhebliche Herausforderungen dar. Für visuelle ZfP-Techniken, wie die in dieser Studie verwendete Risslumineszenz-Methode, ist die Automatisierung der Fehlererkennung durch KI-gestützte Computer-Vision-Systeme ein logischer nächster Schritt. Findet dies als kontinuierliche Überwachung statt, entwickeln sich solche Systeme zu einer KI-gestützten SHM-Methode. Die Implementierung dieser Technologien in sicherheitskritischen Anwendungen erfordert jedoch einen robusten Nachweis ihrer Zuverlässigkeit, der durch diese Weiterentwicklung an Komplexität gewinnt.

Während etablierte Standards für Probability of Detection (POD)-Studien in ZfP-Systemen existieren, fehlen entsprechende Normen für SHM-Systeme, insbesondere solche, die KI einsetzen. Obwohl statistische Methoden für SHM-Systeme verfügbar sind, um Datenabhängigkeiten zu berücksichtigen, ist ein ganzheitlicher POD-Ansatz notwendig, um beeinflussende und variable Störfaktoren umfassend zu adressieren und die Alterung des SHM-Systems zu berücksichtigen.

Diese Arbeit demonstriert Fortschritte bei der Entwicklung einer maßgeschneiderten Methodik, die auf die spezifischen Anforderungen zugeschnitten ist, um die Zuverlässigkeit eines KI-gestützten SHM-Systems auf Basis der Risslumineszenz nachzuweisen. Dieser Ansatz integriert die Prinzipien von POD-Studien mit statistischen Methoden für SHM-Systeme, um eine robuste Zuverlässigkeitsbewertung zu gewährleisten. Er verdeutlicht die erheblichen Herausforderungen, die bereits bei der kontinuierlichen Überwachung einer relativ einfachen visuellen Methode wie der Risslumineszenz auftreten.

Keywords: Risslumineszenz, ZfP, SHM, PoD, Zuverlässigkeit, KI, Computer Vision

BAM
Bundesanstalt für Materialforschung und -prüfung

Sicherheit in Technik und Chemie

27.05.2025

ZUVERLÄSSIGKEITSBASIERTER TECHNOLOGIETRANSFER AM BEISPIEL RISSLUMINESZENZ

Von der ZfP im Labor zum KI-basierten SHM-System für die Industrie – Statusbericht

Paul Gerards-Wünsche, Falk Hille

BAM Fachbereich 7.2 - Ingenieurbau

www.bam.de

1

Vorstellung Risslumineszenz (Riluminati)

Aufbau und Funktionsweise

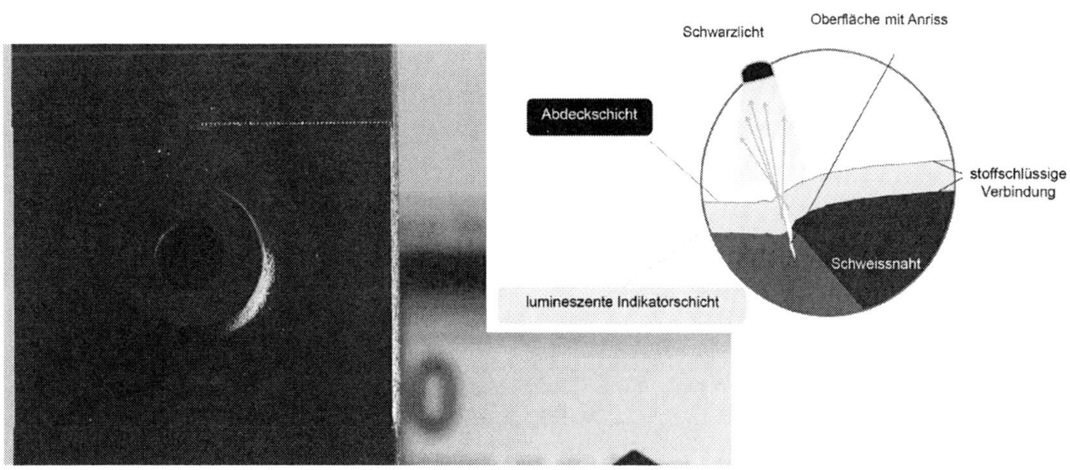

2

Risslumineszenz
Übergeordnete, langfristige Zielstellung

BAM

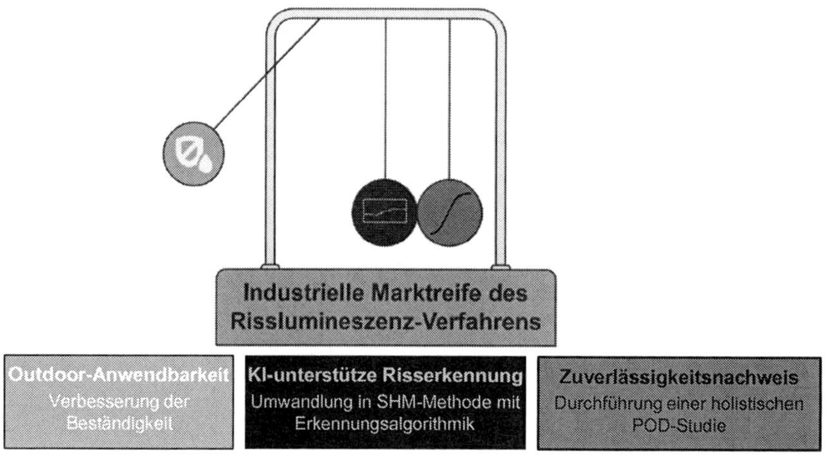

3

Risslumineszenz als SHM-Methode
Bestandteile des Monitoring-Systems

BAM

- Riluminati-Beschichtung
- Kamera (mit Edge-Funktion)
- Automatisierte Risserkennung
- UV-Leuchte
- Energieversorgung und Dateninfrastruktur
- Verfahrmechanismus / Drohne

4

Rissdetektion von ZfP-Verfahren

BAM

Prüfverfahren	Rissbreite [mm]	Risslänge 2c [mm]	Risstiefe a, 2a [mm]	Bemerkung	Anwendungsgrenzen
Sichtprüfung	0,1	2	-	bei sauberer Oberfläche und optischen Hilfsmitteln	komplizierte Geometrie, mangelnder Kontrast
Eindringprüfung	0,01	1	0,5	Werkstoffunabhängiger Einsatz	poröse Werkstoffe, verstopfte Risse, Öffng. zur Oberfläche notwendig, raue Oberfläche
Magnetpulverprüfung	0,001	1	0,1	bei feinen Rissen an der Oberfläche und dicht darunter	nur Ferromagnetika, abhängig von Magnetisierung, Rautiefe, Beleuchtung
Wirbelstromprüfung	0,01	1	0,1	hohe Prüfgeschwindigkeiten realisierbar	nur elektrische Leiter, begrenzte Eindringtiefe, lokale Anwendung
Potentialsondenprüfung	0,01	2	0,2	Risstiefenmessung, ohne Einfluss der Rissbreite, gemittelte Rissgrößen	nur elektrische Leiter, kein elektrischer Kontakt der Rissflanken
Ultraschallprüfung	0,001	1	1	für Innen- und Oberflächenfehler, beliebige Bauteildicke	Ergebnis abhängig von akustischen Werkstoffeigenschaften, komplizierte Bauteil- und Fehlergeometrie
Durchstrahlungsprüfung	0,1	1	2% der Wanddicke	für Innen- und Oberflächenfehler, berührungslose Prüfung	begrenzte Bauteildicke, Strahlenschutz beachten

Quelle: FKM 2018

5

Qualifizierungsmethodik

Komplexität KI-gestützer SHM-Methoden

BAM

- Risslumineszenz-spezifische Qualifizierungsmethodik:
  - Kopplung der Komplexität von SHM-Systemen sowie KI-basierter Algorithmik in etablierte Qualifizierungsmethoden für ZfP-Methoden
- Für SHM-Methoden gibt es erste Ansätze
  - Veröffentlichung SAE ARP6821 (Feb. 2025)
  - Erhöhter Schwierigkeitsgrad durch Service Life des SHM-Systems selbst
  - Infragestellung einer einmaligen Zuverlässigkeitsaussage für alternde SHM-Systeme
- Zertifizierung robuster KI-Systeme schreitet voran (z.B. ISO/IEC TR 24029-1:2021)

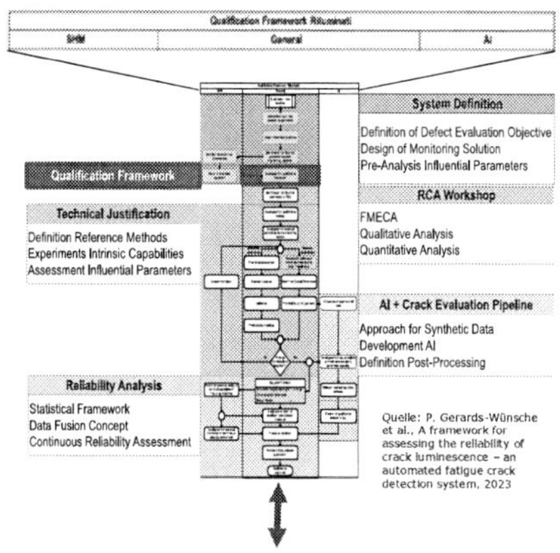

Quelle: P. Gerards-Wünsche et al., A framework for assessing the reliability of crack luminescence – an automated fatigue crack detection system, 2023

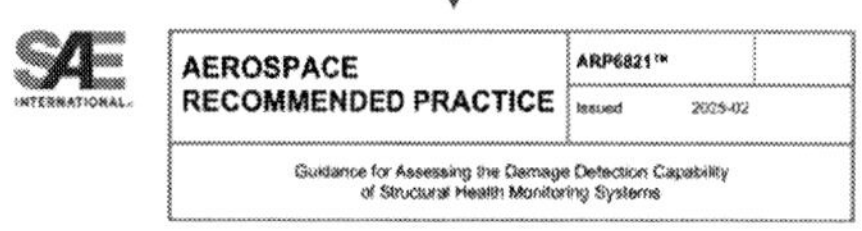

6

Modulares Modell für NDE 4.0

Erweiterung um (KI-)Algorithmen

BAM

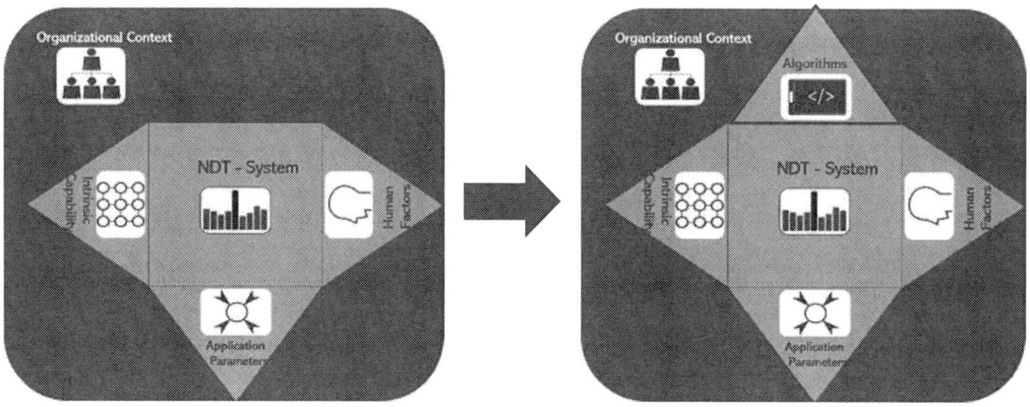

Quelle: V. Rentala, „POD Evaluation: The Key Performance Indicator for NDE 4.0", 2022

7

Fehler-Ursachen-Analyse

1.FMEA-Workshop

BAM

- 1.FMEA Workshop mit Fokus auf die Risslumineszenz-Beschichtung
- Knapp 40 Fehlermodi allein hierfür ermittelt
- Weiterer Workshop zur Analyse des Systemdesigns (Hardware & Software) nötig
- Ableiten möglicher Gegenmaßnahmen
- Untersuchungen essenzieller Störquellen und Unsicherheiten

Experten für Fehler-Ursachen-Analyse			
Menschliche Faktoren	ZfP-Zuverlässigkeit	ZfP-Prüfer für PT und MPI	Chemische Prozesstechnik
Strukturanalyse	KI / Computer Vision	Automatisierungs-technik	Systementwickler

Einbindung weiterer Experten in Folge-Workshops

8

Untersuchungen der Intrinsischen Fähigkeiten

BAM

Ermüdungsversuche – Design of Experiments

- Auswahl verfügbarer Probengeometrien (z.T. begrenzte Stückzahl)
- 2 mm Bleche für Untersuchungen großer Stückzahl (nötig für statistische Auswertung)

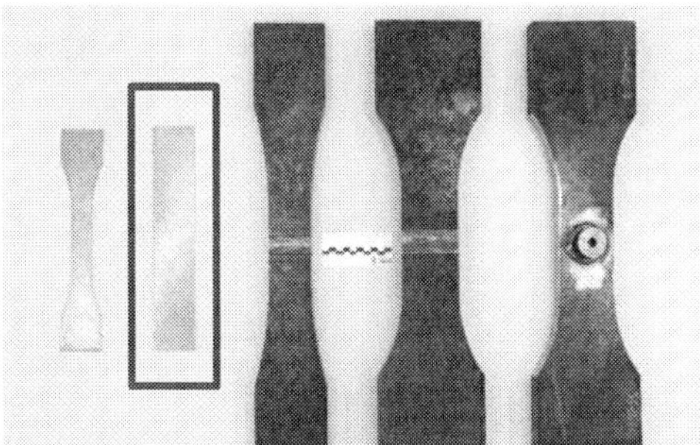

Kleinproben (Blechdicken: 0.8 mm, 2 mm, 25 mm Stumpfnaht, 40mm Stumpfnaht, 40 mm Schweißbuchse)

Großbauteil von Offshore-Windenergieanlagen (Jacketknoten 1:1)

9

Untersuchungen der Intrinsischen Fähigkeiten

BAM

Ermüdungsversuche – Design of Experiments

- Berücksichtigung ermittelter Störfaktoren
- Z.B. Maßgeschneiderte Verdunklungsvorhänge für Ermüdungsprüfstände

Prüfstände mit Verdunklungsvorhängen

10

Untersuchungen der Intrinsischen Fähigkeiten

BAM

Ermüdungsversuche – Design of Experiments

- Ermittlung der Vergleichsverfahren:
  - **Rastlinien** für Validierung aller Versuchsaufbauten (wenige Risslängen)
  - Materialebene: **DCPD** (als Dauermonitoring, BAM 9.4)
  - Bauteilebene: Thermografie (als Dauermonitoring, BAM 8.3)
- Weitere Monitoring-Verfahren getestet:
  - ACPD, GMR-Arrays, Rissfortschritts-DMS

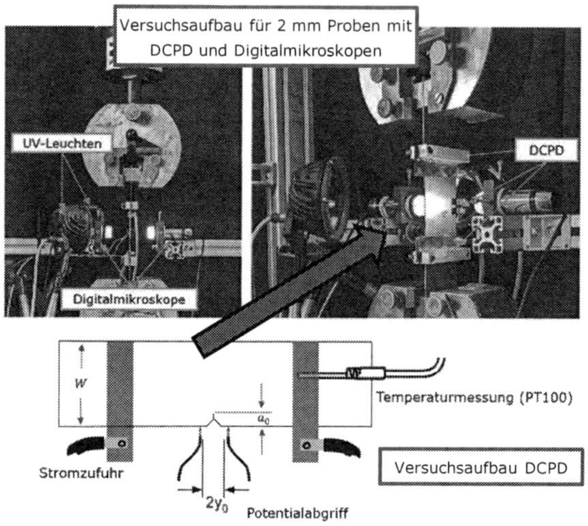

11

Untersuchungen der Intrinsischen Fähigkeiten

BAM

Ermüdungsversuche – 2 mm Bleche

- 2 mm Proben (SENT) mit angeschwungenem Anriss, um Randeffekte zu vermeiden
- Kopplung von Prüfmaschine mit DCPD-Monitoring und Messcomputer für Rastlinien-Erzeugung

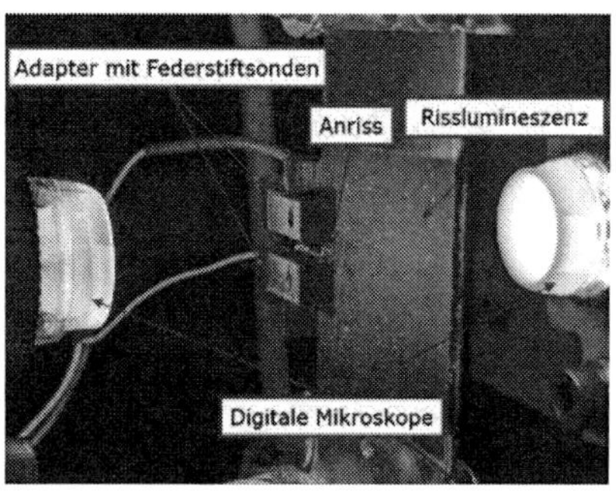

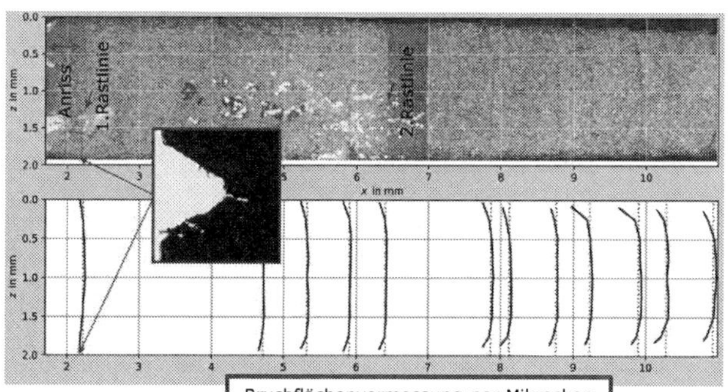

12

Ermüdungsversuche – Kleinproben (t = 40 mm)

BAM

Rissslumineszenz, DMS, Ultraschall (BAM 8.4), Thermografie (BAM 8.3)

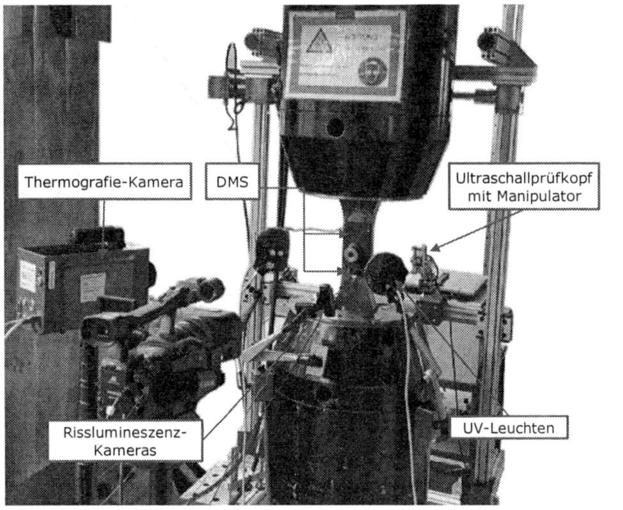

Ablauf

- Rissinitiierungsdetektion
  - ➢ Risslumineszenz, DMS
- Risslänge und -fortschritt
  - ➢ DMS, Risslumineszenz, Thermografie
- Risstiefe
  - ➢ Ultraschall (für Riluminati nicht relevant)
- Validierung
  - ➢ Rastlinien

13

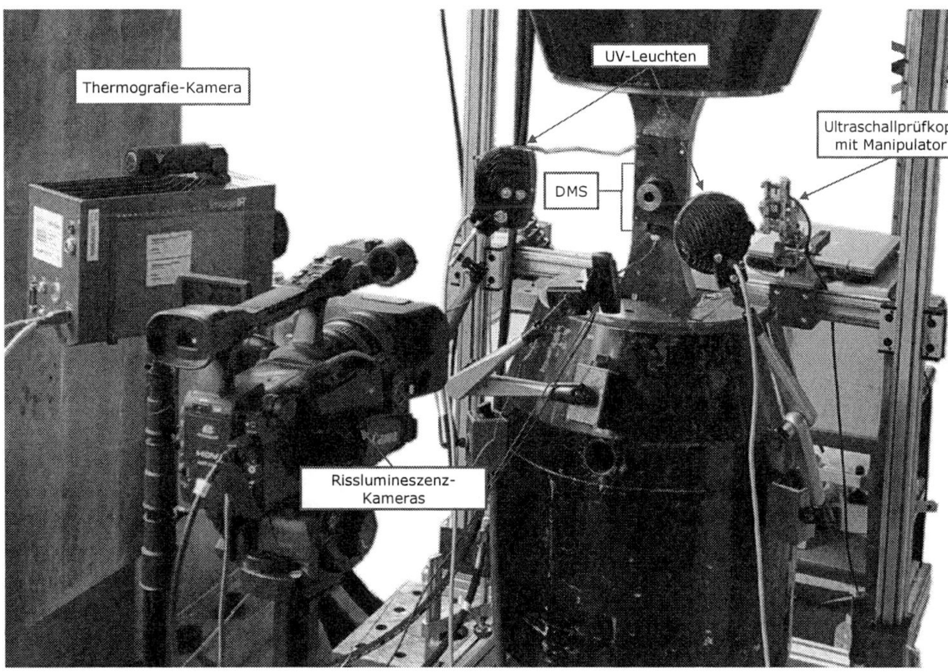

14

40 mm Knochenprobe mit Schweißbuchse

Risslumineszenz, DMS, Thermografie

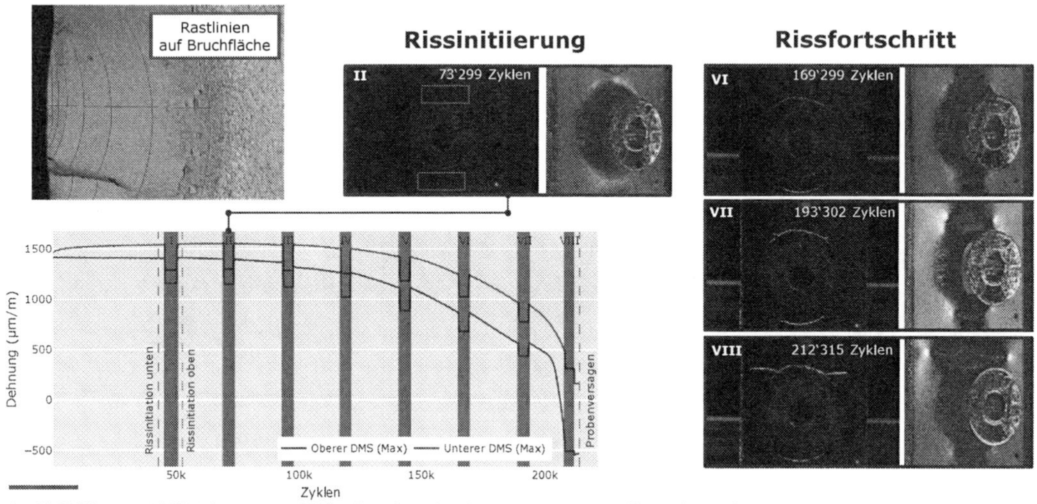

15

Technical Justification

Bauteilebene

- Rissdetektion und –bemessung beim Anschwingen von Austauschkomponenten

16

Technical Justification

Bauteilebene

BAM

- Monitoring während Ermüdungsversuchen von 1:1, 1:1.5 und 1:3.3 Jacket-Knoten (Offshore-Windenergieanlagen)

17

Technical Justification

Abklärungen von Einflussfaktoren

BAM

Messung:
- Blickwinkel und Abstand der Kamera/UV-Leuchte durch Automatisierung
- Störungen
  - Sonneneinfluss
  - Nässe
  - Künstliche Lichtbedingungen
  - Kopplung von Faktoren
  - ...

18

Technical Justification

Abklärungen zur Chloridbeständigkeit

BAM

- Chloridbeständigkeit konnte erhöht werden
- Alleine nicht nutzbar für Offshore-Anwendungen
- Einstufung der Korrosions-beständigkeit (z.B. C2/3) durch Zertifizierungsverfahren als nächster Schritt
- Erste Untersuchungen zur Kopplung der Risslumineszenz mit Offshore-typischen Korrosionsschutzsystemen zeigen Komplexität der individuellen mech. Eigenschaften der Beschichtungen

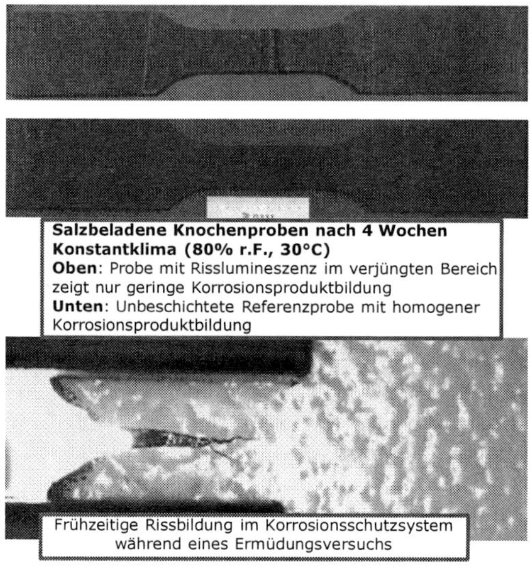

Salzbeladene Knochenproben nach 4 Wochen Konstantklima (80% r.F., 30°C)
Oben: Probe mit Risslumineszenz im verjüngten Bereich zeigt nur geringe Korrosionsproduktbildung
Unten: Unbeschichtete Referenzprobe mit homogener Korrosionsproduktbildung

Frühzeitige Rissbildung im Korrosionsschutzsystem während eines Ermüdungsversuchs

19

Genereller Umgang mit der Degradation

Untersuchungen zur Modellierung der Faktoren

BAM

- Riluminati Lifetime Model wäre v.a. relevant für die zusätzliche Funktion als einfacher Korrosionsschutz
- **Riluminati Service Life Model** ist relevant für die POD Studie
- Minimale **Time to Failure** von Applikationszeitpunkt bis zu dem Zeitpunkt, an dem ein Einflussparameter einen kritischen Schwellwert überschreitet

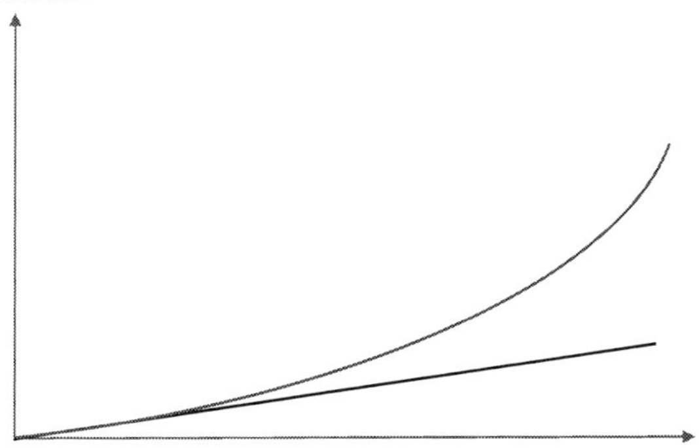

20

Auslagerung auf Umspannstation Meerwind Süd|Ost

Degradation mit und ohne Korrosionsschutz

BAM

- Untersuchungen im Feld zu gekoppelten Degradationsprozessen
- Kombination mit Korrosionsschutz-systemen anderer Industrien (z.B. Stahlbrücken, Krananlagen, Maschinenbau außen)

Prüfaufbau am Vorbild von Vorgängerstudien (Foto: ©EnBW)

MR CHEMIE NDT-MATERIALS

WindMW

brocolor®

Umspannstation des Windparks Meerwind Süd|Ost des Unternehmens WindMW nahe Helgoland (Foto: ©WindMW)

21

Ausblick

BAM

- Auswertung der Ermüdungsversuche
- Erzeugung der POD nach Meeker (Random Effects Modell)
- Durchführung weiterer Untersuchungen zu Messvariabilität und –unsicherheit
- Modellierung dieser und Integration in Erzeugung eines robusten Trainingsdatensatzes für die KI-basierte Algorithmik
- Einbindung der Modelle in POD mittels MAPOD-Ansatz

22

projektträger jülich

Vielen Dank für Ihre Aufmerksamkeit!

Gefördert durch:

Bundesministerium für Wirtschaft und Klimaschutz

aufgrund eines Beschlusses

23

ZfP 4.0 für Produktion und Service in der Luftfahrt

Holger SPECKMANN [1], Jonas HOLTMANN[2], Aswin HARIDAS [3]
Meng LIU[4], Stefan NEUMANN[5]
[1] Testia GmbH, Bremen, Deutschland, holger.speckmann@testia.com
[2] Testia GmbH, Bremen, Deutschland, jonas.holtmann@testia.com
[3] Testia GmbH, Bremen, Deutschland, aswin.haridas@testia.com
[4] Testia GmbH, Ottobrunn, Deutschland, meng.liu@testia.com
[5] Testia GmbH, Bremen, Deutschland, stefan.neumann@testia.com

Kurzfassung

Die Automatisierung der zerstörungsfreien Prüfung und hier insbesondere die automatisierte Auswertung stellt eine der herausragenden Möglichkeiten im Bereich der Digitalisierung und der künstlichen Intelligenz (KI) dar.

Der Vortrag beschäftigt sich mit den ersten technischen Umsetzungen im Bereich des Machine Learnings in der Luftfahrt, erläutert die Nutzung von Computer/Machine Vision Lösungen im Bereich der visuellen Prüfung und wagt einen Ausblick in die umfangreiche Anwendung von Lösungen der KI.

Es wird dargestellt, wie die Analyse von Produktions Parametern (Prozess Monitoring), z.B. im Bereich der Additiven Fertigung, die nachgeschaltete zerstörungsfreie Prüfung reduzieren wird und sich hier bezüglich der Wirtschaftlichkeit positiv auf neue Fertigungstechnologien auswirken wird.

Neben den technischen Ansätzen wird das Thema der Zulassung von KI Lösungen und deren Herausforderungen in der Luftfahrt dargestellt.

Keywords: ZfP4.0, Luftfahrt, Aerospace, Digitalisierung; Automatisierung, Künstliche Intelligenz (KI), Machine Learning, Machine Vision, Produktion, Service, Prozess Monitoring

ZfP 4.0 für Produktion und Service in der Luftfahrt

Holger Speckmann
Jonas Holtmann
Aswin Haridas
Stefan Neumann
Meng Liu

Inhalt

- ZfP 4.0 - Was ist darunter zu verstehen?
- Begrifflichkeiten
- ZfP 4.0 in der Produktion von Luftfahrt Teilen
  - Bildgebende Verfahren verringern die Auswertezeiten (UT + CT)
  - Automatisierte Auswertung und Bewertung
  - Visuelle Prüfung ergänzt EN4179 Methoden
- Prozesskontrolle anstelle von (nachgeschalteter) ZfP
- ZfP 4.0 im Service von Flugzeugen, Hubschraubern, etc.
  - Structural Health Monitoring
- Zulassung und Qualifikation von automatisierter Auswertung unter Verwendung der KI

ZfP 4.0 - Was ist darunter zu verstehen?

Was meint ZfP 4.0?

Quelle: dgzfp.de - ZfP 4.0 aus der Sicht der ZfP-Community: Umfrageergebnisse, Herausforderungen und Perspektiven
Veröffentlicht 23. April 2021

Wieso 4.0?

„Industrie 4.0“ übertragen auf „ZfP 4.0“

Phase 1:	Phase 2:	Phase 3:	Phase 4:
Oberflächliche Prüfung (Sichtprüfung)	Prüfung im Inneren des Materials (Ultraschall, Röntgen, etc.)	Digitalisierte Messmethoden	Nachverfolgung über alle Prozessschritte hinweg (Digital Twin, Machine Learning, Remote Inspection, Künstliche Intelligenz, etc.)

Diese 4 Phasen sind vereinfacht! Reale Entwicklung ist kontinuierlich.

TESTIA
AN AIRBUS COMPANY

Begrifflichkeiten

Begrifflichkeiten

TESTIA
AN AIRBUS COMPANY

Digitalisierung:
Bezeichnet die Umwandlung analoger Informationen, Prozesse oder Geschäftsmodelle in digitale Formate, mit dem Ziel, effizienter, schneller und automatisierter zu arbeiten

Machine Learning:
Ist ein Teilgebiet der Künstlichen Intelligenz (KI), bei dem Computer aus Daten lernen, ohne explizit programmiert zu sein, wie sie bestimmte Aufgaben lösen sollen.

Computer/Machine Vision:
Ist ein Teilgebiet der Künstlichen Intelligenz (KI), das es Computern ermöglicht, Bilder und Videos "zu sehen", zu analysieren und zu interpretieren, ähnlich wies es das menschliche Auge und Gehirn tut.

Künstliche Intelligenz (KI):
Bezeichnet Systeme, die menschliche Intelligenz nachahmen können, als z.B. lernen, planen, Probleme lösen, Sprache verstehen oder sogar kreativ sein.

TESTIA
AN AIRBUS COMPANY

ZfP 4.0 in der Produktion von Flugzeugteilen

Fully Automated Inspection Systems (e.g. for UT)

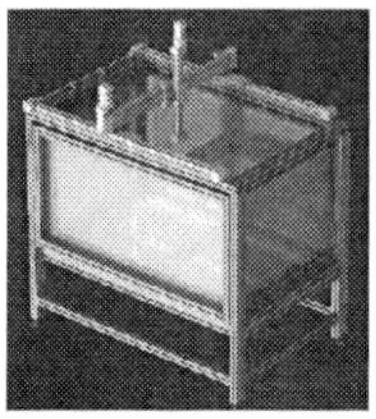

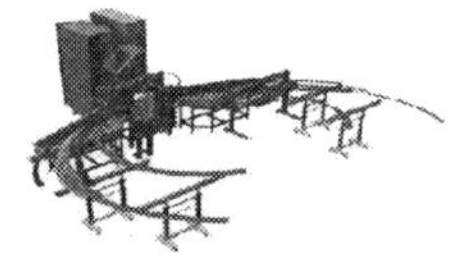

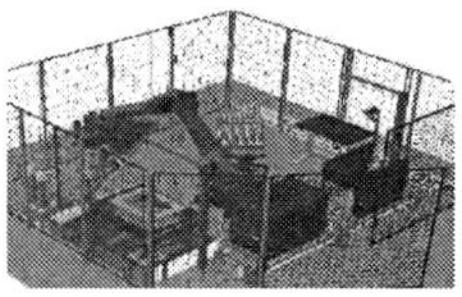

Automated Defect Recognition (ADR) for UT

Data processing software: example NDTkit UT

- Full Harmonization of the UT data analysis process
- Fast Diagnosis
- Assisted defect detection
- Automated reporting

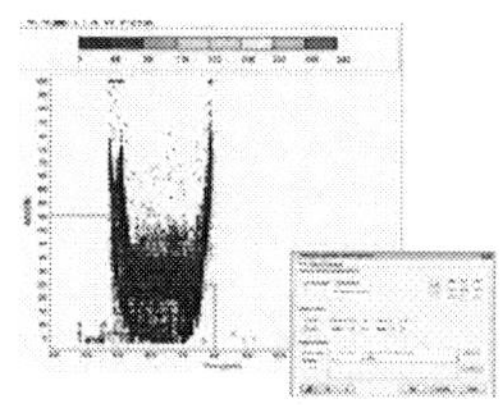

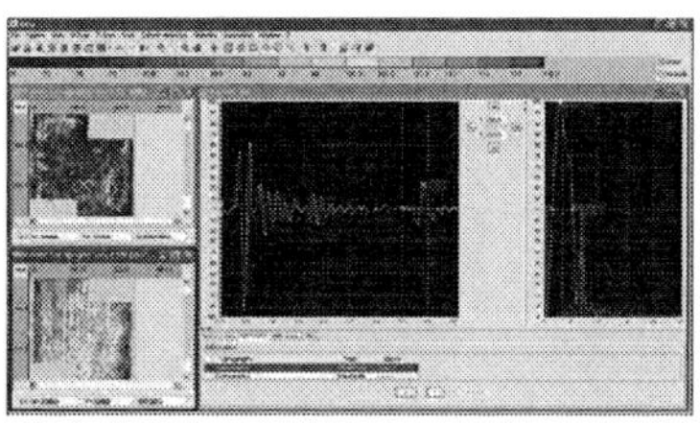

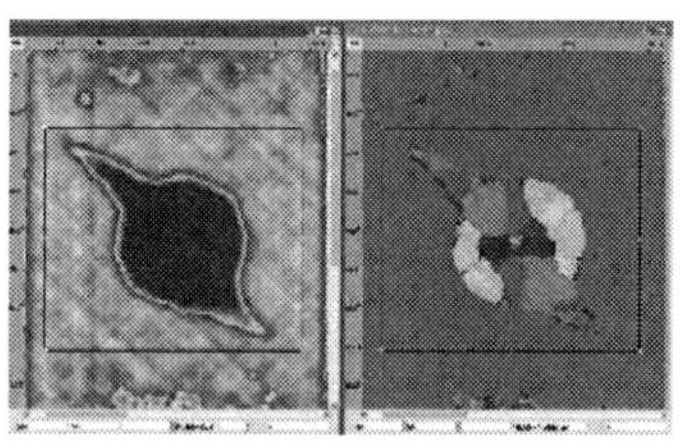

Computer Tomography (CT) - Change to Automation

TESTIA
AN AIRBUS COMPANY

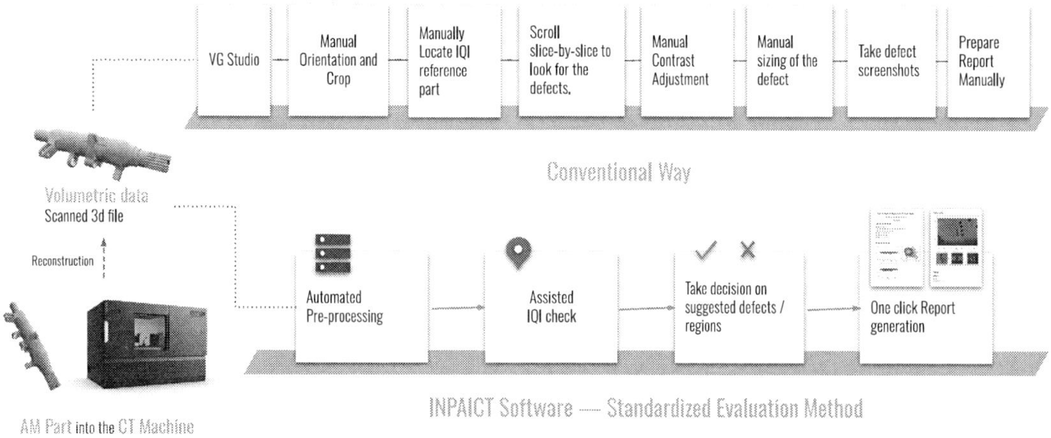

11

INPAICT

TESTIA
AN AIRBUS COMPANY

- Assisted evaluation of Computed Tomography scans
- Drastic reduction in evaluation time and cost
- First step for implementation of AI in aerospace
- Degree of assistance and responsibility of algorithm can be increase incrementally while gaining trust

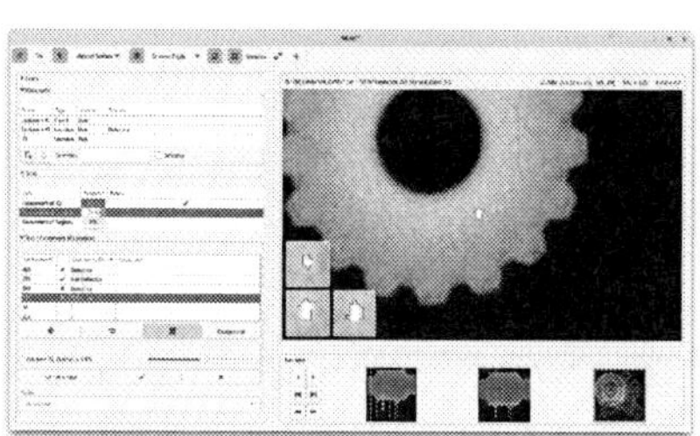

Automated Defect Recognition (ADR) for CT

ATESTIA AN AIRBUS COMPANY

- … by Artificial Intelligence (AI)

- Fully automated detection of defects in CT-Data
- > 80 % time saving on evaluation
- Inspector assistance (Evaluation of defects)
- Automated report generation

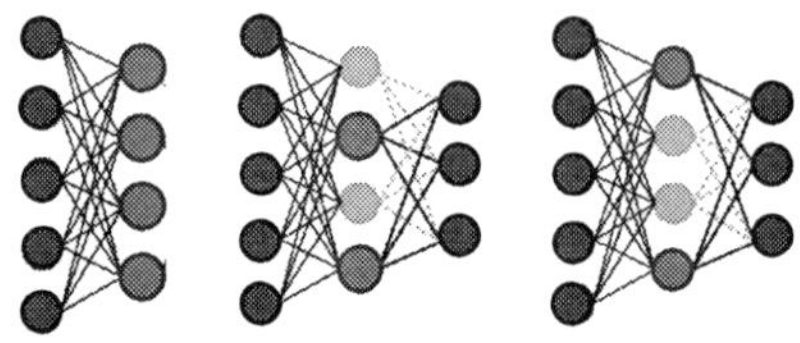

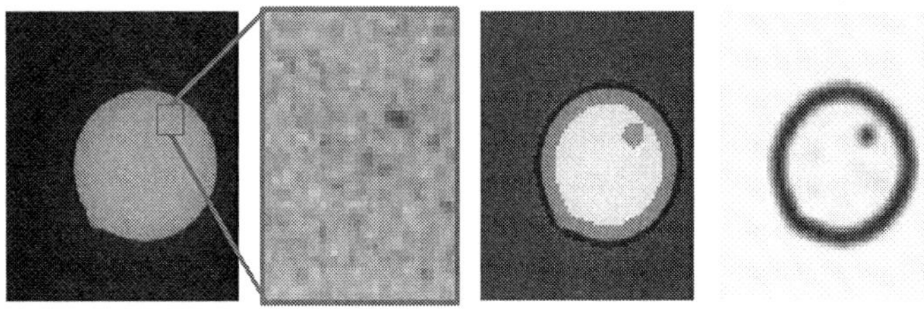

Visual Inspection - What is Machine Vision

ATESTIA AN AIRBUS COMPANY

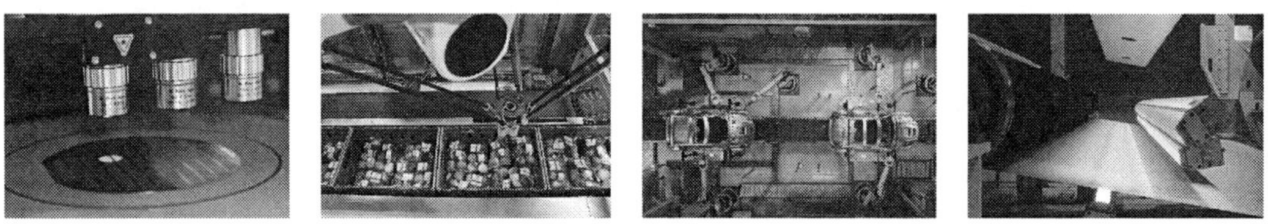

Machine vision (**MV**), as a **systems engineering** discipline, integrates existing technologies and methods to provide imaging/point cloud-based automatic inspection and analysis to solve **real world problems**.

Ref. https://www.infactory-solutions.net/what-is-machine-vision/

Large-Scale-Machine-Vision / Visual Inspection Automation

TESTIA
AN AIRBUS COMPANY

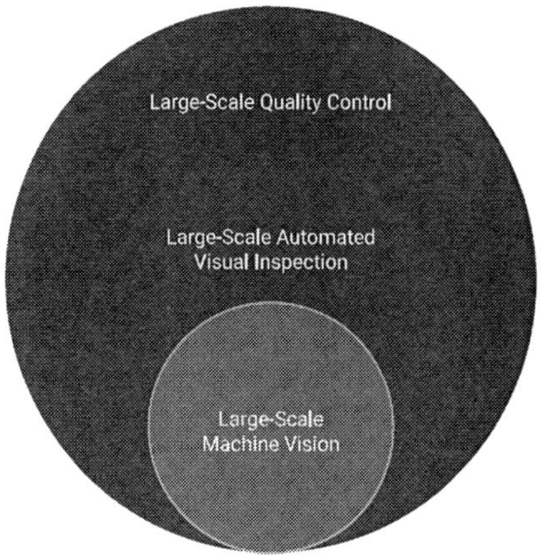

Large-Scale Quality Control (**LSQC**) - the comprehensive methodology to ensure aircraft quality with automation.

Large-Scale Automated Visual Inspection (**LSAVI**) - the collective non-destructive testing method using mainly optical sensors.

Large-Scale Machine vision (**LSMV**) - the core technology of inspection automation using optical sensors.

TESTIA
AN AIRBUS COMPANY

Prozesskontrolle anstelle von (nachgeschalteter) zfP

AM Monitoring

TESTIA
AN AIRBUS COMPANY

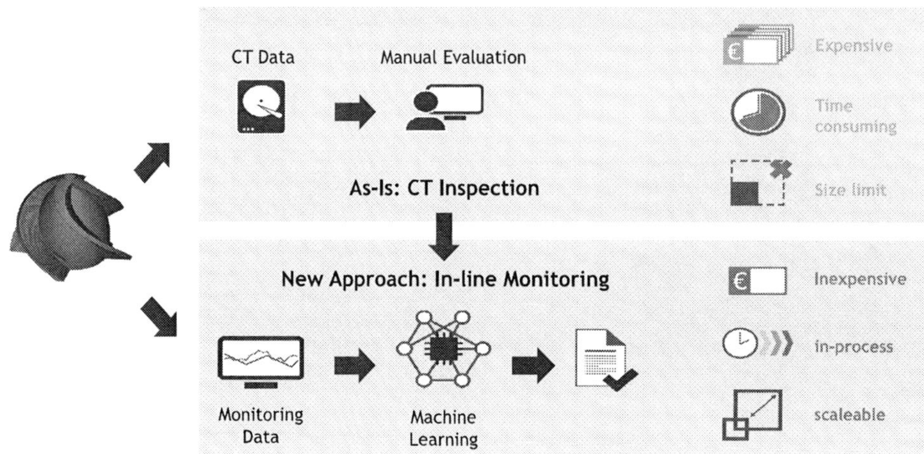

AM Monitoring

TESTIA
AN AIRBUS COMPANY

- Automated in-line and offline defect detection
- Solution can detect anomalies during printing
- Defect visualization
- Containerized application for easy scaling to production needs
- Machine Learning solution for highly complex data interpretation
- Automatic defect alerts

Vision: In-line defect healing by feedback loop to 3D-printing machine

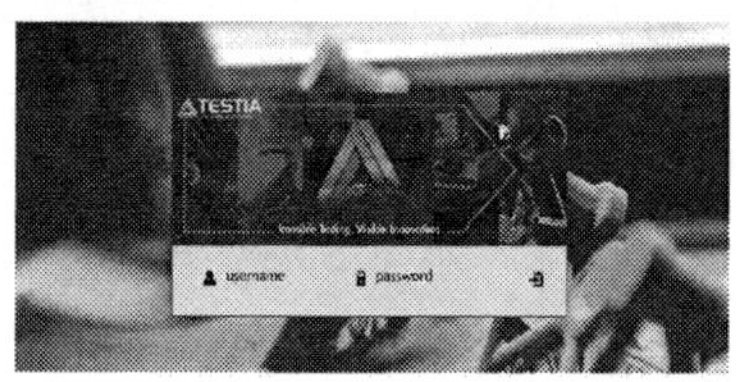

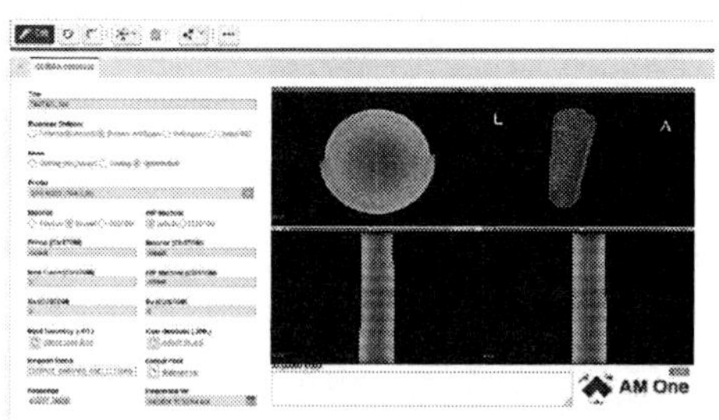

In-Line Vacuum Monitoring

- Simple installation between vacuum pump and part
- Fast identification of leakages in vacuum setups through combined analysis of gas flow and vacuum level
- More efficient bagging of composite parts through customizable vacuum control functionality

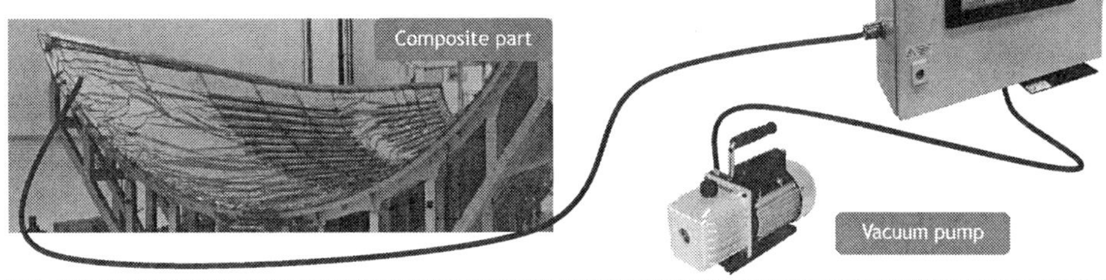

In-line Infusion Monitoring

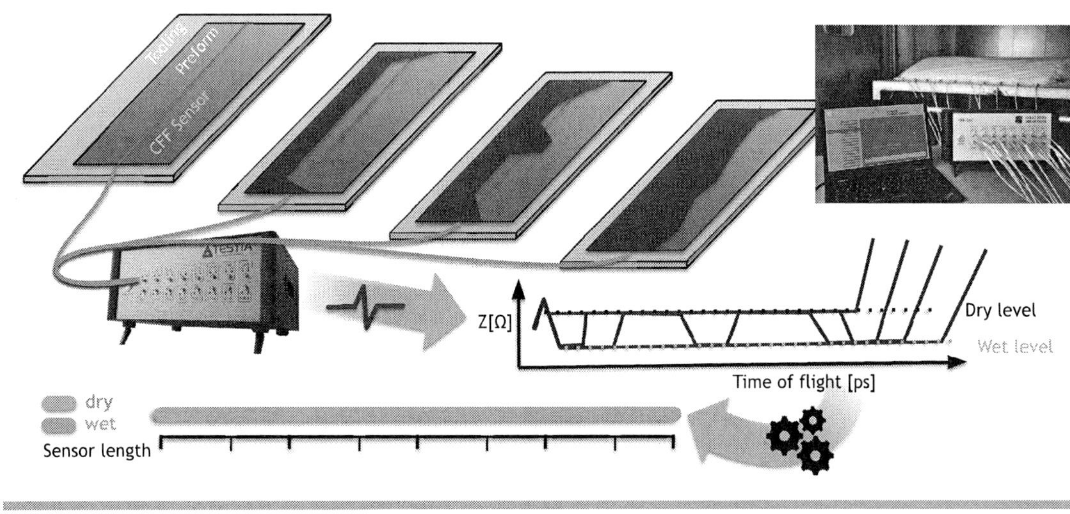

Composite monitoring solutions for the entire process chain

TESTIA
AN AIRBUS COMPANY

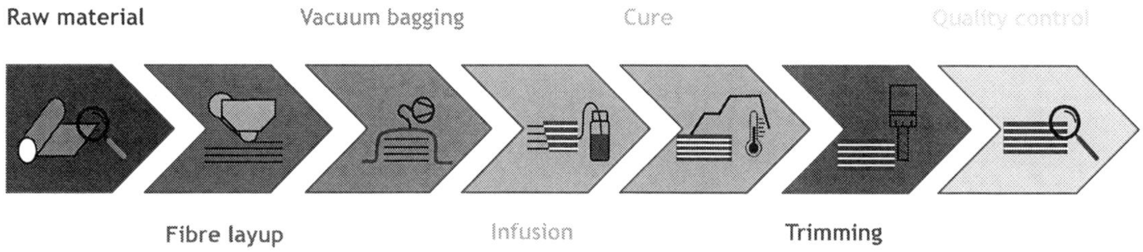

TESTIA
AN AIRBUS COMPANY

ZfP 4.0 im Service

NDT 4.0 Solution for Service

TESTIA AN AIRBUS COMPANY

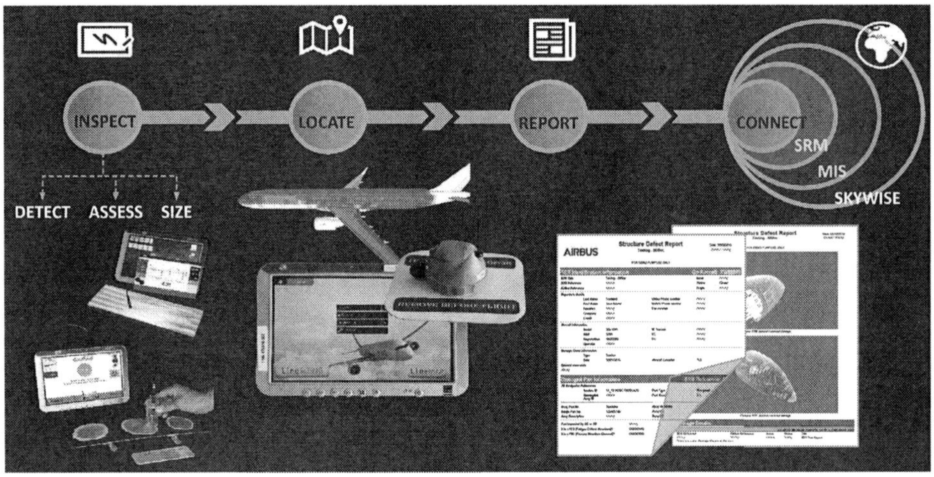

Autonomous SHM (ASHM) for Early Warning

IPR INNOVATIVE PRODUCTS RESOURCES TESTIA AN AIRBUS COMPANY

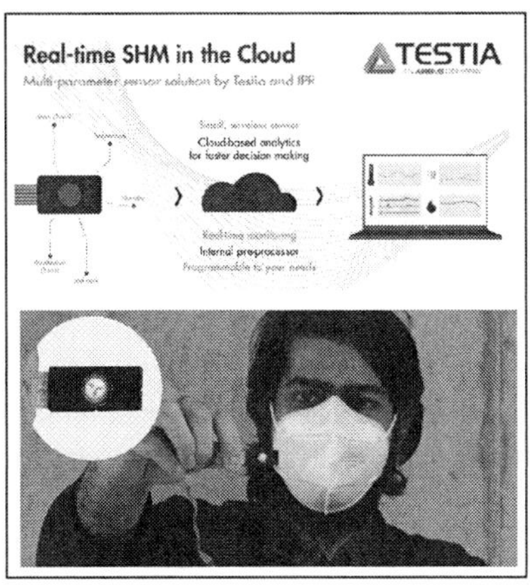

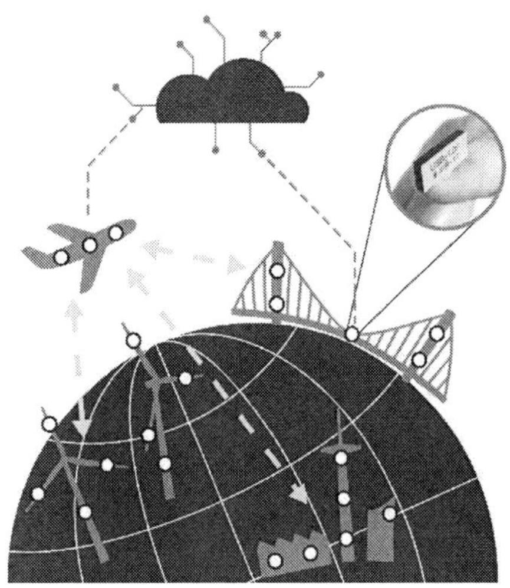

Of-The-Shelf System - Data Collection, Fusion, Analysis

TESTIA
AN AIRBUS COMPANY

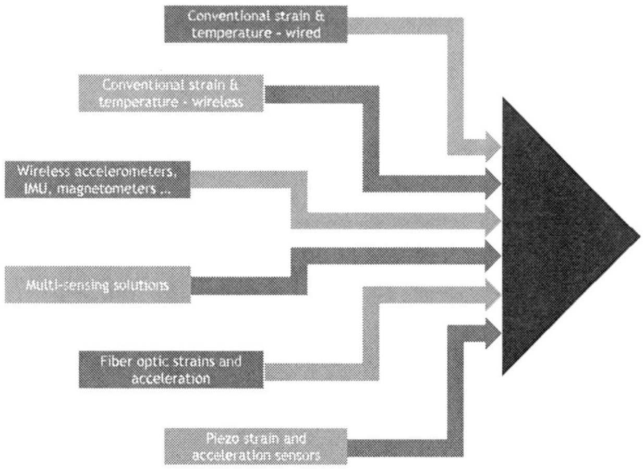

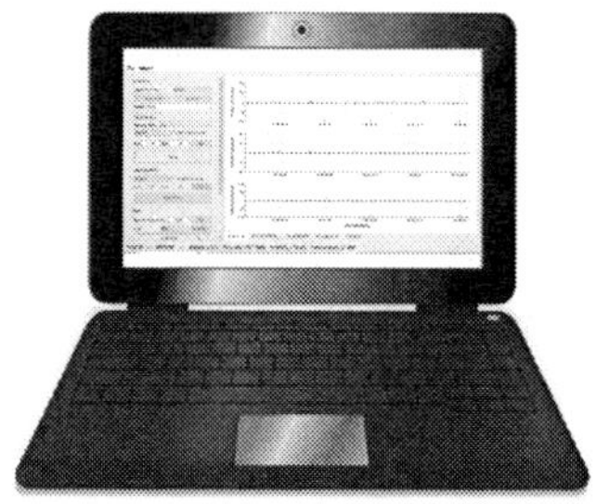

Time Synchronized data from all sensors available on your cloud

Augmented Reality (AR) - Installation & Training

TESTIA
AN AIRBUS COMPANY

- Our AR based solutions aim to reduce overall time for SHM integration & increase efficiency
  - Projecting installation locations & supporting documents, online support, automatic image capture & report generation.
  - First implementations shows time savings of 40%
- Visualize data in realtime using AR (assisting digital twin implementations).
- Standardize SHM trainings using AR tools.

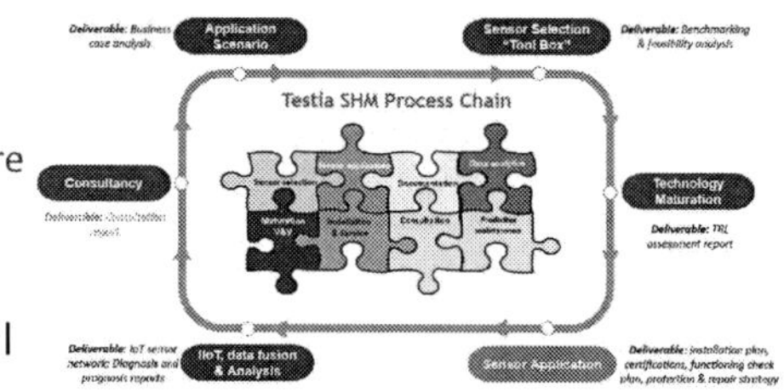

Microsoft, 2023, "Information about HoloLens 2," Accessed: 2025-04-24

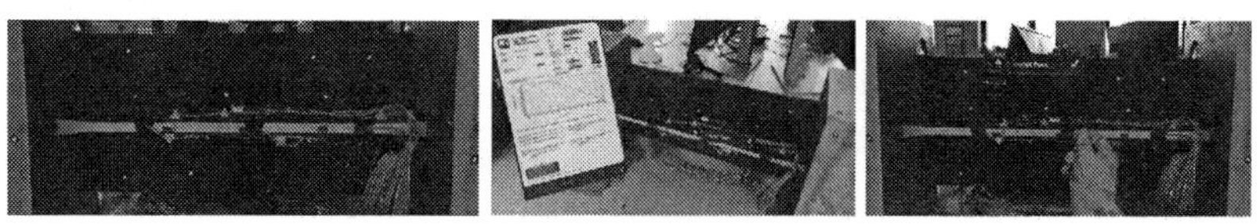

TESTIA
AN AIRBUS COMPANY

Zulassung und Qualifikation von automatisierter Auswertung unter Verwendung von KI

Certification of NDT Systems implementing AI - EASA AI Roadmap

TESTIA
AN AIRBUS COMPANY

- Vision: certifiable, human-centric AI to boost aviation safety & efficiency
- Three general AI levels
  - **Level 1**: Assistance to Human - Decisions taken by end user based on support from AI system
    - Level 1A: Human Augmentation
    - Level 1B: **Human Cognitive Assistance in Decision and Action Selection**
      - Provides automation support to decision making
      - AI system can preset subset of alternatives / selected action solely by end user / AI system has no capability of decision making / end user maintains full authority
  - **Level 2**: Human-AI Teaming - **Partial release of authority to system**, but under full oversight of end user, who remains accountable for operations
  - **Level 3**: Advanced Automation - **AI based system is given full authority** to make and implement decisions.

TESTIA
AN AIRBUS COMPANY
Zusammenfassung

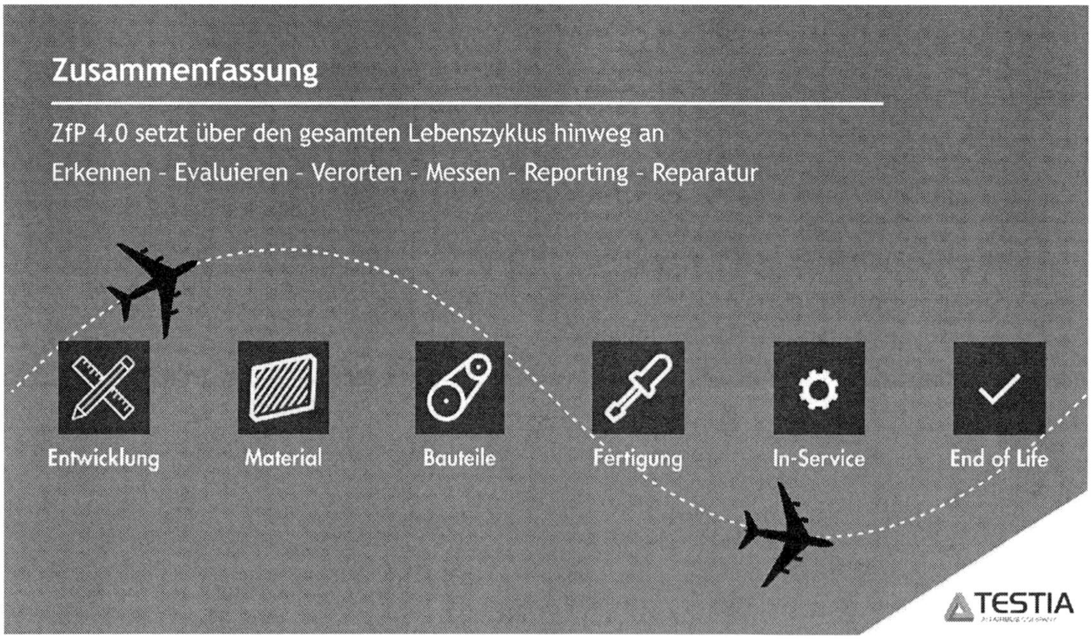
Zusammenfassung
ZfP 4.0 setzt über den gesamten Lebenszyklus hinweg an
Erkennen - Evaluieren - Verorten - Messen - Reporting - Reparatur
Entwicklung
Material
Bauteile
Fertigung
In-Service
End of Life
TESTIA

AUTHOR INDEX